Stochastic Mechanics
Random Media
Signal Processing and Image Synthesis
Mathematical Economics and Finance
Stochastic Optimization
Stochastic Control
Stochastic Models in Life Sciences

Stochastic Modelling and Applied Probability

(Formerly:
Applications of Mathematics)

58

Stochastic Modelling and Applied Probability
formerly: Applications of Mathematics

---

1 Fleming/Rishel, **Deterministic and Stochastic Optimal Control** (1975)
2 Marchuk, **Methods of Numerical Mathematics** (1975, 2nd. ed. 1982)
3 Balakrishnan, **Applied Functional Analysis** (1976, 2nd. ed. 1981)
4 Borovkov, **Stochastic Processes in Queueing Theory** (1976)
5 Liptser/Shiryaev, **Statistics of Random Processes I: General Theory** (1977, 2nd. ed. 2001)
6 Liptser/Shiryaev, **Statistics of Random Processes II: Applications** (1978, 2nd. ed. 2001)
7 Vorob'ev, **Game Theory: Lectures for Economists and Systems Scientists** (1977)
8 Shiryaev, **Optimal Stopping Rules** (1978)
9 Ibragimov/Rozanov, **Gaussian Random Processes** (1978)
10 Wonham, **Linear Multivariable Control: A Geometric Approach** (1979, 2nd. ed. 1985)
11 Hida, **Brownian Motion** (1980)
12 Hestenes, **Conjugate Direction Methods in Optimization** (1980)
13 Kallianpur, **Stochastic Filtering Theory** (1980)
14 Krylov, **Controlled Diffusion Processes** (1980)
15 Prabhu, **Stochastic Storage Processes: Queues, Insurance Risk, and Dams** (1980)
16 Ibragimov/Has'minskii, **Statistical Estimation: Asymptotic Theory** (1981)
17 Cesari, **Optimization: Theory and Applications** (1982)
18 Elliott, **Stochastic Calculus and Applications** (1982)
19 Marchuk/Shaidourov, **Difference Methods and Their Extrapolations** (1983)
20 Hijab, **Stabilization of Control Systems** (1986)
21 Protter, **Stochastic Integration and Differential Equations** (1990)
22 Benveniste/Métivier/Priouret, **Adaptive Algorithms and Stochastic Approximations** (1990)
23 Kloeden/Platen, **Numerical Solution of Stochastic Differential Equations** (1992, corr. 3rd printing 1999)
24 Kushner/Dupuis, **Numerical Methods for Stochastic Control Problems in Continuous Time** (1992)
25 Fleming/Soner, **Controlled Markov Processes and Viscosity Solutions** (1993)
26 Baccelli/Brémaud, **Elements of Queueing Theory** (1994, 2nd. ed. 2003)
27 Winkler, **Image Analysis, Random Fields and Dynamic Monte Carlo Methods** (1995, 2nd. ed. 2003)
28 Kalpazidou, **Cycle Representations of Markov Processes** (1995)
29 Elliott/Aggoun/Moore, **Hidden Markov Models: Estimation and Control** (1995)
30 Hernández-Lerma/Lasserre, **Discrete-Time Markov Control Processes** (1995)
31 Devroye/Györfi/Lugosi, **A Probabilistic Theory of Pattern Recognition** (1996)
32 Maitra/Sudderth, **Discrete Gambling and Stochastic Games** (1996)
33 Embrechts/Klüppelberg/Mikosch, **Modelling Extremal Events for Insurance and Finance** (1997, corr. 4th printing 2003)
34 Duflo, **Random Iterative Models** (1997)
35 Kushner/Yin, **Stochastic Approximation Algorithms and Applications** (1997)
36 Musiela/Rutkowski, **Martingale Methods in Financial Modelling** (1997, 2nd. ed. 2005)
37 Yin, **Continuous-Time Markov Chains and Applications** (1998)
38 Dembo/Zeitouni, **Large Deviations Techniques and Applications** (1998)
39 Karatzas, **Methods of Mathematical Finance** (1998)
40 Fayolle/Iasnogorodski/Malyshev, **Random Walks in the Quarter-Plane** (1999)
41 Aven/Jensen, **Stochastic Models in Reliability** (1999)
42 Hernandez-Lerma/Lasserre, **Further Topics on Discrete-Time Markov Control Processes** (1999)
43 Yong/Zhou, **Stochastic Controls. Hamiltonian Systems and HJB Equations** (1999)
44 Serfozo, **Introduction to Stochastic Networks** (1999)
45 Steele, **Stochastic Calculus and Financial Applications** (2001)
46 Chen/Yao, **Fundamentals of Queuing Networks: Performance, Asymptotics, and Optimization** (2001)
47 Kushner, **Heavy Traffic Analysis of Controlled Queueing and Communications Networks** (2001)
48 Fernholz, **Stochastic Portfolio Theory** (2002)
49 Kabanov/Pergamenshchikov, **Two-Scale Stochastic Systems** (2003)
50 Han, **Information-Spectrum Methods in Information Theory** (2003)

(continued after References)

Peter Kotelenez

Stochastic Ordinary and Stochastic Partial Differential Equations

Transition from Microscopic to Macroscopic Equations

Springer

Author

Peter Kotelenez
Department of Mathematics
Case Western Reserve University
10900 Euclid Ave.
Cleveland, OH 44106–7058
USA
pxk4@cwru.edu

ISBN 978-0-387-74316-5 e-ISBN 978-0-387-74317-2
DOI: 10.1007/978-0-387-74317-2

Library of Congress Control Number: 2007940371

Mathematics Subject Classification (2000): 60H15, 60H10, 60F99, 82C22, 82C31, 60K35, 35K55, 35K10, 60K37, 60G60, 60J60

Printed on acid-free paper.

9 8 7 6 5 4 3 2 1

springer.com

Koty

-

To Lydia

Contents

Introduction

The present volume analyzes mathematical models of time-dependent physical phenomena on three levels: *microscopic, mesoscopic*, and *macroscopic*. We provide a rigorous derivation of each level from the preceding level and the resulting mesoscopic equations are analyzed in detail. Following Haken (1983, Sect. 1.11.6) we deal, "*at the microscopic level, with individual atoms or molecules, described by their positions, velocities, and mutual interactions. At the mesoscopic level, we describe the liquid by means of ensembles of many atoms or molecules. The extension of such an ensemble is assumed large compared to interatomic distances but small compared to the evolving macroscopic pattern... . At the macroscopic level we wish to study the corresponding spatial patterns.*" Typically, at the macroscopic level, the systems under consideration are treated as spatially continuous systems such as fluids or a continuous distribution of some chemical reactants, etc. In contrast, on the microscopic level, Newtonian mechanics governs the equations of motion of the individual atoms or molecules.[1] These equations are cast in the form of systems of deterministic coupled nonlinear oscillators. The mesoscopic level[2] is probabilistic in nature and many models may be faithfully described by stochastic ordinary and stochastic partial differential equations (SODEs and SPDEs),[3] where the latter are defined on a continuum. The macroscopic level is described by time-dependent partial differential equations (PDE's) and its generalization and simplifications.

In our mathematical framework we talk of particles instead of atoms and molecules. The *transition from the microscopic description to a mesoscopic (i.e., stochastic) description* requires the following:

- Replacement of spatially extended particles by point particles
- Formation of small clusters (ensembles) of particles (if their initial positions and velocities are similar)

[1] We restrict ourselves in this volume to "classical physics" (cf., e.g., Heisenberg (1958)).

[2] For the relation between nanotechnology and mesoscales, we refer to Roukes (2001).

[3] In this volume, mesoscopic equations will be identified with SODEs and SPDEs.

- Randomization of the initial distribution of clusters where the probability distribution is determined by the relative sizes of the clusters
- "*Coarse graining*," i.e., representation of clusters as cells or boxes in a grid for the positions and velocities

Having performed all four simplifications, the resulting description is still governed by many deterministic coupled nonlinear oscillators and, therefore, a simplified *microscopic* model.

Given a probability distribution for the initial data, it is possible, through scaling and similar devices, to proceed to the mesoscopic level, governed by SODEs and SPDEs, as follows:

- Following Einstein (1905), we consider the substance under investigation a "solute," which is immersed in a medium (usually a liquid) called the "solvents." Accordingly, the particles are divided into two groups: (1) *Large particles*, i.e., the solute particles; (2) *small particles*, the solvent particles.
- Neglect the interaction between small particles.
- Consider first the interaction between large and small particles. To obtain the Brownian motion effect, increase the initial velocities of the small particles (to infinity). Allow the small particles to escape to infinity after having interacted with the large particles for a macroscopically small time. This small time induces a partition of the time axis into small time intervals. In each of the small time intervals the large particles are being displaced by the interaction with clusters of small particles. Note that the vast majority of small particles have previously not interacted with the large particles and they disappear to infinity after that time step. (Cf. Figs. 1 and 2.) This implies almost independence of the displacements of the large particles in different time intervals and, in the scaling limit, independent increments of the motion of the large particles. To make this rigorous, an infinite system of small particles is needed if the interval size tends to 0 in the scaling limit. Therefore, depending on whether or not friction is included in the equations for the large particles, we obtain that, in the scaling limit, the positions or the velocities of the large particles perform Brownian motions in time.[4] If the positions are Brownian motions, this model is called

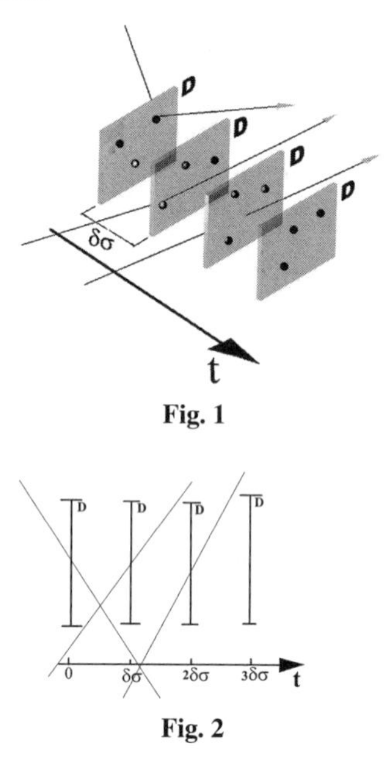

Fig. 1

Fig. 2

[4] The escape to infinity after a short period of interaction with the large particles is necessary to generate independent increments in the limit. This hypothesis seems to be acceptable if for

the Einstein-Smoluchowski model, and if the velocities are Brownian motions, then it is called an Ornstein-Uhlenbeck model (cf. Nelson, 1972).

- The interaction between large particles occurs on a much slower time scale than the interaction between large and small particles and can be included after the scaling limit employing fractional steps.[5] Hence, the positions of the large particles become solutions of a system of SODEs in the Einstein-Smoluchowski model.
- The step from (Einstein-Smoluchowski) SODEs to SPDEs, which is a more simplified mesoscopic level, is relatively easy, if the individual Brownian motions from the previous step are obtained through a Gaussian space–time field, which is uncorrelated in time but spatially correlated. In this case the empirical distribution of the solutions of the SODEs is the solution of an SPDE, independent of the number of particles involved, and the SPDE can be solved in a space of densities, if the number of particles tends to infinity and if the initial particle distribution has a density. The resulting SPDE describes the distribution of matter in a continuum.

The *transition from the mesoscopic SPDEs to macroscopic (i.e., deterministic) PDE's* occurs as follows:

- As the correlation length[6] in the spatially correlated Gaussian field tends to 0 the solutions of the SPDEs tend to solutions of the macroscopic PDEs (as a weak limit).

The mesoscopic SPDE is formally a PDE perturbed by state-dependent Brownian noise. This perturbation is small if the aforementioned correlation length is small. Roughly speaking, the spatial correlations occur in the transition from the microscopic level to the mesoscopic SODEs because the small particles are assumed to move with different velocities (e.g., subject to a Maxwellian velocity distribution). As a result, small particles coming from "far away" can interact with a given large particle "at the same time" as small particles were close to the large particles. This generates a long-range mean-field character of the interaction between small and large particles and leads in the scaling limit to the Gaussian space–time field, which is spatially correlated. Note that the perturbation of the PDE by state-dependent Brownian noise is derived from the microscopic level. We conclude that the correlation length is a result of the discrete spatially extended structures of the microscopic level. Further, on the mesoscopic level, the correlation length is a measure of the strength of the fluctuations around the solutions of the macroscopic equations.

Let $\bar{w}$ denote the average speed of the small particles, $\eta > 0$ the friction coefficient for the large particles. The typical mass of a large particle is $\approx \frac{1}{N}$, $N \in \mathbf{N}$, and

spatially extended particles the interparticle distance is considerably greater than the diameter of a typical particle. (Cf. Fig. 1.) This holds for a gas (cf. Lifshits and Pitayevskii (1979), Ch.1, p. 3), but not for a liquid, like water. Nevertheless, we show in Chap. 5 that the qualitative behavior of correlated Brownian motions is in good agreement with the depletion phenomenon of colloids in suspension.

[5] Cf. Goncharuk and Kotelenez (1998) and also our Sect. 15.3 for a description of this method.

[6] Cf. the following Chap. 1 for more details on the correlation length.

$\sqrt{\varepsilon} > 0$ is the correlation length in the spatial correlations of the limiting Gaussian space–time field. Assuming that the initial data of the small particles are coarse-grained into independent clusters, the following scheme summarizes the main steps in the transition from microscopic to macroscopic, as derived in this book:

Microscopic Level: Newtonian mechanics/systems of deterministic coupled nonlinear oscillators

$\Downarrow \quad (\bar{w} \gg \eta \to \infty)$

Mesoscopic Level: SODEs for the positions of N large particles

$\Downarrow \quad (N \to \infty)$

SPDEs for the continuous distribution of large particles

$\Downarrow \quad (\sqrt{\varepsilon} \to 0)$

Macroscopic Level: PDEs for the continuous distribution of large particles

Next we review the general content of the book. Formally, the book is divided into five parts and each part is divided into chapters. The chapter at the end of each part contain lengthy and technical proofs of some of the theorems that are formulated within the first chapter. Shorter proofs are given directly after the theorems. Examples are provided at the end of the chapters. The chapters are numbered consecutively, independent of the parts.

In Part I (Chaps. 1–3), we describe the transition from the microscopic equations to the mesoscopic equations for correlated Brownian motions. We simplify this procedure by working with a space–time discretized version of the infinite system of coupled oscillators. The proof of the scaling limit theorem from Chap. 2 in Part I is provided in Chap. 3. In Part II (Chaps. 4–7) we consider a general system of Itô SODEs[7] for the positions of the large particles. This is called "mesoscopic level A." The driving noise fields are both correlated and independent, identically distributed (i.i.d) Brownian motions.[8] The coefficients depend on the empirical distribution of the particles as well as on space and time. In Chap. 4 we derive existence and uniqueness as well as equivalence in distribution. Chapter 5 describes the qualitative behavior of correlated Brownian motions. We prove that correlated Brownian motions are weakly attracted to each other, if the distance between them is short (which itself can be expressed as a function of the correlation length). We remark that experiments on colloids in suspension imply that Brownian particles at close distance must have a tendency to attract each other since the fluid between them gets *depleted* (cf. Tulpar et al. (2006) as well as Kotelenez et al. (2007)) (Cf. Fig. 4

[7] We will drop the term "Itô" in what follows, as we will always use Itô differentials, unless explicitly stated otherwise. In the alternative case we will consider Stratonovich differentials and talk about Stratonovich SODEs or Stratonovich SPDEs (cf. Chaps. 5, 8, 14, Sects. 15.2.5 and 15.2.6).

[8] We included i.i.d. Brownian motions as additional driving noise to provide a more complete description of the particle methods in SPDEs.

in Chap. 1). Therefore, our result confirms that correlated Brownian motions more correctly describe the behavior of a solute in a liquid of solvents than independent Brownian motions. Further, we show that the long-time behavior of two correlated Brownian motions is the same as for two uncorrelated Brownian motions if the space dimension is $d \geq 2$. For $d = 1$ two correlated Brownian motions eventually clump. Chapter 6 contains a proof of the flow property (which was claimed in Chap. 4). In Chap. 7 we compare a special case of our SODEs with the formalism introduced by Kunita (1990). We prove that the driving Gaussian fields in Kunita's SODEs are a special case of our correlated Brownian motions. In Part III (*mesoscopic level B*, Chaps. 8–13) we analyze the SPDEs[9] for the distribution of large particles. In Chap. 8, we derive existence and strong uniqueness for SPDEs with finite initial mass. We also derive a representation of semilinear (Itô) SPDEs by Stratonovich SPDEs, i.e., by SPDEs, driven by Stratonovich differentials. In the special case of noncoercive semilinear SPDEs, the Stratonovich representation is a first order transport SPDE, driven by Statonovich differentials. Chapter 9 contains the corresponding results for infinite initial mass, and in Chap. 10, we show that certain SPDEs with infinite mass can have homogeneous and isotropic random fields as their solutions. Chapters 11 and 12 contain proofs of smoothness, an Itô formula and uniqueness, respectively. In Chap. 13 we review some other approaches to SPDEs. This section is by no means a complete literature review. It is rather a random sample that may help the reader, who is not familiar with the subject, to get a first rough overview about various directions and models. Part IV (Chap. 14) contains the macroscopic limit theorem and its complete proof. For semi-linear non-coercive SPDEs, using their Stratonovich representations, the macroscopic limit implies the convergence of a first order transport SPDE to the solution of a deterministic parabolic PDE. Part V (Chap. 15) is a general appendix, which is subdivided into four sections on analysis, stochastics, the fractional step method, and frame-indifference. Some of the statements in Chap. 15 are given without proof but with detailed references where the proofs are found. For other statements the proofs are sketched or given in detail.

Acknowledgement

The transition from SODEs to SPDEs is in spirit closely related to D. Dawson's derivation of the measure diffusion for brachning Brownian motions and the resulting field of superprocesses (cf. Dawson (1975)). The author is indebted to Don Dawson for many interesting and inspiring discussions during his visits at Carleton University in Ottawa, which motivated him to develop the particle approach to SPDEs. Therefore, the present volume is dedicated to Donald A. Dawson on the occasion of his 65th birthday.

A first draft of Chaps. 4, 8, and 10 was written during the author's visit of the Sonderforschungsbereich "Diskrete Strukturen in der Mathematik" of the University of

[9] Cf. our previous footnote regarding our nomenclature for SODEs and SPDEs.

Bielefeld, Germany, during the summer of 1996. The hospitality of the Sonderforschungsbereich “Diskrete Strukturen in der Mathematik” and the support by the National Science Foundation are gratefully acknowledged.

Finally, the author wants to thank the Springer-Verlag and its managing editors for their extreme patience and cooperation over the last years while the manuscript for this book underwent many changes and extensions.

Part I
From Microscopic Dynamics to Mesoscopic Kinematics

Chapter 1
Heuristics: Microscopic Model and Space–Time Scales

On a heuristic level, this section provides the following: space–time scales for the interaction of large and small particles; an explanation of independent increments of the limiting motion of the large particles; a discussion of the modeling difference between one large particle and several large particles, suspended in a medium of small particles; a justification of mean-field dynamics. Finally, an infinite system of coupled nonlinear oscillators for the mean-field interaction between large and small particles is defined.

Fig. 3

Fig. 4

To compute the displacement of large particles resulting from the collisions with small particles, it is usually assumed that the large particles are balls with a spatial extension of average diameter $\hat{\varepsilon}_n \ll 1$. Simplifying the transfer of small particles' momenta to the motion of the large particles, we expect the large particles to perform some type of Brownian motion in a scaling limit. A point of contention, within both the mathematical and physics communities, has centered upon the question of whether or not the Brownian motions of several large particles should be spatially *correlated* or *uncorrelated*. The supposition of uncorrelatedness has been the standard for many models. Einstein (1905) assumed uncorrelatedness provided that the large particles were "sufficiently far separated." (Cf. Fig. 3.) For mathematicians, uncorrelatedness is a tempting assumption, since one does not need to specify or justify the choice of a correlation matrix. In contrast, the empirical sciences have known for some time that two large particles immersed in a fluid become attracted to each if their distance is less than some critical parameter. More precisely, it has been shown that the fluid density between two large particles drops when large particles approach each other, i.e., the fluid between the large particles gets "depleted." (Cf. Fig. 4.) Asakura and Oosawa (1954) were probably the first ones to observe this fact. More recent sources are Goetzelmann et al. (1998), Tulpar et al. (2006) and the references therein, as well as Kotelenez et al. (2007). A simple argument to explain *depletion* is that if the large particles get closer together than the diameter of a typical small particle, the space between the large particles must get depleted.[1] Consequently, the *osmotic pressure* around the large particles can no longer be uniform – as long as the overall density of small particles is high enough to allow for a difference in pressure. This implies that, at close distances, large particles have a tendency to attract one another. In particular, they become spatially correlated. It is now clear that the spatial extension of small and large particles imply the existence of a length parameter governing the *correlations* of the Brownian particles. We call this parameter the "correlation length" and denote it by $\sqrt{\varepsilon}$. In particular, depletion implies that two large particles, modeled as Brownian particles, must be correlated at a close distance.

Another derivation of the correlation length, based on the classical notion of the *mean free path*, is suggested by Kotelenez (2002). The advantage of this approach is that correlation length directly depends upon the density of particles in a macroscopic volume and, for a very low density, the motions of large particles are essentially uncorrelated (cf. also the following Remark 1.2).

We obtain, either by referring to the known experiments and empirical observations or to the "mean free path" argument, a correlation length and the exact derivation of $\sqrt{\varepsilon}$ becomes irrelevant for what follows. Cf. also Spohn (1991), Part II, Sect. 7.2, where it is mentioned that random forces cannot be independent because the "suspended particles all float in the *same* fluid."

Remark 1.1. For the case of just one large particle and assuming no interaction (collisions) between the small particles, stochastic approximations to *elastic collisions* have been obtained by numerous authors. Dürr et al. (1981, 1983) obtain an *Ornstein-Uhlenbeck* approximation[2] to the collision dynamics, generalizing a result of Holley (1971) from dimension $d = 1$ to dimension $d = 3$. The mathematical framework, employed by Dürr et al. (loc.cit.), permits the partitioning of the class of small particles into "fast" and "slowly" moving Particles such that "fast" moving particles collide with the large particle only once and "most" particles are moving fast. After the collision they disappear (towards ∞) and new "independent" small particles may collide with the large particle. Sinai and Soloveichik (1986) obtain an *Einstein-Smoluchowski* approximation[3] in dimension $d = 1$ and prove that almost all small particles collide with the large particle only a finite number of times. A similar result was obtained by Szász and Tóth (1986a). Further, Szász and Tóth

[1] Cf. Goetzelmann et al. (loc.cit.).

[2] This means that the limit is represented by an Ornstein-Uhlenbeck process, i.e., it describes the position and velocity of the large particle – cf. Nelson (1972) and also Uhlenbeck and Ornstein (1930).

[3] This means that the limit is a Brownian motion or, more generally, the solution of an ordinary stochastic differential equation only for the position of the large particle – cf. Nelson (loc.cit.).

(1986b) obtain both Einstein-Smoluchowski and Ornstein-Uhlenbeck approximations for the one large particle in dimension $d = 1$.[4] □

As previously mentioned in the introduction, we note that the assumption of single collisions of most small particles with the large particle (as well as our equivalent assumption) should hold for a (rarefied) gas. In such a gas the mean distance between particles is much greater ($\gg$) than the average diameter of a small particle.[5]

From a statistical point of view, the situation may be described as follows: For the case of just one large particle, the fluid around that particle may look *homogeneous* and *isotropic*, leading to a relatively simple statistical description of the displacement of that particle where the displacement is the result of the "bombardment" of this large particle by small particles. Further, whether or not the "medium" of small particles is spatially correlated cannot influence the motion of only one large particle, as long as the medium is homogeneous and, in a scaling limit, the time correlation time δs tends to 0.[6] The resulting mathematical model for the motion of a single particle will be a diffusion, and the spatial homogeneity implies that the diffusion matrix is constant. Such a diffusion is a Brownian motion.

In contrast, if there are at least two large particles and they move closely together, the fluid around each of them will no longer be homogeneous and isotropic. In fact, as mentioned before, the fluid between them will get depleted. (Cf. Fig. 4.) Therefore, the forces generated by the collisions and acting on two different large particles become statistically correlated if the large particles move together closer than the critical length $\sqrt{\varepsilon}$.

Remark 1.2. Kotelenez (2002, Example 1.2) provides a heuristic "coarse graining" argument to support the derivation of a mean-field interaction in the mesoscale from collision dynamics in the microscale. The principal observation is the following: Suppose the mean distance between particles is much greater ($\gg$) than the average diameter of a small particle. Let $\bar{w}$ be the (large) *average speed* of the small particles, and define the *correlation time* by

$$\delta s := \frac{\sqrt{\varepsilon}}{\bar{w}}.$$

Having defined the correlation length $\sqrt{\varepsilon}$ and the *correlation time* δs, one may, in what follows, assume the small particles to be point particles.

To define the *space–time scales*, let $\mathbf{R}^d$ be partitioned into small cubes, which are parallel to the axes. The cubes will be denoted by $(\bar{r}^\lambda]$, where $\bar{r}^\lambda$ is the center of the cube and $\lambda \in \mathbf{N}$. These cubes are open on the left and closed on the right (in the sense of d-dimensional intervals) and have side length $\delta r \approx \frac{1}{n}$, and the origin 0 is the center of a cell. δr is a *mesoscopic length unit.* The cells and their centers will be used to *coarse-grain* the motion of particles, placing the particles within a cell

[4] Cf. also Spohn (loc.cit.).

[5] Cf. Lifshits and Pitaeyevskii (1979), Ch. 1, p. 3.

[6] Cf. the following (1.1).

at the midpoint. Moreover, the small particles in a cell will be grouped as *clusters* starting in the same midpoint, where particles in a cluster have similar velocities.

Suppose that small particles move with different velocities. Fast small particles coming from "far away" can collide with a given large particle at approximately the same time as slow small particles that were close to the large particle before the collision. If, in repeated microscopic time steps, collisions of a given small particle with the same large particle are negligible, then in a mesoscopic time unit $\delta\sigma$, the collision dynamics may be replaced by long-range mean field dynamics (cf. the aforementioned rigorous results of Sinai and Soloveichik, and Szász and Tóth (loc.cit.) for the case of one large particle). Dealing with a wide range of velocities, as in the Maxwellian case, and working with discrete time steps, a long range force is generated. The correlation length $\sqrt{\varepsilon}$ is preserved in this transition. Thus, we obtain the time and spatial scales

$$\left.\begin{array}{c} \delta s \ll \delta\sigma \ll 1 \\ \delta\rho \ll \delta r \approx \dfrac{1}{n} \ll 1. \end{array}\right\}. \tag{1.1}$$

$\delta\rho$ is the average distance between small particles in $(\bar{r}^{\lambda}]$ and the assumption that there are "many" small particles in a typical cell $(\bar{r}^{\lambda}]$ implies $\delta\rho \ll \delta r$. If we assume that the empirical velocity distribution of the small particles is approximately Maxwellian, the aforementioned mean field force from Example 1.2 in Kotelenez (loc.cit.) is given by the following expression:

$$m\bar{G}_{\varepsilon,M}(r-q) \approx m(r-q)\left(\tfrac{2}{\mathrm{d}\varepsilon}\right)^{\frac{1}{2}}\sqrt{D}\eta_n \frac{1}{(\pi\varepsilon)^{\frac{d}{4}}}\bar{\mathrm{e}}^{\frac{-|r-q|^2}{2\varepsilon}}. \tag{1.2}$$

D is a positive diffusion coefficient, m the mass of a cluster of small particles, and η_n is a friction coefficient for the large particles. r and q denote the positions of large and small particles, respectively. □

A rigorous derivation of the replacement of the collision dynamics by mean-field dynamics is desirable. However, we need not "justify" the use of mean-field dynamics as a coarse-grained approximation to collision dynamics: there are mean-field dynamics on a microscopic level that can result from long range potentials, like a Coulomb potential or a (smoothed) Lenard-Jones potential. Therefore, in Chaps. 2 and 3 we work with a fairly general mean-field interaction between large and small particles and the only scales needed will be[7]

$$\delta\sigma = \frac{1}{n^d} \ll 1, \quad \delta r = \frac{1}{n} \ll 1. \tag{1.3}$$

The choice of $\delta\sigma$ follows from the need to control the variance of sums of independent random variables and its generalization in Doob's inequality. With this

[7] We assume that, without loss of generality, the proportionality factors in the relations for δr and $\delta\sigma$ equal 1.

choice, $\delta\sigma$ becomes a normalizing factor at the forces acting on the large particle motion.[8]

Consider the mean-field interaction with forcing kernel $G_\varepsilon(q)$ on a space–time continuum. Suppose there are N large particles and infinitely many small particles. The position of the ith large particle at time t will be denoted $r^i(t)$ and its velocity $v^i(t)$. The corresponding position and velocity of the λth small particle with be denoted $q^\lambda(t)$ and $w^\lambda(t)$, respectively. $\hat{m}$ is the mass of a large particle, and m is the mass of a small particle. The empirical distributions of large and small particles are (formally) given by

$$\mathcal{X}_N(\mathrm{d}r, t) := \hat{m} \sum_{j=1}^{N} \delta_{r^j(t)}(\mathrm{d}r), \qquad \mathcal{Y}(\mathrm{d}q, t) := m \sum_{\lambda,} \delta_{q^\lambda(t)}(\mathrm{d}q).$$

Further, $\eta > 0$ is a friction parameter for the large particles. Then the interaction between small and large particles can be described by the following infinite system of coupled nonlinear oscillators:

$$\left.\begin{aligned}
\frac{\mathrm{d}}{\mathrm{d}t} r^i(t) &= v^i(t), \quad r^i(0) = r_0^i,\\
\frac{\mathrm{d}}{\mathrm{d}t} v^i(t) &= -\eta v^i(t) + \frac{1}{\hat{m}m} \int G_\varepsilon(\eta, r^i(t) - q)\mathcal{Y}(\mathrm{d}q, t), \quad v^i(0) = v_0^i,\\
\frac{\mathrm{d}}{\mathrm{d}t} q^\lambda(t) &= w^\lambda(t), \quad q^\lambda(0) = q_0^\lambda,\\
\frac{\mathrm{d}}{\mathrm{d}t} w^\lambda(t) &= \frac{1}{\hat{m}m} \int G_\varepsilon(\eta, q^\lambda(t) - r)\mathcal{X}_N(\mathrm{d}r, t), \quad w^\lambda(0) = w_0^\lambda.
\end{aligned}\right\} \tag{1.4}$$

In (1.4) and in what follows, the integration domain will be all of $\mathbf{R}^d$, if no integration domain is specified.[9]

We do not claim that the infinite system (1.4) and the empirical distributions of the solutions are well defined. Instead of treating (1.4) on a space–time continuum, we will consider a suitable space–time coarse-grained version of (1.4).[10] Under suitable assumptions,[11] we show that the positions of the large particles in the space–time coarse-grained version converge toward a system of correlated Brownian motions.[12]

[8] Cf. (2.2).

[9] $G_\varepsilon(\eta, r^i(t) - q)$ has the units $\frac{\ell}{T^2}$ (length over time squared).

[10] Cf. (2.8) in the following Chap. 2.

[11] Cf. Hypothesis 2.2 in the next chapter.

[12] This result is based on the author's paper (Kotelenez, 2005a).

Chapter 2
Deterministic Dynamics in a Lattice Model and a Mesoscopic (Stochastic) Limit

The evolution of a space–time discrete version of the Newtonian system (1.4) is analyzed on a fixed (macroscopic) time interval $[0, \hat{t}]$ *(cf. (2.9)). The interaction between large and small particles is governed by a twice continuously differentiable odd* $\mathbf{R}^d$*-valued function* G.[1] *We assume that all partial derivatives up to order 2 are square integrable and that* $|G|^m$ *is integrable for* $1 \le m \le 4$, *where "integrable" refers to the Lebesgue measure on* $\mathbf{R}^d$. *The function* G *will be approximated by odd* $\mathbf{R}^d$*-valued functions* G_n *with bounded supports (cf. (2.1)). Existence of the space–time discrete version of (1.4) is derived employing coarse graining in space and an Euler scheme in time. The mesoscopic limit (2.11) is a system stochastic ordinary differential equation (SODEs) for the positions of the large particles. The SODEs are driven by Gaussian standard space–time white noise that may be interpreted as a limiting centered number density of the small particles. The proof of the mesoscopic limit theorem (Theorem 2.4) is provided in Chap. 3.*

Hypothesis 2.1 – *Coarse Graining*

- Both single large particles and clusters of small particles, being in a cell $(\bar{r}^\lambda]$,[2] are moved to the midpoint $\bar{r}^\lambda$.
- There is a partitioning of the velocity space

$$\mathbf{R}^d = \cup_{\iota \in \mathbf{N}} B_\iota,$$

and the velocities of each cluster take values in exactly one B_ι where, for the sake of simplicity, we assume that all B_ι are small cubic d-dimensional intervals (left open, right closed), all with the same volume $\le \frac{1}{n^d}$. □

[1] With the exception of Chaps. 5 and 14, we suppress the possible dependence on the correlation length $\sqrt{\varepsilon}$.

[2] Recall from Chap. 1 that $\mathbf{R}^d$ is partitioned into small cubes, $(\bar{r}^\lambda]$, which are parallel to the axes with center $\bar{r}^\lambda$. These cubes are open on the left and closed on the right (in the sense of d-dimensional intervals) and have side length $\delta r = \frac{1}{n}$. n will be the scaling parameter.

Let $\hat{m}$ denote the mass of a large particle and m denote the mass of a cluster of small particles. Set

$$\mathcal{Y}_n(\mathrm{d}q,t) := m\sum_{\lambda,\iota}\delta_{\bar{q}_n(t,\lambda,\iota)}(\mathrm{d}q), \qquad \mathcal{X}_{N,n}(\mathrm{d}r,t) := \hat{m}\sum_{j=1}^{N}\delta_{\bar{r}_n^j(t)}(\mathrm{d}r),$$

where $\bar{r}_n^j(t)$ and $\bar{q}_n(t,\lambda,\iota)$ are the positions at time t of the large and small particles, respectively. "–" means that the midpoints of those cells are taken, where the particles are at time t. E.g., $\bar{r}_n^j(t) = \bar{r}^{\tilde{\lambda}}$ if $r_n^j(t) \in (\bar{r}^{\tilde{\lambda}}]$. $\mathcal{Y}_n$ and $\mathcal{X}_{N,n}$ are called the "empirical measure processes" of the small and large particles, respectively. The labels λ, ι in the empirical distribution $\mathcal{Y}_n$ denote the (cluster of) small particle(s) that started at $t=0$ in $(\bar{r}^{\lambda}]$ with velocities from B_ι.

Let "$\vee$" denote "max." The average speed of the small particles will be denoted $\bar{w}_n$ and the friction parameter of the large particles η_n. The assumptions on the most important parameters are listed in the following

Hypothesis 2.2

$$\eta_n = n^{\tilde{p}}, \quad d > \tilde{p} > 0,$$
$$m = n^{-\zeta}, \quad \zeta \ge 0,$$
$$\bar{w}_n = n^{p}, \quad p > (4d+2)\vee(2\tilde{p}+2\zeta+2d+2). \qquad \square$$

Let $K_n \ge 1$ be a sequence such that $K_n \uparrow \infty$ and $C_b(0)$ be the closed cube in $\mathbf{R}^d$, parallel to the axes, centered at 0 and with side length $b > 0$. Set

$$G_n(q) := \begin{cases} n^d \displaystyle\int_{(\bar{r}^{\lambda}]} G(r)\mathrm{d}r, & \text{if} \quad q \in (\bar{r}^{\lambda}] \text{ and } (\bar{r}^{\lambda}] \subset C_{K_n}(0), \\ 0, & \text{if} \quad q \in (\bar{r}^{\lambda}] \text{ and } (\bar{r}^{\lambda}] \text{ is not a subset of } C_{K_n}(0). \end{cases} \tag{2.1}$$

$|C|$ denotes the Lebesgue measure of a Borel measurable subset C of $\mathbf{R}^k$, $k \in \{d, d+1\}$. Further, $|r|$ denotes the Euclidean norm of $r \in \mathbf{R}^d$ as well as the distance in $\mathbf{R}$. For a vector-valued function F, F_ℓ is its ℓth component and we define the sup norms by

$$|||F_\ell||| := \sup_q |F_\ell(q)|, \quad \ell = 1,\ldots,d, \qquad |||F||| := \max_{\ell=1,..,d} |||F_\ell|||.$$

Let $\wedge$ denote "minimum" and $m \in \{1,\ldots,4\}$. For $\ell = 1,\ldots,d$, by a simple estimate and Hölder's inequality (if $m \ge 2$)

$$\left.\begin{aligned} \sum_{\lambda}|G_{n,\ell}(r-\bar{r}^{\lambda})|^m &= n^d\int |G_{n,\ell}(r-q)|^m\mathrm{d}q \\ &\le (n^d K_n^d |||G_\ell|||^m) \wedge \left(n^d\int |G_\ell(q)|^m\mathrm{d}q\right). \end{aligned}\right\} \tag{2.2}$$

Let $\bar{r}$ the midpoint of an arbitrary cell. Similar to (2.2)

$$\sum_{\lambda} 1_{\{|G_n(\bar{r}-\bar{r}^{\lambda})|>0\}} \le K_n^d n^d. \tag{2.3}$$

Using the oddness of G we obtain

$$\sum_{\lambda} G_n(\bar{r} - \bar{r}^{\lambda}) = 0. \tag{2.4}$$

All time-dependent functions will be constant for $t \in [k\delta\sigma, (k+1)\delta\sigma)$. For notational convenience we use $t, s, u \in \{l\delta\sigma : l \in \mathbf{N}, l\delta\sigma \le \hat{t}\}$, etc. for time, if it does not lead to confusion, and if $t = l\delta\sigma$, then $t- := (l-1)\delta\sigma$ and $t+ := (l+1)\delta\sigma$. We will also, when needed, interpret the time-discrete evolution of positions and velocities as jump processes in continuous time by extending all time-discrete quantities to continuous time. In this extension all functions become cadlag (continuous from the right with limits from the left).[3] More precisely, let $\mathbf{B}$ be some topological space. The space of $\mathbf{B}$-valued cadlag functions with domain $[0, \hat{t}]$ is denoted $D([0, \hat{t}]; \mathbf{B})$.[4] For our time discrete functions or processes $f(\cdot)$ this extension is defined as follows:

$$\bar{f}(t) := f(l\,\delta\sigma), \ \text{if}\ t \in [l\,\delta\sigma, (l+1)\delta\sigma)\,.$$

Since this extension is trivial, we drop the "bar" and use the old notation both for the extended and nonextended processes and functions.

The velocity of the ith large particle at time s, will be denoted $v_n^i(s)$, where $v_n^i(0) = 0\ \forall i$. $w_{0,n}^{\lambda,\iota} \in B_\iota$ will be the initial velocity of the small particle starting at time 0 in the cell $(\bar{r}^{\lambda}]$, $\iota \in \mathbf{N}$. Note that, for the infinitely many small particles, the resulting friction due to the collision with the finitely many large particles should be negligible. Further, in a dynamical Ornstein-Uhlenbeck type model with friction η_n the "fluctuation force" must be governed by a function $\tilde{G}_n(\cdot)$. The relation to a Einstein-Smoluchowski diffusion is given by

$$\tilde{G}_n(r) \approx \eta_n G_n(r),$$

as $\eta_n \longrightarrow \infty$.[5] This factor will disappear as we move from a second-order differential equation to a first-order equation (cf. (2.9) and (3.5)). To simplify the calculations, we will work right from the start with $\eta_n G_n(\cdot)$.

We identify the clusters in the small cells with velocity from B_ι with random variables. The empirical distributions of particles and velocities in cells define, in a

---

[3] From French (as a result of the French contribution to the theory of stochastic integration): f est "*continue á droite et admet une limite á gauche*."

[4] Cf. Sect. 15.1.6 for more details. $D([0, \hat{t}]; \mathbf{B})$ is called the "*Skorokhod space of* **B**-*valued cadlag functions*." If the cadlag functions are defined on $[0, \infty)$ we denote the Skorokhod space by $D([0, \infty); \mathbf{B})$.

[5] Cf., e.g., Nelson (1972) or Kotelenez and Wang (1994).

canonical way, probability distributions. The absence of interaction of the small material particles among themselves leads to the assumption that their initial positions and velocities are independent if modelled as random variables.

Let $\alpha \in (0,1)$ be the expected average volume (in a unit cube) occupied by small particles (for large n) and assume that the initial "density" function $\varphi_n(q)$ for the small particles satisfies:

$$0 \le \varphi_n(q) \le \alpha n^d.$$

We need a state to describe the outcome of finding no particle in the cell (the "empty state"). Let $\diamond$ denote this empty state and set $\hat{\mathbf{R}}^d := \mathbf{R}^d \cup \{\diamond\}$. A suitable metric can be defined on $\hat{\mathbf{R}}^d$ as follows:

$$\left.\begin{aligned} &\rho(r-q) := |r-q| \wedge 1 \ \text{ if } r,q \in \mathbf{R}^d\,; \\ &\hat{\rho}(r-q) := \begin{cases} \rho(r-q), & \text{if} \quad r,q \in \mathbf{R}^d, \\ 1, & \text{if} \quad r \in \mathbf{R}^d \text{ and if } q = \diamond. \end{cases} \end{aligned}\right\} \tag{2.5}$$

The Borel sets in $\mathbf{R}^d$ and in $\hat{\mathbf{R}}^d$ will be denoted by $\mathcal{B}^d$ and $\hat{\mathcal{B}}^d$, respectively. Set

$$\Omega := \{\hat{\mathbf{R}}^{\mathbf{d}} \times \mathbf{R}^{\mathbf{d}}\}^{\mathbf{N}}.$$

The velocity field of the small particles is governed by a strictly positive probability density $\psi(\cdot)$ on $\mathbf{R}^d$, which we rescale as follows:

$$\psi_n(w) := \frac{1}{n^{pd}}\psi\left(\frac{w}{n^p}\right).$$

Define the initial velocities of a cluster of small particles starting in the cell $(\bar{r}^\lambda]$ as random variables with support in some B_ι:

$$w_{0,n}^{\lambda,\iota} \sim \psi_n(w) 1_{B_\iota}(w) \frac{1}{\int_{B_\iota} \psi_n(w)\,\mathrm{d}w}.^{6} \tag{2.6}$$

Further, let

$$\mu_{\lambda,\iota,n} := \int_{(\bar{r}^\lambda]} \varphi_n(q)\,\mathrm{d}q \int_{B_\iota} \psi_n(w)\,\mathrm{d}w$$

be the probability of finding a small particle at $t=0$ in $(\bar{r}^\lambda]$ and with velocity from B_ι. Define random variables $\hat{\zeta}_n^{\lambda,\iota}$ for the initial positions of the small particles as follows:

$$\hat{\zeta}_n^{\lambda,\iota} := \begin{cases} \bar{r}^\lambda, & \text{with probability (w.p.)} \quad \mu_{\lambda,\iota,n}, \\ \diamond, & \text{w.p.} \quad 1-\mu_{\lambda,\iota,n}. \end{cases} \tag{2.7}$$

Denote by $\hat{\mu}_{n,\lambda,\iota}$ and $\nu_{n,\lambda,\iota}$ the distributions of $\hat{\zeta}_n^{\lambda,\iota}$ and $w_{0,n}^{\lambda,\iota}$, respectively. Set

$$\omega_{\lambda,\iota} := \hat{q}^{\lambda,\iota} \times w^{\lambda,\iota} \in \hat{\mathbf{R}}^d \times \mathbf{R}^d$$

6 "$\sim$" here denotes "is distributed."

and

$$P_{n,\lambda,\iota} := \hat{\mu}_{n,\lambda,\iota} \otimes \nu_{n,\lambda,\iota}.$$

We assume the initial positions and velocities to be independent, i.e., we define the initial joint probability distribution of positions and velocities on Ω to be the product measure:

$$P_n := \otimes_{\lambda \in \mathbf{N}} \otimes_{\iota \in \mathbf{N}} P_{n,\lambda,\iota}. \tag{2.8}$$

Formally, the coarse-grained particle evolution in the mesoscale is described by the following Euler scheme:

$$\left.\begin{aligned} r_n^i(t) &= r_n^i(0) + \sum_{s \le t} v_n^i(s)\delta\sigma, \\ v_n^i(s) &= \sum_{0<u\le s} \exp[-\eta_n(s-u)]\frac{1}{\hat{m}m}\int \eta_n G_n(\bar{r}_n^i(u-)-q)\mathcal{Y}_n(dq,u)\,\delta\sigma, \\ q_n(s,\lambda,\iota) &= \bar{r}^\lambda + w_{0,n}^{\lambda,\iota}s + \frac{1}{\hat{m}m}\sum_{u<s}\sum_{v\le u}\int \eta_n G_n(\bar{q}_n(v,\lambda,\iota)-r)\mathcal{X}_{N,n}(dr,v)(\delta\sigma)^2 \\ &\qquad \text{if } \hat{\zeta}_n^{\lambda,\iota} = \bar{r}^\lambda . \end{aligned}\right\} \tag{2.9}$$

Note that the empirical measure processes of the small particles is not a priori finite on bounded sets. In particular, we do not know whether or not infinitely many small particles interact with a given large particle at a given time u. Therefore, we must show existence of the coarse-grained particle model:

Proposition 2.1. *The Euler scheme (2.9) is defined for all $s \ge 0$ a.s.*

The Proof will be provided at the end of this chapter.

We list the remaining hypotheses:

Hypothesis 2.3

Set $\alpha_{n,\lambda} := \int_{(\bar{r}^\lambda]} \varphi_n(q)\,dq$, $\bar{\alpha}_n := \sup_\lambda \alpha_{n,\lambda}$, $\underline{\alpha}_n := \inf_\lambda \alpha_{n,\lambda}$. Let $\delta \in (0, \frac{1}{7})$ and assume

$$n^{d+\delta}(|\alpha - \bar{\alpha}_n| + |\alpha - \underline{\alpha}_n|) \longrightarrow 0. \quad \text{as } n \longrightarrow \infty.$$

□

The speed of convergence assumption will be needed in Lemmas 3.2 and 3.3. The particular choice of δ allows for a simple formulation of the ratios in Hypothesis 2.2 (cf. also Remark 2.7).

Hypothesis 2.4

$\{r_n^1(0),\ldots,r_n^N(0)\}$ and $\{\hat{\zeta}_n^{\lambda,\iota}, w_{n,0}^{\lambda,\iota} : \lambda,\iota \in \mathbf{N}\}$ are independent, where $(r_n^1(0),\ldots, r_n^N(0))$ are the initial positions of the large particles. □

We now describe the possible scaling limit, as $n \to \infty$. An important component of this limit is standard Gaussian space–time white noise $w(dq, dt)$.[7] Suppose the Borel subset A of $\mathbf{R}^d$ has finite Lebesgue measure $|A|$. Denote by $\mathcal{N}(0, |A|)$ the normal distribution with mean 0 and variance $|A|$. Following Walsh (1986), Chap. 1, we have the following.

Definition 2.2. – *Standard Gaussian Space–Time White Noise*
A standard Gaussian Space–Time white noise $w(dq, dt, \omega)$ based on the Lebesgue measure in $\mathbf{R}^d \times \mathbf{R}_+$ is a finitely additive random signed measure on the sets Borel sets A in $\mathbf{R}^d \times \mathbf{R}_+$ of finite Lebesgue measure $|A|$ such that

- $\int_0^\infty \int 1_A(q,t) w(dq, dt, \cdot) \sim \mathcal{N}(0, |A|)$;
- if $A \cap B = \emptyset$, then $\int_0^\infty \int 1_A(q,t) w(dq, dt, \cdot)$ and $\int_0^\infty \int 1_B(q,t) w(dq, dt, \cdot)$ are independent. □

Remark 2.3.

- Walsh's theory of stochastic integration with respect to $w(dq, dt)$ and, more generally, with respect to martingale measures is based on Itô's approach.[8] We show in Chap. 4 ((4.14) and (4.15)) that stochastic integrals in finite dimensional space, driven by $w(dq, dt)$ are equivalent to sums of stochastic (Itô) integrals, driven by infinitely many i.i.d. Brownian motions.
- A description of $w(dq, dt)$ as a generalized Gaussian space–time field is provided in Sects. 15.2.2 and 15.2.4. Of special importance are (15.69) and (15.126). In fact, we show in (15.126) that the finitely additive signed measure $w(dq, dt, \omega)$ has a generalized (Radon-Nikodym) derivative, $\frac{\partial^{d+1}}{\partial s \partial q_1 \ldots \partial q_d} \bar{\hat{w}}(\cdot, \cdot, \omega)$, which is a Schwarz distribution[9] over $\mathbf{R}^{d+1}$, i.e., in the space of Schwarz distributions over $\mathbf{R}^{d+1}$
$$\frac{\partial^{d+1}}{\partial s \partial q_1 \ldots \partial q_d} \bar{\hat{w}}(q, s) dq\, ds \sim w(dq, ds). \tag{2.10}$$
- A modeling interpretation of $w(dq, dt)$ in terms of the (centered) occupation measure (cf. (2.16) (3.15)) is provided in Remark 3.10 at the end of Chap. 3. It

[7] $w(dq, ds)$ is the space–time generalization of the time increments of a scalar valued standard Brownian motion $\beta(ds)$. If restricted to only positive coordinates $r_k \geq 0,\ k = 1, \ldots, d$ the integrated version, $\int_0^t \int_0^{r_1} \ldots \int_0^{r_d} w(dq, dt)$, is called the Brownian sheet. Cf. also (15.124).

[8] The most important features of Itô integration are presented in Sect. 15.2.5.

[9] Here "$\sim$" means "equivalent in distribution" and $\bar{\hat{w}}(\cdot, \cdot)$ is a suitably defined Brownian sheet on $\mathbf{R}^d \times [0, \infty)$. We refer to our analysis of $w(dq, dt)$ in Sect. 15.2.4, in particular, (15.126). Our space–time white noise $w(dq, dt)$ and $\bar{\hat{w}}(\cdot, \cdot)$ can be identified as $\mathcal{S}'$-valued random fields, i.e., as a random Schwarz distributions. If we assume that our space–time white noise was defined as integration with respect to the Brownian sheet $\bar{\hat{w}}(\cdot, \cdot)$, (2.10) can be strengthened to $\frac{\partial^{d+1}}{\partial s \partial q_1 \ldots \partial q_d} \bar{\hat{w}}(q, s, \omega) dq ds \equiv w(dq, ds, \omega)$ a.s. and we can drop the "$\bar{\hat{}}$" over w. In the terminology of Gel'fand and Vilenkin (1964), Chap. III.1.3, $\frac{\partial^{d+1}}{\partial s \partial q_1 \ldots \partial q_d} \bar{\hat{w}}(\cdot, \cdot)$ may be called a "unit random field." We refer to our Sect. 15.1.3 for a presentation of Schwarz distributions and some of their properties.

follows, in particular, that $\frac{\partial^{d+1}}{\partial s \partial q_1 \ldots \partial q_d} \bar{\hat{w}}(\cdot, \cdot)$ may be interpreted as an approximation to the centered "number density" of the small particles in the mesoscopic scaling limit. □

Hypothesis 2.5

- Suppose there are $\mathbf{R}^d$-valued square integrable random variables $(r^1(0), \ldots, r^N(0))$ such that
$$\left(r_n^1(0), \ldots, r_n^N(0)\right) \Longrightarrow \left(r^1(0), \ldots, r^N(0)\right), \quad \text{as } n \longrightarrow \infty,$$
where "$\Longrightarrow$" denotes weak convergence.[10]
- $\sup\limits_n \sum\limits_{i=1}^{N} E|r_n^i(0)| < \infty$.
- There is a finite constant $c > 0$ such that the function G satisfies the following Lipschitz condition:
$$\sum_{j=1,\ldots,d} \int (G_j(r_1 - q) - G_j(r_2 - q))^2 \mathrm{d}q \le c\rho(r_1 - r_2)$$
for all r_1 and r_2 in $\mathbf{R}^d$, where G_j is the jth component of G.
- There is a scalar standard Gaussian white noise, $w(\mathrm{d}q, \mathrm{d}s)$, on $\mathbf{R}^d \times \mathbf{R}_+$, defined on the same probability space as $(r^1(0), \ldots, r^N(0))$ such that $(r^1(0), \ldots, r^N(0))$ and $w(\mathrm{d}q, \mathrm{d}s)$ are independent. □

Consider the stochastic integral equations:

$$r^i(t) = r^i(0) + \sqrt{\alpha} \int_0^t \int G(r^i(u) - q) w(\mathrm{d}q, \mathrm{d}u), \quad i = 1, \ldots, N. \tag{2.11}$$

By the assumptions on G, in addition to Hypothesis 2.5, (2.11) has a unique solution and is a Markov process in $\mathbf{R}^{Nd}$.[11] Moreover, we show in Chap. 5 that the solutions of (2.11) are correlated Brownian motions. The following mesoscopic limit theorem is the main result of Part I. It establishes the Einstein-Smoluchowski model as an approximation to the evolution of the positions of the large particles.

Theorem 2.4. – *Mesoscopic Limit Theorem*

Under Hypotheses 2.1–2.5

$$(r_n^1(\cdot), \ldots, r_n^N(\cdot)) \Longrightarrow (r^1(\cdot), \ldots r^N(\cdot))$$
$$\text{in } D([0, \hat{t}]; \mathbf{R}^{\mathrm{d}N}), \text{ as } n \longrightarrow \infty,$$

where $(r^1(\cdot), \ldots, r^N(\cdot))$ are the unique solutions of (2.11) *and* $(r_n^1(\cdot), \ldots, r_n^N(\cdot))$ *are the solutions of the Euler scheme (2.9).*

[10] For the definition of weak convergence we refer to Sect. 15.2.1, (15.56).

[11] Cf. Kotelenez (1995b) and Theorems 4.5 and 4.7 in the following Chap. 4.

The Proof will be provided in Chap. 3. □

Remark 2.5. Formally,[12] we may replace the stochastic integrator, $w(\mathrm{d}q, \mathrm{d}s)$, in (2.11) by integrating with respect to the limiting centered number density, i.e.,

$$r^i(t) = r^i(0) + \sqrt{\alpha}\int_0^t \int G(r^i(u) - q)\frac{\partial^{d+1}}{\partial u \partial q_1 \dots \partial q_d} w(q, u, \omega)\mathrm{d}q\mathrm{d}u, i = 1, \dots, N. \tag{2.12}$$

Hence, we could say that the displacement of the large particles in the mesoscopic scaling limit is driven by the limiting centered number density of the small particles. Of course, it is not clear whether the right-hand side in (2.12) is well-defined since $r^i(u)$ depends on the events, generated by the increments of $w(\mathrm{d}q, \mathrm{d}s)$ until time u.[13] Under an additional assumption on G[14], we show in Proposition 15.60 that the Itô and Stratonovich integrals coincide for the right-hand side of (2.11). Although this is encouraging, it does not suffice to give a rigorous meaning to (2.12), unless the integrand were independent of $w(\mathrm{d}q, \mathrm{d}s)$. In such a case, e.g., if the integrand were deterministic, the above representation could become rigorous as a Stieltjes integral (cf. Sect. 15.1.5). □

We state the obvious changes in the model and the limit theorem for the case of a system without friction. (2.9) must be replaced by

$$\left.\begin{aligned} r_n^i(t) &= r_n^i(0) + \sum_{s \le t} v_n^i(s)\delta\sigma, \\ v_n^i(s) &= \sum_{0 < u \le s} \frac{1}{\hat{m}m} \int G_n(\bar{r}_n^i(u-) - q)\mathcal{Y}_n(\mathrm{d}q, u)\delta\sigma, \\ q_n(s, \lambda, \iota) &= \bar{r}^\lambda + w_{0,n}^{\lambda,\iota} s + \frac{1}{\hat{m}m}\sum_{u<s}\sum_{v \le u} \int G_n(\bar{q}_n(v, \lambda, \iota) - r)\mathcal{X}_{N,n}(\mathrm{d}r, v)(\delta\sigma)^2, \end{aligned}\right\} \tag{2.13}$$

The limiting equation becomes

$$\left.\begin{aligned} r^i(t) &= r^i(0) + \int_0^t v^i(u)\mathrm{d}u, \\ v^i(t) &= \sqrt{\alpha}\int_0^t \int G(r^i(u) - q)w(\mathrm{d}q, \mathrm{d}u), \\ & i = 1, \dots, N. \end{aligned}\right\} \tag{2.14}$$

Finally, we replace Hypothesis 2.2 by

[12] Cf. (15.126).

[13] Cf. Sects. 15.2.5. and 15.2.6 for a discussion of Itô and Stratonovich integrals.

[14] Cf. (15.195).

Hypothesis 2.6

$$m = n^{-\zeta}, \quad \zeta \geq 0,$$
$$\bar{w}_n = n^p, \quad p > (4d + 2) \vee (2\zeta + 2d + 2). \qquad \square$$

We then obtain

Theorem 2.6. *Under Hypotheses 2.1, 2.3–2.5, and 2.6*

$$(r_n^1(\cdot), v_n^1(\cdot), \ldots, r_n^N(\cdot), v_n^N(\cdot)) \Longrightarrow (r^1(\cdot), v^1(\cdot), \ldots, r^N(\cdot), v^N(\cdot))$$
$$\text{in } D([0, \hat{t}]; \mathbf{R}^{2dN}), \text{ as } n \longrightarrow \infty,$$

where $(r^1(\cdot), v^1(\cdot), \ldots, r^N(\cdot), v^N(\cdot))$ *are the unique solutions of* (2.14) *and* $(r_n^1(\cdot), v_n^1(\cdot), \ldots, r_n^N(\cdot), v_n^N(\cdot))$ *are the solutions of the Euler scheme (2.13).*

The proof is somewhat simpler than the proof of Theorem 2.4 and can be obtained by doing the appropriate changes in the proof of Theorem 2.4. □

Remark 2.7.

- Hypotheses 2.4 and 2.5 are self-explanatory and do not require any further comments.
- Hypothesis 2.3 is trivially satisfied for $\varphi \equiv \alpha n^d$. The following example appears often in equilibrium statistical mechanics as one factor of a Gibbs distribution if U is the potential for the interaction force. It also satisfies Hypothesis 2.3.
  Let $U \geq 0$ be smooth with bounded support and set
  $$\varphi_n(q) := \alpha n^d \exp\left(\frac{-1}{n^{2p}\hat{m}} \int U(q - r)\mathcal{X}_{N,n}(\mathrm{d}r)\right) 1_{(\bar{r}^\lambda]}(q),$$
  where $n^p = \bar{w}_n$ is the average speed (cf. Hypothesis 2.2).
- We can also derive an Ornstein-Uhlenbeck model, i.e., a version of (2.14) with a fixed friction coefficient, by making suitable adjustments in the model and Hypothesis 2.2. □

Proof of Proposition 2.1

Suppose that up to time $t-$ we have determined the particle evolution according to (2.9) and that for $u \leq t-$ the empirical measure process $\mathcal{Y}(\mathrm{d}q, u)$ is a.s. finite on bounded sets. Assume, without loss of generality, $t- < 1$. It follows that $\max_{0 \leq u \leq t-} \max_{i=1,\ldots,N} |r_n^i(u)| \leq \tilde{K}_{n,t-}$ a.s., where $\tilde{K}_{n,t-}$ is some finite positive number (which depends on ω). The assumption is true, of course, for $t- = 0$.

Since by induction assumption the empirical distributions $\mathcal{X}_N(\mathrm{d}r, u)$, $u \leq t-$, are a.s defined, we can define the positions of the small particles at t by

$$q_n(t,\lambda,\iota) := \begin{cases} \bar{r}^\lambda + w_{0,n}^{\lambda,\iota} t \\ + \frac{1}{\hat{m}m} \sum_{u \le t-} \sum_{v \le u} \int \eta_n G_n(\bar{q}_n(v,\lambda,\iota) - r) \mathcal{X}_N(\mathrm{d}r, v)(\delta\sigma)^2, \\ \quad \text{if } \hat{\zeta}_n^{\lambda,\iota} = \bar{r}^\lambda ; \\ \diamond \ , \text{ if } \hat{\zeta}_n^{\lambda,\iota} = \diamond. \end{cases} \tag{2.15}$$

The velocity field $v_n^i(t)$ can be defined if $\mathcal{Y}(\mathrm{d}q,t)$ is a.s. finite on $C_{K_n}(\bar{r}_n^i(t-))$[15] for $i = 1,\ldots,n$. The analysis of this problem simplifies if we restrict $\mathcal{Y}(\mathrm{d}q,t)$ to cells and then "paste" the results together. We define the "occupation measure"[16] associated with (2.15) as follows:

$$I_A^n(t) := \frac{1}{m}\int_A \mathcal{Y}_n(\mathrm{d}q,t) = \sum_{\lambda,\iota} 1_A(\bar{q}_n(t,\lambda,\iota)). \tag{2.16}$$

We partition $\mathbf{N}\times\mathbf{N}$, the set of indices for the positions and velocities, as follows: Choose an integer $\bar{K}_{n,t} > \tilde{K}_{n,t-}$ (a more precise relation will be determined in (2.19)). We decompose the velocity indices:

$$\mathbf{J}_{n,t} := \{\iota : B_\iota \subset C_{\bar{K}_{n,t}}(0)\}, \ \ \mathbf{J}_{n,t}^\perp := \mathbf{N}\setminus \mathbf{J}_{n,t}.$$

We then choose a positive integer $\hat{K}_{n,t} > \bar{K}_{n,t} + \tilde{K}_{n,t-} + K_n$, and we decompose the position indices:

$$\mathbf{L}_{n,t} := \{\lambda : (\bar{r}^\lambda] \subset C_{\hat{K}_{n,t}}(0)\}, \ \ \mathbf{L}_{n,t}^\perp := \mathbf{N}\setminus \mathbf{L}_{n,t}.$$

We may, without loss of generality, assume that for $s \le t-$ the sequences of constants $\{\tilde{K}_{n,s}\}$, $\{\bar{K}_{n,s}\}$ and $\{\hat{K}_{n,s}\}$ have been defined and are monotone increasing in s. Altogether we partition the pairs of indices into 3 sets:

$$\mathbf{N}\times\mathbf{N} = (\mathbf{N}\times\mathbf{J}_{\mathbf{n,t}}^\perp) \cup (\mathbf{L}_{\mathbf{n,t}}\times\mathbf{J}_{\mathbf{n,t}}) \cup (\mathbf{L}_{\mathbf{n,t}}^\perp\times\mathbf{J}_{\mathbf{n,t}}). \tag{2.17}$$

The first set contains labels for all possible positions and all "large" velocities. The second set is finite and the third one is chosen in such a way that small particles starting in those positions with labels for the "small" velocities from $\mathbf{J}_{n,t}$ cannot reach the cube with side length $\tilde{K}_{n,t-} + K_n$ during a time of length 1. Hence, those particles cannot interact with the large particles during that time interval (since the large particles are at time points $u \le t-$ in $C_{\tilde{K}_{n,t-}}$). Next we decompose the occupation measure $I_A^n(t)$ into the sum of three different measures, restricting the summation for each measure to the sets of indices from the decomposition (2.17):

$$I_A^n(t) = I_{A,\mathbf{N}\times\mathbf{J}_{\mathbf{n,t}}^\perp}^n(t) + I_{A,\mathbf{L}_{n,t}\times\mathbf{J}_{n,t}}^n(t) + I_{A,\mathbf{L}_{n,t}^\perp\times\mathbf{J}_{n,t}}^n(t), \tag{2.18}$$

where $A := (\bar{r}^\lambda]$ for some λ is a subset of the cube with side length $\tilde{K}_{n,t-} + K_n$. By the choice of the indices, $I_{A,\mathbf{L}_{n,t}^\perp\times\mathbf{J}_{n,t}}^n(t) = 0$ and $I_{A,\mathbf{L}_{n,t}\times\mathbf{J}_{n,t}}^n(t)$ is finite.

---

[15] K_n is defined in (2.1).

[16] $\frac{1}{|A|} I_A^n(t)$ is the "number density."

Therefore, all we must show is that $I^n_{A,\mathbf{N}\times\mathbf{J}^{\perp}_{\mathbf{n},\mathbf{t}}}(t) < \infty$ a.s. for all $A \subset C_{\tilde{K}_{n,t-}+K_n}(0)$. Since $\tilde{K}_{n,t-} < \infty$ a.s., we may, again without loss of generality, assume $1 \le \tilde{K}_n := \tilde{K}_{n,t-} < \infty$ is deterministic (partitioning Ω into sets where the bounds can be chosen deterministic). Hence, we may also assume that $\bar{K}_n := \bar{K}_{n,t} > \tilde{K}_n$ and $\mathbf{J}^{\perp}_n := \mathbf{J}^{\perp}_{n,t}$ are deterministic. Define

$$\left.\begin{aligned} \bar{c} &:= (|||G|||N+2)(\hat{t}+1), \\ c_n &:= \bar{c}K_n^d, \\ \tilde{c}_n &:= c_n(\tilde{K}_n \vee n^{\tilde{p}+\zeta}), \\ \bar{K}_n &:= 4\tilde{c}_n n^d, \end{aligned}\right\} \tag{2.19}$$

where we may assume that $\forall \iota$ either $B_\iota \subset C_{\bar{K}_n}(0)$ or $B_\iota \cap C_{\bar{K}_n}(0)$ has Lebesgue measure 0. The choice of ratios between $\bar{K}_n$, $\tilde{K}_n$, and K_n is motivated by the estimates in Lemmas 3.1 and 3.3 of the next section. Let $\hat{n}_1$ be an integer such that $\hat{n}_1 \ge 3$. To complete the proof of Proposition 2.1 we derive the following Lemmas 2.8 and 2.10 in addition to Corollary 2.9.

Lemma 2.8. *Let $u, s \le t$. Suppose $\exists \tilde{\omega} \in \Omega$, $\iota \in \mathbf{J}^{\perp}_n$ and u such that $q_n(u, \lambda, \iota, \tilde{\omega}) \in C_{\tilde{c}_n}(0)$. Then $\forall s \ne u$, $\forall \omega \in \Omega$ and $\forall n \ge \hat{n}_1$*

$$q_n(s, \lambda, \iota, \omega) \notin C_{2\tilde{c}_n}(0).$$

Proof. (i) Suppose, without loss of generality, $\hat{\zeta}_n^{\lambda,\iota} = \bar{r}^{\lambda}$ in (2.15) and abbreviate:

$$H_n(t, \lambda, \iota) := n^{\tilde{p}+\zeta} \frac{1}{\hat{m}} \sum_{s<t} \sum_{u \le s} \int G_n(\bar{q}_n(u, \lambda, \iota) - r)\mathcal{X}_N(\mathrm{d}r, u)(\delta\sigma)^2.$$

Using the abbreviations from (2.19), it follows that for $u \ne s$

$$|H_n(s, \lambda, \iota, \omega) - H_n(u, \lambda, \iota, \omega)| \le |s-u|\bar{c}n^{\tilde{p}+\zeta} \tag{2.20}$$

and, consequently,

$$|q_n(s, \lambda, \iota, \omega) - q_n(u, \lambda, \iota, \omega)| \ge |s-u| \left[\left| w_{0,n}^{\lambda,\iota}(\omega) \right| - \bar{c}n^{\tilde{p}+\zeta} \right] > 3\tilde{c}_n. \tag{2.21}$$

Thus, for $n \ge \hat{n}_1$ and $\forall \omega \in \Omega$

$$q_n(s, \lambda, \iota, \omega) \in C_{2\tilde{c}_n}(0) \text{ for at most one } s \le t.$$

Since $C_{\tilde{c}_n}(0) \supset C_{\tilde{K}_n+K_n}(0)$, we obtain for $n \ge \hat{n}_1$, $\iota \in \mathbf{J}^{\perp}_n$ and $\omega \in \Omega$

$$\max_{s \le t} |H_n(s, \lambda, \iota, \omega)| \le \bar{c}n^{\tilde{p}+\zeta}\delta\sigma. \tag{2.22}$$

(ii) Let $\omega, \tilde{\omega} \in \Omega$ and $u \ne s$. We have

$$\begin{aligned} &|q_n(s, \lambda, \iota, \omega) - q_n(u, \lambda, \iota, \tilde{\omega})| \\ &\ge |q_n(s, \lambda, \iota, \omega) - q_n(u, \lambda, \iota, \omega)| - |q_n(u, \lambda, \iota, \omega) - q_n(u, \lambda, \iota, \tilde{\omega})| \\ &= I - II. \end{aligned}$$

By (2.21) $I > 3\tilde{c}_n$. Further, by Hypothesis 2.1, (2.22) and by the fact that, for $n \geq \hat{n}_1$, $\delta\sigma = \frac{1}{n^d} \leq \frac{1}{9}$ we conclude that

$$II \leq 2\bar{c}n^{\tilde{p}+\zeta}n^{-d} + \frac{2}{n} \leq \tilde{c}_n.$$

Therefore,

$$|q_n(s, \lambda, \imath, \omega) - q_n(u, \lambda, \imath, \tilde{\omega})| > 2\tilde{c}_n.$$

□

Corollary 2.9.

$$\begin{aligned} &\forall n \geq \hat{n}_1, \forall \lambda, \quad \forall \imath \in \mathbf{J}_n^{\perp}, \quad \forall \omega, \quad \forall s \in (0, t], \forall A = (\bar{a}] \subset C_{\tilde{c}_n}(0): \\ &1_A(q_n(s, \lambda, \imath, \omega)) = 1_A(\bar{r}^{\lambda} + w_{0,n}^{\lambda,\imath}(\omega)s)1_{\{\hat{\zeta}_n^{\lambda,\imath}(\omega) = \bar{r}^{\lambda}\}}. \end{aligned}$$

Proof. Note that on the set $\{\hat{\zeta}_n^{\lambda,\imath} = \bar{r}^{\lambda}\}$

$$q_n(s, \lambda, \imath) = \bar{r}^{\lambda} + w_{0,n}^{\lambda,\imath,}s + H_n(s, \lambda, \imath).$$

The assumption

$$1_A(q_n(s, \lambda, \tilde{\omega})) \neq 0 \text{ for some } \tilde{\omega} \text{ and some } A \subset C_{\tilde{c}_n}(0)$$

implies by Lemma 2.8

$$q_n(u, \lambda, \imath, \omega) \notin C_{2\tilde{c}_n}(0) \;\; \forall u \neq s, \;\; \forall \omega.$$

The definition of the constants implies for $\forall n \geq \hat{n}_1$:

$$|\bar{q}_n(u, \lambda, \imath, \omega) - r| \geq 2\tilde{c}_n - \tfrac{1}{2n} - \tilde{K}_n > \tilde{c}_n \tag{2.23}$$

for $r \in \cup_{v<t}\text{supp}(\mathcal{X}_N(\cdot, (v)))$, where "supp" denotes the support of the measure. Thus, $\bar{q}_n(u, \lambda, \imath, \omega) - r \notin C_{K_n}(0) \;\; \forall \omega \;\; \forall r \in \cup_{v<t}\text{supp}(\mathcal{X}_N(\cdot, (v)))$, whence for $u < s$

$$G_n(\bar{q}_n(u, \lambda, \imath, \omega) - r) = 0 \;\; \forall(\omega, u < s, \;\; r \in \cup_{v<t}\text{supp}(\mathcal{X}_N(\cdot, (v, \omega)))). \tag{2.24}$$

This implies

$$H_n(s, \lambda, \imath, \omega) = 0 \;\; \forall(\omega, s \leq t). \tag{2.25}$$

□

Recall the definition of $\bar{\alpha}$ and $\underline{\alpha}$ from Hypothesis 2.3. Corollary 2.9 and Lemma 2.8 imply

Lemma 2.10. $\forall A \subset C_{\tilde{c}_n}(0), \quad \forall s \in (0, t]$

$$\int_{\{|x| > \frac{\tilde{K}_n}{n^{\tilde{p}}}\}} \psi_n(x)\mathrm{d}x\underline{\alpha}_n \leq EI^n_{A,\mathbf{N}\times\mathbf{J}_\mathbf{n}^{\perp}}(s) \leq \int_{\{|x| > \frac{\tilde{K}_n}{n^{\tilde{p}}}\}} \psi_n(x)\mathrm{d}x\bar{\alpha}_n.$$

Proof. By Corollary 2.9 we have the representation

$$I^n_{A,\mathbf{N}\times\mathbf{J}^\perp_\mathbf{n}}(s) = \sum_{\lambda,\iota} 1_A\left(\bar{r}^\lambda + w^{\lambda,\iota}_{0,n}s\right) 1_{\{\hat{\zeta}^{\lambda,\iota}_n = \bar{r}^\lambda\}}.$$

Denoting by $\bar{a}$ the midpoint of the cell A,

$$|q_n(s,\lambda,\iota) - \bar{a}| \le \tfrac{1}{2n} \Longleftrightarrow |sw^{\lambda,\iota}_{0,n} + \bar{r}^\lambda - \bar{a}| \le \tfrac{1}{2n} \Longleftrightarrow w^{\lambda,\iota}_{0,n} \in C_{\frac{1}{2ns}}(\tfrac{-\bar{r}^\lambda + \bar{a}}{s}).$$

The Lebesgue measure does not charge the boundaries of the cubes. Therefore, using equivalent representations of the indicator function and change of variables $x := sw$, we obtain

$$\begin{aligned} EI^n_{A,\mathbf{N}\times\mathbf{J}^\perp_\mathbf{n}}(s) &= \sum_\lambda \sum_{\iota\in\mathbf{J}^\perp_n} P(\{q_n(s,\lambda,\iota) \in A\}) \\ &= \sum_\lambda \sum_{\iota\in\mathbf{J}^\perp_n} \int 1_{B_\iota}(w) 1_{C_{\frac{1}{2sn}}\left(-\frac{\bar{r}^\lambda}{s}\right)}\left(w - \frac{\bar{a}}{s}\right) \psi_n(w) \mathrm{d}w\alpha_{n,\lambda} \\ &= \frac{1}{s^d}\sum_\lambda \int 1_{C_{\frac{1}{2n}}(-\bar{r}^\lambda + \bar{a})}(x)(1 - 1_{C_{s\bar{K}_n}(0)}(x))\psi_n\left(x\frac{1}{s}\right) \mathrm{d}x\alpha_{n,\lambda} \\ &= \frac{1}{s^d}\sum_\lambda \int 1_{(-\bar{r}^\lambda + \bar{a}]}(x)(1 - 1_{C_{s\bar{K}_n}(0)}(x))\psi_n\left(x\frac{1}{s}\right) \mathrm{d}x\alpha_{n,\lambda}. \end{aligned}$$

Taking sup and inf with regard to $\alpha_{n,\lambda}$ allows us to incorporate the summation over λ into the integral and use the obvious relation:

$$\sum_\lambda 1_{(-\bar{r}^\lambda + \bar{a}]}(x) = 1_{\cup_\lambda(-\bar{r}^\lambda + \bar{a}]}(x) \equiv 1.$$

Change of variables finishes the proof of Lemma 2.10. □

Lemma 2.10 implies in particular that a.s. $I^n_{A,\mathbf{N}\times\mathbf{J}^\perp_\mathbf{n}}(t) < \infty$ for any $A \subset C_{\bar{K}_n + K_n}(0) \subset C_{\tilde{c}_n}(0)$. Hence, by (2.18) and the arguments given thereafter, $\mathcal{Y}(\mathrm{d}q,t)$ is a.s. finite on $C_{K_n}(\bar{r}^i_n(t-))$ for $i = 1,\ldots,n$. Altogether we obtain Proposition 2.1 from Lemmas 2.8, 2.10, and Corollary 2.9. □

We conclude this chapter with some comments on the model.

Remark 2.11. Here are some arguments to support the initial discretization of time and space.

- Time
  The collision dynamics should (in the microscale) be described by straight lines (the free motion between collisions). The discrete time approach reflects this by using an "average" time interval. The "information" the particle carries after some collision will be a time-delayed one at the next collision. An Euler scheme for the mean field dynamics in discrete time preserves this feature. The delays

in time become infinitesimally small in the macroscale (as in Itô differentials). If, however, for some other microscopic interaction time delays would be inadequate, we need to interpolate between two "collisions," i.e., between future and past, to obtain an adequate continuous time description. For correlation Brownian motions, this could result in correction terms (s., e.g., the classical papers of Wong and Zakai (1965) and Stratonovich (1964) as well as our Sect. 15.6.5).

- Space
  The discretization in space is an example of "coarse graining" and it simplifies the calculations. Second, this discretization allows us to model the observation that, in the framework of classical mechanics, one cannot precisely determine the position and the velocity of a given particle – as a result of "measurement errors" (cf., e.g., Heisenberg (1958)). We incorporate this observation into our model as follows: Instead of saying our particle is at time t in the position $q(t)$ and has the velocity $w(t)$, we say it is in the cell $(r^\lambda]$ and its velocity is from the cell B_ι (s. Chap. 2 for the precise definitions). We then model the "measurement errors" by random variables, where each random variable takes values only in some small cell. In particular, the clusters of small particles that are initially in a given cell $(r^\lambda]$ with approximately the same initial velocities from the cell B_ι become one small particle, placed at the midpoint $\bar{r}^\lambda$, and its velocity is randomly distributed over the cell B_ι. We use the material distributions of the small particles in a cell $(r^\lambda]$ to define the probability that the cell does contain the "random particle" with velocities from B_ι (cf. (2.6)). Finally, we make the independence assumptions about the initial positions and velocities (justifying this assumption by assuming no interaction...), and we have a relatively simple standard probabilistic set-up that we can work with. □

Remark 2.12.

- The relation of our initial distribution to the more common Poisson random measure for the particles or particles and velocities can be described as follows (cf., e.g., Dürr et al. (1981, 1983), Sinai and Soloveichik (1986), Spohn (1991), Szász and Tóth (1986, 1987) for the use of the Poisson random measure): Assume for simplicity $B_\iota \in \{(\bar{r}^\lambda] : \lambda \in \mathbf{N}\}\ \forall \iota$ and $\varphi_n(q) = \alpha n^d$. Let A and B cubes in $\mathbf{R}^d$, which can be represented as finite unions of cells $(\bar{r}^\lambda]$ and set

$$\tilde{\mu}_n(\omega, A \times B) := \#\{(\lambda, \iota) := \left(\hat{\zeta}_n^{\lambda,\iota}, w_{n,0}^{\lambda,\iota}\right) \in A \times B\}.$$

Since $\tilde{\mu}_n(\omega, A \times B) \le n^{2d}|A||B|$ we see immediately that $\tilde{\mu}_n(\omega, A \times B)$ is not a Poisson random measure (say, on the grid). However, in many other aspects it is similar to the Poisson random measure for positions and velocities. In particular, we have independence in sets (A, B) and $(\tilde{A}, \tilde{B})$, if $(A, B) \cap (\tilde{A}, \tilde{B}) = \emptyset$. We obviously have the representation

$$\{\omega : \tilde{\mu}_n(\omega, A \times B) = k\} = \left\{\omega : \sum_{\lambda,\iota} 1_{(\hat{\zeta}_n^{\lambda,\iota}, w_{n,0}^{\lambda,\iota}) \in A \times B}(\omega) = k\right\}.$$

Abbreviate $\gamma_p := \alpha \nu_n(B_{\iota_p})$. If $k \in \mathbf{N}$ and $k \leq l := n^d|A|$ and $m := n^d|B|$

$$P\{\omega : \tilde{\mu}_n(\omega, A \times B) = k\} = \sum_{k_1+\cdots+k_m=k} \prod_{p=1}^{m} \left[\binom{l}{k_p} \gamma_p^{k_p} (1-\gamma_p)^{l-k_p} \right].$$

Therefore, we may call our initial distribution a "multinomial random measure." Note that in our framework time and spatial scales as well as mass and average velocities are functions of n. If we fix the average velocity $\bar{w}_n \gg 1$ then, for sets B not far from the origin, the density in ν_n is almost constant. Hence, the γ_p are approximately equal to $\approx cn^{-d}$, where c is a constant. We obtain that the multinomial distribution is close to the usual binomial distribution, i.e.,

$$\{\tilde{\mu}_n(\omega, A \times B) = k\} \approx \binom{n^{2d}|A||B|}{k} (cn^{-d})^k (1 - cn^{-d})^{n^{2d}|A||B|-k}.$$

As a result, we could evoke the usual Poisson limit theorem and obtain (for B close to the origin) that our distribution can be approximated by a Poisson random measure with intensity measure $\alpha n^d |A| \times \nu_n(B)$.

- Typically, in many scaling limits the forces must be scaled as well. Kotelenez (2002, Example 1.2) derives a long-range smooth force through a heuristic scaling limit argument in the passage from the microscale to the mesoscale (cf. Remark 1.2 and (1.2)). A scaling of the type $n^b G_n(n(r-q))$, where $b \geq 0$, would make the action of the force local in the limit. Such a result would be inconsistent with the observation made in Remark 1.2. In other words, if we want to capture the long-range effects of bombardments of the large particles by small particles that move with different velocities, we ought to avoid the localizing scaling in the argument of G_n. In Example 1.3 of Kotelenez (loc.cit.), this long range effect of different velocities is compounded by the presence of a long-range interaction potential. Accounting for these long-range effects requires summing up over potentially infinitely many contributions and converting sums into integrals (cf. (2.2)). The choice $\delta\sigma$ from (1.1) provides the correct normalizing factor for this approach. Our approach generates a global Brownian medium that preserves the long-range effects of the original particle interaction in the correlation operator.
- A simpler version of the Einstein-Smoluchowski equations than (2.11) was shown to be the limit of the so-called Orstein-Uhlenbeck model of Brownian motion by I'lin and Khasminskii (1964) and Nelson (1972). This was generalized by Kotelenez and Wang (1994) to the case described by (2.11). Note that in these results the Ornstein-Uhlenbeck model was a second order stochastic differential equation where the velocity was a Wiener process (I'lin and Khasminskii (loc.cit.) and Nelson (loc.cit), and in Kotelenez and Wang (loc.cit) the velocity was driven by Gaussian space–time white noise. These results were only a reduction of a second order stochastic equation to a first order SODE (called Einstein-Smoluchowski) as a consequence of a very large friction η_n and, therefore, describe the transition from stochastic dynamics to stochastic kinematics.

In contrast, in the derivation presented here, no stochastic effects are assumed, i.e., the motion of both large and small particles is, by assumption, entirely deterministic. Only with regard to initial conditions we assume randomness and independence. However, because of the very large velocities of the small particles and their independent starts, the changes in the velocities of the large particles become approximately independent. Scaling yields the stochastic effects for the large particles, i.e., the independence in the changes in time. For η_n fixed this would result in an Ornstein-Uhlenbeck model, which was the starting point in the analysis by I'lin and Khasminskii as well as Nelson (loc.cit) to derive the simpler Einstein-Smoluchowski equations. Therefore, in our derivation, we have joined two steps (for the slowly moving large particles) into one:

(i) Transition from deterministic dynamics to stochastic dynamics, as $n \longrightarrow \infty$.
(ii) Transition from stochastic dynamics to stochastic kinematics, as $\eta_n \longrightarrow \infty$.

□

Chapter 3
Proof of the Mesoscopic Limit Theorem

Let $\mathbf{D}_n \subset \mathbf{N}$, $r(u)$ be some cadlag process and $J^n(u)$ some (nice) occupation measure process with support in the cells. Assume that both $r(u)$ and $J^n(u)$ are constant for $u \in [(l-1)\delta\sigma, l\delta\sigma)$, $l \in \mathbf{N}$. Abbreviate

$$\Lambda(t, \mathbf{D}_n, r, J^n) := \frac{1}{\hat{m}} \sum_{u \le t} \sum_{\lambda \in \mathbf{D}_n} G_n(\bar{r}(u-) - \bar{r}^\lambda) J^n_{[\bar{r}^\lambda)}(u)\delta\sigma. \tag{3.1}$$

Further, $\Lambda(ds, \ldots)$ denotes the (Stieltjes) integrator, defined by the increments of $\Lambda(s, \ldots)$.

Using these abbreviations, we plug $v_n^i(s)$ into the equation for $r_n^i(t)$ in (2.9) and replace integration with respect to $\mathcal{Y}(\mathrm{d}q, u)$ by integration over the sum of occupation measures, $I^n(u)$, in accordance with (2.16). We then change the order of the summation in the resulting double sum. By Hypothesis 2.2 and the summation formula for geometric sums, we obtain that the evolution of the large particles in (2.9) may be described by

$$r_n^i(t) = r_n^i(0) + \Lambda(t, \mathbf{N}, r_n^i, I^n) - \int_0^t \exp[-\eta_n(t-s)]\Lambda(\mathrm{d}s, \mathbf{N}, r_n^i, I^n). \tag{3.2}$$

Further, let $\tilde{K}_n > 0$ such that $\sum\limits_n \frac{1}{\tilde{K}_n} < \infty$. By the uniform boundedness of the first moments of $r_n^i(0)$ (cf. Hypothesis 2.5) and Chebyshev's inequality, we obtain that for all $\tilde{\zeta} > 0$ there is an $n(\tilde{\zeta})$ such that

$$\begin{aligned}
& P(\cup_{n \ge n(\tilde{\zeta})}\{\omega \in \Omega : \max_{i=1,\ldots,N} |r_n^i(0,\omega)| \ge \tilde{K}_n\}) \\
& \le \sum_{n \ge n(\tilde{\zeta})} P(\{\omega \in \Omega : \max_{i=1,\ldots,N} |r_n^i(0,\omega)| \ge \tilde{K}_n\}) \\
& \qquad \le \sum_{n \ge n(\tilde{\zeta})} \frac{\max_{i=1,\ldots,N} E|r_n^i(0)|}{\tilde{K}_n} \\
& \qquad\qquad \le \sum_{n \ge n(\tilde{\zeta})} \frac{\text{const}}{\tilde{K}_n} \le \tilde{\zeta}.
\end{aligned}$$

Hence, we may, without loss of generality, assume that $\forall n \in \mathbf{N}$:

$$\left.\begin{array}{c}\sup\limits_{\omega\in\Omega} \max_{i=1,\ldots,N} |r_n^i(0,\omega)| \le \frac{\tilde{K}_n}{4},\\ \text{where } \tilde{K}_n := n^{d+2\delta} \text{ and } 0 < \delta < \frac{1}{7}.\end{array}\right\} \tag{3.3}$$

For the support of G_n we assume: $K_n := n^{\frac{\delta}{d}}$.

The choice $\tilde{K}_n$ follows from the need to control certain error terms that are of the order of n^d (cf. (3.18)) and the choice $0 < \delta < \frac{1}{7}$ makes the choice of p in the representation of the average speed simpler.[1] Summarizing the previous definitions and referring to Hypothesis 2.2 for the definition of $\tilde{p}$ and ζ, the constants from (2.19) now have the following values:

$$\left.\begin{array}{l}0 < \delta < \frac{1}{7},\\ \tilde{K}_n := n^{d+2\delta} \quad \text{(bound in (3.3) and (3.8))},\\ K_n := n^{\frac{\delta}{d}} \quad \text{(for the support of } G_n\text{)},\\ \bar{c} := (|||G|||N + 2)(\hat{t} + 1),\\ c_n := \bar{c}n^{\delta},\\ \tilde{c}_n := \bar{c}n^{(d+3\delta)\vee(\tilde{p}+\zeta+\delta)} \quad \text{(cf. (3.6) and (3.1))},\\ \bar{K}_n := 4\bar{c}n^{(2d+3\delta)\vee(\tilde{p}+\zeta+\delta+d)}.\end{array}\right\} \tag{3.4}$$

It follows from integration by parts in Stieltjes integrals that

$$\sup_{t\le\hat{t}}\left|\int_0^t \exp[-\eta_n(t-s)]\Lambda(ds,\mathbf{D}_n,r_n^i,J^n)\right| \le 2\sup_{t\le\hat{t}}|\Lambda(t,\mathbf{D}_n,r_n^i,J^n)|. \tag{3.5}$$

Consequently, bounds on $\Lambda(t,\mathbf{D}_n,r_n^i,J^n)$ imply bounds on $\int_0^t \exp[-\eta_n(t-s)]\Lambda$ $(ds,\mathbf{D}_n,r_n^i,J^n)$ and, therefore, most of the following analysis will be focused on the asymptotics of $\Lambda(t,\mathbf{D}_n,r_n^i,J^n)$.

Recall the definition of the constants $\tilde{K}_n$ and $\tilde{c}_n$ from (3.4). Define the following sequence of stopping times:

$$\left.\begin{array}{c}T_n := \inf\left\{s \le \hat{t} : \max_{i=1,..,N} |\Lambda(t,\mathbf{C}_n,r_n^i,I^n)| \ge \frac{\tilde{K}_n-1}{4}\right\},\\ \text{where } \mathbf{C}_n := \{\lambda : (\bar{r}^{\lambda}] \subset C_{\tilde{c}_n}(0)\}.\end{array}\right\} \tag{3.6}$$

Denote integration against measures from a set of time dependent empirical measure processes $\mathcal{Z}(s)$, $s \in [0,t-]$, by $\mathcal{Z}(dr,[0,t-])$. Set for $0 \le s \le \hat{t}$

$$\left.\begin{array}{l}\tilde{r}_n^i(s) := r_n^i(s\wedge T_n),\\ \tilde{\mathcal{X}}_N(dr,\cdot) := \hat{m}\sum\limits_{i=1}^N \delta_{\tilde{r}_n^i(\cdot)}(dr),\\ \tilde{q}_n(t,\lambda,\iota) := q_n(t,\lambda,\iota,\tilde{\mathcal{X}}_N(dr,[0,t-])),\\ \tilde{I}_A^n(t) := \sum\limits_{\lambda,\iota} 1_A(\tilde{q}_n(t,\lambda,\iota)).\end{array}\right\} \tag{3.7}$$

---

[1] Cf. Hypothesis 2.2, Lemmas 3.2–3.3 and (3.10).

Here $q_n(t, \lambda, \imath, \tilde{\mathcal{X}}_N(dr, [0, t-]))$ is the solution of the third recursive scheme in (2.9) with $\mathcal{X}_N(dr, [0, t-])$ being replaced by $\tilde{\mathcal{X}}_N(dr, [0, t-])$.[2] As a consequence of (3.3) and (3.5) we obtain

$$\sup_{\omega\in\Omega}\sup_{s\leq\hat{t}}\max_{i=1,..,N}|r_n^i(s\wedge T_n)| < \tilde{K}_n \quad \forall n \in \mathbf{N}. \tag{3.8}$$

The stopping implies that we may proceed as in the derivation of Proposition 2.1 with deterministic constants $\bar{K}_n > \tilde{K}_n$ and deterministic sets of labels $\mathbf{J}_n^\perp := \mathbf{N}\setminus\mathbf{J_n}$, which are now all independent of t as well. The precise definitions of $\tilde{K}_n$,K_n and $\bar{K}_n$ are given in (3.4). The decomposition of the occupation measure $\tilde{I}_A^n$ into the sum of two different measures follows the pattern of (2.18). One of the measures is driven by the "small" velocities ($\imath \in \mathbf{J}_n$) and the other one by the "large" velocities $\imath \in \mathbf{J}_n^\perp := \mathbf{N} \setminus \mathbf{J}_n$. We no longer need to partition the position indices. Using the abbreviations $\tilde{I}_A^{n,\perp} := \tilde{I}^n_{A,\mathbf{N}\times\mathbf{J_n^\perp}}$ and $\tilde{I}^n_{A,\mathbf{J}_n} := \tilde{I}^n_{A,\mathbf{N}\times\mathbf{J_n}}$, we have

$$\tilde{I}_A^n(s) = \tilde{I}^n_{A,\mathbf{J}_n}(s) + \tilde{I}_A^{n,\perp}(s), \tag{3.9}$$

Lemmas 2.8 and 2.10 as well as Corollary 2.9 carry immediately over to $\tilde{I}_A^{n,\perp}(t)$ and $\tilde{q}_n(t, \lambda, \imath)$ for $\imath \in \mathbf{J}_n^\perp$. Lemma 2.8 and Corollary 2.9 imply

Lemma 3.1. *$\forall A \subset C_{\tilde{c}_n}(0)$ $\tilde{I}_A^{n,\perp}(s)$ (from (3.9)) is independent in $s = k\,\delta\sigma$, $s \leq \hat{t}$, and $\tilde{I}_A^{n,\perp}(\cdot)$ does not depend on the states $\tilde{r}_n^i$, $i = 1, \ldots, N$.*

Proof. Let $u \neq s$. For $t \in \{s, u\}$ set

$$E_t := \cup\left\{\lambda \in \mathbf{N}, \imath \in \mathbf{J}_n^\perp : \tilde{q}_n(t, \lambda, \imath, \omega) \in C_{\tilde{c}_n}(0) \text{ for at least one } \omega\right\}.$$

We represent $\tilde{I}_A^{n,\perp}(t)$ as the sum over all indices from E_t since the other indices would not change the values of the occupation measures. Therefore, by Corollary 2.9

$$\tilde{I}_A^{n,\perp}(t,\omega) = \sum_{(\lambda,\imath)\in E_t} 1_A\left(\bar{r}^\lambda + w_{0,n}^{\lambda,\imath}(\omega)t\right) 1_{\{\hat{\zeta}_n^{\lambda,\imath}(\omega)=\bar{r}^\lambda\}}.$$

First of all, this shows the independence with respect to the states $\tilde{r}_n^i$, $i = 1, \ldots, N$. Further, by Lemma 2.8 and Corollary 2.9,

$$E_s \cap E_u = \emptyset.$$

$\tilde{I}_A^{n,\perp}(s)$ and $\tilde{I}_A^{n,\perp}(u)$ as sums of independent families of random variables are independent. □

---

[2] Cf. also (2.15).

Lemma 3.2.

$$n^{d+\delta}\left[\max_{A\subset C_{\bar{c}_n}(0)}\max_{\delta\sigma\le s\le t}|EI^n_{A,\mathbf{N}\times\mathbf{J}_n^{\perp}}(s)-\alpha|\right]\longrightarrow 0,^3$$

as $n\longrightarrow\infty$.

The *Proof* easily follows from Lemma 2.10 and (2.19), (3.3), (3.4) in addition to Hypotheses 2.2 and 2.3. □

The following Lemma 3.3 provides a bound from above for $\tilde{I}^n_{A,\mathbf{J}_n}(s)$, showing that the contribution of $\tilde{I}^n_{A,\mathbf{J}_n}(s)$ to the time evolution of the large particles becomes negligible in the limit. In other words, we need to consider only the "large velocities" from $\mathbf{J}_n^{\perp}$. In what follows recall that

$$s>0\Longleftrightarrow s\ge\delta\sigma=n^{-d},$$

since u,s,t take only discrete time values in $(0,\hat{t}]$, which are multiples of $\delta\sigma$.

Lemma 3.3. *There is a finite constant $\hat{c}$, depending on $\psi, G, N, \hat{t}$, such that $\forall A\subset C_{\bar{c}_n}(0)$, $\forall s\in(0,\hat{t}]$ and arbitrary constant $L>1$:*

$$\sum_{\lambda}\sum_{\iota\in\mathbf{J}_n}P(\{\tilde{q}_n(s,\lambda,\iota)\in A\})\le\hat{c}n^{([(4d+6\delta)\vee(2\bar{p}+2\zeta+2d+2\delta)]-p)d}\bar{\alpha}_n,^4$$

Proof. Let us denote by $\check{Q}_n(s,\lambda,\iota)$ the restriction of the random variable $\tilde{Q}_n(s,\lambda,\iota)$ to those ω, where $\zeta_n^{\lambda,\iota}=\bar{R}^{\lambda}$.

(i) Similarly to the proof of Lemma 2.8, we abbreviate

$$H_n\left(t,\lambda,w_{0,n}^{\lambda,\iota}\right):=n^{\bar{p}+\zeta}\frac{1}{\hat{m}}\sum_{s<t}\sum_{u\le s}\int G_n\left(\bar{\tilde{q}}_n(u,\lambda,\iota)-r\right)\tilde{\mathcal{X}}_N(\mathrm{d}r,u)(\delta\sigma)^2.$$

As in the proof of Lemma 2.10 we obtain

$$\|\bar{\tilde{q}}_n(s,\lambda,\iota)-\bar{a}\|\le\frac{1}{2n}\Longleftrightarrow w_{0,n}^{\lambda,\iota}\in C_{\frac{1}{2ns}}\left(\frac{-\bar{r}^{\lambda}-H_n\left(s,\lambda,w_{0,n}^{\lambda,\iota}\right)+\bar{a}}{s}\right).$$

Set

$$\hat{H}_n(s,\lambda,w_{0,n}^{\lambda,\iota}):=H_n\left(s,\lambda,w_{0,n}^{\lambda,\iota}\right)-\bar{a}.$$

We can easily show (by induction) that

$$w\longmapsto H_n(s,\lambda,w)$$

---

[3] Recall that α is the expected average volume (in a unit cube) occupied by small particles (for large n). Cf. also Hypothesis 2.3.

[4] Cf. Hypothesis 2.3 for the definition of $\bar{\alpha}_n$.

is $\mathcal{B}^d - \mathcal{B}^d$-measurable for all n, λ, s and, therefore, the same holds for $\hat{H}_n(s, \lambda, w)$. Thus,

$$
\begin{aligned}
I_\lambda &:= \sum_{\iota \in \mathbf{J}_n} P(\{\breve{q}_n(s, \lambda, \iota) \in A\}) \\
&= \frac{1}{s^d n^{pd}} \int 1_{C_{\frac{1}{2n}}(-\bar{r}^\lambda)} \left(x + \hat{H}_n\left(s, \lambda, \frac{x}{s}\right)\right) 1_{C_{s\bar{K}_n}(0)}(x) \psi\left(x \frac{1}{n^p s}\right) \mathrm{d}x \alpha_{n,\lambda} \\
&\leq \frac{|||\psi|||}{s^d n^{pd}} \int 1_{C_{\frac{1}{2n}}(-\bar{r}^\lambda)} \left(x + \hat{H}_n\left(s, \lambda, \frac{x}{s}\right)\right) 1_{C_{s\bar{K}_n}(0)}(x) \mathrm{d}x \alpha_{n,\lambda}.
\end{aligned}
$$

$\bar{a} \in C_{\tilde{c}_n}(0)$ implies: $\forall s \leq t, \forall \lambda, \ |\hat{H}_n(s, \lambda, w)| \leq \bar{c} n^{\tilde{p}+\zeta} + \tilde{c}_n - 1 \leq 2\tilde{c}_n - 1 = 2\bar{c} n^{(d+3\delta)\vee(\tilde{p}+\zeta+\delta)} - 1$.[5] The product of the two indicator functions in the integral in the right-hand side of the above equation will be 0 if $|\bar{r}^\lambda| > s\bar{K}_n + 2\tilde{c}_n - 1 + \frac{1}{2n}$. Hence, we can estimate the integral of the product of the two indicator functions by $s^d \bar{K}_n^d \times \int 1_{C_{\frac{1}{2n}}(-\bar{r}^\lambda)}(x) 1_{C_{s\bar{K}_n + 2\tilde{c}_n}(0)}(x) \mathrm{d}x n^d$. Altogether, we obtain

$$
I_\lambda \leq \frac{|||\psi|||}{n^{pd}} \int 1_{C_{\frac{1}{2n}}(-\bar{r}^\lambda)}(x) 1_{C_{s\bar{K}_n + 2\tilde{c}_n}(0)}(x) \mathrm{d}x \bar{\alpha}_n \bar{K}_n^d n^d,
$$

where $\int 1_{C_{\frac{1}{2n}}(-\bar{r}^\lambda)}(x) 1_{C_{s\bar{K}_n + 2\tilde{c}_n}(0)}(x) \mathrm{d}x n^d$ is an upper bound for the indicator function $1_{\{|\bar{r}^\lambda| \leq s\bar{K}_n + 2\tilde{c}_n - 1 + \frac{1}{2n}\}}$.

Hence,

$$
\sum_\lambda \sum_{\iota \in \mathbf{J}_n} P(\{\breve{q}_n(s, \lambda, \iota) \in A\}) \leq \frac{|||\psi|||}{n^{pd}} \int_{C_{s\bar{K}_n + 2\tilde{c}_n}(0)} \mathrm{d}x [4\tilde{c}_n n^d]^d n^d \overline{\alpha}_n.
$$

The last integral is estimated above by $\left[\frac{\hat{t}\bar{K}_n + 2\tilde{c}_n}{n^p}\right]^d$. Recalling the abbreviations from (3.4), we finish the proof of Lemma 3.3. □

Note that by the assumption on φ_n we have $\bar{\alpha}_n \leq \alpha$. Since by Hypothesis 2.2

$$
n^{d+\delta} n^{([(4d+6\delta)\vee(2\tilde{p}+2\zeta+2d+2\delta)]-p)d} \longrightarrow 0 \text{ as } n \longrightarrow \infty,
$$

we obtain

$$
n^{d+\delta} \left[\max_{\lambda \in \mathbf{C}_n} \max_{\delta\sigma \leq s \leq \hat{t}} E\tilde{I}^n_{(\bar{r}^\lambda], \mathbf{J}_n}(s)\right] \longrightarrow 0, \text{ as } n \longrightarrow \infty, \tag{3.10}
$$

where $\mathbf{C}_n$ was defined in (3.6). With the decomposition (3.9) we obtain

$$
n^{d+\delta} |E\tilde{I}^n_A(s) - \alpha| \leq n^{d+\delta} |E\tilde{I}^n_{A,\mathbf{J}_n}(s)| + n^{d+\delta} |E\tilde{I}^{n,\perp}_A(s) - \alpha|.
$$

Hence, (3.10) in addition to Lemma 3.2 implies

---

[5] Cf. (2.19) and (3.4).

$$n^{d+\delta}\left[\max_{\lambda\in\mathbf{C}_n}\max_{\delta\sigma\le s\le\hat{t}}\left|E\tilde{I}^n_{(\bar{r}^\lambda]}(s)-\alpha\right|\right]\longrightarrow 0,\ \text{as } n\longrightarrow\infty. \tag{3.11}$$

It follows that there is an

$$\hat{n}_2\ge\hat{n}_1$$

such that for all $n\ge\hat{n}_2$

$$\max_{\lambda\in\mathbf{C}_n}\max_{\delta\sigma\le s\le\hat{t}}E\tilde{I}^n_{(\bar{r}^\lambda]}(s)\le\alpha+1. \tag{3.12}$$

For $\lambda\notin\mathbf{C}_n$ and $|\bar{r}|<\tilde{K}_n$ we have $|\bar{r}^\lambda-\bar{r}|>n^{d+2\delta}(\bar{c}-1)>n^{d+2\delta}$ (cf. (3.4) and (3.6)). We conclude

$$G_n\left(\bar{\tilde{r}}^i_n(u-)-\bar{r}^\lambda\right)=0\quad\forall u\le\hat{t},\forall n\ge\hat{n}_2,\ \ \forall\lambda\notin\mathbf{C}_n. \tag{3.13}$$

Therefore, we obtain

$$\Lambda\left(t\wedge T_n,\mathbf{N},r^i_n,I^n\right)\equiv\Lambda\left(t\wedge T_n,\mathbf{N},\tilde{r}^i_n,\tilde{I}^n\right)\equiv\Lambda\left(t\wedge T_n,\mathbf{C}_n,\tilde{r}^i_n,\tilde{I}^n\right), \tag{3.14}$$

where the first relation is trivial.

In what follows, we show that we may work, without loss of generality, with centered occupation measures as defined by

$$\tilde{I}^{n,\perp,c}_A(t):=\tilde{I}^{n,\perp}_A(t)-E\tilde{I}^{n,\perp}_A(t)\quad\forall t. \tag{3.15}$$

We obtain

$$\left.\begin{aligned}\tilde{r}^i_n(\cdot)=r^i_n(0)+\Lambda(\cdot,\mathbf{C}_n,\tilde{r}^i_n,\tilde{I}^{n,\perp,c})\\ -\int_0^t\exp[-\eta_n(\cdot-s)]\Lambda\left(ds,\mathbf{C}_n,\tilde{r}^i_n,\tilde{I}^{n,\perp,c}\right)+U^{i,n}_1(\cdot),\end{aligned}\right\} \tag{3.16}$$

where

$$\left.\begin{aligned}U^{i,n}_1(t):=\Lambda\left(t\wedge T_n,\mathbf{C}_n,\tilde{r}^i_n,\tilde{I}^n\right)-\Lambda\left(t,\mathbf{C}_n,\tilde{r}^i_n,\tilde{I}^{n,\perp,c}\right)\\ -\int_0^t\exp\left[-\eta_n(t-s)\right]\left[\Lambda\left(d(s\wedge T_n),\mathbf{C}_n,\tilde{r}^i_n,\tilde{I}^n\right)-\Lambda\left(ds,\mathbf{C}_n,\tilde{r}^i_n,\tilde{I}^{n,\perp,c}\right)\right].\end{aligned}\right\} \tag{3.17}$$

Lemma 3.4.

$$E\sup_{t\le\hat{t}}\left|\Lambda\left(t,\mathbf{C}_n,\tilde{r}^i_n,\tilde{I}^n\right)-\Lambda\left(t,\mathbf{C}_n,\tilde{r}^i_n,\tilde{I}^{n,\perp,c}\right)\right|$$
$$\longrightarrow 0,\quad as\ n\longrightarrow\infty\,.$$

Proof.

(i) By (2.3), (3.4), and (3.10)

$$\begin{aligned}&E\sum_{t\le\hat{t}}\left|\Lambda\left(t,\mathbf{C}_n,\tilde{r}^i_n,\tilde{I}^n\right)-\Lambda\left(t,\mathbf{C}_n,\tilde{r}^i_n,\tilde{I}^{n,\perp}\right)\right|\\ &\quad\le\hat{t}\frac{|||G|||}{\hat{m}}n^{d+\delta}\sup_{\lambda\in\mathbf{C}_n}\max_{\delta\sigma\le t\le\hat{t}}E\tilde{I}^n_{(\bar{r}^\lambda],\mathbf{J}_n}(t)\\ &\quad\longrightarrow 0,\quad\text{as } n\longrightarrow\infty.\end{aligned}$$

(ii) $\sum_{\lambda\in\mathbf{C}_n} G_n\left(\tilde{\bar{r}}^i_n(t-)-\bar{r}^\lambda\right) E\tilde{I}^{n,\perp}_{(\bar{r}^\lambda]}(t)\delta\sigma \;=\; \sum_{\lambda\in\mathbf{C}_n} G_n\left(\tilde{\bar{r}}^i_n(t-)-\bar{r}^\lambda\right)\left[E\tilde{I}^{n,\perp}_{(\bar{r}^\lambda]}(t)-\alpha\right]$, employing (2.4) in addition to (3.13). Hence, by (2.3), (3.4), and Lemma 3.2 for $\tilde{I}^{n,\perp} - \tilde{I}^{n,\perp,c} = E\tilde{I}^{n,\perp}$

$$
\begin{aligned}
&E\sup_{t\le\hat{t}}\left|\Lambda\left(t,\mathbf{C}_n,\tilde{r}^i_n,\tilde{I}^{n,\perp}\right)-\Lambda\left(t,\mathbf{C}_n,\tilde{r}^i_n,\tilde{I}^{n,\perp,c}\right)\right|\\
&\le \hat{t}\frac{|||G|||}{\hat{m}}n^{d+\delta}\left[\sup_{\lambda\in\mathbf{C}_n}\max_{\delta\sigma\le s\le\hat{t}}\left|E\tilde{I}^{n,\perp}_{(\bar{r}^\lambda]}(s)-\alpha\right|\right]\\
&\longrightarrow 0 \text{ as } n\longrightarrow\infty,
\end{aligned}
$$

which completes the proof. □

We obtain for $n \ge \hat{n}_2$ by Chebyshev's inequality in addition to (2.3), (3.3), (3.4), and (3.12):

$$
\left.
\begin{aligned}
&P\left\{\max_{i=1,\ldots,N}\sup_{0\le t\le\hat{t}}\left|\Lambda\left(t,\mathbf{C}_n,\tilde{r}^i_n,\tilde{I}^n\right)-\Lambda\left(t\wedge T_n,\mathbf{C}_n,\tilde{r}^i_n,\tilde{I}^n\right)\right|>0\right\}\\
&\quad\le P\left\{\max_{i=1,\ldots,N}\sup_{0\le t\le\hat{t}}\left|\Lambda\left(t,\mathbf{C}_n,\tilde{r}^i_n,\tilde{I}^n\right)\right|\ge\frac{\tilde{K}_n-1}{4}\right\}\\
&\quad\le\frac{8|||G|||\hat{t}(\alpha+1)}{n^\delta}\\
&\quad\longrightarrow 0,\ \text{as } n\longrightarrow\infty.
\end{aligned}
\right\}
\tag{3.18}
$$

Corollary 3.5.

$$
\max_{i=1,\ldots,N}\sup_{t\le\hat{t}}|U^{i,n}_1(t)|\longrightarrow 0 \quad \textit{stochastically, as } n\longrightarrow\infty.
$$

Proof.

(i) By Lemma 3.4 in addition to (3.18)

$$
\begin{aligned}
&\sup_{t\le\hat{t}}\left|\Lambda\left(t\wedge T_n,\mathbf{C}_n,\tilde{r}^i_n,\tilde{I}^n\right)-\Lambda\left(t,\mathbf{C}_n,\tilde{r}^i_n,\tilde{I}^{n,\perp,c}\right)\right|\\
&\qquad\longrightarrow 0 \text{ stochastically, as } n\longrightarrow\infty.
\end{aligned}
$$

Let us write $\Lambda(s)$, etc. and suppress the dependence on the other parameters. Similar to (3.5)

$$
\begin{aligned}
&\left\|\int_0^t \exp[-\eta_n(t-s)][\Lambda(\mathrm{d}(s\wedge T_n))-\Lambda(\mathrm{d}s)]\right\|\\
&\qquad\le\sup_{t\le\hat{t}}2\|\Lambda(t\wedge T_n)-\Lambda(t)\|,
\end{aligned}
$$

and the proof is complete. □

We are now ready to apply the existing tools of stochastic analysis, in particular, limit theorems as developed by Kurtz and Protter (1998). In view of the previous estimates, we will assume $n \ge \hat{n}_2$. Following the scheme in Kurtz and Protter (loc.cit.)

we show first that the driving term in $\Lambda(\cdot, \mathbf{C}_n, \tilde{r}^i_n, \tilde{I}^{n,\perp,c})$ converges to some type of Brownian noise.[6] To this end we introduce filtrations σ-algebras. We use the common notation $\sigma\{\cdot\}$ to denote the σ-algebra generated by the the quantities in the braces. Set

$$\left.\begin{aligned} \mathcal{G}_{n,s} &:= \sigma\{1_A(\tilde{q}_n(s,\lambda,\iota,\cdot)), (\lambda,\iota) \in \mathbf{N} \times \mathbf{J}^{\perp}_n, A \subset C_{\tilde{c}_n}(0)\},\\ \mathcal{F}_{n,s-} &:= \sigma\{\tilde{r}^i_n(u), 0 \le u \le s-, i = 1,\ldots,N\},\\ \widehat{\mathcal{F}}_{n,s} &= \sigma\{\mathcal{F}_{n,s-}, \mathcal{G}_{n,u}, u \le s\}. \end{aligned}\right\} \tag{3.19}$$

Again, as in the proof of Lemma 3.1, Lemma 2.8, and Corollary 2.9 imply that $\mathcal{G}_{n,s}$ and $\mathcal{G}_{n,t}$ are independent if $t \neq s$. Further, we have

Proposition 3.6. $\Lambda(\cdot, \mathbf{C}_n, \tilde{r}^i_n, \tilde{I}^{n,\perp,c})$ *is a square integrable mean zero* $\widehat{\mathcal{F}}_{n,\cdot}$*-martingale.*[7]

Proof.

(i) Let $A := (\bar{r}^\lambda] \subset C_{\tilde{c}_n}(0)$ for some λ. We obtain that both $\tilde{I}^{n,\perp}_A(s)$ and $\tilde{I}^{n,\perp,c}_A(s)$ are $\mathcal{G}_{n,s}$-measurable. Hence, $\Lambda\left(\cdot, \mathbf{C}_n, \tilde{r}^i_n, \tilde{I}^{n,\perp,c}\right)$ is $\widehat{\mathcal{F}}_{n,\cdot}$-adapted.

(ii) Further, by independence and the mean zero property, using the definition in (3.7), and by (3.12) for $n \ge \hat{n}_2$

$$E\left(\left[\tilde{I}^{n,\perp,c}_A(s)\right]^2\right) \le E\tilde{I}^n_A(s) \le \alpha + 1.$$

Therefore, $\Lambda(\cdot, \mathbf{C}_n, \tilde{r}^i_n, \tilde{I}^{n,\perp,c})$ is square integrable.

(iii) Finally,

$$\widehat{\mathcal{F}}_{n,s} = \widehat{\mathcal{F}}_{n,s-} \otimes \mathcal{G}_{n,s} \ \forall s,$$

where for two σ-algebras $\mathcal{A}, \mathcal{B}$, $\mathcal{A} \otimes \mathcal{B}$ is the product σ-algebra. Hence, using conditional expectations, we see that $\Lambda(\cdot, \mathbf{C}_n, \tilde{r}^i_n, \tilde{I}^{n,\perp,c})$ is a square integrable mean zero $\widehat{\mathcal{F}}_{n,s}$-martingale. □

Denote by $L_{2,\widehat{\mathcal{F}}_{n,\cdot-}}([0,\hat{t}] \times \mathbf{R}^d \times \Omega)$ the space of all real-valued processes f, which are constant in $t \in [l\delta\sigma, (l+1)\delta\sigma)$, $l \in \mathbf{N} \cup \{0\}$, and depend on $q \in \mathbf{R}^d$ in addition to the dependence on ω such that

(i) f is square integrable on $[0,\hat{t}] \times \mathbf{R}^d \times \Omega$ with respect to $\mathrm{d}t \times \mathrm{d}q \times \mathrm{d}P$;
(ii) $f(s,\cdot,\cdot)$ is $\mathcal{B}^d \otimes \widehat{\mathcal{F}}_{n,s-}$-measurable.

Setting $\hat{G}_n(r,q) := G_n(\bar{r} - \bar{r}^\lambda)$ for $q \in (\bar{r}^\lambda]$, we obtain $\hat{G}_{n,j}(\bar{\tilde{r}}^i_n(\cdot-),\cdot) \in L_{2,\widehat{\mathcal{F}}_{n,\cdot-}}([0,\hat{t}] \times \mathbf{R}^d \times \Omega)$, and

[6] Cf. also the end of our Sect. 15.2.4.

[7] We refer to Sect. 15.2.3 for the definition and properties of martingales.

$$\left.\begin{array}{c}\Lambda\left(t, \mathbf{C}_n, \tilde{r}_n^i, \tilde{I}^{n,\perp,c}\right) \\ = \sum_{s \le t} \frac{1}{\hat{m}} \int \hat{G}_n\left(\bar{\tilde{r}}_n^i(s-), q\right) \left[\sum_{\lambda \in \mathbf{C}_n} 1_{(\bar{r}^\lambda]}(q) n^d \tilde{I}^{n,\perp,c}_{(\bar{r}^\lambda]}(s)\right] \mathrm{d}q\, \delta\sigma. \end{array}\right\} \quad (3.20)$$

Formally, this representation is the integration of elements from $L_{2,\widehat{\mathcal{F}}_{n,\cdot-}}([0,\hat{t}] \times \mathbf{R}^d \times \Omega)$ against the time increments of a sum of step functions, where by the restriction to $\mathbf{C}_n$ the sum is finite (cf. (3.6)). To indicate that the summation is restricted to $\mathbf{C}_n$, we will in what follows use the symbol $\widehat{\sum}$ instead of just $\sum$.

We define an infinite dimensional integrator $M_n(s)$ by setting for $f \in L_{2,\widehat{\mathcal{F}}_{n,\cdot-}}([0,\hat{t}] \times \mathbf{R^d} \times \Omega)$

$$\int_0^t \langle f(s-), dM_n(s)\rangle := \sum_{s \le t} \int f(s-,q) \left[\sum_{\lambda \in \mathbf{C}_n} 1_{(\bar{r}^\lambda]}(q) n^d \tilde{I}^{n,\perp,c}_{(\bar{r}^\lambda]}(s)\right] \mathrm{d}q \delta\sigma. \tag{3.21}$$

We obtain

$$\Lambda_j\left(t, \mathbf{C}_n, \tilde{r}_n^i, \tilde{I}^{n,\perp,c}\right) = \int_0^t < \hat{G}_{n,j}\left(\bar{\tilde{r}}_n^i(s-), \cdot\right), \mathrm{d}M_n(s) > .$$

In what follows, $[\cdot,\cdot]$ will denote the mutual quadratic variation of square integrable martingales. Let $f, g \in L_{2,\widehat{\mathcal{F}}_{n,\cdot-}}([0,\hat{t}] \times \mathbf{R}^d \times \Omega)$. For $k \in \{f, g\}$ set

$$\widehat{k}_n(q) := \begin{cases} n^d \int_{(\bar{r}^\lambda]} k(q)\mathrm{d}q, & \text{if} \quad q \in (\bar{r}^\lambda] \text{ and } (\bar{r}^\lambda] \subset C_{\tilde{c}_n}(0), \\ 0, & \text{if} \quad q \in (\bar{r}^\lambda] \text{ and } (\bar{r}^\lambda] \text{ is not a subset of } C_{\tilde{c}_n}(0). \end{cases} \tag{3.22}$$

Since, for a pure jump process, the quadratic variation is the sum of the squares of the jumps, we obtain

$$\left.\begin{array}{c}\left[\int_0^t \langle f(s-), \mathrm{d}M_n(s)\rangle, \int_0^t \langle g(s-), \mathrm{d}M_n(s)\rangle\right] \\ = \sum_{s \le t} \widehat{\sum_A} \widehat{\sum_B} \widehat{f}_n(s-,\bar{a}) \widehat{g}_n(s-,\bar{b}) I_A^{n,\perp,c}(s) \tilde{I}_B^{n,\perp,c}(s) (\delta\sigma)^2. \end{array}\right\} \quad (3.23)$$

To obtain bounds on the quadratic variation, we first need some estimates on the evolution of the small particles. If $A \cap B = \emptyset$, then

$$\left.\begin{array}{c} 1_A(\tilde{q}_n(s,\lambda,\imath,\omega)) 1_B(\tilde{q}_n(s,\lambda,\imath,\omega)) = 0 \ \forall \omega; \\ E 1_A^c(\tilde{q}_n(s,\lambda,\imath)) 1_B^c(\tilde{q}_n(s,\lambda,\imath)) = -E 1_A(\tilde{q}_n(s,\lambda,\imath)) E 1_B(\tilde{q}_n(s,\lambda,\imath)). \end{array}\right\} \quad (3.24)$$

Recall that $s = l\delta\sigma$ for some $l \in \mathbf{N}$. From Hypothesis 2.1, (2.6), (2.7), and Corollary 2.9 in addition to the independence of $\hat{\zeta}_n^{\lambda,\imath}$ and $w_{0,n}^{\lambda,\imath}$, we obtain

$$E 1_A(\tilde{q}_n(s,\lambda,\imath)) \le \frac{|||\psi|||}{n^{d(p+1)}}. \tag{3.25}$$

Hence, by (3.12) in addition to (3.24) and (3.25) and the definitions of $\hat{n}_2$ before (3.12) we have for $n \geq \hat{n}_2$ and $A \cap B = \emptyset$

$$\left| E\left(\tilde{I}_A^{n,\perp,c}(s)\tilde{I}_B^{n,\perp,c}(s)\right)\right| \leq \frac{|||\psi|||}{n^{d(p+1)}}(\alpha+1). \tag{3.26}$$

For the case $A = B$, we now improve the estimate in step (ii) of the proof of Proposition 3.6, using (3.25):

$$\left.\begin{array}{c} E\left[\left(\tilde{I}_A^{n,\perp,c}(s)\right)^2\right] \\ \in \left(E\tilde{I}_A^{n,\perp}(s) - \frac{|||\psi|||}{n^{d(p+1)}}(\alpha+1), E\tilde{I}_A^{n,\perp}(s) + \frac{|||\psi|||}{n^{d(p+1)}}(\alpha+1)\right). \end{array}\right\} \tag{3.27}$$

Hence, by Lemma 3.2 we obtain for $A = (\bar{r}^\lambda] \subset C_{\tilde{c}_n}(0)$

$$E\left(\tilde{I}_A^{n,\perp,c}(s)\right)^2 \longrightarrow \alpha, \quad \text{as } n \longrightarrow \infty. \tag{3.28}$$

Let us use the following simplifying notation:

$$\tilde{q}_n(t,\lambda,\imath,\perp) := \begin{cases} \tilde{q}_n(t,\lambda,\imath), \text{ if } \imath \notin \mathbf{J}_n, \\ \diamond, \qquad\qquad \text{if } \imath \in \mathbf{J}_n; \end{cases}$$

$$1_A^c(\tilde{q}_n(s,\lambda,\imath,\perp)) := 1_A(\tilde{q}_n(s,\lambda,\imath,\perp)) - E1_A(\tilde{q}_n(s,\lambda,\imath,\perp));$$

$$h(c,s,\lambda,\imath) := \widehat{\sum_A} \hat{f}_n(s-,\bar{a})1_A^c(\tilde{q}_n(s,\lambda,\imath,\perp))\delta\sigma.$$

Note that $h(c,s,\lambda,\imath) = 0$ for $\imath \in \mathbf{J}_n$. Apparently,

$$\langle f(s-), \mathrm{d}M_n(s)\rangle = \sum_{\lambda,\imath} h(c,s,\lambda,\imath).$$

We obtain the (discrete time) increment of the quadratic variation

$$[\langle f(s-), \mathrm{d}M_n(s)\rangle] = \left(\sum_{\lambda,\imath} h(c,s,\lambda,\imath)\right)^2.$$

Suppose $(\lambda,\imath) \neq (\tilde{\lambda},\tilde{\imath})$. The independence of $\widehat{\mathcal{F}}_{n,s-}$ and $\mathcal{G}_{n,s}$ (cf. step (iii) in the proof of Proposition 3.6) in addition to the independence in $\lambda,\imath$ and the mean zero property of $1_A^c(\tilde{q}_n(s,\lambda,\imath))$ yields

$$E(h(c,s,\lambda,\imath)h(c,s,\tilde{\lambda},\tilde{\imath})|\widehat{\mathcal{F}}_{n,s-}) = 0.$$

Therefore,

$$\begin{aligned} E\left[\left(\sum_{\lambda,\imath} h(c,s,\lambda,\imath)\right)^2 |\widehat{\mathcal{F}}_{n,s-}\right] &= \sum_{\lambda,\imath} E\left[(h(c,s,\lambda,\imath))^2|\widehat{\mathcal{F}}_{n,s-}\right] \\ &= (\delta\sigma)^2\widehat{\sum_A}\left(\hat{f}_n(s-,\bar{a})\right)^2\sum_{\lambda,\imath} E\left(1_A^c(\tilde{q}_n(s,\lambda,\imath,\perp))\right)^2 \end{aligned}$$

$$+(\delta\sigma)^2\widehat{\sum_A}\hat{f}_n(s-,\bar{a})\widehat{\sum_B}\hat{f}_n(s-,\bar{b})1_{A\cap B=\emptyset}\sum_{\lambda,\imath} E1_A^c(\tilde{q}_n(s,\lambda,\imath,\perp))\,1_B^c(\tilde{q}_n(s,\lambda,\imath,\perp)).$$

Let $|\widehat{\sum}|$ be the number of nontrivial terms in $\widehat{\sum}$. By (3.6) the summation is restricted to the set of all $A = [\bar{r}^{\tilde{\lambda}})$ for some $\tilde{\lambda}$ such that $[\bar{r}^{\tilde{\lambda}}) \subset C_{\tilde{c}_n}$. Therefore, by (3.4)

$$\left|\widehat{\sum}\right| \le \tilde{c}_n^d n^d = \bar{c}^d n^{[(d+3\delta)\vee(\tilde{p}+\zeta+\delta)]d} n^d.$$

Using (3.24), (3.25), (3.12), the Cauchy-Schwarz inequality in the summation "$\widehat{\sum}$," (1.3) and the definition of $\widehat{f}_n$, the double sum of mixed terms in $\widehat{\sum}$ can be estimated from above as follows:

$$\begin{aligned} &|(\delta\sigma)^2 \widehat{\sum_A} \widehat{f}_n(s-,\bar{a}) \widehat{\sum_B} \widehat{f}_n(s-,\bar{b}) 1_{A\cap B=\emptyset} \sum_{\lambda,\iota} E 1_A^c(\tilde{q}_n(s,\lambda,\iota,\perp)) 1_B^c(\tilde{q}_n(s,\lambda,\iota,\perp))| \\ &\quad \le (\alpha+1)|||\psi|||(\delta\sigma)\bar{c}^d n^{d([(d+3\delta)\vee(\tilde{p}+\zeta+\delta)]-p)} \int (f(s-,q))^2 \mathrm{d}q. \end{aligned}$$

Next, by (3.24) and Lemma 3.2

$$(\delta\sigma)^2 \widehat{\sum} \left(\widehat{f}_n(s-,\bar{a})\right)^2 \sum_{\lambda,\iota} E\left(1_A^c(\tilde{q}_n(s,\lambda,\iota,\perp))\right)^2 \in (x,y),$$

where $x = (\delta\sigma)^2 \widehat{\sum}\left(\widehat{f}_n(s-,\bar{a})\right)^2 \left[\alpha - o(n^{d+\delta})\right]\left(1 - |||\psi|||n^{-d(p+1)}\right)$ and $y = (\delta\sigma)^2 \widehat{\sum}\left(\widehat{f}_n(s-,\bar{a})\right)^2 \left[\alpha + o(n^{d+\delta})\right](1 + |||\psi|||n^{-d(p+1)})$ with nonnegative functions $o(\cdot)$. Finally, let

$$\hat{n}_3 \ge \hat{n}_2 \quad \text{and such that } \forall n \ge \hat{n}_3: \quad 0 \le o(n^{d+\delta}) \le \tfrac{\alpha}{3} \text{ and } |||\psi|||n^{-d(p+1)} < \tfrac{1}{4}.$$

Before continuing, let us introduce an abbreviated notation for $L_2(\mathbf{R}^d, \mathrm{d}r)$, the real-valued functions on $\mathbf{R}^d$ that are square integrable with respect to the Lebesgues measures $\mathrm{d}r$.

$$\left.\begin{aligned} &\mathbf{H}_0 := L_2(\mathbf{R}^d, \mathrm{d}r), \quad \text{and for } f, g \in \mathbf{H}_0: \\ &\langle f, g\rangle_0 := \int f(q)g(q)\mathrm{d}q, \quad |f|_0 := \sqrt{< f, f >_0}. \end{aligned}\right. \tag{3.29}$$

Thus, for all $n \ge \hat{n}_3$ and $\forall s \in (0, \hat{t}]$

$$\left.\begin{aligned} &\delta\sigma|\widehat{f}_n(s-)|_0^2 \frac{\alpha}{2} \\ &< (\delta\sigma)^2 \widehat{\sum} \left(\widehat{f}_n(s-,\bar{a})\right)^2 \sum_{\lambda,\iota} E\left(1_A^c(\tilde{q}_n(s,\lambda,\iota,\perp))\right)^2 \\ &< \delta\sigma\left|\widehat{f}_n(s-)\right|_0^2 2\alpha \le \delta\sigma\|f(s-)\|_0^2 2\alpha. \end{aligned}\right\} \tag{3.30}$$

We obtain altogether for $n \ge \hat{n}_3$ from the previous estimates and (3.30)[8]

[8] Here, the duality $< f(s-), \mathrm{d}M_n(s) >$ is identical to the scalar product $< f(s-), \mathrm{d}M_n(s) >_0$. However, by Proposition 3.7, $M_n(\cdot)$ will converge to a distribution-valued process and for certain distributions the scalar product $< \cdot, \cdot >_0$ may be extended to a duality between "test functions" and "distributions." The notation $< f(s-), \mathrm{d}M_n(s) >$ indicates this fact. We refer to Sect. 15.1.3 for more details.

$$\left.\begin{array}{r}E\left[\left(\sum_{\lambda,\iota} h(c,t,\lambda,\iota)\right)^2 |\widehat{\mathcal{F}}_{n,s-}\right] \\ = E\left([\langle f(s-), dM_n(s)\rangle] \,|\widehat{\mathcal{F}}_{n,s-}\right) \\ \le (\alpha+1)\delta\sigma \int (f(s-,q))^2 \mathrm{d}q \left[1 + |||\psi|||\bar{c}^d n^{\mathrm{d}([(d+3\delta)\vee(\tilde{p}+\zeta+\delta)]-p)}\right].\end{array}\right\} \quad (3.31)$$

We also need an estimate of the "quadratic variation" of the quadratic variation $[< f(s-), \mathrm{d}M_n(s) >]$, which, for a pure jump process, is the sum of the 4th powers of the jump sizes. Moreover, this estimate is used for a law of large numbers (LLN) argument for the quadratic variation of functionals of M_n. Note that the $\widehat{f}_n(s-, \bar{a}_i)$ are measurable with respect to $\widehat{\mathcal{F}}_{n,s-}$ and the $1_{A_i}{}^c(\tilde{q}_n(s,\lambda,\iota,\perp))$ are independent of $\widehat{\mathcal{F}}_{n,s-}$. Therefore, we have

$$E\left(\left(\sum_{\lambda,\iota} h(c,s,\lambda,\iota)\right)^4 |\widehat{\mathcal{F}}_{n,s-}\right)$$

$$= (\delta\sigma)^4 \widehat{\sum_{A_1}}\cdots\widehat{\sum_{A_4}}\prod_{i=1}^4 \widehat{f}_n(s-,\bar{a}_i) E\prod_{i=1}^4\left(\sum_{\lambda,\iota} 1^c_{A_i}(\tilde{q}_n(s,\lambda,\iota,\perp))\right).$$

The mean zero property of $1_{A_i}{}^c(\tilde{q}_n(s,\lambda,\iota,\perp))$ and independence in λ,ι yields:

$$\left.\begin{array}{l} E\prod_{i=1}^4\left(\sum_{\lambda,\iota} 1_{A_i}{}^c(\tilde{q}_n(s,\lambda,\iota,\perp))\right) = \sum_{\lambda,\iota} E\prod_{i=1}^4 1_{A_i}{}^c(\tilde{q}_n(s,\lambda,\iota,\perp)) \\ + \sum_{(\lambda,\iota)\neq(\tilde{\lambda},\tilde{\iota})} E\left(\prod_{i\in\{1,2\}} 1_{A_i}{}^c(\tilde{q}_n(s,\lambda,\iota,\perp))\right) E\left(\prod_{j\in\{3,4\}} 1_{A_j}{}^c(\tilde{q}_n(s,\tilde{\lambda},\tilde{\iota},\perp))\right) \\ + \sum_{(\lambda,\iota)\neq(\tilde{\lambda},\tilde{\iota})} E\left(\prod_{i\in\{1,3\}} 1_{A_i}{}^c(\tilde{q}_n(s,\lambda,\iota,\perp))\right) E\left(\prod_{j\in\{2,4\}} 1_{A_j}{}^c(\tilde{q}_n(s,\tilde{\lambda},\tilde{\iota},\perp))\right) \\ + \sum_{(\lambda,\iota)\neq(\tilde{\lambda},\tilde{\iota})} E\left(\prod_{i\in\{1,4\}} 1_{A_i}{}^c(\tilde{q}_n(s,\lambda,\iota,\perp))\right) E\left(\prod_{j\in\{2,3\}} 1_{A_j}{}^c(\tilde{q}_n(s,\tilde{\lambda},\tilde{\iota},\perp))\right).\end{array}\right\} \quad (3.32)$$

In the expansion of $E((\sum_{\lambda,\iota} h(c,t,\lambda,\iota))^4|\widehat{\mathcal{F}}_{n,s-})$ the resulting terms with $(\lambda,\iota)\neq(\tilde{\lambda},\tilde{\iota})$ are estimated as follows, where we show the computations only for the first of the three terms, since the other terms can obviously be estimated in the same way.

$$
\left.\begin{aligned}
&(\delta\sigma)^4 \sum \sum_{(\lambda,\iota)\neq(\tilde{\lambda},\tilde{\iota})} \widehat{\sum_{A_1}}\widehat{\sum_{A_2}} \prod_{i\in\{1,2\}} \widehat{f}_n(s-,\bar{a}_i) E\left(\prod_{i\in\{1,2\}} 1^c_{A_i}(\tilde{q}_n(s,\lambda,\iota,\perp))\right)\\
&\quad\times \widehat{\sum_{A_3}}\widehat{\sum_{A_4}} \prod_{j\in\{3,4\}} \widehat{f}_n(s-,\bar{a}_j) E\left(\prod_{j\in\{3,4\}} 1^c_{A_j}(\tilde{q}_n(s,\tilde{\lambda},\tilde{\iota},\perp))\right)\\
&\qquad\qquad \leq \left(E\left(\left(\sum_{\lambda,\iota} h(c,t,\lambda,\iota)\right)^2 |\widehat{\mathcal{F}}_{n,s-}\right)\right)^2\\
&\qquad\qquad = (E([\langle f(s-), \mathrm{d}M_n(s)\rangle]|\widehat{\mathcal{F}}_{n,s-}))^2.
\end{aligned}\right\}
\tag{3.33}
$$

Next, the other case for the fourth moment can be written as follows:

$$
\widehat{\sum_{A_1,A_2,A_3,A_4}} \sum_{\lambda,\iota} (\delta\sigma)^4 \prod_{i=1}^4 \widehat{f}_n(s-,\bar{a}_i) E \prod_{i=1}^4 [1_{A_i}(\tilde{q}_n(s,\lambda,\iota,\perp)) - E1_{A_i}(\tilde{q}_n(s,\lambda,\iota,\perp))].
$$

Multiplying out, we group the terms into different cases. One case contains terms of the form $E\prod_{i=1}^4 1_{A_i}(\tilde{q}_n(s,\lambda,\iota,\perp))$. However, this term is only nontrivial if $A_1 = A_2 = A_3 = A_4$, because $\tilde{q}_n(s,\lambda,\iota,\perp)$ cannot be in two disjoint sets at the same time. Therefore, instead of quadruple summation, we have only the summation over A_1, for example. In the other cases, the quadruple summation also reduces accordingly, except in the case $E\prod_{i=1}^4 E1_{A_i}(\tilde{q}_n(s,\lambda,\iota,\perp))$. If we get more than one factor $E1_{A_i}(\tilde{q}_n(s,\lambda,\iota,\perp))$ in those products we can use estimate (3.25) as a bound on this factor. By (3.12) we obtain for $n \geq \hat{n}_3$

$$
\begin{aligned}
&\widehat{\sum_{A_1,A_2,A_3,A_4}} \sum_{\lambda,\iota} (\delta\sigma)^4 \prod_{i=1}^4 \widehat{f}_n(s-,\bar{a}_i) E \prod_{i=1}^4 1^c_{A_i}(\tilde{q}_n(s,\lambda,\iota,\perp))\\
&\qquad \leq (\alpha+1) \sum_{a_1} |\widehat{f}_n(s-,\bar{a}_1))|^4 (\delta\sigma)^4\\
&\quad +(\alpha+1) \widehat{\sum_{a_k,a_l}} |\widehat{f}_n(s-,\bar{a}_k)|^3 |\widehat{f}_n(s-,\bar{a}_l)| \cdot |||\psi||| n^{-d(p+1)} (\delta\sigma)^4\\
&+(\alpha+1) \widehat{\sum_{a_k,a_l,a_{\tilde{k}}}} |\widehat{f}_n(s-,\bar{a}_k)|^2 |\widehat{f}_n(s-,\bar{a}_l)\widehat{f}_n(s-,\bar{a}_{\tilde{k}})| \cdot |||\psi|||^2 n^{-2d(p+1)} (\delta\sigma)^4\\
&\quad +(\alpha+1) \widehat{\sum_{\{a_{k_l},l=1,..,4\}}} \prod_{i=1}^4 |\widehat{f}_n(s-,\bar{a}_{k_i})|^4 |||\psi|||^3 n^{-3d(p+1)} (\delta\sigma)^4.
\end{aligned}
$$

Converting the summation into integration,

$$\left.\begin{aligned}
\widehat{\sum_{A_1,A_2,A_3,A_4}} \sum_{\lambda,\iota} (\delta\sigma)^4 \prod_{i=1}^{4} \widehat{f}_n(s-,\bar{a}_i) E \prod_{i=1}^{4} 1_{A_i}{}^{c}(\tilde{q}_n(s,\lambda,\iota,\perp))& \\
\leq (\delta\sigma)^4(\alpha+1)n^d \int |\widehat{f}_n(s-,q)|^4 dq& \\
+(\delta\sigma)^4 |||\psi|||(\alpha+1)n^{d-dp}\int |\widehat{f}_n(s-,q)|^3 dq \int |\widehat{f}_n(s-,q)| dq& \\
+(\delta\sigma)^4 |||\psi|||^2(\alpha+1)n^{d-2dp}\int |\widehat{f}_n(s-,q)|^2 dq \left(\int |\widehat{f}_n(s-,q)| dq\right)^2& \\
+(\delta\sigma)^4 |||\psi|||^3(\alpha+1)n^{d-3dp}\left(\int |\widehat{f}_n(s-,q)| dq\right)^4.&
\end{aligned}\right\} \quad (3.34)$$

From (3.33) and (3.34) in addition to (3.31) we obtain

$$\left.\begin{aligned}
E\left(\left(\sum_{\lambda,\iota} h(c,t,\lambda,\iota)\right)^4 |\widehat{\mathcal{F}}_{n,s-}\right) &= E\left(([\langle f(s-), dM_n(s)\rangle])^2 |\widehat{\mathcal{F}}_{n,s-}\right) \\
&\leq (\delta\sigma)^2 c(\alpha+1)^2 \sum_{1\leq i,j\leq 4}\left(\int |\widehat{f}_n|^i(s-,q)dq\right)^j,
\end{aligned}\right\} \quad (3.35)$$

where $c > 0$ is some finite constant, which depends only on the powers $i = 1,\ldots,4$, $|||\psi|||$ and the constant $\bar{c}$ from (3.4). Finally, from the calculations in (3.32) and (3.33), we obtain for $n \geq \hat{n}_3$ by the estimate in (ii) of the proof of Proposition 3.6.

$$E\left(\tilde{I}_A^{n,\perp,c}(s)\right)^4 \leq c(\alpha+1)^2. \quad (3.36)$$

The previous computations were a direct analysis of the quadratic variation of Λ_j. To apply the results by Kurtz and Protter (loc. cit.) we need to separate a driving term in $\Lambda(\cdot,\mathbf{C}_n,\tilde{r}_n^i,\tilde{I}^{n,\perp,c})$, which depends only on $\tilde{I}^{n,\perp,c}$ and not on the "state" $\tilde{r}_n^i$. Let us describe this term in a convenient way.

Let $\hat{\mathbf{N}} := \mathbf{N}\cup\{0\}$ denote the nonnegative integers. The multiindices from $\hat{\mathbf{N}}^d$ will be denoted in bold face. Choose the normalized Hermite functions $\{\phi_{\mathbf{k}}\}_{\mathbf{k}\in\hat{\mathbf{N}}^d}$ as a complete orthonormal system (CONS) in $\mathbf{H}_0$.[9] Specializing to $f(s,\cdot,\omega) := \phi_{\mathbf{k}}$, we set

$$m_{\mathbf{k},n}(t) := \int_0^t \langle\phi_{\mathbf{k}}, dM_n(s)\rangle = \sum_{s\leq t}\int \phi_{\mathbf{k}}(q)\left[\sum_{\lambda\in\mathbf{C}_n} 1_{(\bar{r}^\lambda]}(q)n^d\tilde{I}^{n,\perp,c}_{(\bar{r}^\lambda]}(s)\right] dq\,\delta\sigma. \quad (3.37)$$

As before, $<\cdot,\cdot>$ is a duality between generalized functions and smooth functions which extends the usual scalar product $<\cdot,\cdot>_0$ on $\mathbf{H}_0$. Obviously, $m_{\mathbf{k},n}(\cdot)$

[9] The precise definition and the proof of completeness are provided in Sect. 15.1.3, (15.11) and Proposition 15.8.

are mean zero square integrable martingales and do not depend on the state $\bar{\tilde{r}}^i_n(s-)$. Therefore, they are a perfect candidate for the martingale central limit theorem.[10] Since $|\phi_{\mathbf{k}}|_0 = 1$, we obtain from (3.31) and Doob's inequality[11] that for $n \geq \hat{n}_3$ and for all $\mathbf{k}$

$$E \sup_{t\leq \hat{t}} (m_{\mathbf{k},n}(t))^2 \leq 4(\alpha+1)\hat{t}\left[1 + |||\psi|||\bar{c}^d n^{d([(d+3\delta)\vee(\tilde{p}+\zeta+\delta)]-p)}\right]. \quad (3.38)$$

A Fourier expansion yields

$$\Lambda(t, \mathbf{C}_n, \tilde{r}^i_n, \tilde{I}^{n,\perp,c}) = \frac{1}{\hat{m}}\sum_{s\leq t}\sum_{\mathbf{k}}\int \hat{G}_n(r,q)\phi_{\mathbf{k}}(q)\mathrm{d}q\, m_{\mathbf{k},n}(\mathrm{d}s). \quad (3.39)$$

(3.23) becomes

$$[m_{\mathbf{k},n}(t), m_{\mathbf{l},n}(t)] = \sum_{s\leq t}\widehat{\sum_A\sum_B}\widehat{\phi}_{\mathbf{k},n}(\bar{a})\widehat{\phi}_{\mathbf{l},n}(\bar{b})\tilde{I}^{n,\perp,c}_A(s)\tilde{I}^{n,\perp,c}_B(s)(\delta\sigma)^2. \quad (3.40)$$

By Lemma 3.1, the right-hand side of (3.40) is a sum of independent random variables (in the time variable s!). Referring to (3.4) for the definition of $\tilde{c}_n$, we note that by (1.3), (3.26), and (3.6), the notational convention (3.22) and the Cauchy-Schwarz inequality[12]

$$\begin{aligned}
&\left|\widehat{\sum_A\sum_B} 1_{A\cap B=\emptyset}\widehat{\phi}_{\mathbf{k},n}(\bar{a})\widehat{\phi}_{\mathbf{l},n}(\bar{b})(\delta\sigma)E\tilde{I}^{n,\perp,c}_A(s)\tilde{I}^{n,\perp,c}_B(s)\right| \\
&\leq \frac{n^d}{n^{p(d+1)}}|||\psi|||(\alpha+1)\int|\phi_{\mathbf{k}}|(x)1_{C_{\tilde{c}_n}(0)}(x)dx\int|\phi_{\mathbf{l}}|(x)1_{C_{\tilde{c}_n}(0)}(x)\mathrm{d}x \\
&\leq \frac{n^d}{n^{p(d+1)}}|||\psi|||(\alpha+1)\tilde{c}^d_n.
\end{aligned}$$

We obtain from the above inequality in addition to the scaling assumption (1.3), (3.4), (3.3), (3.4), and (3.28), Hypothesis 2.2 and Lebesgue's dominated convergence theorem (applied to $\int\phi_{\mathbf{k},n}(q)\phi_{\mathbf{l},n}(q)\mathrm{d}q$)

$$\sum_{s\leq t}\widehat{\sum_A\sum_B}\widehat{\phi}_{\mathbf{k},n}(\bar{a})\widehat{\phi}_{\mathbf{l},n}(\bar{b})(\delta\sigma)^2E\tilde{I}^{n,\perp,c}_A(s)\tilde{I}^{n,\perp,c}_B(s)\longrightarrow \alpha t\delta_{\mathbf{k},\mathbf{l}}, \quad (3.41)$$

where $\delta_{\mathbf{k},\mathbf{l}} = 1$, if $\mathbf{k} = \mathbf{l}$ and $\delta_{\mathbf{k},\mathbf{l}} = 0$ otherwise. As before, we use the following simplifying notation:

$$\tilde{h}(c,s,\lambda,\iota,\mathbf{k}) := \sum_A\widehat{\phi}_{\mathbf{k},n}(\bar{a})1^c_A(\tilde{q}_n(s,\lambda,\iota,\perp))\delta\sigma, \quad \text{as } n\longrightarrow\infty\,.$$

[10] Cf. Theorem 15.39 in Sect. 15.2.3.

[11] Cf. Theorem 15.32 in our Sect. 15.2.3.

[12] Proposition 15.4 in Sect. 15.1.2.

Set $c(\phi_{\mathbf{k}}) := \sup_n \sum_{1\le i,j\le 4} (\int |\widehat{\phi}_{\mathbf{k},n}|^i(q)dq)^j$ and note that $c(\phi_{\mathbf{k}}) < \infty$. We use independence and the mean zero property in addition to (3.35) and the fact that the summation $\sum_{s\le t}$ has $\frac{t}{\delta\sigma}$ terms. Therefore, the variance of (3.40) satisfies for $n \ge \hat{n}_3$:

$$\left.\begin{aligned} &E([m_{\mathbf{k},n}(t), m_{\mathbf{l},n}(t)] - E[m_{\mathbf{k},n}(t), m_{\mathbf{l},n}(t)])^2 \\ &\le \sum_{s\le t} E(\sum_{\lambda,\iota} \tilde{h}(c,s,\lambda,\iota,\mathbf{k}))^2 \left(\sum_{\lambda,\iota} \tilde{h}(c,s,\lambda,\iota,\mathbf{l})\right)^2 \\ &\le \frac{1}{2}\sum_{s\le t}\left[E\left(\left(\mathrm{d}\left[m_{\mathbf{k},n}(s)\right]\right)^2\right) + E\left(\left(\mathrm{d}\left[m_{\mathbf{l},n}(s)\right]\right)^2\right)\right] \\ &\le \delta\sigma t c(\alpha+1)^2 \frac{1}{2}[c(\phi_{\mathbf{k}}) + c(\phi_{\mathbf{l}})] \quad \text{by (3.34)}, \\ &\longrightarrow 0, \text{ as } n \longrightarrow \infty . \end{aligned}\right\} \tag{3.42}$$

Consequently, we obtain the weak law of large numbers from Lemma 3.1 in addition to (3.31), (3.41), and (3.42):

$$\left.\begin{aligned} \forall \mathbf{k}, \mathbf{l}, \forall t \ge 0 \ \ [m_{\mathbf{k},n}(t), m_{\mathbf{l},n}(t)] \longrightarrow t\alpha\delta_{\mathbf{k},\mathbf{l}} \\ \text{in probability, as } n \longrightarrow \infty. \end{aligned}\right\} \tag{3.43}$$

It is required to check a jump size condition in the martingale central limit theorem. Recall that we extended our time-discrete quantities to continuous time in a canonical way, where the processes are considered cadlag jump processes.[13] By (3.37)

$$m_{\mathbf{k},n}(t) - m_{\mathbf{k},n}(t-) = \sum_A \widehat{\phi}_{\mathbf{k},n}(\bar{a}) \tilde{I}_A^{n,\perp,c}(t)\delta\sigma \qquad \forall \mathbf{k}.$$

Hence, by (3.42)

$$\left.\begin{aligned} &\left\{E\left[\sup_{t\le\hat{t}} |m_{\mathbf{k},n}(t) - m_{\mathbf{k},n}(t-)|\right]\right\}^2 \\ &\le \left\{\sum_{t\le\hat{t}} E(\sum_{\lambda,\iota} \tilde{h}(c,t,\lambda,\iota,\mathbf{k}))^4\right\}^{\frac{1}{2}} \\ &\le \sqrt{c}(\alpha+1)\sqrt{c(\phi_{\mathbf{k}})}\sqrt{\delta\sigma\hat{t}} \\ &\longrightarrow 0\,, \text{ as } n \longrightarrow \infty. \end{aligned}\right\} \tag{3.44}$$

Let $\mathbf{k}_1, \ldots, \mathbf{k}_j, \ldots$ be an enumeration of our multiindices. Further, let

$$\beta_{\mathbf{k}_1}(\cdot), \ldots, \beta_{\mathbf{k}_j}(\cdot), \ldots$$

be independent identically distributed (i.i.d.) real-valued standard Brownian motions. By (3.43) and (3.44) we may employ the martingale central limit theorem (Theorem 15.39):

[13] Cf. the procedure at the beginning of Chap. 2.

Proposition 3.7.

$$\forall j \in \mathbf{N} \ (m_{\mathbf{k}_1,n}(\cdot), \ldots, m_{\mathbf{k}_j,n}(\cdot)) \Longrightarrow (\sqrt{\alpha}\beta_{\mathbf{k}_1}(\cdot), \ldots, \sqrt{\alpha}\beta_{\mathbf{k}_j}(\cdot)), \quad as\ n \longrightarrow \infty.$$

□

The process $(\sqrt{\alpha}\beta_{\mathbf{k}_1}(\cdot), \ldots, \sqrt{\alpha}\beta_{\mathbf{k}_j}(\cdot), \ldots)$ is cylindrical on the space of square summable sequences, i.e., for each t the distribution of $(\sqrt{\alpha}\beta_{\mathbf{k}_1}(t), \ldots, \sqrt{\alpha}\beta_{\mathbf{k}_j}(t), \ldots)$ is (only) a finitely additive measure on that sequence space. Therefore,

$$W(\cdot) := \sum_{\mathbf{k}\in\hat{\mathbf{N}}^d} \beta_{\mathbf{k}}(\cdot)\phi_{\mathbf{k}} \tag{3.45}$$

is a standard cylindrical Brownian motion in $\mathbf{H_0}$, i.e., its covariance operator on $\mathbf{H}_0$ is the identity operator. Further, without loss of generality, we may assume that $W(\cdot)$ from (3.45) has representation[14]

$$W(t) \equiv \sum_{\mathbf{k}\in\hat{\mathbf{N}}^d} \int_0^t \int \phi_{\mathbf{k}}(q) w(\mathrm{d}q, \mathrm{d}u)\phi_{\mathbf{k}}. \tag{3.46}$$

We have

$$M_n(\cdot) = \sum_{\mathbf{k}\in\hat{\mathbf{N}}^d} m_{\mathbf{k},n}(\cdot)\phi_{\mathbf{k}}. \tag{3.47}$$

We now analyze the "covariance" operator

$$\Theta_n(s) := \frac{1}{\delta s} E|[M_n]|(s) - |[M_n]|(s-), \tag{3.48}$$

where $|[M_n]|(\cdot)$ is the tensor quadratic variation associated with $E[\langle f, \mathrm{d}M_n(s)\rangle, \langle g, \mathrm{d}M_n(s)\rangle]$, where $f, g \in \mathbf{H}_0$.[15] Note that the increments $[\langle f, \mathrm{d}M_n(s)\rangle, \langle g, \mathrm{d}M_n(s)\rangle]$ are independent of the "past" $\widehat{\mathcal{F}}_{n,s-}$. Therefore, we may use the conditional expectation instead of the unconditional one. Recalling (3.23) and the notational convention (3.22), $\Theta_n(s)$ is defined by

$$\left.\begin{array}{l} E([\langle f, \mathrm{d}M_n(s)\rangle, \langle g, \mathrm{d}M_n(s)\rangle]|\widehat{\mathcal{F}}_{n,s-}) \\ = \langle \Theta_n(s) f, g\rangle_0 \delta\sigma = \langle \Theta_n(s)\widehat{f}_n, \widehat{g}_n\rangle_0 \delta\sigma. \end{array}\right\} \tag{3.49}$$

Apparently, $\Theta_n(s)$ is nonnegative (i.e., it has a nonnegative spectrum); it is symmetric and bounded (by (3.30) and $\|\widehat{f}\|_0 \le \|f\|_0$). Additionally, $\Theta_n(s)$ has the finite dimensional subspace $\mathbf{H}_{0,n}$ as its range, where $\mathbf{H}_{0,n} := \{\widehat{f}_n : f \in \mathbf{H}_0\}$. Therefore, it is compact and has a discrete real-valued spectrum. Note that this spectrum is deterministic by the independence of the increments $\langle f, \mathrm{d}M_n(s)\rangle$ and $\langle g, \mathrm{d}M_n(s)\rangle$ of

[14] It follows from Sect. 15.2.2 that both (3.45) and the right-hand side of (3.46) define $\mathbf{H}_0$-valued standard cylindrical motions. In particular, both processes are identically distributed. Cf. (15.69). Cf. also (4.14) and (4.29) in Chap. 4.

[15] Cf. Sect. 15.2.5, (15.177). For continuous martingales $\Theta_n(s)$ would be defined as the covariance of $\frac{dM_n(s)}{ds}$ on $\mathbf{H}_0$. Cf. Sect. 15.2.3, Definition 15.43.

$\widehat{\mathcal{F}}_{n,s-}$. Consequently, denoting by $\mu_k(s)$ the eigenvalues of $\Theta_n(s)$, we obtain from (3.29) for $n \geq \hat{n}_3$

$$\max_{0<s\leq \hat{t}} \sup_k \mu_k(s) \leq 2\alpha. \tag{3.50}$$

It is well known[16] that $\Theta_n(s)$ is symmetric and nonnegative in addition to (3.50) implies

$$\left.\begin{array}{l} \sup\limits_{0\leq s\leq \hat{t}} \|\Theta_n(s)\|_{\mathcal{L}(\mathbf{H_0})} \leq 2\alpha, \\ \text{and } 0 \leq \langle \Theta_n(s) f, f\rangle_0 \leq 2\alpha \langle f, f\rangle_0 \ \forall f \in \mathbf{H}_0, \ \forall n \in \mathbf{N}\ 0 \leq s \leq \hat{t}. \end{array}\right\} \tag{3.51}$$

Next, set

$$\mathcal{J}_n := \{\tilde{f} := (f_1, \ldots, f_{\mathrm{d}N})^T : f_i \in L_{2,\widehat{\mathcal{F}}_{n,\cdot-}}[0,\hat{t}] \times \Omega \quad \text{and } f_i(\cdot) \text{ is cadlag }\},$$

where $L_{2,\widehat{\mathcal{F}}_{n,\cdot-}}([0,\hat{t}] \times \Omega)$ is the space of those elements from $L_{2,\widehat{\mathcal{F}}_{n,\cdot-}}([0,\hat{t}] \times \mathbf{R}^d \times \Omega)$, which do not depend on $q \in \mathbf{R}^d$. Further, let $\mathcal{B}_0$ the Borel σ-algebra on $\mathbf{H}_0$. Set

$$\mathcal{A}_n := \{\tilde{f} : \exists \tilde{\phi}_1, \ldots, \tilde{\phi}_L \in \mathbf{H}_0,\ L \in \mathbf{N}\text{such that } \tilde{f}(s) := \sum_{k=1}^{L} \tilde{f}^k(s)\tilde{\phi}_k,\ \tilde{f}(\cdot) \in \mathcal{J}_n\}.$$

Further, set

$$\mathcal{A}_{n,0} := \{\tilde{f} \in \mathcal{A}_n \quad \text{such that } |\tilde{f}(t,\cdot,\omega)|_{0,\mathrm{d}N} \leq 1\},$$

where $|h|^2_{0,\mathrm{d}N} := \sum_{j=1,\ldots,\mathrm{d}N} |h_j|_0^2$ for an $\mathbf{R}^{\mathrm{d}N}$-valued square integrable function $h = (h_1, \ldots, h_{\mathrm{d}N})$. The stochastic integral of the components of $\tilde{f} \in \mathcal{A}_n$ with respect to M_n has the representation[17]

$$\int_0^t \langle f_i(q,s-), \mathrm{d}M_n(s)\rangle = \sum_{k=1}^{L} \int_0^t f_i^k(s-)\langle \tilde{\phi}_k, \mathrm{d}M_n(s)\rangle,\ i = 1, \ldots, \mathrm{d}N. \tag{3.52}$$

Proposition 3.8.

$$\limsup_{c\longrightarrow\infty} \sup_{f\in\mathcal{A}_{n,0}} \sup_n P\left\{\sup_{t\leq\hat{t}} \sum_{i\in\{1,\ldots,\mathrm{d}N\}} \left|\int_0^t \langle f_i(q,s), \mathrm{d}M_n(s)\rangle\right| > c\right\} = 0,$$

and, consequently, the sequence M_n is uniformly tight in the sense of Definition 15.53 in Sect. 15.2.5.

[16] cf., e.g., Kato (1976), Chap. Vol. 2, (2.4). The conclusion (3.51) does not depend on the discreteness of the spectrum.

[17] Cf. (15.176).

Proof. $\sum_{i\in\{1,\ldots,dN\}} \int_0^t < F_i(q,s-), \mathrm{d}M_n(s) > |$ is a submartingale. Hence, by the submartingale inequality[18] for $c > 0$ and the Cauchy-Schwarz inequality

$$
\begin{aligned}
&P\left\{\sup_{t\le\hat{t}} \sum_{i\in\{1,\ldots,dN\}} \left|\int_0^t \langle f_i(\cdot), s), \mathrm{d}M_n(s)\rangle\right| > c\right\} \\
&\quad\le \sum_{i\in\{1,\ldots,dN\}} \frac{1}{c} E\left|\int_0^{\hat{t}} \langle f_i(\cdot), \mathrm{d}M_n(s)\rangle\right| \\
&\quad\le \sum_{i\in\{1,\ldots,dN\}} \frac{1}{c}\sqrt{E\left(\int_0^{\hat{t}} \langle f_i(\cdot), \mathrm{d}M_n(s)\rangle\right)^2}
\end{aligned}
$$

By the rules for Itô integrals,[19] in addition to (3.49), (3.22), and (3.51),

$$
\begin{aligned}
E\left(\int_0^{\hat{t}} \langle f_i(\cdot), \mathrm{d}M_n(s)\rangle\right)^2 &= E\int_0^{\hat{t}} \langle f_i(\cdot), \mathrm{d}M_n(s)\rangle^2 \\
&= E\int_0^{\hat{t}} E(\langle f_i(\cdot), \mathrm{d}M_n(s)\rangle^2 | \widehat{\mathcal{F}}_{n,s-}) \\
&= \sum_{s\le\hat{t}} E\langle \Theta_n(s)\widehat{f_i}(\cdot), \widehat{f_i}(\cdot)\rangle\delta\sigma \\
&\le 2\alpha\sum_{s\le\hat{t}} E|f_i(\cdot)|_0^2\delta\sigma.
\end{aligned}
$$

Again by the Cauchy-Schwarz inequality and the assumption $\sum_{i\in\{1,\ldots,dN\}} |f_i(\cdot)|_0^2 \le 1$, the preceding steps imply

$$
\begin{aligned}
&P\left\{\sup_{t\le\hat{t}}\{1,\ldots,\mathrm{d}N\}| \int_0^t \langle f_i(\cdot), s), \mathrm{d}M_n(s) > | > c\right\} \\
&\le \frac{(\mathrm{d}N)}{c} 2\alpha\hat{t} \longrightarrow 0, \quad \text{as } c\longrightarrow\infty.
\end{aligned}
\tag{3.53}
$$

□

Now we set for $j = 1,\ldots,d$ (cf. (3.17))

$$
U_{2,j}^{i,n}(\cdot) := -\int_0^{\cdot} \exp[-\eta_n(\cdot - s)]\Lambda_j\left(\mathrm{d}s, \mathbf{C}_n, \tilde{R}_n^i, \tilde{I}^{n,\perp,c}\right). \tag{3.54}
$$

Although $U_{2,j}^{i,n}(t)$ in the last representation is a stochastic convolution integral, we may nevertheless apply the Itô formula[20] to the second factor of the process

[18] Theorem 15.32, (i).

[19] By a cadlag generalization of (15.151) – cf. Metivier and Pellaumail (1980) (Sects. 3.2, 4.2).

[20] For the Itô formula, cf. Sect. 15.2.5, Theorem 15.50.

$$\exp[-\eta_n t]\int_0^t \exp[\eta_n s]\mathrm{d}m_{i,j}(s),$$

where we use the abbreviation

$$m_{i,j}(s) := \Lambda_j(s, \mathbf{C}_n, \tilde{R}_n^i, \tilde{I}^{n,\perp,c}).$$

For a step function cadlag process $f(s)$, constant in $u \in ((k-1)\delta\sigma, k\,\delta\sigma)$ and evaluated at $s = k\,\delta\sigma,\ \ k = 0, 1, 2, \ldots,$ $f(s-)$ is the limit from the left. Again, by the cadlag generalization of (15.151) (cf. Metivier and Pellaumail (loc.cit.)),

$$|U_{2,j}^{i,n}(t)|^2 \le 2\int_0^t \exp[-2\eta_n(t-s)]\int_0^{s-}\exp[-\eta_n(s-u)]\mathrm{d}m_{i,j}(u)\mathrm{d}m_{i,j}(s) \\ +\exp[-2\eta_n t]\int_0^t \exp[2\eta_n s]\mathrm{d}[m_{i,j}(s)].$$

We apply the Itô formula again along with an estimate for the quadratic variation from Metivier and Pellaumail (loc.cit.)

$$\left(\int_0^t \exp[2\eta_n s]\mathrm{d}[m_{i,j}(s)]\right)^2 \le 2\int_0^t \exp[2\eta_n s]\int_0^{s-}\exp[2\eta_n u]\mathrm{d}[m_{i,j}(u)]\mathrm{d}[m_{i,j}(s)] \\ +\int_0^t \exp[4\eta_n s]\mathrm{d}[[m_{i,j}(s)]],$$

where $\mathrm{d}[[m_{i,j}(s)]]$ is the increment of the "quadratic variation" of the quadratic variation. Abbreviate

$$z_n(t) := 2\int_0^t \exp[-4\eta_n(t-s)]\int_0^{s-}\exp[-2\eta_n(s-u)]\,\mathrm{d}\,[m_{i,j}(u)]\mathrm{d}[m_{i,j}(s)] \\ +\int_0^t \exp[-4\eta_n(t-s)]\,\mathrm{d}\,[[m_{i,j}(s)]].$$

Thus,

$$\exp[-2\eta_n t]\int_0^t \exp[2\eta_n s]\mathrm{d}[m_{i,j}(s)] \le \sqrt{z_n(t)}.$$

Altogether, we obtain

$$\left.\begin{array}{c} |U_{2,j}^{i,n}(t)|^2 \\ \le 2\displaystyle\int_0^t \exp[-2\eta_n(t-s)]\int_0^{s-}\exp[-\eta_n(s-u)]\mathrm{d}m_{i,j}(u)\mathrm{d}m_{i,j}(s) + \sqrt{z_n(t)}. \end{array}\right\} \tag{3.55}$$

Next, by Theorem 15.42 of our Sect. 15.2.3 there is a $c(\hat{t}) < \infty$ such that

$$\left.\begin{array}{l} E\displaystyle\sup_{0\le t\le \hat{t}}\left|2\int_0^t \exp\left[-2\eta_n(t-s)\right]\int_0^{s-}\exp\left[-\eta_n(s-u)\right]\mathrm{d}m_{i,j}(u)\mathrm{d}m_{i,j}(s)\right| \\ \quad\le c(\hat{t})\left[E\left(\displaystyle\int_0^{\hat{t}}\left(2\int_0^{s-}\exp\left[-\eta_n(s-u)\right]\mathrm{d}m_{i,j}(u)\right)\mathrm{d}m_{i,j}(s)\right)^2\right]^{\frac{1}{2}}. \end{array}\right\} \tag{3.56}$$

Recall the representation of $m_{i,j}(s)$ after (3.21). By (3.31) and (2.2) there is an $\hat{n}_4 \geq \hat{n}_3$ such that for $n \geq \hat{n}_4$

$$E([\mathrm{d}m_{i,j}(s)]|\widehat{\mathcal{F}}_{n,s-}) \leq \delta\sigma c_{2,G,\alpha}, \tag{3.57}$$

where $c_{2,G,\alpha} := \max_{1\leq j\leq d} |G_j|_0^2 2(\alpha+1) < \infty$. Therefore, using conditional mathematical expectations, by the Itô formula

$$\left.\begin{aligned} &E\left(\int_0^{\hat{t}} (2\int_0^{s-} \exp[-\eta_n(s-u)]\,\mathrm{d}m_{i,j}(u)\mathrm{d}m_{i,j}(s)\right)^2 \\ &= E\left(\int_0^{\hat{t}} (2\int_0^{s-} \exp[-\eta_n(s-u)]\,\mathrm{d}m_{i,j}(u))^2 E\left(\mathrm{d}\left[m_{i,j}(s)\right]|\widehat{\mathcal{F}}_{n,s-}\right)\right) \\ &\qquad \leq 4\sum_{s\leq\hat{t}}\sum_{0<u<s} \exp[-2\eta_n(s-u)](c_{2,G,\alpha})^2(\delta\sigma)^2. \end{aligned}\right\} \tag{3.58}$$

Recalling that $u = l\,\delta\sigma,\ l \in \mathbf{N}$, the summation formula for geometric sums yields

$$\sum_{0<u<s} \exp[-2\eta_n(s-u)]\delta\sigma \leq \frac{(1-\exp[-2\eta_n s])\delta\sigma}{\exp[2\eta_n\,\delta\sigma]-1}.$$

Further,

$$\frac{\delta\sigma}{\exp[2\eta_n\,\delta\sigma]-1} = \frac{\delta\sigma}{2\eta_n \displaystyle\int_0^{\delta\sigma} \exp[2\eta_n u]\mathrm{d}u} \leq \frac{1}{2\eta_n}.$$

Since $\sum_{s\leq\hat{t}} \delta\sigma = \hat{t}$ the calculations (3.56)–(3.58) in addition to the assumptions on G imply

$$\left.\begin{aligned} &E\sup_{0<t\leq\hat{t}}\left|2\int_0^t \exp[-2\eta_n(t-s)]\int_0^{s-} \exp[-\eta_n(s-u)]\mathrm{d}m_{i,j}(u)\mathrm{d}m_{i,j}(s)\right| \\ &\qquad \leq 2c(\hat{t})c_{2,G,\alpha}\sqrt{\frac{\hat{t}}{2\eta_n}}. \end{aligned}\right\} \tag{3.59}$$

Let us now show that the last term in (3.55) tends to 0. Since the integrands and the integrator increments are positive,

$$\begin{aligned} &\sup_{t\leq\hat{t}} 2\int_0^t \exp[-4\eta_n(t-s)]\int_0^{s-} \exp[-2\eta_n(s-u)]\mathrm{d}[m_{i,j}(u)]\mathrm{d}[m_{i,j}(s)] \\ &\qquad \leq 2\int_0^{\hat{t}}\int_0^{s-} \exp[-2\eta_n(s-u)]\mathrm{d}[m_{i,j}(u)]\mathrm{d}[m_{i,j}(s)]. \end{aligned}$$

Therefore, proceeding as in the derivation of (3.58) and (3.59), using conditional expectations and the summation formula for geometric sums,

$$
\begin{aligned}
&E\sup_{t\le\hat t} 2\int_0^t \exp[-4\eta_n(t-s)]\int_0^{s-}\exp[-2\eta_n(s-u)]\mathrm{d}[m_{i,j}(u)][\mathrm{d}m_{i,j}(s)]\\
&\quad\le 2E\int_0^{\hat t}\int_0^{s-}\exp[-2\eta_n(s-u)]\mathrm{d}[m_{i,j}(u)]\mathrm{d}E\left([m_{i,j}(s)]|\widehat{\mathcal{F}}_{n,s-}\right)\\
&\qquad\le 2(c_{2,G,\alpha})^2\frac{\hat t}{2\eta_n}.
\end{aligned}
$$

Finally, from the positivity of the increments of the quadratic variation in addition to (3.35) and the homogeneity and integrability of $|G|^m, m = 1,\ldots,4$, we obtain for $n \ge \hat n_4$

$$
\begin{aligned}
E\sup_{t\le\hat t}\int_0^t \exp[-4\eta_n(t-s)]\mathrm{d}[[m_{i,j}(s)]] &\le E\int_0^{\hat t}\mathrm{d}[[m_{i,j}(s)]]\\
&=\int_0^{\hat t} E(E(\mathrm{d}[[m_{i,j}(s)]]|\widehat{\mathcal{F}}_{n,s-}))\le \delta\sigma\hat t c(4,G,\alpha),
\end{aligned}
$$

where $c(4,G,\alpha) := c(\alpha+1)^2\sup_n \sum_{1\le k,l\le 4}(\int |G_j|^k(q)|\mathrm{d}q)^l) < \infty$. Hence, for $n \ge \hat n_4$ we obtain the following estimate for the second term in (3.53):

$$
E\sup_{t\le\hat t}\sqrt{z_n(t)} \le \sqrt{2(c_{2,G,\alpha})^2\frac{\hat t}{2\eta_n}+\delta\sigma\hat t c(4,G,\alpha)}. \tag{3.60}
$$

So, (3.55) in addition to (3.59) and (3.60) implies

$$
\left.
\begin{aligned}
&\max_{\mathrm{j}=1,\ldots,\mathrm{d},\mathrm{i}=1,\ldots,\mathrm{N}} E\sup_{t\le\hat t}|U_{2,j}^{i,n}(t)|^2\\
&\le c(\hat t)2c_{2,G,\alpha}\sqrt{\frac{\hat t}{2\eta_n}}+\sqrt{2(c_{2,G,\alpha})^2\frac{\hat t}{2\eta_n}+\delta\sigma\hat t c(4,G,\alpha)}\\
&\qquad\longrightarrow 0,\quad \text{as } n\longrightarrow\infty\,.
\end{aligned}
\right\} \tag{3.61}
$$

Now set

$$
U_j^{i,n}(t) := r_{j,n}^i(0)+U_{1,j}^{i,n}(t)+U_{2,j}^{i,n}(t),\quad i=1,\ldots,N,\ j=1,\ldots,\mathrm{d},
$$

where $U_{1,j}^{i,n}(t)$ was defined in (3.17) and shown in Corollary 3.5 to converge stochastically to 0 uniformly in $t \le \hat t$. Denote the strong dual of $\mathbf{H}_0$ by $\mathbf{H}_0'$ and identify $\mathbf{H}_0'$ with $\mathbf{H}_0$. Let $\mathcal{S}'$ be the Schwarz space of tempered distributions over $\mathbf{R}^d$, and let $\mathcal{H}_{-\gamma}$ be a Hilbert distribution subspace of $\mathcal{S}'$ such that the imbedding[21]

$$
\mathbf{H}_0' \subset \mathcal{H}_{-\gamma} \quad \text{is Hilbert–Schmidt.}
$$

[21] Cf. Sect. 15.1.3, in particular, (15.32) and (15.36).

By Hypotheses 2.4 and 2.5, Corollary 3.5, (3.61) and a standard argument about weak convergence and stochastic convergence to 0[22], and by Theorem 15.54 (Sect. 15.2.5) we obtain

$$\left.\begin{aligned}&((U^{1,n}(\cdot),\ldots,U^{N,n}(\cdot)),M_n(\cdot)) \Longrightarrow ((r^1(0),\ldots,r^N(0)),W(\cdot))\\ &\qquad\text{in } \mathbf{R}^{dN}\times D([0,\hat{t}];\mathcal{H}_{-\gamma}),\ \text{as } n\longrightarrow\infty.\end{aligned}\right\} \tag{3.62}$$

We now write the solution of (3.16) as the solution of the stochastic integral equation

$$\tilde{r}_n^i(t) = \int_0^t \left\langle \hat{G}_n\left(\tilde{r}_n^i(s-)\right), \mathrm{d}M_n\right\rangle + U^{i,n}(t),\ \ i=1,\ldots,N, \tag{3.63}$$

where $\langle\cdot,\cdot\rangle$ is the duality between the (cylindrical $\mathbf{H}_0$)-valued sequence of martingales M_n and the $\mathbf{H}_0^d$-valued function $\hat{G}_n(r^i_{M_n}(s-),\cdot)$. Expression (3.62) is of the type that is considered in Sect. 15.2.5 (Theorem 15.54). By the identification of $\mathbf{H}_\mathbf{0}'$ with $\mathbf{H}_\mathbf{0}'$ the components of $\hat{G}_n(r,\cdot)$ are elements of $\mathbf{H}_0$. Recalling the notation introduced before (3.20), we evoke the dominated convergence theorem, the uniform continuity and integrability of $|G_j|^m, m=1,\ldots,4$, and we obtain for an arbitrary $c<\infty$ and all $j=1,\ldots,d$

$$\sup_{\{|R|\le c\}} |G_{n,j}(r-\cdot)-G_j(r-\cdot)|_0^2 \longrightarrow 0,\ \ \text{as } n\longrightarrow\infty. \tag{3.64}$$

This, in addition to the properties of G, implies the condition (15.181) in Sect. 15.2.5.

The preceding calculations in addition to Theorem 15.54 imply the main lemma of this section:

Lemma 3.9. *Under Hypotheses 2.1–2.6*

$$(\tilde{r}_n^1(\cdot),\ldots,\tilde{r}_n^N(\cdot)) \Longrightarrow (r^1(\cdot),\ldots,r^N(\cdot))$$

$$\text{in } D([0,\hat{t}];\mathbf{R}^{dN}),\ \text{as } n\longrightarrow\infty,$$

where $(r^1(\cdot),\ldots,r^N(\cdot))$ are the unique solutions of the stochastic integral equation (2.11). □

Proof of Theorem 2.4

We show that we may replace $\left(\tilde{r}_n^1(\cdot),\ldots,\tilde{r}_n^N(\cdot)\right)=\left(r_n^1(\cdot\wedge T_n),\ldots,r_n^N(\cdot\wedge T_n)\right)$ in Lemma 3.9 by $\left(r_n^1(\cdot),\ldots,r_n^N(\cdot)\right)$, where the stopping times T_n were defined in (3.6).

[22] Cf. Corollary 15.25 in Sect. 15.2.1.

(i) Note that by a theorem of Levy the $R^i(\cdot) - R^i(0)$ are $\mathbf{R}^d$-valued Brownian motions.[23] The initial conditions are bounded in probability. Therefore, we have

$$\left.\begin{array}{l} P\left\{\sup_{0\le t\le \hat{t}} \sum_{i=1,\dots,d} |r^i(t)| \ge L\right\} \le \sum_{i=1}^{d} P\left\{\sup_{0\le t\le \hat{t}} |r^i(t)| \ge \frac{L}{d}\right\} \\ \le \sum_{i=1}^{d} P\left\{\sup_{0\le t\le \hat{t}} |r^i(t) - r^i(0)| \ge \frac{L}{2d}\right\} + \sum_{i=1}^{d} P\left\{|r^i(0)| \ge \frac{L}{2d}\right\} \\ \longrightarrow 0, \text{ as } L \longrightarrow \infty. \end{array}\right\} \tag{3.65}$$

We restrict the metric $\mathrm{d}_{D,\mathbf{R}^{\mathrm{d}N}}(X, Y)$ on the Skorokhod space $D([0, \infty); \mathbf{R}^{\mathrm{d}N})$[24] to $D([0, \hat{t}]; \mathbf{R}^{\mathrm{d}N})$. We observe that for any $L > 0$

$$\left\{x(\cdot) \in D([0, \hat{t}]; \mathbf{R}^{\mathrm{d}N}) : \sup_{t\le \hat{t}} |x(t)|_{\mathrm{d}N} \ge L\right\}$$

are closed subsets of the metric space $(D([0, \hat{t}]; \mathbf{R}^{\mathrm{d}N}), d_{D,\mathbf{R}^{\mathrm{d}N}})$. From the definition of the stopping times T_n in (3.6) we obtain

$$P\{T_n < \hat{t}\} \le P\left\{\sup_{t\le \hat{t}} \max_{i=1,\dots,N} |\Lambda(t, \mathbf{C}_n, \tilde{r}^i_n, \tilde{\mathcal{I}}^n)| \ge \frac{\tilde{K}_n - 1}{4}\right\}.$$

Set $r(\cdot) := (r^1(\cdot), \dots, r^N(\cdot))$ and similarly for $\tilde{r}_n(\cdot)$. Theorem 15.24 in Sect. 15.2.1 in addition to Lemma 3.4, Lemma 3.9, and (3.62) implies

$$\left.\begin{array}{l} \limsup_{n\longrightarrow\infty} P\left\{\sup_{t\le \hat{t}} \max_{i=1,\dots,N} \left|\Lambda\left(t, \mathbf{C}_n, \tilde{r}^i_n, \tilde{\mathcal{I}}^n\right)\right| \ge L\right\} \\ \quad \le \limsup_{n\longrightarrow\infty} P\left\{\sup_{t\le \hat{t}} |\tilde{r}_n(t)|_{\mathrm{d}N} \ge \frac{L}{2}\right\} \\ \quad + \limsup_{n\longrightarrow\infty} P\left\{\sup_{t\le \hat{t}} |U^n(t)|_{\mathrm{d}N} \ge \frac{L}{2}\right\} \\ \le P\left\{\sup_{0\le t\le \hat{t}} |r(t)|_{\mathrm{d}N} \ge \frac{L}{2}\right\} + P\left\{|r(0)|_{\mathrm{d}N} \ge \frac{L}{2}\right\}. \end{array}\right\} \tag{3.66}$$

Altogether we obtain from the calculations (3.63) through (3.66):

$$\limsup_{n\longrightarrow\infty} P\{T_n < \hat{t}\} = 0. \tag{3.67}$$

---

[23] Cf. Sect. 15.2.3, Theorem 15.37, and Proposition 5.2 in Chap. 5. The joint $\mathbf{R}^{\mathrm{d}N}$-valued process is not Brownian and, by Definition 5.3, is a family of N correlated Brownian motions.

[24] Cf. Sect. 15.1.6, (15.53).

(ii) Let d_p denote the Prohorov metric on the space of probability measures on the complete separable metric space $(D_{\mathbf{R}^{dN}}[0,\hat{t}], d_{D,\mathbf{R}^{dN}})$.[25] Denote the probability distributions of $r_n(\cdot)$ and $r(\cdot)$ on $(D([0,\hat{t}];\mathbf{R}^{dN}), d_{D,\mathbf{R}^{dN}})$ by P_n and $\bar{P}$, respectively. We can easily show that for any $\bar{\delta} > 0$ there is an $n(\bar{\delta})$ such that for all $n \geq n(\bar{\delta})$

$$d_p(P_n, \bar{P}) < \bar{\delta}. \tag{3.68}$$

(iii) Finally, note that weak convergence is equivalent to the convergence in the Prohorov metric.[26] □

Remark 3.10. – On Space–Time White Noise

Recall that we constructed the "occupation measure" $\tilde{I}_A^n(t)$ as a positive random measure by counting the number of small particles that are in A at time t. Further, we showed that summing up the centered occupation measure ((3.15)) in time from 0 to t essentially determined an infinite dimensional square integrable mean zero martingale, $M_n(\cdot)$, (cf. (3.37)) which, by the martingale central limit theorem, was shown to converge to a (scalar) standard cylindrical $\mathbf{H}_0-$valued Brownian motion.[27] Considering space and time as variables in the occupation measure, the approximate independence of the occupation measure in space and time was a result of the independent starts of the small particles. In the scaling limit the centered occupation measure is approximated by the scalar valued standard Gaussian white noise $w(dq, dt)$ or, equivalently, by $\frac{\partial^{d+1}}{\partial s \partial q_1 \ldots \partial q_d}\bar{\tilde{w}}(\cdot,\cdot)$[28] as the limit of the centered number density of the small particles. Consequently, we may interpret the independence in space and time of the limiting generalized random field $w(dq, dt)$ the result of independent starts and large velocities of the underlying medium of small particles. □

[25] Section 15.2.1, (15.57).

[26] Section 15.2.1, Theorem 15.21.

[27] Proposition 3.7 and (3.45). Cf. also (5.2).

[28] Cf. our comments to (2.10) in Chap. 2.

Part II
Mesoscopic A: Stochastic Ordinary Differential Equations

Chapter 4
Stochastic Ordinary Differential Equations: Existence, Uniqueness, and Flows Properties

The principal steps in defining "mesoscopic" equations for the positions of particles in $\mathbf{R}^d$ are as follows: We first construct solutions of stochastic ordinary differential equations in the sense of Itô (SODEs)[1] *such that solutions $r(t, \tilde{\mathcal{Y}}, q)$ depend "nicely" on some measure-valued process $\tilde{\mathcal{Y}}$ and on the initial condition q. To proceed from SODEs to stochastic partial differential equations (SPDEs) for the mass distribution of particles, governed by the* flow $q \mapsto r(t, \tilde{\mathcal{Y}}, q)$, *we proceed in a second step as follows: Suppose $r(t, \tilde{\mathcal{Y}}, q)$ is measurable in q with respect to the Borel σ-algebra $\mathcal{B}^d$ of $\mathbf{R}^d$ and integrable with respect to some initial measure μ_0. If $\mu_0(\mathbf{R}^d) < \infty$, we can define a finite measure-valued process*[2]

$$\mathcal{Y}(t)(\mathrm{d}r) := \int \delta_{(r(t,\tilde{\mathcal{Y}},q))}(\mathrm{d}r)\mu_0(\mathrm{d}q). \tag{4.1}$$

The process $\mathcal{Y}(t)$ is the solution of a bilinear SPDE (cf. Chap. 8, (8.25)). If, in addition to the previous step, the input process $\tilde{\mathcal{Y}}(\cdot)$ equals $\mathcal{Y}(\cdot)$ from the left-hand side of (4.1), then $\mathcal{Y}(t)$ is the solution of a semilinear or quasilinear SPDE.[3] *For this case we use the notation $\mathcal{X}(t)$ or $\mathcal{X}_\varepsilon(t)$ instead of $\mathcal{Y}(t)$.*

4.1 Preliminaries

We introduce the necessary notation and definitions. Recall from (2.5) the definition of the metric on $\mathbf{R}^d$

$$\rho(r-q) := |r-q| \wedge 1,$$

---

[1] Cf. the following (4.9) and (4.10).

[2] (4.1) is by definition the image of the initial measure μ_0 under the "flow" $q \mapsto r(t, \tilde{\mathcal{Y}}, q)$, i.e., $\int_A \mathcal{Y}(t)(\mathrm{d}r) = \int \delta_{(r(t,\tilde{\mathcal{Y}},q))}(A)\mu_0(\mathrm{d}q) = \mu_0(r_\varepsilon^{-1}(t, \tilde{\mathcal{Y}}, s, q))(A)$, where $A \in \mathcal{B}^d$. Cf. also (8.50) in Chap. 8.

[3] Cf. Chap. 8, (8.26).

where $r, q \in \mathbf{R}^d$, $|r-q|$ is the Euclidean distance on $\mathbf{R}^d$ and "$\wedge$" denotes "minimum." The particle distributions will be measures. Set

$$\mathbf{M}_f := \left\{ \mu : \mu \text{ is a finite Borel measure on } \mathbf{R}^d \right\}. \tag{4.2}$$

Let γ_f be the Wasserstein distance on $\mathbf{M}_f$. For the precise definition and properties we refer to Sect. 15.1.4, (15.38) and Proposition 15.9. Here we note that $(\mathbf{M}_f, \gamma_f)$ is a complete and separable metric space. Further, we note that γ_f is the restriction of a norm[4] on a space of distributions to the cone of nonnegative finite Borel measures and that

$$\gamma_f(\mu) = \mu(\mathbf{R}^d). \tag{4.3}$$

By $\mathbf{M}_\infty$ we denote the σ-finite Borel measures μ on $\mathbf{R}^d$, i.e., μ is σ-finite. For the description of those $\mu \in \mathbf{M}_\infty$, which have a density with respect to the Lebesgue measure, we choose the following (standard) weight function making the constants integrable over $\mathbf{R}^d$ (cf. Sect. 15.2.8):

$$\varpi(r) = (1 + |r|^2)^{-\gamma}, \quad \gamma > \frac{d}{2}. \tag{4.4}$$

In what follows we write $\varpi\mu$ etc. to denote the finite measure that is represented as μ with density ϖ. Set

$$\mathbf{M}_{\infty,\varpi} := \left\{ \mu \in \mathbf{M}_\infty : \int \varpi(q)\mu(\mathrm{d}q) < \infty \right\} \tag{4.5}$$

where the integration is taken over $\mathbf{R}^d$.[5] If we define for $\mu, \nu \in \mathbf{M}_{\infty,\varpi}$

$$\gamma_\varpi(\mu - \nu) := \gamma_f(\varpi(\mu - \nu))$$

we obtain that $(\mathbf{M}_{\infty,\varpi}, \gamma_\varpi)$ is isometrically isomorphic to $(\mathbf{M}_f, \gamma_f)$. Therefore, $(\mathbf{M}_{\infty,\varpi}, \gamma_\varpi)$ is also a complete separable metric space (and even a separable Fréchet space).

The stochastic set-up is provided as follows: $(\Omega, \mathcal{F}, \mathcal{F}_t, P)$ is a stochastic basis with right continuous filtration. All our stochastic processes are assumed to live on Ω and to be $\mathcal{F}_t$-adapted (including all initial conditions in stochastic ordinary differential equations (SODEs) and stochastic partial differential equations (SPDEs)). Moreover, the processes are assumed to be $\mathrm{d}P \otimes \mathrm{d}t$-measurable, where $\mathrm{d}t$ is the Lebesgue measure on $[0, \infty)$.[6]

Let $w_\ell(\mathrm{d}r, \mathrm{d}t)$ be i.i.d. real-valued space–time white noises on $\mathbf{R}^d \times \mathbf{R}_+$, $\ell = 1, \ldots, d$ (cf. Definition 2.2). Set $w(p,t) := (w_1(p,t), \ldots, w_d(p,t))^T$, where "$T$"

---

[4] It follows that $(\mathbf{M}_f, \gamma_f)$ is a separable Fréchet space. Cf. Sect. 15.1.1.

[5] Recall from (1.4) our convention not to indicate the integration domain when integrating over $\mathbf{R}^d$.

[6] Cf., e.g., Metivier and Pellaumail (1980), Ch. 1.2., and Dellacherie (1975), Ch. III.2.

denotes the transpose. Finally, $\{\beta^{\perp,n}(\cdot)\}_{n\in\mathbf{N}}$ is a family of i.i.d. standard $\mathbf{R}^d$-valued Brownian motions, which is independent of $\{w_\ell\}_{\ell=1,..,d}$.[7]

Notation

(i) If $(\mathbf{K}, d_{\mathbf{K}})$ is some metric space with a metric $d_{\mathbf{K}}$, then $C([s,T];\mathbf{K})$ is the space of continuous $\mathbf{K}$-valued functions, defined on $[s,T]$.
$L_{0,\mathcal{F}_s}(\mathbf{K})$ is the space of $\mathbf{K}$-valued $\mathcal{F}_s$-measurable random variables, and $L_{0,\mathcal{F}}(C([s,T];\mathbf{K}))$ is the space of $\mathbf{K}$-valued adapted and $dt\otimes dP$-measurable processes with sample paths in $C([s,T];\mathbf{K})$ a.s.

(ii) Let $p\ge 1$ and e be a fixed element of $(\mathbf{K}, d_{\mathbf{K}})$. $L_{p,\mathcal{F}_s}(\mathbf{K})$ and $L_{p,\mathcal{F}}(C([s,T];\mathbf{K}))$ are those elements in $L_{0,\mathcal{F}_s}(\mathbf{K})$ and $L_{0,\mathcal{F}}(C([s,T];\mathbf{K}))$, respectively, with finite pth moments of $d_{\mathbf{K}}(\xi,e)$, where for the processes $\sup\limits_{s\le t\le T} d^p_{\mathbf{K}}(\xi(t),e)$ has to be integrable.

(iii) For $p\in\{0\}\cup[1,\infty)$ $L_{p,\mathcal{F}}(C([s,\infty);\mathbf{K}))$ is the space of those processes $\xi(\cdot)$ such that $\xi(\cdot\wedge T)\in L_{p,\mathcal{F}}(C([s,T];\mathbf{K}))$ $\forall T>s$.

Definition 4.1.

- *A random variable $\tau:\Omega\longrightarrow[0,\infty]$ is called a "stopping time" if for every $t\in[0,\infty)$ $\{\omega:\tau(\omega)\le t\}\subset\mathcal{F}_t$.*
- *Let $(\mathbf{K}_\ell, d_{\mathbf{K},\ell})$ and $(\tilde{\mathbf{K}}_k, d_{\tilde{\mathbf{K}},k})$ be metric spaces with metrics $d_{\mathbf{K},\ell}$, $d_{\tilde{\mathbf{K}},k}$ and $\xi_\ell(t)$ and $\mu_{s,k}$, $\ell=1,\ldots,m$, $k=1,\ldots,\tilde m$ be elements from $L_{0,\mathcal{F}}(C([s,T];\mathbf{K}_\ell))$ and $L_{0,\mathcal{F}}(C([s,T];\tilde{\mathbf{K}}_k))$, respectively. We will call a sequence of stopping times $\tau_n\in[s,\infty]$ localizing stopping times for $\xi_\ell(\cdot,\omega)$ and $\mu_{s,k}$, $\ell=1,\ldots,m$, $k=1,\ldots,\tilde m$, if*

$$
\begin{aligned}
&\tau_n(\omega) := \tau_n(\omega,\xi_\ell,\mu_{s,k},\ell=1,\ldots,m,k=1,..,\tilde m)\\
:= &\inf\{t\in[s,\infty]: \max_{\ell=1,\ldots,m}\max_{k=1,\ldots,\tilde m} d_{\mathbf{K},\ell}(\xi_\ell(t,\omega),e)\vee d_{\tilde{\mathbf{K}},k}(\mu_{s,k},\tilde e)\ge n\},
\end{aligned}
$$

and

$$
\forall T>0\ \ P\{\omega:\tau_n(\omega)<T\}\downarrow 0,\ \ as\ n\longrightarrow\infty. \qquad \square
$$

Let τ_n be a sequence of localizing stopping times for $\xi_\ell(\cdot,\omega)$ and set for $p\ge 1$. Set

$$
\begin{aligned}
L_{\mathrm{loc},p,\mathcal{F}}(C([s,T];\mathbf{K})) := \Bigg\{&\xi(\cdot)\in L_{0,\mathcal{F}}(C([s,T];\mathbf{K})):\\
&\times\ \exists\ \text{"l.s.t."}(\tau_n:=\tau_n(\xi))\text{s.t. } E\Big(\sup_{s\le t\le T\wedge\tau_n} d^p_{\mathbf{K}}(\xi(t),e)\Big)<\infty\Bigg\}\ \ \forall n,
\end{aligned}
$$

[7] The i.i.d. standard Brownian motions,$\{\beta^{\perp,n}(\cdot)\}_{n\in\mathbf{N}}$, are being used to obtain more general SODEs and SPDEs. However, the main focus in this volume are on stochastic equations driven by

$$\{w_\ell\}_{\ell=1,..,d}$$

$$L_{\text{loc},p,\mathcal{F}}(C((s,T];\mathbf{K})) := \Bigg\{ \xi(\cdot) \in \mathcal{K}_{[s,T]} :$$

$$\times\ \exists\ \text{“l.s.t.”}(\tau_n := \tau_n(\xi))\text{s.t. } E\left(\sup_{s\le t\le T\wedge\tau_n} d_K^p(\xi(t),e)1_{\{\tau_n>s\}}\right) < \infty \Bigg\}\quad \forall n,$$

$$L_{\text{loc},p,\mathcal{F}_s}(\mathbf{K}) := \{\eta \in L_{0,\mathcal{F}_s}(\mathbf{K}) :\})) :$$
$$\times\ \exists\ \text{“l.s.t.”}(\tau_n := \tau_n(\eta))\text{s.t. } \eta 1_{\tau_n>s} \in L_{p,\mathcal{F}_s}(\mathbf{K})\big\}\ \ \forall n\big\}$$

where “l.s.t.” means “localizing stopping time.” If the particular level n at which we stop or truncate the random maps is irrelevant (as it will be most of the time), we drop the subscript n and call τ a “localizing” stopping time. This means that for each level n there is such a τ with the properties, described above.

Let us make some comments on this definition:

Remark 4.2.

- In most cases, our metric spaces will be either vector spaces with a norm or positive cones of vector spaces, i.e., the set of elements of a vector space that is invariant with respect to multiplications with nonnegative real numbers, e.g., $\mathbf{M}_f$. In those cases the metric will be the restriction of a norm to the positive cones (e.g., γ_f) and we may choose $e := \mathbf{0}$, i.e., the “null” element.
- For $\mu_s \in L_{0,\mathcal{F}_s}(\mathbf{M}_f)$ or $L_{0,\mathcal{F}_s}(\mathbf{M}_{\infty,\varpi})$ we have $\forall p \ge 1 : \mu_s 1_{\{\tau_n>s\}} \in L_{p,\mathcal{F}_s}(\mathbf{M}_f)$ or $\in L_{p,\mathcal{F}_s}(\mathbf{M}_{\infty,\varpi})$, respectively. A similar statement holds for the processes.
- Assume for simplicity $m = \tilde{m} = 1$ and $\mathbf{K} = \tilde{\mathbf{K}}$ and d_K a norm with $e := \mathbf{0}$. Denoting
$$\Omega_n := \{\omega : d_K(\xi(s,\omega)) < n\}$$
we have
$$P\{\Omega_n\} \uparrow 1,\ \text{as}\ n \longrightarrow \infty.$$
Therefore, setting
$$\tau_n(\omega) := \inf\{t \in [s,\infty] :\ d_K(\xi(t,\omega)) \ge n\},$$
it follows that
$$\tau_n(\omega)_{|\Omega\backslash\Omega_n} = s, \quad \Omega_n = \{\omega : \tau_n > s\}.$$
On $\Omega\backslash\Omega_n$, however, $d_K(\xi(s),e)$ does not have to be integrable, whereas in $L_{\text{loc},p,\mathcal{F}}(C([s,T];\mathbf{K}))$ also the pth moment at the initial time s must be integrable. So, $L_{\text{loc},p,\mathcal{F}}(C((s,T];\mathbf{K}))$ strictly contains $L_{\text{loc},p,\mathcal{F}}(C([s,T];\mathbf{K}))$. The above set Ω_n is $\mathcal{F}_s$-measurable. Therefore, $\xi(\cdot)1_{\{\tau_n>s\}} \in L_{\text{loc},p,\mathcal{F}}(C([s,T];\mathbf{K}))$. Consequently, we may work with pth moments estimates for processes that may not be integrable at the initial time point s. □

Let us now return to our approach to mesoscopic equations. The definition (4.1) implies *mass conservation*, i.e., $\mathcal{Y}(t,\omega,\mathbf{R}^d) = \mu_0(\omega,\mathbf{R}^d)$. Hence, we have in this case $\gamma_f(\mathcal{Y}(t,\omega)) = \mathcal{Y}(t,\omega,\mathbf{R}^d) = \mu_0(\omega,\mathbf{R}^d) = \gamma_f(\mu_0(\omega)) < \infty$. Therefore, bounds on the initial measure also yield bounds for the measure uniformly in $t \geq 0$, if we work with $\mathbf{M}_f$-valued initial conditions. The situation with $\mathbf{M}_{\infty,\varpi}$-valued initial conditions is not so straightforward. Although we might think of $\int_A \varpi(q)\mu(dq)$ as a finite measure on the Borel sets even if $\mu(\mathbf{R}^d) = \infty$, we do not have in general $\int \varpi(q)\mathcal{Y}(t,dq) = \int \varpi(q)\mu_0(dq)$ – even if $\mu_0(\mathbf{R}^d) = \mathcal{Y}(t,\mathbf{R}^d) < \infty$. The problem is that the weight function ϖ "distorts" the mass distribution. However, if we assume $E\gamma_{\infty,\varpi}(\mu_0) < \infty$, we can find an important subspace of $L_{0,\mathcal{F}}(C([s,\infty);\mathbf{M}_{\infty,\varpi}))$ where this assumption implies $E \sup_{0\leq t\leq T} \gamma_{\infty,\varpi}(\mu(t)) < \infty$.

To this end, let Ψ denote the set of adapted measurable flows φ from $[0,T]\times\mathbf{R}^d$ into $\mathbf{R}^d$, which satisfy the following conditions:

(i) $\varphi(0,q) \equiv q$.
(ii) $t \longmapsto \varphi(t,q)$ is continuous a.s. $\forall q$ and measurable in $q \in \mathbf{R}^d$.
(iii) $\forall q \quad \varphi(t,q)$ is a square integrable semimartingale. If $b(t,q)$ and $m(t,q)$ are the process of bounded variation and the martingale in the Doob-Meyer decomposition, respectively, both $b(t,q)$ and the mutual quadratic variations[8] of the one-dimensional martingale components, $[m_k(t,q), m_\ell(t,q)]$ are differentiable with respect to t and the derivatives are bounded uniformly in $\omega \in \Omega, q \in \mathbf{R}^d$ and $t \in [0,T]$ for all $T > 0$.

Clearly, condition (iii) implies for $\varphi \in \Psi$

$$\sup_q E \sup_{0\leq s\leq T} |\varphi(s,q) - q|^2 < \infty.$$

Proposition 4.3.

(a) Suppose $\mu_0 \in \mathcal{M}_{\infty,\varpi}$ *and* $\varphi \in \Psi$. *Then*

$$t \longmapsto \mu(\cdot) := \mu_0 \circ \varphi^{-1}(\cdot) \in L_{0,\mathcal{F}}(C([0,\infty);\mathbf{M}_{\infty,\varpi})). \tag{4.6}$$

(b) Suppose, in addition, that for $p \geq 1 \;\; E\gamma_\varpi^{p}(\mu_0) < \infty$. *Then*

$$E \sup_{0\leq t\leq T} \gamma_\varpi^{p}(\mu(t)) \leq c_T E\gamma_\varpi^{p}(\mu_0) < \infty, \tag{4.7}$$

where the finite constant c_T *depends only on* T *and the bounds of the derivatives of the characteristics of the semimartingale.*

Equation (4.6) states that the boundedness of the moment at the initial time implies $\mu(\cdot) \in L_{p,\mathcal{F}}(C([s,T];\mathbf{M}_{\infty,\varpi}))$.

The *proof* is given in Sect. 15.2.8. □

Notation

All constants will be denoted by $c_F, c_{\mathcal{J}}, c_{\mathcal{J},T}$, etc. and are assumed to be nonnegative and finite.

[8] Cf. Sects. 15.2.3 and 15.2.5, in particular, (15.149) and (15.92).

4.2 The Governing Stochastic Ordinary Differential Equations

Existence and uniqueness are proved for a class of SODEs that are driven by both space–time white noise, $w(\mathrm{d}q, \mathrm{d}s)$, and by finitely many independent Brownian motions. The coefficients may depend on measure processes as parameters or on the empirical measure processes for the solutions.

In what follows let $(\mathbf{M}, \gamma) \in \{(\mathbf{M}_f, \gamma_f), (\mathbf{M}_{\infty,\varpi}, \gamma_{\varpi})\}$ and let $\mathcal{M}_{d\times d}$ denote the $d \times d$ matrices over $\mathbf{R}$. We now define the coefficients for the stochastic ordinary differential equations (SODEs):

$$\left.\begin{aligned} &F : \mathbf{R}^d \times \mathbf{M} \times \mathbf{R} \to \mathbf{R}^d;\\ &\mathcal{J} : \mathbf{R}^d \times \mathbf{R}^d \times \mathbf{M} \times \mathbf{R} \to \mathcal{M}_{d\times d};\\ &\sigma_n^{\perp} : \mathbf{R}^d \times \mathbf{M} \times \mathbf{R} \to \mathcal{M}_{d\times d}, \quad n \in \mathbf{N}. \end{aligned}\right\} \tag{4.8}$$

We consider two types of stochastic ordinary differential equations:

$$\left.\begin{aligned} \mathrm{d}r^i(t) &= F(r^i(t), \tilde{\mathcal{Y}}(t), t)\mathrm{d}t\\ &\quad + \int \mathcal{J}(r^i(t), p, \tilde{\mathcal{Y}}(t), t) w(\mathrm{d}p, \mathrm{d}t) + \sigma_i^{\perp}(r^i(t), \tilde{\mathcal{Y}}(t), t)\mathrm{d}\beta^{\perp,i}(t)\\ r^i(s) &= r_s^i \in L_{2,\mathcal{F}_s}(\mathbf{R}^d), \;\; \tilde{\mathcal{Y}} \in L_{\mathrm{loc},2,\mathcal{F}}(C((s,T];\mathbf{M})), \;\; i = 1,\ldots,N, \end{aligned}\right\} \tag{4.9}$$

where $L_{2,\mathcal{F}_s}(\mathbf{R}^d)$ is the space of $\mathbf{R}^d$-valued, $\mathcal{F}_s$-measurable random variables r_s such that $E|r_s|^2 < \infty$. (4.9) describes the motion of a system of diffusing particles in a random environment (represented by $\tilde{\mathcal{Y}}$, w_ℓ, $\ell = 1,\ldots,d$, $\beta^{\perp,i}$, $i = 1,\ldots,N$.). The $\mathbf{R}^{Nd}$-valued process with d-dimensional components $r^i(t)$ will be denoted $r_N(t)$.

$$\left.\begin{aligned} \mathrm{d}r_N^i(t) &= F(r_N^i(t), \mathcal{X}_N(t), t)\mathrm{d}t\\ &\quad + \int \mathcal{J}(r_N^i(t), p, \mathcal{X}_N(t), t) w(\mathrm{d}p, \mathrm{d}t) + \sigma_i^{\perp}(r_N^i(t), \mathcal{X}_N(t), t)\mathrm{d}\beta^{\perp,i}(t),\\ r_N^i(s) &= r_s^i \in L_{2,\mathcal{F}_s}(\mathbf{R}^d), \; i = 1,\ldots N, \; \mathcal{X}_N(t) := \sum_{i=1}^N m_i \delta_{r_N^i(t)}, \;\; m_i \geq 0. \end{aligned}\right\} \tag{4.10}$$

Expression (4.10) represents the motion of N interacting and diffusing particles in the random environment (represented by $w_\ell, \ell = 1,\ldots,d, \beta^{\perp,i}, i = 1,\ldots,N$). δ_{r^i} is the point measure concentrated in r^i. $\mathcal{X}_N(t)$ is the *empirical measure process* associated with (4.10). The general measure process $\tilde{\mathcal{Y}}(t)$ in (4.9) has been replaced by the empirical measure process of the solution of (4.10). Therefore, the $N \cdot d$-vector-valued solution of (4.10) depends on N, whereas in (4.9) it is independent of N.

In what follows, γ is assumed to be an arbitrary metric from $\{\gamma_f, \gamma_{\varpi}\}$ and μ is in the corresponding spaces $\mathbf{M}_f$ and $\mathbf{M}_{\infty,\varpi}$, respectively.

Hypothesis 4.1[9]

Suppose $(r_\ell, \mu_\ell, t) \in \mathbf{R}^d \times \mathbf{M} \times \mathbf{R}$, $\ell = 1, 2$. Let c_F, $c_{\mathcal{J},\sigma^\perp}$, $c_{F,\mathcal{J},\sigma^\perp} \in (0, \infty)$. Assume global Lipschitz and boundedness conditions:

$$\left.\begin{aligned}
&(a)\quad |F(r_1,\mu_1,t) - F(r_2,\mu_2,t)| \\
&\qquad \le c_F\{(\gamma(\mu_1) \vee \gamma(\mu_2))\rho(r_1 - r_2) + \gamma(\mu_1 - \mu_2)\}, \\
&\qquad \sum_{k,\ell=1}^{d} \Big[\int (\mathcal{J}_{k\ell}(r_1,p,\mu_1,t) - \mathcal{J}_{k\ell}(r_2,p,\mu_2,t))^2 \mathrm{d}p \\
&\qquad\qquad + \sup_{n\in\mathbf{N}} |\sigma^\perp_{n,k\ell}(r_1,\mu_1,t) - \sigma^\perp_{n,k\ell}(r_2,\mu_2,t)|^2\Big] \\
&\qquad \le c^2_{\mathcal{J},\sigma^\perp}\{(\gamma^2(\mu_1) \vee \gamma^2(\mu_2))\rho^2(r_1 - r_2) + \gamma^2(\mu_1 - \mu_2)\}; \\
&(b)\quad |F_\varepsilon(r,\mu,t)|^2 + \sum_{k,\ell=1}^{d} \Big\{\int \mathcal{J}^2_{\varepsilon,k\ell}(r,p,\mu,t)\mathrm{d}p + \sup_{n\in\mathbf{N}} |\sigma^\perp_{n,k\ell}(r,\mu,t)\Big\} \\
&\qquad \le c_{F,\mathcal{J},\sigma^\perp}.
\end{aligned}\right\} \tag{4.11}$$

The constants $c_F, c_{\mathcal{J}}$, and $c_{F,\mathcal{J},\sigma^\perp}$ in (4.11) may also depend on the space dimension d. □

Alternatively, if we drop the boundedness assumption on the coefficients in (4.11), we need to impose a linear growth condition in addition to corresponding Lipschitz conditions from (4.11) in terms of the Euclidean metric:

$$\left.\begin{aligned}
&(a)\quad |F(r_1,\mu_1,t) - F(r_2,\mu_2,t)| \\
&\qquad \le c_F\{\gamma(\mu_1) \vee \gamma(\mu_2)|r_1 - r_2| + \gamma(\mu_1 - \mu_2)\}, \\
&\qquad \sum_{k,\ell=1}^{d} \Big[\int (\mathcal{J}_{k\ell}(r_1,p,\mu_1,t) - \mathcal{J}_{k\ell}(r_2,p,\mu_2,t))^2 \mathrm{d}p \\
&\qquad\qquad + \sup_{n\in\mathbf{N}} |\sigma^\perp_{n,k\ell}(r_1,\mu_1,t) - \sigma^\perp_{n,k\ell}(r_2,\mu_2,t)|^2\Big] \\
&\qquad \le c^2_{\mathcal{J},\sigma^\perp}\{(\gamma^2(\mu_1) \vee \gamma^2(\mu_2))|r_1 - r_2|^2 + \gamma^2(\mu_1 - \mu_2)\}; \\
&(b)\quad |F(r,\mu,t)| \le c_F(1 + |r|), \\
&\qquad |\sum_{k,\ell=1}^{d} \Big\{\int \mathcal{J}^2_{k\ell}(r,p,\mu,t)\mathrm{d}p + \sup_{n} |\sigma^\perp_{n,k\ell}(r,\mu,t)|^2\Big\} \le c^2_{\mathcal{J},\sigma^\perp}(1 + |r|^2).
\end{aligned}\right\} \tag{4.12}$$

Note that we only need to construct solutions on sets of ω with large probability. Working either with $\mathbf{M}_f$ or $\mathbf{M}_{\infty,\varpi}$, boundedness of the norm of the initial measure implies boundedness for all $t \in [0, T]$. Therefore, we will be able to use stopping time techniques to enforce uniformly bounded Lipschitz constants on the coefficients F, $\mathcal{J}$, and $\sigma^\perp_n$. In fact, with only minor modifications, we may assume that

[9] For some SODEs with singular interaction cf. Skorohod (1996).

the coefficients also depend on ω and for given input variables are $\mathcal{F}_t$-adapted. This is standard in stochastic analysis (cf., e.g., Liptser and Shiryayev (1974) or Kurtz and Protter (1996)). However, the dependence of our coefficients on empirical measures requires more notation than in the traditional treatment of SODEs. To avoid cumbersome notation, the analysis in this volume is restricted to coefficients that depend on ω only through solutions of the SODEs (the Markovian case!) or explicit input processes.

In Sect. 4.4 (i.e., at the end of this section), we provide a number of examples of coefficients $F, \mathcal{J}$ satisfying Hypothesis 4.1. Examples for $\sigma^{\perp}$ can be obtained from the examples for F.

In this section we will use both metrics and conditions (4.11) and (4.12). In the derivation of SPDEs we will work with the bounded metric $\rho(\cdot - \cdot)$. This allows us to work directly in the case of finite measures with the metric γ_f and, employing the Kantorovich-Rubinstein theorem for some important estimates, we may use the Wasserstein metric (15.38)/(15.39). With the exception of Chaps. 9 and 14, the corresponding theorems, based on the Euclidean metric and the additional linear growth assumption, can be obtained after minor modifications from the proofs based on the bounded metric as long as we work with $\mathbf{M}_f$.[10]

Remark 4.4.

- Borkar (1984) was probably the first author to use Brownian white noise as a driving term for stochastic ordinary differential equations (SODEs). The step from SODEs for the position of particles to the empirical measure and related stochastic partial differential equations (SPDEs) was introduced by the author in his papers (1995a, b).
- Recall from (3.29) that $\mathbf{H}_0$ is the space of measurable functions on $\mathbf{R}^d$, which are square integrable with respect to the Lebesgue measure with norm $|\cdot|_0$ and scalar product $< \cdot, \cdot >_0$. Let $\{\phi_n\}_{n \in N}$ be a complete orthonormal system (CONS) in $\mathbf{H}_0$ and define an $\mathcal{M}_{d \times d}$-valued function $\widehat{\phi}_n(\cdot)$ by

$$\widehat{\phi}_n := \begin{pmatrix} \phi_n \; 0 \; \ldots\ldots. \; 0 \\ 0 \; \phi_n \; \ldots\ldots \; 0 \\ \ldots\ldots\ldots\ldots\ldots\ldots\ldots\ldots \\ \ldots\ldots\ldots\ldots\ldots\ldots\ldots\ldots \\ \\ \ldots\ldots\ldots\ldots\ldots\ldots\ldots\ldots \\ 0 \; 0 \; \ldots\ldots. \; \phi_n \end{pmatrix}, \tag{4.13}$$

  i.e., $\widehat{\phi}_n(\cdot)$ is a $d \times d$ matrix-valued function whose entries on the main diagonal are all $\phi_n(\cdot)$ and whose other entries are all 0. Set

$$\beta^n(t) := \int_0^t \int \widehat{\phi}_n(p) w(\mathrm{d}p, \mathrm{d}s). \tag{4.14}$$

---

[10] The analysis of $\mathbf{M}_{\infty,\varpi}$ processes and solutions of SPDEs in Chap. 9 is based on Proposition 4.3 and, in the present setting, requires bounded coefficients.

Then the $\beta^n(\cdot)$ are i.i.d. standard $\mathbf{R}^d$-valued Brownian motions (or Wiener processes).[11] Moreover, for any $\tilde{\mathcal{Y}} \in L_{\text{loc},2,\mathcal{F}}(C((s,T];\mathbf{M}))$ and $r(\cdot) \in L_{0,\mathcal{F}}(C([0,\infty);\mathbf{R}^d))$ (the space of $\mathbf{R}^d$-valued adapted continuous processes)

$$\int \mathcal{J}(r(t),p,\tilde{\mathcal{Y}}(t),t)w(\mathrm{d}p,\mathrm{d}t) = \sum_{n=1}^{\infty} \int \mathcal{J}(r(t),p,\tilde{\mathcal{Y}}(t),t)\widehat{\phi}_n(p)\mathrm{d}p\mathrm{d}\beta^n(t). \tag{4.15}$$

The right-hand side of (4.15) defines the increment of an $\mathbf{R}^d$-valued square integrable continuous martingale $m(\cdot)$.[12] In particular, (4.11) implies for the mutual quadratic variation of the one-dimensional components of $m(t) := m(r(\cdot),\tilde{\mathcal{Y}}(\cdot),t)$ the following estimate:

$$[m_k(t), m_\ell(t)] \le c_{\mathcal{J}} t. \tag{4.16}$$

□

Notation

Let $\mathcal{G}_{s,t}$ (resp. $\mathcal{G}_t$) be the σ-algebra generated by $w(\mathrm{d}p,\mathrm{d}u)$ between s and t (resp. 0 and t) for $t \ge s$. By analogy, for $t \ge s$, $\mathcal{G}_{s,t}^{\perp}$ and $\mathcal{G}_t^{\perp}$ are the σ-algebras generated by $\{\mathrm{d}\beta^{\perp,n}(u)\}_{n\in\mathbf{N}}$ between s and t and 0 and t, respectively. The cylinder set filtration on $C([s,\infty);\mathbf{M})$ is denoted $\mathcal{F}_{\mathbf{M},s,t}$. We write $\mathcal{F}_{\mathbf{M},t}$ if $s=0$. Here, we used again the abbreviation: $\mathbf{M} \in \{\mathbf{M}_f, \mathbf{M}_{\infty,\varpi}\}$ etc. Completed σ-algebras will be denoted with a bar on top of the σ-algebra, e.g. $\bar{\mathcal{G}}_{s,t}$.

Denote a solution of (4.9), if it exists, by $r_N(t,\tilde{\mathcal{Y}},r_{N,s},s)$. Further, if f is a stochastic process on $[s,\infty)$ with values in some metric space, we set for $t \ge s$

$$(\pi_{s,t}f)(u) := f(u \wedge t), \ (u \ge s).$$

In the following theorem $(\mathbf{M},\gamma) \in \{(\mathbf{M}_f,\gamma_f),(\mathbf{M}_{\infty,\varpi},\gamma_\varpi)\}$.

Theorem 4.5. *Assume either (4.11) or (4.12) and $\tilde{\rho}(\cdot) \in \{\rho(\cdot), |\cdot|\}$. Then*

(1) To each $s \ge 0$, $r_s^k \in L_{2,\mathcal{F}_s}(\mathbf{R}^d)$, $\tilde{\mathcal{Y}} \in L_{\text{loc},2,\mathcal{F}}(C((s,T];\mathbf{M}))$ (4.9) has a unique solution $r^k(\cdot,\tilde{\mathcal{Y}},r_s^k,s) \in L_{\text{loc},2,\mathcal{F}}(C([s,T];\mathbf{R}^d))$.

(2) Let $\tilde{\mathcal{Y}}_i \in L_{\text{loc},2,\mathcal{F}}(C((s,T];\mathbf{M}))$ and $r_{s,i}^k \in L_{2,\mathcal{F}_s}(\mathbf{R}^d)$, $i=1,2$. Then for any $T \ge s$ and any stopping time $\tau \ge s$, which is localizing for $\tilde{\mathcal{Y}}_i$, $i=1,2$,

$$\left.\begin{aligned} &E \sup_{s\le t\le T\wedge\tau} \tilde{\rho}^2(r^k(t,\tilde{\mathcal{Y}}_1,r_{s,1}^k,s) - r^k(t,\tilde{\mathcal{Y}}_2,r_{s,2}^k,s))1_{\{\tau>s\}} \\ &\le c_{T,F,\mathcal{J},\sigma,\tilde{\mathcal{Y}},\tau}\{E(\tilde{\rho}^2(r_{s,1}^k - r_{s,2}^k)1_{\{\tau>s\}}) \\ &\quad + E\int_s^{T\wedge\tau} (\gamma^2(\tilde{\mathcal{Y}}_1(u) - \tilde{\mathcal{Y}}_2(u))1_{\{\tau>s\}}\mathrm{d}u\}. \end{aligned}\right\} \tag{4.17}$$

[11] Cf. Sect. 15.2.2.

[12] The statement follows from Doob's inequality (Theorem 15.32, Part (b)) and the fact that the terms in the right-hand side of (4.15) are uncorrelated martingales, employing (15.92).

Further, with probability 1 *uniformly in* $t \in [s, \infty)$

$$r^k(t, \tilde{\mathcal{Y}}, r^k_{s,1}, s) \equiv r^k(t, \pi_{s,t}\tilde{\mathcal{Y}}, r^k_{s,1}, s). \tag{4.18}$$

(3) For any $N \in \mathbf{N}$ *there is a* $\mathbf{R}^{\mathrm{d}N}$*-valued map in the variables* $(t, \omega, \mu(\cdot), r_N, s)$, $0 \le s \le t < \infty$ *such that for any fixed* $s \ge 0$

$$\bar{r}_N(\cdot, \ldots, \cdot, s) : \Omega \times C([s, T]; \mathbf{M}) \times \mathbf{R}^{\mathrm{d}N} \to C([s, T]; \mathbf{R}^{\mathrm{d}N}),$$

and the following holds:

(i) For any $t \in [s, T]$ $\bar{r}_N(t, \cdot, \ldots, \cdot, s)$ *is* $\overline{\mathcal{G}}_{s,t} \otimes \overline{\mathcal{G}}^{\perp,[1,N]}_{s,t} \otimes \mathcal{F}_{\mathbf{M},s,t} \otimes \mathcal{B}^{\mathrm{d}N} - \mathcal{B}^{\mathrm{d}N}$*-measurable.*

(ii) The i*th* d*-vector of* $\bar{r}_N = (\bar{r}^1, \ldots, \bar{r}^i, \ldots \bar{r}^N)$ *depends only on the* i*th* d*-vector initial condition* $r^i_s \in L_{2,\mathcal{F}_s}(\mathbf{R}^d)$ *and the Brownian motion* $\beta^{\perp,i}(\cdot)$*, in addition to its dependence on* $w(\mathrm{d}q, \mathrm{d}t)$ *and* $\tilde{\mathcal{Y}} \in L_{2,\mathcal{F}}(C([s, T]; \mathbf{M}))$*, and with probability 1 (uniformly in* $t \in [s, \infty)$*)*

$$\bar{r}^i_N(t, \cdot, \tilde{\mathcal{Y}}, r^i_s, s) \equiv r^i(t, \tilde{\mathcal{Y}}, r^i_s, s), \tag{4.19}$$

where the right-hand side of (4.19) is the i*th* d*-dimensional component of the solution of (4.9).*

(iii) If $u \ge s$ *is fixed, then with probability* 1 *(uniformly in* $t \in [u, \infty)$*)*

$$\bar{r}_N(t, \cdot, \pi_{u,t}\tilde{\mathcal{Y}}, \bar{r}_N(u, \cdot, \pi_{s,u}\tilde{\mathcal{Y}}, r_{N,s}, s), u) \equiv \bar{r}_N(t, \cdot, \pi_{s,t}\tilde{\mathcal{Y}}, r_{N,s}, s). \tag{4.20}$$

Proof. Statements (1) and (2) are standard, and we prove them, without loss of generality, on $[0, T]$ for fixed $T > 0$, $s = 0$ and $N = 1$. Whenever possible, we suppress the dependence on $s = 0$ and the particle index $i = 1$ in our notation. Moreover, by our global Lipschitz (and linear growth) assumptions we may, without loss of generality, assume that the coefficients are time independent.

Further, our stopping times are localizing, i.e., for each $\delta > 0$ and each $b > 0$ there is a stopping time τ and an $\mathcal{F}_0$-measurable set Ω_b such that

$$P\left\{ \sup_{0 \le t \wedge \tau \le T} \gamma(\tilde{\mathcal{Y}}(t, \omega)) 1_{|\Omega_b} \le b \right\} \ge 1 - \delta.$$

Therefore, we may, without loss of generality, assume that

$$\operatorname*{ess\,sup}_{\omega} \sup_{0 \le t \le T \wedge \tau} \gamma(\tilde{\mathcal{Y}}(t)) := c_{\mathcal{Y}} < \infty.$$

(i) Let $q_\ell(\cdot \wedge \tau) \in L_{2,\mathcal{F}}(C([0, \infty); \mathbf{R}^d))$ and set

$$\tilde{q}_\ell(t) := q_\ell(0) + \int_0^t F(q_\ell(s), \tilde{\mathcal{Y}}_\ell(s))\mathrm{d}s + \int_0^t \int \mathcal{J}(q_\ell(s), p, \tilde{\mathcal{Y}}_\ell(s)) w(\mathrm{d}p, \mathrm{d}s)$$
$$+ \int_0^t \sigma_1^\perp(q(s), \tilde{\mathcal{Y}}_\ell(s))\mathrm{d}\beta^{\perp,1}(s))$$

whose right-hand side is a well-defined Itô integral, driven by a semimartingale.[13] Observe that for a constant $c \geq 1$ and $a \geq 0$

$$a \wedge c \leq (a \wedge 1)c.$$

Therefore, working in the following steps $\tilde{\rho}(\cdot) \in \{\rho(\cdot), |\cdot|\}$, we assume that, without loss of generality, the constants in (4.11) are greater or equal to 1.

(ii) From the preceding estimate and $c_F \geq 1$ we obtain

$$|F(q_1(s), \tilde{\mathcal{Y}}_1(s)) - F(q_2(s), \tilde{\mathcal{Y}}_2(s))| \leq c_F \tilde{\rho}(F(q_1(s), \tilde{\mathcal{Y}}_1(s)) - F(q_2(s), \tilde{\mathcal{Y}}_2(s)))$$

Hence,

$$\left.\begin{array}{l}
\tilde{\rho}\left(\int_0^{t\wedge\tau} F(q_1(s), \tilde{\mathcal{Y}}_1(s))\mathrm{d}s - \int_0^{t\wedge\tau} F(q_2(s), \tilde{\mathcal{Y}}_2(s))\mathrm{d}s\right) \\
\leq \left|\int_0^{t\wedge\tau} F(q_1(s), \tilde{\mathcal{Y}}_1(s))\mathrm{d}s - \int_0^{t\wedge\tau} F(q_2(s), \tilde{\mathcal{Y}}_2(s))\mathrm{d}s\right| \\
\leq \int_0^{t\wedge\tau} |F(q_1(s), \tilde{\mathcal{Y}}_1(s)) - F(q_2(s), \tilde{\mathcal{Y}}_2(s))|\mathrm{d}s \\
\leq \int_0^{t\wedge\tau} c_F \tilde{\rho}\left(F(q_1(s), \tilde{\mathcal{Y}}_1(s)) - F(q_2(s), \tilde{\mathcal{Y}}_2(s))\right)\mathrm{d}s \\
\leq c_F \left\{\int_0^{t\wedge\tau} c_{\tilde{\mathcal{Y}}} \tilde{\rho}(q_1(s) - q_2(s))\mathrm{d}s + \int_0^{t\wedge\tau} \gamma(\tilde{\mathcal{Y}}_1(s) - \tilde{\mathcal{Y}}(s))\mathrm{d}s\right\} \\
\text{(by (4.11), (a), or (4.12) , (a))}
\end{array}\right\} \quad (4.21)$$

where $c_{\tilde{\mathcal{Y}}} \geq \text{ess}\sup_{\omega} \sup_{0\leq t<\infty} \gamma(\tilde{\mathcal{Y}}_1(t\wedge\tau,\omega)) \sup_{\omega} \sup_{0\leq t<\infty} \gamma(\tilde{\mathcal{Y}}_2(t\wedge\tau,\omega))$.

(iii) Similarly, by (4.11) and the independence of w_ℓ and $\beta^{\perp,1}$

$$\left.\begin{array}{l}
\sum_{k,\ell=1}^{d}\left[\int_0^{t\wedge\tau}\int \{\mathcal{J}_{k\ell}(q_1(s), p, \tilde{\mathcal{Y}}_1(s)) - \mathcal{J}_{k\ell}(q_2(s), p, \tilde{\mathcal{Y}}_2(s))\} w(\mathrm{d}p, \mathrm{d}s)\right] \\
\quad + \sum_{k,\ell=1}^{d}\left[\int_0^{t\wedge\tau} \{\sigma^{\perp}_{1,k\ell}(q_1(s), \tilde{\mathcal{Y}}_1(s)) - \sigma^{\perp}_{1,k\ell}(q_2(s), \tilde{\mathcal{Y}}_2(s))\}\mathrm{d}\beta^{\perp,1}(s)\right] \\
\leq c_{\mathcal{J},\sigma}\left\{\int_0^{t\wedge\tau} c^2_{\tilde{\mathcal{Y}}} \tilde{\rho}^2(q_1(s) - q_2(s))\mathrm{d}s + \int_0^{t\wedge\tau} \gamma^2(\tilde{\mathcal{Y}}_1(s) - \tilde{\mathcal{Y}}_2(s))\mathrm{d}s\right\}.
\end{array}\right\} \quad (4.22)$$

[13] Cf. Sect. 15.2.5.

(iv) By the Cauchy–Schwarz and Doob inequalities[14]

$$
\begin{aligned}
E \sup_{0\le t\le T} \tilde{\rho}^2(\tilde{q}_\ell(t\wedge\tau)) &\le 3E\tilde{\rho}^2(q(0)) + 3T\int_0^{T\wedge\tau} c_F\tilde{\rho}^2|F(q(s),\tilde{\mathcal{Y}}(s))|^2\mathrm{d}s \\
&\quad +3\sum_{k,\ell=1}^d E\Bigg[\int_0^{T\wedge\tau}\Bigg\{\Big[\int \mathcal{J}_{k\ell}(q(s),p,\tilde{\mathcal{Y}}(s))w(\mathrm{d}p,\mathrm{d}s)\Big] \\
&\qquad +\Big[\sigma^\perp_{1,k\ell}(q(s),\tilde{\mathcal{Y}}(s))\mathrm{d}\beta^{\perp,1}(s)\Big]\Bigg]\Bigg\} \\
&\le 3E\tilde{\rho}^2(q(0)) + c_{F,\mathcal{J},\sigma,T,\tau}E\int_0^{T\wedge\tau}(1+\tilde{\rho}^2(q(s)))\mathrm{d}s \\
&\quad \text{(by (4.11) (b) or (4.12) (b))} \\
&\le 3E\tilde{\rho}^2(q(0)) + c_{F,\mathcal{J},\sigma,T,\tau}E\int_0^{T}(1+\tilde{\rho}^2(q(s\wedge\tau)))\mathrm{d}s
\end{aligned}
$$

The Gronwall inequality[15] implies

$$
E \sup_{0\le t\le T} \tilde{\rho}^2(\tilde{q}_\ell(t\wedge\tau)) \le \tilde{c}_{F,\mathcal{J},\sigma,T,\tau}E\tilde{\rho}^2(q(0)), \tag{4.23}
$$

whence $\tilde{q}_\ell(\cdot\wedge\tau)\in L_{2,\mathcal{F}}(C([0,\infty);\mathbf{R}^d))$.

(v) (4.21)–(4.23), the Cauchy-Schwarz and Doob inequalities imply

$$
\left.
\begin{aligned}
&E \sup_{0\le t\le T} \tilde{\rho}^2(\tilde{q}_1(t\wedge\tau)-\tilde{q}_2(t\wedge\tau)) \\
&\le c_{F,\mathcal{J},\sigma,T}\Bigg\{E\tilde{\rho}^2(q_1(0)-q_2(0)) + E\int_0^T c^2_{\tilde{\mathcal{Y}}}\tilde{\rho}^2(q_1(s\wedge\tau)-q_2(s\wedge\tau))\mathrm{d}s \\
&\qquad +E\int_0^{T\wedge\tau}\gamma^2(\tilde{\mathcal{Y}}_1(s)-\tilde{\mathcal{Y}}_2(s))\mathrm{d}s\Bigg\}.
\end{aligned}
\right\} \tag{4.24}
$$

(vi) Choosing first $\tilde{\mathcal{Y}}_1\equiv\tilde{\mathcal{Y}}_2$, the existence of a continuous solution is derived by the standard Picard-Lindelöf procedure, defining iteratively:

$$
\begin{aligned}
q_{n+1}(t) &:= q(0) + \int_0^t F(q_n(s),\tilde{\mathcal{Y}}_\ell(s))\mathrm{d}s + \int_0^t\int \mathcal{J}(q_n(s),p,\tilde{\mathcal{Y}}_\ell(s))w(\mathrm{d}p,\mathrm{d}s) \\
&\quad + \int_0^t \sigma_1^\perp(q_n(s),\tilde{\mathcal{Y}}_\ell(s))\mathrm{d}\beta^{\perp,1}(s) \\
q_0(t) &:\equiv q(0).
\end{aligned}
$$

By (4.23) and (4.24) $q_n(\cdot)$ converge uniformly on the compact interval $[0,T]$ in mean square towards a solution of (4.9). This follows from the contraction

[14] Cf. Proposition 15.4 in Sect. 15.1.2 and Theorem 15.32 in Sect. 15.2.3.

[15] Cf. Sect. 15.1.2, Proposition 15.5

mapping principle.[16] The uniqueness follows from (4.24) and Gronwall's inequality. Having thus established the existence of unique (continuous) solutions $r(\cdot, \tilde{\mathcal{Y}}_\ell, r_{0,\ell})$, $\ell = 1, 2$, (4.17) follows from (4.24) and Gronwall's inequality.

(vii) The relation (4.18) follows immediately from the construction.

(viii) Statement 3 is proved in Sect. 6. □

Remark 4.6. By Part (3) of Theorem 4.4 and the boundedness of the coefficients F, $\mathcal{J}$, and σ_i the map

$$q \mapsto \psi^i(t, q) := \bar{r}^i_N(t, \tilde{\mathcal{Y}}, q) \tag{4.25}$$

is in Ψ from Proposition 4.3 for arbitrary input process $\tilde{\mathcal{Y}}$ and $i = 1, \ldots, N$. □

Next, we consider the $\mathbf{R}^{dN}$-valued system of coupled SODEs (4.10). Since for each ω the initial measure is a finite sum of point measures, it is finite. Therefore,

$$\mathcal{X}_N(s) := \sum_{i=1}^{N} m_i \delta_{r^i(s)}, \in L_{0,\mathcal{F}_s}(\mathbf{M}).$$

Further, a solution of (4.10), if it exists, preserves the initial mass, i.e., $\mathcal{X}_N(\cdot, \mathbf{R}^d) \equiv \sum_{i=1}^{N} m_i$, where $\mathcal{X}_N(t) := \sum_{i=1}^{N} m_i \delta_{r^i(t)}$. Therefore, we may take $\tilde{\mathcal{Y}}(t) := \mathcal{X}_N(t) := \sum_{i=1}^{N} m_i \delta_{r^i(t)}$ in Theorem 4.4. We endow $\mathbf{R}^{dN}$ with the metric

$$\rho_N(r_N, q_N) := \max_{1 \le i \le N} \rho(r_i, q_i),$$

where $r_N := (r^1, \ldots, r^N)$, $q_N := (q_1, \ldots, q_N) \in \mathbf{R}^{dN}$.

Theorem 4.7. Assume (4.11) or (4.12) in addition to $\mathcal{X}_N(s) \in L_{0,\mathcal{F}_s}(\mathbf{M}_f)$. Then, to each initial condition $r_N(s) \in L_{0,\mathcal{F}_s}(\mathbf{R}^{dN})$ (4.10) has a unique solution $r_N(\cdot, r_N(s)) \in L_{0,\mathcal{F}}(C([s, \infty); \mathbf{R}^{dN}))$, which is a Markov process on $\mathbf{R}^{dN}$.

Proof.

(i) Assume, without loss of generality, $s = 0$. Let $b > 0$ and choose a localizing stopping time $\tau > 0$ for $\mathcal{X}_N(0)$ such that

$$0 \le \mathcal{X}_N(0, \omega, \mathbf{R}^d) 1_{\{\tau_n > 0\}} \le b \quad \text{a.s}$$

Since $1_{\{\tau_n > 0\}}$ is $\mathcal{F}_0$-measurable, we may therefore, without loss of generality, assume

[16] Cf., Theorem 15.2 in Sect. 15.1.1. Picard-Lindelöf iterative approximations of the solutions of SODEs are standard in stochastic analysis – cf., e.g., Gikhman and Skorokhod (1982), Chap. IV.1, or Protter (2004), Chap. V.3. The argument for SODEs is essentially the same as for ODE's. For the latter case, cf. Coddington and Levinson, 1955, Chap. 1.3.

$$0 \le \mathcal{X}_N(0,\omega,\mathbf{R}^d) \le b \quad \text{a.s.} \tag{4.26}$$

Define recursively for $i = 1, \ldots, N$

$$q^i_{n+1}(t) := q^i(0) + \int_0^t F(q^i_n(s), \mathcal{Y}_n(s))\mathrm{d}s + \int_0^t \int \mathcal{J}(q^i_n(s), p, \mathcal{Y}_n(s))w(\mathrm{d}p, \mathrm{d}s)$$
$$+ \int_0^t \sigma_i^{\perp}(q^i_n(s), \mathcal{Y}_n(s))\mathrm{d}\beta^{i,\perp}(s),$$

where now

$$\mathcal{Y}_n(s) := \sum_{i=1}^N m_i \delta_{q^i_n(s)}.$$

(ii) We now proceed, following the pattern of the proof of Theorem 4.5.[17] We compare two steps in the recursive scheme, indexed by 1 and 2 (instead of n and m).

$$\begin{aligned}
\gamma_f(\mathcal{Y}_1(s), \mathcal{Y}_2(s)) &= \sup_{\|f\|_{L,\infty}\le 1} \int f(r)(\mathcal{Y}_1(s) - \mathcal{Y}_2(s)) \\
&= \sup_{\|f\|_{L,\infty}\le 1} \sum_{i=1}^N m_i \left[f(q^i_1(s)) - f(q^i_2(s))\right] \\
&= \sup_{\|f\|_{L,\infty}\le 1} \sum_{i=1}^N m_i \frac{[f(q^i_1(s)) - f(q^i_2(s))]}{\rho(q^i_1(s), q^i_2(s))} \rho(q^i_1(s), q^i_2(s)) \\
&\le \sum_{i=1}^N m_i \rho(q^i_1(s), q^i_2(s)) \\
&\le b\rho_N(q_{1,N}(s), q_{2,N}(s)),
\end{aligned}$$

i.e.,

$$\gamma_f(\mathcal{Y}_1(s), \mathcal{Y}_2(s)) \le b\rho_N(q_{1,N}(s), q_{2,N}(s)). \tag{4.27}$$

(iii) Using estimates (4.21)–(4.24) in addition to (4.27) and letting "tilde" indicate the next step, we obtain that for $i = 1, \ldots, N$

$$\begin{aligned}
&E \sup_{0\le t\le T\wedge\tau} \rho^2(\tilde{q}^i_1(t), \tilde{q}^i_2(t)) \\
&\le c_{F,\mathcal{J},T}\{E\rho^2(q_1(0), q_2(0)) + E\int_0^{T\wedge\tau} c_b \rho_N^2(q_{1,N}(s), q_{2,N}(s)\mathrm{d}s\}.
\end{aligned}$$

(iv) Summing up from 1 to N and using

$$\max_{i=1,\ldots,N} (\rho^2(\tilde{q}^i_1(t), \tilde{q}^i_2(t)) \le \sum_{i=1}^N (\rho^2(\tilde{q}^i_1(t), \tilde{q}^i_2(t))$$

---

[17] To simplify the notation we provide the argument only for the bounded metric $\rho(\cdot)$ under the assumption (4.11). The corresponding steps, based on (4.12) can be obtained as in Theorem 4.5.

in addition to Gronwall's inequality, the contraction mapping principle implies the existence of a unique continuous solution.

(v) The Markov property follows from the more general Proposition 15.64 in Sect. 15.2.7. □

4.3 Equivalence in Distribution and Flow Properties for SODEs

We derive equivalence of the distributions of the solutions of (4.9) if we replace the space–time white noise $w(\cdot,\cdot)$ by a space–time field with the same distribution. This analysis is motivated by different space–time white noise fields $w(\cdot,\cdot)$ as a result of shifts and rotations (cf. the following Chap. 5, (5.23)–(5.25), and Chap. 10. Finally, we analyze the backward flow for (4.9).

For the analysis of equivalent distributions it is convenient to have a nice state space for $w(\mathrm{d}p, \mathrm{d}t)$ and the infinite sequence of i.i.d. Brownian motions $\{\beta^{\perp,n}(\cdot)\}$. The representation (4.14)/(4.15) suggests for fixed t a state space on which we can define spatial Gaussian standard white noise as a countably additive Gauss measure. To generalize (3.45) to $\mathbf{R}^d$-valued processes let $\mathbf{H}_0^d$ be the Cartesian product $\mathbf{H}_0 \times \ldots \times \mathbf{H}_0$ (d times). Endowed with the scalar product $<\cdot,\cdot>_{0,d}$ as the sum of the corresponding d scalar products $<\cdot,\cdot>_0$, $\mathbf{H}_0^d$ is a real separable Hilbert space. The usual definition of standard cylindrical Brownian motions[18] implies that

$$\left.\begin{aligned} W(t) &:= \sum_{n=1}^{\infty} \widehat{\phi}_n(\cdot) \int_0^t \int \widehat{\phi}_n(p) w(\mathrm{d}p, \mathrm{d}s) = \sum_{n=1}^{\infty} \widehat{\phi}_n(\cdot) \beta^n(t), \\ W^{\perp}(t) &:= \sum_{n=1}^{\infty} \widehat{\phi}_n(\cdot) \beta^{\perp,n}(t) \end{aligned}\right\} \tag{4.28}$$

define $\mathbf{H}_0^d$-valued standard cylindrical Brownian motions. It follows from Sects. 15.2.2 and 15.2.4 that

$$W(t) = \frac{\partial^d}{\partial r_1 \ldots \partial r_d} \int_0^t w(\cdot, \mathrm{d}s), \tag{4.29}$$

where the derivatives are taken in the distributional sense.[19] To regularize $W(\cdot)$ and $W^{\perp}(\cdot)$, we need to choose an appropriate Hilbert distribution space. The precise definition and analysis are provided in Sect. 15.1.3 in the framework of the well known Schwarz space $\mathcal{S}'$ of tempered distributions. We show in Sect. 15.1.3 ((15.32)–(15.36)) that there is a Hilbert distribution space $\mathbf{H}_0 \subset \mathcal{H}_{-\gamma} \subset \mathcal{S}'$ with $\gamma > d$ such that, imbedded into $\mathcal{H}_{-\gamma}$, the $\mathbf{H}_0$-valued standard Brownian motion,

[18] Cf. Sect. 15.2.2.

[19] Cf. (2.10), (15.69), (15.126), and (15.127).

defined by (3.45) becomes an $\mathcal{H}_{-\gamma}$-valued regular Brownian motion. More precisely, the γ is chosen such that

$$\sum_{n=1}^{\infty} \|\phi_n\|_{-\gamma}^2 < \infty, \tag{4.30}$$

which means that the imbedding $\mathbf{H}_0 \subset \mathcal{H}_{-\gamma}$ is Hilbert-Schmidt.[20] As in the definition of $\mathbf{H}_0^d$, we define $\mathcal{H}_{-\gamma} \times \ldots \times \mathcal{H}_{-\gamma}$ (d times) and let $\langle \cdot, \cdot \rangle_{-\gamma,d}$ be the sum of the corresponding d scalar products $\langle \cdot, \cdot \rangle_{-\gamma}$. Setting

$$(\mathbf{H}_{w,d}, \langle \cdot, \cdot \rangle_{w,d}) := (\mathcal{H}_{-\gamma}^d, \langle \cdot, \cdot \rangle_{-\gamma,d}), \tag{4.31}$$

the imbedding

$$\mathbf{H}_0^d \subset \mathbf{H}_{w,d}, \tag{4.32}$$

is Hilbert-Schmidt. Hence, imbedded into $\mathbf{H}_{w,d}$, $W(\cdot)$ and $W^{\perp}(\cdot)$ become $\mathbf{H}_{w,d}$-valued regular Brownian motions. In $\mathbf{H}_w$, $W(\cdot)$ and $W^{\perp}(\cdot)$ are mean zero square integrable continuous martingales. By Doob's inequality for $\tilde{W} \in \{W, W^{\perp}\}$

$$E\left\{\sup_{0 \le t \le T} \|\tilde{W}(t)\|_{w,d}^2\right\} \le 4d \sum_{n=1}^{\infty} \|\phi_n\|_{-\gamma}^2 T < \infty, \tag{4.33}$$

recalling that $\hat{\phi}_n \beta^n(t) = (\phi_n \beta_1^n(t), \ldots, \phi_n \beta_d^n(t))^T$ and similarly for $\beta^{n,\perp}(\cdot)$. Consequently, the sums in (4.28), imbedded into $\mathbf{H}_{w,d}$, converge uniformly on compact intervals in mean square. It follows that

$$W(\cdot),\ W^{\perp}(\cdot) \in L_{0,\mathcal{F}}(C([0,\infty); \mathbf{H}_{w,d})). \tag{4.34}$$

It is more convenient to analyze probability distributions on the "canonical" probability space. Set

$$\Omega := C([0,\infty); \mathbf{H}_{w,d} \times \mathbf{H}_{w,d} \times \mathbf{M}) \times \mathbf{R}^{\mathrm{d}N}. \tag{4.35}$$

Notation

(i) Let $\mathcal{F}_{\mathbf{H}_{w,d} \times \mathbf{H}_{w,d} \times \mathbf{M},s,t}$ denote the σ-algebra of cylinder sets in $C([s,t]; \mathbf{H}_{w,d} \times \mathbf{H}_{w,d} \times \mathbf{M})$ and $\bar{\mathcal{F}}_{\mathbf{H}_{w,d} \times \mathbf{H}_{w,d} \times \mathbf{M},s,t}$ denotes the completion of $\mathcal{F}_{\mathbf{H}_{w,d} \times \mathbf{H}_{w,d} \times \mathbf{M},s,t}$ with respect to the underlying probability measure P. $\bar{\mathcal{F}}_{\mathbf{H}_{w,d} \times \mathbf{H}_{w,d} \times \mathbf{M},s,t} \otimes \mathcal{B}^{\mathrm{d}N}$ is the filtration on the canonical probability space. Finally we set $\bar{\mathcal{F}} := \bar{\mathcal{F}}_{\mathbf{H}_{w,d} \times \mathbf{H}_{w,d} \times \mathbf{M},0,\infty}$.

(ii) $\beta^{\perp,[j,k]}$ is the $\mathbf{R}^{(k+1-j)d}$-dimensional standard Brownian motion with d-dimensional components $\beta^{\perp,i}$, $i = j, \ldots, k$.

(iii) If f and g are random variables with values in some measurable space, we write

$$f \sim g,$$

if f and g have the same distribution.

[20] Cf. (15.35)–(15.36).

Hypothesis 4.2

Suppose for $s \geq 0$ there are two sets of random variables $(w_i, W_i^{\perp}, \mathcal{Y}_i, r_{s,i})$, $i = 1, 2$, on $(\Omega, \mathcal{F}, P)$ such that

(1) w_i are two space–time white noises with W_i in $L_{0,\mathcal{F}}(C([0,\infty); \mathbf{H}_{w,d}))$ such that in the (weak) representation (4.28) $W_i(\cdot, t) = \sum_{n=1}^{\infty} \hat{\phi}_n(\cdot)\beta^{n,i}(t)$, where $\{\beta^{n,i}(\cdot)\}_{n\in\mathbf{N}}$ are two families of independent $\mathbf{R}^d$-valued standard Brownian motions.
(2) $\{\beta^{n,\perp,i}\}_{\{n\in\mathbf{N}, i=1,2\}}$ are two sequences of i.i.d. $\mathbf{R}^d$-valued standard Brownian motions, independent of the white noises w_i, with representations $W^{\perp,i}$ in $L_{0,\mathcal{F}}(C([0,\infty); \mathbf{H}_{w,d}))$.
(3) $\tilde{\mathcal{Y}}_i \in L_{\mathrm{loc},2,\mathcal{F}}(C([s,T]; \mathbf{M}))$ and for $T \geq t \geq u \geq s$

$$\pi_{u,t}\tilde{\mathcal{Y}}_1 \sim \pi_{u,t}\tilde{\mathcal{Y}}_2.$$

(4) $r_{N,i,s} \in L_{2,\mathcal{F}_s}(\mathbf{R}^{\mathrm{d}N})$ and

$$r_{N,1,s} \sim r_{N,2,s}.$$

Finally, suppose

$$\begin{aligned}(W_1, W_1^{\perp}, \tilde{\mathcal{Y}}_1, r_{N,1,s}) &\sim (W_2, W_2^{\perp}, \tilde{\mathcal{Y}}_2 r_{N,2,s}) \\ &\text{in } C([s,T]; \mathbf{H}_{w,d} \times \mathbf{H}_{w,d} \times \mathbf{M} \times \mathbf{R}^{\mathrm{d}N}).\end{aligned} \tag{4.36}$$

□

We denote by $\bar{r}_N(\cdot, w_i, \beta^{\perp,[1,N]}, \tilde{\mathcal{Y}}_i, r_{N,i,s}, s)$ the unique continuous solutions of (4.9) with input variables $(w_i, \beta^{\perp,i,[1,N]}, \tilde{\mathcal{Y}}_i, r_{N,i,s})$, $i = 1, 2$, with properties from Part (3) of Theorem 4.5.

Theorem 4.8. *Assuming Hypothesis 4.2,*

$$\begin{aligned} r_N\ (\cdot, w_1, \beta^{\perp,1,[1,N]}, \tilde{\mathcal{Y}}_1, r_{N,1,s}, s) &\sim r_N(\cdot, w_2, \beta^{\perp,2,[1,N]}, \\ &\tilde{\mathcal{Y}}_2, r_{N,2,s}, s) \ \textit{in}\ C[s,T]; \mathbf{R}^{\mathrm{d}N}). \end{aligned} \tag{4.37}$$

In particular, for any $t \geq s$

$$r_N(t, w_1, \beta^{\perp,1,[1,N]}, \tilde{\mathcal{Y}}_1, r_{N,1,s}, s) \sim r_N(t, w_2, \beta^{\perp,2,[1,N]}, \tilde{\mathcal{Y}}_2, r_{N,2,s}, s). \tag{4.38}$$

Proof. *By the results of Theorem 4.5, there is a solution map*

$$\bar{\bar{r}}_N \colon C([0,T]; \mathbf{H}_{w,d} \times \mathbf{H}_{w,d} \times \mathbf{M}) \times \mathbf{R}^{Nd} \to C([0,T]; \mathbf{R}^{Nd})$$

such that for all $t \in [0,T]$

$$(\pi_t\, \bar{\bar{r}})(\cdot,\cdot,\cdot,\cdot) \text{ is } \bar{\mathcal{F}}_{\mathbf{H}_{w,d}\times\mathbf{H}_{w,d}\times\mathbf{M},0,t} \otimes \mathcal{B}^d - \mathcal{F}_{\mathbf{R}^{Nd},0,t}\text{-measurable,} \tag{4.39}$$

where $\mathcal{F}_{\mathbf{R}^{Nd},0,t}$ is the σ-algebra of cylinder sets on $C([0,t]; \mathbf{R}^{Nd})$. Moreover, in $C([0,t]; \mathbf{R}^{Nd})$

$$\overset{=}{r}_N(\cdot, \pi.W_i, \pi.\beta^{\perp,i,[1,N]}, \pi.\tilde{\mathcal{Y}}_i, r_{N,0}) \sim \bar{r}_N(\cdot, w_i, \beta^{\perp,i,[1,N]}, \tilde{\mathcal{Y}}_i, , r_{N,0}), \quad i \equiv 1, 2, \tag{4.40}$$

(cf. Ikeda and Watanabe (1981), Chap. IV). Expressions (4.36), (4.39), and (4.40) imply (4.37) and (4.38). □

The SODE (4.9) describes the random "forward" flow. Next, we follow the procedure in Ikeda and Watanabe (1981), Chap. V, to construct the random "backward" flow. Fix $T > s$ and consider

$$\left.\begin{aligned} \mathrm{d}r^i(t) &= -F(r^i(t), \tilde{\mathcal{Y}}(T+s-t), T+s-t)\mathrm{d}t \\ &\quad + \int \mathcal{J}(r^i(t), p, \tilde{\mathcal{Y}}(T+s-t), T+s-t)\check{w}(\mathrm{d}p, \mathrm{d}t) \\ &\quad + \sigma_i^{\perp}(r^i(t), p, \tilde{\mathcal{Y}}(T+s-t), T+s-t)\mathrm{d}\check{\beta}^{\perp,i}(t), \\ &\quad\quad r^i(s) = r^i \text{ (deterministic)}, t \in [s, T], i = 1, \ldots, N, \end{aligned}\right\} \tag{4.41}$$

where

$$\left.\begin{aligned} \check{w}(\mathrm{d}p, t) &:= w(\mathrm{d}p, T-t) - w(\mathrm{d}p, T), \\ \check{\beta}^{\perp,i}(t) &:= \beta^{\perp,i}(T-t) - \beta^{\perp,i}(T). \end{aligned}\right\} \tag{4.42}$$

Replacing $F(r, \tilde{\mathcal{Y}}(t), t)$ by $-F(r, \tilde{\mathcal{Y}}(T+s-t), T+s-t)$, $\mathcal{J}(r, p, \tilde{\mathcal{Y}}(t), t)$ by $\mathcal{J}(r, p, \tilde{\mathcal{Y}}(T+s-t), T+s-t)$, $\tilde{\sigma}_i^{\perp}(r, \tilde{\mathcal{Y}}(t), t)$, $\sigma_i^{\perp}(r, \tilde{\mathcal{Y}}(T+s-t), T+s-t)$, w by $\check{w}$ and $\beta^{\perp,i}$ by $\check{\beta}^{\perp,i}$, Theorem 4.5 implies the existence of a unique Itô-solution of (4.41). Moreover, setting $\overline{\mathcal{G}}_t := \check{\mathcal{G}}_t := \overline{\sigma}(\check{W}(u), u \le t)$ and $\overline{\mathcal{G}}_t^{\perp,[j,k]} := \check{\mathcal{G}}_t^{\perp,[j.k]} := \overline{\sigma}(\{\check{\beta}^{\perp,i}(u), u \le t, i = j, \ldots, k\}$ (where $\overline{\sigma}(\cdot)$ is the completed σ-algebra), this solution can be represented through the $\check{\mathcal{G}}_t \otimes \check{\mathcal{G}}_t^{\perp,[1,N]} \otimes \mathcal{F}_{\mathbf{M},t} \otimes \mathcal{B}^N - \mathcal{B}^{Nd}$-measurable map $\check{r}_N(t, \cdot, \cdot, s, \check{w}, \check{\beta}^{\perp,[1,N]})$ of Part 3 of Theorem 4.5. The measurability properties of $\check{r}$ allow us to define

$$\xi_N \mapsto \check{r}_N(\cdot, \tilde{\mathcal{Y}}, \xi_N, s, \check{w}, \check{\beta}^{\perp,[1,N]}) \tag{4.43}$$

for any $\mathbf{R}^N$-valued random variable ξ_N such that for deterministic $\xi_N = r_N$ $\check{r}_N(\cdot, \tilde{\mathcal{Y}}, r_N, s, \check{w}, \check{\beta}^{\perp,[1,N]})$ is the unique Itô-solution of (4.41) with $\check{r}_N(s, \tilde{\mathcal{Y}}, r, s, \check{w}, \check{\beta}^{\perp,[1,N]}) = r_N$.

Remark 4.9.

(i) In general, $\check{r}_N(\cdot, \tilde{\mathcal{Y}}, \xi_N, s, \check{w}, \check{\beta}^{\perp,[1,N]})$ cannot be interpreted as an Itô-solution of (4.41) with initial condition ξ_N at time s, because ξ_N can be anticipating with respect to $\check{w}, \check{\beta}^{\perp,[1,N]}$.

(ii) Since in (4.43) we do not need the measurability in μ, we may always assume $\tilde{\mathcal{Y}} \in L_{0,\mathcal{F}}(C([s, T]; \mathbf{M}))$, etc. and assume that $\check{r}_N(t, \tilde{\mathcal{Y}}, \cdot, s, \check{w}, \check{\beta}^{\perp,[1,N]})$ is measurable with respect to (ω, r). □

We show that (4.41) is indeed the SODE for the backward flow, under an additional smoothness assumption on the coefficients of (4.9). To this end, we introduce the following notation, which will also be used in the following sections:

Notation

For $m \in \mathbf{N}$ let $C_b^m(\mathbf{R}^d, \mathbf{R})$ be the space of m times continuously differentiable bounded real valued functions on $\mathbf{R}^d$, where all derivatives up to order m are bounded, $C_0^m(\mathbf{R}^d, \mathbf{R})$ is the subspace of $C_b^m(\mathbf{R}^d, \mathbf{R})$ whose elements and derivatives vanish at infinity and $C_c^m(\mathbf{R}^d, \mathbf{R})$ the subspace of $C_0^m(\mathbf{R}^d, \mathbf{R})$, whose elements have compact support. For $m = 0$ we drop the superscript in $C_b^m(\mathbf{R}^d, \mathbf{R})$, etc.

To describe the order of differentiability of $f : \mathbf{R}^d \to \mathbf{R}$ we introduce the following notation on multiindices:

$$\mathbf{n} = (n_1, \ldots, n_d) \in (\mathbf{N} \cup \{0\})^d.$$

$|\mathbf{n}| := n_1 + \ldots + n_d$. If $\mathbf{n} = (n_1, \ldots, n_d)$ and $\mathbf{m} = (m_1, \ldots, m_d)$, then $\mathbf{n} \le \mathbf{m}$ iff $n_i \le m_i$ for $i = 1, \ldots, d$. $\mathbf{n} < \mathbf{m}$ iff $\mathbf{n} \le \mathbf{m}$ and $|\mathbf{n}| < |\mathbf{m}|$. For arbitrary $\mathbf{n}, \mathbf{m}$ we define

$$\mathbf{n} - \mathbf{m} := (k_1, \ldots, d_d),$$

where $k_i = (n_i - m_i) \vee 0$, $i = 1, \ldots, d$, $\mathbf{n} = (n_1, \ldots, n_d)$, and $\mathbf{m} = (m_1, \ldots, m_d)$. Differential operators with respect to the space variables are usually denoted by $\partial^{\mathbf{j}}$, $\mathbf{j} \in (\mathbf{N} \cup \{0\})^d$, e.g., for $\mathbf{j} = (j_1, \ldots, j_d)$ we write

$$\partial^{\mathbf{j}} f(r) = \frac{\partial^{\mathbf{j}}}{(\partial r_{j_1})^{j_1} \ldots (\partial r_{j_d})^{j_d}} f(r).$$

However, if there are several groups of space variables, like (r, p, q) etc., we will write $\partial_r^{\mathbf{j}}$, $\partial_q^{\mathbf{j}}$, etc. to indicate the variable on which $\partial^{\mathbf{j}}$ is acting. Partial derivatives of first and second order may also be denoted by ∂_k, $\partial_{k\ell}^2$, resp. $\partial_{k,r}$, $\partial_{k\ell,r}^2$ (indicating again in the latter cases the relevant space variables). If $f \in C_b^m(\mathbf{R}^d, \mathbf{R})$ we set

$$|||f|||_m := \max_{|\mathbf{j}| \le m} \sup_{r \in \mathbf{R}^d} |\partial^{\mathbf{j}} f(r)|. \tag{4.44}$$

.

Abbreviate for $u \in [0, T]$

$$\left.\begin{array}{ll}
\tilde{F}_k(r, u) := F_k(r, \tilde{\mathcal{Y}}(u), u), & k = 1, \ldots, d; \\
\tilde{\mathcal{J}}_{k\ell}(r, p, u) := \mathcal{J}_{k\ell}(r, p, \tilde{\mathcal{Y}}(u), u) & k, \ell = 1, \ldots, d; \\
\tilde{\sigma}_{n,k\ell}^{\perp}(r, u) := \sigma_n^{\perp}(r, \tilde{\mathcal{Y}}(u), u), & k, \ell = 1, \ldots, d, n, \ell = 1, \ldots, N.
\end{array}\right\} \tag{4.45}$$

Suppose that for some $m \ge 1$

$$\max_{1 \le k, \ell \le d} ess \sup_{\omega \in \Omega,\, 0 \le u \le T} \{|||\tilde{F}_k(\cdot, u, \omega)|||_m + \max_{|\mathbf{j}| \le m+1} \left\| \sum_{\ell=1}^d \int (\partial_r^{\mathbf{j}} \tilde{\mathcal{J}})_{k\ell}^2(\cdot, p, u, \omega) \mathrm{d}p \right\| + |||\tilde{\sigma}_{n,k\ell}^{\perp}(\cdot, u, \omega)|||_{m+1}\} < \infty. \tag{4.46}$$

Remark 4.10. For homogeneous diffusion coefficients (and $\sigma_n^{\perp} \equiv 0, n = 1, \ldots, N$) condition (4.46) is similar to (8.35) on $\tilde{D}$, defined in (8.34) If, in addition, $\sigma_n^{\perp}$ is not identically $0, n = 1, .., N$, then (4.46) is similar to (8.35) and (8.75), respectively (cf. Chap. 8). Indeed, assuming $|\mathbf{j}| \leq m + 1$,

$$\int \left(\sum_{\ell=1}^{d} (\partial_{rr}^{\mathbf{j}} \tilde{\mathcal{J}}_{k\ell})(r, p, u)(\partial_{rr}^{\mathbf{j}} \tilde{\mathcal{J}}_{k\ell})(r, p, u) \mathrm{d}p \right.$$

$$= \partial_r^{\mathbf{j}} (\partial_q^{\mathbf{j}} \int \sum_{\ell=1}^{d} \tilde{\mathcal{J}}_{k\ell})(r, p, u) \tilde{\mathcal{J}}_{k\ell} \left. \vphantom{\sum_{\ell=1}^{d}} \right) (q, p, u) \mathrm{d}p_{|q=r}$$

$$= \partial_r^{\mathbf{j}} \partial_q^{\mathbf{j}} (-1)^{|\mathbf{j}|} \tilde{D}_{kk}(r - q, u)_{|q=r}$$

□

Theorem 4.11. *Fix $T > s$ and assume (4.46) with $m \geq 1$ in addition to (4.11). Then, with probability 1,*

$$\bar{r}_N(T-t, \tilde{\mathcal{Y}}, r_N, s, w, \beta^{\perp,[1,N]}) = \check{r}(t, \tilde{\mathcal{Y}}, \bar{r}(T-s, \tilde{\mathcal{Y}}, r, s, w, \beta^{\perp,[1,N]}), \check{w}, \check{\beta}^{\perp,[1,N]}), \tag{4.47}$$

uniformly in $r_N \in \mathbf{R}^N$ and $t \in [s, T]$, where the left- hand side in (4.47) is the (ω, r)-measurable version of the solution of the "forward" SODE (4.9), and the right-hand side is the measurable version (in (ω, r)) of the "backward" SODE (4.41).

Proof. (6.33), (6.38), and (6.39) in Chap. 6 imply (4.47). □

4.4 Examples

Since the inclusion of a time dependence in the coefficients does not cause any additional difficulty, it suffices to provide examples of time-independent coefficients. Further, we note that examples of coefficients, depending on finite measures $\mu(dr)$ can be immediately generalized to examples, depending on $\varpi(r)\mu(\mathrm{d}r)$, where $\mu \in \mathbf{M}_{\infty,\varpi}$. Therefore, it suffices to provide examples of coefficients F and $\mathcal{J}$ that depend on finite measures. We support this claim analyzing at the end a simple example of a diffusion kernel depending on measures from $\mathbf{M}_{\infty,\varpi}$.

We use the following abbreviation:

$$\mathbf{W}_{0,1,1} := L_1(\mathbf{R}^d; \mathrm{d}r)$$

where the right-hand side is the space of real valued functions on $\mathbf{R}^d$ which are integrable with respect to the Lebesgue measure $\mathrm{d}r$. The usual L_1-norm will be denoted $\| \cdot \|_{0,1,1}$.

Example 4.12. • (Fin.,F) Suppose there is a sequence of kernels $K_n := (K_{1n},\ldots,K_{dn})\colon \mathbf{R}^d \longrightarrow \mathbf{R}^d$ such that

(i) $K_{\ell n} \in C_b^1(\mathbf{R}^d,\mathbf{R}) \cap \mathbf{W}_{0,1,1}$ for $n \in \mathbf{N}\cup\{0\}$, $\ell = 1,\ldots,d$;
(ii) $\sup\limits_{n\in\mathbf{N}\cup\{0\}} \max_{\ell=1,\ldots,d}\{\||K_{\ell n}|\|_1 + \|K_{\ell n}\|_{L,\infty} + \int |K_{\ell n}|(r)\mathrm{d}r\} =: c_K < \infty.$

Further, for $f \in \mathbf{W}_{0,1,1}$ and $\mu \in \mathbf{M}_f$ we define

$$f * \mu^{*n}(r) := \begin{cases} \int f(r-p)\mu^{*n}(\mathrm{d}p), & \text{if } n \geq 1\,, \\ f(r), \text{ if } n = 0, \end{cases}$$

where

$$\int f(r-p)\mu^{*n}(\mathrm{d}p) := \int\cdots\int f(r-(p_1+\cdots+p_n))\mu(\mathrm{d}p_1)\cdots\mu(\mathrm{d}p_n).$$

Let two finite measures μ_1 and μ_2 be given, and suppose p_n is a sequence of nonnegative numbers such that

$$\sum_{n=1}^{\infty} p_n n(\gamma_f(\mu_1)\vee\gamma_f(\mu_2))^n \leq c_F(\gamma_f(\mu_1)\vee\gamma_f(\mu_2)) < \infty. \tag{4.48}$$

Set

$$F(r,\mu_i) := \sum_{n=0}^{\infty} p_n K_n * \mu_i^{*n}(r) \;\; i = 1,2. \tag{4.49}$$

We verify the Lipschitz assumption:

$$\begin{aligned} |F(r_1,\mu_1) - F(r_2,\mu_1)| &\leq \sum_{\ell=1}^{d}\sum_{n=0}^{\infty} p_n\Big|\int\cdots\int (K_{n\ell}(r_1-(p_1+\cdots+p_n))) \\ &\quad -K_{n\ell}(r_2-(p_1+\cdots+p_n))\mu_1(\mathrm{d}p_1)\cdots\mu_1(\mathrm{d}p_n)\Big| \\ &\leq \mathrm{d}c_K\sum_{n=0}^{\infty} p_n\rho(r_1,r_2)\gamma_f(\mu_1)^n \end{aligned}$$

Further,

$$\begin{aligned} |F(r_2,\mu_1) - F(r_2,\mu_2)| &\leq \sum_{\ell=1}^{d}\sum_{n=0}^{\infty} p_n\Big|\int\cdots\int K_{n\ell}(r_2-(p_1+\cdots+p_n)) \\ &\quad \times(\mu_1(\mathrm{d}p_1)\cdots\mu_1(\mathrm{d}p_n) - \mu_2(\mathrm{d}p_1)\cdots\mu_2(\mathrm{d}p_n)\Big| \end{aligned}$$

$$\leq \sum_{\ell=1}^{d}\sum_{n=0}^{\infty} p_n \Big| \int \cdots \int K_{n\ell}(r_2 - (p_1 + \cdots + p_n))$$

$$\times [(\mu_1(\mathrm{d}p_1) - \mu_2(\mathrm{d}p_1))\mu_1(\mathrm{d}p_2)\cdots \mu_1(\mathrm{d}p_n)$$

$$+ \sum_{\ell=1}^{d}\sum_{n=0}^{\infty} p_n \Big| \int \cdots \int K_{n\ell}(r_2 - (p_1 + \cdots + p_n))\mu_2(\mathrm{d}p_1)$$

$$\times (\mu_1(\mathrm{d}p_2)\cdots \mu_1(\mathrm{d}p_n) - \mu_2(\mathrm{d}p_2)\cdots \mu_2(\mathrm{d}p_n)|$$

$$\vdots$$

$$\leq \sum_{\ell=1}^{d}\sum_{n=0}^{\infty} p_n \sum_{k=1}^{n} \Big| \int \cdots \int K_{n\ell}(r_2 - (p_1 + \cdots + p_n))$$

$$\times [\mu_2(\mathrm{d}p_1)\cdots \mu_2(\mathrm{d}p_{k-1})(\mu_1(\mathrm{d}p_k) - \mu_2(\mathrm{d}p_k))\mu_1(\mathrm{d}p_{k+1})\cdots \mu_1(\mathrm{d}p_n)|$$

with the usual interpretation of the factors if $k = 1$ or $k = n$

$$\leq \sum_{\ell=1}^{d}\sum_{n=0}^{\infty} p_n \gamma_f(\mu_1, \mu_2)\|K_{n\ell}\|_{L,\infty,1}(\gamma_f(\mu_1) \vee \gamma_f(\mu_2))^{n-1}$$

$$\leq \mathrm{d}c_K \sum_{n=0}^{\infty} p_n \gamma_f(\mu_1, \mu_2)(n-1)(\gamma_f(\mu_1) \vee \gamma_f(\mu_2))^{n-1}.$$

Applying the triangle inequality the previous two calculations yield

$$|F(r_1, \mu_1) - F(r_2, \mu_2)| \leq c_F((\gamma_f(\mu_1) \vee \gamma_f(\mu_2))\rho(r_1, r_2) + \gamma_f(\mu_1, \mu_2)). \quad (4.50)$$

Hence, $F(r, \mu)$ satisfies the assumptions on F with $\gamma = \gamma_f$.

Let $\mathbf{H}_2$ denote the Sobolev space square integrable real-valued functions on $\mathbf{R}^d$, which are twice differentiable in the generalized sense such that their first and second partial derivatives are in $\mathbf{H}_0$. The Hilbert norm on $\mathbf{H}_2$ is denoted by $\|\cdot\|_2$. Further, $\|\cdot\|_L$ denotes the Lipschitz norm on the real-valued functions with domain $\mathbf{R}^d$.[21] □

Example 4.13. • (Fin.,$\mathcal{J}$) Suppose there is a sequence of kernels $\Gamma_n := (\Gamma_{k\ell n}) : \mathbf{R}^d \longrightarrow \mathcal{M}_{d\times d}$ such that

(i) $\Gamma_{k\ell n} \in C_b^2(\mathbf{R}^d, \mathbf{R}) \cap \mathbf{H}_2$ for $n \in \mathbf{N} \cup \{0\}$, $\ell = 1, \ldots, d$;
(ii) $\sup_{n\in\mathbf{N}\cup\{0\}} \max_{k,\ell=1,\ldots,d}\{|||\Gamma_{k\ell n}|||^2 + \|\Gamma_{k\ell n}\|_L^2 + \|\Gamma_{k\ell n}\|_2\} =: c_\Gamma < \infty$.

Let two finite measures μ_1 and μ_2 be given, and suppose q_n is a sequence of nonnegative numbers such that

[21] Cf. Sect. 15.1.4, (15.37).

$$\sum_{n=1}^{\infty} q_n(n+1)((\gamma_f(\mu_1) \vee \gamma_f(\mu_2))^n \le c_{\mathcal{J}}((\gamma_f(\mu_1) \vee \gamma_f(\mu_2)) < \infty.$$

Then

$$\mathcal{J}(r, p, \mu_i) := \sum_{n=0}^{\infty} q_n \Gamma_n * \mu_i^{*n}(r-p), \quad i = 1, 2, \tag{4.51}$$

is homogeneous in the position coordinates and satisfies the assumptions on $\mathcal{J}$ with respect to γ_f.

Indeed,

$$\begin{aligned}
&\sum_{k,\ell=1}^{d} \int (\mathcal{J}_{k\ell}(r_1, p, \mu_1) - \mathcal{J}_{k\ell}(r_2, p, \mu_1))^2 \mathrm{d}p \\
&\le \sum_{k,\ell=1}^{d} \sum_{n,\tilde{n}}^{d} q_n (\gamma_f(\mu_1))^n n q_{\tilde{n}} (\gamma_f(\mu_1))^{\tilde{n}} \tilde{n} \sup_{\zeta,\tilde{\zeta}} \\
&\quad \left\{ \left(\int (\Gamma_{k\ell n}(r_1 - p + \zeta) - \Gamma_{k\ell n}(r_2 - p + \zeta))^2 \mathrm{d}p \right)^{\frac{1}{2}} \right. \\
&\quad \left. \left(\int (\Gamma_{k\ell n}(r_1 - p + \tilde{\zeta}) - \Gamma_{k\ell n}(r_2 - p + \tilde{\zeta}))^2 \mathrm{d}p \right)^{\frac{1}{2}} \right\} \\
&\le d^2 c_{\mathcal{J}}^2 \gamma_f^2(\mu_1) \rho^2(r_1, r_2).
\end{aligned}$$

By shift invariance of the Lebesgue measure,

$$\begin{aligned}
&\int\int \left[\int \Gamma_{k\ell n}(r - p - \tilde{p}) \Gamma_{k\ell n}(r - p - \hat{p}) \mathrm{d}p \right] (\mu_1 - \mu_2)(\mathrm{d}\tilde{p})(\mu_1 - \mu_2)(\mathrm{d}\hat{p}) \\
&= \int\int \left[\int \Gamma_{k\ell n}(\hat{p} - \tilde{p} - p) \Gamma_{k\ell n}(-p) \mathrm{d}p \right] (\mu_1 - \mu_2)(\mathrm{d}\tilde{p})(\mu_1 - \mu_2)(\mathrm{d}\hat{p}).
\end{aligned}$$

The function

$$F(\hat{p}) := \int\int \Gamma_{k\ell n}(\hat{p} - \tilde{p} - p) \Gamma_{k\ell n}(-p) \mathrm{d}p (\mu_1 - \mu_2)(\mathrm{d}\tilde{p})$$

is Lipschitz continuous and bounded. Indeed,

$$\begin{aligned}
G(\tilde{p}) := &\sup_{\{\hat{p}_2 \neq \hat{p}_1, |\hat{p}_2 - \hat{p}_1| \le 1\}} \frac{1}{\rho(\hat{p}_2 - \hat{p}_1)} \int \left[\Gamma_{k\ell n}(\hat{p}_2 - \tilde{p} - p) \right. \\
&\left. - \Gamma_{k\ell n}(\hat{p}_1 - \tilde{p} - p) \right] \Gamma_{k\ell n}(-p) \mathrm{d}p
\end{aligned}$$

is Lipschitz continuous and bounded. A Lipschitz constant can be computed as follows:

$$
\begin{aligned}
\left|\frac{G(\tilde{p}_2)-G(\tilde{p}_1)}{\rho(\tilde{p}_2-\tilde{p}_1)}\right| &\le \sup_{\{\hat{p}_2\neq\hat{p}_1,|\hat{p}_2-\hat{p}_1|\le 1\}} \frac{1}{\rho(\hat{p}_2-\hat{p}_1)\rho(\tilde{p}_2-\tilde{p}_1)} \\
&\quad\times\left|\int \Gamma_{k\ell n}(\hat{p}_2-\tilde{p}_2-p)-\Gamma_{k\ell n}(\hat{p}_1-\tilde{p}_2-p)-\Gamma_{k\ell n}(\hat{p}_2-\tilde{p}_1-p)\right. \\
&\quad\left.+\Gamma_{k\ell n}(\hat{p}_1-\tilde{p}_1-p)\Gamma_{k\ell n}(-p)\mathrm{d}p\right| \\
&= \sup_{\{\hat{p}_2\neq\hat{p}_1,|\hat{p}_2-\hat{p}_1|\le 1\}}\left|\sum_{i,j=1}^{d}\frac{(\hat{p}_{2,j}-\hat{p}_{1,j})(\tilde{p}_{2,j}-\tilde{p}_{1,j})}{\rho(\hat{p}_2-\hat{p}_1)\rho(\tilde{p}_2-\tilde{p}_1)}\int_0^1\int_0^1 \mathrm{d}\alpha\,\mathrm{d}\beta\right. \\
&\quad\left.\times\int \partial_{i,j}^2\Gamma_{k\ell n}(\hat{p}_1+\beta(\hat{p}_2-\hat{p}_1)-\tilde{p}_1-\alpha(\tilde{p}_2-\tilde{p}_1)-p)\Gamma_{k\ell n}(-p)\mathrm{d}p\right| \\
&\le \sum_{i,j=1}^{d}\int_0^1\int_0^1 \mathrm{d}\alpha\,\mathrm{d}\beta\left|\int \partial_{i,j}^2\Gamma_{k\ell n}(\hat{p}_1+\beta(\hat{p}_2-\hat{p}_1)\right. \\
&\quad\left.-\tilde{p}_1-\alpha(\tilde{p}_2-\tilde{p}_1)-p)\Gamma_{k\ell n}(-p)\mathrm{d}p\right| \\
&\le \sum_{i,j=1}^{d}\int_0^1\int_0^1 \mathrm{d}\alpha\,\mathrm{d}\beta\left(\int (\partial_{i,j}^2\Gamma_{k\ell n})^2(\hat{p}_1+\beta(\hat{p}_2-\hat{p}_1)-\tilde{p}_1\right. \\
&\quad\left.-\alpha(\tilde{p}_2-\tilde{p}_1)-p)\mathrm{d}p\right)^{\frac{1}{2}}\left(\int \Gamma_{k\ell n}^2(-p)\mathrm{d}p\right)^{\frac{1}{2}}
\end{aligned}
$$

(by the Cauchy-Schwarz inequality)

$$
= \sum_{i,j=1}^{d}\left(\int (\partial_{i,j}^2\Gamma_{k\ell n})^2(-p)\mathrm{d}p\right)^{\frac{1}{2}}\left(\int \Gamma_{k\ell n}^2(-p)\mathrm{d}p\right)^{\frac{1}{2}}
$$

(by the shift invariance of the Lebesgue measure $\mathrm{d}p$)

$$
\le d\|\Gamma_{k\ell n}\|_2\|\Gamma_{k\ell n}\|_0
$$

$\left(\text{by the Cauchy-Schwarz inequality with respect to } \sum_{i,j=1}^{d}\right)$.

Hence,

$$
\left|\frac{F(\hat{p}_2)-F(\hat{p}_1)}{\rho(\hat{p}_2,\hat{p}_1)}\right| \le d\|\Gamma_{k\ell n}\|_2\|\Gamma_{k\ell n}\|_0\gamma_f(\mu_1-\mu_2).
$$

Altogether, we obtain:

$$
\left|\int\int\left[\int \Gamma_{k\ell n}(r-p-\tilde{p})\Gamma_{k\ell n}(r-p-\hat{p})\mathrm{d}p\right](\mu_1-\mu_2)(\mathrm{d}\tilde{p})(\mu_1-\mu_2)(\mathrm{d}\hat{p})\right|
$$

$$
\le d\|\Gamma_{k\ell n}\|_2\|\Gamma_{k\ell n}\|_0\gamma_f^2(\mu_1-\mu_2).
$$

Using this bound in addition to the triangle inequality, we obtain, repeating the calculations for (Fin.,F), the global Lipschitz property:

$$
\begin{aligned}
&\sum_{k,\ell=1}^{d} \int (\mathcal{J}_{k\ell}(r_2, p, \mu_1) - \mathcal{J}_{k\ell}(r_1, p, \mu_2))^2 \, \mathrm{d}p \\
&\le c^2_{\mathcal{J},d}((\gamma_f(\mu_1) \vee \gamma_f(\mu_2))^2 \rho^2(r_2 - r_1) + \gamma_f^2(\mu_1 - \mu_2)),
\end{aligned}
\tag{4.52}
$$

where again the Lipschitz constant at $\gamma_f(\mu_2, \mu_1)$ may be chosen independent of the mass of the measures. □

Example 4.14. (Infin.,$\mathcal{J}$) Let us analyze a version of the above example for $n = 1$ and $\mu \in \mathbf{M}_{\infty,\varpi}$.

Let $\Gamma = (\Gamma_{k\ell}) \in \mathcal{M}_{d\times d}$ with its one-dimensional components $\Gamma_{k\ell} \in C_b^2(\mathbf{R}^d, \mathbf{R}) \cap \mathbf{H}_2$. Set

$$
\tilde{\mathcal{J}}(r, p, \mu) := \int \Gamma(r - p - q)\varpi(q)\mu(\mathrm{d}q) \tag{4.53}
$$

Abbreviate

$$
D_{k\ell}(r_2 - r_1) := \sum_{m=1}^{d} \int \Gamma_{km}(r_2 - p)\Gamma_{\ell m}(r_1 - p)\mathrm{d}p
$$

and note that $D_{k\ell}(r_2 - r_1) = D_{k\ell}(r_1 - r_2)$, i.e, $D_{k\ell}(\cdot)$ is even. Further, $D_{k\ell}(\cdot) \in C_b^2(\mathbf{R}^d, \mathbf{R})$. Then

$$
\begin{aligned}
&\sum_{m=1}^{d} \int (\tilde{\mathcal{J}}_{km}(r_2, p, \mu) - \tilde{\mathcal{J}}_{km}(r_1, p, \mu))^2 \mathrm{d}p \\
&\quad = \sum_{m=1}^{d} \int\int\int (\Gamma_{km}(r_2 - p - q) - \Gamma_{km}(r_1 - p - q))(\Gamma_{\ell m}(r_2 - p - \tilde{q}) \\
&\quad\quad -\Gamma_{\ell m}(r_1 - p - \tilde{q}))\mathrm{d}p\varpi(q)\mu(\mathrm{d}q)\varpi(\tilde{q})\mu(\mathrm{d}\tilde{q}) \\
&\quad = \int\int \left[2D_{k\ell}(\tilde{q} - q) - 2D_{k\ell}(\tilde{q} - q + r_2 - r_1)\right] \varpi(q)\mu(\mathrm{d}q)\varpi(\tilde{q})\mu(\mathrm{d}\tilde{q})
\end{aligned}
$$

Since $D_{k\ell}(\cdot)$ is even and in $C_b^2(\mathbf{R}^d, \mathbf{R})$ it follows that

$$
|D_{k\ell}(\tilde{q} - q) - D_{k\ell}(\tilde{q} - q + r_2 - r_1)| \le c^2_{\mathcal{J}} \rho^2(r_2 - r_1).
$$

This implies in particular that $D_{k\ell}$ is a bounded Lipschitz function. Hence,

$$
\sum_{m=1}^{d} \int (\tilde{\mathcal{J}}_{km}(r_2, p, \mu) - \tilde{\mathcal{J}}_{km}(r_1, p, \mu))^2 \mathrm{d}p \le c^2_{\mathcal{J}} \gamma^2_{\varpi}(\mu)\rho^2(r_2, r_1).
$$

Further,

$$
\begin{aligned}
&\sum_{m=1}^{d}\int(\tilde{\mathcal{J}}_{km}(r,p,\mu_2)-\tilde{\mathcal{J}}_{km}(r,p,\mu_1))^2\mathrm{d}p\\
&=\sum_{m=1}^{d}\int\int\int(\Gamma_{km}(r-p-q)(\Gamma_{\ell m}(r-p-\tilde{q})\mathrm{d}p\varpi(q)\varpi(\tilde{q})(\mu_2(\mathrm{d}q)\\
&\quad-\mu_1(\mathrm{d}q))(\mu_2(\mathrm{d}\tilde{q})-\mu_1(\mathrm{d}\tilde{q}))\\
&=\int\int D_{k\ell}(\tilde{q}-q)\varpi(q)(\mu_2(\mathrm{d}q)-\mu_1(\mathrm{d}q))\varpi(\tilde{q})(\mu_2(\mathrm{d}\tilde{q})-\mu_1(\mathrm{d}\tilde{q}))\\
&\le c_{\mathcal{J}}^2\gamma_{\varpi}^2(\mu_2,\mu_1).
\end{aligned}
$$

Altogether,

$$
\begin{aligned}
\sum_{m=1}^{d}\int(\tilde{\mathcal{J}}_{km}(r_2,p,\mu_2)-\tilde{\mathcal{J}}_{km}(r_1,p,\mu_1))^2\mathrm{d}p &\le c_{\mathcal{J}}^2((\gamma_{\varpi}(\mu_1)\wedge\gamma_{\varpi}(\mu_2))\\
&\quad\times\rho^2(r_2-r_1)+\gamma_{\varpi}^2(\mu_2-\mu_1)).
\end{aligned}
\tag{4.54}
$$

Recall that because of the mass conservation we may assume that $\gamma_f(\tilde{\mathcal{Y}}(t,\omega))=\gamma_f(\tilde{\mathcal{Y}}(0,\omega))$ if $\gamma_f(\tilde{\mathcal{Y}}(0,\omega))\le c<\infty$. Hence, we easily see that $\mathcal{J}(r,\mu)$ satisfies the assumption (4.11), provided we work with $L_{\mathrm{loc},2,\mathcal{F}}(C((s,T];\mathbf{M}_f))$ and assume that the total mass of the measures $\tilde{\mathcal{Y}}(0,\omega)$ is bounded.

Restricting the $\tilde{\mathcal{Y}}(\cdot)\in L_{\mathrm{loc},2,\mathcal{F}}(C((s,T];\mathbf{M}_{\varpi}))$ to measure processes generated by flows in the sense of Proposition 4.3 allows us to use stopping time techniques to enforce uniform Lipschitz bounds on the coefficients F, $\mathcal{J}$ and $\sigma^{\perp}$ (cf. the proof of Proposition 4.3 in Sect. 15.2.8). □

Chapter 5
Qualitative Behavior of Correlated Brownian Motions

5.1 Uncorrelated and Correlated Brownian Motions

We define correlated Brownian motions and compare their space–time correlations with uncorrelated Brownian motions. Further, some simple properties of correlated Brownian motions are derived.

Recall the N-particle motion described by (2.11) where we may incorporate the coefficient α into the definition of G_ε. By Theorems 4.5 or 4.7, (2.11) has a unique solution for every adapted and square integrable initial condition $r^i(0)$, $i = 1, \ldots, N$. Let us now generalize (2.11) to the following system of SODEs in integral form, which is a special case of equations (4.9) and (4.10).

$$r(t, r_0^i) = r_0^i + \int_0^t \int \Gamma_\varepsilon(r(s, r_0^i), q, t) w(\mathrm{d}q, \mathrm{d}t), \quad i = 1, \ldots, N. \tag{5.1}$$

As in (4.9) and (4.10), $w(\mathrm{d}q, \mathrm{d}t)$ is a d-dimensional space–time white noise and $\Gamma_\varepsilon(r, q)$ is an $\mathcal{M}_{d\times d}$-valued function such that Hypothesis 4.1 holds with Γ_ε replacing $\mathcal{J}$. Hypothesis 4.1 guarantees the existence and uniqueness of the solution of (5.1). In addition, we assume that the kernel function Γ_ε depends on a correlation length parameter $\varepsilon > 0$. Let us denote the square integrable continuous martingales from the right-hand side of (5.1) by

$$m_\varepsilon(t, r_0^i) := \int_0^t \int \Gamma_\varepsilon(r(u, r_0^i), q) w(\mathrm{d}q, \mathrm{d}u), \quad i = 1, \ldots, N. \tag{5.2}$$

We may use the increments of $m_\varepsilon(t, r_0^i)$ as a stochastic perturbation of a deterministic ODE. Of particular interest are the special cases of (4.9) or (4.10), if we replace $\mathcal{J}$ by Γ_ε and assume $\sigma^\perp \equiv 0$. We may simplify the set of SODEs even further, assuming the initial conditions to be deterministic points r_0^i at time $t = 0$ and $r_0^i \neq r_0^j$, if $i \neq j$. To be specific, let us consider (4.10). We may then rewrite the resulting SODE in the following equivalent form:

$$\left.\begin{array}{l} \mathrm{d}r_N^i(t) = F(r_N^i(t), \mathcal{X}_N(t), t)\mathrm{d}t + m_\varepsilon(\mathrm{d}t, r_0^i), \\[2ex] r_N^i(0) = r_0^i, \; i = 1, \ldots N, \; \mathcal{X}_N(t) := \displaystyle\sum_{i=1}^N m_i \delta_{r_N^i(t)}, \quad m_i \in \mathbf{R} \end{array}\right\} \tag{5.3}$$

where the unperturbed ODE is cast in the form

$$\left.\begin{aligned}&\frac{\partial}{\partial t} r_N^i(t) = F(r_N^i(t), \mathcal{X}_N(t), t),\\ &r_N^i(0) = r_0^i,\ i = 1, \dots N,\ \mathcal{X}_N(t) := \sum_{i=1}^N m_i \delta_{r_N^i(t)},\quad m_i \in \mathbf{R}.\end{aligned}\right\} \tag{5.4}$$

In a more traditional approach, the stochastic perturbation of (5.4) is modeled by i.i.d. Brownian motions $\beta(\cdot, r_0^i)$, $i = 1, \dots, N$, with variance parameter $D > 0$. The initial condition r_0^i specifies the Brownian motion that perturbs the motion of the ith particle, yielding the following system of SODEs

$$\left.\begin{aligned}&\mathrm{d}r_N^{\perp i}(t) = F(r_N^{\perp i}(t), \mathcal{X}_N(t), t)\mathrm{d}t + \beta(\mathrm{d}t, r_0^i),\\ &r_N^i(0) = r_0^i,\ i = 1, \dots N,\ \mathcal{X}_N(t) := \sum_{i=1}^N m_i \delta_{r_N^i(t)},\quad m_i \in \mathbf{R}.\end{aligned}\right\} \tag{5.5}$$

Remark 5.1. There seem to be three reasons for the use of (5.5).

- The first reason is the relatively simple space–time correlations of the noise $\beta(\mathrm{d}t, r_0^i)$.
- The second reason is related to the deterministic behavior of $\mathcal{X}_N(t)$, as $N \longrightarrow \infty$ if the masses $m_i \approx \frac{1}{N}$. We refer the reader to Chap. 14 for more details on the macroscopic limit problem and will focus here our attention on the space–time correlations.
- The third reason is historical precedence: Einstein (1905) derives, in his first paper on Brownian motion, the diffusion equation with constant diffusion coefficient $\times \Delta$, assuming that "*neighboring particles are sufficiently separated.*" On the Basis of Einstein's arguments, scientists, more or less tacitly, assume that the Brownian particles *are sufficiently separated.*[1] □

Following the analysis of both generalized and classical random processes and random fields by Gel'fand and Vilenkin (1964), we choose test functions from the Schwarz space of infinitely often differentiable and rapidly decreasing real-valued functions on $\mathbf{R}$, $\mathcal{S}(\mathbf{R})$, for the time parameter.[2] The duality between functions from $\mathcal{S}(\mathbf{R})$ and its dual is denoted by $(\cdot, \cdot)$. Note that the deterministic equation (5.4) is perturbed by the time-derivatives of the Brownian motions $\beta(\cdot, r_0)$, where we are dropping the superscript "i" at the initial condition, since the initial conditions can vary on the continuum $\mathbf{R}^d$. We choose $\varphi, \psi \in \mathcal{S}(\mathbf{R})$ and we obtain from (15.1120/(15.113) in Sect. 15.2.4

[1] It is obvious that two Brownian particles at a close distance must be correlated. One argument to support this claim is the depletion phenomenon. The main goal of this chapter is to define a model of correlated Brownian motions, which at large distance are approximately uncorrelated but at distances less or equal $\sqrt{\varepsilon}$ show attractive behavior. Under appropriate assumptions (cf. (5.10) and Sect. 5.3) we show that $m_\varepsilon(\cdot, r_0)$, $r_0 \in \mathbf{R}^d$, may serve as a model for correlated Brownian motions with the desired properties.

[2] We refer the reader to our Sect. 15.1.3 for a description of the Schwarz space.

$$\left.\begin{aligned}&E\left(\varphi,\left(\frac{\mathrm{d}}{\mathrm{d}t}\right)\beta_k(\cdot,r_0)\right)\left(\psi,\left(\frac{\mathrm{d}}{\mathrm{d}t}\right)\beta_\ell(\cdot,\tilde{r}_0)\right)\\&=\int\varphi(t)\psi(t)\mathrm{d}t\,D\delta_{0,d}(r_0-\tilde{r}_0)\\&=\int_0^\infty\int_0^\infty\varphi(t)\psi(s)\delta_0(t-s)\mathrm{d}t\,\mathrm{d}s\,D\delta_{0,d}(r_0-\tilde{r}_0).\end{aligned}\right\}\tag{5.6}$$

Here, $\delta_0(t-s)$ and $\delta_{0,d}(r_0-\tilde{r}_0)$ are the delta-functions with support in 0 in $\mathbf{R}$ and $\mathbf{R}^d$, respectively. We refer to Lemma 15.45 in Sect. 15.2.4, which shows that for $k=1,\ldots,d$ and $r_0\in\mathbf{R}^d$ the generalized random processes $\frac{\mathrm{d}}{\mathrm{d}t}\beta_k(\cdot,r_0,\omega)$ are with probability 1 elements of $\mathcal{S}'(\mathbf{R})$, i.e., of the strong dual of $\mathcal{S}(\mathbf{R})$. Recall the definition of the tensor product between the bilinear functionals $\delta_0(t-s)$ and $\delta_{0,d}(r_0-\tilde{r}_0)$, denoted $\delta_{0,d}(r_0-\tilde{r}_0)\otimes\delta_0(t-s)$. Denoting the space–time correlations by $\overline{Cov}_{d+1}$, we obtain

$$\overline{Cov}_{d+1}=D\delta_{0,d}(r_0-\tilde{r}_0)\otimes\delta_0(t-s).\tag{5.7}$$

The derivation (5.7) begs the question concerning the space–time correlations of the stochastic perturbations in (5.3). We recall the definition of the tensor quadratic variation of a vector-valued martingale, as given by Metivier and Pellaumail (1980), Sect. 2.3. In case of finite-dimensional martingales the tensor quadratic variation reduces to the mutual quadratic variation of all components of the vector-valued martingale. The tensor quadratic variation of the $\mathbf{R}^{2d}$-valued continuous square integrable martingale $(m_\varepsilon(\cdot,r_0),m_\varepsilon(\cdot,\tilde{r}_0))$ is defined by

$$\left.\begin{aligned}&[m_{\varepsilon,k}(t,r_0),m_{\varepsilon,\ell}(t,\tilde{r}_0)]=\int_0^t\int\sum_{m=1}^d\Gamma_{\varepsilon,km}(r(u,r_0),q)\Gamma_{\varepsilon,\ell m}(r(u,\tilde{r}_0),q)\mathrm{d}q\,\mathrm{d}u,\\&\qquad k,\ell=1,\ldots,d,r_0,\tilde{r}_0\in\mathbf{R}^d.\end{aligned}\right\}\tag{5.8}$$

Expression (5.8) has a derivative in t. It follows from Itô's formula that this derivative serves as the diffusion matrix in an equation for the mass distribution associated with (5.1):

$$\left.\begin{aligned}&D_{\varepsilon,k\ell}(r,\tilde{r}):=\int\sum_{m=1}^d\Gamma_{\varepsilon,km}(r,q)\Gamma_{\varepsilon,\ell m}(\tilde{r},q)\mathrm{d}q,\\&\text{where } r(t,r_0)=r,\quad r(t,\tilde{r}_0)=\tilde{r}.\end{aligned}\right\}\tag{5.9}$$

The diffusion matrix in $\mathcal{M}_{(2d)\times(2d)}$ consists of 4 $d\times d$ matrices, which describe the spatial pair correlations between the motions of two given particles, indexed by r_0 and $\tilde{r}_0$, at a given time.

Of special interest are the block diagonal $d\times d$ matrices, i.e., when $r_0=r_0$ or $\tilde{r}_0=\tilde{r}_0$, in particular, for the case of spatially homogeneous diffusion coefficients, i.e., when the diffusion coefficients satisfy

$$D_{\varepsilon,k\ell}(r,q)=D_{\varepsilon,k\ell}(r-q)\quad\forall r,q.\tag{5.10}$$

If $r_0 = \tilde{r}_0$ and the diffusion matrix is homogeneous, the uniqueness of the solutions implies that the tensor quadratic variation for one given particle is a constant matrix times t. We remark that the class of equations (5.1) with resulting spatially homogeneous diffusion matrix includes those equations where the kernel function Γ_ε is itself spatially homogeneous, i.e., where

$$\Gamma_\varepsilon(r,q) = \Gamma_\varepsilon(r-q) \quad \forall r,q. \tag{5.11}$$

Indeed, assuming (5.11), the shift invariance of the Lebesgue measure yields

$$\left.\begin{aligned} D_{\varepsilon,k\ell}(r(u.r_0), r(u,\tilde{r}_0)) &= \int \sum_{m=1}^{d} \Gamma_{\varepsilon,km}(r(u,r_0)-q)\Gamma_{\varepsilon,\ell m}(r(u,\tilde{r}_0)-q)\mathrm{d}q \\ &= \int \sum_{m=1}^{d} \Gamma_{\varepsilon,km}(r(u,r_0)-r(u,\tilde{r}_0)-q)\Gamma_{\varepsilon,\ell m}(-q)\mathrm{d}q \\ &=: D_{\varepsilon,k\ell}(r(u,r_0)-r(u,\tilde{r}_0)). \end{aligned}\right\} \tag{5.12}$$

However, we show in Chap. 7 that there are also spatially inhomogeneous $\Gamma_\varepsilon(r,q)$ with spatially homogeneous diffusion matrix. The key observation for the remainder of this section is the following.

Proposition 5.2. *Suppose the diffusion matrix associated with (5.1) is spatially homogeneous. Then, for each r_0 $M(\cdot, r_0)$, defined by (5.2), is a d-dimensional Brownian motion with diffusion matrix*

$$\begin{pmatrix} D_{\varepsilon,11}(0) & D_{\varepsilon,12}(0) & \dots & D_{\varepsilon,1d}(0) \\ D_{\varepsilon,21}(0) & D_{\varepsilon,22}(0) & \dots & D_{\varepsilon,2d}(0) \\ & & \vdots & \\ D_{\varepsilon,d1}(0) & D_{\varepsilon,d2}(0) & \dots & D_{\varepsilon,dd}(0) \end{pmatrix} \tag{5.13}$$

Proof. The proposition is a direct consequence of a d-dimensional generalization of Paul Levy's theorem.[3] □

Henceforth we assume that the diffusion matrix, associated with (5.1), is spatially homogeneous, i.e., that (5.10) holds. If certain properties hold for the more general case of a possibly inhomogeneous diffusion matrix such as Propositions 5.4, 5.5, and Corollary 5.6, we will state this explicitly.[4]

The computation of the spatial correlations for $\frac{\mathrm{d}}{\mathrm{d}t}\beta(\cdot, r_0)$ is easy because of the independence assumptions. To derive the corresponding space–time correlations for $\frac{\mathrm{d}}{\mathrm{d}t}m_\varepsilon(\cdot, r_0)$, we treat r_0 as a variable and, referring to the calculations in Sect. 15.2.4, we have that $\frac{\mathrm{d}}{\mathrm{d}t}m_\varepsilon(\cdot,\cdot,\omega) \in \mathcal{S}'(\mathbf{R}^{d+1})$, the dual of the infinitely often differentiable, rapidly decreasing real-valued functions on $\mathbf{R}^{d+1}$. Choose $\varphi_d, \psi_d \in \mathcal{S}(\mathbf{R}^d)$ and $\varphi, \psi \in \mathcal{S}(\mathbf{R})$. We show in (15.113)/(15.119) of Sect. 15.2.4 that, using the tensor quadratic variation of the martingale $(m_\varepsilon(\cdot, r_0), m_\varepsilon(\cdot, \tilde{r}_0)$, we obtain the following random covariance for the space–time random field $\frac{\mathrm{d}}{\mathrm{d}t}m_\varepsilon(\cdot,\cdot)$

[3] Cf. Theorem 15.37 in Sect. 15.2.3.

[4] Cf. Remark 5.7.

$$
\begin{aligned}
&Cov_{d+1,\omega}\left[(\varphi_d\varphi, \frac{\mathrm{d}}{\mathrm{d}t}m_\varepsilon(\cdot,\cdot))(\psi_d\psi, \frac{\mathrm{d}}{\mathrm{d}t}m_\varepsilon(\cdot,\cdot))\right] \\
&= \int\int\int_0^\infty\int_0^\infty \varphi_d(r_0)\psi_d(\tilde{r}_0)\varphi(t)\psi(s)\delta_0(t-s) \\
&\int\sum_{m=1}^d \Gamma_{\varepsilon,km}(r(t,r_0,\omega),q)\Gamma_{\varepsilon,\ell m}(r(s,\tilde{r}_0,\omega),q)\mathrm{d}q\,\mathrm{d}s\,\mathrm{d}t\,\mathrm{d}r_0\,\mathrm{d}\tilde{r}_0.
\end{aligned}
$$

Hence, employing (5.9) and (5.10) in addition to the (15.119), the random covariance is an $\mathcal{M}_{d\times d}$-valued bilinear functional on $\mathcal{S}(\mathbf{R}^{d+1})$

$$
Cov_{d+1,\omega} = \delta_{t-s}\otimes D_\varepsilon(r(t,r_0,\omega) - r(s,\tilde{r}_0,\omega)). \tag{5.14}
$$

The space–time correlations for $m_\varepsilon(\cdot,\cdot)$, $\overline{Cov_{d+1}}$, are obtained by taking the mathematical expectation in (5.14):

$$
\overline{Cov_{d+1}} = \delta_{t-s}\otimes E(D_\varepsilon(r(t,r_0) - r(s,\tilde{r}_0))). \tag{5.15}
$$

Observe that the joint motion of $(m_\varepsilon(\cdot,r_0), m_\varepsilon(\cdot,\tilde{r}_0))$ for $r_0\neq\tilde{r}_0$ is not Gaussian, because the $2d\times 2d$-diffusion matrix is not constant in the space variable. We are now ready to formulate the main concept of this section:

Definition 5.3. *Suppose the diffusion matrix associated with (5.1) is spatially homogeneous. The N d-dimensional martingales $m_\varepsilon(\cdot,r^i)$, defined by (5.2), $i=1,\dots,N$, are called "N d-dimensional correlated Brownian motions with start in 0 and the pair space–time correlation is given by (5.15). If, in addition, the initial conditions in (5.1) are deterministic, the solutions $r(\cdot,r_0^i), i = 1,\dots N$, are called "N d-dimensional correlated Brownian motions with start in $(r_0^1,\dots,r_0^N)$."* □

In the following proposition we allow adapted random initial conditions. We then show that particles, performing correlated Brownian motions and starting with ositive probability at the same point, will stick together with the same probability.

Proposition 5.4. *Let $A := \{\omega : r_0^i(\omega) = r_0^j(\omega)\}$ and assume $P(A) > 0$. Then,*

$$
r(\cdot,r_0^i)1_A \equiv r(\cdot,r_0^j)1_A \quad \textit{with probability 1.} \tag{5.16}
$$

Proof. (i) Set $r_{A,0} := r_0^i 1_A$ and let $r_A(t,r_{A,0})$ be the unique solution of the stochastic integral equation

$$
r_A(t,r_{A,0}) = r_{A,0} + \int_0^t\int \Gamma_\varepsilon(r_A(s,r_{A,0}),q)w(\mathrm{d}p,\mathrm{d}s) - \int_0^t\int \Gamma_\varepsilon(0,q)w(\mathrm{d}p,\mathrm{d}s)1_{\Omega\setminus A}, \tag{5.17}
$$

Suppose $m\in\{i,j\}$. Then,

$$
\begin{aligned}
r(t,r_0^m)1_A &= r_{A,0} + \int_0^t\int \Gamma_\varepsilon(r(s,r_0^m),q)w(\mathrm{d}p,\mathrm{d}s)1_A \\
&= r_{A,0} + \int_0^t\int \Gamma_\varepsilon(r(s,r_0^m)1_A,q)w(\mathrm{d}p,\mathrm{d}s)1_A
\end{aligned}
$$

(since the integral is 0 for $\omega\notin A$ we may change the values of Γ for those ω)

$$= r_{A,0} + \int_0^t \int \Gamma_\varepsilon(r(s, r_0^m)1_A, q)w(\mathrm{d}p, \mathrm{d}s) - \int_0^t \int \Gamma_\varepsilon(r(s, r_0^m)1_A, q)w(\mathrm{d}p, \mathrm{d}s)1_{\Omega\setminus A}$$

$$= r_{A,0} + \int_0^t \int \Gamma_\varepsilon(r(s, r_0^m)1_A, q)w(\mathrm{d}p, \mathrm{d}s) - \int_0^t \int \Gamma_\varepsilon(0, q)w(\mathrm{d}p, \mathrm{d}s)1_{\Omega\setminus A}$$

(since we now may change the values of Γ for $\omega \in A$) .

By the uniqueness of the solution of (5.1) we conclude that with probability 1 $r(\cdot, r_0^i)1_A \equiv r_A(\cdot, r_{A,0}) \equiv r(\cdot, r_0^j)1_A$. □

Next, let us show that the large particles solving (5.1) and starting at different positions, will never collide.

Proposition 5.5.[5]

Suppose $i \neq j$ and

$$P\{\omega : |r_0^i(\omega) - r_0^j(\omega)| = 0\} = 0.$$

Then

$$P\{\cup_{t\geq 0}\{\omega : |r(t, \omega, r_0^i) - r(t, \omega, r_0^j)| = 0\}\} = 0. \tag{5.18}$$

Proof. Set

$$\tau := \inf\{t > 0 : |r(t, r_0^i) - r(t, r_0^j)| = 0\},$$

where, as customary, we set $\tau = \infty$ if the set in the right-hand side is empty. τ is a stopping time.[6] We need to show $\tau = \infty$ a.s. Abbreviate

$$g(r(t, r_0^i), r(t, r_0^j), p)$$

$$:= \begin{cases} |r(t, r_0^i) - r(t, r_0^j)|^{-2}(r(t, r_0^i) - r(t, r_0^j)) \cdot (\Gamma_\varepsilon(r(t, r_0^i), p) - \Gamma_\varepsilon(r(t, r_0^j), p)), & \text{if } t < \tau, \\ 0, & \text{if } t \geq \tau \text{ and } \tau < \infty\,. \end{cases}$$

Note that the assumptions on Γ_ε imply a global Lipschitz condition. Therefore, there is also a finite constant c_1 such that

$$\int_0^t \int g^2(r(s, r_0^i), r(s, r_0^j), p)\mathrm{d}p\, \mathrm{d}s \leq c_1 t \;\forall t.$$

Hence, employing Walsh (1986), Chap. 2, we obtain that

$$\int_0^t \int g(r(s, r_0^i), r(s, r_0^j), p)w(\mathrm{d}p, \mathrm{d}s)$$

is a continuous adapted square integrable martingale whose quadratic variation is bounded by $c_1 t$. Again, by the assumptions on Γ_ε, there is a finite constant c_2 such

[5] The proof is an adaptation of a proof by Krylov (2005) to our setting. An alternative proof was provided by Dawson (1993) under the assumption $E|r_0^i - r_0^j|^{-2} < \infty$.

[6] Cf. Liptser and Shiryayev (1974), Chap. 1.3, Lemma 1.11.

that for the increments of the mutual quadratic variations the following estimate holds:

$$\sum_{k,\ell=1}^{d} \mathrm{d}\Big[\int (\Gamma_{\varepsilon,k}(r(t,r_0^i),p) - \Gamma_{\varepsilon,k}(r(t,r_0^j),p))w(\mathrm{d}p,\mathrm{d}t),$$

$$\int (\Gamma_{\varepsilon,\ell}(r(t,r_0^i),p) - \Gamma_{\varepsilon,\ell}(r(t,r_0^j),p))w(\mathrm{d}p,\mathrm{d}t)\Big]$$

$$= \sum_{k,\ell=1}^{d} \int (\Gamma_{\varepsilon,k}(r(t,r_0^i),p) - \Gamma_{\varepsilon,k}(r(t,r_0^j) - p))(\Gamma_{\varepsilon,\ell}(r(t,r_0^i),p)$$
$$-\Gamma_{\varepsilon,\ell}(r(t,r_0^j),p))\mathrm{d}p\,\mathrm{d}t$$
$$\le c_2|r(t,r_0^i) - r(t,r_0^j)|^2\mathrm{d}t.$$

Hence,

$$f(r(t,r_0^i),r(t,r_0^j)) := \begin{cases} \sum_{k,\ell=1}^{d} \int \frac{(\Gamma_{\varepsilon,k}(r(t,r_0^i),p)-\Gamma_{\varepsilon,k}(r(t,r_0^j),p))(\Gamma_{\varepsilon,\ell}(r(t,r_0^i),p)-\Gamma_{\varepsilon,\ell}(r(t,r_0^j),p))}{|r(t,r_0^i)-r(t,r_0^j)|^2}\mathrm{d}p, \\ \qquad \text{if } t < \tau, \\ 0, \qquad \text{if } t \ge \tau \text{ and } \tau < \infty, \end{cases}$$

is a bounded, $\mathrm{d}t \otimes \mathrm{d}P$-measurable adapted process. The Itô formula yields

$$\mathrm{d}|r(t,r_0^i) - r(t,r_0^j)|^2 = |r(t,r_0^i) - r(t,r_0^j)|^2$$
$$\int g(r(t,r_0^i),r(t,r_0^j),p)w(\mathrm{d}p,\mathrm{d}s) + |r(t,r_0^i)$$
$$-r(t,r_0^j)|^2 f(r(t,r_0^i),r(t,r_0^j))\mathrm{d}t. \tag{5.19}$$

Consequently, $|r(t,r_0^i) - r(t,r_0^j)|^2$ is the solution of a bilinear SODE, driven by a "nice" semi-martingale. We verify the following representation by Itô's formula:

$$\left.\begin{array}{l} |r(t\wedge\tau,r_0^i) - r(t\wedge\tau,r_0^j)|^2 \\ = |r_0^i - r_0^j|^2 \exp\Big[\int_0^{t\wedge\tau}\int g(r(s,r_0^i),r(s,r_0^j),p)w(\mathrm{d}p,\mathrm{d}s) \\ \qquad + \int_0^{t\wedge\tau}(f(r(s,r_0^i),r(s,r_0^j)) \\ \qquad -\frac{1}{2}\int g^2(r(s,r_0^i),r(s,r_0^j),p)\mathrm{d}p)\mathrm{d}s\Big]. \end{array}\right\} \tag{5.20}$$

We obtain from the assumption and (5.20) that $\tau = \infty$ a.s. □

Propositions 5.4 and 5.5 together imply the following:

Corollary 5.6. *Let* $A := \{\omega : r^i(0,\omega) = r^j(0,\omega)\}$ *and assume* $1 > P(A) > 0$. *Then* $\forall t \geq 0$

$$r(t,\omega,r_0^i) = r(t,\omega,r_0^j) \quad \text{if } \omega \in A, \text{ and } r(t,\omega,r_0^i) \neq r(t,\omega,r_0^j)\ \forall t \text{ if } \omega \notin A. \tag{5.21}$$

□

Remark 5.7. Propositions 5.4, 5.5, and Corollary 5.6 also hold for the solutions of (5.1) with inhomogeneous diffusion matrix, since we have not used the homogeneity of the diffusion matrix in the proofs. □

5.2 Shift and Rotational Invariance of $w(\mathrm{d}q, \mathrm{d}t)$

We prove shift and rotational invariance of $w(\mathrm{d}q, \mathrm{d}t)$.

Denote by $\mathcal{O}(d)$ the orthogonal $d \times d$ matrices over $\mathbf{R}^d$, i.e., $\mathcal{O}(d) := \{Q \in \mathcal{M}_{d\times d} : \det(Q) = \pm 1\}$. Let $h \in \mathbf{R}^d$ and $Q \in \mathcal{O}(d)$. Let A be a bounded Borel set in $\mathbf{R}^d$ and define the spatially shifted (resp. rotated) space–time white noise by

$$\left.\begin{aligned} \int_A w_{-h}(\mathrm{d}q,\mathrm{d}s) &:= \int_{A+h} w(\mathrm{d}q,\mathrm{d}s),\\ \int_A w_{Q^{-1}}(\mathrm{d}q,\mathrm{d}s) &:= \int_{QA} w(\mathrm{d}q,\mathrm{d}s). \end{aligned}\right\} \tag{5.22}$$

Recall that

$$\hat{w}(r,t) := \int_0^t \int_0^{r_d} \cdots \int_0^{r_1} w(\mathrm{d}q,\mathrm{d}s)$$

is called a "*Brownian sheet*," if $r_i \geq 0,\ i = 1,\ldots,d$. Although we will not need the notion of the Brownian sheet in what follows, we remark that for $r, r-h$ and $Q^{-1}r$ vectors with nonnegative components (5.22) may be written as transformations of the Brownian sheet:[7]

$$\left.\begin{aligned} \hat{w}_{-h}(r,t) &:= \hat{w}(r-h,t),\\ \hat{w}_{Q^{-1}}(r,t) &:= \hat{w}(Q^{-1}r,t). \end{aligned}\right\} \tag{5.23}$$

Let h^* denote the shift operator on $\mathbf{H}_0$ defined through

$$(h^* f)(r) := f(r+h),$$

where $f \in \mathbf{H}_0$. Similarly, we define the rotation operator Q^* on $\mathbf{H}_0$ through

$$(Q^* f)(r) := f(Qr).$$

[7] We refer to Sect. 15.2.4, in particular, to (15.125) for an extension of the Brownian sheet to $\mathbf{R}^d \times [0,\infty)$.

Let $\{\phi_n\}$ be a CONS in $\mathbf{H}_0$ and define $\mathbf{H}_0$-valued standard cylindrical Brownian motions $W_{-h}(t)$, $W_{Q^{-1}}(t)$ by:

$$\left.\begin{aligned} W_{-h}(t) &:= \sum_{n\in\mathbf{N}} \phi_n(\cdot) \int \phi_n(q) w_{-h}(\mathrm{d}q, t), \\ W_{Q^{-1}}(t) &:= \sum_{n\in\mathbf{N}} \phi_n(\cdot) \int \phi_n(q) w_{Q^{-1}}(\mathrm{d}q, t). \end{aligned}\right\} \tag{5.24}$$

Note that

$$\int 1_A(q) w_{-h}(\mathrm{d}q, t) = \int 1_{\{A+h\}}(q) w(\mathrm{d}q, t) = \int 1_A(q-h) w(\mathrm{d}q, t),$$

and

$$\int 1_A(q) w_{Q^{-1}}(\mathrm{d}q, t) = \int 1_{QA}(q) w(\mathrm{d}q, t) = \int 1_A(Q^{-1}q) w(\mathrm{d}q, t).$$

Therefore,

$$\begin{aligned} \sum_{n\in\mathbf{N}} \phi_n(\cdot) \int \phi_n(q) w_{-h}(\mathrm{d}q, t) &= \sum_{n\in\mathbf{N}} \phi_n(\cdot) \int \phi_n(q-h) w(\mathrm{d}q, t) \\ &= \sum_{n\in\mathbf{N}} (h^*\phi_n)(\cdot - h) \int \phi_n(q-h) w(\mathrm{d}q, t); \end{aligned}$$

$$\begin{aligned} \sum_{n\in\mathbf{N}} \phi_n(\cdot) \int \phi_n(q) w_{Q^{-1}}(\mathrm{d}q, t) &= \sum_{n\in\mathbf{N}} \phi_n(\cdot) \int \phi_n(Q^{-1}q) w(\mathrm{d}q, t) \\ &= \sum_{n\in\mathbf{N}} (Q^*\phi_n)(Q^{-1}(\cdot)) \int \phi_n(Q^{-1}q) w(\mathrm{d}q, t). \end{aligned}$$

Choose a suitable Hilbert distribution space $\mathcal{H}_{-\gamma}$ as in Sect. 15.1.3 (15.36), such that the imbedding

$$\mathbf{H}_0 \subset \mathcal{H}_{-\gamma}.$$

is Hilbert-Schmidt.[8] Set

$$\mathbf{H}_w := \mathcal{H}_{-\gamma}, \quad \text{(for some arbitrary fixed) } \gamma > \mathrm{d}$$

Then, $W_{-h}(t)$, $W_{Q^{-1}}(t)$, and $W(t)$, are regular $\mathbf{H}_w$-valued Brownian motions with continuous sample paths. Let us now show that the systems $\{(h^*\phi)_n\}$ and $\{(Q^*\phi)_n\}$ are complete orthonormal systems (CONS) in $\mathbf{H}_0$.

[8] Cf. also Chap. 4 (4.31).

$$\int (Q^*\phi_n)(q)(Q^*\phi_m)(q)\mathrm{d}q = \int \phi_n(Qq)\phi_m(Qq)\mathrm{d}q$$
$$= \int \phi_n(q)\phi_m(q)|\det(Q^{-1})|\mathrm{d}q = \delta_{n,m},$$

because $|\det(Q^{-1})| = 1$. The completeness follows from

$$0 =< f, \phi_n(Q(\cdot)) >_0 \quad \forall n \text{ iff } 0 =< f(Q^{-1}(\cdot)), \phi_n(\cdot) >_0 \quad \forall n$$

$$\text{iff a.e. } 0 = f(Q^{-1}(q)) \text{ iff a.e. } 0 = f(q), \quad \text{because } Q^{-1} \text{ is a bijection.}$$

Therefore, for $f \in \mathbf{H}_0$

$$< f, W_{Q^{-1}}(t) >_0 = \sum_{n\in\mathbf{N}} \int \phi_n(q)f(q)\mathrm{d}q \int \phi_n(q)w_{Q^{-1}}(\mathrm{d}q, t)$$
$$= \sum_{n\in\mathbf{N}} \int \phi_n(q)f(q)\mathrm{d}q \int \phi_n(Q^{-1}q)w(\mathrm{d}q, t)$$
$$= \sum_{n\in\mathbf{N}} \int \phi_n(Q^{-1}q)f(Q^{-1}q)|\det(Q)|\mathrm{d}q \int \phi_n(Q^{-1}q)w(\mathrm{d}q, t)$$
$$= \sum_{n\in\mathbf{N}} \int \phi_n(Q^{-1}q)(Q^{-1*}f)(q)\mathrm{d}q \int \phi_n(Q^{-1}q)w(\mathrm{d}q, t)$$
$$< (Q^{-1*}f), W(t) >_0 .$$

But

$$< (Q^{-1*}f), W(t) >_0 \sim \mathcal{N}(0, |Q^{-1*}f|_0^2) = \mathcal{N}(0, |f|_0^2),$$

because $|Q^{-1*}f|_0^2 = |f|_0^2$. Similarly for $\{(h^*\phi_n)\}$. As a consequence we obtain

Proposition 5.8.

$$W_{Q^{-1}}(\cdot) \sim W_{-h}(\cdot) \sim W(\cdot) \;\; in \;\; C([0,\infty); \mathbf{H}_w), \tag{5.25}$$

i.e., all three processes are different versions of the standard cylindrical process $W(\cdot)$ in $\mathbf{H}_0$. □

5.3 Separation and Magnitude of the Separation of Two Correlated Brownian Motions with Shift-Invariant and Frame-Indifferent Integral Kernels

We restrict our analysis to two correlated Brownian motions, $r(\cdot, r_0^i)$, $i = 1, 2$, as described by (2.9). We obtain Markov and SODE representations for the separation $r(\cdot, r_0^2) - r(\cdot, r_0^1)$ and, under the additional assumption of frame-indifference of *the kernel G_ε, also Markov and SODE representations for the magnitude of the separation, $|r(\cdot, r_0^2) - r(\cdot, r_0^1)|$.*

Setting $G_\varepsilon = \Gamma_{\varepsilon,\cdot 1}$ and assuming that the other column vectors of Γ_ε are identically 0 in addition to spatial homogeneity of G_ε, (5.1) reduces to (2.11), where we now use the notation $w(\mathrm{d}q, \mathrm{d}t)$ to denote a scalar valued space–time white noise.

We denote the solutions of (2.11) with initial conditions r_0^i, $i = 1, 2$, and driving noise $w(\mathrm{d}r, \mathrm{d}t)$ by $r^1(\cdot) := r(\cdot, r_0^1, w)$ and $r^2(\cdot) := r(\cdot, r_0^2, w))$, respectively. For $h \in \mathbf{R}^d$ and $i = 1, 2$, we see that

$$\left.\begin{aligned} r^i(t) + h &= r^i(0) + h + \int_0^t \int G_\varepsilon(r^i(u) + h - q + h) w(\mathrm{d}q, \mathrm{d}u) \\ &= r^i(0) + h + \int_0^t \int G_\varepsilon(r^i(u) + h - q) w_{-h}(\mathrm{d}q, \mathrm{d}u) \end{aligned}\right\} \tag{5.26}$$

Hence, the left-hand side of (5.26) is the solution $(r(\cdot, r_0^1 + h, w_{-h}), r(\cdot, r_0^2 + h, w_{-h}))$. Further, the assumptions on G_ε imply that, by Part (3) of Theorem 4.5, the solutions of (2.11) have a version that is measurable in all parameters, i.e., in the initial conditions and in the noise process, considered as a distribution-valued Wiener process.[9] This implies by Theorem 4.8 for $N = 2$ that

$$(r(\cdot, r_0^1, w) + h, r(\cdot, r_0^2, w) + h) \sim (r(\cdot, r_0^1 + h, w), r(\cdot, r_0^2 + h, w)).$$

To obtain a similar statement for rotations, we follow Truesdell and Noll (1965), Sect. 17,[10] and call functions $\vartheta : \mathbf{R}^d \longrightarrow \mathbf{R}$, $G : \mathbf{R}^d \longrightarrow \mathbf{R}^d$, and $A : \mathbf{R}^d \longrightarrow \mathcal{M}_{d\times d}$ *"frame-indifferent"* if $\forall r \in \mathbf{R}^d$, $Q \in \mathcal{O}(d)$

$$\left.\begin{aligned} \vartheta(Qr) &= \vartheta(r), \\ G(Qr) &= QG(r) \\ A(Qr) &= QA(r)Q^T. \end{aligned}\right\} \tag{5.27}$$

To describe the structure of matrix-valued frame-indifferent functions, we first denote the subspace spanned by r, $\{r\}$, and the subspace orthogonal to $\{r\}$, we denote $\{r\}^\perp$. We define the projection operators

$$P(r) := \frac{rr^T}{|r|^2}, \quad P^\perp(r) := I_d - P(r), \tag{5.28}$$

where rr^T is the matrix product between the column vector r and the row vector r^T, resulting in the $d \times d$ matrix with entries $r_k r_\ell$, $k, \ell = 1, \ldots, d$. Assuming A is symmetric, we show in Sect. 15.2.8, Theorem 15.71, that these functions are frame-indifferent if and only if there are scalar functions $\alpha, \beta, \lambda, \lambda^\perp : \mathbf{R}_+ \longrightarrow \mathbf{R}$ such that $\forall r \in \mathbf{R}^d$, $Q \in \mathcal{O}(d)$

---

[9] Cf. (5.24) and (5.25).

[10] Cf. also Kotelenez et al. (2007).

$$\left.\begin{aligned} \vartheta(r) &\equiv \alpha(|r|^2), \\ G(r) &\equiv \beta(|r|^2)r, \\ A(r) &\equiv \lambda(|r|^2)P(r) + \lambda^{\perp}(|r|^2)P^{\perp}(r). \end{aligned}\right\} \tag{5.29}$$

Example 5.9. Consider the kernel, derived in (1.2) from a Maxwellian velocity field

$$\left.\begin{aligned} &G_{\varepsilon,M}(r) := \bar{c}_{\varepsilon} r \ \exp\left(-\tfrac{|r|^2}{2\varepsilon}\right) \\ &\left(\text{where} \quad \bar{c}_{\varepsilon} := \sqrt{\alpha D}(\pi\varepsilon)^{\frac{d}{4}}\left(\tfrac{2}{d\varepsilon}\right)^{\frac{1}{2}}\frac{1}{(2\pi\varepsilon)^{\frac{d}{2}}}\right). \end{aligned}\right\} \tag{5.30}$$

Clearly, $G_{\varepsilon,M}$ is both shift-invariant and, by (5.27), also frame-indifferent.

Now suppose G_{ε} is frame-indifferent. Then

$$\left.\begin{aligned} Qr^i(t) &= Qr^i(0) + \int_0^t \int G_{\varepsilon}(Qr^i(u) - Qq)w(dq, du) \\ &= Qr^i(0) + \int_0^t \int G_{\varepsilon}(Qr^i(u) - q)w_{Q^{-1}}(dq, du) \end{aligned}\right\} \tag{5.31}$$

Summarizing and using the same argument for the rotation as for the shift, we obtain by Theorem 4.8:

Proposition 5.10.

$$\left.\begin{aligned} &(r(\cdot, r_0^1, w) + h, r(\cdot, r_0^2, w) + h) \sim (r(\cdot, r_0^1 + h, w), r(\cdot, r_0^2 + h, w)), \\ &\qquad\qquad \textit{and, if } G_{\varepsilon} \textit{ is frame-indifferent,} \\ &(Qr(\cdot, r_0^1, w), Qr(\cdot, r_0^2, w)) \sim (r(\cdot, Qr_0^1, w), r(\cdot, Qr_0^2, w)). \end{aligned}\right\} \tag{5.32}$$

where the pair processes are considered as $C([0,\infty); \mathbf{R}^{2d})$*-valued random variables.* □

Note that (5.32) is only correct if we shift both r^1 and r^2 by the same d-dimensional vector h, or rotate them by the same orthogonal matrix Q. Abbreviate $\hat{r} := (r^1, r^2) \in \mathbf{R}^{2d}$ with d-dimensional coordinates r^1 and r^2, respectively. Set

$$D_{\varepsilon,k\ell,ij}(\hat{r}) := \int G_{\varepsilon,k}(r^i - q)G_{\varepsilon,\ell}(r^j - q)dq, \quad k, \ell = 1, .., d, \quad i, j = 1, 2. \tag{5.33}$$

By Proposition 15.72 of Sect. 15.2.8, $D_{\varepsilon,k\ell,ij}$ is symmetric for each fixed (i, j). The generator, $A_{\varepsilon,\mathbf{R}^{2d}}$, of the Markov pair process $\hat{r}(\cdot) := (r^1(\cdot), r^2(\cdot))$ is given by

$$A_{\varepsilon,\mathbf{R}^{2d}} := \frac{1}{2}\sum_{k,\ell=1}^{d}\sum_{i,j=1}^{2} D_{\varepsilon,k\ell,ij}(\hat{r})\frac{\partial^2}{\partial r_k^i \partial r_\ell^j}. \tag{5.34}$$

A core for this generator is, e.g., $C_0^2(\mathbf{R}^{2d}, \mathbf{R})$, the space of twice continuously differentiable real-valued functions on $\mathbf{R}^{2d}$, which vanish at infinity.[11]

[11] Cf. Ethier and Kurtz (loc.cit.), Sect. 8.2, Theorem 2.5.

We are interested in local effects of the diffusion coefficient on the $|r^2 - r^1|$. An initial step in this direction is change of coordinates. First, we employ the shift invariance to obtain a Markov representation for $r(\cdot, r_0^2) - r(\cdot, r_0^1)$.

Let I_d be the identity matrix in $\mathbf{R}^d$ and define the following unitary matrix $\mathcal{A}$ in $\mathbf{R}^{2d}$:

$$\mathcal{A} := \frac{1}{\sqrt{2}} \begin{pmatrix} I_d & -I_d \\ I_d & I_d \end{pmatrix}. \tag{5.35}$$

We then define:

$$\begin{pmatrix} r^1 \\ r^2 \end{pmatrix} = \mathcal{A} \begin{pmatrix} u \\ v \end{pmatrix}. \tag{5.36}$$

In (5.36) $u, v, r^1, r^2 \in \mathbf{R}^d$. It follows

$$\left.\begin{aligned} u &= \frac{1}{\sqrt{2}}(r^1 + r^2) \\ v &= \frac{1}{\sqrt{2}}(r^2 - r^1). \end{aligned}\right\} \tag{5.37}$$

Let B be a Borel set in $\mathbf{R}^d$. Set

$$\Gamma_B := \mathcal{A}(\mathbf{R}^d \times B)$$

and

$$\hat{r}(u, v) := \begin{pmatrix} r^1(u, v) \\ r^2(u, v) \end{pmatrix},$$

where the coordinate functions are determined by the linear map (5.35). We wish to compute the probability, $P(t, \hat{r}, \Gamma_B)$, for the two-particle motion to be at time t in Γ_B, having started in $\hat{r}$. Note that $r^i(t) = r(t, r^i),\ i = 1, 2$, i.e., $r^i(t)$ are the solutions of (2.11) with start in $r^i,\ i = 1, 2$. By (5.36)

$$\left.\begin{aligned} P(t, \hat{r}, \Gamma_B) &= P\left(\left(\tfrac{1}{\sqrt{2}}(r(t, r^1) + r(t, r^2), \tfrac{1}{\sqrt{2}}(r(t, r^2) - r(t, r^1)))\right) \in \mathbf{R}^d \times B\right) \\ &= P\left(\tfrac{1}{\sqrt{2}}(r(t, r^2) - r(t, r^1)) \in B\right). \end{aligned}\right\} \tag{5.38}$$

Let $h \in \mathbf{R}^d$ and $\hat{h} := \begin{pmatrix} h \\ h \end{pmatrix}$. By (5.32), (5.36), and (5.37)

$$\left.\begin{aligned} P(t, \hat{r} + \hat{h}, \Gamma_B) &= P(\tfrac{1}{\sqrt{2}}(r(t, r^2) + h - (r(t, r^1) + h) \in B) \\ &= P(\tfrac{1}{\sqrt{2}}(r(t, r^2) - r(t, r^1)) \in B) = P(t, \hat{r}, \Gamma_B). \end{aligned}\right\} \tag{5.39}$$

Therefore, the probability in (5.38) depends only the starts $q := \frac{1}{\sqrt{2}}(r^2 - r^1)$ and, following Dynkin (1965), Chap. 10.6, we may define the "*marginal*" transition probability distribution

$$\bar{P}(t,a,B) := P(t,\hat{r},\Gamma_B)\Big|_{\left\{\frac{1}{\sqrt{2}}(r^2-r^1)=a\right\}}. \tag{5.40}$$

Next, let $f \in C_0^2(\mathbf{R}^{2d},\mathbf{R})$ and denote

$$\tilde{f}(u,v) := f(r^1(u,v), r^2(u,v)).$$

To calculate the action of the generator on f in the new coordinates u and v from (5.37) we need to express the derivatives with respect to (r^1,r^2) in terms of derivatives with respect to (u,v). Expression (5.36) implies

$$\left.\begin{aligned}
&\frac{\partial^2}{\partial r_k^1 \partial r_\ell^1} f + \frac{\partial^2}{\partial r_k^2 \partial r_\ell^2} f = \frac{\partial^2}{\partial u_k \partial u_\ell}\tilde{f} + \frac{\partial^2}{\partial v_k \partial v_\ell}\tilde{f};\\
&\frac{\partial^2}{\partial r_k^1 \partial r_\ell^2} f = \frac{1}{2}\left[\frac{\partial^2}{\partial u_k \partial v_\ell}\tilde{f} - \frac{\partial^2}{\partial v_k \partial u_\ell}\tilde{f} + \frac{\partial^2}{\partial u_k \partial u_\ell}\tilde{f} - \frac{\partial^2}{\partial v_k \partial v_\ell}\tilde{f}\right].
\end{aligned}\right\} \tag{5.41}$$

Suppose $\tilde{f}$ does not depend on u. Using the notation r and q instead of v, we then define

$$(T_{\varepsilon,\mathbf{R}^d}(t)\tilde{f})(q) := \int \tilde{f}(r)\bar{P}(t,q,\mathrm{d}r). \tag{5.42}$$

Let $T_{\varepsilon,\mathbf{R}^{2d}}(t)$ denote the Markov semigroup of operators generated by the transition probabilities $P(t,\hat{r},\cdot)$ and denote its generator by $A_{\varepsilon,\mathbf{R}^{2d}}$. By definition,

$$(T_{\varepsilon,\mathbf{R}^d}(t)\tilde{f})(q) = (T(t)_{\varepsilon,\mathbf{R}^{2d}} f)(\hat{r})\Big|_{\left\{\frac{1}{\sqrt{2}}(r^2-r^1)=q\right\}}. \tag{5.43}$$

Therefore, we calculate the time derivative of the left-hand side of (5.43) in terms of the generator $A_{\varepsilon,\mathbf{R}^{2d}}$. The result defines an operator $A_{\varepsilon,\mathbf{R}^d}$ acting on those $f \in C_0^2(\mathbf{R}^{2d},\mathbf{R})$ for which $\tilde{f}$ does not depend on u. Note that $D_{\varepsilon,k\ell,11} = D_{\varepsilon,k\ell,22}$ and $D_{\varepsilon,k\ell,12} = D_{\varepsilon,k\ell,21}$. Taking into account that $\tilde{f}$ does not depend on u, we obtain from (5.41) and (5.43)

$$\left.\begin{aligned}
(A_{\varepsilon,\mathbf{R}^d}\tilde{f})(q) &:= \frac{\partial}{\partial t}(T_{\varepsilon,\mathbf{R}^d}(t)\tilde{f})(q)|_{t=0}\\
&= \frac{\partial}{\partial t}(T(t)_{\varepsilon,\mathbf{R}^{2d}} f)(\hat{r})_{|\frac{1}{\sqrt{2}}(r^2-r^1)=q}|_{t=0}\\
&= \frac{1}{2}\sum_{k,\ell=1}^{d} \bar{D}_{\varepsilon,k\ell}(\sqrt{2}q)\left(\frac{\partial^2}{\partial q_k \partial q_\ell}\tilde{f}\right)(q),
\end{aligned}\right\} \tag{5.44}$$

where we abbreviated

$$\left.\begin{aligned}
\bar{D}_{\varepsilon,k\ell}(\sqrt{2}q) &:= [\tilde{D}_{\varepsilon,k\ell}(0) - \tilde{D}_{\varepsilon,k\ell}(\sqrt{2}q)],\\
\tilde{D}_{\varepsilon,k\ell}(\sqrt{2}q) &:= \int G_{\varepsilon,k}(q-\tilde{q})G_{\varepsilon,\ell}(-\tilde{q})\mathrm{d}\tilde{q}, \quad k,\ell = 1,\ldots,d.
\end{aligned}\right\} \tag{5.45}$$

Definition 5.11. *The Markov process $r_\varepsilon(\cdot, r_0)$ with generator $A_{\varepsilon,\mathbf{R}^d}$ will be called the* "separation process" *for the two correlated Brownian motions $r(\cdot, r_0^2)$ and $r(\cdot, r_0^1)$.* □

Proposition 5.12.

(i) $\bar{D}_{\varepsilon,k\ell}(\sqrt{2}q)$ is nonnegative definite, and its entries of are in $C_b^2(\mathbf{R}^d, \mathbf{R})$.
(ii) $A_{\varepsilon,\mathbf{R}^d}$, defined by (5.44), is the generator of a Feller semigroup on $C_0(\mathbf{R}^d, \mathbf{R})$.
(iii) If, in addition to the previous assumptions on G_ε, the range of G_ε is not a proper subspace of $\mathbf{R}^d$, then for all $q \in \mathbf{R}^d \setminus \{0\}$ $\bar{D}_{\varepsilon,k\ell}(\sqrt{2}q)$ is positive definite and therefore invertible.
(iv) $\bar{D}_{\varepsilon,k\ell}(\sqrt{2}q) = 0\ \forall k, \ell$ if and only if $q = 0$ unless $G_{\varepsilon,k}(\cdot) \equiv 0\ \forall k$.

Proof. Let $(x_1, \ldots, x_d) \in \mathbf{R}^d$. To prove the first part of (i) suppose $(x_1, \ldots, x_d) \neq (0, \ldots, 0)$. Denote the entries of $D_{\varepsilon,k\ell,11}(0)$ by $c_{k\ell}$ and the entries of $D_{\varepsilon,k\ell,12}(\sqrt{2}q)$ by $d_{k\ell}$. Set $A(\tilde{q}) := \sum_{k=1}^{d} x_k G_{\varepsilon,k}(-\tilde{q})$ and $B(\tilde{q}) := \sum_{k=1}^{d} x_k G_{\varepsilon,k}(q - \tilde{q})$. Note that

$$\sum_{k,\ell} c_{k\ell} x_k x_\ell = \int A^2(\tilde{q}) \mathrm{d}\tilde{q}, \quad \sum_{k,\ell} d_{k\ell} x_k x_\ell = \int A(\tilde{q}) B(\tilde{q}) \mathrm{d}\tilde{q}.$$

Further, by change of variables and the shift-invariance of the Lebesgue measure

$$\int B^2(\tilde{q}) \mathrm{d}\tilde{q} = \int A^2(\tilde{q}) \mathrm{d}\tilde{q}.$$

Then

$$\left.\begin{aligned} 0 \le \int (A(\tilde{q}) - B(\tilde{q}))^2 \mathrm{d}\tilde{q} &= \int A^2(\tilde{q}) \mathrm{d}\tilde{q} + \int B^2(\tilde{q}) \mathrm{d}\tilde{q} - 2 \int A(\tilde{q}) B(\tilde{q}) \mathrm{d}\tilde{q} \\ &= 2 \left[\int A^2(\tilde{q}) \mathrm{d}\tilde{q} - \int A(\tilde{q}) B(\tilde{q}) \mathrm{d}\tilde{q} \right]. \end{aligned}\right\} \tag{5.46}$$

This proves that $\bar{D}_{\varepsilon,k\ell}(\sqrt{2}q)$ is nonnegative definite. The second statement in (i) follows directly from our assumptions on G_ε.

The properties of $\bar{D}_{\varepsilon,k\ell}(\sqrt{2}q)$ imply that $A_{\varepsilon,\mathbf{R}^d}$ is the generator of a Feller semigroup on $C_0(\mathbf{R}^d, \mathbf{R})$.[12]

Finally, we have equality in (5.46) if, and only if, $A(\cdot) \equiv B(\cdot)$. Hence, $A(\cdot)$ is a periodic function with period q. This contradicts the integrability of $G_{\varepsilon,k}(\cdot)$ on $\mathbf{R}^d$ unless $A(\cdot) \equiv 0$. Since by assumption the range of G_ε cannot be restricted to a proper subspace of $\mathbf{R}^d$ it follows that $x_i = 0$ for all $i = 1, \ldots, d$.

$\bar{D}_\varepsilon(\sqrt{2}q) = \mathbf{0}$ (the matrix whose entries are all 0) if and only if $(\tilde{D}_\varepsilon(0) - \tilde{D}_\varepsilon(\sqrt{2}q))x \cdot y = 0\ \ \forall x = (x_1, .., x_d)^\perp, y = (y_1, .., y_d)^\perp \in \mathbf{R}^d$. Choosing $x = y = (0, ..0, 1, 0.., 0)^\perp$ where the entry 1 is at the kth coordinate, the arguments of step (iii) imply that $A(\cdot) \equiv G_{\varepsilon,k}(\cdot)$ is a periodic function with period q. This contradicts the integrability of $G_{\varepsilon,k}(\cdot)$ on $\mathbf{R}^d$ unless $A(\cdot) \equiv 0$. □

[12] Cf. Ethier and Kurtz (loc.cit.), Sect. 8.2, Theorem 2.5.

Remark 5.13. The preceding calculations show that, assuming shift-invariance of the kernel G_ε, the Markov semigroup $T_{\varepsilon,\mathbf{R}^{2d}}(t)$, defined on $C_0(\mathbf{R}^{2d}, \mathbf{R})$, has a nontrivial invariant subspace. After the rotation (5.35)–(5.37) this subspace can be written as $C_0(\mathbf{R}^d, \mathbf{R})$, and $T_{\varepsilon,\mathbf{R}^{2d}}(t)$, restricted to this subspace, defines a Feller-Markov semigroup $T_{\varepsilon,\mathbf{R}^d}(t)$. The corresponding Markov process is equivalent in distribution to $r(t) := \frac{1}{\sqrt{2}}(r(t,r^2) - r(t,r^1))$. We may, therefore, call the system

$$\left.\begin{aligned} u(t) &= \frac{1}{\sqrt{2}}(r(t,r^1) + r(t,r^2)) \\ v(t) &= \frac{1}{\sqrt{2}}(r(t,r^2) - r(t,r^1)) \end{aligned}\right\} \tag{5.47}$$

"partially decoupled in distribution." □

The assumptions imply that there is a unique function $\sigma_\varepsilon(q)$ with values in the nonnegative definite matrices such that

$$\left.\begin{aligned} &\sum_j \sigma_{\varepsilon,kj}(q)\sigma_{\varepsilon,\ell j}(q) = \bar{D}_{\varepsilon,k\ell}(\sqrt{2}q), \quad k,\ell = 1,\ldots,d; \\ &\qquad \sum_{k,j} |\sigma_{\varepsilon,kj}(r) - \sigma_{\varepsilon,kj}(q)| \le c|r-q| \\ &\qquad \text{(for some finite constant } c\text{)}, \end{aligned}\right\} \tag{5.48}$$

i.e., the nonnegative definite square root of $(\bar{D}_{\varepsilon,k\ell}(\sqrt{2}q))$ is Lipschitz continuous.[13] Therefore, the separation process, generated by $A_{\varepsilon,\mathbf{R}^d}$, can be represented as the unique solution $r(\cdot,q)$ of the stochastic differential equation in the sense of Itô:

$$\left.\begin{aligned} dr(t) &= \sum_j \sigma_{\varepsilon,\cdot j}(r)\beta_j(dt), \\ r(0) &:= q, \end{aligned}\right\} \tag{5.49}$$

where the β_j are i.i.d. one-dimensional standard Brownian motions, $j = 1,\ldots,d$. For the solution we have a.s. $r(\cdot,q) \in C([0,\infty);\mathbf{R}^d)$.[14]

In addition to the shift-invariance we henceforth also assume that our kernel G_ε is frame-indifferent .

We first analyze the matrix function r $\mapsto$ $\tilde{D}_\varepsilon(\sqrt{2}r)$ from (5.45), which may be written as

$$\tilde{D}_\varepsilon\left(\sqrt{2}q\right) := \int G_\varepsilon\left(\frac{1}{2}q - \tilde{q}\right) G_\varepsilon^T\left(-\frac{1}{2}q - \tilde{q}\right) d\tilde{q}.$$

By the frame-indifference assumption and (5.29)

13 Cf. Stroock and Varadhan (1979), Chap. 5.2, Theorem 5.2.3. Cf. also our representation (5.60) for frame-indifferent $\bar{D}_\varepsilon$).

14 Cf. also our representation (5.68) for frame-indifferent $\bar{D}_\varepsilon$.

$$\tilde{D}_\varepsilon(\sqrt{2}r) = \int (q+r)(q-r)^T \beta_\varepsilon(|q+r|^2)\beta_\varepsilon(|q-r|^2)\mathrm{d}q \tag{5.50}$$

Since $\int (rq^T - qr^T)\beta_\varepsilon(|q+r|^2)\beta_\varepsilon(|q-r|^2)\mathrm{d}q = 0$ (5.50) reduces to

$$\tilde{D}_\varepsilon(\sqrt{2}r) = \int qq^T \beta_\varepsilon(|q+r|^2)\beta_\varepsilon(|q-r|^2)\,\mathrm{d}q - rr^T \int \beta_\varepsilon(|q+r|^2)\beta_\varepsilon(|q-r|^2)\,\mathrm{d}q. \tag{5.51}$$

Define a matrix function of r by

$$A_\varepsilon(r) := \int qq^T \beta_\varepsilon(|q+r|^2)\beta_\varepsilon(|q-r|^2)\,\mathrm{d}q \tag{5.52}$$

and a scalar function of r by

$$\phi_\varepsilon(r) := \int \beta_\varepsilon(|q+r|^2)\beta_\varepsilon(|q-r|^2)\,\mathrm{d}q. \tag{5.53}$$

Thus,

$$\tilde{D}_\varepsilon(\sqrt{2}r) = A_\varepsilon(r) - \phi_\varepsilon(r)rr^T. \tag{5.54}$$

Let $Q \in \mathcal{O}(d)$. $|q+r| = |Qq + Qr|$ and $|q-r| = |Qq - Qr|$ in addition to a change-in-variables $\tilde{q} := Qq$ yield the following:

$$QA_\varepsilon(r)Q^T = A_\varepsilon(Qr) \text{ and } \phi_\varepsilon(r) = \phi_\varepsilon(Qr), \tag{5.55}$$

that is, they are frame-indifferent in the sense of (5.27). By (5.29) $A_\varepsilon(r)$ and $\phi_\varepsilon(r)$ must have the form

$$A_\varepsilon(r) = \lambda_\varepsilon(|r|^2)P(r) + \lambda_{\perp,\varepsilon}(|r|^2)P^\perp(r) \text{ and } \phi_\varepsilon(r) = \gamma_\varepsilon(|r|^2) \tag{5.56}$$

for scalar functions

$$\lambda_\varepsilon, \lambda_{\perp,\varepsilon}, \gamma_\varepsilon : \mathbf{R}_+ \to \mathbf{R};\quad \xi \mapsto \lambda_\varepsilon(\xi), \lambda_{\perp,\varepsilon}(\xi), \gamma_\varepsilon(\xi).$$

At $r = 0$

$$\lambda_\varepsilon(0) = c_\varepsilon > 0 \qquad \text{and} \qquad \lambda_{\perp,\varepsilon}(0) = c_\varepsilon. \tag{5.57}$$

Moreover, it follows from the definitions in (5.51) and (5.52), together with our basic regularity hypotheses, that

$$\lim_{|r|\to\infty} \lambda_\varepsilon(|r|^2) = 0, \quad \lim_{|r|\to\infty} \lambda_{\perp,\varepsilon}(|r|^2) = 0 \text{ and } \lim_{|r|\to\infty} |r|^2\gamma_\varepsilon(|r|^2) = 0. \tag{5.58}$$

Combining, we see that $\tilde{D}_\varepsilon$ has the special form

$$\tilde{D}_\varepsilon(\sqrt{2}r) = \lambda_{\perp,\varepsilon}(|r|^2)P^\perp(r) + (\lambda_\varepsilon(|r|^2) - |r|^2\gamma_\varepsilon(|r|^2))P(r). \tag{5.59}$$

Putting it all together, we see that the pair diffusion matrix, $\bar{D}_\varepsilon(\sqrt{2}r)$, of (5.45) must have the special form

$$\bar{D}_\varepsilon(\sqrt{2}r) = \alpha_{\perp,\varepsilon}(|r|^2)P^\perp(r) + \alpha_\varepsilon(|r|^2)P(r), \tag{5.60}$$

$$\left.\begin{aligned}\alpha_{\perp,\varepsilon}(|r|^2) &= c_\varepsilon - \lambda_{\perp,\varepsilon}(|r|^2)\\ \alpha_\varepsilon(|r|^2) &= c_\varepsilon - \lambda_\varepsilon(|r|^2) + |r|^2\gamma_\varepsilon(\tfrac{1}{2}|r|^2).\end{aligned}\right\} \tag{5.61}$$

Hence

$$\lim_{|r|\to\infty} \bar{D}_\varepsilon(\sqrt{2}r) = c_\varepsilon I_d. \tag{5.62}$$

Recall that $\bar{D}_\varepsilon(\sqrt{2}r)$ is nonnegative definite by Proposition 5.12. Choose first a unit vector $\bar{z} \in \mathbf{R}^d$ that is perpendicular to r. Then from (5.60)

$$\alpha_{\perp,\varepsilon}(|r|^2) = (\bar{D}_\varepsilon(\sqrt{2}r)\bar{z}) \cdot \bar{z} \geq 0 \;\; \forall r \neq 0. \tag{5.63}$$

Similarly, choosing $\bar{r} := \frac{1}{|r|}r$, we obtain

$$\alpha_\varepsilon(|r|^2) = (\bar{D}_\varepsilon(\sqrt{2}r)\bar{r}) \cdot \bar{r} \geq 0 \;\; \forall r \neq 0. \tag{5.64}$$

By Part (iv) of Proposition 5.12,

$$\alpha_\varepsilon(|r|^2) + \alpha_{\perp,\varepsilon}(|r|^2) > 0 \quad \forall r \neq 0. \tag{5.65}$$

Finally, by (5.63) and (5.64),

$$\text{Whenever } \; \bar{D}_\varepsilon(\sqrt{2}r) \; \text{ is positive definite, } \alpha_\varepsilon(|r|^2) > 0 \text{ and } \alpha_{\perp,\varepsilon}(|r|^2) > 0. \tag{5.66}$$

Remark 5.14.

(i) The condition in Part (iii) of Proposition 5.12 guarantees that $\bar{D}_\varepsilon(\sqrt{2}r)$ is positive definite $\forall r \neq 0$.
(ii) If, in addition to the assumptions of Proposition 5.12, we suppose that the kernel G_ε is frame-indifferent, then we obtain a simple representation of the nonnegative square root of $\bar{D}_\varepsilon(\sqrt{2}r)$:[15]

$$\bar{D}_\varepsilon^{\frac{1}{2}}(\sqrt{2}r) = \sqrt{\alpha_{\perp,\varepsilon}(|r|^2)}P^\perp(r) + \sqrt{\alpha_\varepsilon(|r|^2)}P(r). \tag{5.67}$$

Consequently, the stochastic ordinary differential equation (5.49) for the separation simplifies as well:

$$\left.\begin{aligned}\mathrm{d}r(t) = \sqrt{\alpha_{\perp,\varepsilon}(|r|^2)}P^\perp(r)\tilde{\beta}(\mathrm{d}t) + \sqrt{\alpha_\varepsilon(|r|^2)}P(r)\tilde{\beta}(\mathrm{d}t),\\ r(0) := r_0,\end{aligned}\right\} \tag{5.68}$$

where $\tilde{\beta}(\cdot)$ is an $\mathbf{R}^d$-valued standard Brownian motion. □

Proposition 5.15. In addition to the assumptions of Proposition 5.12 suppose that the kernel G_ε is frame-indifferent, and let $r(t,q)$ be the solution of (5.49) (resp. of (5.68)). Then

$$x(t,a) := |r(t,q)|, \;\; x(0,a) = |q|, \tag{5.69}$$

[15] Cf. (5.48) and (5.60).

is equivalent in distribution to a one-dimensional Feller diffusion. Its generator is an extension of the differential operator

$$\left.\begin{aligned} (A_{\varepsilon,\mathbf{R}}\varphi)(x) &= \frac{1}{2}\left((d-1)\alpha_{\perp,\varepsilon}(x^2)\frac{1}{x}\frac{\mathrm{d}\varphi(x)}{\mathrm{d}x} + \alpha_\varepsilon(x^2)\frac{\mathrm{d}^2\varphi(x)}{(\mathrm{d}x)^2}\right), \\ \varphi'(0) &= 0 \end{aligned}\right\} \tag{5.70}$$

where $\varphi \in C^2(\mathbf{R}; \mathbf{R})$.

Proof.

(i) If $q, \tilde{q} \in \mathbf{R}^d$ and $|q| = |\tilde{q}|$ then there is a $Q \in \mathcal{O}(d)$ such that

$$\tilde{q} = Qq.$$

On the other hand, we may suppose

$$q = r_0^2 - r_0^1, \quad \tilde{q} = \tilde{r}_0^2 - \tilde{r}_0^1.$$

Hence,

$$\tilde{r}_0^2 - \tilde{r}_0^1 = Qr_0^2 - Qr_0^1.$$

Then, by Proposition 5.10, (5.49), (5.39), and (5.40),

$$r(t,q) \sim r(t, r_0^2, w) - r(t, r_0^1, w) \sim r(t, Qr_0^2, w) - r(t, Qr_0^1, w) \sim r(t, \tilde{q}). \tag{5.71}$$

Let B be a Borel subset of $\mathbf{R}$ and $|\cdot|^{-1}B$ denote the inverse image of the map $r \longrightarrow |r|$. (5.71) and the preceding considerations imply by Dynkin (loc.cit.) that

$$P_{\mathbf{R}}(t, x, B)_{|\{x:=|q|\}} := P\{r(t,q) \in |\cdot|^{-1}B\} \tag{5.72}$$

is a family of transition probabilities of a Markov diffusion in $\mathbf{R}_+$.

(ii) The boundary condition follows from the fact that our "test functions" $\tilde{f}(r) = \varphi(|r|)$ are even.

(iii) Next, we compute the action of the differential operator in (5.44) on functions $\tilde{f}$ of the form $\tilde{f}(q) = \varphi(|q|)$, where $\varphi \in C^2(\mathbf{R}; \mathbf{R})$. As a preliminary, compute the matrix $\nabla_q \nabla_q^T \tilde{f}$ whose components are given by $\frac{\partial^2}{\partial q_i \partial q_j}\varphi(|q|)$. This computation yields

$$(\nabla_q \nabla_q^T \tilde{f})(q) = \frac{\varphi'(|q|)}{|q|}\left(I_d - \frac{qq^T}{|q|^2}\right) + \varphi''(|q|)\frac{qq^T}{|q|^2}, \tag{5.73}$$

provided $q \neq 0$. In terms of the projections from (5.28), (5.73) is cast in the form

$$(\nabla\nabla^T \tilde{f})(q) = \frac{\varphi'(|q|)}{|q|}P^\perp(q) + \varphi''(|q|)P(q). \tag{5.74}$$

Using the diffusion matrix $\bar{D}_\varepsilon(\sqrt{2}q)$ in the form of (5.60), the partial differential operator of (5.44) applied to φ reduces to the ordinary differential operator

$$\left.\begin{aligned}(A_{\varepsilon,\mathbf{R}}\varphi(|q|)) &= \tfrac{1}{2}tr\left(\bar{D}_\varepsilon(\sqrt{2}q)(\nabla\nabla^T\tilde{f})(q)\right)\\ &= \tfrac{1}{2}\left((d-1)\alpha_{\perp,\varepsilon}(|q|^2)\frac{\varphi'(|q|)}{|q|} + \alpha_\varepsilon(|q|^2)\varphi''(|q|)\right).\end{aligned}\right\} \tag{5.75}$$

This yields the generator for the magnitude process, defined in (5.70). The Feller property of the diffusion follows because the semigroup $\bar{T}_\varepsilon(t)$ for the separation process is Feller and the semigroup for the magnitude of the separation process is obtained by a restriction of $\bar{T}_\varepsilon(t)$ to functions $\varphi(|q|)$ from the preceding proof. □

Definition 5.16. *The real-valued Markov process $x(t,y)$, generated by $A_{\varepsilon,\mathbf{R}}$ from (5.70) will be called the* "magnitude of the separation process" *or, simply, the* "magnitude process." □

Remark 5.17.

(i) Changing to spherical coordinates

$$r = (r_1,\ldots,r_d) \longrightarrow (\tfrac{r_1}{|r|},\ldots,\tfrac{r_d}{|r|},|r|), \tag{5.76}$$

the preceding calculations show that the magnitude of the separation process, $|r(t)|$, is partially decoupled in distribution (cf. Remark 5.13).

(ii) By (5.58) and (5.61) $\forall\varepsilon > 0$ and large x

$$(A_{\varepsilon,\mathbf{R}}\varphi)(x) \approx \frac{c_\varepsilon}{2}(d-1)\frac{1}{x}\frac{\mathrm{d}\varphi(x)}{\mathrm{d}x} + \frac{c_\varepsilon}{2}\frac{\mathrm{d}^2\varphi(x)}{(\mathrm{d}x)^2}, \tag{5.77}$$

This is for $\frac{c_\varepsilon}{2} = 1$ the generator of the Bessel process associated with a d-dimensional standard Brownian motion $\beta(\cdot)$.[16] In other words, for large x the magnitude process behaves like the Bessel process of the Brownian motion $\sqrt{\frac{c_\varepsilon}{2}}\beta(\cdot)$. We will, therefore, call the magnitude process with generator (5.70) a "*Bessel-type process*."

(iii) From (5.70) we obtain that the magnitude of the separation process is equivalent in distribution to the solution of the following stochastic ordinary differential equation:

$$\left.\begin{aligned}\mathrm{d}x &= \frac{1}{2}(d-1)\frac{\alpha_{\perp,\varepsilon}(x^2)}{x}\mathrm{d}t + \sqrt{\alpha_\varepsilon(x^2)}\beta(\mathrm{d}t),\\ x(0) &= x_0,\end{aligned}\right\} \tag{5.78}$$

where $\beta(\cdot)$ is a one-dimensional standard Brownian motion and where, for the interpretation of the magnitude, we assume $x_0 > 0$. □

[16] Cf. Dynkin, loc. cit., Chap. X, (10.87).

5.4 Asymptotics of Two Correlated Brownian Motions with Shift-Invariant and Frame-Indifferent Integral Kernels

We apply criteria from stochastic analysis of one-dimensional diffusions to obtain a proof of the long-time behavior of the magnitude process both for $d=1$ and for $d \geq 2$.[17]

As the case $d=1$ is relatively simple, we will first focus on $d\geq 2$ and then provide the argument for $d=1$. The case $d=2$ and $d\geq 3$ require different assumptions on the kernel G_ε.

Lemma 5.18.[18] *For $x > 0$ define functions*

$$\psi(x) \in \{\log(x+e), 1(x)\}, \tag{5.79}$$

where $1(x) \equiv 1$, $\log(\cdot)$ *is the natural logarithm and* $\log(e) = 1$. *Suppose*

$$\int |G_{k\ell}(-p)|^2 \psi^2(|p|)\mathrm{d}p < \infty \;\; \forall k, \ell. \tag{5.80}$$

Then

$$\int |G_{k\ell}(q-p)G_{k\ell}(-p)|\mathrm{d}p\,\psi(|q|) \longrightarrow 0, \;\; \text{as } |q| \longrightarrow \infty \; \forall k, \ell. \tag{5.81}$$

Proof. In what follows we assume $\psi(x) = \log(x+e)$. The proof for $\psi(x) \equiv 1$ follows the same pattern. Since the functions $G_{k\ell}(\cdot)\log(|\cdot|+e)$ are continuous and square integrable, we obtain

$$|G_{k\ell}(q-p)|\log(|q-p|+e) \longrightarrow 0, \;\; \text{as } |q| \longrightarrow \infty \;\; \forall p \in \mathbf{R}^d.$$

Further, for given $\varepsilon > 0$ there is an $N = N(\varepsilon) > 0$ such that

$$\int 1_{\{|p|\geq N\}} G^2_{k\ell}(-p)\log^2(|p|+e)\mathrm{d}p < \frac{\varepsilon^2}{16[\max_{k,\ell}\int G^2_{k\ell}(-p)\log^2(|p|+e)\mathrm{d}p + 1]} =: \tilde{\varepsilon}^2.$$

Note that

$$\begin{aligned}\log(|q|+e) &\leq \log(|q-p|+e) + \log(|p|+e)\\ &\leq 2[\log(|q-p|+e) \vee \log(|p|+e)]\\ &\leq 2\log(|q-p|+e)\cdot\log(|p|+e).\end{aligned}$$

---

[17] In Sect. 5.8 we obtain the same results, on the basis of theorems of Friedman (1975), without the assumption of frame-indifference. However, the proofs under the assumption of frame-indifference are clearer and simpler. Therefore, we have included this subsection for the convenience of the reader.

[18] We remark that Lemma 5.18 and the following Corollary 5.19 do not require frame-indifference.

This implies

$$\log(|q|+e) \le 2\log(|q-p|+e)\cdot\log(|p|+e). \tag{5.82}$$

Hence,

$$\left.\begin{aligned}
&\sup_q \int 1_{\{|p|\ge N\}}|G_{k\ell}(q-p)G_{k\ell}(-p)|\mathrm{d}p\log(|q|+e)\\
&\le \sup_q 2\sqrt{\int 1_{\{|p|\ge N\}}G^2_{k\ell}(q-p)\log^2(|q-p|+e)\mathrm{d}p}\\
&\quad\sqrt{\int 1_{\{|p|\ge N\}}G^2_{k\ell}(-p)\log^2(|p|+e)\mathrm{d}p}\\
&< \sup_q 2\sqrt{\int G^2_{k\ell}(q-p)\log^2(|q-p|+e)\mathrm{d}p\cdot\tilde{\varepsilon}} = 2\\
&\quad\sqrt{\int G^2_{k\ell}(-p)\log^2(|p|+e)\mathrm{d}p\cdot\tilde{\varepsilon}} \le \frac{\varepsilon}{2}.
\end{aligned}\right\} \tag{5.83}$$

Employing first (5.82) and then Lebesgue's dominated convergence theorem, we obtain that there is an $|q|(N(\varepsilon)) > 0$ such that

$$\left.\begin{aligned}
&\int 1_{\{|p|<N\}}|G_{k\ell}(q-p)G_{k\ell}(-p)|\mathrm{d}p\log(|q|+e)\\
&\le \int 1_{\{|p|<N\}}|G_{k\ell}(q-p)\log(|q-p|+e)G_{k\ell}(-p)\log(|p|+e)|\mathrm{d}p\\
&< \frac{\varepsilon}{2},\quad \forall q \text{ whenever } |q| \ge |q|(N(\varepsilon)).
\end{aligned}\right\} \tag{5.84}$$

(5.83) and (5.84) together imply (5.81). □

Corollary 5.19. *Suppose (5.80). Then, with $\psi(\cdot)$ from (5.79),*

$$|\bar{D}_{\varepsilon,k\ell}(\sqrt{2}q) - D_{\varepsilon,k\ell,11}(0)|\psi(|q|), \longrightarrow 0 \ \textit{as}\ |q| \longrightarrow \infty\ \textit{for}\ k,\ell = 1,\ldots,d. \tag{5.85}$$

Set

$$\zeta_0 := \inf\{\zeta \ge 0 : \alpha_\varepsilon(x) > 0 \ \text{ whenever } x \ge \zeta\}. \tag{5.86}$$

□

Remark 5.20.

We note that by (5.57), (5.58), and (5.61), $0 \le \zeta_0 < \infty$. If, in addition, the conditions of Part (iii) of Proposition 5.12 hold, (5.66) implies that $\zeta_0 = 0$. □

Recall in the following theorem that for $d \ge 2$ $x(t,x_0) \sim |r(t,r^2) - r(t,r^1)|_{\{x_0=|r^2-r^1|\}}$, where $r(\cdot,r^i)$, $i = 1,2$, are solutions of (2.11). For $d = 1$, $r(t,q) \sim (r(t,r^2) - r(t,r^1))_{\{q=r^2-r^1\}}$

Theorem 5.21.[19]

Suppose (5.80) with $\psi(x) \equiv 1$ if $d \neq 2$ and $\psi(x) = \log(x+e)$ if $d = 2$. Further, suppose that the diffusion matrix with entries $D_{\varepsilon,k\ell}(0)$ is positive definite in addition the the conditions of Proposition 5.12, Part (iii). Finally, assume for $d = 2$ that $\alpha_\varepsilon(\xi) > 0 \;\forall \xi > 0$. Then the following holds:

(i) $\{0\}$ is an attractor for the $r(\cdot)$, if $d = 1$, i.e.,[20]

$$P\{\lim_{t\to\infty} r(t,q) = 0\} = 1, \quad \text{if } d = 1 \text{ and } q \neq 0. \tag{5.87}$$

(ii) For $d = 2$ and $x_0 > 0$ let $0 < a < b$ and $x(\cdot, x_0)$ is the solution of (5.78). Then

$$P\{x(t,x_0) \text{ hits } (a,b) \text{ at a sequence of times increasing to } \infty\} = 1, \quad \text{if } d = 2, \tag{5.88}$$

i.e., the solution of (5.78) is recurrent. A similar statement holds for $x_0 < 0$.

(iii) For $d \geq 3$ and $x_0 \neq 0$ the solution of (5.78) is transient, i.e,

$$P\{\lim_{t\to\infty} |x(t,x_0)| = \infty\} = 1. \tag{5.89}$$

As a consequence, if $d = 1$ the two Brownian particles $r(\cdot, r^i)$, $i = 1,2$, will eventually clump. Further, if $d = 2$, the two Brownian particles $r(\cdot, r^i)$, $i = 1,2$, will attract and repel each other infinitely often and, if $d \geq 3$ the distance between the particles will tend to ∞ with probability 1, as $t \longrightarrow \infty$.

Proof. Case $d = 1$.

Our coefficient σ is Lipschitz continuous and $\sigma(0) = 0$. Therefore, $\sigma_\varepsilon^2(x) \leq (\text{const})x^2$. Hence, for $y > 0$

$$\int_0^y \frac{1}{\sigma_\varepsilon^2(x)}\,\mathrm{d}x = \infty. \tag{5.90}$$

This result implies (5.87).[21]

Case $d \geq 2$.

Assuming $\alpha_\varepsilon(\xi) > 0 \;\forall \xi \geq \rho \geq 0$ the following functional (5.91) was used by Gikhman and Skorokhod[22] to study the asymptotic behavior of solutions of one-dimensional stochastic ordinary differential equations.[23]

$$s(x,\rho) := \int_\rho^x \exp\left[-\int_\rho^y \frac{(d-1)\alpha_{\perp,\varepsilon}(z^2)}{z\alpha_\varepsilon(z^2)}\mathrm{d}z\right]\mathrm{d}y. \tag{5.91}$$

[19] We refer to Remark 5.34 at the end of Sect. 5.8 for a comparison with the separation of independent Brownian motions.

[20] This result was first derived by Kotelenez (2005b).

[21] Cf. Ikeda and Watanabe (1981), Chap. 6.3, Theorem 3.1; Skorohod (1987), Chap. 1.3, Theorem 19 and the references therein.

[22] Cf.Gikhman and Skorokhod (1968), Chap. 4, Sect. 16.

[23] Cf. also Skorokhod (loc.cit.), Chap. 1.3. For the notation – cf. Ikeda and Watanabe (loc.cit.), Chap. VI.3.

We must show that

$$\left.\begin{array}{ll} s(x,\xi) \longrightarrow \pm\infty, \text{ as } x \longrightarrow \pm\infty & \text{if } d=2 \\ \\ s(x,\xi) \longrightarrow s^{\pm}, \text{ as } x \longrightarrow \pm\infty & \text{if } d>2, \end{array}\right\} \tag{5.92}$$

where $-\infty < s^- \le s^+ < \infty$. The statement then follows from Theorem 3.1 in Ikeda and Watanabe (loc.cit.).

(i) We verify that for any numbers x, ρ, ξ

$$s(x,\xi) = s(\rho,\xi) + \exp\left[-\int_{\xi}^{\rho} \frac{(d-1)\alpha_{\perp,\varepsilon}(u^2)}{u\alpha_{\varepsilon}(u^2)}\,\mathrm{d}u\right] s(x,\rho), \tag{5.93}$$

whence we may choose appropriate values for ρ and prove (5.92) for $s(x,\rho)$ instead of $s(x,\xi)$.
Further, for $0<\rho<x$

$$s(-x,-\rho) = -s(x,\rho). \tag{5.94}$$

Therefore, we may, in what follows, restrict ourselves to the case $0<\rho<x$.

(ii) Suppose $d=2$. Let $\bar{z}$ be a unit vector in $\mathbf{R}^2$, which is perpendicular to q. By (5.45) and (5.59)

$$\lambda_{\perp,\varepsilon}(|q|^2) = \lambda_{\perp,\varepsilon}(|q|^2)P^{\perp}(q)\bar{z}\cdot\bar{z} = \int G_{\varepsilon}(q-\tilde{q})G_{\varepsilon}^{\perp}(-\tilde{q})\mathrm{d}\tilde{q}\bar{z}\cdot\bar{z}. \tag{5.95}$$

Hence, by (5.81),

$$|\lambda_{\perp,\varepsilon}(|q|^2)|\log(|q|+e) \longrightarrow 0, \quad \text{as } |q|\longrightarrow\infty. \tag{5.96}$$

Similarly, we set $\bar{q} := \frac{1}{|q|}q$. Then, by (5.45) and (5.59),

$$\begin{aligned} \lambda_{\varepsilon}(|q|^2) - |q|^2\gamma_{\varepsilon}(|q|^2) &= (\lambda_{\varepsilon}(|q|^2) - |q|^2\gamma_{\varepsilon}(|q|^2))P(q)\bar{q}\cdot\bar{q} \\ &= \int G_{\varepsilon}(q-\tilde{q})G_{\varepsilon}^{T}(-\tilde{q})\mathrm{d}\tilde{q}\bar{q}\cdot\bar{q}. \end{aligned} \tag{5.97}$$

Again, by (5.81),

$$|\lambda_{\varepsilon}(|q|^2) - |q|^2\gamma_{\varepsilon}(|q|^2)|\log(|q|+e) \longrightarrow 0, \quad \text{as } |q|\longrightarrow\infty. \tag{5.98}$$

Abbreviate

$$\mu_{\varepsilon}(x^2) := \lambda_{\varepsilon}(x^2) - x^2\gamma_{\varepsilon}(x^2). \tag{5.99}$$

Then, by (5.61),

$$\frac{\alpha_{\perp,\varepsilon}(x^2)}{\alpha_{\varepsilon}(x^2)} = 1 + \frac{[\mu_{\varepsilon}(x^2) - \lambda_{\perp,\varepsilon}(x^2)]c_{\varepsilon}^{-1}}{1-\mu_{\varepsilon}(x^2)c_{\varepsilon}^{-1}}. \tag{5.100}$$

By (5.96) and (5.98) there is a $\rho > 0$ such that for all $x \geq \rho$

$$\frac{[\mu_\varepsilon(x^2) - \lambda_{\perp,\varepsilon}(x^2)]c_\varepsilon^{-1}}{1 - \mu_\varepsilon(x^2)c_\varepsilon^{-1}} \frac{\log(x+e)}{\log(x+e)} \leq \frac{\frac{1}{2}}{\frac{1}{2}\log(x+e)} = \frac{1}{\log(x+e)}. \tag{5.101}$$

Hence, for $x \geq \rho (> 0)$

$$\frac{\alpha_{\perp,\varepsilon}(x^2)}{x\alpha_\varepsilon(x^2)} \leq \frac{1}{x} + \frac{1}{x\ \log(x+e)} \leq \frac{1}{x} + \frac{1}{x\ \log(x)}. \tag{5.102}$$

Integrating from ρ to $y \geq \rho$:

$$\int_\rho^y \left(\frac{1}{u} + \frac{1}{u\ \log(u)}\right) du = \log\left(\frac{y}{\rho}\right) + \log\left(\frac{\log(y)}{\log(\rho)}\right), \tag{5.103}$$

whence

$$\exp\left(-\int_\rho^y \left(\frac{1}{u} + \frac{1}{u\log(u)}\right) du\right) = \frac{\rho}{y}\cdot\frac{\log(\rho)}{\log(y)}. \tag{5.104}$$

Thus,

$$\exp\left(-\int_\rho^y \left(\frac{(\alpha_{\perp,\varepsilon}(u^2)}{u\alpha_\varepsilon(u^2)}\right) du\right) \geq \frac{\rho}{y}\cdot\frac{\log(\rho)}{\log(y)}. \tag{5.105}$$

This implies

$$s(x,\rho) \geq \int_\rho^x \frac{\rho}{y}\cdot\frac{\log(\rho)}{\log(y)} dy = \rho\log(\rho)\log\left(\frac{\log(x)}{\log(\rho)}\right) \longrightarrow \infty \text{ as } x \longrightarrow \infty, \tag{5.106}$$

which finishes the proof for $d = 2$.

(iii) If for $\zeta_0 > 0$ from (5.86) $\alpha_\varepsilon(\zeta_0) = 0$ then, by (5.65) $\alpha_{\perp,\varepsilon}(\zeta_0) > 0$, i.e., at the point ζ_0 (5.78) has a positive drift and no diffusion. Consequently, it is increasing at that point. Since we are claiming transience for $d > 2$, we may for $d > 2$, without loss of generality, assume that for all $x \geq \zeta_0$ $\alpha_\varepsilon(x^2) > 0$. So, let $d \geq 3$ and $0 < \delta < \frac{1}{2}$. By (5.58) and (5.61) there is a $\rho > 0$ such that for $z > \rho > 0$

$$\frac{(d-1)(1+\delta)}{x} \geq \frac{(d-1)\alpha_{\perp,\varepsilon}(x^2)}{x\alpha_\varepsilon(x^2)} \geq \frac{(d-1)(1-\delta)}{x}. \tag{5.107}$$

Then

$$\begin{aligned} s(x,\rho) &\leq \int_\rho^x \exp\left[-\int_\rho^y \frac{(d-1)(1-\delta)}{z} dz\right] dy \\ &= \int_\rho^x \left(\frac{\rho}{y}\right)^{(d-1)(1-\delta)} dy \leq \int_\rho^x \left(\frac{\rho}{y}\right)^{(2-2\delta)} dy \\ &= \frac{\rho^{(2-2\delta)}}{2\delta - 1}\left[x^{(2\delta-1)} - \rho^{(2\delta-1)}\right] \longrightarrow \frac{\rho^{(2-2\delta)}}{1-2\delta}\rho^{(2\delta-1)}, \quad \text{as } x \longrightarrow \infty. \end{aligned}$$

Since $s(x, \rho)$ is monotone increasing in x, the previous steps imply (5.92) for $d > 2$.

(iv) The statements about the distance of the particles are a simple consequence of the recurrence and transience properties, respectively. □

5.5 Decomposition of a Diffusion into the Flux and a Symmetric Diffusion

Starting with Van Kampen's definition of the flux of a one-dimensional diffusion, we develop an alternative mathematical definition of a flux for Markovian SODEs. On the basis of the concept of symmetric Dirichlet forms, we first decompose a quadratic form, defined by the generator of the SODEs, into a symmetric Dirichlet form and a nonsymmetric form. The flux is defined through the nonsymmetric form.

Van Kampen (1983), Chap. VIII.1, considers one-dimensional diffusions with drift $a(x)$ and quadratic variation $b(x) \geq 0\ \forall x$. Suppose that both coefficients are bounded and continuous with bounded continuous derivatives. The Fokker-Planck equation[24] for such diffusions is

$$\frac{\partial P(x,t)}{\partial t} = -\frac{\partial}{\partial x}\big[a(x)P(x,t)\big] + \frac{1}{2}\frac{\partial^2}{\partial x^2}\big[b(x)P(x,t)\big], \tag{5.108}$$

where $P(x,t)$ is the probability density at time t in the point x. Van Kampen (loc.cit.) then defines the "*probability flux*" of the one-dimensional diffusion (5.108)

$$J(x,t) := a(x)P(x,t) - \frac{1}{2}\frac{\partial}{\partial x}\big[b(x)P(x,t)\big]. \tag{5.109}$$

Setting $\sigma(x) := \sqrt{b(x)}$ and choosing a one-dimensional standard Brownian motion $\beta(\cdot)$, we obtain the stochastic differential equation for the diffusion (5.108):

$$\left.\begin{aligned} \mathrm{d}x &= a(x)\mathrm{d}t + \sigma(x)\beta(\mathrm{d}t), \\ x(0) &= x_0. \end{aligned}\right\} \tag{5.110}$$

To better understand the meaning of the flux, consider first the case with no drift. Let us stick, for the time being, to van Kampen's notation of the probability density $P(x,t)$. We discretize the space into densely packed sites, denoted x, y, etc. The flux can be obtained as follows:

Fix a point $x \in \mathbf{R}$ and consider the diffusion between x and its right neighbor, $x + \delta x$ between some time t and $t + \delta t$. Then the rate of probability flow from x to $x + \delta x$ during a short time interval δt is proportional to $\sigma^2(x)P(x,t)\delta t$ and the rate of probability flow from $x + \delta x$ to x is proportional to $\sigma^2(x+\delta x)P(x+\delta x,t)\delta t$. The difference is the net flow from x to $x + \delta x$. A formal calculation leads to the definition of the flux (as the instantaneous rate of change of the net flow):[25]

---

[24] Cf. Sect. 15.2.7, in particular, (15.236).

[25] Cf. Van Kampen, loc.cit. Chap. 8.1, (1.3).

$$J(x,t) := -\frac{1}{2}\frac{\partial}{\partial x}(b(x)P(x,t)). \tag{5.111}$$

Further, if the probability density near x is approximately constant at time t and δx is "small", then the sign of the flux completely depends on the derivative of the diffusion coefficient $b(x)$, and the derivative of $b(x)$ describes the "*bias of the diffusion*" in one or the other direction. We, therefore, obtain that a one-dimensional diffusion with spatially dependent diffusion coefficient and no drift has a natural bias, depending on the sign of the derivative of the diffusion coefficient. In contrast, if

$$\sigma^2(x)P(x,t) = \sigma^2(x+\delta x)P(x+\delta x,t), \tag{5.112}$$

then the (probability) flux between these two points equals 0, and we conclude that the solute (i.e., the diffusing substance) between the sites x and $x+\delta x$ is in (local) equilibrium. If (5.112) holds between all neighboring sites, the diffusing substance is in (global) equilibrium.

The problem, however, is that the probability flux (5.109) cannot be explicitly computed, unless the probability density is stationary. For dimensions $d>1$, even in the stationary case, we usually cannot compute the solution of (5.109). Let us, therefore, consider the fundamental solution of the Fokker-Planck equation, $p(t,x,y)$, which is the probability density at y at time $t>0$, where at $t=0$ the probability distribution is the point measure $\delta_x(A)$ (and its density, in the distributional sense, the Dirac delta function with support in x). Let us first consider the case $b(z) \equiv \bar{b} > 0$. The diffusion then is a Brownian motion with transition probability density

$$p(t,x,y) := \frac{1}{\sqrt{2\pi\bar{b}t}}\exp\left[-\frac{(y-x)^2}{2\bar{b}t}\right].$$

Apparently,

$$p(t,x,y) = p(t,y,x) \ \forall t>0, \ \ \forall x,y \in \mathbf{R}. \tag{5.113}$$

Hence, if the transition probability density of a solute with transition probability satisfies (5.113), the transition from site x to site y during a fixed time interval t is the same as the transition from site y to site x during the same time interval t. We call the solute (whose probability distribution is the solution of a diffusion equation, similarly to (5.108)) in "(global) equilibrium" if (5.113) holds. Observe that for nonzero drift and constant diffusion coefficient, the deviation from equilibrium (or bias) is completely described by the sign of the drift. It seems to be natural to incorporate the bias from the nonconstant diffusion coefficient into the drift and to define an additional stochastic motion that is unbiased. To this end let us consider the *Dirichlet form* corresponding to (5.110).[26] Let A be the generator of the Markov process, defined through (5.110) or (5.108) and f and g be in the domain of A such that their first and second partial derivatives are continuous and in $\mathbf{H}_0 := L_2(\mathbf{R}, dr)$. A is given by

[26] Cf. Fukushima (1980) as well as Ma and Röckner (1992).

$$(Af)(x) := \left(a(x)\frac{\partial}{\partial x}f\right)(x) + \left(b(x)\frac{\partial^2}{(\partial x)^2}f\right)(x). \tag{5.114}$$

Assume, for simplicity, that the coefficients in (5.109) are bounded and continuous and that $b(x)$ has a bounded and continuous derivative. Suppose also $b(x) > 0\ \forall x$. Then A is closable in $\mathbf{H}_0$, and its closure defines a Markov semigroup on $\mathbf{H}_0$ as well. By a slight abuse of notation, we denote this closure also A. Set

$$\mathcal{E}(f, g) := \langle -Af, g\rangle_0, \tag{5.115}$$

where, as before, $\langle\cdot,\cdot\rangle_0$ is the L_2-scalar product. Integration by parts in the diffusion term and rearranging the terms yields

$$\mathcal{E}(f, g) = \mathcal{E}_1(f, g) + \mathcal{E}_2(f, g), \tag{5.116}$$

where

$$\left.\begin{aligned} \mathcal{E}_1(f, g) &= -\int\left[a(x)\left(\frac{\partial}{\partial x}f\right)(x) - \frac{1}{2}\left(\frac{\partial}{\partial x}b(x)\right)\left(\frac{\partial}{\partial x}f\right)(x)\right]g(x)\mathrm{d}x, \\ \mathcal{E}_2(f, g) &:= \frac{1}{2}\int b(x)\left(\frac{\partial}{\partial x}f\right)(x)\left(\frac{\partial}{\partial x}g\right)(x)\mathrm{d}x. \end{aligned}\right\} \tag{5.117}$$

$\mathcal{E}_2(f, g)$ is a closable *symmetric Dirichlet form*,[27] and since $b(x) > 0\ \forall x$ (by assumption) the domain of its closure is the standard Hilbert-Sobolev space $\mathbf{H}_1$ whose vectors are those $f \in \mathbf{H}_0$, which have square integrable derivatives (in the generalized sense). Consequently, there is a self-adjoint nonpositive operator A_2 defined through the form $\mathcal{E}_2(f, g)$. The form $\mathcal{E}_1(f, g)$ is, of course, not symmetric and defines in a unique way a first order partial differential operator A_1. More precisely, we have

$$\left.\begin{aligned} \langle -A_1 f, g\rangle_0 &= \mathcal{E}_1(f, g), \\ \langle -A_2 f, g\rangle_0 &= \mathcal{E}_2(f, g) \end{aligned}\right\} \tag{5.118}$$

for sufficiently smooth f and g. For appropriately chosen f the generators are given by

$$\left.\begin{aligned} (A_1)f(y) &:= a(y)\left(\frac{\partial}{\partial y}f\right)(y) - \frac{1}{2}\left(\frac{\partial}{\partial y}b(y)\right)\left(\frac{\partial}{\partial y}f\right)(y); \\ (A_2)f(z) &:= \frac{1}{2}\left(\frac{\mathrm{d}}{\mathrm{d}z}\left(b(z)\frac{\mathrm{d}}{\mathrm{d}z}f\right)\right)(z), \end{aligned}\right\} \tag{5.119}$$

where the latter generator is in *divergent form* and where we employed the simple identiy

$$\frac{1}{2}\left(\frac{\mathrm{d}}{\mathrm{d}z}b\right)(z)\left(\frac{\mathrm{d}}{\mathrm{d}z}f\right)(z) + \frac{1}{2}b(z)\left(\frac{\mathrm{d}^2}{(\mathrm{d}z)^2}f\right)(z) = \frac{1}{2}\left(\frac{\mathrm{d}}{\mathrm{d}z}\left(b(z)\frac{d}{dz}f\right)\right)(z).$$

[27] Cf. Fukushima (loc.cit.), Example 1.2.1.

A_2 is the generator of a symmetric Markov semigroup of self-adjoint operators on $\mathbf{H}_0$, $T(2,t)$. Further, this semigroup that defines a semigroup of transition probabilities on $C_b(\mathbf{R},\mathbf{R})$. A_1 itself generates a strongly continuous group, $T(1,t)$ both on $\mathbf{H}_0$ and on $C_b(\mathbf{R},\mathbf{R})$. In terms of semigroups and generators we may consider A_1 a relatively bounded perturbation of A_2, and the Markov semigroup $T(t)$ with generator A can be obtained through the Trotter product formula, applied to $T(1,t)$ and $T(2,t)$.[28] We call $T(t)$ the "*Trotter product*" of $T(1,t)$ and $T(2,t)$. Recalling the abbreviation

$$\sigma(z) := \sqrt{b(z)}$$

and observing that

$$\frac{1}{2}\frac{\mathrm{d}}{\mathrm{d}y}(\sigma^2(y)) = \left(\frac{\mathrm{d}}{\mathrm{d}y}\sigma(y)\right)\sigma(y),$$

the (Markovian) group $T(1,t)$ is generated by the solutions of the following ordinary differential equation (ODE)

$$\left.\begin{aligned} \frac{\mathrm{d}y}{\mathrm{d}t} &= a(y) - \left(\frac{\mathrm{d}}{\mathrm{d}y}\sigma(y)\right)\sigma(y) \\ y(0) &= y_0, \end{aligned}\right\} \tag{5.120}$$

and the (Markovian) semigroup $T(2,t)$ is generated by the solution of the stochastic ordinary differential equation (SODE)

$$\left.\begin{aligned} \mathrm{d}z &= (\tfrac{\mathrm{d}}{\mathrm{d}z}\sigma(z))\sigma(z)\mathrm{d}t + \sqrt{\sigma(z)}\mathrm{d}\beta, \\ z(0) &= z_0. \end{aligned}\right\} \tag{5.121}$$

Alternatively to the Trotter product formula, we may apply the fractional step method[29] to show that the mixture of the solutions of (5.120) and (5.121) converges under suitable assumptions on the initial conditions to the solution of (5.110). We call the resulting limit the "*fractional step product*" of the two equations (5.120) and (5.121). With this terminology (5.110) is the fractional step product of (5.120) and (5.121). Moreover, if the probability density is constant near the site x van Kampen's probability flux is proportional to the right-hand side of the ODE (5.120). We, therefore, call the right-hand side of (5.120) the "*(local) flux*" of the diffusion (5.110), and most of the time we will not explicitly mention the attribute "*local*".[30]

Our assumptions imply that the family of transition probabilities for (5.116)–(5.119), $P(t,x,A)$, $P_1(t,x,A)$, $P_2(t,x,A)$, have densities $p(t,x,y)$, $p_1(t,x,y)$, $p_2(t,x,y)$. For $P_2(t,x,A)$ this means[31]

$$P_2(t,x,A) := \int 1_A(y)p_2(t,x,y)\mathrm{d}y \quad \forall \text{ Borel set } A \subset \mathbf{R} \text{ and } \forall x \in \mathbf{R}. \tag{5.122}$$

---

[28] Cf. Davies (1980), Chap. 4.3 and Chap. 3.4.

[29] Cf. Sect. 15.3.

[30] Cf. the following (5.130) and Definition 5.23 for the usefulness of this approach versus (5.109).

[31] Cf. Stroock and Varadhan (1979), Sect. 9.1, Theorem 9.1.9.

The smoothness assumptions on the coefficient $b(z)$ implies that $p_2(t, x, y)$ is continuous in x and y for all $t > 0$[32] The symmetry of the Dirichlet form $\mathcal{E}_2(f, g)$ implies that the semigroup, defined through the transition probabilities $P_2(t, x, A)$, is symmetric in $\mathbf{H}_0$[33] Employing (5.122), the symmetry, change of variables $x \leftrightarrow y$ and Fubini's theorem imply

$$\int\int f(x)p_2(t,x,y)g(y)\mathrm{d}y\mathrm{d}x = \int\int f(x)p_2(t,y,x)g(y)\mathrm{d}y\mathrm{d}x \;\; \forall f, g \in \mathbf{H}_0. \tag{5.123}$$

We, therefore, obtain

$$p_2(t,x,y) = p_2(t,y,x) \;\; \forall t > 0, \; \forall x, y \in \mathbf{R}. \tag{5.124}$$

We conclude that the diffusion, defined by $\mathcal{E}_2(f, g)$, is in equilibrium, which is the same as to say that the flux equals 0.

Finally, we note that we may rewrite (5.110) into Stratonovich SODE:

$$\left.\begin{aligned} \mathrm{d}x &= \left[a(x) - \left(\frac{\partial}{\partial x}\sigma\right)(x)\sigma(x)\right]\mathrm{d}t + \left[\left(\frac{\partial}{\partial x}\sigma\right)(x)\sigma(x)\mathrm{d}t + \sigma(x)\beta(\mathrm{d}t)\right] \\ &= \left[a(x) - \left(\frac{\partial}{\partial x}\sigma\right)(x)\sigma(x)\right]\mathrm{d}t + \left[\frac{1}{2}\left(\frac{\partial}{\partial x}\sigma\right)(x)\sigma(x)\mathrm{d}t + \sigma(x)\circ\beta(\mathrm{d}t)\right], \\ &\qquad x(0) = x_0, \end{aligned}\right\} \tag{5.125}$$

where $\circ$ denotes the Stratonovich differential.[34] We observe that the Stratonovich differential needs to be corrected as well in order to describe an unbiased stochastic motion.

The derivation of the flux for the d-dimensional case may be easily achieved by decomposing the stochastic motion into an equilibrium motion (via a symmetric Dirichlet form) and a flux (via a nonsymmetric first order form). A generalization of (5.49) is

$$\left.\begin{aligned} \mathrm{d}r &= a(r)\mathrm{d}t + \sum_j \sigma_{\cdot,j}(r)\beta_j(\mathrm{d}t), \\ r(0) &= r_0. \end{aligned}\right\} \tag{5.126}$$

Here $a(r)$ is an $\mathbf{R}^d$-valued drift and $\sigma(r)$ is a symmetric matrix. We assume that both coefficients are bounded with bounded continuous derivatives. Now $\mathbf{H}_0$ denotes $L_2(\mathbf{R}^d, \mathrm{d}r)$ and the scalar product $< \cdot, \cdot >_0$ is the corresponding L_2-scalar product. We define the *divergence of the matrix* $D(r)$ by[35]

$$-\frac{1}{2}\mathrm{div}(D(r)) = -\frac{1}{2}\left(\sum_{\ell=1}^{d}\frac{\partial D_{1\ell}(r)}{\partial r_\ell}, \ldots, \sum_{\ell=1}^{d}\frac{\partial D_{d\ell}(r)}{\partial r_\ell}\right)^T = -\frac{1}{2}\sum_{\ell=1}^{d}\frac{\partial D_{\cdot\ell}(r)}{\partial r_\ell}. \tag{5.127}$$

[32] Cf. Ladyzhenskaya et al. (1967), Sect. IV.11.

[33] Cf. Fukushima (loc.cit.), Sect. 1, (1.4.12).

[34] Cf. also Sect. 15.2.6.

[35] Cf. Gurtin (1981), Sect. II.5.

Besides the above modifications, we may use the same notation as in the one-dimensional case with corresponding assumptions and obtain

$$
\left.\begin{aligned}
(Af)(r) = a(r)\cdot(\nabla f) + \frac{1}{2}\sum_{i,j} D_{ij}(r)\left(\frac{\partial^2}{\partial r_i \partial r_j} f\right)(r)\\
\mathcal{E}(f,g) := < -Af, g >_0,\\
\mathcal{E}_1(f,g) = -\int \left[a(r) - \frac{1}{2}\mathrm{div}(D(r))\right]\cdot(\nabla f)(r)g(r)\mathrm{d}r,\\
\mathcal{E}_2(f,g) := \int \frac{1}{2}\sum_{i,j} D_{ij}(r)\left(\frac{\partial}{\partial r_i} f\right)(r)\left(\frac{\partial}{\partial r_j} g\right)(r)\mathrm{d}r,\\
\mathcal{E}(f,g) = \mathcal{E}_1(f,g) + \mathcal{E}_2(f,g),
\end{aligned}\right\} \tag{5.128}
$$

where A in (5.128) is the generator of the Markov process, associated with (5.126). As in one dimension, the existence of a transition probability density $p_2(t,q,r)$ for the symmetric Markov diffusion, defined by $\mathcal{E}_2(f,g)$, implies the symmetry of the densities:

$$
p_2(t,q,r) = p_2(t,r,q) \;\; \forall t > 0, \; \forall q, r \in \mathbf{R}^d. \tag{5.129}
$$

Moreover, the symmetry of the densities obviously implies symmetry of the Markov diffusion semigroups in $\mathbf{H}_0$. In other words, if the Markov transition probabilities have densities, then symmetry of the Markov diffusion is equivalent to (5.129). Hence, we arrive at the following decomposition of (5.126) into two differential equations, corresponding to the decomposition of $\mathcal{E}(f,g)$ into the non-symmetric and symmetric parts $\mathcal{E}_1(f,g)$ and $\mathcal{E}_2(f,g)$. Setting

$$
Fl(q) := a(q) - \tfrac{1}{2}\mathrm{div}(D(q)), \} \tag{5.130}
$$

we obtain the deterministic ODE

$$
\frac{\mathrm{d}q}{\mathrm{d}t} = Fl(q), \quad q(0) = q_0 \tag{5.131}
$$

and the symmetric SODE

$$
\mathrm{d}p = -Fl(p)\mathrm{d}t + \sum_j \sigma_{\cdot,j}(p)\beta_j(\mathrm{d}t), \quad p(0) = p_0. \tag{5.132}
$$

Definition 5.22. *$Fl(r)$ in (5.130) is called the "(d-dimensional) flux" for the diffusion (5.126).* □

Let $q(t)$ be the solution of the ODE in (5.131). Then

$$
\frac{\mathrm{d}|q(t)|^2}{\mathrm{d}t} = 2Fl(q(t))\cdot q(t). \tag{5.133}
$$

We conclude that $|q(t)|$ is decreasing or increasing dependent on whether the sign in (5.133) is negative or positive. Further, the motion of the solution $r(t)$ of the symmetric Itô equation in (5.132) is unbiased. Consequently, the motion of (5.126) has a preferred tendency towards the origin or away from the origin, depending on the sign of the right hand of (5.133).[36] Since for the separation of the two particles a motion toward the origin describes an attractive phenomenon and the reverse motion implies repulsion, we arrive at the following.

Definition 5.23. *A point $r \in \mathbf{R}^d$ is called* "attractive" *if $Fl(r) \cdot r < 0$. The point r is called* "repulsive" *if $Fl(r) \cdot r > 0$.* □

In (5.126) and in Definition 5.23 we see that the decomposition of (5.110) into (5.120) and (5.121) gives us an explicit way to describe the inherent bias in a spatially dependent diffusion (with or without drift) through (5.120). Consequently, the use of the flux from (5.120) has an advantage over the use of the probability flux (5.109). Obviously, the same comments hold for the d-dimensional case, if we generalize (5.109) to d dimensions.

5.6 Local Behavior of Two Correlated Brownian Motions with Shift-Invariant and Frame-Indifferent Integral Kernels

We derive the flux for the separation and also for the magnitude of the separation and define attractive and repulsive points. Under the assumption of frame-indifference, we obtain the radial flux that characterized attractive and repulsive points through their distance from the origin.

Theorem 5.24. The SODE (5.68) for the separation $r(t,q)$ is obtained as the fractional step product of the following two differential equations:

$$\left.\begin{aligned} \frac{dq}{dt}(t) &= \left\{ \frac{1}{2}\frac{(d-1)}{|q|^2}[\alpha_{\perp,\varepsilon}(|q|^2) - \alpha_\varepsilon(|q|^2)] - (\alpha'_\varepsilon)(|q|^2) \right\} q, \\ q(0) &:= q_0, \end{aligned}\right\} \tag{5.134}$$

and

$$\left.\begin{aligned} dr(t) = &- \left\{ \frac{1}{2}\frac{(d-1)}{|r|^2}[\alpha_{\perp,\varepsilon}(|r|^2) - \alpha_\varepsilon(|r|^2)] - (\alpha'_\varepsilon)(|r|^2) \right\} r dt \\ &+ \left[\sqrt{\alpha_{\perp,\varepsilon}(|r|^2)} P^\perp(r) + \sqrt{\alpha_\varepsilon(|r|^2)} P(r) \right] \tilde{\beta}(dt), \\ r(0) := r_0. \end{aligned}\right\} \tag{5.135}$$

[36] This is true for short times if the distribution near the point of investigation is approximately constant. A divergence theorem argument, provided in (15.165)–(15.167), supports this conclusion.

The right-hand side of (5.134) is the flux of (5.68), and (5.135) is the symmetric diffusion.

Proof. Recall that by (5.130) the flux for (5.68) is just the divergence of the matrix $\bar{D}_\varepsilon$. Employing the representation (5.60), we first compute

$$\frac{\partial \alpha_{\perp,\varepsilon}(|r|^2)}{\partial r_\ell} = (\alpha'_{\perp,\varepsilon})(|r|^2)2r_\ell, \quad \frac{\partial \alpha_\varepsilon(|r|^2)}{\partial r_\ell} = (\alpha'_\varepsilon)(|r|^2)2r_\ell. \tag{5.136}$$

Further,

$$\frac{\partial}{\partial r_\ell}\left(\frac{r_k r_\ell}{|r|^2}\right) = \frac{r_k}{|r|^2} + \frac{r_l \delta_{k\ell}}{|r|^2} - \frac{2r_k r_\ell^2}{|r|^4}. \tag{5.137}$$

Observe that

$$\left.\begin{aligned}
&\sum_{\ell=1}^{d}(\alpha'_{\perp,\varepsilon})(|r|^2)2r_\ell\left[\delta_{k\ell} - \frac{r_k r_\ell}{|r|^2}\right] = 0;\\
&\sum_{\ell=1}^{d}(\alpha'_\varepsilon)(|r|^2)2r_\ell\frac{r_k r_\ell}{|r|^2} = 2(\alpha'_\varepsilon)(|r|^2)r_k;\\
-&\sum_{\ell=1}^{d}\alpha_{\perp,\varepsilon}(|r|^2)\left[\frac{r_k}{|r|^2}(1+\delta_{k\ell}) - \frac{2r_k r_\ell^2}{|r|^4}\right] = -\alpha_{\perp,\varepsilon}(|r|^2)\frac{(d-1)r_k}{|r|^2};\\
&\sum_{\ell=1}^{d}\alpha_\varepsilon(|r|^2)\left[\frac{r_k}{|r|^2}(1+\delta_{k\ell}) - \frac{2r_k r_\ell^2}{|r|^4}\right] = \alpha_\varepsilon(|r|^2)\frac{(d-1)r_k}{|r|^2}.
\end{aligned}\right\} \tag{5.138}$$

Recalling (5.130), we obtain from the preceding calculations for the kth component of the d-dimensional flux $Fl(r)$

$$Fl_k(r) = \left\{\frac{1}{2}\frac{(d-1)}{|r|^2}\left[\alpha_{\perp,\varepsilon}(|r|^2) - \alpha_\varepsilon(|r|^2)\right] - (\alpha'_\varepsilon)(|r|^2)\right\} r_k. \tag{5.139}$$

The statement about the fractional step product follows from Goncharuk and Kotelenez (1998). □

Observe that, under the frame-indifference assumption, the flux is parallel to $\frac{1}{|r|}r$. Therefore, if a point r is attractive or repulsive then all points on the boundary ∂B of a ball B, centered at the origin with radius $|r|$, are either all attractive or all repulsive. In other words, attractive and repulsive domains can be completely characterized by the distance $|r|$ to the origin. (Note that the outward normal at $r \in \partial B$ is $\frac{1}{|r|}r$.) In this case we change coordinates and define the *radial flux* for (5.68) through

$$\bar{Fl}(x) := \frac{1}{2}\left\{\frac{(d-1)}{x}\left[\alpha_{\perp,\varepsilon}(x^2) - \alpha_\varepsilon(x^2)\right] - \left(\frac{\mathrm{d}}{\mathrm{d}x}\alpha_\varepsilon\right)(x^2)\right\}, \quad x := |r|. \tag{5.140}$$

Then we have

Proposition 5.25. *The sphere* $\{r : |r| = x\}$ *is attractive for (5.68) if* $\bar{F}l(x) < 0$, *and it is repulsive if* $\bar{F}l(x) > 0$.

It remains to discuss the relation of Proposition 5.25 to the one-dimensional Itô SODE (5.78) for the radial component $|r(t)|$ *of the solution of (5.68). As a preliminary, we analyze the symmetric diffusion (5.135) and its Dirichlet form* $\mathcal{E}_2(f, g)$.

Let $A(S^{d-1})$ *be the surface measure of the unit sphere in* $\mathbf{R}^d$ *and define a measure* m *on the Borel sets of* $[0, \infty)$:

$$m(\mathrm{d}x) := A(S^{d-1})x^{d-1}\mathrm{d}x. \tag{5.141}$$

□

Proposition 5.26.

(i) *The transition probabilities,* $P_2(t, q, A)$, *of the Markov diffusion (5.135) have densities* $p_2(t, q, r)$ *such that for any* $Q_1, Q_2 \in \mathcal{O}(d)$

$$p_2(t, Q_1q, Q_2r) = p_2(t, q, r) \quad \forall t > 0, q, r \in \mathbf{R}^d. \tag{5.142}$$

(ii) *Restrict* $\mathcal{E}_2(f, g)$ *to (smooth and integrable)* $\tilde{f}(r) := \varphi(|r|)$ *and* $\tilde{g}(r) := \psi(r)$. *Then*

$$\mathcal{E}_2(\varphi, \psi) = \frac{1}{2}\int_0^\infty \alpha_\varepsilon(x^2)\varphi'(x)\psi'(x)m(\mathrm{d}x). \tag{5.143}$$

Proof.

(i) With the obvious modifications (5.71) and (5.72) carry over to the diffusion (5.135). Hence,

$$P_2(t, Qq, A) = P_2(t, q, A) \ \forall Q \in \mathcal{O}(d) \quad \forall t > 0, q, \in \mathbf{R}^d \ \forall A \in \mathcal{B}^d.$$

Stroock and Varadhan (loc.cit.) and Ladyženskaja et al. (loc.cit.) guarantee the existence of a density, continuous in q, r for $t > 0$. Therefore, the previous equation implies for the densities

$$p_2(t, Qq, r) = p_2(t, q, r) \ \forall Q \in \mathcal{O}(d) \quad \forall t > 0, q, r \in \mathbf{R}^d.$$

By symmetry in q and r we obtain the same statement with respect to r, i.e.,

$$p_2(t, q, Qr) = p_2(t, q, r) \ \forall Q \in \mathcal{O}(d) \quad \forall t > 0, q, r \in \mathbf{R}^d.$$

We now apply another element from $\mathcal{O}(d)$ to the coordinate q and obtain (5.142).

(ii) Recalling the definition (5.28) of $P(r)$ and $P^\perp(r)$, the generator, A_2 for (5.135) is given by

$$\left.\begin{aligned}(A_2f)(r) = &-\left\{\frac{1}{2}\frac{(d-1)}{|r|^2}[\alpha_{\perp,\varepsilon}(|r|^2) - \alpha_\varepsilon(|r|^2)] - (\alpha_\varepsilon')(|r|^2)\right\} r \cdot \nabla f(r)\\ &+\frac{1}{2}\sum_{k,\ell=1}^{d}\left[\alpha_{\perp,\varepsilon}(|r|^2)\left(\delta_{k\ell} - \frac{r_kr_\ell}{|r|^2}\right) + \alpha_\varepsilon(|r|^2)\frac{r_kr_\ell}{|r|^2}\right]\left(\frac{\partial^2}{\partial_k\partial_\ell}f\right)(r),\end{aligned}\right\} \tag{5.144}$$

where f is sufficiently smooth and integrable. As before, we consider only (smooth) functions $\tilde{f}(r) = \varphi(|r|), \tilde{g}(r) := \psi(|r|)$. Note that

$$\left.\begin{aligned} &\nabla\varphi(|r|) = (\varphi)'(|r|)\frac{1}{|r|}r, \quad \nabla\psi(|r|) = (\psi)'(|r|)\frac{1}{|r|}r; \\ &\frac{\partial^2}{\partial r_k \partial r_\ell}\varphi(|r|) = (\varphi')(|r|)\left[\frac{\delta_{k\ell}}{|r|} - \frac{r_k r_\ell}{|r|^3}\right] + (\varphi'')(|r|)\frac{r_k r_\ell}{|r|^2}; \\ &\sum_{k,\ell=1}^{d}\left[r_k r_\ell \delta_{k\ell} - \frac{r_k^2 r_\ell^2}{|r|^2}\right] = 0; \\ &\sum_{k,\ell=1}^{d}\left[\delta_{k\ell} - \frac{r_k r_\ell}{|r|^2}\right]^2 = d-1. \end{aligned}\right\} \tag{5.145}$$

We, therefore, obtain

$$\left.\begin{aligned} &\mathcal{E}_2(\varphi,\psi) \\ &= \frac{1}{2}\int\left\{\frac{d-1}{|r|}[\alpha_{\perp,\varepsilon}(|r|^2) - \alpha_\varepsilon(|r|^2)] - 2(\alpha_\varepsilon')(|r|^2)\right\}\varphi'(|r|)\psi(|r|)\mathrm{d}r \\ &\quad -\frac{1}{2}\int\frac{d-1}{|r|}\alpha_{\perp,\varepsilon}(|r|^2)\varphi'(|r|)\psi(|r|)\mathrm{d}r - \frac{1}{2}\int\alpha_\varepsilon(|r|^2)\varphi''(|r|)\psi(|r|)\mathrm{d}r \\ &= -\frac{1}{2}\int\left\{\frac{d-1}{|r|}\alpha_\varepsilon(|r|^2) + 2(\alpha_\varepsilon')(|r|^2)\right\}\varphi'(|r|)\psi(|r|)\mathrm{d}r \\ &\quad -\frac{1}{2}\int\alpha_\varepsilon(|r|^2)\varphi''(|r|)\psi(|r|)\mathrm{d}r. \end{aligned}\right\} \tag{5.146}$$

Since all the functions in the last line of (5.146) depend only on $|r|$, we may change the measure according to (5.141) and obtain

$$\begin{aligned} \mathcal{E}_2(\varphi,\psi) &= -\frac{1}{2}\int_0^\infty \frac{d-1}{x}\alpha_\varepsilon(x^2)\varphi'(x)\psi(x)m(\mathrm{d}x) \\ &\quad -\frac{1}{2}\int_0^\infty\left[\frac{\mathrm{d}}{\mathrm{d}x}(\alpha(x^2)\varphi'(x))\right]\psi(x)m(\mathrm{d}x) \end{aligned} \tag{5.147}$$

Finally, we integrate by parts and obtain (5.143). □

By (5.142)

$$\bar{p}_2(t,|q|,|r|) := p_2(t,q,r). \tag{5.148}$$

is a continuous function of the arguments $t, x = |q|, y = |r|$, which is symmetric in x and y.

Corollary 5.27. *Let $r(t, r_0)$ be the solution of (5.135). Then $x(r, x_0)_{x_0=|r_0|} = |r(t, r_0)|$ is equivalent in distribution to a one-dimensional Markov diffusion with probability density*

$$p_{2,\mathbf{R}}(t, x, y) := A(S^{d-1})\bar{p}_2(t, x, y)y^{d-1} \ \forall t > 0, x, y \in \mathbf{R}_+, \tag{5.149}$$

where $A(S^{d-1})$ is the surface measure of the d-dimensional unit sphere (cf. (5.141)). Moreover, this density is symmetric with respect to $m(\mathrm{d}x)$ from (5.141). The Dirichlet form for this diffusion is given by (5.143). It is equivalent in distribution to the solution of the Itô SODE

$$\mathrm{d}x = -\bar{F}l(x)\mathrm{d}t + \sqrt{\alpha_\varepsilon(x^2)}\beta(\mathrm{d}t), \ \ x(0) = x_0, \tag{5.150}$$

where $\bar{F}l(x)$ is the radial flux from (5.140). Finally, if $q(t, q_0)$ is the solution of (5.134), *then $y(t, y_0)_{y_0=|q_0|} = |q(t, q_0)|$ can be obtained as the solution of the ODE*

$$\frac{\mathrm{d}y}{\mathrm{d}t} = \bar{F}l(y), \quad y(0) = y_0. \tag{5.151}$$

Proof. That the radial part of (5.135) is itself equivalent to a Markov diffusion on the positive real line follows exactly like the corresponding statement Proposition 5.15. To prove (5.149), consider the probability

$$P(|r(t, q)| \in [a, b]) = A(S^{d-1}) \int_a^b \bar{p}_2(t, |q|, y)y^{d-1}\mathrm{d}y.$$

Dividing both sides of the last equality by $b - a$ and assuming $b \downarrow a$ we obtain (5.149). To check the symmetry let φ, ψ be "nice" functions, defined on $[0, \infty)$. Then

$$\begin{aligned}\int_0^\infty \int_0^\infty \varphi(y)p_{2,\mathbf{R}}(t, x, y)\mathrm{d}y\psi(x)m(\mathrm{d}x) = [A(S^{d-1})]^2 \\ \int_0^\infty \int_0^\infty \varphi(y)\psi(x)\bar{p}_2(t, x, y)y^{d-1}x^{d-1}\mathrm{d}y\mathrm{d}x.\end{aligned} \tag{5.152}$$

The symmetry of $\bar{p}_2(t, x, y)$ in x, y in addition to (5.152) implies the symmetry of $p_{2,\mathbf{R}}(t, x, y)$ with respect to the initial distribution $m(\mathrm{d}x)$. It is obvious that (5.143) is the Dirichlet form for the radial diffusion $|r(t, q)|$ and that the representation (5.150)/(5.151) holds. □

Theorem 5.28. [37]
Consider the two correlated Brownian motions $r(t, r_0^1)$ and $r(t, r_0^2)$ as solutions of (2.11). Suppose

$$\frac{\alpha_{\perp,\varepsilon}(x^2)}{\alpha_\varepsilon(x^2)} \leq 1 \ \forall x > 0. \tag{5.153}$$

[37] We refer to Remark 5.34 at the end of Sect. 5.8 for a comparison with the separation of independent Brownian motions.

Then there is an interval (a,b) with $a \geq 0$ such that the magnitude of the separation $|r(t,r^1(0)) - r(t,r^2(0)| \in (a,b)$ has a bias to decrease, i.e., the two Brownian particles become attracted to each other if their distance is in (a,b).

Proof. Note that $\alpha_\varepsilon(x^2)$ equals 0 at $x = 0$. It is nonnegative by (5.64) and that there is an interval, where $\alpha_\varepsilon(x^2) > 0$. Hence, it must have a positive derivative on some interval (a,b). This, in addition to the assumption (5.153) and (5.140), implies that $\bar{F}l(x) < 0$ in (a,b). □

5.7 Examples and Additional Remarks

Based on the definition of the flux, attractive zones are explicitly computed for the Maxwellian kernel. The flux of the magnitude of d-dimensional separation process, which is based on the d-dimensional Lebesgue measure, is compared with a one-dimensional flux, based on the one-dimensional Lebesgue measure. These two fluxes are shown to be different. We also provide a derivation of the flux, employing the divergence theorem.

Example 5.29. Consider the Maxwellian kernel $G_{\varepsilon,M}(r) = \bar{c}_\varepsilon r \exp\left(-\frac{|r|^2}{2\varepsilon}\right)$ (cf. 5.30). The mutual quadratic variations are as follows:

$$
\left.
\begin{array}{l}
D_{\varepsilon,k\ell,12}(\sqrt{2}r) := \left\{
\begin{array}{lll}
-c_{\varepsilon,d} \exp\left(-\dfrac{|r|^2}{2\varepsilon}\right) r_k r_\ell, & \text{if} & k \neq \ell, \\[2ex]
-c_{\varepsilon,d} \exp\left(-\dfrac{|r|^2}{2\varepsilon}\right)\left(-\dfrac{\varepsilon}{2} + r_k^2\right), & \text{if} & k = \ell.
\end{array}
\right. \\
\qquad\qquad \text{with } c_{\varepsilon,d} := \bar{c}_\varepsilon^2 (\pi\varepsilon)^{\frac{d}{2}} .
\end{array}
\right\}
\tag{5.154}
$$

Indeed, recalling the definition (5.33),

$$
D_{k\ell,12}(\hat{r}) = \int G_{\varepsilon,k}(r_1 - q) G_{\varepsilon,\ell}(r_2 - q) \mathrm{d}q
$$

The first observation is that for $i = 1, 2$

$$
\left.
\begin{aligned}
G_{\varepsilon,\ell,i}(r_i - q) &= \bar{c}_\varepsilon (r_i - q)_\ell \exp\left(-\frac{|r_i - q|^2}{2\varepsilon}\right) \\
&= -\bar{c}_\varepsilon \varepsilon \partial_\ell \exp\left(-\frac{|r_i - q|^2}{2\varepsilon}\right).
\end{aligned}
\right\}
\tag{5.155}
$$

Hence,

$$
\left.
\begin{aligned}
D_{k\ell,12}(\hat{r}) &= (\bar{c}_\varepsilon \varepsilon)^2 \int \partial_k \exp\left(-\frac{|r_1 - q|^2}{2\varepsilon}\right) \partial_\ell \exp\left(-\frac{|r_2 - q|^2}{2\varepsilon}\right) dq \\
&= (\bar{c}_\varepsilon \varepsilon)^2 \int \frac{\partial}{\partial q_k} \exp\left(-\frac{|r_1 - q|^2}{2\varepsilon}\right) \frac{\partial}{\partial q_\ell} \exp\left(-\frac{|r_2 - q|^2}{2\varepsilon}\right) dq \\
&= -(\bar{c}_\varepsilon \varepsilon)^2 \int \exp\left(-\frac{|r_1 - q|^2}{2\varepsilon}\right) \frac{\partial^2}{\partial q_k \partial q_\ell} \exp\left(-\frac{|r_2 - q|^2}{2\varepsilon}\right) dq \\
&\qquad \text{(integrating by parts)} \\
&= -(\bar{c}_\varepsilon \varepsilon)^2 \int \exp\left(-\frac{|r_1 - q|^2}{2\varepsilon}\right) \times \begin{cases} \frac{(r_2-q)_k}{\varepsilon} \frac{(r_2-q)_\ell}{\varepsilon} \exp\left(-\frac{|r_2-q|^2}{2\varepsilon}\right) dq, & \text{if} \quad k \neq \ell, \\ \left(-\frac{1}{\varepsilon} + \frac{(r_2-q)_k^2}{\varepsilon^2}\right) \exp\left(-\frac{|r_2-q|^2}{2\varepsilon}\right) dq, & \text{if} \quad k = \ell. \end{cases}
\end{aligned}
\right\}
\tag{5.156}
$$

Using shift invariance, we obtain for $k \neq \ell$

$$
D_{k\ell,12}(\hat{r}) = -\bar{c}_\varepsilon^2 \int \exp\left(-\frac{|r_1 - r_2 - q|^2 + |q|^2}{2\varepsilon}\right) q_k q_\ell \, dq
$$

and for $k = \ell$

$$
D_{k\ell,12}(\hat{r}) = -\bar{c}_\varepsilon^2 \int \exp\left(-\frac{|r_1 - r_2 - q|^2 + |q|^2}{2\varepsilon}\right) (-\varepsilon + q_k^2) \, dq.
$$

We use

$$
\exp\left(-\frac{|r - q| + |q|^2}{2\varepsilon}\right) = \exp\left(-\frac{|r|^2}{4\varepsilon}\right) \exp\left(-\frac{|q - \frac{1}{2}r|^2}{\varepsilon}\right). \tag{5.157}
$$

Hence, we obtain for $k \neq \ell$

$$
D_{k\ell,12}(\hat{r}) = -\bar{c}_\varepsilon^2 2 \exp\left(-\frac{|r_1 - r_2|^2}{4\varepsilon}\right) \int \exp\left(-\frac{|\frac{1}{2}(r_1 - r_2) - q|^2}{\varepsilon}\right) q_k q_\ell \, dq
$$

and for $k = \ell$

$$
D_{k\ell,12}(\hat{r}) = -\bar{c}_\varepsilon^2 \exp\left(-\frac{|r_1 - r_2|^2}{4\varepsilon}\right) \int \exp\left(-\frac{|\frac{1}{2}(r_1 - r_2) - q|^2}{\varepsilon}\right) (-\varepsilon + q_k^2) dq.
$$

The previous steps, in addition to a simple integration, imply

$$
D_{k\ell,12}(\hat{r}) := \begin{cases} -\bar{c}_\varepsilon^2 (\pi \varepsilon)^{\frac{d}{2}} \exp\left(-\frac{|r_1-r_2|^2}{4\varepsilon}\right) \frac{(r_1-r_2)_k (r_1-r_2)_\ell}{4}, & \text{if} \quad k \neq \ell, \\ -(\bar{c}_\varepsilon^2 (\pi \varepsilon)^{\frac{d}{2}} \exp\left(-\frac{|r_1-r_2|^2}{4\varepsilon}\right) \left(-\frac{\varepsilon}{2} + \frac{(r_1-r_2)_k^2}{4}\right), & \text{if} \quad k = \ell. \end{cases} \tag{5.158}
$$

Recalling $\sqrt{2}r = r_1 - r_2$, this yields (5.154).

Hence,

$$\bar{D}_{\varepsilon,k\ell}(\sqrt{2}r) := \begin{cases} \frac{c_{\varepsilon,d}}{2} \exp\left(-\frac{|r|^2}{2\varepsilon}\right) r_k r_\ell, & \text{if } k \neq \ell, \\ \varepsilon \frac{c_{\varepsilon,d}}{2}\left[1 - \exp\left(-\frac{|r|^2}{2\varepsilon}\right)\right] + \frac{c_{\varepsilon,d}}{2} \exp\left(-\frac{|r|^2}{2\varepsilon}\right) r_k^2\Big)\Big], & \text{if } k = \ell. \end{cases} \tag{5.159}$$

The diffusion matrix $\bar{D}_\varepsilon(\sqrt{2}r)$ is clearly frame-indifferent. Employing the representation (5.60), we obtain

$$\left.\begin{aligned} \alpha_{\perp,\varepsilon}(|r|^2) &= \hat{c}_\varepsilon\left(1 - \exp\left[-\frac{|r|^2}{2\varepsilon}\right]\right), \\ \alpha_\varepsilon(|r|^2) &= \hat{c}_\varepsilon\left(1 - \exp\left[-\frac{|r|^2}{2\varepsilon}\right] + \frac{|r|^2}{\varepsilon}\right)\exp\left[-\frac{|r|^2}{2\varepsilon}\right], \end{aligned}\right\} \tag{5.160}$$

and, therefore,

$$\frac{\mathrm{d}}{\mathrm{d}x}\alpha_\varepsilon(x^2) = \frac{\hat{c}_\varepsilon}{\varepsilon}x\left[3 - \frac{x^2}{\varepsilon}\right]\exp\left[-\frac{x^2}{2\varepsilon}\right].$$

So, the radial flux from (5.140) is

$$\bar{F}l(x) = -\frac{\hat{c}_\varepsilon}{\varepsilon}x\ \exp\left[-\frac{x^2}{2\varepsilon}\right]\left[d + 2 - \frac{x^2}{\varepsilon}\right] \tag{5.161}$$

We obtain that the interval $(0, \sqrt{\varepsilon(d+2)})$ is attractive for two correlated Brownian motions with Maxwellian kernel $G_{\varepsilon,M}(r) = \bar{c}_\varepsilon r\ \exp\left(-\frac{|r|^2}{2\varepsilon}\right)$.

We remark that, for the Maxwellian case, the divergence of the matrix $\bar{D}_\varepsilon(\sqrt{2}r)$ may also be computed directly without recourse to the representation (5.60), and we obtain

$$\mathrm{div}\bar{D}_\varepsilon(\sqrt{2}r) \cdot \frac{1}{|r|}r = \bar{F}l(|r|). \qquad \square$$

Remark 5.30. So far we have focused on the d-dimensional origin of the radial diffusion $|r(t,q)|$. Using the d-dimensional Lebesgue measure as the equilibrium measure for $r(t,q)$ from (5.68) we have derived $m(\mathrm{d}x) = A(S^{d-1})x^{d-1}\,\mathrm{d}x$ in (5.141) as the equilibrium measure for $|r(t,q)|$ and determined the radial flux (5.140). On the other hand, we could have decomposed (5.78) right away into two differential equations, following the pattern of (5.120)/(5.121):

$$\left.\begin{aligned} \hat{F}l(x) &:= \frac{1}{2}\left\{\frac{(d-1)}{x}\alpha_{\perp,\varepsilon}(x^2) - \left(\frac{\mathrm{d}}{\mathrm{d}x}\alpha_\varepsilon\right)(x^2)\right\}; \\ \frac{\mathrm{d}y}{\mathrm{d}t} &= \hat{F}l(y), \quad y(0) = y_0; \\ \mathrm{d}z &= -\hat{F}l(z)\mathrm{d}t + \sqrt{\alpha_\varepsilon(z^2)}\beta(\mathrm{d}t) \quad z(0) = z_0. \end{aligned}\right\} \tag{5.162}$$

The difference between the radial flux $\bar{F}l(x)$ and the one-dimensional flux $\hat{F}l(x)$ is the term $-\frac{1}{2}\frac{(d-1)}{x}\alpha_\varepsilon(x^2)$. Recall that $\mathcal{E}_2(\varphi,\psi)$ from (5.143) has representation (5.147). (In fact, we derived (5.143) from (5.147).) The Dirichlet form $\mathcal{E}_1(\varphi,\psi)$, generated by (5.151), is given by

$$\begin{aligned}\mathcal{E}_1(\varphi,\psi) = \int_0^\infty \frac{1}{2}\Bigg\{ &\frac{(d-1)}{x}\Big[\alpha_{\perp,\varepsilon}(x^2) - \alpha_\varepsilon(x^2)\Big] \\ &- \left(\frac{\mathrm{d}}{\mathrm{d}x}\alpha_\varepsilon\right)(x^2)\Bigg\}\varphi'(x)\psi(x)m(\mathrm{d}x)\end{aligned} \tag{5.163}$$

So,

$$\left.\begin{aligned} &\mathcal{E}_1(\varphi,\psi) + \mathcal{E}_2(\varphi,\psi) \\ &= \int_0^\infty \frac{1}{2}\left\{\frac{(d-1)}{x}\alpha_{\perp,\varepsilon}(x^2) - \left(\frac{\mathrm{d}}{\mathrm{d}x}\alpha_\varepsilon\right)(x^2)\right\}\varphi'(x)\psi(x)m(\mathrm{d}x) \\ &\quad -\frac{1}{2}\int_0^\infty\left[\frac{\mathrm{d}}{\mathrm{d}x}(\alpha(x^2)\varphi'(x))\right]\psi(x)m(\mathrm{d}x) \\ &=: \hat{\mathcal{E}}_1(\varphi,\psi) + \hat{\mathcal{E}}_2(\varphi,\psi). \end{aligned}\right\} \tag{5.164}$$

In this derivation two other Dirichlet forms, $\hat{\mathcal{E}}_1(\varphi,\psi)$ and $\hat{\mathcal{E}}_2(\varphi,\psi)$, emerge, corresponding to decomposition (5.162) of (5.78). An important observation is that the second form is not symmetric with respect to $m(\mathrm{d}x)$. However, if we now replace in the right-hand side of (5.164) $m(\mathrm{d}x)$ from (5.141) by the Lebesgue measure $\mathrm{d}x$, then the second form becomes symmetric, and the first form contains the flux for (5.78) as a one-dimensional SODE with Lebesgue measure as the equilibrium measure. This last representation of the flux is equivalent to van Kampen's probability flux, if the density near the point x is constant, as we mentioned before. Summarizing, we remark that, for frame-indifferent diffusions in $\mathbf{R}^d$, $d > 1$, the analysis of the radial part of that diffusion has to take into account the "correct" equilibrium measure in $\mathbf{R}^d$ and then compute the corresponding equilibrium measure on $[0,\infty)$ (in our case $m(\mathrm{d}x)$). A careless application of the generator of the radial part, without computing the corresponding equilibrium measure on $[0,\infty)$, can lead to an incorrect definition of the flux.[38] □

Additional Comments

(i) Einstein (1905), Chap. 4, (2), derives his famous formula for the diffusion coefficient D for the Brownian motion of one particle by assuming some force K and equilibrium between the diffusion and the force K. In a kinematic stochastic model,

[38] Cf. also a fluid mechanics argument in our Additional Comment (ii).

the force becomes a drift coefficient. Therefore, Einstein's equilibrium condition is equivalent to the condition that the flux equals 0 (with respect to some equilibrium measure). In our approach, the Lebesgue measure dr, i.e., the uniform initial distribution, is the equilibrium measure, and the flux describes local deviations from that equilibrium.

(ii) It is interesting in its own right to generalize (5.109), using the divergence theorem. We use, without loss of generality, the same notation as in (5.127).[39] Let $X(t, r)$ denote the density of the solute at location $r \in \mathbf{R}^d$ and time t. Let $\nu := (\nu_1, \ldots, \nu_d)$ be a unit vector in $\mathbf{R}^d$. The pair (r, ν) determines an oriented plane in $\mathbf{R}^d$ through r with orienting normal ν.

Consider a small right cylindrical "pill box" $\mathcal{P}(r, \nu)$, with axis ν passing through the point r. One end of the cylinder is located at r and the other end is located at $r + \lambda\nu$, $0 < \lambda \ll 1$. We now interpret the transition probabilities for the diffusion of $r(t)$ as the distribution of some matter. The matter flowing out of the pill box $\mathcal{P}(r, \nu)$ in a small time step through its boundary $\partial\mathcal{P}(r, \nu)$ is described by the following surface integral

$$\int_{\partial\mathcal{P}(r,\nu)} \frac{1}{2}\bar{D}_\varepsilon(\sqrt{2}\tilde{r})\tilde{\nu}(\tilde{r})X(t, r) \cdot \nu(\tilde{r})\mathrm{d}A(\tilde{r}),$$

where $\mathrm{d}A(\tilde{r})$ is the surface measure, $\tilde{\nu}(\tilde{r})$ is the outward unit normal at $\tilde{r} \in \partial\mathcal{P}(r, \nu)$ and $X(t, r)$ is the density. Then, using the Divergence Theorem[40], we have

$$\int_{\partial\mathcal{P}(r,\nu)} \frac{1}{2}\bar{D}_\varepsilon(\sqrt{2}\tilde{r})\tilde{\nu}(\tilde{r})X(t, \tilde{r}) \cdot \nu\,\mathrm{d}A(\tilde{r}) = \int_{\mathcal{P}(r,\nu)} \frac{1}{2}\mathrm{div}(\bar{D}_\varepsilon(\sqrt{2}\tilde{r})\nu(\tilde{r})X(t, \tilde{r}))\mathrm{d}\tilde{r}. \tag{5.165}$$

Next, we suppose that the length, λ, and the area of the cross section are both small and that the ratio of the length to the cross-sectional area is also small. This implies that the contribution through the lateral surface is small in comparison with that for the ends. We arrive at the following observation: the flow of (probability) matter out of the pill-box through the lateral surfaces becomes negligible. As a result, we obtain essentially the difference of the probability "mass," moving in a small time step in the direction of the normal ν, minus the probability "mass," moving in the direction of $-\nu$. Note that $\tilde{\nu}(\tilde{r}) \equiv \pm\nu$ for $\tilde{r}$ in the cross-sectional areas. Therefore, dividing the volume integral by the volume of the pill box, we define the *surface flux density* $J_\nu(r, t)$ as the instantaneous rate of change in the net probability flow through a unit area of plane at r in the ν direction by

$$J_\nu(r, t) := -\frac{1}{2}\nabla\cdot((\bar{D}_\varepsilon(\sqrt{2}r)X(t, r))\nu) = -\frac{1}{2}(\mathrm{div}(\bar{D}_\varepsilon(\sqrt{2}r)X(t, r)))\cdot\nu. \tag{5.166}$$

In terms of the divergence of the diffusion matrix, $\mathrm{div}\bar{D}_\varepsilon$, the surface flux density is

---

[39] We follow arguments, provided by M. Leitman – cf. Kotelenez et al. (2007).

[40] Cf. Gurtin, loc.cit.

$$J_\nu(r,t) = -\frac{1}{2}X(t,r)\mathrm{div}(\bar{D}_\varepsilon(\sqrt{2}r)\cdot\nu) - \frac{1}{2}\nabla_r X(t,r)\cdot(\bar{D}_\varepsilon(\sqrt{2}r)\cdot\nu). \quad (5.167)$$

Again, if $X(t,r)$ is approximately constant at time t near r the sign of the flux will be completely determined by the divergence of the matrix $\bar{D}_\varepsilon(\sqrt{2}r)$.

We note that for $d = 1$ and no drift, the surface flux density is exactly van Kampen's probability flux (5.109). For $d > 1$ and no drift the d-dimensional flux density and $X(t,r)$ approximately constant near r the d-dimensional flux density is our d-dimensional flux $Fl(r)$ from (5.130)/(5.139), which we used to derive the radial flux $\bar{Fl}(x)_{|x=|r|}$ in (5.140). In other words, Leitman's application of the divergence theorem provides an independent proof of our claim that $\bar{Fl}(x)_{|x=|r|}$ is the correct flux for the radial motion, $|r(t)|$, of a frame-indifferent diffusion $r(t)$ (cf. our Remark 5.30).

(iii) Friedman (loc.cit.), Chap. 9.4, calls the flux $Fl(x)$ in (5.130) the "*Fichera drift*." Following Friedman, loc.cit., the Fichera drift plays a role in trapping diffusion in domain B if the "normal diffusion" to ∂B at r equals 0, i.e., if

$$\sum_{k,\ell=1}^{d} D_{k\ell}(\sqrt{2}r)\nu_k\nu_\ell = 0. \quad (5.168)$$

Employing the factorization as in (5.48) of $D_{k\ell}(\sqrt{2}r)$, we obtain that (5.168) is equivalent to

$$\left.\begin{aligned} &\sum_{j=1}^{d}\left(\sum_{k=1}^{d}\sigma_{kj}(r)\nu_k\right)^2 = 0, \quad \text{if } r\in\partial B \\ &\Longleftrightarrow \\ &\sum_{k=1}^{d}\sigma_{kj}(r)\nu_k = 0 \;\; \forall j\in\{1,..,d\}, \quad \text{if } r\in\partial B\,. \end{aligned}\right\} \quad (5.169)$$

Using (5.48) and assuming (5.168) and that the coefficients in (5.49) have continuous bounded partial derivatives, we obtain

$$\begin{aligned} &-\frac{c}{2}\sum_{k=1}^{d}\sum_{\ell=1}^{d}\frac{\partial D_{k\ell}(\sqrt{2}r)}{\partial r_\ell}\nu_k \\ &= -\frac{c}{2}\sum_{k=1}^{d}\sum_{\ell=1}^{d}\sum_{j=1}^{d}\left(\frac{\partial\sigma_{kj}(r)}{\partial r_\ell}\right)\sigma_{\ell j}(r)\nu_k - \frac{c}{2}\sum_{k=1}^{d}\sum_{\ell=1}^{d}\sum_{j=1}^{d}\sigma_{kj}(r)\frac{\partial\sigma_{\ell j}(r)}{\partial r_\ell}\nu_k \\ &= -\frac{c}{2}\sum_{k=1}^{d}\sum_{\ell=1}^{d}\sum_{j=1}^{d}\left(\frac{\partial\sigma_{kj}(r)}{\partial r_\ell}\right)\sigma_{\ell j}(r)\nu_k, \end{aligned}$$

since

$$-\frac{1}{2}\sum_{k=1}^{d}\sum_{\ell=1}^{d}\sum_{j=1}^{d}\sigma_{kj}(r)\frac{\partial\sigma_{\ell j}(r)}{\partial r_\ell}\nu_k = -\frac{1}{2}\sum_{\ell=1}^{d}\sum_{j=1}^{d}\frac{\partial\sigma_{\ell j}(r)}{\partial r_\ell}\left(\sum_{k=1}^{d}\sigma_{kj}(r)\nu_k\right) = 0$$

(by (5.168)).

If we write the SODE, corresponding to (5.49), in Stratonovich form (indicated by "∘" in the stochastic differential,[41] we obtain that $r(t)$ is the solution of

$$\left.\begin{aligned} \mathrm{d}r(t) = -\frac{1}{2}\sum_{\ell=1}^{d}\sum_{j=1}^{d}\left(\frac{\partial \sigma_{\cdot j}(r)}{\partial r_\ell}\right)\sigma_{\ell j}(r)\mathrm{d}t + \sum_j \sigma_{\cdot j}(r)\circ \beta_j(\mathrm{d}t), \\ r(0) := q. \end{aligned}\right\} \tag{5.170}$$

$-\frac{1}{2}\sum_{\ell=1}^{d}\sum_{j=1}^{d}(\frac{\partial \sigma_{\cdot j}(r)}{\partial r_\ell})\sigma_{\ell j}(r)$ is the so-called Stratonovich correction term, which we have to add if we want to transform an Itô equation into an equivalent Stratonovich equation. The above derivations imply the following:

In addition to the previous assumptions on the domain B and the coefficients of (5.49), assume that the normal diffusion to ∂B at r equals 0. Then the Fichera drift (which is the same as our flux) equals the Stratonovich correction term, i.e.,

$$-\frac{1}{2}\sum_{\ell=1}^{d}\frac{\partial \bar{D}_{\cdot\ell}(r)}{\partial r_\ell} = -\frac{1}{2}\sum_{\ell=1}^{d}\sum_{j=1}^{d}\left(\frac{\partial \sigma_{\cdot j}(r)}{\partial r_\ell}\right)\sigma_{\ell j}(r). \tag{5.171}$$

(iv) The result of Theorem 5.21 (and the following more general Theorem 5.34) imply in particular that, under the assumption of frame-indifference and smooth kernels G_ε, the separation of two correlated Brownian motions cannot be "trapped" within a (small) distance for all $t > 0$ if $d \geq 2$. For more general d-dimensional diffusions Friedman (loc.cit., Sect. 9.5) provides sufficient conditions for a solution to be "trapped" in a given domain. This begs the question whether there are diffusion matrices for the separation, as defined in (5.45), that can force the separation to remain in a ball with center 0 for all $t > 0$. The investigation of this interesting problem goes beyond the scope of this book.

(v) Suppose that $\beta_\varepsilon(x) > 0 \ \forall x$ (cf. (5.29)). Further suppose that $\beta_\varepsilon(\cdot)$ is twice continuously differentiable, such that $\forall x \quad \beta'_\varepsilon(x) \leq 0$ and $(\log(\beta_\varepsilon(x)))'' \leq 0$. Following Kotelenez et al (loc.cit.) we call such a function "*logarithmically concave*." For the proof of the recurrence for $d = 2$ of the separation in Theorem 6.1 of Kotelenez et al., it is assumed that $\beta_\varepsilon(\cdot)$ is logarithmically concave instead of our integrability condition (5.80). First note that for logarithmically concave $\beta_\varepsilon(\cdot)$ we obtain (5.153), i.e.,[42]

$$0 \leq \frac{\alpha_{\perp,\varepsilon}(x^2)}{\alpha_\varepsilon(x^2)} \leq 1.$$

Hence, for logarithmically concave $\beta_\varepsilon(\cdot)$ the recurrence proof for $d = 2$ in Theorem 5.28 is simpler. We wish to show that the assumption of logarithmically concavity implies the integrability condition (5.80) with $\psi(x) = \log(x + e)$, if we have already integrability with $\psi)(\cdot) \equiv 1$. Set

[41] Cf. Sect. 15.2.6.

[42] Recall that this is our assumption in Theorem 5.28.

$$\gamma_\varepsilon(x) := \log(\beta_\varepsilon(x)).$$

The differentiability assumptions imply

$$\gamma_\varepsilon'(x) \le 0, \gamma_\varepsilon''(x) \le 0 \;\; \forall x.$$

Hence by the integrability there must be an $x_0 > 0$ and $\delta > 0$ such that

$$\forall x \ge x_0 \;\; \gamma_\varepsilon(x) \le -\delta,$$

whence

$$\forall x \ge x_0 \;\; \gamma_\varepsilon(x) \le -\delta - a(x - x_0).$$

This is equivalent to

$$\forall x \ge x_0 \;\; \beta_\varepsilon(x) \le \exp[-\delta - a(x - x_0)]. \tag{5.172}$$

Since for continuous functions $G_{k\ell}$ we must check the integrability condition (5.80) only outside a bounded ball, (5.172) implies (5.80).

(vi) Theorem 5.28 states that there is an attractive zone for two Brownian particles at close distances. This is consistent with the depletion effect. The fact that, under our assumptions, the two Brownian particles never hit and for $d > 2$ eventually move apart by Theorem 5.21 is consistent with the existence of a strong repulsive (electrostatic) force at distance ≈ 0. Both the attractive behavior for distances near 0 and the "repulsive" behavior at distance ≈ 0 are in good agreement with empirical data about colloids in fluids.[43]

5.8 Asymptotics of Two Correlated Brownian Motions with Shift-Invariant Integral Kernels

This subsection contains generalizations of Sect. 5.4 without the assumption of frame indifference.

Friedman (1975), Chap. 9, provides conditions for transience and recurrence of d-dimensional diffusions, represented by SODEs, as described by our equation (5.49). In what follows, we will adjust those conditions for the dimension $d = 2$ and $d \ge 3$ to our setting.

Since the coefficients in (5.49) are Lipschitz and bounded, the Lipschitz and linear growth condition, A.1 in Friedman (loc.cit), Chap. 9.1 holds for (5.49). Condition B.1 in Friedman (loc.cit.), Chap. 9.5, reduces in our setting to the nondegeneracy of $D_{\varepsilon,k\ell}(\sqrt{2}q)$, for which we have given a sufficient condition in Proposition 5.12, Part (iii).

Condition B.2 in Friedman (loc.cit.), Chap. 9.5, we can simplify in our setting as follows: Consider the function $R(q) = |q|$, which is infinitely often continuously

[43] Cf. Tulpar, John Y. (2006), and the references therein as well as Kotelenez et al. (loc.cit.).

differentiable for $q \neq 0$. For $q \neq 0$ denote by $D_q|q|$ is the directional derivative of the function $R(q) = |q|$ at q in the direction of q. If $q \neq 0$, we calculate

$$\left.\begin{array}{l} \dfrac{\partial}{\partial q_k}|q| = \dfrac{q_k}{|q|}; \\ \dfrac{\partial^2}{\partial q_k \partial q_\ell}|q| := \begin{cases} \frac{1}{|q|} - \frac{q_k^2}{|q|^2}, & \text{if} \quad k = \ell\,, \\ -\frac{q_k q_\ell}{|q|^2}, & \text{if} \quad k \neq \ell; \end{cases} \\ D_q|q| \equiv 1. \end{array}\right\} \tag{5.173}$$

Set for $q \neq 0$

$$B(q) := \frac{1}{2} \sum_{k,\ell=1}^{d} \bar{D}_{\varepsilon,k\ell}(\sqrt{2}q) \frac{\partial^2}{\partial q_k \partial q_\ell}|q|,$$

$$P(q) := \frac{1}{2} \sum_{k,\ell=1}^{d} \bar{D}_{\varepsilon,k\ell}(\sqrt{2}q) \frac{\partial}{\partial q_k}|q| \frac{\partial}{\partial q_\ell}|q|,$$

and for $x > 0$

$$\epsilon(x) := \frac{1}{\log x}.$$

Lemma 5.31. *There is a $\delta_0 > 0$ such that $0 < |q| \leq \delta_0$ implies*

$$\frac{B(|q|)}{|q|} \geq \frac{P(|q|)}{|q|^2}(1 + \epsilon(|q|)). \tag{5.174}$$

Proof. By (5.173)

$$B(|q|) = \frac{1}{2|q|} \sum_{k=1}^{d} \bar{D}_{\varepsilon,kk}(\sqrt{2}q) - P(|q|),$$

whence (5.174) is equivalent to

$$\frac{1}{2} \sum_{k=1}^{d} \bar{D}_{\varepsilon,kk}(\sqrt{2}q) \geq P(|q|)[|q| + (1 + \epsilon(|q|))]. \tag{5.175}$$

Note that

$$\bar{D}_{\varepsilon,kk}(\sqrt{2}q) \longrightarrow \bar{D}_{\varepsilon,kk}(0) > 0, \quad \text{as } |q| \downarrow 0 \text{ for } k = 1, \ldots, d. \tag{5.176}$$

Further,[44]

$$\epsilon(x) \longrightarrow -\infty, \text{ as } x \downarrow 0, \tag{5.177}$$

and for $0 < x < y < 1$

[44] (5.177) is Condition (5.3) in Friedman (loc.cit.), Chap. 9.5.

$$\left.\begin{aligned}&\int_x^y \frac{1}{u}\exp\left[\int_u^y \frac{\epsilon(v)}{v}dv\right]du\\&=\int_x^y \frac{1}{u}\exp\left[\log\log(v)|_u^y\right]du\\&=\int_x^y \frac{1}{u}\frac{\log y}{\log u}du \longrightarrow \infty, \text{ as } x\downarrow 0.\end{aligned}\right\}\tag{5.178}$$

We obtain from (5.173) in addition to Proposition 5.12 that $P(q) > 0$ for all $q \neq 0$. Hence, by (5.176) and (5.177), for sufficiently small $|q|$ the left-hand side in (5.175) is positive, whereas the right-hand side is negative.

Remark 5.32. Expression (5.43) implies Condition B.2 in Friedman (loc.cit.), Chap. 9.5. Further, assuming the matrix with entries $D_{\varepsilon,k\ell,11}(0)$ is positive definite, (5.85) implies Friedman's Condition A.3' if the dimension $d \geq 3$ and $\psi(x) \equiv 1$. Friedman's Condition A.4' holds if $d = 2$ and $\psi(x) = \log(x+e)$. (Friedman, loc.cit. Chap. 9.2, pp. 200 and 202), taking into account the "Remark" on p. 202. □

We are now ready to derive the announced generalization of Theorem 5.21. We recall that in the following theorem $r(t,q) \sim (r(t,r^2) - r(t,r^1))_{\{q=r^2-r^1\}}$, where $r(\cdot,r^i)$, $i = 1,2$, are solutions of (2.11).

Theorem 5.33. *Consider the SODE (5.49). Suppose the matrix with entries $D_{\varepsilon,k\ell,11}(0)$ is positive definite in addition to the conditions of Proposition 5.12, Part (iii). Further, suppose (5.80) with $\psi(x) \equiv 1$ if $d \neq 2$ and $\psi(x) = \log(x+e)$ if $d = 2$. Let $q \neq 0$. Then*

- *$\{0\}$ is an attractor for the $r(\cdot)$, if $d = 1$, i.e.,*

$$P\{\lim_{t\to\infty} r(t,q) = 0\} = 1, \quad \text{if } d = 1 \tag{5.179}$$

- *$r(\cdot,q)$ is recurrent, if $d = 2$, i.e., for any ball $B_\gamma(p) := \{\tilde{q} : |\tilde{q} - p| \leq \gamma\}$, $p \in \mathbf{R}^d$, $\gamma > 0$,*

$$P\{r(t,q) \text{ hits } B_\gamma(p) \text{ at a sequence of times increasing to } \infty\} = 1, \quad \text{if } d = 2 \tag{5.180}$$

- *$r(\cdot,q)$ is transient, if $d \geq 3$, i.e.,*

$$P\{\lim_{t\to\infty} |r(t,q)| = \infty\} = 1, \quad \text{if } d \geq 3 \tag{5.181}$$

As a consequence, if $d = 1$ the two Brownian particles $r(\cdot,r^i)$, $i = 1,2$, will eventually clump. Further, if $d = 2$, the two Brownian particles $r(\cdot,r^i)$, $i = 1,2$, will attract and repel each other infinitely often and, if $d \geq 3$ the distance between the particles will tend to ∞ with probability 1, as $t \longrightarrow \infty$.

Proof. Case $d = 1$ has been shown in Theorem 5.21.

The cases $d \geq 2$ follow from Friedman (1975), Chap. 9, as shown in the previous steps (cf. our Remark 5.32). For $d = 2$ we employ Friedman's Theorem 6.1 in Chap. 9 and the Remark at the end of that section, and the case $d \geq 3$ follows from Friedman's Theorem 5.3 in Chap. 9. □

Remark 5.34.

- Consider, as in the traditional approach, two i.i.d. standard $\mathbf{R}^d$-valued Brownian motions, $\beta^1(t)$ and $\beta^2(t)$ to describe the positions of two Brownian particles. The difference, $\beta((t) := \beta^2(t) - \beta^1(t)$, is again a d-dimensional Brownian motion, which for fixed t is distributed acoording to $\mathcal{N}(0, 2tI_d)$. For $d = 1, 2$ $\beta(\cdot)$ is recurrent, and for $d = 3$ it is transient. Therefore, the only difference in the long-time behavior between our model of convolutions of space–time Brownian noise with a homogeneous kernel $G_\varepsilon(r)$ and the traditional one appears in dimension $d = 1$.
- However, we have seen in Theorem 5.28 that, under the assumption of frame-indifference, for short times the behavior of our correlated Brownian motions is different from the traditional one also for $d > 1$, because they show attractive and repulsive phenomena, depending on the distance of the particles. We expect a similar phenomenon also without the assumption of frame-indifference. □

Chapter 6
Proof of the Flow Property

We first derive a version of the solution of (4.9), which is measurable in all parameters (Sect. 6.1). In Sect. 6.2 smoothness results are derived for the forward and backward flows.

6.1 Proof of Statement 3 of Theorem 4.5

(i) For this proof we may, without loss of generality, assume $N = 1$ and that $\sigma_i^{\perp} \equiv 0 \ \forall i$. We adjust the classical proof of the Markov property for certain SODEs to our setting.[1] Let "diameter" be defined as usual for metric spaces $(\mathbf{B}, d_B)$, i.e., for a Borel set $A \subset \mathbf{B}$ we set

$$\text{diam}\,(A) := \sup_{b,\tilde{b}\in A} d_B(b, \tilde{b}).$$

$C^1([0,T];\mathbf{M})$ is the space of continuously differentiable functions from $[0,T]$ into $\mathbf{M}$, where $\mathbf{M} \in \{\mathbf{M}_f, \mathbf{M}_{\infty,\varpi}\}$ and recall that both spaces are separable complete metric spaces.[2] Note that $C^1([0,T];\mathbf{M})$ is dense in $C([0,T];\mathbf{M})$, which itself is separable.[3] Therefore, for each $n \in \mathbf{N}$ we define a countable decomposition of $C([0,T];\mathbf{M})$ into disjoint Borel sets D_ℓ^n of diameter $\le 3^{-n}$, $\ell, n \in \mathbf{N}$, such that each D_ℓ^n contains an $\eta_\ell^n \in C^1([0,T];\mathbf{M})$. By the separability of $\mathbf{R}^d$, there is a sequence of countable decompositions $\{E_k^m\}_{k\in\mathbf{N}}$, where E_k^m are nonempty Borel sets of diameter $\le 3^{-m}$ for all $k \in \mathbf{N}$ and $m \in \mathbf{N}$. The distance is measured in the Euclidean norm. In each E_k^m we choose an arbitrary but fixed element r_k^m, $k, m \in \mathbf{N}$. Finally, we decompose $[0,T)$ into intervals $\left[s_j^N, s_{j+1}^N\right)$, $j = 0, 1, \ldots$, where $s_j^N := \left(3^{-N} j\right) \wedge T$. Now we define maps

[1] Cf., e.g., Dynkin, 1965, Chap. VI, §2.

[2] Cf. Sect. 15.1.4, Proposition 15.9, and (15.45), respectively.

[3] Cf. Treves, 1967, Chap. 44, Theorem 44.1.

$$\left.\begin{array}{l} f_m : \mathbf{R}^d \to \mathbf{R}^d,\ f_m(r) = r_k^m, \text{ if } r \in E_k^m; \\ h_n : C([0,T];\mathbf{M}) \to C([0,T];\mathbf{M}),\ \ h_n(\eta) = \eta_\ell^n \in C^1([0,T];\mathbf{M}) \cap D_\ell^n, \text{ if } \eta \in D_\ell^n; \\ g_N : [0,T) \to [0,T),\ \ g_N(s) = s_{j-1}^N, \text{ if } s \in \left[s_{j-1}^N, s_j^N\right); \\ k, \ell, j = 1, 2, \ldots \end{array}\right\} \tag{6.1}$$

Let $S_N([0,T];\mathbf{M})$ be the space of $\mathbf{M}$-valued functions on $[0,T]$, which are constant on $[s_{j-1}^N, s_j^N)$, $j = 1, 2, \ldots$ and set

$$CS_N([0,T];\mathbf{M}) := C([0,T];\mathbf{M}) \cup \bigcup_{N=1}^{\infty} S_N([0,T];\mathbf{M}).$$

We endow $CS_N([0,T];\mathbf{M})$ with the metric $|||\eta_1 - \eta_2|||_{\gamma,T} := \sup_{0\le t\le T} \gamma(\eta_1(t) - \eta_2(t))$, where we recall that $\gamma \in \{\gamma_f, \gamma_\varpi\}$, depending on whether $\mathbf{M} = \mathbf{M}_f$ or $\mathbf{M} = \mathbf{M}_\varpi$. We then define

$$\left.\begin{array}{l} \pi_N : C([0,T];\mathbf{M}) \to S_N([0,T];\mathbf{M}) \subset CS_N([0,T];\mathbf{M}), \\ \eta(\cdot) \mapsto \pi_N\eta(\cdot) := \eta(g_N(\cdot)). \end{array}\right\} \tag{6.2}$$

Let $\mathcal{B}_{C([0,t];\mathbf{M})}$ and $\mathcal{F}_{\mathbf{M},t}$ be the σ-algebras of the Borel sets and the cylinder sets on $C([0,t];\mathbf{M})$, respectively, $t \le T$. The corresponding σ-algebras on $CS_N([0,t];\mathbf{M})$ will be denoted $\mathcal{B}_{CS_N([0,t];\mathbf{M})}$ and $\hat{\mathcal{F}}_{\mathbf{M},t}$, respectively. By the separability of $\mathbf{M}$ and the definition of $|||\cdot|||_{\gamma,T}$ we have for all $t \le T$ [4]

$$\left.\begin{array}{r} \mathcal{B}_{C([0,t];\mathbf{M})} = \mathcal{F}_{\mathbf{M},t}, \\ \mathcal{B}_{CS_N([0,t];\mathbf{M})} = \hat{\mathcal{F}}_{\mathbf{M},t}. \end{array}\right\} \tag{6.3}$$

The continuity of the imbedding

$$C([0,T];\mathbf{M}) \subset CS_N([0,T];\mathbf{M})$$

implies

$$\mathcal{B}_{CS_N([0,t];\mathbf{M})} \cap C([0,t];\mathbf{M}) = \mathcal{B}_{C([0,t];\mathbf{M})}, \tag{6.4}$$

where the left-hand side of (6.4) is the trace of $\mathcal{B}_{CS_N([0,t];\mathbf{M})}$ in $C([0,t];\mathbf{M})$. Since for $\pi_t := \pi_{0,t}$

$$\pi_t\mathcal{F}_{\mathbf{M},T} = \mathcal{F}_{\mathbf{M},t}, \qquad \pi_t\hat{\mathcal{F}}_{\mathbf{M},T} = \hat{\mathcal{F}}_{\mathbf{M},t}$$

we obtain

$$D_{k,t}^n := \pi_t D_k^n \in \mathcal{F}_{\mathbf{M},t}, \qquad D_{k,t}^{n,N} := \pi_t\pi_N D_k^n \in \hat{\mathcal{F}}_{\mathbf{M},t}. \tag{6.5}$$

[4] Cf., e.g., Kuo, 1975, Chap. 1, §3, whose proof can be easily generalized to our setting.

(ii) Let $0 \le s \le u \le t$. By (4.18) and the uniqueness of the solution of (4.9)

$$r(t, \pi_{s,t}\tilde{\mathcal{Y}}, r_s, s) = r(t, \pi_{u,t}\tilde{\mathcal{Y}}, r(u, \pi_{s,u}\tilde{\mathcal{Y}}, r_s, s), u). \tag{6.6}$$

Hence, the solution of (4.9) is also well defined for measurable and adapted input processes with trajectories in $S_N([0,T]; \mathbf{M})$ and (4.18) holds for step function valued processes $\tilde{\mathcal{Y}}$[5] as well. Denote by $\tilde{r}(\cdot, \omega, \pi_N h_n \eta, f_m(r), g_N(s))$ the unique continuous (Itô-)solution of (4.9) with fixed deterministic input $(\pi_N h_n(\eta), f_m(r), g_N(s))$. By (4.18) with probability 1 (uniformly in $t \in [g_N(s), T]$)

$$\tilde{r}(t, \omega, \pi_N h_n(\eta), f_m(r), g_N(s)) \equiv \tilde{r}(t, \omega, \pi_t \pi_N h_n(\eta), f_m(r), g_N(s)). \tag{6.7}$$

Let $B \in \mathcal{B}^d$. By adaptedness

$$A_{t,\pi_N h_n(\eta), f_m(r), g_N(s)} := \{\omega : \tilde{r}(t, \omega, \pi_N h_n(\eta), f_m(r), g_N(s)) \in B\} \in \mathcal{G}_t. \tag{6.8}$$

Set

$$\tilde{r}^{-1}(t, \cdot, \cdot, \cdot, \cdot)(B) := \bigcup_{k,\ell,j} A_{t,\pi_N \eta_\ell^n, r_k^m, s_j^N} \times D_{\ell,t}^{n,N} \times E_k^m \times \left[s_{j-1}^N, s_j^N\right)$$

(the subindex t at $D_\ell^{n,N}$ is justified by (6.7) for the inverse image). Then, by (6.5)

$$\tilde{r}^{-1}(t, \cdot, \cdot, \cdot, \cdot)(B) \in \mathcal{G}_t \otimes \hat{\mathcal{F}}_{\mathbf{M},t} \otimes \mathcal{B}^d \otimes \mathcal{B}_{[0,t]}. \tag{6.9}$$

(iii) We now employ (4.17) and the Borel-Cantelli lemma to remove step by step the discretization of our input variables, extending the proof given by Dynkin[6] to our setting. The first step is the limit $m \to \infty$, and we may directly copy Dynkin's proof. This step yields the existence of a function

$$\approx{r}: 0 \le s \le t \le T \times \Omega \times CS_N([0,T]; \mathbf{M}) \times \mathbf{R}^d \to \mathbf{R}^d$$

and sets $\tilde{\Omega}_{\tilde{\eta},r,s}$ with $P(\tilde{\Omega}_{\tilde{\eta},r,s}) = 1$ such that uniformly in $t \in [g_N(s), T]$

$$\overset{\approx}{r}(t, \omega, \pi_N h_n(\eta), f_m(r), g_N(s)) :\equiv \left\{ \begin{array}{l} \tilde{r}(t, \omega, \pi_N h_n(\eta), f_m(r), g_N(s)), \text{ if } \omega \in \tilde{\Omega}_{\eta,r,s}, \\ 0, \text{ if } \omega \notin \tilde{\Omega}_{\tilde{\eta},r,s}), \end{array} \right\} \tag{6.10}$$

and for any ω, η, r, s uniformly on $[g_N(s), T]$

$$\lim_{m\to\infty} \overset{\approx}{r}(\cdot, \omega, \pi_N h_n(\eta), f_m(r), g_N(s)) = \overset{\approx}{r}(\cdot, \omega, \pi_N h_n(\eta), r, g_N(s)). \tag{6.11}$$

---

[5] We can define the solution of (4.9) recursively on intervals $\left[s_j^N, s_{j+1}^N\right)$, $j = 0, 1, \ldots$ and patch them together. Alternatively, we could have included step functions $\tilde{\mathcal{Y}}$ as input processes into (4.9), which would have lead to the same result.

[6] Cf. Dynkin (loc. cit.), Chap. XI, §2, (11.44) and (11.45).

Expression (6.10) and (6.11) imply that both $\tilde{\tilde{r}}\,(t,\cdot,\pi_N h_n(\cdot), f_m(\cdot), g_N(\cdot))$ and $\tilde{\tilde{r}}\,(t,\cdot,\pi_N h_n(\cdot),\cdot, g_N(\cdot))$ are $\overline{\mathcal{G}}_t \otimes \hat{\mathcal{F}}_{\mathbf{M},t} \otimes \mathcal{B}^d \otimes \mathcal{B}_{[0,t]} - \mathcal{B}^d$ measurable. Further,

$$\left.\begin{array}{l} \tilde{\tilde{r}}\,(t,\cdot,\pi_N h_n(\eta), r_{g_N(s)}, g_N(s)) \text{ is a solution of (4.9) for any} \\ r_{g_N(s)} \in L_{2,\mathcal{F}_{g_N(s)}},\ \eta \in C([0,T];\mathbf{M}). \end{array}\right\} \tag{6.12}$$

(iv) In this step we show that $\tilde{\tilde{r}}\,(t,\cdot,\pi_N h_n(\tilde{\mathcal{Y}}), r_{g_N(s)}, g_N(s))$ is a solution of (4.9) for $r_{g_N(s)} \in L_{2,\mathcal{F}_{g_N(s)}}$, $\tilde{\mathcal{Y}} \in L_{2,\mathcal{F}}(C([0,T];\mathbf{M}))$. Without loss of generality, let $g_N(s) = 0$. Since on $[0, s_1^N)$ $\pi_N h_n(\tilde{\mathcal{Y}})(s) = h_n(\tilde{\mathcal{Y}})(0)$ is $\mathcal{F}_0$-measurable, we may again copy Dynkin's proof to show that $\tilde{\tilde{r}}\,(t,\cdot,\pi_N h_n(\tilde{\mathcal{Y}}), r_0, 0)$ is a solution of (4.9) on $[0, s_1^N)$ and, by a continuity argument, also on $[0, s_1^N]$. Set $r_{s_1^N} := \lim_{t\uparrow s_1^N} \tilde{\tilde{r}}\,(t,\cdot,\pi_N h_n(\tilde{\mathcal{Y}}), r_0, 0)$. As before, we conclude that

$$\tilde{\tilde{r}}\,(t,\cdot,\pi_N h_n(\tilde{\mathcal{Y}}), r_{s_1^N}) \text{ solves (4.9) on } \left[s_1^N, s_2^N\right].$$

Set

$$\Omega_{k_0} := \left\{\omega : \pi_N h_n(\tilde{\mathcal{Y}})(0) = \eta_{k_0}^n(0)\right\},\ \Omega_{k_1} := \left\{\omega : \pi_N h_n\left(\tilde{\mathcal{Y}}\right)\left(s_1^N\right) = \eta_{k_1}^n\left(s_1^N\right)\right\}$$

Since $\Omega_{k_0} \in \mathcal{F}_0$ and $\Omega_{k_1} \in \mathcal{F}_{s_1^N}$, a.s. for $t \in \left[s_1^N, s_2^N\right]$

$$\begin{aligned} &\tilde{\tilde{r}}\left(t,\cdot,\pi_N h_n\left(\tilde{\mathcal{Y}}\right), r_{s_1^N}\right) 1_{\Omega_{k_0}\cap\Omega_{k_1}} \\ &= \tilde{\tilde{r}}\left(t,\cdot,\pi_N \eta_{k_1}^n\ \tilde{\tilde{r}}\left(s_1^N,\cdot,\pi_N \eta_{k_0}^n, r_0, 0\right), s_1^N\right) 1_{\Omega_{k_0}\cap\Omega_{k_1}} \\ &= \tilde{\tilde{r}}\left(t,\cdot,\pi_N \eta_{k_0,k_1}^n, r_0, 0\right) 1_{\Omega_{k_0}\cap\Omega_{k_1}} \end{aligned}$$

with $\eta_{k_0,k_1}^n \in CS_N([0,T];\mathbf{M})$ such that $\pi_N \eta_{k_0,k_1}^n(0) = \eta_{k_0}^n(0)$ and $\pi_N \eta_{k_0,k_1}^n(s_1^N) = \eta_{k_1}^n(s_1^N)$. The second identity derives from the unique continuation property of solutions of (4.9) and (6.12). So, we obtain a.s. for $t \in [s_1^N, s_2^N]$

$$\tilde{\tilde{r}}\left(t,\cdot,\pi_N h_n\left(\tilde{\mathcal{Y}}\right), r_{s_1^N}, s_1^N\right) 1_{\Omega_{k_0}\cap\Omega_{k_1}} = \tilde{\tilde{r}}\left(t,\cdot,\pi_N h_n\left(\tilde{\mathcal{Y}}\right), r_0, 0\right) 1_{\Omega_{k_0}\cap\Omega_{k_1}}.$$

Since $\Omega = \bigcup_{k_0,k_1} \Omega_{k_0} \cap \Omega_{k_1}$, this implies that a.s. for any $t \in \left[s_1^N, s_2^N\right]$

$$\tilde{\tilde{r}}\left(t,\cdot,\pi_N h_n\left(\tilde{\mathcal{Y}}\right), r_{s_1^N}, s_1^N\right) \equiv \tilde{\tilde{r}}\left(t,\cdot,\pi_N h_n\left(\tilde{\mathcal{Y}}\right), r_0, 0\right),$$

whence $\tilde{\tilde{r}}\,(\cdot,\cdot,\pi_N h_n(\tilde{\mathcal{Y}}), r_0, 0)$ is a solution of (4.9) on $[0, s_2^N]$. Repeating this procedure a finite number of times finishes step (iv).

(v) We verify that for any n and k

$$\left.\begin{array}{l}(\alpha)\displaystyle\int_0^T \gamma\left(\pi_N\eta_k^n(s)-\eta_k^n(s)\right)\mathrm{d}s \le c_T\left\|\frac{\partial}{\partial t}\eta_k^n\right\|_{\gamma,T}^2 \mathrm{e}^{-N}\\ \text{and for any } s,\eta, \text{ and } r\\ (\beta)E|r(s,\eta,r,g_N(s))-r|^2\le c_T 3^{-N}.\end{array}\right\} \tag{6.13}$$

Hence, as in step (iv), employing (4.17) and the Borel–Cantelli lemma, we obtain a function

$$\overset{=}{r}: 0\le s\le t\le T\times\Omega\times CS_N([0,T];\mathbf{M})\times\mathbf{R}^d\to\mathbf{R}^d$$

and sets $\overset{\approx}{\Omega}_{\eta,r,s}$ with $P\left(\overset{\approx}{\Omega}_{\eta,r,s}\right)=1$ such that uniformly in $t\in[g_N(s),T]$

$$\overset{=}{r}\,(t,\omega,\pi_N h_n(\eta),f_N(r),g_N(s)) :\equiv \left\{\begin{array}{l}\overset{\approx}{r}\,(t,\omega,\pi_N h_n(\eta),f_N(r),g_N(r)),\ \text{if}\ \omega\in\overset{\approx}{\Omega}_{\eta,r,s}\\ 0,\ \text{if}\ \omega\notin\overset{\approx}{\Omega}_{\eta,r,s};\end{array}\right\} \tag{6.14}$$

and for any ω,η,r,s uniformly in $t\in[g_N(s),T]$

$$\lim_{N\to\infty}\overset{=}{r}\,(\cdot,\omega,\pi_N h_n(\eta),f_N(r),g_N(s))=\overset{=}{r}\,(\cdot,\omega,h_n(\eta),r,s). \tag{6.15}$$

Clearly, $\overset{=}{r}$ has the same measurability properties as $\overset{\approx}{r}$. Moreover, setting $\tilde{\Omega}:=\bigcap_{\ell,k,j}\overset{\approx}{\Omega}_{\eta_\ell^n,r_k^N,s_j^N}$ we have $P(\tilde{\Omega})=1$, whence $\overset{=}{r}\,(\cdot,\cdot,\pi_N h_n(\tilde{\mathcal{Y}}),f_N(r_{g_N(s)}),g_N(s))$ is a solution of (4.9) (being a.s. equal to $\overset{\approx}{r}\,(\cdot,\cdot,\pi_N h_n(\tilde{\mathcal{Y}}),f_N(r_{g_N}(s)),g_N(s)))$. Next, we may suppose $E|r_{g_N(s)}|^2\to E|r_s|^2$, as $N\to\infty$. Using the usual truncation technique we may, without loss of generality, assume ess $\sup_\omega\sup_{0\le t\le T}\gamma\left(\tilde{\mathcal{Y}}(t,\omega)\right)<\infty$. We then choose a subsequence $\tilde{N}(N)\to\infty$ such that

$$E\int_0^T\gamma^2\left(\pi_{\tilde{N}}h_n\left(\tilde{\mathcal{Y}}(t)\right)-h_n\left(\tilde{\mathcal{Y}}(t)\right)\right)\mathrm{d}t\le c_T 3^{-N}. \tag{6.16}$$

The choice of $\tilde{N}(N)$ implies that a.s. uniformly on $[s,T]$

$$\overset{=}{r}\,(\cdot,\cdot,\pi_{\tilde{N}}h_n(\tilde{\mathcal{Y}}),f_{\tilde{N}}(r_{g_{\tilde{N}}(s)}),g_N(s))\to r(\cdot,h_n(\tilde{\mathcal{Y}}),r_s,s), \tag{6.17}$$

the solution of (4.9). (6.14) and (6.17) imply that a.s.

$$\overset{=}{r}\left(\cdot,\cdot,h_n(\tilde{\mathcal{Y}}),r_s,s\right)\equiv r\left(\cdot,h_n(\tilde{\mathcal{Y}}),r_s,s\right),$$

i.e., $\overset{=}{r}\left(\cdot,\cdot,h_n(\tilde{\mathcal{Y}}),r_s,s\right)$ solves (4.9).

(vi) We now proceed as before. By (4.17) and the Borel–Cantelli lemma we obtain

$$\bar{r} : 0 \le s \le t \le T \times CS_N([0,T];\mathbf{M}) \times \mathbf{R}^d \times [0,t] \to \mathbf{R}^d$$

and sets $\overline{\Omega}_{\tilde{\eta},r,s}$ with $P\left(\overline{\Omega}_{\tilde{\eta},r,s}\right) = 1$ such that uniformly in $t \in [g_N(s), T]$

$$\bar{r}(t,\omega,h_N(\eta), f_N(r), g_N(s)) := \begin{Bmatrix} \bar{\bar{r}}\,(t,\omega,h_N(\eta), f_N(r), g_N(s)), & \text{if } \omega \in \overline{\Omega}_{\eta,r,s}, \\ 0, \ \text{if } \omega \notin \overline{\Omega}_{\eta,r,s}; & \end{Bmatrix} \tag{6.18}$$

and for any ω, η, r, s uniformly in $t \in [g_N(s), T]$

$$\lim_{N\to\infty} \bar{r}(\cdot,\omega,h_N(\eta), f_N(r), g_N(s)) \equiv \bar{r}(\cdot,\omega,\eta,r,s). \tag{6.19}$$

This implies the $\overline{\mathcal{G}}_t \otimes \hat{\mathcal{F}}_{\mathbf{M},t} \otimes \mathcal{B}^d \otimes \mathcal{B}_{[0,t]} - \mathcal{B}^d$-measurability of $\bar{r}(t,\cdot,\cdot,\cdot,\cdot)$ and, by a simple modification of the previous steps, also statement 3)(i). Further, as before, $\bar{r}(t,\cdot,h_N(\tilde{\mathcal{Y}}), f_N(r_{g_N(s)}), g_N(s))$ is a solution of (4.9).
Again, by (4.17) and the Borel-Cantelli lemma, we obtain that $\bar{r}(t,\cdot,\tilde{\mathcal{Y}},r_s,s)$ is a solution of (4.9). □

6.2 Smoothness of the Flow

Let $(\mathbf{B}, \|\cdot\|_B)$ be some Banach space with norm $\|\cdot\|_B$. A $\mathbf{B}$-valued random field is by definition a $\mathcal{B}^d \otimes \mathcal{F} - \mathcal{B}_B$-measurable map from $\mathbf{R}^d \times \Omega$ into $\mathbf{B}$, where $\mathcal{B}_\mathbf{B}$ is the Borel σ-algebra on $\mathbf{B}$.

Definition 6.6. *Let $p \ge 1$. A $\mathbf{B}$-valued random field is said to be an element of $L_{p,loc}(\mathbf{R}^d \times \Omega; \mathbf{B})$, if for any $L < \infty$*

$$E\int_{|q|\le L} \|f(q)\|_B^p \, dq < \infty. \qquad \square$$

Let $\bar{r}(\cdot,q) := \bar{r}(\cdot,\tilde{\mathcal{Y}},s,q,w)$ be the measurable version of the solution of (4.9) with $\bar{r}(s,q) = q$, as derived in Theorem 4.5 under the Hypothesis 1.

Lemma 6.3.

(I) For any $p \ge 2$ and $T > s$, $\bar{r}(\cdot,\cdot) \in L_{p,\mathrm{loc}}(\mathbf{R}^d \times \Omega; C([s,T];\mathbf{R}^d)$.
(II) Let $q, \tilde{q} \in \mathbf{R}^d$. Then, for any $p \ge 2$ and $T > s$

$$E\sup_{s\le t\le T} |\bar{r}(t,q) - \bar{r}(t,\tilde{q})|^p \le c_{T,F,\mathcal{J},p}|q-\tilde{q}|^p. \tag{6.20}$$

Proof.

(i) Theorem 4.5 implies that $\bar{r}(\cdot,\cdot)$ is $C([s,T];\mathbf{R}^d)$-valued random field for any $T > s$. The integrability with respect to (q,ω) follows from the boundedness and integrability assumptions on the coefficients F and $\mathcal{J}$.

(ii) (6.20) follows from our assumptions and the Cauchy-Schwarz and Burkholder-Davis-Gundy inequalities.[7] □

Endow $\mathcal{M}_{d\times d}$, the space of $d \times d$ matrices over $\mathbf{R}$, with the Euclidean metric of $\mathbf{R}^{d^2}$ and let δ denote the Fréchet derivative.

Lemma 6.4. *Suppose the coefficients of (4.9) satisfy condition (4.46) with $m \geq 1$ in addition to condition (4.11). Then, $\bar{r}(t,\cdot)$ is at least once differentiable with respect to q_k in the generalized sense, $k = 1, \ldots, d$, and for any $p \geq 2$, $T > s$*

$$\varphi(\cdot,\cdot) := \frac{\partial}{\partial r}\bar{r}(\cdot,\cdot) \in L_{p,\mathrm{loc}}(\mathbf{R}^d \times \Omega; C([s,T]; \mathcal{M}_{d\times d})). \tag{6.21}$$

Moreover, for any $q \in \mathbf{R}^d$, $\varphi(\cdot,q)$ is the Itó-solution of the following bilinear $\mathcal{M}_{d\times d}$-valued SODE:

$$\mathrm{d}\varphi(t,q) = \left\{ \begin{array}{l} (\delta F)(\bar{r}(t,q),t)\varphi(t,q)\mathrm{d}t + \sum_{\ell=1}^{d} \int (\delta \mathcal{J}_\ell)(\bar{r}(t,q),p,t)\varphi(t,q)w_\ell(\mathrm{d}p,\mathrm{d}t) \\ \varphi(s,q) = I_d \text{ for all } q \in \mathbf{R}^d, \end{array} \right\} \tag{6.22}$$

where I_d is the identity matrix in $\mathcal{M}_{d\times d}$.

Proof. We adjust the proof of Theorem 12, Chap. 4, of Gihman and Skorohod (1982) to our setting.

(i) Assumption (4.46) ($m \geq 1$) implies that for each $q \in \mathbf{R}^d$ (6.22) has a unique solution $\varphi(\cdot,q) \in C([s,T];\mathcal{M}_{d\times d})$ a.s. Moreover, for each $p \geq 2$ there is a finite $c_{T,F,\mathcal{J},p,d}$ such that

$$\sup_{q\in\mathbf{R}^d} E \sup_{s\leq t\leq T} |\varphi(t,q)|^p \leq c_{T,F,\mathcal{J},p,d}. \tag{6.23}$$

(ii) First, we need some notation. For $q \in \mathbf{R}^d$ and $\triangle_j \in \mathbf{R}$ set

$$q_{+\triangle_j} := (q_1,\ldots,q_{j-1},q_j+\triangle_j,q_{j+1},\ldots,q_d)^T.$$

Now let $f : \mathbf{R}^d \to \mathbf{R}$. Set

$$\partial_{\triangle,j} f(q) = \frac{1}{\triangle_j}(f(q_{+\triangle_j}) - f(q)).$$

Next, assume f to be differentiable and define the "quotient" map for f, $Q_\triangle f : \mathbf{R}^{2d} \to \mathbf{R}^d$, as follows:

$$(r,\hat{r}) \mapsto (Q_\triangle f)_k(r,\tilde{r}) = \left\{ \begin{array}{ll} \frac{f(\tilde{r}_1,\ldots,\tilde{r}_{k-1},r_k,\ldots r_d) - f(\tilde{r}_1,\ldots,\tilde{r}_k,r_{k+1},\ldots r_d)}{r_k-\tilde{r}_k}, & \text{if } r_k \neq \tilde{r}_k \\ \partial_k f(\tilde{r}_1,\ldots,\tilde{r}_k,r_{k+1},\ldots,r_d), & \text{if } r_k = \tilde{r}_k, \end{array} \right\} \tag{6.24}$$

[7] Cf. for the latter inequality Sect. 15.2.3, Theorem 15.40.

$k = 1, \ldots, d$, where $f(\tilde{r}_1, \ldots, \tilde{r}_j, r_{j+1}, \ldots, r_d)$ means the first j coordinates are those of $\tilde{r}$ and the last $d - j$ coordinates are those of r. Moreover, if $j = 0$, then we have $f(r_1, \ldots r_d)$ and for $j = d$ we obtain $f(\tilde{r}_1, \ldots, \tilde{r}_d)$. Recalling the abbreviations (4.43), we obtain by assumption

$$(Q_\triangle \tilde{F}_i)_k(\bar{r}(u, q_{+\triangle_j}), \bar{r}(u, q), u),$$
$$(Q_\triangle \tilde{\mathcal{J}}_{i\ell})_k(\bar{r}(u, q_{+\triangle_j}), \tilde{p}, \bar{r}(u, q), \tilde{p}, u)$$

are well defined for $i, j, \ell, k = 1, \ldots d$. They are also adapted processes and (for fixed $\triangle_j$) measurable in (u, q, ω) (as a consequence of the measurability and adaptedness of $\bar{r}(\cdot, \cdot)$ with respect to (u, q, ω) and the continuous differentiability of F and $\mathcal{J}$). We must verify that these "quotients" are also "nicely" integrable. By Taylor's formula

$$\left.\begin{aligned} &(Q_\triangle \tilde{\mathcal{J}}_{i\ell})_k(\bar{r}(u, q_{+\triangle_j}), \tilde{p}, \bar{r}(u, q), \tilde{p}, u) \\ &= \partial_{k,r} \tilde{\mathcal{J}}_{i\ell}(\tilde{r}, \tilde{p}, u) + \tfrac{1}{2}\partial^2_{kk,r} \tilde{\mathcal{J}}_{i\ell}(\hat{r}, \tilde{p}, u)(\bar{r}_k(u, q_{+\triangle_j}) - \bar{r}_k(u, q)) \end{aligned}\right\} \tag{6.25}$$

where

$$\tilde{r} = (\bar{r}_1(u, q), \ldots, \bar{r}_k(u, q), \bar{r}_{k+1}(u, q_{+\triangle_j}), \ldots, \bar{r}_d(u, q_{+\triangle_j})) =: \tilde{r}(u, q, j, k)$$

and

$$\begin{aligned} \hat{r} &= (\bar{r}_1(u, q), \ldots, \bar{r}_{k-1}(u, q), \theta(\bar{r}_k(u, q_{+\triangle_j}) \\ &\quad - \bar{r}_k(u, q)) + \bar{r}_k(u, q), \bar{r}_{k+1}(u, q_{+\triangle_j}), \ldots, \bar{r}_d(u, q_{+\triangle_j})) \\ &=: \hat{r}(u, q, j, k) \end{aligned}$$

with $\theta \in (0, 1)$. By (4.46)

$$\int \left(\partial^2_{kk,r} \tilde{\mathcal{J}}_{i\ell}\right)^2 (r, \tilde{p}, u) \mathrm{d}\tilde{p}_{|r=\hat{r}(u,q,j,k)} \le d_{1,T} \tag{6.26}$$

uniformly in i, k, ℓ, j, ω, r and $u \le T$. Equation (6.26) allows us to reduce the integrability properties of $(Q_\triangle \tilde{\mathcal{J}}_{i\ell})_k(\bar{r}(u, q_{+\triangle_j}), \tilde{p}, \bar{r}(u, q), \tilde{p}, u)$ to those of $\partial_k \tilde{\mathcal{J}}_{i\ell}(\tilde{r}(u, q, j), \tilde{p})$. In particular, (6.25) and (6.26) imply

$$\left.\begin{aligned} &\int \left\{(Q_\triangle \tilde{\mathcal{J}}_{i\ell})_k(\bar{r}(u, q_{+\triangle_j}), \tilde{p}, \bar{r}(u, q), \tilde{p}, u) - \partial_k \tilde{\mathcal{J}}_{i\ell}(\tilde{r}(u, q, j), \tilde{p}, u)\right\}^2 \mathrm{d}\tilde{p} \\ &\le d_{1,T} |\bar{r}_k(u, q_{+\triangle_j}) - \bar{r}_k(u, q)|^2 \end{aligned}\right\} \tag{6.27}$$

for $i, j, k, \ell \in \{1, \ldots, d\}$, where we used the abbreviations, introduced after (6.25). With the same abbreviations, we obtain immediately

$$\left.\begin{aligned} &|(Q_\triangle \tilde{F}_i)_k(\bar{r}(u, q_{+\triangle_j}), \bar{r}(u, q), u)) - \partial_k \tilde{F}_i(\tilde{r}(u, q, j), u)| \\ &\le d_{3,T} |\bar{r}(u, q_{+\triangle_j}) - \bar{r}(u, q)|. \end{aligned}\right\} \tag{6.28}$$

(iii) Now we define the $\mathcal{M}_{d\times d}$-valued process $\eta_\triangle(t,q)$ by

$$\eta_{\triangle,i,j}(t,q) = \partial_{\triangle,j}\bar{r}_i(t,q), \tag{6.29}$$

where $\partial_{\triangle,j}$ acts on the jth component of q. Then, with $\tilde{\mathcal{J}}_{\cdot\ell}$ being the ℓth column of $\tilde{\mathcal{J}}$,

$$\left.\begin{aligned}\eta_\triangle(t,q) = I &+ \int_s^t (Q_\triangle \tilde{F})(\bar{r}(u,q_{+\triangle_j}),\bar{r}(u,q),u)\eta_\triangle(u,q)\mathrm{d}u\\ &+\sum_{\ell=1}^d \int_s^t\int (Q_\triangle\tilde{\mathcal{J}}_{\cdot\ell})(\bar{r}(u,q_{+\triangle_j}),\tilde{p},\bar{r}(u,q),\tilde{p},u)\eta_\triangle(u,q)w_\ell(\mathrm{d}\tilde{p},\mathrm{d}u),\end{aligned}\right\} \tag{6.30}$$

which may be rewritten as an integral equation associated with (6.22) plus two error terms. Setting

$$\psi_\triangle(t,q) := \varphi(t,q) - \eta_\triangle(t,q),$$

we have

$$\left.\begin{aligned}\psi_\triangle(t,q) = &\int_s^t (\delta\tilde{F})(\bar{r}(u,q),u)\psi_\triangle(u,q)\mathrm{d}q\\ &+\sum_{\ell=1}^d\int_s^t\int(\delta\tilde{\mathcal{J}}_{\cdot\ell})(\bar{r}(u,q),\tilde{p},u)\psi_\triangle(u,q)\tilde{w}_\ell(\mathrm{d}\tilde{p},\mathrm{d}u)\\ &+\int_s^t D_F(\bar{r}(u,q),\triangle,u)\mathrm{d}u\\ &+\sum_{\ell=1}^d\int_s^t\int D_{\mathcal{J}\cdot\ell}(\bar{r}(u,q),\tilde{p},\triangle,u)w_\ell(\mathrm{d}\tilde{p},\mathrm{d}u),\end{aligned}\right\} \tag{6.31}$$

where the error terms are defined by

$$\begin{aligned}D_F(\bar{r}(u,q),\triangle,u) := ((Q_\triangle\tilde{F})&(\bar{r}(u,q_{+\triangle_j}),\bar{r}(u,q),u)\\ &-\delta\tilde{F}(\tilde{r}(u,q,j),u))\eta_\triangle(u,q),\end{aligned}$$

and

$$\begin{aligned}D_{\mathcal{J}\cdot\ell}(\bar{r}(u,q),\tilde{p},\triangle,u) := ((Q_\triangle\tilde{\mathcal{J}}_{\cdot\ell})&(\bar{r}(u,q_{+\triangle_j}),\tilde{p},\bar{r}(u,q),\tilde{p},u)\eta_\triangle(u,q)\\ &-\delta\tilde{\mathcal{J}}_{\cdot\ell}(\bar{r}(u,q),\tilde{p},u))\eta_\triangle(u,q).\end{aligned}$$

(iv) By the Burkholder-Davis-Gundy inequality, in addition to the usual inequality for Riemann-Lebesgues integrals and (6.20), (6.27), and (6.28), we obtain

$$E\sup_{s\le t\le T}|\psi_\triangle(t,q)|^p \le c_{T,F,\mathcal{J},p}|(\triangle_1,\ldots,\triangle_d)^T|^p. \tag{6.32}$$

Letting $\triangle_i \longrightarrow 0$ for $i = 1, \ldots, d$ implies that

$$\varphi(\cdot, \cdot) = \frac{\partial}{\partial r}\bar{r}(\cdot, \cdot).$$

The measurability of φ with respect to q follows from the measurability of $\eta_\triangle(\cdot, \cdot)$ in q, since by (6.32) $\eta_\triangle(t, q)$ approximates $\varphi(t, q)$ uniformly in q, and so, a fortiori, also in $L_{p,\mathrm{loc}}(\mathbf{R}^d \times \Omega; C([s, T]; \mathbf{R}^d)$, where the local integrability is trivial. □

Next, we set

$$\tilde{\mathcal{J}}_M(r, p, u) := \sum_{n=1}^{M} \int \tilde{\mathcal{J}}(r, q, u)\hat{\phi}_n(q)\mathrm{d}q\hat{\phi}_n(p), \tag{6.33}$$

with $\hat{\phi}_n$ from (4.13). This implies[8]

$$\int \tilde{\mathcal{J}}_M(r, p, u)w(\mathrm{d}p, \mathrm{d}u) = \sum_{n=1}^{M} \sigma_n(r, u)\mathrm{d}\beta_n(u), \tag{6.34}$$

where $\sigma_n(r, u) := \int \mathcal{J}(r, q, u)\phi_n(q)\mathrm{d}q$. Hence, (4.9) with $\mathcal{J}_M$ instead of $\mathcal{J}$, becomes an SODE driven by finitely many Brownian motions. We see that the validity of the conditions (4.11) and (4.46) on $\mathcal{J}$ entails the validity of the same conditions for $\mathcal{J}_M$. Hence, Theorem 4.5 and Lemmas 6.3 and 6.4 hold for (4.9), if we replace $\mathcal{J}$ by $\mathcal{J}_M$. Moreover, all these results also carry over to the SODE for the backward flow (4.41) associated with $(\check{\mathcal{J}}_M, \check{r})$ and $\check{w}(\mathrm{d}p, t)$ defined by (4.42) for fixed $T > 0$, and $t \in [0, T]$. Therefore, we may, without loss of generality, focus in what follows on the forward equation. We denote the measurable version of the solutions of the forward flow, associated with $\mathcal{J}_M$, by

$$\bar{r}_M(\cdot, q) := \bar{r}_M(\cdot, \mathcal{Y}, s, q, w).$$

If $\mathbf{B}$ is some metric space, then $L_0(\Omega; \mathbf{B})$ is by definition the space of $\mathbf{B}$-valued random variables. Fix $T > s$.

Lemma 6.5. *Let $m \geq 1$ and suppose (4.46). Let $\mathbf{j}$ and $\boldsymbol{\ell}$ be multiindices with $|\mathbf{j}| \leq m$ and $|\boldsymbol{\ell}| \leq m - 1$. Then for any $p \geq 2$*

$$\partial^{\mathbf{j}}\bar{r}_M \in L_{p,\mathrm{loc}}(\mathbf{R}^d \times \Omega; C([s, T]; \mathbf{R}^d)) \tag{6.35}$$

and

$$\partial^{\boldsymbol{\ell}}\bar{r}_M \in L_0(\Omega; C(\mathbf{R}^d \times [s, T]; \mathbf{R}^d). \tag{6.36}$$

[8] Cf. (4.14), (4.15), and (4.28).

Proof. Expression (6.35) for $|\mathbf{j}| \le 1$ is a consequence of Lemma 6.4. For $|\mathbf{j}| > 1$ we can (repeatedly) apply the procedure of the proof of Lemma 6.4 to SODEs in increasing dimensions. Therefore, to prove (6.35) for $m = 2$, we consider the $\mathbf{R}^{d^2+d}$-valued SODE for $(\bar{r}, \varphi)$ and obtain a bilinear SODE for $\nabla((\bar{r}, \varphi)^T)$ on $\mathcal{M}_{(d^2+d)\times(d^2+d)}$ (∇ now being the $(d^2 + d)$-dimensional gradient). Expression (6.36) follows from a straightforward generalization of Lemma 2.1 and Propositions 2.1 and 2.2 in Chap. V of Ikeda and Watanabe (1981). □

Next, let

$$\check{r}_M(\cdot, q) = \check{r}(\cdot, \mathcal{Y}, s, q, \check{w})$$

denote the measurable version of the solutions of the backward flows associated with $\mathcal{J}_M$. The solution of (6.22) associated $\check{r}_M$ will be denoted by $\check{\varphi}_M$. We show that $\check{r}_M$ indeed defines the flow inverse to $\bar{r}_M$, following the procedure of Ikeda and Watanabe (loc. cit) and approximate $\bar{r}_M$ and $\check{r}_M$ by solutions of suitably chosen random ODEs (Wong-Zakai approximations), for which the invertibility will follow from uniqueness.

Remark 6.6. For time-independent coefficients Ikeda and Watanabe (loc. cit.) provide nice approximations of Itô SODEs by random ODEs. An analysis of the proof of the approximation theorems in Ikeda and Watanabe (Chap. V, 2, Chap. VI, 7) shows the following: No explicit use of Stratonovich SODEs is made. Instead, random ODEs with different coefficients (containing the "correction" term) approximate the Itô-SODEs. The independence of time and chance of the coefficients is not necessary – the "correction" term has the same form for time (and chance)-dependent coefficients and for those that do not depend on t or ω (cf. the following (6.37)).[9] □

We now define the coefficients for the drift of the forward random ODE, resp. of the backward random ODE:

$$\left.\begin{aligned} F_i^+(r,t) &= F_i(r,t) - \frac{1}{2}\sum_{n=1}^{M}\sum_{j,k=1}^{d} \left(\partial_k \sigma_{n,ij}(r,t)\right)\sigma_{n,kj}(r,t)) \text{ (forward)},\\ F_i^-(r,t) &= F_i(r,T-t) + \frac{1}{2}\sum_{n=1}^{M}\sum_{j,k=1}^{d} \left(\partial_k \sigma_{n,ij}(r,T-t)\right)\sigma_{n,kj}(r,T-t) \text{ (backward)}. \end{aligned}\right\} \tag{6.37}$$

Lemma 6.7. *Suppose (4.46) with $m \ge 1$. Then F_i^+ and F_i^- are globally Lipschitz with Lipschitz constants independent of t and ω.*

[9] Cf. Sect. 15.2.6.

Proof.

(i)

$$
\begin{aligned}
&|(\partial_k \sigma_{n,ij}(r,t))\sigma_{n,kj}(r,t) - (\partial_k \sigma_{n,ij}(q,t))\sigma_{n,kj}(q,t)| \\
&\le |\partial_k \sigma_{n,ij}(r,t) - \partial_k \sigma_{n,ij}(q,t)| \cdot |\sigma_{n,kj}(r,t)| \\
&\quad + |\sigma_{n,kj}(r,t) - \sigma_{n,kj}(q,t)| \cdot |\partial_k \sigma_{n,ij}(q,t)| \\
&=: A_{1,n}(t,r,q) + A_{2,n}(t,r,q).
\end{aligned}
$$

(ii)

$$
\sum_{n=1}^{M} A_{1,n}(t,r,q) \le \left(\sum_{n=1}^{\infty} \sigma_{n,kj}^2(r,t)\right)^{\frac{1}{2}} \left(\sum_{n=1}^{\infty} (\partial_k \sigma_{n,ij}(r,t) - \partial_k \sigma_{n,ij}(q,t)\Big)^2 \right)^{\frac{1}{2}}.
$$

Further,

$$
\sum_{n=1}^{\infty} \sigma_{n,kj}^2(r,t) = \int \tilde{\mathcal{J}}_{kj}^2(r,p,t)\,\mathrm{d}\hat{p} \le d_{1,T}
$$

uniformly in $t \in [s,T]$ and ω by (4.14)/(4.15). Next,

$$
\begin{aligned}
(\partial_k \sigma_{n,ij}(r,t) - \partial_k \sigma_{n,ij}(q,t))^2 &= \left(\int (\partial_k \tilde{\mathcal{J}}_{ij}(r,\tilde{p},t) - \partial_k \tilde{\mathcal{J}}_{ij}(q,\tilde{p},t))\phi_n(\tilde{p})\mathrm{d}\tilde{p} \right)^2 \\
&= \left(\sum_{\ell=1}^{d} \int_{q_\ell}^{r_\ell} \int \partial_{k\ell}^2 \tilde{\mathcal{J}}_{ij}((q,r,\alpha)_\ell, \tilde{p}, t)\phi_n(\tilde{p})\mathrm{d}\tilde{p}\,\mathrm{d}\alpha \right)^2
\end{aligned}
$$

(where $(q,r,\alpha)_\ell = (q_1,\ldots,q_{\ell-1},\alpha,r_{\ell+1},\ldots r_d)$ with the usual interpretation for $\ell = 1$ and $\ell = d$, cf. (6.24))

$$
\le d|r-q| \cdot \sum_{\ell=1}^{d} \int_{\overline{q}_\ell}^{\overline{r}_\ell} \left(\int \partial_{k\ell}^2 \tilde{\mathcal{J}}_{ij}((q,r,\alpha)_\ell, \tilde{p}, t)\phi_n(\tilde{p})d\tilde{p} \right)^2 \mathrm{d}\alpha
$$

by the Cauchy-Schwarz inequality, where $\overline{r}_\ell = r_\ell, \overline{q}_\ell = q_\ell$ if $r_\ell \ge q_\ell$ and $\overline{r}_\ell = q_\ell, \overline{q}_\ell = r_\ell$ if $q_\ell > r_\ell$). Summing up from $n = 1,\ldots,\infty$ implies

$$
\begin{aligned}
&\sum_{n=1}^{\infty} \left(\partial_k \sigma_{n,ij}(r,t) - \partial_k \sigma_{n,ij}(q,t)\right)^2 \\
&\le d|r-q| \cdot \sum_{\ell=1}^{d} \int_{\overline{q}_\ell}^{\overline{r}_\ell} \int \left(\partial_{k\ell}^2 \tilde{\mathcal{J}}_{ij}((q,r,\alpha)_\ell, \tilde{p}, t)\right)^2 \mathrm{d}\tilde{p}\,\mathrm{d}\alpha \\
&\le d_{1,T} d^2 |r-q|^2
\end{aligned}
$$

by (4.46).

Hence,

$$\sum_{n=1}^{M} \big(A_{1,n}(t,r,q)\big) \le \tilde{d}_{1,T} d|r-q|. \tag{6.38}$$

uniformly in t and ω (and M).

(iii)

$$\sum_{n=1}^{\infty} \left(\partial_k \sigma_{n,ij}\right)^2 (r,t) = \int \left(\partial_k \tilde{\mathcal{J}}_{ij}(r,\tilde{p},t)\right)^2 \mathrm{d}\tilde{p} \le d_{1,T}.$$

Repeating the derivation of (6.38) (with the obvious changes)

$$\sum_{n=1}^{M} \big|A_{2,n}(t,r,q)\big| \le \tilde{d}_{1,T} d|r-q| \tag{6.39}$$

(uniformly in t, ω and M).

(iv) Since F_i, $i = 1 \ldots d$, are globally Lipschitz by assumption, the lemma is proven. □

Lemma 6.8. *Under the assumption (4.46), $m \ge 1$, for any $T > s$ with probability 1*

$$\overline{r}_M(T-t,r) = \check{r}_M(t,\overline{r}_M(T,r)) \tag{6.40}$$

uniformly in $r \in \mathbf{R}^d$ and $t \in [s,T]$.

Proof.

(i) We take the Wong-Zakai approximations to the SODEs for $\overline{r}_M$ and $\check{r}_M$ with piecewise linear approximations $\beta_{n,L}(t)$ of β_n, $n = 1,\ldots,M$, $L \in \mathbf{N}$, where

$$\left.\begin{aligned}\beta_{n,L}(t) = L\left\{\left(\frac{j+1}{L}-t\right)\beta_n\left(\frac{j}{L}\right)+\left(t-\frac{j}{L}\right)\beta_n\left(\frac{j+1}{L}\right)\right\}\\ \text{if } \frac{j}{L} \le t \le \frac{j+1}{L}, \quad j = 0,1,\ldots.\end{aligned}\right\} \tag{6.41}$$

By Lemma 6.7, the Wong-Zakai approximations of the forward and backward flows have unique solutions $\overline{r}_{M,L}(t,r)$ and $\check{r}_{M,L}(t,r)$, respectively. Moreover, $\overline{r}_{M,L}(T-t,r)$, the solution of the approximate forward random ODE, evaluated at $T-t$, satisfies the random ODE

$$\frac{\mathrm{d}r(t)}{\mathrm{d}t} = -F^+(r(t),T-t) - \sum_{n=1}^{M} \sigma_n(r(t),T-t)\frac{\mathrm{d}\beta_{n,L}}{\mathrm{d}t}(t), \tag{6.42}$$

which is the backward random ODE for $\check{r}_{M,L}(t,\cdot)$. By uniqueness (6.40) follows, if we replace $\overline{r}_M$ and $\check{r}_M$ by $\overline{r}_{M,L}$ and $\check{r}_{M,L}$, respectively, uniformly in $r \in \mathbf{R}^d$ and $t \in [s,T]$.

(ii) Next, we check that we may (trivially) generalize Lemma 7.2 in Ikeda and Watanabe (loc. cit.), Chap. VI, to our equations with time – and ω – dependent coefficients, which implies for any multiindex $\mathbf{j}$ with $|\mathbf{j}| \le m$), any $N \in \mathbf{N}$, $p \ge 2$ and any $\tilde{r}_{M,L} \in \{\bar{r}_{M,L}, \check{r}_{M,L}\}$.

$$\sup_{|r|\le N} \sup_{L} E \sup_{s\le t\le T} |\partial^{\mathbf{j}} \tilde{r}_{M,L}(t,r)|^p < \infty \tag{6.43}$$

The same estimates (without L) hold for $\bar{r}_M(\cdot, r)$ and $\check{r}_M(\cdot, r)$ and their partial derivatives up to order m by (6.35). Hence, we truncate all involved processes using stopping times $\sigma_{k,L}$ and σ_k such that for $\mathbf{j}$ with $|\mathbf{j}| \le m$ and $(\tilde{r}_{M,L}, \tilde{r}_M) \in \{(\bar{r}_{M,L}, \bar{r}_M), (\check{r}_{M,L}, \check{r}_M)\}$

$$\sup_{s\le t\le T\wedge\sigma_{k,L}} |\partial^{\mathbf{j}} \tilde{r}_{M,L}(t,r)| + \sup_{s\le t\le T\wedge\sigma_k} |\partial^{\mathbf{j}} \tilde{r}_M(t,r)| \le K < \infty, \tag{6.44}$$

whence

$$\left.\begin{aligned}
&\sup_{s\le t\le T\wedge\sigma_{k,L}\wedge\sigma_k} |\partial^{\mathbf{j}} \tilde{r}_{M,L}(t,r) - \partial^{\mathbf{j}} \tilde{r}_M(t,r)|^p \\
&\le (2K)^{p-2} \cdot \sup_{s\le t\le T\wedge\sigma_{k,L}\wedge\sigma_k} |\partial^{\mathbf{j}} \tilde{r}_{M,L}(t,r) - \partial^{\mathbf{j}} \tilde{r}_M(t,r)|^2, \\
&\sup_{T\wedge\sigma_{k,L}\wedge\sigma_k\le t\le T} |\partial^{\mathbf{j}} \tilde{r}_{M,L}(t,r) - \partial^{\mathbf{j}} \tilde{r}_M(t,r)|^p \\
&\le \frac{2}{K} \sup_{s\le t\le T} (|\partial^{\mathbf{j}} \tilde{r}_{M,L}(t,r)| + |\partial^{\mathbf{j}} \tilde{r}_M(t,r))|)^{p+1}.
\end{aligned}\right\} \tag{6.45}$$

The estimates (6.43) and (6.45) imply that for any $N \in \mathbf{N}$[10]

$$\left.\begin{aligned}
&\sup_{|r|\le N} E \sup_{T\wedge\sigma_{k,L}\wedge\sigma_k\le t\le T} |\partial^{\mathbf{j}} \tilde{r}_{M,L}(t,r) - \partial^{\mathbf{j}} \tilde{r}_M(t,r)|^p \\
&\quad\le \frac{2}{K} \sup_{|r|\le N} E \sup_{s\le t\le T} (|\partial^{\mathbf{j}} \tilde{r}_{M,L}(t,r)| + |\partial^{\mathbf{j}} \tilde{r}_M(t,r))|)^{p+1} \\
&\quad\to 0, \text{ as } K \to \infty.
\end{aligned}\right\} \tag{6.46}$$

(iii) We check that Theorem 7.2, Ch. VI, in Ikeda and Watanabe (loc. cit.) generalizes to our equations, which implies by (6.45) and (6.46) for any multiindex $\mathbf{j}$ with $|\mathbf{j}| \le m$, $N \in \mathbf{N}$ and $p \ge 2$

$$\sup_{|r|\le N} E \sup_{s\le t\le T} |\partial^{\mathbf{j}} \tilde{r}_{M,L}(t,r) - \partial^{\mathbf{j}} \tilde{r}_M(t,r)|^p \to 0, \text{ as } L \to \infty, \tag{6.47}$$

which, of course, implies for any $N \in \mathbf{N}$, $|\mathbf{j}| \le m$ and $p \ge 2$

$$E \int_{|r|\le N} \sup_{s\le t\le T} |\partial^{\mathbf{j}} \tilde{r}_{M,L}(t,r) - \partial^{\mathbf{j}} \tilde{r}_M(t,r)|^p \, dr \to 0, \text{ as } L \to \infty. \tag{6.48}$$

[10] Cf. Ikeda and Watanabe (loc. cit.), Chap. V, Proof of Lemma 2.1 therein.

(iv) Choose $p > \mathrm{d} \vee 2$ (where "$\vee$" denotes "maximum"). Then (6.48) implies by the Sobolev imbedding theorem[11] that for any $N \in \mathbf{N}$ and $|\mathbf{j}| \le m-1$

$$E \sup_{|r|\le N} \sup_{s\le t\le T} |\partial^{\mathbf{j}}\tilde{r}_{M,L}(t,r) - \partial^{\mathbf{j}}\tilde{r}_M(t,r)| \to 0, \text{ as } L \to \infty. \tag{6.49}$$

(v) (6.40) follows now from steps (i)–(iv) and (6.49). □

Lemma 6.9. *Assume (4.46), $m \ge 1$, and let $p \ge 2$. Then, for any $(\tilde{r}_M, \tilde{r}) \in \{(\bar{r}_M, \bar{r}), (\check{r}_M, \check{r})\}$ and any $L < \infty$*

$$\sup_{|q|\le L} E \sup_{s\le t\le T} |\tilde{r}_M(t,q) - \tilde{r}(t,q)|^p \to \infty, \text{ as } M \to \infty. \tag{6.50}$$

Proof.

(i) Assume, without loss of generality, $(\tilde{r}_M, \tilde{r}) = (\bar{r}_M, \bar{r})$. By the Burkholder-Davis-Gundy and Hölder inequalities

$$\left.\begin{aligned}
&E \sup_{s\le t\le T} |\bar{r}(t,q) - \bar{r}_M(t,q)|^p\\
&\le c_{T,p}\Bigg\{\int_s^T E|F(\bar{r}(u,q),u) - F(\bar{r}_M(u,q),u)|^p\,\mathrm{d}u\\
&\qquad + \int_s^T E\left\{\sum_{n=1}^M |\sigma_n(\bar{r}(u,q),u) - \sigma_n(\bar{r}_M(u,q),u)|^2\right\}^{\frac{p}{2}} \mathrm{d}u\\
&\qquad + \int_s^T E\left\{\sum_{n=M+1}^\infty |\sigma_n(\bar{r}(u,q),u)|^2\right\}^{\frac{p}{2}} \mathrm{d}u\Bigg\}\\
&=: c_{T,p}\left\{\sum_{i=1}^3 A_{i,M}(T)\right\}.
\end{aligned}\right\} \tag{6.51}$$

(ii)

$$\begin{aligned}
A_{2,M}(T) &\le \int_s^T E\left\{\sum_{n=1}^\infty |\sigma_n(\bar{r}(u,q),u) - \sigma_n(\bar{r}_M(u,q),u)|^2\right\}^{\frac{p}{2}} \mathrm{d}u\\
&\le c_{\mathcal{J},p}\int_s^T E|\bar{r}(u,q) - \bar{r}_M(u,q)|^p\,\mathrm{d}u
\end{aligned}$$

by assumption (4.11) on $\tilde{\mathcal{J}}(r,p,u)$.

(iii) Since $\sum_{n=1}^\infty |\sigma_n(\bar{r}(u,q),u)|^2 = \sum_{k,\ell=1}^d \int \mathcal{J}_{k\ell}^2(\bar{r}(u,q),\tilde{p},u)^2\,\mathrm{d}\tilde{p} < \infty$ uniformly in u, ω, and q the monotone convergence theorem implies that for any $\varepsilon > 0$ there is an $M(\varepsilon)$ such that for all $M \ge M(\varepsilon)$

$$A_{3,M}(T) \le \varepsilon.$$

[11] Cf., e.g., Triebel (1978), Chap. 4.6.1.

(iv)

$$A_{1,M}(T) \leq c_{F,p} \int_s^T E|\overline{r}(u,q) - \overline{r}_M(u,q)|^p \, du$$

by the Lipschitz condition on F.

(v) Hence, for all $M \geq M(\varepsilon)$

$$E \sup_{s \leq t \leq T} |\overline{r}(t,q) - \overline{r}_M(t,q)|^p \leq \tilde{c}_{T,F,\mathcal{J},p}\varepsilon. \tag{6.52}$$

□

Let $\overline{\varphi}(t,q)$ and $\check{\varphi}(t,q)$ be the solutions of (6.22) associated with the forward flow $\overline{r}(t,q)$ and backward flow $\check{r}(t,q)$, respectively.

Lemma 6.10. *Assume (4.46), $m \geq 1$, and let $p \geq 2$. Then for any $(\tilde{\varphi}_M, \tilde{\varphi}) \in \{(\overline{\varphi}_M, \overline{\varphi}), (\check{\varphi}_M, \check{\varphi})\}$ and any $L < \infty$*

$$\sup_{|q| \leq L} E \sup_{s \leq t \leq T} |\tilde{\varphi}_M(t,q) - \tilde{\varphi}(t,q)|^p \to \infty, \text{ as } M \to \infty. \tag{6.53}$$

Proof.

(i) Assume without loss of generality $(\tilde{\varphi}_M, \tilde{\varphi}) = (\overline{\varphi}_M, \overline{\varphi})$. Set $\psi_M := \overline{\varphi} - \overline{\varphi}_M$. Then

$$\begin{aligned}
\psi_M(t,q) = & \int_s^t \delta\tilde{F}(\overline{r}(u,q),u)\psi_M(u,q)\mathrm{d}u \\
& + \sum_{\ell=1}^d \int_s^t \int \delta\tilde{\mathcal{J}}_{M,\cdot\ell}(\overline{r}(u,q),\tilde{p},u)\psi_M(u,q) w_\ell(\mathrm{d}\tilde{p},\mathrm{d}u) \\
& + \int_s^t (\delta\tilde{F}(\overline{r}_M(u,q),u) - \delta\tilde{F}(\overline{r}(u,q),u))(\psi_M(u,q) - \overline{\varphi}(u,q))\mathrm{d}q \\
& + \sum_{\ell=1}^d \int_s^t \int (\delta\tilde{\mathcal{J}}_{M,\cdot\ell}(\overline{r}_M(u,q),\tilde{p},u) \\
& - \delta\tilde{\mathcal{J}}_{M,\cdot\ell}(\overline{r}(u,q),\tilde{p},u))(\psi_M(u,q) - \overline{\varphi}(u,q)) w_\ell(\mathrm{d}\tilde{p},\mathrm{d}u) \\
& + \sum_{\ell=1}^d \sum_{n=M+1}^\infty \int_s^t \delta\sigma_{n,\cdot\ell}(\overline{r}(u,q),u)\mathrm{d}\beta_{n,\ell}(u) \\
= & \sum_{i=1}^5 B_{i,M}(t).
\end{aligned}$$

(ii) By the assumptions on $\mathcal{J}$ we obtain as in the previous proof of the previous lemma (step iv) (cf. also the proof of Lemma 6.7) that for any $p \geq 2$ and $\varepsilon > 0$

$$\sup_q E \sup_{s \leq t \leq T} |B_{5,M}(t,q)|^p \leq \varepsilon$$

provided $M \geq M(\varepsilon)$.

(iii) We easily see that for any $p \geq 2$ and $L < \infty$

$$\sup_M \sup_{|q| \leq L} E \sup_{s \leq t \leq T} |\tilde{\varphi}_M(t,q)|^p < \infty. \tag{6.54}$$

Therefore, the boundedness and continuity of all coefficients in addition to Lemma 6.9 imply (6.53). □

Corollary 6.11. *Assume (4.46) with $m \geq 1$. Then, for any $(\tilde{r}_M, \tilde{r}) \in \{(\bar{r}_M, \bar{r}), (\check{r}_M, \check{r})\}$, $L < \infty$ and $|\mathbf{j}| \leq m - 1$*

$$E \sup_{|q| \leq L} \sup_{s \leq t \leq T} |\partial^{\mathbf{j}} \tilde{r}_M(t,q) - \partial^{\mathbf{j}} \tilde{r}(t,q)| \to 0, \text{ as } M \to \infty. \tag{6.55}$$

Moreover, for any multiindex $\mathbf{j}$ with $|\mathbf{j}| \leq m - 1$

$$\partial^{\mathbf{j}} \tilde{r} \in L_0(\Omega; C(\mathbf{R}^d \times [s,T]; \mathbf{R}^d)). \tag{6.56}$$

Proof. Expression (6.55) follows in the same way as (6.49). Expression (6.56) follows from (6.55) and (6.36). □

Chapter 7
Comments on SODEs: A Comparison with Other Approaches

7.1 Preliminaries and a Comparison with Kunita's Model[1]

We briefly review Kunita's generalization of the classical Itô SODEs, driven by Brownian noise. Correlation functionals for Gaussian random fields are shown to have a representation as convolutions of certain kernels with Gaussian space–time white noise (Proposition 7.1). Under a global Lipschitz assumption, it follows that Kunita's SODEs for Brownian flows are a special case of our SODEs (4.10).

In view of (4.15), we may interpret the present definition of SODEs as Itô SODEs driven by infinitely many i.i.d. Brownian motions. SODEs, driven by finitely many i.i.d. Brownian motions have been the standard Brownian model in stochastic analysis for a number of decades.[2] In addition, the driving noise in SODEs have been generalized to martingales and semimartingales.[3]

A significant step beyond this standard model was accomplished by Kunita (1990). Kunita[4] defines $C(\mathbf{R}^d; \mathbf{R}^{\mathbf{k}})$-valued Brownian motion $S(\cdot, \cdot)$ as a continuous independent increment process. If we evaluate $S(\cdot, \cdot)$ at the spatial points $r_1, \ldots, r_N$ we obtain an $\mathbf{R}^{Nk}$-valued independent increment process $(S(r_1, \cdot), \ldots, S(r_N, \cdot))$ with continuous sample paths. This is equivalent to the statement that $(S(r_1, \cdot), \ldots, S(r_N, \cdot))$ is a Gaussian process with independent increments.[5] If we, in addition, assume that the process has mean 0 and stationary increments, then the process is an $\mathbf{R}^{Nk}$-valued "Brownian motion" in the traditional sense.[6] The

[1] The comparison with Kunita's approach is taken from the appendix in Kotelenez and Kurtz (2006).

[2] Cf., e.g., Gihman and Skorohod (1968, 1982), Arnold (1973), or Ikeda and Watanabe (1981).

[3] In addition to some of the previously mentioned monographs, cf., e.g., Metivier and Pellaumail (1980) and Protter (2004) and the references therein.

[4] Kunita (1990), (loc.cit.), Sect. 3.1.

[5] Cf. Gikhman and Skorokhod (1971), Chap. III.5, Theorem 5.

[6] Note that the covariance may not be $I_{Nk}t$ in our definition of Brownian motion, where I_{Nk} is the identity matrix on $\mathbf{R}^{Nk}$. If the covariance matrix is $I_{Nk}t$, the Brownian motion is called "*standard*." Cf. also Sects. 15.2.2 and 15.2.4.

existence of a $C(\mathbf{R}^d; \mathbf{R}^\mathbf{k})$-valued Brownian motion $S(\cdot, \cdot)$ is equivalent to the existence of a Gaussian random field with mean

$$E[S(r,t)] = \int_0^t b(r,u)\mathrm{d}u, \quad \mathrm{Cov}(S(r,t), S(q,s)) = \int_0^{t\wedge s} a(r,q,u)\mathrm{d}u, \quad (7.1)$$

subject the assumptions that $b(r,u)$ is jointly continuous in all variables and Hölder continuous in r and $a(r,q,u)$ is jointly continuous in all variables and Hölder continuous in (r,q).[7] The proof of the existence is based on Kolmogorov's existence theorem and the existence of a continuous version follows from Kolmogorov's continuity theorem.[8] Kunita calls $b(r,u)$ and $a(r,q,u)$ the "*infinitesimal mean*" and "*infinitesimal covariance*" of the random field, respectively. Following Kunita's notation we define the Lipschitz norms:

$$\left.\begin{aligned} \|b(t)\|_{0,1} &:= \sup_{r\neq q} \frac{|b(r,t)-b(q,t)|}{|r-q|} \\ \|a(t)\|_{0,1} &:= \sup_{r\neq q, \tilde{r}\neq\tilde{q}} \frac{|a(r,\tilde{r},t)-a(q,\tilde{r},t)-a(r,\tilde{q},t)+a(q,\tilde{q}.t)|}{|r-q||\tilde{r}-\tilde{q}|}, \end{aligned}\right\} \quad (7.2)$$

Under the additional assumptions

$$\left.\begin{aligned} &|b(r,t)| \le c(1+|r|), \quad \|a(r,q,t)\| \le c(1+|r|)(1+|q|) \ \text{(linear growth)}, \\ &\sup_{t\ge 0}[\|b(t)\|_{0,1} + \|a(t)\|_{0,1}] \le c \quad \text{(globally Lipschitz)}, \end{aligned}\right\} \quad (7.3)$$

Kunita[9] constructs a "*forward Brownian flow*" $\varphi_{s,t}(r)$ as the unique solution of a stochastic ordinary differential equation driven by the Gaussian field (or random function) $S(r,t)$:

$$\varphi_{s,t}(r) = r + \int_s^t S(\varphi_{s,u}(r), \mathrm{d}u). \quad (7.4)$$

where in (7.4) the stochastic integral must be interpreted in the sense of Itô. To better understand the formalism involved, let us look at a special case of such a Gaussian space–time random field:

$$\tilde{S}(r,t) := \int_0^t F(r,u)\mathrm{d}u + \int_0^t \sum_{i=1}^N \sigma_i^\perp(r,u)\mathrm{d}\beta^{\perp,i}(u),$$

where $k = d$ and all coefficients and the Brownian motions are as in (4.10), except they do not depend on the empirical distribution of the solutions of (4.10) (and

[7] Cf. Kunita (loc.cit., Sect. 4.2).

[8] Cf. Gikhman and Skorokhod, (1971), Sect. III.1 and Sect. III.5, where the Gaussian field is called "Gaussian random function," for a proof. Cf. also Ibragimov (1983) for a generalization of Kolmogorov's continuity theorem.

[9] Kunita (1990), (loc.cit.), Sect. 4.2, Theorems 4.2.5 and 4.2.8.

$\mathcal{J} \equiv 0$). In this case, (7.4) becomes exactly a special case of (4.10 (with $\mathcal{J} \equiv 0$), and the Itô differential $S(r(u,s,q), \mathrm{d}u)$ in (4.10) is the usual (Itô) semimartingale differential

$$\tilde{S}(r(u,s,q), \mathrm{d}u) = F(r(u,s,q),u)\mathrm{d}u + \sum_{n=1}^{N} \sigma_i^{\perp}(r(u,s,q),u)\mathrm{d}\beta^{\perp,i}(u).$$

Let us now assume that the coefficients F and $\mathcal{J}$ in our equation (4.10) depend only on the position r and the time t, which we call the "bilinear case."[10] Further, assume $\sigma_i^{\perp}(\cdot,\cdot) \equiv 0 \; \forall i$. Then

$$\hat{S}(r,t) := \int_0^t F(r,u)\mathrm{d}u + \int_0^t \int \mathcal{J}(r,p,u)w(\mathrm{d}p,\mathrm{d}u) \tag{7.5}$$

is a Gaussian space–time random field (or Gaussian random function) with continuous sample paths. Further, (4.10) becomes the stochastic integral equation

$$r(t,s,r) = r + \int_s^t F(r(u,s,r),u)\mathrm{d}u + \int_s^t \int \mathcal{J}(r(u,s,r),p,u)w(\mathrm{d}p,\mathrm{d}u), \tag{7.6}$$

which may be written as

$$r(t,s,r) = r + \int_s^t \hat{S}(r(u,s,r),\mathrm{d}u). \tag{7.7}$$

We obtain that

$$\left.\begin{aligned} \hat{b}(r,t) &:= F(r,t) \\ \hat{a}(r,q,t) &:= \int (\mathcal{J}(r,p,t)\mathcal{J}^T(q,p,t))\,\mathrm{d}p \end{aligned}\right\} \tag{7.8}$$

are the infinitesimal mean and infinitesimal covariance, respectively.

Our linear growth assumption (4.12) is similar to Kunita's Condition (A.2),[11] except we do not assume the coefficient to be continuous in t. Since we solve our SODEs on a finite time interval, our assumption of uniform boundedness in t is the same as in Kunita's Theorem 4.2.5, and it may be easily removed by working with localizing stopping times. Further, let us comment on the Lipschitz assumption in Kunita.[12] Since this is an assumption on the components of the matrices involved we may, without loss of generality, assume that the coefficients are real valued. Our assumption (4.12) then implies in the terminology of Kunita that

$$|a(r,r,t) + a(q,q,t) - 2a(r,q,t)| \le c|r-q|^2, \tag{7.9}$$

[10] The term "bilinear" refers to the fact that the associated SPDE for the empirical distribution will be bilinear as long as the coefficients of the underlying SODE do not depend on the empirical distribution of the solutions (cf. Chap. 8, (8.26)). The SODE (7.6) needs, of course, not be bilinear.

[11] Kunita (1990), (loc.cit.), Sect. 4.2.

[12] Kunita (1990), (loc.cit.), Theorem 4.2.5.

which is apparently weaker than Kunita's continuity assumption on $a(r, q, t)$ in addition to the global Lipschitz assumption (7.3).

It remains to show that each Gaussian random field (random function), considered by Kunita may be represented as in (7.5). Since a Gaussian random field (random function) is uniquely determined by its mean and covariance, we have just to show that the class of infinitesimal means and covariances from (7.8) contains those considered by Kunita.

The case for the mean is trivial. Therefore, without loss of generality, we may assume that the mean is 0. Let us fix t and first consider the case of real-valued random fields. Set $\mathbf{B} := C(\mathbf{R}^d; \mathbf{R})$ and endow $\mathbf{B}$ with the topology of uniform convergence on bounded sets. Let $\mathcal{B}$ be the Borel σ-algebra on $\mathbf{B}$. Given $a(r, q, t)$ from (7.1), Kolmogorov's theorem implies that the Gaussian random field with covariance $a(r, q, t)$ has sample paths in $C(\mathbf{R}^{d+1}; \mathbf{R})$.[13] Hence, for fixed t, the Gauss measure μ on the space of functions from $\mathbf{R}^d$ with values in $\mathbf{R}$ is supported by $\mathbf{B}$, endowed with $\mathcal{B}$.[14] Thus, we have the existence of a (continuous) Gaussian random field $\xi(r, t)$ such that

$$\forall t \quad a(r, q, t) = E(\xi(r, t)\xi(q, t)) = \int_{\mathbf{B}} (b(r, t)b(q, t))\mu(\mathrm{d}b(\cdot, t). \tag{7.10}$$

Observe that $\mathbf{B}$ is a Polish space. Hence, any measure on $(\mathbf{B}, \mathcal{B})$ is a Radon measure.[15] Therefore, by Corollary 3.2.7 in Chap. 3.2 of Bogachev (1997), $L_2(\mathbf{B}, \mathcal{B}, \mathrm{d}\mu)$ is separable. The separability of $L_2(\mathbf{B}, \mathcal{B}, \mathbf{d}\mu)$ implies that

$$L_2(\mathbf{B}, \mathcal{B}, \mathrm{d}\mu) \cong L_2(\mathbf{R}^d, \mathcal{B}^d, \mathrm{d}r) \quad \text{(isometrically isomorphic)}\,. \tag{7.11}$$

Note for the special case of a separable Banach space $\mathbf{B}$, the separability of $L_2(\mathbf{B}, \mathcal{B}, \mathrm{d}\mu)$ follows from the Itô–Wiener–Chaos decomposition (by construction a countable complete orthonormal system of eigenfunctions for $L_2(\mathbf{B}_N, \mathcal{B}_N, du_N)$.[16] The relation (7.11) immediately generalizes to $\mathbf{R}^d$-valued Gaussian random fields. Thus, there are Borel measurable $\mathbf{R}^d$ valued functions $\mathcal{J}(r, p)$, which are Borel measurable in p and continuous in r such that[17]

$$\forall t, r, q \quad a(r, q, t) = E(\xi(r, t)\xi^T(q, t)) = \int_{\mathbf{R}^d} \mathcal{J}(r, p, t)\mathcal{J}^T(q, p, t)\mathrm{d}p. \tag{7.12}$$

If $\mathcal{J}(r, p, t) = \mathcal{J}(r - p, t)$, i.e., if the kernel is spatially homogeneous, then also $a(r, q, t) = a(r - q, t)$. For spatially homogeneous kernels $\mathcal{J}(r, t)$ is called the

---

[13] Cf. Ibragimov (1983).

[14] It is actually supported by an even smaller space of more regular functions. Recall the classical Wiener space and the regularity of the standard Brownian motion expressed by Levy's modulus of continuity.

[15] Cf. Bourbaki (1977), Chap. IX, Sect. 3.3.

[16] Cf. Watanabe (1984), Sect. 1.2. Cf. also Gikhman and Skorokhod (1971), Chap. VIII, and for newer results on the Malliavian calculus, Üstünel (1995) and the references therein.

[17] The observation, leading to (7.12), is essentially due to Dorogovtsev (2004a).

convolution root of $a(r,t)$.[18] We extend this term also to the inhomogeneous case and show that there are "many" spatially homogeneous $a(r,t)$ whose convolution root is necessarily (spatially) inhomogeneous.[19]

Since Gaussian fields are uniquely determined by their mean and their correlation operator we have

Proposition 7.1.

$$S(\cdot,\cdot) \sim \int_0^{\cdot} F(\cdot,u)\mathrm{d}u + \int_0^{\cdot}\int \mathcal{J}(\cdot,p,u)w(\mathrm{d}p,\mathrm{d}u). \tag{7.13}$$

□

Hence, (7.11) allows us to represent Kunita's Gaussian random fields directly as the sum of a deterministic integral and a stochastic integral, where the latter is driven by a standard Gaussian space–time white noise, as in our Chaps. 2 and 4. Recalling Kunita's stronger regularity assumptions on the mean and covariance, it follows that the Gaussian random fields that drive the stochastic differential equations of Kunita (loc.cit.) are essentially a special case of the fields given in (7.5), which themselves are a special case of the fields, driving the SODEs in Chap. 4.

Remark 7.2. Suppose that the covariance of the Gaussian space–time random field is given by the right-hand side of (7.12). By Theorem 1 of Gikhman and Skorokhod (loc.cit.), Chap. 4.5, there is an orthogonal random measure $\hat{w}(\mathrm{d}p,\mathrm{d}t)$ with the Lebesgue measure as its "structure measure." $\hat{w}(\mathrm{d}p,\mathrm{d}t)$ is measurable with respect to the random variables generated by the linear combinations of $S(r_i,t_i), i = 1,\ldots,N, N\in\mathbf{N}$, such that $S(t,x)$ has the following representation:

$$S(\cdot,\cdot) = \int_0^{\cdot} F(\cdot,u)\mathrm{d}u + \int_0^{\cdot}\int \mathcal{J}(\cdot,p,u)\hat{w}(\mathrm{d}p,\mathrm{d}u) \text{ a.s..} \tag{7.14}$$

Since $S(r,t)$ is Gaussian, the random measure $\hat{w}(\mathrm{d}p,\mathrm{d}t)$ is Gaussian and, in view of the orthogonality, it follows that $\hat{w}(\mathrm{d}p,\mathrm{d}t)$ is (another) standard Gaussian space–time white noise. □

Remark 7.3. Let us specialize the analysis to the case of homogeneous mean zero Gaussian fields. Since the vector-valued case can be easily derived from the analysis of scalar fields and t can be kept fixed, we assume that, without loss of generality, the correlation functionals are real valued and independent of t. For $f\in C(\mathbf{R}^d;\mathbf{R})$ define the norm:[20]

$$|||f|||_L := |||f||| + \sup_{r\neq 0}\frac{|f(0)-f(r)|}{|r|^2} \tag{7.15}$$

and set

$$C_L(\mathbf{R}^d;\mathbf{R}) := \{f\in C(\mathbf{R}^d;\mathbf{R}) : |||f|||_L < \infty\}. \tag{7.16}$$

---

[18] Cf. Ehm et al. (2004) and the references therein.

[19] Cf. Remark 7.3 and Example 7.6.

[20] Cf. (7.9).

We verify that class of real-valued homogeneous and autonomous a from (7.1), satisfying(7.2)/(7.3), has the following properties:

$$\left.\begin{array}{ll}\text{(i)} & a \in C_L(\mathbf{R}^d; \mathbf{R});\\ \text{(ii)} & a(0) \geq |a(r)| \ \ \forall r;\\ \text{(iii)} & a(r) = a(-r) \ \ \forall r.\end{array}\right\} \tag{7.17}$$

We denote the class of functions with properties (i)–(iii) from (7.17) by $\tilde{C}_L$. Next, let $\Gamma \in \mathbf{H}_0$. Then,

$$\int \Gamma(r-p)\Gamma(-p)\mathrm{d}p \in C_0(\mathbf{R}^d; \mathbf{R}), \tag{7.18}$$

i.e., in the space of continuous functions that vanish at infinity.[21] Let us call the mapping from $\mathbf{H}_0$ into $C_0(\mathbf{R}^d; \mathbf{R})$ an s-convolution (since for the convolution we would have to evaluate the second integrand at p rather than at $-p$). Thus, our result implies that there are "many" homogeneous correlation functionals a resulting from spatially inhomogeneous s-convolution-type products of functions $\mathcal{J}(r, p)$. □

7.2 Examples of Correlation Functions

Three examples of correlation functionals are analyzed.

Before giving examples convolution roots, related to Fourier transforms, let us prove a simple fact about odd integrable functions f.

Proposition 7.4. *Let $f \in L_1(\mathbf{R}^d; \mathrm{d}r)$ such that f is odd, i.e., $f(p) = -f(-p)$ a.s. Then*

$$\int f(p)\mathrm{d}q = 0. \tag{7.19}$$

Proof. For $d = 1$ this is trivial. If $d > 1$ we divide $\mathbf{R}^d$ into 2^d infinite squares as follows: each of the squares has as one-dimensional boundaries either $\{-\infty, 0\}$ or $\{0, \infty\}$ and we add linear manifolds of dimension less than d, i.e., where at least one of the coordinates equals 0 instead of an interval for the boundary of the squares. Choose one of the squares and call it B_+. Then there is exactly one other square, called B_- such that $p \in B_+$ if and only if $-p \in B_-$. Note that by change of variables $p \leftrightarrow -p$ in addition to the oddness of f

$$\int_{B_-} f(p)\mathrm{d}p = -\int_{B_+} f(p)\mathrm{d}p.$$

Since the integration over the linear manifolds of dimension less than d equals 0, we obtain from the preceding equation

[21] Cf., e.g., Folland (1984), Chap. 8.2, (8.8) Proposition.

$$\int f(p)\mathrm{d}p = \sum_B \left[\int_{B_+} f(p)\mathrm{d}q + \int_{B_-} f(p)\mathrm{d}p \right] = 0.$$

□

Example 7.5. Let $\varphi(p) \in L_1(\mathbf{R}^d, \mathrm{d}r)$ an even and nonnegative function. Set

$$a(r) := \int \cos(r \cdot p)\varphi(p)\mathrm{d}p, \tag{7.20}$$

i.e., $a(r)$ is the Fourier transform of a finite measure with a density. We verify that $a(\cdot) \in \tilde{C}_L$.[22] Further, evoking the Riemann-Lebesgue Theorem[23], we obtain that $a(\cdot) \in \tilde{C}_L \cap C_0(\mathbf{R}^d; \mathbf{R})$. Set

$$\Gamma(r,p) := [\cos(r \cdot p) + \sin(r \cdot p)]\sqrt{\varphi(p)}. \tag{7.21}$$

Since

$$\cos(r \cdot p)\cos(q \cdot p) = \tfrac{1}{2}[\cos((r-q)\cdot p) + \cos((r+q)\cdot p)];$$
$$\sin(r \cdot p)\sin(q \cdot p) = \tfrac{1}{2}[\cos((r-q)\cdot p) - \cos((r+q)\cdot p)]$$

we obtain

$$\begin{aligned} &[\cos(r \cdot p) + \sin(r \cdot p)][\cos(q \cdot p) + \sin(q \cdot p)] \\ &= [\cos((r-q)\cdot p) + \cos(r \cdot p)\sin(q \cdot p) + \sin(r \cdot p)\cos(q \cdot p) \end{aligned}$$

The last two functions are odd in p. Since $\varphi(\cdot)$ is even and integrable it follows from Proposition 7.4 that

$$\int [\cos(r \cdot p)\sin(q \cdot p) + \sin(r \cdot p)\cos(q \cdot p)]\varphi(p)\mathrm{d}p = 0.$$

So,

$$\int \Gamma(r,p)\Gamma(q,p)\mathrm{d}p = \int \cos((r-q)\cdot p)\varphi(p)\mathrm{d}p = a(r-q). \tag{7.22}$$

□

Example 7.6. Set

$$a(r) := 1 + \cos(r \cdot b), \quad r, b \in \mathbf{R}^d, b \neq 0. \tag{7.23}$$

Obviously, $a(r) \notin C_0(\mathbf{R}^d; \mathbf{R})$. Further,

$$a(r) = \int \exp(ir \cdot \lambda)\left[\frac{1}{2}\delta_{-b}(\mathrm{d}\lambda) + \delta_0(\mathrm{d}\lambda) + \frac{1}{2}\delta_b(\mathrm{d}\lambda)\right], \tag{7.24}$$

i.e., $a(r)$ is the Fourier transform of the symmetric spectral measure

---

[22] $\tilde{C}_L$ has been defined by (7.17).

[23] Cf. Bauer (1968), Chap. VIII, Sect. 48.

$$\mu(\mathrm{d}\lambda) := 2\pi \left[\frac{1}{2}\delta_{-b}(\mathrm{d}\lambda) + \delta_0(\mathrm{d}\lambda) + \frac{1}{2}\delta_b(\mathrm{d}\lambda) \right]. \tag{7.25}$$

Hence $a(r)$ defines a correlation functional of a homogeneous random field.[24] As before we employ the trigonometric identity

$$\cos((r-q)\cdot b) = \cos(r\cdot b)\cos(q\cdot b) + \sin(r\cdot b)\sin(q\cdot b) \tag{7.26}$$

and note that

$$\begin{aligned} \int & [\cos(r\cdot\lambda)\sin(q\cdot\lambda) + \sin(r\cdot\lambda)\cos(q\cdot\lambda)] \\ & \times \left[\frac{1}{2}\delta_{-b}(\mathrm{d}\lambda) + \delta_0(\mathrm{d}\lambda) + \frac{1}{2}\delta_b(\mathrm{d}\lambda) \right] = 0, \end{aligned} \tag{7.27}$$

because the integrand is odd and the measure is symmetric. We obtain that

$$\begin{aligned} a(r-q) = \int & [\cos(r\cdot\lambda) + \sin(r\cdot\lambda)]\,[\cos(q\cdot\lambda) + \sin(q\cdot\lambda)] \\ & \times \left[\frac{1}{2}\delta_{-b}(\mathrm{d}\lambda) + \delta_0(\mathrm{d}\lambda) + \frac{1}{2}\delta_b(\mathrm{d}\lambda) \right], \end{aligned} \tag{7.28}$$

Equation (7.28) represents $a(r-q)$ as an s-convolution with respect to the spectral measure. Our goal, however, is to represent $a(r-q)$ by an s-convolution with respect to the Lebesgue measure, as in (7.8). Abbreviate

$$\begin{aligned} C_1(0) &:= \left\{ p \in \mathbf{R}^d : -\tfrac{1}{2} \le p_i \le \tfrac{1}{2},\ i = 1,\ldots,d \right\}, \\ B_+ &: \left\{ p \in \mathbf{R}^d : p_i \ge \tfrac{1}{2}\ \forall i \right\}, \\ B_- &: \left\{ p \in \mathbf{R}^d : p_i \le -\tfrac{1}{2}\ \forall i \right\}, \end{aligned} .$$

Let $h(p) \ge 0$ be an even and integrable function such that

$$h(p) := \left\{ \begin{array}{l} 1,\ \text{if } p \in C_1(0), \\ 0,\ \text{ if there are } i, j \text{such that } -\dfrac{1}{2} \le p_i \le \dfrac{1}{2} \text{ and } |p_j| > \dfrac{1}{2}, \\ \ge 0\ ,\ \text{if} p \in B_+ \cup B_- \text{ and such that } \displaystyle\int_{B_+} h(p)\mathrm{d}p = \frac{1}{2} = \int_{B_-} h(p)\mathrm{d}p. \end{array} \right\} \tag{7.29}$$

Define an odd function $\phi(p)$ by

$$\phi(p) := \begin{cases} -1, & \text{if } \quad p \in B_-, \\ 1, & \text{if } \quad p \in B_+, \\ 0, & \text{otherwise.} \end{cases} \tag{7.30}$$

[24] Cf. Yaglom (1957) or Gel'fand and Vilenkin (1964).

Set

$$\Gamma(r,p) := [\cos(r\cdot b\phi(p)) + \sin(r\cdot b\phi(p))]\sqrt{h(p)}. \tag{7.31}$$

As $\phi(p) = 0\ \forall p \in C_1(0)$, we have $\Gamma(r,p) = 1\ \forall r$ and $\forall p \in C_1(0)$. It follows that

$$\int_{C_1(0)} \Gamma(r,p)\Gamma(q,p)\mathrm{d}p = \int_{C_1(0)} h(p)\mathrm{d}p = 1.$$

Further,

$$\begin{aligned}
&\int_{B_-} \Gamma(r,p)\Gamma(q,p)\mathrm{d}p \\
&\quad = [\cos(-r\cdot b) + \sin(-r\cdot b)]\,[\cos(-q\cdot b) + \sin(-q\cdot b)] \int_{B_-} h(p)\mathrm{d}p \\
&\quad = \frac{1}{2}\,[\cos(-r\cdot b) + \sin(-r\cdot b)]\,[\cos(-q\cdot b) + \sin(-q\cdot b)]\,,
\end{aligned}$$

$$\begin{aligned}
&\int_{B_+} \Gamma(r,p)\Gamma(q,p)\mathrm{d}p \\
&\quad = [\cos(r\cdot b) + \sin(r\cdot b)]\,[\cos(q\cdot b) + \sin(q\cdot b)] \int_{B_+} h(p)\mathrm{d}p \\
&\quad = \frac{1}{2}\,[\cos(r\cdot b) + \sin(r\cdot b)]\,[\cos(q\cdot b) + \sin(q\cdot b)]
\end{aligned}$$

and

$$\int_{\mathbf{R}^d \setminus (B_+ \cup B_- \cup C_1(0))} \Gamma(r,p)\Gamma(q,p)\mathrm{d}p = 0.$$

Adding up these results and employing the trigonometric identity (7.26) yields

$$a(r-q) = \int \Gamma(r,p)\Gamma(q,p)\mathrm{d}p, \tag{7.32}$$

i.e., $a(r-q)$ has representation (7.8) with inhomogeneous kernel $\Gamma(r,p)$. □

Example 7.7. For $d = 1$ we may take $b = 1$ and obtain

$$a(x) := 1 + \cos(x). \tag{7.33}$$

Since $a(x) = a(|x|)$ the associated process is also isotropic. □

Part III
Mesoscopic B: Stochastic Partial Differential Equations

Chapter 8
Stochastic Partial Differential Equations: Finite Mass and Extensions

8.1 Preliminaries

The variational and the semigroup approaches are reviewed and the particle approach is introduced. In conclusion, we define continuum and macroscopic limits.

Generalizing Pardoux's variational approach to SPDEs,[1] Krylov and Rozovsky (1979) consider a class of SPDEs of the following type

$$\left.\begin{aligned} \mathrm{d}X &= A(t,X)\mathrm{d}t + B(t,X)\mathrm{d}W + \mathrm{d}Z, \\ X(0) &= X_0. \end{aligned}\right\} \tag{8.1}$$

$A(t, X)$ is an unbounded closed operator on some Hilbert space $\mathbf{H}$. Further, in the variational frame-work there are two Banach spaces $\mathbf{B}$ and $\mathbf{B}'$ such that

$$\mathbf{B} \subset \mathbf{H} = \mathbf{H}' \subset \mathbf{B}' \tag{8.2}$$

with dense continuous imbeddings, where "$'$" denotes the (strong) dual space. The spaces $\mathbf{B}$ and $\mathbf{B}'$ are chosen such that $A(t, \cdot)$ may be considered a bounded operator from $\mathbf{B}$ into $\mathbf{B}'$. $W(\cdot)$ is a $\mathbf{K}$-valued regular Brownian motion, i.e., its covariance operator, Q_W, is nuclear where $\mathbf{K}$ is another (separable) Hilbert space.[2] $B(t, X)$ is a linear operator from $\mathbf{K}$ into a class of unbounded operators on $\mathbf{H}$. In the most important examples $A(t, X) = A(t)X$ is a linear parabolic second-order operator (which may depend on ω), and the unboundedness of $B(t, X)$ is of the order of a first-order linear operator. $Z(t, \omega)$ is just a state independent $\mathbf{H}$-valued process. We do not wish to present all the specific details and assumptions, except for the "coercivity."[3] This is, by definition, the assumption that

$$\langle 2A(t,X), X\rangle + \mathrm{Trace}\{\mathrm{B(t,X)Q_W B'(t,X)}\} \le -\gamma_1\|\mathrm{X}\|^2_{\mathbf{B}} + \gamma_2\|\mathrm{X}\|^2_{\mathbf{H}} + \mathrm{const.} \tag{8.3}$$

[1] Cf. Pardoux (1975).

[2] Cf. Sect. 15.2.2.

[3] Krylov and Rozovsky (1979) call condition (8.3) "coercivity of the pair."

Here, $\langle \cdot, \cdot \rangle$ is the duality between $\mathbf{B}$ and $\mathbf{B}'$, extending the scalar product $\langle \cdot, \cdot \rangle_{\mathbf{H}}$ on $\mathbf{H}$. γ_1 and γ_2 are positive constants, and the norms in (8.3) are the norms on $\mathbf{B}$ and $\mathbf{H}$, respectively. We call $\int_0^t \mathrm{Trace\,B(s,X)Q_W B'(s,X)ds}$ the trace of the *tensor quadratic variation* of the martingale noise.[4] The key in this assumption is the term $-\gamma_1 \|X\|_{\mathbf{B}}^2$ which allows a smoothing result for the solutions, i.e., starting in $\mathbf{H}$ the solution may be in $\mathbf{B}$ for $t > 0$. This assumption is satisfied if, e.g., $A = \Delta$ and $BX\mathrm{d}W = \nabla X \cdot \mathrm{d}W$, where $W(\cdot) = (W_1(\cdot), \ldots, W_d(\cdot))^{\mathrm{T}}$ and the components $W_i(\cdot)$ are $\mathbf{H}$-valued regular Brownian motions, i.e., whose covariance operators, Q_{W_i}, are nuclear (cf. Definition 15.1 of Chap. 15). An important consequence of (8.3) is that it satisfies the assumptions of the Zakai equation.[5] Consider first the deterministic case, i.e., replace the stochastic term by a deterministic perturbation $f(\cdot) \in L_2([0,T];\mathbf{B}')$ and suppose that $\mathbf{B}$ is itself a Hilbert space. Under these assumptions, Lions and Magenes[6] show that the solution $X(\cdot)$ of the corresponding parabolic problem with $X(0) = \mathbf{0}$ lives in $L_2([0,T];\mathbf{B}) \cap C([0,T];\mathbf{H})$. Therefore, the action of the operator is smoothing the solution. The SPDE (8.1) is obtained by replacing the deterministic perturbation by the martingale term $\int_0^t B(s,X)\mathrm{d}W$. Since the additive term Z in (8.1) is just an "external" perturbation, suppose, in what follows, $Z(\cdot) \equiv 0$. To explain the notion of a solution in the variational context of Pardoux and of Krylov and Rozovsky we introduce the following notation, similar to DaPrato and Zabczyk (1992):

Notation

$L_{p,\mathcal{F}}([s,T] \times \Omega; \mathbf{B})$, $p \in [1,\infty) \cup \{0\}$, is the set of all $Y(\cdot,\cdot) : [s,T] \times \Omega \to \mathbf{B}$, such that $Y(\cdot,\cdot)$ is $\mathrm{d}t \otimes \mathrm{d}P$-measurable, adapted to $\mathcal{F}_t$, and p-integrable with respect to $\mathrm{d}t \otimes \mathrm{d}P$ on $[s,T] \times \Omega$.[7]

If a solution of (8.1) exists for $X_0 \in L_{2,\mathcal{F}_0}(\mathbf{H})$, then Pardoux (loc. cit.) as well as Krylov and Rozovsky (loc. cit.) show the analog to the result of Lions and Magenes, namely that

$$X(\cdot, X_0) \in L_{0,\mathcal{F}}(C([0,T];\mathbf{H})) \cap L_{2,\mathcal{F}}([0,T] \times \Omega; \mathbf{B}), \tag{8.4}$$

where we chose $p = 2$. Strictly speaking, (8.4) means that the solution $X(\cdot, X_0)$ has two versions, one with continuous paths in the space $\mathbf{H}$ and the other with merely measurable paths in the smoother space $\mathbf{B}$. It is important to note that the $\mathbf{B}$-valued version of this solution is smoother than the initial condition.

Returning to the deterministic framework of Lions and Magenes (loc. cit.), it follows in the time invariant case that A generates an analytic semigroup $T(t)$ (both on $\mathbf{B}$ and on $\mathbf{H}$).[8] Further, Tanabe (1979), Chap. 5, provides conditions which

[4] Cf. Sects. 15.2.3 and 15.2.4.

[5] Expression (13.23) in Chap. 13.2. Cf. Zakai (1969) and Pardoux (1979).

[6] Cf. Lions and Magenes (1972), Chap. 3, Sect. 4, (4.3), (4.4), Theorem 4.1.

[7] For $p = 0$, we define the pth moment to be equal to $\int_s^T E(\|Y(u)\|_{\mathbf{B}} \wedge 1)\mathrm{d}u$.

[8] Cf. Tanabe (1979), Chap. 3.6, Theorem 3.6.1.

guarantee the existence of a two-parameter group, $U(t,s)$, generated by $A(t)$.[9] Then, employing "variation of constants," (8.1) is transformed into the (stochastic) integral equation

$$X(t) = U(t,0)X_0 + \int_0^t U(t,s)X(s)\mathrm{d}s + \int_0^t U(t,s)B(s,X(s))\mathrm{d}W(s) + \int_0^t U(t,s)\mathrm{d}Z(s). \tag{8.5}$$

A solution of (8.5) is called a "mild solution" of (8.1). Assuming coercivity of the pair, A being independent of t and a Lipschitz condition on B and a suitable Z, DaPrato (1982) obtains the unique mild solution of (8.5) for $X_0 \in L_{2,\mathcal{F}_0}(\mathbf{H})$.

The Itô formula provides the simplest key to understanding the coercivity condition (8.3). Suppose, for the time being, that (8.5) has a strong solution in $\mathbf{H}$ ("strong" in the sense of PDEs), i.e., a solution which is in the domain of $A(\cdot)$, considered as an unbounded operator on $\mathbf{H}$. By the Itô formula[10]

$$\begin{aligned}\|X(t)\|_{\mathbf{H}}^2 = \|X(0)\|_{\mathbf{H}}^2 + \int_0^t \{\langle 2A(s,X(s)), X(s)\rangle_{\mathbf{H}} \\ + \mathrm{Trace}\{B(s,X(s))Q_W B^*(s,X(s))\}\}ds \\ + \text{(mean zero) martingale}(t).\end{aligned} \tag{8.6}$$

Employing the a priori estimate (8.3) and taking mathematical expectations,

$$\begin{aligned}E\|X(t)\|_{\mathbf{H}}^2 \le E\|X(0)\|_{\mathbf{H}}^2 + \int_0^t \left[-\gamma_1 E\|X(s)\|_{\mathbf{B}}^2 + \gamma_2 E\|X(s)\|_{\mathbf{H}}^2\right]\mathrm{d}s + t(\text{const.}) \\ \Longleftrightarrow \int_0^t E\|X(s)\|_{\mathbf{B}}^2 \le \frac{1}{\gamma_1}\left\{E\|X(0)\|_{\mathbf{H}}^2 + \gamma_2\int_0^t E\|X(s)\|_{\mathbf{H}}^2\mathrm{d}s \right. \\ \left. - E\|X(t)\|_{\mathbf{H}}^2 + t(\text{const.})\right\},\end{aligned} \tag{8.7}$$

i.e., we obtain an estimate of the solution in the norm $\|\cdot\|_{\mathbf{B}}$ in terms of the solution of the weaker norm $\|\cdot\|_{\mathbf{H}}$. In the variational approach (for $p = 2$), the notion of a strong solution is weakened to the assumption that $X(\cdot) \in L_{2,\mathcal{F}}([0,T]\times\Omega;\mathbf{B})$. Accordingly, the Itô formula must be proved under those weakened assumptions to provide a priori estimates.[11]

To illustrate our particle approach to SPDEs, let us present a simple example from 2D fluid mechanics.[12] Assume first that the fluid is ideal. Set

$$g(|r|) := \frac{1}{2\pi}\log|r|, \quad \nabla^{\perp} := \left(-\frac{\partial}{\partial r_2}, \frac{\partial}{\partial r_2}\right)^{\mathrm{T}}.$$

[9] Cf. Sect. 15.2.3, Definition 15.41, and Sect. 15.2.7, Proposition 15.64.

[10] Cf. Metivier and Pellaumail (1980), Chap. 2.3.7.

[11] Cf. Krylov and Rozovsky (loc. cit.), Theorem 3.1.

[12] Cf. Marchioro and Pulvirenti (1982).

Further, for $\eta > 0$, let g_η be a smooth approximation of g if $\eta \to 0$ and set

$$K_\eta(r) := \nabla^\perp g_\eta(|r|).$$

Consider the system of N coupled ordinary differential equations

$$\left.\begin{aligned} \frac{\mathrm{d}}{\mathrm{d}t} r^i(t) &= \sum_{j=1}^N a_j K_\eta(r^i - r^j) = \int K_\eta(r^i - q)\mathcal{Y}_{\eta,N}(t, \mathrm{d}q), \\ r^i(0) &= r_0^i, \ \ i = 1, \ldots, N, \ \ \mathcal{Y}_{\eta,N}(t) = \sum_{i=1}^N a_i \delta_{r^i(t)}. \end{aligned}\right\} \tag{8.8}$$

where $a_j \in \mathbf{R}$ are the (positive and negative) intensities of the rotational motion of the fluid and $\mathcal{Y}_{\eta,N}(t)$ is the empirical process, generated by the solutions of the ODEs.

It is well known[13] that $\mathcal{Y}_\eta(t)$ is an approximate solution of the (first-order quasilinear) Euler equation

$$\left.\begin{aligned} \frac{\partial}{\partial t} Y(t,r) &= -\nabla(Y(t,r) \int \nabla^\perp g(|r-q|) Y(t,q)\mathrm{d}q), \\ Y(0,r) &= Y_0(r), \end{aligned}\right\} \tag{8.9}$$

provided the initial empirical measure $\sum_{i=1}^N a_i \delta_{r^i(0)} \approx Y_0(r)$.

Next, assume that the fluid is viscous with viscosity coefficient $\nu > 0$. In this case, the Euler equation is replaced by a (semilinear second-order) Navier–Stokes equation for the vorticity distribution:

$$\left.\begin{aligned} \frac{\partial}{\partial t} X(t,r) &= \nu \triangle X(t,r) - \nabla(X(t,r) \int \nabla^\perp g(|r-q|) X(t,q)\mathrm{d}q), \\ X(0,r) &= X_0(r). \end{aligned}\right\} \tag{8.10}$$

A natural question is how to generalize the approximation of (8.9) through (8.8) to the viscous case. Chorin[14] replaces the system of ODEs (8.8) by the system of SODEs:

$$\left.\begin{aligned} \mathrm{d}r^i(t) &= \int K_\eta(r^i - q)\mathcal{X}_{\eta,N}(t, \mathrm{d}q)\mathrm{d}t + \sqrt{2\nu}\beta^{\perp,i}(\mathrm{d}t), \\ r^i(0) &= r_0^i, \ \ i = 1, \ldots, N, \ \ \mathcal{X}_{\eta,N}(t) = \sum_{i=1}^N a_i \delta_{r^i(t)}, \end{aligned}\right\} \tag{8.11}$$

[13] Cf., e.g., Marchioro and Pulvirenti (loc. cit.).

[14] Cf. Chorin (1973) and the references therein.

where $\{\beta^{\perp,i}\}$ are $\mathbf{R}^2$-valued i.i.d. standard Brownian motions (cf. (4.10)). Marchioro and Pulvirenti (loc. cit.) obtain in a rigorous limit theorem

$$\begin{aligned}&\mathcal{X}_{\eta,N}(\cdot) \longrightarrow X(\cdot), \text{ as } N \longrightarrow \infty \text{ and } \eta = \eta(N) \to 0\\ &\text{(suitably coupled with the limit behaviors of } N)\end{aligned} \tag{8.12}$$

in a weak topology, provided that the initial measures $\mathcal{X}_{\eta,N}(0)$ converge to X_0. We call a limit of the type (8.12), where the limit is the solution of a deterministic partial differential equation, a "macroscopic limit."

The use of ODEs and SODEs in 2D fluid mechanics to approximately solve the Euler equation and the Navier–Stokes equation, respectively, is known as the *point vortex* method.

Besides the obvious difference of approximations of a first-order and a second-order PDE,[15] there is another feature which the approximations $\mathcal{Y}_{\eta,N}$ and $\mathcal{X}_{\eta,N}$ do not share. Let φ be a smooth test function on $\mathbf{R}^2$ with compact support. Let $\langle\cdot,\cdot\rangle$ denote the duality between smooth functions and distributions, extending the L_2-scalar product $\langle\cdot,\cdot\rangle_0$.[16] The chain rule and integration against the empirical process implies

$$\begin{aligned}&\frac{\mathrm{d}}{\mathrm{d}t}\langle\mathcal{Y}_{\eta,N}(t),\varphi\rangle = \left\langle\mathcal{Y}_{\eta,N}(t), \nabla\varphi\cdot\int K_\eta(\cdot-q)\mathcal{Y}_{\eta,N}(t,\mathrm{d}q)\right\rangle, \text{ which is equivalent to}\\ &\frac{\partial}{\partial t}\mathcal{Y}_{\eta,N}(t) = -\nabla\cdot(\mathcal{Y}_{\eta,N}(t)\textstyle\int K_\eta(\cdot-q)\mathcal{Y}_{\eta,N}(t,\mathrm{d}q)),\end{aligned} \tag{8.13}$$

i.e., the empirical process $\mathcal{Y}_{\eta,N}$ is a weak solution of an approximation of the Euler-type equation (cf. (8.9)). Unfortunately, the empirical process $\mathcal{X}_{\eta,N}$ is not a weak solution of a Navier–Stokes type equation. The method fails as a result of the choice of different Brownian motions $\beta^{\perp,i}$ as perturbations of different point vortices (or particles). Therefore, Kotelenez (1995a) replaces in (8.11) the independent Brownian motions by correlated Brownian motions $\int G_\varepsilon(r^i-q)w(\mathrm{d}q,\mathrm{d}t)$, assuming $\int G_\varepsilon(r-q)G_\varepsilon^{\mathrm{T}}(r-q)\mathrm{d}q = I_d$ $(d=2)$. Following this procedure, we obtain a system of SODEs driven by correlated Brownian motions

$$\left.\begin{aligned}&\mathrm{d}r^i(t) = \sum_{j=1}^N a_j K_\eta(r_\varepsilon^i - r_\varepsilon^j)\mathrm{d}t + \sqrt{2\nu}\int G_\varepsilon(r_\varepsilon^i - q)w(\mathrm{d}q,\mathrm{d}t),\\ &r_\varepsilon^i(0) = r_0^i,\ \ i=1,\ldots,N,\ \ \mathcal{X}_{\eta,\varepsilon,N}(t) = \sum_{i=1}^N a_i\delta_{r_\varepsilon^i(t)}.\end{aligned}\right\} \tag{8.14}$$

We may now apply the Itô formula (i.e., the extended chain rule) and obtain, similarly to (8.13), an SPDE as an approximation to the Navier–Stokes equation (8.10).

[15] The following parabolic second-order SPDE (8.14) can be written as a first-order transport SPDE, if we replace the Itô differential $w(\mathrm{d}q,\mathrm{d}t)$ by the Stratonovich differential $w(\mathrm{d}q,\circ\mathrm{d}t)$. Cf. Theorem 8.25 in Sect. 8.6.

[16] Here $\mathbf{H}_0$ denotes $L_2(\mathbf{R}^2,\mathrm{d}r)$.

$$
\left.
\begin{aligned}
\mathrm{d}\langle \mathcal{X}_{\eta,\varepsilon,N}(t),\varphi\rangle &= \Big[\Big\langle \mathcal{X}_{\eta,\varepsilon,N}(t), \nu\triangle\varphi + \nabla\varphi\cdot\int K_\eta(\cdot - q)\mathcal{X}_{\eta,\varepsilon,N}(t,\mathrm{d}q)\Big\rangle\Big]\mathrm{d}t\\
&+\Big\langle \mathcal{X}_{\eta,\varepsilon,N}(t), \nabla\varphi\cdot\sqrt{2\nu}\int G_\varepsilon(\cdot - q)w(\mathrm{d}q,\mathrm{d}t)\Big\rangle, \text{ which is equivalent to}\\
\mathrm{d}\mathcal{X}_{\eta,\varepsilon,N}(t) &= \nu\triangle\mathcal{X}_{\eta,\varepsilon,N}(t) - \nabla\cdot(\mathcal{X}_{\eta,\varepsilon,N}(t)\int K_\eta(\cdot - q)\mathcal{X}_{\eta,\varepsilon,N}(t,\mathrm{d}q))\\
&-\nabla\cdot(\mathcal{X}_{\eta,\varepsilon,N}(t)\sqrt{2\nu}\int G_\varepsilon(\cdot - q)w(\mathrm{d}q,\mathrm{d}t)).
\end{aligned}
\right. \tag{8.15}
$$

Further, it is shown in Kotelenez (1995a) that for fixed $\varepsilon > 0$, fixed $\eta > 0$ and $N \longrightarrow \infty$, the empirical processes $\mathcal{X}_{\eta,\varepsilon,N}(t,\mathrm{d}q))$ tend toward a smooth solution of (8.15), provided the initial distribution tends to a smooth initial distribution. Here and in what follows, "smooth" solution means that the solution has a density with respect to the Lebesgue measure, and that the density itself is a solution of (8.15).[17] This is in contrast to the macroscopic behavior of the empirical process $\mathcal{X}_{\eta,N}(t,\mathrm{d}q)$ of (8.12).

The example from 2D fluid mechanics may be generalized to a system of diffusing and interacting particles where, for most purposes, the real-valued intensities are replaced by nonnegative weights, the masses of the individual particles. Similarly to the drift in (8.11), the diffusion may also depend on the empirical process, which leads to quasilinear PDEs or SPDEs. If the Brownian motions are chosen uncorrelated as in (8.11), macroscopic limits have been obtained by numerous authors.[18] Note that in those macroscopic limit theorems, the original particle system is governed by a system of interacting ODEs (as in (8.8)). The diffusion is added such that each particle is driven by an independent standard Brownian motion (times a diffusion coefficient $\mathcal{J}_0(r,\mu,\ldots) \in \mathcal{M}_{d\times d}$). Let us use the same notation as in (8.8)–(8.12) for the relevant quantities but omit, without loss of generality, the dependence on the smoothing parameter η (which is not needed in most of the models mentioned above). Choose a twice continuously differentiable function φ on $\mathbf{R}^d$ and denote now by $\langle\cdot,\cdot\rangle$ the duality between measures and continuous functions. Under nontrivial assumptions on the coefficients, Itô's formula yields the following (incremental) quadratic variations[19] for the duality between test functions and the empirical process, associated with the interacting and diffusion particle system:

$$
\left.
\begin{aligned}
\mathrm{d}[\langle\mathcal{X}_{0.N}(t),\varphi\rangle] &= \sum_{i=1}^{N}\frac{1}{N^2}\sum_{k,\ell=1}^{d}(\partial_k\varphi)(r^i_{0,N}(t))(\partial_\ell\varphi)(r^i_{0,N}(t))\\
&\qquad D_{0,k,\ell}(r^i_{0,N}(t),r^i_{0,N}(t),\mathcal{X}_{0,N}(t),t)\mathrm{d}t\\
&= O_\varphi\left(\frac{1}{N}\right)\mathrm{d}t,
\end{aligned}
\right\} \tag{8.16}
$$

[17] Cf. also Theorem 8.6.

[18] Cf. Oelschläger (1984, 1985), Gärtner (1988), and the references therein.

[19] Cf. Sects. 15.2.3 and 15.2.4. We may consider the left-hand side of (8.16), the tensor quadratic variation of $\mathcal{X}_{0.N}(t)$, in weak form.

where

$$D_0(r^i, r^j, \mu, t) := \begin{cases} \mathcal{J}_0(r^i, \mu, t)\mathcal{J}_0^T(r^j, \mu, t), & \text{if} \quad i = j, \\ \mathbf{0}, & \text{if} \quad i \neq j. \end{cases}$$

Here, $\mathbf{0} \in \mathcal{M}_{d\times d}$ is the matrix with all entries being equal to 0 and the subscript "0" indicates that the correlation length equals 0, i.e., the processes are driven by uncorrelated Brownian motions.[20] Clearly, we have N terms in the sum (8.16), divided by N^2, which suggests a generalization of Chebyshev inequality estimates in the law of large numbers for independent random variables with suitable moment conditions. Thus, one can show that $\mathcal{X}_{0,N}(\cdot) \Longrightarrow X(\cdot)$, as $N \longrightarrow \infty$, where now $X(\cdot)$ is the solution of a quasilinear parabolic partial differential equation (PDE) of McKean–Vlasov type.[21]

We return now to the main theme of this chapter, namely the derivation of SPDEs from particle distributions, generalizing (8.14)/(8.15). The stochastic Navier–Stokes equation (8.15) is generalized to a class of quasilinear SPDEs by the author in Kotelenez (1995b), where the quasilinearity appears if, as in (4.10), the diffusion kernel $\mathcal{J}_\varepsilon(r, q, \mathcal{X}_{\varepsilon,N}(\cdot))$ depends also on the empirical process. Suppose that $\sigma_i^\perp \equiv 0\ \forall i$ in (4.10). Then the motion of the particles in (4.10) is perturbed only by correlated Brownian motions, where we now explicitly use the subscript ε for the correlation length. Choose, as in (8.16), a twice continuously differentiable function φ on $\mathbf{R}^d$. Itô's formula yields the following (incremental) quadratic variations for duality between test functions and the empirical process $X_{\varepsilon,N}(t)$, associated with the interacting and diffusion particle system:

$$\left.\begin{aligned} \mathrm{d}[\langle \mathcal{X}_{\varepsilon,N}(t), \varphi\rangle] &= \sum_{i,j=1}^{N} \frac{1}{N^2} \sum_{k,\ell=1}^{\mathrm{d}} (\partial_k\varphi)(r^i_{\varepsilon,N}(t))(\partial_\ell\varphi)(r^j_{\varepsilon,N}(t)) \\ &\qquad D_{\varepsilon,i,j,k,\ell}(r^i_{\varepsilon,N}, r^j_{\varepsilon,N}, \mathcal{X}_{\varepsilon,N}(t), t)\mathrm{d}t \\ &= O_{\varepsilon,\varphi}(1)\,\mathrm{d}t, \end{aligned}\right\} \tag{8.17}$$

where

$$D_\varepsilon(r^i, r^j, \mu, t) := \int \mathcal{J}_\varepsilon(r^i_\varepsilon, q, \mu, t)\mathcal{J}_\varepsilon^T(r^j_\varepsilon, q, \mu, t)\mathrm{d}q \quad \forall\, i, j = 1, \ldots, N.$$

In difference from (8.16), we now have N^2 terms in the sum (8.17), divided by N^2, and the noise does not disappear in the limit as $N \longrightarrow \infty$. As a consequence, the limit $N \longrightarrow \infty$ is a stochastic space–time field and solves a corresponding SPDE.[22] Observe that the limit $N \longrightarrow \infty$ is, most of the time, performed to obtain a continuum model. In the case of (8.17), this limit is an SPDE which can be solved

[20] Cf. Chap. 14, in particular, (14.1)–(14.3).

[21] Cf. (14.5)/(14.6) in Chap. 14 of this book as well as Oelschläger (1984, 1985) and Gärtner (1988). Cf. also (8.72).

[22] Vaillancourt (1988) obtains similar SPDEs in the limit $N \longrightarrow \infty$ for a somewhat different choice of the diffusion.

on the space of densities (with respect to the Lebesgue measure, Theorems 8.3 and 8.6). Therefore, we will call the limit $N \longrightarrow \infty$ a "continuum limit." Summarizing the previous observations, the continuum limit of the empirical process (provided such a limit exists) is the solution of a *macroscopic* PDE (i.e., deterministic), if each particle is perturbed by its own independent Brownian motion (times diffusion coefficients). The continuum limit, however, is the solution of a *mesoscopic* SPDE, if the particle system is perturbed by correlated Brownian motions. Returning to the problem of macroscopic behavior of the particle distribution, we mention that by our Theorem 14.2 for the case of nonnegative masses we obtain a macroscopic limit as $\varepsilon \longrightarrow 0$.

Summarizing, we may conclude that the use of correlated Brownian motions allows us to perform the *transition to a macroscopic model* in two steps:

1. We first perform a *continuum limit*, which is described by an SPDE, defined on a suitable space of densities. This SPDE is a PDE plus a stochastic term, representing the interaction of the (large) particles with the medium of small particles.[23] Further, the stochastic term contains the correlation length $\varepsilon > 0$ as a parameter, and this parameter "measures" the distance to a macroscopic (i.e., deterministic) PDE.
2. We then perform a *macroscopic limit* as $\varepsilon \longrightarrow 0$. In this limit the solution of the SPDE tends to the solution of a PDE.

In view of our results of Chap. 5, the above procedure allows us to work with the SPDE as a more realistic continuum model, where the stochastic perturbation is small. This perturbation, derived from the interaction of small and large particles, describes in the continuum model the error which we commit when simplifying a discrete microscopic model of atoms and molecules by a continuum model for their distribution. We will refrain here from a discussion of whether or not point vortices in viscous fluid should be driven by uncorrelated or correlated Brownian motions. As we shall see later in Remark 8.13, the SPDEs obtained for the empirical processes of (4.19) are noncoercive, if we assume $\sigma_i^{\perp} \equiv 0 \; \forall i$ in (4.10) (cf. also Theorem 8.25 and the relation between the stochastic Navier–Stokes equation, driven by Itô differentials, and a stochastic Euler-type equation, driven by Stratonovich differentials).

Kurtz and Xiong (1999) extend the author's particle approach to SPDEs to include quasilinear SPDEs of coercive type.[24] This approach combines (8.11) and (8.14) into one SODE (as in our (4.9) or (4.10)), and Kurtz and Xiong obtain an SPDE in the continuum limit as $N \longrightarrow \infty$. Using this generalization, Kurtz and Xiong (2001) solve the Zakai equation by particle methods. In Sect. 8.4, we analyze a special case of the Kurtz and Xiong result, namely where the mass is conserved.[25]

---

[23] We recall that this interaction is observable and deterministic on a microscopic level (cf. Chaps. 1 and 2).

[24] Cf. Remark 8.14.

[25] For other approaches and results on SPDEs, we refer the reader to Chap. 13.

8.2 A Priori Estimates

The general conditions of Theorem 4.5 on initial values and input processes will be assumed throughout the rest of this section. A priori estimates for measure valued SPDEs are derived, based on Hypothesis 4.1.

We assume that the initial mass is finite in the a priori estimates for the empirical processes for (4.9) and (4.10) . As in Chap. 4, we suppress a possible dependence of the kernels $\mathcal{J}$ on the correlation length parameter ε for the rest of this section. Set

$$\mathbf{M}_{f,d} := \{\mu \in \mathbf{M}_f : \mu \text{ is finite sum of point measures}\}$$

and

$$L_{0,\mathcal{F}_s}(\mathbf{M}_{f,d}) := \left\{ \mathcal{X}_s : \mathcal{X}_s = \mathcal{X}_{s,N} := \sum_{i=1}^{N} m_i \delta_{r_s^i} \right. \\ \left. \text{with weights and initial positions adapted, } N \in \mathbf{N} \right\}. \tag{8.18}$$

In what follows, $\mu_{s,1}$ and $\mu_{s,2} \in L_{0,\mathcal{F}_s}(\mathbf{M}_{f,d})$ are the initial distributions of the particles, described in Theorem 4.5 and $\tilde{\mathcal{Y}}, \tilde{\mathcal{Y}}_i \in L_{loc,2,\mathcal{F}}(C((s,T];\mathbf{M}_f))$, $i = 1,2$ be arbitrary and fixed.

Lemma 8.1. *Assume Hypothesis 4.1. Let $\mu_i(t)$, $i = 1,2$, be the empirical measures, generated by the solutions $r(t,\tilde{\mathcal{Y}},r_{s,i})$ of (4.9) with initial distributions $\mu_{s,i}$, and let τ be a localizing stopping time for $\tilde{\mathcal{Y}}$, $\mu_{s,1}$ and $\mu_{s,2}$. Then there is a constant $c_{T,F,\mathcal{J},\tau}$ such that*

$$E\left(\sup_{s\le t\le T\wedge\tau} \gamma_f^2(\mu_1(t)-\mu_2(t))1_{\{\tau>s\}} \right) \le c_{T,F,\mathcal{J},\tau} E(\gamma_f^2(\mu_{s,1}-\mu_{s,2})1_{\{\tau>s\}}). \tag{8.19}$$

Proof. Suppose, without loss of generality, that $\tilde{\mathcal{Y}} \in L_{loc,2,\mathcal{F}}(C([s,T];\mathbf{M}_f))$ and that the initial distributions are nonrandom (employing first conditional expectation in the following estimates conditioned on the events from $\mathcal{F}_s$) with total mass $\bar{m}_i$, $i = 1,2$. Then, by (15.43)[26]

$$\begin{aligned} &E \sup_{s\le t\le T\wedge\tau} \gamma_f^2(\mu_1(t)-\mu_2(t)) \\ &\le E \sup_{s\le t\le T\wedge\tau} \left[(\bar{m}_1\wedge\bar{m}_2)\,\tilde{\gamma}_1\left(\tfrac{\mu_1(t)}{\bar{m}_1}-\tfrac{\mu_2(t)}{\bar{m}_2}\right) + |\bar{m}_1-\bar{m}_2| \right]^2. \end{aligned}$$

[26] The following derivation employs definitions and notation from Sect. 15.1.4; e.g., $C(\mu,\nu)$ denotes the set of joint distributions Q on the Borel sets of $\mathbf{R}^{2d}$ for two probability measures, μ and ν on $\mathcal{B}^d$.

By (15.40)

$$
\begin{aligned}
&E \sup_{s\le t\le T\wedge\tau}\left[(\bar m_1\wedge\bar m_2)\tilde\gamma_1\left(\frac{\mu_1(t)}{\bar m_1}-\frac{\mu_2(t)}{\bar m_2}\right)\right]^2\\
&=(\bar m_1\wedge\bar m_2)^2 E \sup_{s\le t\le T\wedge\tau}\ \inf_{Q_1\in C\left(\frac{\mu_{s,1}}{\bar m_1},\frac{\mu_{s,2}}{\bar m_2}\right)}\left[\int\int \varrho(r(t,\tilde{\mathcal Y},q)-r(t,\tilde{\mathcal Y},\tilde q))Q_1(\mathrm dq,\mathrm d\tilde q)\right]^2\\
&\le(\bar m_1\wedge\bar m_2)^2 E \sup_{s\le t\le T\wedge\tau}\left[\int\int \varrho(r(t,\tilde{\mathcal Y},q)-r(t,\tilde{\mathcal Y},\tilde q))Q_1(\mathrm dq,\mathrm d\tilde q)\right]^2
\end{aligned}
$$

(where $Q_1\in C\left(\frac{\mu_{s,1}}{\bar m_1},\frac{\mu_{s,2}}{\bar m_2}\right)$ is arbitrary)

$$
\begin{aligned}
&=(\bar m_1\wedge\bar m_2)^2 E \sup_{s\le t\le T\wedge\tau}\int\int\int\int \varrho(r(t,\tilde{\mathcal Y},q)-r(t,\tilde{\mathcal Y},\tilde q))\varrho(r(t,\tilde{\mathcal Y},p)\\
&\quad -r(t,\tilde{\mathcal Y},\tilde p))Q_1(\mathrm dq,\mathrm d\tilde q)Q_1(\mathrm dp,\mathrm d\tilde p).
\end{aligned}
$$

By the Cauchy–Schwarz inequality along with the usual properties of integrals and suprema in addition to (4.17) and the fact that the initial values and distributions are deterministic we obtain

$$
\begin{aligned}
&E \sup_{s\le t\le T\wedge\tau}\int\int\int\int \varrho(r(t,\tilde{\mathcal Y},q)-r(t,\tilde{\mathcal Y},\tilde q))\varrho(r(t,\tilde{\mathcal Y},p)\\
&\quad -r(t,\tilde{\mathcal Y},\tilde p))Q_1(\mathrm dq,\mathrm d\tilde q)Q_1(\mathrm dp,\mathrm d\tilde p)\\
&\le\int\int\int\int\sqrt{E \sup_{s\le t\le T\wedge\tau}\varrho^2(r(t,\tilde{\mathcal Y},q)-r(t,\tilde{\mathcal Y},\tilde q))}\\
&\quad\sqrt{E \sup_{s\le t\le T\wedge\tau}\varrho^2(r(t,\tilde{\mathcal Y},p)-r(t,\tilde{\mathcal Y},\tilde p))}\,Q_1(\mathrm dq,\mathrm d\tilde q)Q_1(\mathrm dp,\mathrm d\tilde p)\\
&\le\tilde c_{T,F,\mathcal J,\tau}\int\int\int\int\sqrt{\varrho^2(q-\tilde q)}\sqrt{\varrho^2(p-\tilde p)}\,Q_1(\mathrm dq,\mathrm d\tilde q)Q_1(\mathrm dp,\mathrm d\tilde p)\\
&=\tilde c_{T,F,\mathcal J,\tau}\int\int\int\int\varrho(q-\tilde q)\varrho(p-\tilde p)Q_1(\mathrm dq,\mathrm d\tilde q)Q_1(\mathrm dp,\mathrm d\tilde p)\\
&=\tilde c_{T,F,\mathcal J,\tau}\tilde\gamma_1^2\left(\frac{\mu_{s,1}}{\bar m_1}-\frac{\mu_{s,2}}{\bar m_2}\right)
\end{aligned}
$$

(since $Q_1\in C\left(\frac{\mu_{s,1}}{\bar m_1},\frac{\mu_{s,2}}{\bar m_2}\right)$ was arbitrary).

Altogether,

$$
E \sup_{s\le t\le T\wedge\tau}\left[(\bar m_1\wedge\bar m_2)\tilde\gamma_1\left(\frac{\mu_1(t)}{\bar m_1}-\frac{\mu_2(t)}{\bar m_2}\right)\right]^2\le\tilde c_{T,F,\mathcal J,\tau}(\bar m_1\wedge\bar m_2)^2\tilde\gamma_1^2\left(\frac{\mu_{s,1}}{\bar m_1}-\frac{\mu_{s,2}}{\bar m_2}\right).
$$

We add to the previous estimates $|\bar{m}_1 - \bar{m}_2|^2$ on both sides, multiply the result by 2 and employ $2a^2 + 2b^2 \le 2(a+b)^2$ for nonnegative a, b. Then (15.43) implies (8.19) (with $c_{T,F,\mathcal{J},\tau} := 4\tilde{c}_{T,F,\mathcal{J},\tau}$). □

Recall that, according to Definition 4.1, a stopping time τ is called "localizing" for initial measures $\mu_{s,k}$, $k = 1, \ldots, m$, and measure-valued input processes $\tilde{\mathcal{Y}}_\ell(t)$, $\ell = 1, \ldots, n$, if there is a finite constant c such that

$$\sum_{k=1}^{m} \gamma_f(\mu_{s,k}) 1_{\{\tau > s\}} + \sum_{\ell=1}^{n} \gamma_f(\tilde{\mathcal{Y}}_\ell(t \wedge \tau)) 1_{\{\tau > s\}} \le c \quad \text{a.s.}$$

Lemma 8.2. *Assume Hypothesis 4.1. Let $\mu(t, \tilde{\mathcal{Y}}_i, \mu_{s,i})$, $i = 1, 2$, be the empirical measures, generated by the solutions $r(t, \tilde{\mathcal{Y}}_i, r_{s,i})$ of (4.9) with initial distributions $\mu_{s,1}$ and $\mu_{s,2}$, respectively. Further, let τ be a localizing stopping time for $\tilde{\mathcal{Y}}_i$, $\mu_{s,i}$ and $\mu_{s,2}$, $i = 1, 2$. Then there is a constant $c_{T,F,\mathcal{J},\tau}$ such that*

$$\left.\begin{aligned} &E\left(\sup_{s \le t \le T \wedge \tau} \gamma_f^2(\mu(t, \tilde{\mathcal{Y}}_1, \mu_{s,1}) - \mu(t, \tilde{\mathcal{Y}}_2, \mu_{s,2})) 1_{\{\tau > s\}}\right) \\ &\quad \le 2c_{T,F,\mathcal{J},\tau} E(\gamma_f^2(\mu_{s,1} - \mu_{s,2}) 1_{\{\tau > s\}}) \\ &\qquad + 2c_{T,F,\mathcal{J},\tau} E\left([\mu_{s,1}(\mathbf{R}^d)]^2 \int_s^{T \wedge \tau} \gamma_f^2(\tilde{\mathcal{Y}}_1(u) - \tilde{\mathcal{Y}}_2(u)) \mathrm{d}u 1_{\{\tau > s\}}\right). \end{aligned}\right\} \tag{8.20}$$

Proof. As in the proof of Lemma 8.1, we may, without loss of generality, assume that the initial distributions are deterministic and that the $\tilde{\mathcal{Y}}_\ell \in L_{loc,2,\mathcal{F}}(C([s,T]; \mathbf{M}_f))$, $\ell = 1, 2$.

(i)

$$\begin{aligned} &\gamma_f(\mu(t, \tilde{\mathcal{Y}}_1, \mu_{s,1}) - \mu(t, \tilde{\mathcal{Y}}_2, \mu_{s,2})) \\ &\quad \le \gamma_f(\mu(t, \tilde{\mathcal{Y}}_1, \mu_{s,1}) - \mu(t, \tilde{\mathcal{Y}}_2, \mu_{s,1})) + \gamma_f(\mu(t, \tilde{\mathcal{Y}}_2, \mu_{s,1}) - \mu(t, \tilde{\mathcal{Y}}_2, \mu_{s,2})) \\ &\quad =: \sum_{i=1}^{2} I_i(t). \end{aligned}$$

(ii)

$$\begin{aligned} &E \sup_{s \le t \le T \wedge \tau} \gamma_f^2(\mu(t, \tilde{\mathcal{Y}}_1, \mu_{s,1}) - \mu(t, \tilde{\mathcal{Y}}_2, \mu_{s,1})) \\ &\quad = E \sup_{s \le t \le T \wedge \tau} \sup_{\|f\|_{L,\infty} \le 1} \left[\int (f(r(t, \tilde{\mathcal{Y}}_1, q)) - f(r(t, \tilde{\mathcal{Y}}_2, q))) \mu_{s,1}(\mathrm{d}q)\right]^2 \\ &\quad \le E \sup_{s \le t \le T \wedge \tau} \left|\int (\rho(r(t, \tilde{\mathcal{Y}}_1, Z_1, q) - r(t, \tilde{\mathcal{Y}}_2, Z_1, q)) \mu_{s,1}(\mathrm{d}q)\right|^2 \\ &\quad \le \int \mu_{s,1}(dq) \int E \sup_{s \le t \le T \wedge \tau} \rho^2(r(t, \tilde{\mathcal{Y}}_1, q) - r(t, \tilde{\mathcal{Y}}_2, q)) \mu_{s,1}(\mathrm{d}q) \end{aligned}$$

(by the Cauchy–Schwarz inequality)

$$\leq [\mu_{s,1}(\mathbf{R}^d)]^2 \tilde{c}_{T,F,\mathcal{J},\tau} E \int_s^{T\wedge\tau} \gamma^2(\tilde{\mathcal{Y}}_1(u) - \tilde{\mathcal{Y}}_2(u))\mathrm{d}u$$

(by (4.17) and the assumption that the initial distributions are in $L_{0,\mathcal{F}_s}(\mathbf{M}_{f,d})$).

(iii) Applying (8.19) to the term $I_2(t)$ completes the proof. □

8.3 Noncoercive SPDEs

Assuming Hypothesis 4.1, we obtain results for noncoercive SPDEs on the space of finite measures as well as the existences of smooth solutions and extensions of function-valued SPDEs on weighted Sobolev spaces. The general conditions of Theorem 4.5 on initial values and input processes will be assumed throughout the rest of this section.

Suppose throughout this Sect. 8.3

Hypothesis 8.1

$$\sigma_i^{\perp} \equiv 0 \quad \forall i \;\text{ in (4.9)/(4.10).}$$ □

Further, assume first that the initial mass is finite. Recall our notational convention $\mathbf{M} \in \{\mathbf{M}_f, \mathbf{M}_{\infty,\varpi}\}$ and $L_{loc,p,\mathcal{F}}(C((s,T];\mathbf{M})) \in \{L_{loc,p,\mathcal{F}}(C((s,T];\mathbf{M}_f)), L_{loc,p,\mathcal{F}}(C((s,T];\mathbf{M}_{\infty,\varpi}))\}$. The empirical process associated with (4.9) is given by

$$\mathcal{Y}_N(t) := \sum_{i=1}^N m_i \delta_{r(t,r_s^i)}, \tag{8.21}$$

where $r(t,r_s^i)$ is the solution of (4.9) starting at r_s^i with input process $\tilde{\mathcal{Y}} \in L_{loc,2,\mathcal{F}}(C((s,T];\mathbf{M}))$ (cf. Theorem 4.5). The random weights $m_i = m_i(\omega)$ are ≥ 0 and $\mathcal{F}_s$-measurable. We note that

$$\left.\begin{array}{l}\mathcal{Y}_N \in L_{loc,p,\mathcal{F}}(C((s,T];\mathbf{M}_f)) \;\forall p\geq 1,\\[1ex] \text{and, if the } m_i \text{ are deterministic, then}\\[1ex] \mathcal{Y}_N \in L_{loc,p,\mathcal{F}}(C([s,T];\mathbf{M}_f)) \;\forall p\geq 1.\end{array}\right\} \tag{8.22}$$

The following derivation of estimates of solutions of (8.25), resp. (8.26), depends on bounds on the initial distributions. Therefore, we will often replace the absolute expectation by the conditional one, conditioned on the events from $\mathcal{F}_s$.[27] It follows

[27] Cf. also the proofs of Lemmas 8.1 and 8.2.

that in the conditional expectation the initial distribution is deterministic and we may assume the second relation in (8.22) instead of the more general first relation in that formula.

Recall from (4.15) the abbreviation

$$m(r(\cdot), \tilde{\mathcal{Y}}(\cdot), t) := \int_0^t \int \mathcal{J}(r(s), p, \tilde{\mathcal{Y}}(s), s) w(\mathrm{d}q, \mathrm{d}s).$$

Generalizing the procedure in Chap. 5, we replace the processes $r(\cdot), \tilde{\mathcal{Y}}(\cdot)$ and $q(\cdot), \tilde{\mathcal{Y}}(\cdot)$ in $m(\cdot, \cdot, t)$ by fixed (deterministic!) elements $(r, \mu), (q, \mu) \in \mathbf{M} \times \mathbf{R}^d$, i.e., we consider the Gaussian space–time fields

$$m(r, \mu, t) := \int_0^t \int \mathcal{J}(r, p, \mu, s) w(\mathrm{d}p, \mathrm{d}s), \quad m(q, \mu, t)$$
$$:= \int_0^t \int \mathcal{J}(q, p, \mu, s) w(\mathrm{d}p, \mathrm{d}s).$$

We then define the generalization of (5.9), setting[28]

$$\tilde{D}_{k\ell}(\mu, r, q, t) := \sum_{j=1}^{d} \int \mathcal{J}_{kj}(r, p, \mu, t) \mathcal{J}_{\ell j}(q, p, \mu, t) \mathrm{d}p,$$
$$D_{k\ell}(\mu, r, t) := \tilde{D}_{k\ell}(\mu, r, r, t),$$

and obtain, by the Cauchy–Schwarz inequality,

$$\left. \begin{aligned} & |\tilde{D}_{k\ell}(\mu, r, q, t)| \\ & \leq \sqrt{\sum_{j=1}^{d} \int \mathcal{J}_{kj}^2(r, p, \mu, t) \mathrm{d}p} \sqrt{\sum_{j=1}^{d} \int \mathcal{J}_{\ell j}^2(q, p, \mu, t) \mathrm{d}p} \\ & = \sqrt{D_{kk}(\mu, r, t)} \sqrt{D_{\ell\ell}(\mu, q, t)}. \end{aligned} \right\} \tag{8.23}$$

Take a test function $\varphi \in C_c^3(\mathbf{R}^d, \mathbf{R})$ and recall that $\langle \cdot, \cdot \rangle$ denotes the duality between measures and continuous test functions with compact supports (extending

[28] Observe that $\tilde{D}_{k\ell}(\mu, r, q, t) = \frac{\mathrm{d}}{\mathrm{d}t}[m_k(r, \mu, t), m_\ell(q, \mu, t)]$. In Chap. 5 and Sect. 15.2.4 we analyze the case of correlated Brownian motions. In that case $\mathcal{J}$ does not depend on the input process $\tilde{\mathcal{Y}}(\cdot)$, $F(\cdot, \cdot, \cdot) \equiv 0$, and the diffusion matrix is spatially homogeneous. The derivative of the mutual tensor quadratic variation is called the "random covariance" of the correlated Brownian motions $r(\cdot), q(\cdot)$ which are solutions of (5.1). (Cf. also Chap. 5, (5.8) and (5.9).) In fact, the derivation of (15.121) can immediately be generalized to the case of input process dependent diffusion kernels $\mathcal{J}$. The only difference is that for this case the marginals $r(\cdot)$ and $q(\cdot)$ are no longer Brownian motions.

the $\mathbf{H}_0$ inner product). Let $r^j(t)$ be the solutions of (4.9) with the initial conditions r_s^j, $j = 1, \ldots, N$ and $\tilde{\mathcal{Y}} \in L_{loc,2,\mathcal{F}}(C((s,T];\mathbf{M}))$ and observe that under Hypothesis 4.1 the right-hand side of (8.23) is bounded. Itô's formula yields

$$
\begin{aligned}
\langle \mathrm{d}\mathcal{Y}_N(t), \varphi\rangle &= \mathrm{d}\langle \mathcal{Y}_N(t), \varphi\rangle = \sum_{j=1}^{N} m_j \mathrm{d}\varphi(r^j(t)) \\
&= \sum_{j=1}^{N} m_j (\nabla\varphi)(r^j(t)) \cdot F(r^j(t), \tilde{\mathcal{Y}}, t)\mathrm{d}t \\
&\quad + \sum_{j=1}^{N} m_j (\nabla\varphi)(r^j(t)) \cdot \int \mathcal{J}(r^j(t), p, \tilde{\mathcal{Y}}, t) w(\mathrm{d}p, \mathrm{d}t) \\
&\quad + \sum_{j=1}^{N} m_j \tfrac{1}{2} \sum_{k,\ell=1}^{d} (\partial^2_{k\ell}\varphi)(r^j(t)) D_{k\ell}(\tilde{\mathcal{Y}}, r^j(t), t)\mathrm{d}t \\
&= \langle \mathcal{Y}_N(t), (\nabla\varphi)(\cdot) \cdot F(\cdot, \tilde{\mathcal{Y}}, t)\mathrm{d}t\rangle \\
&\quad + \left\langle \mathcal{Y}_N(t), (\nabla\varphi)(\cdot) \cdot \int \mathcal{J}(\cdot, p, \tilde{\mathcal{Y}}, t) w(\mathrm{d}p, \mathrm{d}t)\right\rangle \\
&\quad + \left\langle \mathcal{Y}_N(t), \frac{1}{2} \sum_{k,\ell=1}^{d} (\partial^2_{k\ell}\varphi)(\cdot) D_{k\ell}(\tilde{\mathcal{Y}}, \cdot, t) \right\rangle \mathrm{d}t \\
&\quad \text{(by the definition of the duality between the measure} \\
&\quad\quad \mathcal{Y}_N(t) \text{and the corresponding test functions)} \\
&= \left\langle \frac{1}{2} \sum_{k,\ell=1}^{d} D_{k\ell}(\tilde{\mathcal{Y}}, \cdot, t)\mathcal{Y}_N(t), \partial^2_{k\ell}\varphi \right\rangle \mathrm{d}t \\
&\quad + \sum_{k=1}^{d} \langle \mathcal{Y}_N(t) F_k(\cdot, \tilde{\mathcal{Y}}, t)\mathrm{d}t, \partial_k\varphi\rangle \\
&\quad + \sum_{k=1}^{d} \langle \mathcal{Y}_N(t) \int \mathcal{J}_{k\cdot}(\cdot, p, \tilde{\mathcal{Y}}, t) w(\mathrm{d}p, \mathrm{d}t), \partial_k\varphi\rangle \\
&\quad \text{(by rearranging terms in the duality and changing the order} \\
&\quad\quad \text{of the summation of the brackets } \langle\cdot,\cdot\rangle),
\end{aligned}
$$

i.e.,

$$
\left.
\begin{aligned}
\langle \mathrm{d}\mathcal{Y}_N(t), \varphi\rangle = &\left\langle \frac{1}{2} \sum_{k,\ell=1}^{d} \partial^2_{k\ell}(D_{k\ell}(\tilde{\mathcal{Y}}, \cdot, t)\mathcal{Y}_N(t))\mathrm{d}t, \varphi \right\rangle \\
&- \langle \nabla \cdot (\mathcal{Y}_N(t) F(\cdot, \tilde{\mathcal{Y}}, t))\mathrm{d}t, \varphi\rangle \\
&- \left\langle \nabla \cdot (\mathcal{Y}_N(t) \int \mathcal{J}(\cdot, p, \tilde{\mathcal{Y}}, t) w(\mathrm{d}p, \mathrm{d}t)), \varphi \right\rangle \\
&\text{(integrating by parts in the generalized sense).}
\end{aligned}
\right\}
\tag{8.24}
$$

Employing the linearity of the duality $\langle\cdot,\cdot\rangle$ and the fact that the test functions from $C_c^3(\mathbf{R}^d, \mathbf{R})$ uniquely determine the measure, we obtain that the empirical process associated with (4.9), $\mathcal{Y}_N(t)$, is a *weak solution* of the bilinear SPDE

$$\left.\begin{aligned} d\mathcal{Y} &= \left(\frac{1}{2}\sum_{k,\ell=1}^{d} \partial^2_{k\ell}(D_{k\ell}(\tilde{\mathcal{Y}},\cdot,t)\mathcal{Y}) - \nabla\cdot(\mathcal{Y}F(\cdot,\tilde{\mathcal{Y}},t)\right)dt \\ &\quad -\nabla\cdot(\mathcal{Y}\int \mathcal{J}(\cdot,p,\tilde{\mathcal{Y}},t)w(dp,dt)) \end{aligned}\right\} \tag{8.25}$$

with initial condition $\mathcal{Y}(s) = \sum_{i=1}^{N} m_i\delta_{r^i_s}$. Note that we use "weak solution" as is customary in the area of partial differential equations. Since we do not change the probability space and work strictly with the same stochastic driving term, our solution is a "strong (Itô) solution" in the terminology of stochastic analysis.

The empirical process associated with (4.10) is $\mathcal{X}_N(t)$, which is explicitly defined in (4.10). We verify that $\mathcal{X}_N \in L_{loc,2,\mathcal{F}}(C((s,T];\mathbf{M}_f))$. Replacing $\tilde{\mathcal{Y}}$ by $\mathcal{X}_N$ in the derivation (8.24)/(8.25), we also obtain that the empirical process associated with (4.10), $\mathcal{X}_N(t)$, is a weak solution of the quasilinear SPDE:

$$\left.\begin{aligned} d\mathcal{X} &= \left(\frac{1}{2}\sum_{k,\ell=1}^{d} \partial^2_{k\ell,r}(D_{k\ell}(\mathcal{X},\cdot,t)\mathcal{X}) - \nabla\cdot(\mathcal{X}F(\cdot,\mathcal{X},t))\right)dt \\ &\quad -\nabla\cdot(\mathcal{X}\int \mathcal{J}(\cdot,p,\mathcal{X},t)w(dp,dt)) \end{aligned}\right\} \tag{8.26}$$

with initial condition $\mathcal{X}(s) = \sum_{i=1}^{N} m_i\delta_{r^i_s}$.

We next extend the empirical processes associated with (4.9) and (4.10), respectively, by continuity from discrete initial distributions to general initial distribution. This *extension by continuity* is one of the main tools in the derivation of SPDEs from the empirical distributions of finitely many SODEs.[29] We start with (4.9).

Theorem 8.3. *Under Hypotheses 4.1 and 8.1, the following holds: For any fixed* $\tilde{\mathcal{Y}} \in L_{loc,2,\mathcal{F}}(C((s,T];\mathbf{M}))$ *the map*

$$\mathcal{X}_{s,N} \longmapsto \mathcal{Y}_N(\cdot,\tilde{\mathcal{Y}},\mathcal{X}_{s,N}) \text{ from } L_{0,\mathcal{F}_s}(\mathbf{M}_{f,d}) \text{ into } L_{loc,2,\mathcal{F}}(C((s,T];\mathbf{M}_f))$$
extends uniquely to a map
$$\mathcal{X}_s \longmapsto \mathcal{Y}(\cdot,\tilde{\mathcal{Y}},\mathcal{X}_s) \text{ from } L_{0,\mathcal{F}_s}(\mathbf{M}_f) \text{ into } L_{loc,2,\mathcal{F}}(C((s,T];\mathbf{M}_f)).$$

Further, let $\mathcal{X}_{s,\ell} \in L_{0,\mathcal{F}_s}(\mathbf{M}_f)$, $\tilde{\mathcal{Y}}_\ell \in L_{loc,2,\mathcal{F}}(C((s,T];\mathbf{M}_f))$, $\ell = 1,2$, *and let* τ *be a localizing stopping time for those four random quantities. Then,*

$$\left.\begin{aligned} &E\left(\sup_{s\le t\le T\wedge\tau} \gamma_f^2\left(\mathcal{Y}(t,\tilde{\mathcal{Y}}_1,\mathcal{X}_{s,1}1_{\{\tau>s\}}) - \mathcal{Y}(t,\tilde{\mathcal{Y}}_2,\mathcal{X}_{s,2}1_{\{\tau>s\}})\right)\right) \\ &\le 3c_{T,F,\mathcal{J},\tau}E\left(\gamma_f^2((\mathcal{X}_{s,1}-\mathcal{X}_{s,2})1_{\{\tau>s\}})\right) \\ &\quad +3c_{T,F,\mathcal{J},\tau}E\left([\mu_{s,1}(\mathbf{R}^d)]^2\int_s^{T\wedge\tau} \gamma_f^2(\tilde{\mathcal{Y}}_1(u)-\tilde{\mathcal{Y}}_2(u))1_{\{\tau>s\}}du\right). \end{aligned}\right\} \tag{8.27}$$

[29] Cf. Theorem 15.1 in Sect. 15.1.1.

Moreover, if $m(\omega) := \mathcal{X}_s(\omega, \mathbf{R}^d) = \gamma_f(\mathcal{X}_s(\omega))$ is the random mass at time s, then with probability 1 uniformly in $t \in [s, T]$

$$\mathcal{Y}\left(t, \omega, \tilde{\mathcal{Y}}, \mathcal{X}_s(\omega)\right) \in \mathbf{M}_{m(\omega)}. \tag{8.28}$$

Proof. Working with localizing stopping times, we may, without loss of generality, assume that all quantities are square integrable. The extension from $L_{2,\mathcal{F}_s}(\mathbf{M}_{f,d})$ into $L_{2,\mathcal{F}}(C([s,T];\mathbf{M}_f))$ and (8.27) is a straightforward consequence of (8.20) and the extension by continuity, since $L_{2,\mathcal{F}_s}(\mathbf{M}_{f,d})$ is dense in $L_{2,\mathcal{F}_s}(\mathbf{M}_f)$ with respect to the metric $\sqrt{E\gamma^2(\cdot,\cdot)}$.

Let

$$\tau_n(\omega) := \inf\{t \in [s, \infty] : \gamma_f(\tilde{\mathcal{Y}}(t,\omega)) + \gamma_f(\mathcal{X}(s,\omega)) \geq n\}.$$

Set $\Omega_n := \{\omega : \tau_n > s\}$. Note that for $n \in \mathbf{N}$

$$\mathcal{Y}1_{\Omega_n}(t, \omega, \tilde{\mathcal{Y}}, \mathcal{X}_s(\omega)) \equiv \mathcal{Y}1_{\Omega_n}(t, \omega, \hat{\mathcal{Y}}, \hat{\mathcal{X}}_s(\omega))$$

for all adapted measure processes $\tilde{\mathcal{Y}}, \hat{\mathcal{Y}}$ from $L_{loc,2,\mathcal{F}}(C((s,T];\mathbf{M}))$ and adapted initial measures $\mathcal{X}_s, \hat{\mathcal{X}}_s$ from $L_{0,\mathcal{F}_s}(\mathbf{M}_f)$ whose restrictions to Ω_n satisfy

$$\tilde{\mathcal{Y}}1_{\Omega_n} \equiv \hat{\mathcal{Y}}1_{\Omega_n}, \quad \mathcal{X}_s 1_{\Omega_n} \equiv \hat{\mathcal{X}}_s 1_{\Omega_n}.$$

This is a simple consequence of the construction of the empirical process $\mathcal{Y}$ from finitely many particles and their associated stochastic ordinary differential equations. Further, for $n_1 < n_2$

$$\mathcal{Y}1_{\Omega_{n_1}}\left(t, \omega, \tilde{\mathcal{Y}}, \mathcal{X}_s(\omega)1_{\Omega_{n_2}}\right) = \mathcal{Y}1_{\Omega_{n_1}}\left(t, \omega, \tilde{\mathcal{Y}}1_{\Omega_{n_1}}, \mathcal{X}_s 1_{\Omega_{n_2}} 1_{\Omega_{n_1}}(\omega)\right)$$

Hence, if $n_1 < n_2$, then

$$\left.\begin{aligned} \mathcal{Y}1_{\Omega_{n_1}}\left(t, \omega, \tilde{\mathcal{Y}}, \mathcal{X}_s(\omega)1_{\Omega_{n_2}}\right) &= \mathcal{Y}1_{\Omega_{n_1}}\left(t, \omega, \tilde{\mathcal{Y}}1_{\Omega_{n_1}}, \mathcal{X}_s(\omega)1_{\Omega_{n_1}}\right) \\ &= \mathcal{Y}\left(t, \omega, \tilde{\mathcal{Y}}, \mathcal{X}_s(\omega)1_{\Omega_{n_1}}\right), \end{aligned}\right\} \tag{8.29}$$

because for those ω, where the initial measure is $\mathbf{0}$, the process $\mathcal{Y}(t, \omega, \tilde{\mathcal{Y}}, \mathcal{X}_s(\omega)) \equiv \mathbf{0}$ by mass conservation. Hence, there is an $\mathbf{M}_f$-valued process $\mathcal{Y}(t, , \tilde{\mathcal{Y}}, \mathcal{X}_s)$, whose restriction to any of the Ω_n is the afore-mentioned extension. Clearly, $\mathcal{Y}(t, , \tilde{\mathcal{Y}}, \mathcal{X}_s)$ is an extension from $L_{0,\mathcal{F}_s}(\mathbf{M}_f)$ into $L_{loc,2,\mathcal{F}}(C((s,T];\mathbf{M}_f))$. The uniqueness follows because all the restrictions are unique and the τ_n are localizing for the input processes and the initial measures.

Expression (8.28) is a simple consequence of the conservation of mass. □

We next consider the mesoscopic equation (8.26) associated with the coupled system of SODEs (4.10). We denote the empirical process of the N-particle system by $\mathcal{X}(t, \mathcal{X}_{s,N})$, with $\mathcal{X}_{s,N}$ being its initial state.

Theorem 8.4. *Under Hypotheses 4.1 and 8.1 the following holds: The map*

$$\mathcal{X}_{s,N} \mapsto \mathcal{X}_N(\cdot, \mathcal{X}_{s,N}) \text{ from } L_{0,\mathcal{F}_s}(\mathbf{M}_{f,d}) \text{ into } L_{loc,2,\mathcal{F}}(C((s,T];\mathbf{M}_f))$$
extends uniquely to a map
$$\mathcal{X}_s \mapsto \mathcal{X}(\cdot, \mathcal{X}_s) \text{ from } L_{0,\mathcal{F}_s}(\mathbf{M}_f) \text{ into } L_{loc,2,\mathcal{F}}(C((s,T];\mathbf{M}_f)).$$

Further, let $\mathcal{X}_{s,\ell} \in L_{0,\mathcal{F}_s}(\mathbf{M}_f)$ and a localizing stopping time τ for both initial conditions. Then,

$$E\left(\sup_{s\le t\le T\wedge\tau} \gamma_f^2(\mathcal{X}(t,\mathcal{X}_{s,1}) - \mathcal{X}(t,\mathcal{X}_{s,2})1_{\{\tau>s\}})\right) \le c^2_{T,F,\mathcal{J},\tau} E(\gamma_f^2(\mathcal{X}_{s,1} - \mathcal{X}_{s,2})1_{\{\tau>s\}}). \tag{8.30}$$

Moreover, if $m(\omega) := \mathcal{X}_s(\omega, \mathbf{R}^d)$ is the random mass at time s, then with probability 1 uniformly in $t \in [s,T]$

$$\mathcal{X}(t,\omega,\mathcal{X}_s(\omega)) \in \mathbf{M}_{m(\omega)}. \tag{8.31}$$

Proof. Set $\mathcal{Y} := \mathcal{X}(\cdot, \mathcal{X}_{s,N})$. We obtain $\mathcal{X}(t, \mathcal{X}_{s,N}) = \mathcal{Y}(t, \mathcal{Y}, \mathcal{X}_{s,N})$. Apply (8.27) and Gronwall's inequality. □

Theorem 8.5. *Suppose $\mathcal{X}_s \in L_{0,\mathcal{F}_s}(\mathbf{M})$ in addition to Hypotheses 4.1 and 8.1. Then the following statements hold:*

1. *If $\tilde{\mathcal{Y}} \in L_{loc,2,\mathcal{F}}(C((s,T];\mathbf{M}))$, then $\mathcal{Y}(\cdot, \tilde{\mathcal{Y}}, \mathcal{X}_s)$, the process obtained in Theorem 8.3, is a solution of (8.25).*
2. *Assume also that the conditions of Theorem 4.7 hold. Then $\mathcal{X}(\cdot, \mathcal{X}_s)$, the process obtained in Theorem 8.4, is a solution of (8.26).*

Proof. [30] We must confirm that the solutions satisfy the SPDEs in weak form, as described in the derivation (8.25). We will show (2). The proof of (1) is simpler.

Without loss of generality, let $s = 0$ and $\gamma_f(\mathcal{X}_0(\omega)) \le b < \infty$ a.s. for some finite $b > 0$. Let $\mathcal{X}(t) := \mathcal{X}(t, \mathcal{X}_0)$ be the extension of the empirical process $\mathcal{X}_N(t) := \mathcal{X}(t, \mathcal{X}_{N,0})$, where we assume $E\gamma_f^2(\mathcal{X}_{N,0} - \mathcal{X}_0) \longrightarrow 0$.

Let $\varphi \in C_c^3(\mathbf{R}^d;\mathbf{R})$. Note that $\|\varphi\|_{L,\infty}$, $\|\partial_\ell\varphi\|_{L,\infty}$, and $\|\partial^2_{k,\ell}\varphi\|_{L,\infty}$ are finite for all $k, \ell = 1,\ldots,d$, where $\|\cdot\|_{L,\infty}$ denotes the bounded Lipschitz norm.[31] Substituting on the right-hand side of (8.24) $\mathcal{X}(\cdot)$ for $\mathcal{Y}_N(\cdot)$ and for $\tilde{\mathcal{Y}}(\cdot)$ or $\mathcal{X}_N(\cdot)$ for $\mathcal{Y}_N(\cdot)$ and for $\tilde{\mathcal{Y}}(\cdot)$, the right-hand side of (8.24) is well defined for both substitutions.

(i) $|\langle \mathcal{X}(t) - \mathcal{X}_N(t), \varphi\rangle| \longrightarrow 0$, as $N \longrightarrow \infty$,
since, by Theorem 8.4, we have convergence to 0 if we take the sup over all φ with $\|\varphi\|_{\infty,L} \le 1$. By the same argument, the assumptions imply

[30] For another and simpler proof via a flow representation of the empirical processes cf. Theorem 8.12. However, the present proof is straightforward and can be generalized to the coercive SPDE (8.54).

[31] Cf. (15.37).

$$|\langle \mathcal{X}_0 - \mathcal{X}_{N,0}, \varphi \rangle| \longrightarrow 0, \text{ as } N \longrightarrow \infty.$$

Therefore, we must show that the difference of the right-hand sides of (8.24), evaluated at $\mathcal{X}(t)$ and $\mathcal{X}_N(t)$, respectively, and integrated from 0 to t, converges to 0. Set $h_N(t) := \mathcal{X}(t) - \mathcal{X}_N(t)$. Then

$$\begin{aligned}
&E\left\{\int_0^t \langle h_N(t), (\nabla\varphi)(\cdot) \cdot \int \mathcal{J}(\cdot, p, \mathcal{X}(s), t) w(\mathrm{d}p, \mathrm{d}s)\rangle\right\}^2 \\
&= \sum_{k,\ell,m=1}^{d} \int_0^t E \int\int h_N(t, \mathrm{d}r) h_N(t, \mathrm{d}q) \partial_k \varphi(r) \partial_\ell \varphi(q) \\
&\quad \times \int \mathcal{J}_{km}(r, p, \mathcal{X}(s), t) \mathcal{J}_{\ell m}(q, p, \mathcal{X}(s), t) \mathrm{d}p\, \mathrm{d}s \\
&\longrightarrow 0, \text{ as } N \longrightarrow \infty,
\end{aligned}$$

because, by Hypothesis 4.1, the integral with respect to $\mathrm{d}p$ is bounded by a constant, uniformly in all variables.

(ii) Since $f(r) \equiv 1$ is uniformly Lipschitz,

$$\lim_{N\to\infty} \mathcal{X}_{N,0}(\mathbf{R}^d, \omega) = \mathcal{X}_0(\mathbf{R}^d, \omega).$$

The mass conservation implies (uniformly in t)

$$\lim_{N\to\infty} \mathcal{X}_N(t)(\mathbf{R}^d, \omega) = \mathcal{X}(t)(\mathbf{R}^d, \omega).$$

Hence, we obtain that

$$\mathcal{X}_N(t)(\mathbf{R}^d, \omega) = \mathcal{X}_{0,N}(\mathbf{R}^d, \omega) \le b < \infty \text{ a.s.}$$

Thus,

$$\begin{aligned}
&E\left\{\int_0^t \left\langle \mathcal{X}_N(t), (\nabla\varphi)(\cdot) \cdot \int [\mathcal{J}(\cdot, p, \mathcal{X}(s), t) - \mathcal{J}(\cdot, p, \mathcal{X}_N(s), t)] w(\mathrm{d}p, \mathrm{d}s)\right\rangle\right\}^2 \\
&= \sum_{k,\ell,m=1}^{d} \int_0^t E \int\int \mathcal{X}_N(t, \mathrm{d}r) \mathcal{X}_N(t, \mathrm{d}q) \partial_k \varphi(r) \partial_\ell \varphi(q) \\
&\quad \times \int [\mathcal{J}_{km}(r, p, \mathcal{X}(s), t) - \mathcal{J}_{km}(r, p, \mathcal{X}_N(s), t)] \\
&\quad [\mathcal{J}_{\ell m}(q, p, \mathcal{X}(s), t) - \mathcal{J}_{\ell m}(q, p, \mathcal{X}_N(s), t)] \mathrm{d}p\, \mathrm{d}s \\
&\le \sum_{k,\ell,m=1}^{d} \int_0^t E \int\int \mathcal{X}_N(t, \mathrm{d}r) \mathcal{X}_N(t, \mathrm{d}q) |\partial_k \varphi(r) \partial_\ell \varphi(q)| \\
&\quad \times \sqrt{\int [\mathcal{J}_{km}(r, p, \mathcal{X}(s), t) - \mathcal{J}_{km}(r, p, \mathcal{X}_N(s), t)]^2 \mathrm{d}p} \\
&\quad \sqrt{\int [\mathcal{J}_{\ell m}(q, p, \mathcal{X}(s), t) - \mathcal{J}_{\ell m}(q, p, \mathcal{X}_N(s), t)]^2 \mathrm{d}p}\, \mathrm{d}s
\end{aligned}$$

(by the Cauchy–Schwarz inequality)

$$\leq c_{\mathcal{J}}^2 \sum_{k,\ell,m=1}^{d} \int_0^t E \int\int \mathcal{X}_N(t,\mathrm{d}r)\mathcal{X}_N(t,\mathrm{d}q)|\partial_k\varphi(r)\partial_\ell\varphi(q)|\gamma_f^2(\mathcal{X}(s)-\mathcal{X}_N(s))\mathrm{d}s$$

(by Hypothesis 4.1)

$$\leq c_{\mathcal{J}}^2 \max_{k=1,..,d} |||\partial_k\varphi|||^2 d^2 b^2 \int_0^t E\gamma_f^2(\mathcal{X}(s)-\mathcal{X}_N(s))\mathrm{d}s \longrightarrow 0, \quad \text{as } N \longrightarrow \infty.$$

(iii) For the integral containing the diffusion coefficients we use

$$\begin{aligned}
&|D_{k\ell}(\mu,r,t) - D_{k\ell}(\nu,r,t)|\\
&= \left|\sum_{m=1}^{d}\int[\mathcal{J}_{km}(r,p,\mu,t)\mathcal{J}_{\ell m}(r,p,\mu,t) - \mathcal{J}_{km}(r,p,\nu,t)\mathcal{J}_{\ell m}(r,p,\nu,t)]\mathrm{d}p\right|\\
&\leq \left|\sum_{m=1}^{d}\int[\mathcal{J}_{km}(r,p,\mu,t) - \mathcal{J}_{km}(r,p,\nu,t)]\mathcal{J}_{\ell m}(r,p,\mu,t)\mathrm{d}p\right|\\
&\quad + \left|\sum_{m=1}^{d}\int \mathcal{J}_{km}(r,p,\mu,t)[\mathcal{J}_{\ell m}(r,p,\mu,t) - \mathcal{J}_{\ell m}(r,p,\nu,t)]\mathrm{d}p\right|\\
&\leq \sum_{m=1}^{d}\sqrt{\int[\mathcal{J}_{km}(r,p,\mu,t) - \mathcal{J}_{km}(r,p,\nu,t)]^2\mathrm{d}p}\sqrt{\int \mathcal{J}_{\ell m}^2(r,p,\mu,t)\mathrm{d}p}\\
&\quad + \sum_{m=1}^{d}\sqrt{\int \mathcal{J}_{km}^2(r,p,\mu,t)\mathrm{d}p}\sqrt{\int[\mathcal{J}_{\ell m}(r,p,\mu,t) - \mathcal{J}_{\ell m}(r,p,\nu,t)]^2\mathrm{d}p}
\end{aligned}$$

(by the Cauchy–Schwarz inequality),

whence, by Hypothesis 4.1,

$$|D_{k\ell}(\mu,r,t) - D_{k\ell}(\nu,r,t)| \leq c\gamma_f(\mu-\nu). \tag{8.32}$$

Using (8.32), we prove the convergence to 0 of the corresponding deterministic integrals, following the pattern of step (ii). □

In what follows we focus on "smooth" solutions of (8.25) and (8.26). To this end, we introduce suitable functions spaces.

Notation

Set for $m \in \mathbf{N} \cup \{0\}$

$$\mathbf{B}_{d,1,m} := \{f : \mathbf{R}^d \to \mathbf{R} : \ f \text{ is } \mathcal{B}^d - \mathcal{B}^1 \text{ measurable and } \exists\, \partial^{\mathbf{j}} f,\ |\mathbf{j}| \leq \mathbf{m}\},$$

where $\partial^{\mathbf{j}} f = \frac{\partial^{j_1,\dots,j_d}}{\partial r^{j_1}\dots\partial r^{j_d}} f$ was introduced at the end of Sect. 4.3 (before (4.45)) and these derivatives are taken in the generalized sense. For $p > 0$ and $f \in \mathbf{B}_{d,1,m}$, we set

$$\|f\|^p_{m,p,\Phi} := \sum_{|\mathbf{j}|\le m} \int |\partial^{\mathbf{j}} f|^p(r)\Phi(r)\mathrm{d}r.$$

Let

$$\Phi \in \{1, \varpi\},$$

where $1(r) \equiv 1$, and set

$$\mathbf{W}_{m,p,\Phi} := \{f \in \mathbf{B}_{d,1,m} : \|f\|_{m,p,\Phi} < \infty\}.$$

For the L_2-space, introduced in (3.29) of Sect. 3, we have

$$(\mathbf{H}_0, \|\cdot\|_0) = (\mathbf{W}_{0,2,1}, \|\cdot\|_{0,2,1}).$$

Further,

$$(\mathbf{H}_{0,\varpi}, \|\cdot\|_{0,\varpi}) := (\mathbf{W}_{0,2,\varpi}, \|\cdot\|_{0,2,\varpi})$$

and

$$(\mathbf{H}_m, \|\cdot\|_m) := (\mathbf{W}_{m,2,1}, \|\cdot\|_{m,2,1}).$$

Note that smoothness for the bilinear SPDE (8.25) implies smoothness for the quasilinear SPDE (8.26) (by taking $\mathcal{Y} \equiv \mathcal{X} \in L_{loc,2,\mathcal{F}}(C((s,T];\mathbf{M}_f))$, provided $\mathcal{X}_s \in L_{0,\mathcal{F}_s}(\mathbf{M}_f)$). To make the arguments more transparent we assume that the diffusion coefficients from (8.23) are spatially homogeneous, i.e.,

$$\left.\begin{aligned} \tilde{D}_{k\ell}(\mu, r, q, t) &= \tilde{D}_{k\ell}(\mu, r-q, t),\\ D_{k\ell}(\mu, r, t) &\equiv \tilde{D}_{k\ell}(\mu, 0, t). \end{aligned}\right\} \tag{8.33}$$

We abbreviate

$$\left.\begin{aligned} D_{k\ell}(s) &:= D_{k\ell}(\tilde{\mathcal{Y}}(s), 0, s),\\ F_k(r,s) &:= F_k(r, \tilde{\mathcal{Y}}(s), s),\\ \mathcal{J}(r,p,s) &:= \mathcal{J}(r,p,\tilde{\mathcal{Y}}(s), s),\\ \mathrm{d}m(s) &:= \int \mathcal{J}(\cdot, p, s) w(\mathrm{d}p, \mathrm{d}s),\\ \tilde{D}_{k\ell}(s, r-q) &:= D_{k\ell}(\tilde{\mathcal{Y}}, r-q, s). \end{aligned}\right\} \tag{8.34}$$

Fix $m \in \mathbf{N} \cup \{0\}$, $T > 0$ and suppose there are finite constants $d_{i,T}$, $i = 1,2,3$ such that

$$\left.\begin{aligned} &\max_{1\le k,\ell\le d} \operatorname{ess}\sup_{(t,\omega)\in[0,T]\times\Omega} |||\tilde{D}_{k\ell}(t,\omega)|||_{m+3} \le d_{1,T};\\ &\max_{1\le k,\ell\le d} \operatorname{ess}\sup_{(t,\omega)\in[0,T]\times\Omega} \max_{1\le k\le d} \operatorname{ess}\sup_{(t,\omega)\in[0,T]\times\Omega} |||F_k(t,\omega)|||_{m+1} \le d_{2,T};\\ &\tilde{D}_{k\ell}(s,r) \equiv \tilde{D}_{k\ell}(s,-r)\ \forall k,\ell\ \text{(symmetry of the diffusion)}. \end{aligned}\right\} \tag{8.35}$$

For the definition of the norms $|||\cdot|||_m$ we refer to (4.44). We call m the "smoothness degree of the coefficients" F and $\mathcal{J}$. If we make more differentiability assumptions on the kernels in the examples of Sect. 4.4, (8.35) will hold for those examples. The symmetry assumption follows immediately, if we, e.g., assume that $\mathcal{J}$ is symmetric. A more interesting sufficient condition is obtained, if we assume the generating kernel of $\mathcal{J}$ to be frame indifferent.[32]

Notation

Following the notation of DaPrato and Zabczyk (1992), let $L_{1,\mathcal{F}}([0,T]\times\Omega)$ be the space of $dt\otimes dP$-measurable and adapted processes. Further, we set[33]

$$
\begin{aligned}
&L_{\bar{n}p,\mathcal{F}}([s,T]\times\Omega;\mathbf{W}_{m,p,\Phi})\\
&\quad:=\{\tilde{Y}:[s,T]\times\Omega\mapsto\mathbf{W}_{m,p,\Phi}:\tilde{Y}\text{ is}\mathcal{F}_t\\
&\quad\text{-adapted and }\|\tilde{Y}\|_{m,p,\Phi}^{\bar{n}p}\in L_{1,\mathcal{F}}([0,T]\times\Omega;\mathbf{R})\},\\
&L_{\bar{n}p,\mathcal{F}}((s,T]\times\Omega);\mathbf{W}_{m,p,\Phi})\\
&\quad:=\{\tilde{Y}:\ \tilde{Y}1_{\{\tau>s\}}\in L_{\bar{n}p,\mathcal{F}}([s,T]\times\Omega);\mathbf{W}_{m,p,\Phi})\},\\
&\quad\text{where }\tau\text{is an arbitrary localizing stopping time for}\\
&\quad\quad\tilde{Y}(s)\text{ and }p\in[1,\infty)\cup\{0\}\text{and}\bar{n}\in\mathbf{N}.
\end{aligned}
$$

Theorem 8.6. *Assume that the conditions of part (1) of Theorem 8.5 hold in addition to (8.35) and (8.33), and let $p\in\{2,4,6,\ldots\}$ be an even integer. Suppose that for $\Phi\in\{1,\varpi\}$*

$$Y_s\in L_{\mathrm{loc},p,\mathcal{F}_s}(\mathbf{W}_{m,p,\Phi})\cap L_{0,\mathcal{F}_s}(\mathbf{M}_f),\tag{8.36}$$

where $m\in\mathbf{N}\cup\{0\}$ and $\bar{n}\in\mathbf{N}$. Then the solution of (8.25), derived in Theorem 8.5, is unique and has a density with respect to the Lebesgue measure $Y(\cdot,Y_s):=Y(\cdot,\tilde{\mathcal{Y}},Y_s)$ such that for any $T>s$

$$Y(\cdot,Y_s)\in L_{\bar{n}p,\mathcal{F}}((s,T]\times\Omega;\mathbf{W}_{m,p,\Phi})\cap L_{0,\mathcal{F}}(C([s,T];\mathbf{M}_f)).\tag{8.37}$$

Moreover, there are finite constants $\hat{c}_{m,T,\Phi}:=c_{m,T}(d_1,d_2)<\infty$, depending on the bounds $d_1=d_{1,T}$ and $d_2=d_{2,T}$ from (8.35), and the following estimate holds:

$$\sup_{s\le t\le T}E(\|Y(t)\|_{m,p,\Phi}^p1_{\{\tau>s\}})\le\hat{c}_{m,T,\Phi}E(\|Y(s)\|_{m,p,\Phi}^p1_{\{\tau>s\}}),\tag{8.38}$$

where $\tau\ge s$ is a localizing stopping time for $\tilde{\mathcal{Y}}$ and Y_s.

Finally, if $m\ge 2$,

$$Y(\cdot)\in L_{0,\mathcal{F}}(C([s,T];\mathbf{W}_{m-2,p,\Phi})).\tag{8.39}$$

---

[32] Cf. Example 8.27 at the end of this chapter.

[33] Cf. (8.4).

Proof. The smoothness, (8.38) and (8.39), is proven in Chap. 11. The uniqueness follows from the bilinearity and (8.38). □

Corollary 8.7. *Suppose the positive and negative parts of the initial condition in (8.25), $Y_+(0) := Y(0) \wedge 0$ and $Y_-(0) := Y(0) - Y_+(0)$ are in $L_{loc,p,\mathcal{F}_s}(\mathbf{W}_{m,p,\Phi}) \cap L_{0,\mathcal{F}_s}(\mathbf{M}_f)$. Then there is a unique solution of (8.25), $Y(\cdot, Y(0))$ such that (8.37) and (8.39) hold for $Y(\cdot, Y(0))$.*

Proof. The statement follows from Theorem 8.6 and the bilinearity of (8.25). □

We next derive the Itô formula for $\|Y(t)\|^p_{0,p,\Phi}$ under smoothness assumption on the coefficients and the initial conditions. We add the following abbreviations to our notation with multi-indices.[34]

Notation

$1_k := (0, \ldots, 0, 1, 0, \ldots, 0)$ (i.e., the d-vector where all entries except for the kth entry are 0 and the kth entry is 1. Further, we set

$$\mathcal{L}_{k,\ell,\mathbf{n}}(s) := \partial^{\mathbf{n}} \tilde{D}_{k\ell}(s, r)_{|r=0}. \tag{8.40}$$

The definition of $\tilde{D}_{k\ell}(s, r)$ implies that

$$\tilde{D}_{k\ell}(s, r) = \tilde{D}_{\ell k}(s, -r),$$

whence

$$\mathcal{L}_{k,\ell,\mathbf{n}}(s) = (-1)^{|\mathbf{n}|} \mathcal{L}_{\ell,k,\mathbf{n}}(s). \tag{8.41}$$

Theorem 8.8 (Itô formula). *Let $\Phi \in \{1, \varpi\}$ and assume (8.35) with $m = 0$. Let $Y(\cdot) := Y(\cdot, Y_s) := Y(\cdot, \tilde{\mathcal{Y}}, Y_s)$ be the (smooth) solution of (8.25) and recall the abbreviations from (8.34). Suppose*

$$Y(s) \in L_{2,\mathcal{F}_s}(\mathbf{H}_{0,\varpi}) \quad \textit{and} \quad \tilde{\mathcal{Y}} \in L_{2,\mathcal{F}}(C([0, T]; \mathbf{M})).$$

Then,

$$\left.\begin{aligned}
\int Y^2(t, r)\Phi(r)\mathrm{d}r = &\int Y^2(s, r)\Phi(r)\mathrm{d}r \\
&+ \frac{1}{2}\sum_{k,\ell=1}^{d} \int_s^t \int Y^2(u, r)\partial^2_{k\ell}\Phi(r)\mathrm{d}r\, D_{k\ell}(u)\mathrm{d}u \\
&- \sum_{k,\ell=1}^{d} \int_s^t \int Y^2(u, r)\Phi(r)\mathrm{d}r\, \mathcal{L}_{k,\ell,1_k+1_\ell}(u)\mathrm{d}u \\
&+ \sum_{\ell=1}^{d} \int_s^t \int Y^2(u, r)((\partial_\ell F_\ell)(r, u)\Phi(r) + F_\ell(r, u)\partial_\ell\Phi(r))\mathrm{d}r\, \mathrm{d}u \\
&+ \sum_{\ell=1}^{\mathrm{d}} \int_s^t \int Y^2(u, r)((\mathrm{d}\partial_\ell m_\ell(u, r))\Phi(r) + \mathrm{d}m_\ell(u, r)\partial_\ell\Phi(r))\mathrm{d}r.
\end{aligned}\right\} \tag{8.42}$$

[34] Cf. the end of Sect. 4.3.

Proof is given in Chap. 11.

From (8.42) we obtain the Itô formula in the special case $m = 0, p = 2$, and $\Phi \equiv 1$:

$$\left.\begin{aligned}&\int Y^2(t,r)\mathrm{d}r\\&= \int Y^2(s,r)\mathrm{d}r - \sum_{k,\ell=1}^{d}\int_s^t\int Y^2(u,r)\mathrm{d}r\,\mathcal{L}_{k,\ell,1_k+1_\ell}(u)\mathrm{d}u\\&+\sum_{\ell=1}^{d}\int_s^t\int Y^2(u,r)\partial_\ell F_\ell(r,u)\mathrm{d}r\,\mathrm{d}u + \sum_{\ell=1}^{d}\int_s^t\int Y^2(u,r)\mathrm{d}\partial_\ell m_\ell(u,r))\mathrm{d}r.\end{aligned}\right\} \tag{8.43}$$

□

The smoothness result (Theorem 8.6 and Corollary 8.7) imply the following.

Lemma 8.9. *Let $Y(t,Y_s)$ and $\tilde{Y}(t,\tilde{Y}_s)$ be two solutions of (8.25), whose initial conditions satisfy the assumptions of Corollary 8.7 with $|\mathbf{m}| =: m = 2$, $p = 2$, and smoothness degree $\overline{m} > d+4$. Then there is a sequence of localizing stopping times $s \le \tau_N \longrightarrow \infty$ a.s., as $N \longrightarrow \infty$ such that*

$$\sup_{s\le t<\infty}\max_{|\mathbf{m}|\le 2}[\|\partial^{\mathbf{m}}Y(t\wedge\tau_N,Y_s)\|_{0,\Phi} + \|\partial^{\mathbf{m}}\tilde{Y}(t\wedge\tau_N,\tilde{Y}_s))\|_{0,\Phi}] \le N \ \ a.s. \tag{8.44}$$

Proof. Our assumptions imply that both Y and $\tilde{Y}$ are in $C([0,\infty);\mathbf{W}_{1,2,\Phi})$. □

We now derive uniqueness for the quasilinear SPDE (8.26) under the smoothness assumptions (8.35). Let $\partial_r^{\mathbf{n}}$ be a partial differential operator acting on the variable r in the following formula and set div $F := \sum_{k=1}^{d}\partial_k F_k$. We assume the following (additional) Lipschitz condition on the coefficients of (4.9)/(4.10):

$$\left.\begin{aligned}&\sup_{t\ge 0}|\mathrm{div}\ F(r,\mu_1) - \mathrm{div}\ F(r,\mu_2)| \le c\gamma_\varpi(\mu_1-\mu_2)\\&\sup_{t\ge 0}\sum_{m,\ell=1}^{d}\sum_{|\mathbf{n}|\le 2}\int[\partial_r^{\mathbf{n}}\mathcal{J}_{\ell m}(r,p,\mu_1) - \partial_r^{\mathbf{n}}\mathcal{J}_{\ell m}(r,p,\mu_2)]^2\mathrm{d}p \le c\gamma_\varpi^2(\mu_1-\mu_2).\end{aligned}\right\} \tag{8.45}$$

As we noted after (8.35), Sect. 4.4, Examples, provides coefficients which satisfy (8.45), if we increase the smoothness assumptions on the kernels (cf. (4.49) and (4.51)). Further, assume that μ_i have densities $X_i \in \mathbf{H}_{0,\varpi}$, $i = 1,2$. Identifying the measures with their densities in what follows, the Cauchy–Schwarz inequality implies

$$\begin{aligned}\gamma_\varpi(\mu_1-\mu_2) &\le \sup_{\|f\|_{L,\infty}\le 1}\int|f|(q)\varpi(q)|X_1-X_2|(q)\mathrm{d}q\\&= \int\varpi(q)|X_1-X_2|(q)\mathrm{d}q \le \|\sqrt{\varpi}\|_0\|X_1-X_2\|_{0,\varpi},\end{aligned}$$

i.e.,

$$\gamma_\varpi(\mu_1-\mu_2) \le \|\sqrt{\varpi}\|_0\|X_1-X_2\|_{0,\varpi}. \tag{8.46}$$

If $\mathbf{B}$ is some function space, we denote the cone of functions with nonnegative values by $\mathbf{B}_+$. From (8.46) we obtain

$$\left.\begin{array}{c}\mathbf{H}_{0,\varpi,+} \subset \mathbf{M}_{\infty,\varpi}\,, \\ C([s,T];\mathbf{H}_{0,\varpi,+}) \subset C([s,T];\mathbf{M}_{\infty,\varpi}) \;\forall\; 0 \le s < T < \infty\end{array}\right\} \tag{8.47}$$

with continuous imbeddings. Since

$$\mathbf{M}_{\infty,\varpi} \subset \mathcal{S}',$$

where $\mathcal{S}'$ is the space of tempered distributions over $\mathbf{R}^d$,[35] these imbeddings are also dense.

Lemma 8.10. *Let $Y(t,Z,Y_s)$ and $\tilde{Y}(t,\tilde{Z},\tilde{Y}_s)$ be two solutions of (8.25), whose initial conditions satisfy the assumptions of Theorem 8.6 with $|\mathbf{m}| =: m = 2$, $p = 2$, and smoothness degree $\overline{m} > d+4$ in addition to (8.45) and Hypothesis 4.1 with $\gamma = \gamma_\varpi$. Further, assume that $Z, \tilde{Z} \in L_{loc,2,\mathcal{F}}(C((s,T];\mathbf{H}_{0,\varpi,+})) \cap L_{loc,2,\mathcal{F}}(C([s,T];\mathbf{M}_{\infty,\varpi}))$. Then there are a finite $c_{F,\mathcal{J},\varpi}$ and a sequence of stopping times $\tau_N \ge s$ which are localizing for $Z, \tilde{Z}, Y_s$, and $\tilde{Y}_s$ such that for any $T > s$*

$$\left.\begin{array}{l}\sup\limits_{s\le t\le T} E[\|Y(t\wedge\tau_N,Y_s) - \tilde{Y}(t\wedge\tau_N,\tilde{Y}_s)\|_{0,\varpi}^2 1_{\{\tau_N>0\}}] \\ \le \exp((T-s)c_{F,\mathcal{J},\varpi})N\left[E\left\{\displaystyle\int_s^{T\wedge\tau_N} \|Z(u)-\tilde{Z}(u))\|_{0,\varpi}^2 1_{\{\tau_N>s\}}du\right.\right. \\ \qquad\qquad\qquad\qquad \left.+\|Y_s-\tilde{Y}_s\|_{0,\varpi}^2 1_{\{\tau_N>s\}}\right].\end{array}\right\} \tag{8.48}$$

The *Proof* uses the same ideas as the proof of Theorem 8.8 and is given in Sect. 12. □

Lemma 8.10 implies the following uniqueness theorem for (8.26).

Theorem 8.11. *Suppose the conditions of Theorem 8.6 hold with $|\mathbf{m}| =: m = 2$, $p = 2$, and smoothness degree $\overline{m} > d+4$ in addition to (8.45). Further, suppose the initial condition X_s for (8.26) is in $L_{0,\mathcal{F}_s}(\mathbf{H}_{0,\Phi,+})$, $\Phi \in \{1,\varpi\}$. Then the following holds:*

If there is a solution $X(\cdot,X_s) \in L_{loc,2,\mathcal{F}}(C((s,T];\mathbf{H}_{0,\Phi,+}))$ of (8.26), then this solution is unique.

Proof. Again we may assume that, without loss of generality, the initial condition is in $L_{2,\mathcal{F}_s}((\mathbf{H}_{0,\varpi,+})$. Let there be two solutions $X(\cdot,X_s)$ and $\tilde{X}(\cdot,X_s)$ in $L_{2,\mathcal{F}}(C([s,T];\mathbf{H}_{0,\Phi,+}))$. Then both solutions are also in $L_{2,\mathcal{F}}(C([s,T];\mathbf{H}_{0,\varpi,+}))$, and we may consider both solutions as solutions of (8.25) with input processes X and $\tilde{X}$, respectively. By (8.48) and the Gronwall inequality

$$X(t\wedge\tau_N,X_s) \equiv \tilde{X}(t\wedge\tau_N,X_s).$$

Our assumptions imply that $\tau_N \longrightarrow \infty$ a.s., as $N \longrightarrow \infty$, where N is the bound from (8.44) with $\Phi = \varpi$. □

[35] Cf. (15.32) in Sect. 15.1.3.

We next derive a flow representation for the solutions of (8.25)/(8.26). Consider test functions $\varphi \in C_c^1(\mathbf{R}^d, \mathbf{R})$ and let $\mu \in \mathbf{M}_f$. Then the duality

$$\langle \varphi, \mu \rangle := \int \varphi(r)\mu(\mathrm{d}r). \tag{8.49}$$

is a continuous bilinear form between $(\mathbf{M}_f, \gamma_f)$ and $C_c^1(\mathbf{R}^d, \mathbf{R})(\subset C_{L,\infty}(\mathbf{R}^d; \mathbf{R}))$.

Theorem 8.12. *Let $r(t, \tilde{\mathcal{Y}}, s, q)$ be the solution of (4.9) with "input" $\tilde{\mathcal{Y}}(\cdot)$ and let $r(t, \mathcal{X}, s, q)$ be the solution of (4.10) with "input" $\mathcal{X}(\cdot)$. Both solutions start at s in deterministic q and, without loss of generality, are the versions, measurable in all parameters, as constructed in Statement 3 of Theorem 4.5. Further, let $\mathcal{Y}(t) := \mathcal{Y}(t, \tilde{\mathcal{Y}}, s, \mathcal{Y}(s))$ be the solution of (8.25) with input process $\tilde{\mathcal{Y}}(\cdot) \in L_{loc,2,\mathcal{F}}(C((s,T]; \mathbf{M}))$ and initial value $\mathcal{Y}(s)$ at s and $\mathcal{X}(t) := \mathcal{X}(t, s, \mathcal{X}(s))$ be the solution of (8.26) with initial value $\mathcal{X}(s)$ at s. We then have the following flow representations:*

$$\left.\begin{aligned} \mathcal{Y}(t) &= \int \delta_{(r(t,\tilde{\mathcal{Y}},s,q))}\mathcal{Y}(s, \mathrm{d}q) = \mathcal{Y}(s)(r^{-1}(t, \tilde{\mathcal{Y}}, s, q))(\cdot), \\ \mathcal{X}(t) &= \int \delta_{(r(t,\mathcal{X},s,q))}\mathcal{X}(s, \mathrm{d}q) = \mathcal{X}(s)(r^{-1}(t, \mathcal{X}, s, q))(\cdot). \end{aligned}\right\} \tag{8.50}$$

Moreover, if uniqueness holds, then for $t \geq s \geq 0$[36]

$$\int \delta_{(r(t,\mathcal{X},0,q))}\mathcal{X}(0, \mathrm{d}q) = \int \delta_{(r(t,\mathcal{X},s,q))}\mathcal{X}(s, \mathrm{d}q). \tag{8.51}$$

Further, for $\varphi \in C_c^3(\mathbf{R}^d, \mathbf{R})$, Itô's formula implies[37]

$$\left.\begin{aligned} \langle \varphi, \mathcal{Y}(t, \mathcal{X}_s)\rangle &= \langle \varphi, \mathcal{Y}_s\rangle + \int_s^t \Bigg\langle \frac{1}{2}\sum_{k,\ell=1}^d (\partial_{k\ell}^2\varphi)(r(u, \tilde{\mathcal{Y}}, \cdot))D_{k\ell}(\tilde{\mathcal{Y}}, \cdot, u) \\ &\quad + (\nabla\varphi)(r(u, \tilde{\mathcal{Y}}, \cdot)) \cdot F(\cdot, \tilde{\mathcal{Y}}, u), \mathcal{Y}(s) \Bigg\rangle \mathrm{d}u \\ &\quad + \int_s^t \Big\langle \nabla\varphi(r(u, \mathcal{X}, \cdot)) \cdot \mathcal{J}(\cdot, p, \tilde{\mathcal{Y}}, u), \mathcal{Y}(s)\Big\rangle w(\mathrm{d}p, \mathrm{d}u) \\ &= \Bigg\langle \varphi, \mathcal{Y}_s\rangle + \int_s^t \langle \frac{1}{2}\sum_{k,\ell=1}^d (\partial_{k\ell}^2\varphi)(\cdot)D_{k\ell}(\tilde{\mathcal{Y}}, \cdot, u) \\ &\quad + (\nabla\varphi)(\cdot) \cdot F(\cdot, \tilde{\mathcal{Y}}, u), \mathcal{Y}(u, \tilde{\mathcal{Y}}, \mathcal{Y}(s)) \Bigg\rangle \mathrm{d}u \\ &\quad + \int_s^t \Big\langle \nabla\varphi(\cdot) \cdot \mathcal{J}(\cdot, p, \tilde{\mathcal{Y}}, u), \mathcal{Y}(u, \tilde{\mathcal{Y}}, \mathcal{Y}(s))\Big\rangle w(\mathrm{d}p, \mathrm{d}u). \end{aligned}\right\} \tag{8.52}$$

[36] Cf. (4.1). The formula is similar to formulas applied in Markov processes. Cf., e.g., Dynkin (1965) or Ethier and Kurtz (1986).

[37] The assumption $\varphi \in C_c^3(\mathbf{R}^d, \mathbf{R})$ implies that the first and second partial derivatives of φ are in $C_c^1(\mathbf{R}^d, \mathbf{R}) \subset C_{L,\infty}(\mathbf{R}^d; \mathbf{R})$. Therefore, the duality between those derivatives of the test function and the initial measure is continuous.

Substituting $\mathcal{X}(t, \mathcal{X}(s))$ for $\mathcal{Y}(t)$ and $\tilde{\mathcal{Y}}(t)$ in (8.52), we obtain the analogous statement for the solution of (8.26).[38]

Proof. Without loss of generality we may restrict the proof to the quasilinear SPDE (8.26). Note that $\mathcal{X}(\cdot, \mathcal{X}_s) \in L_{\mathrm{loc},2,\mathcal{F}}(C((s,T]; \mathbf{M}_f))$. Therefore, $\int \delta_{(r(t,\mathcal{X},s,q))}\mathcal{X}(s, \mathrm{d}q)$ is well defined and an element from $\mathbf{M}_f$ for all ω. Starting in just N points, i.e., $\mathcal{X}(s) = \mathcal{X}_N(s)$, we have $\mathcal{X}(t, s, \mathcal{X}_N(s)) = \mathcal{X}_N(t)$, whence (8.50) reduces to the definition of the empirical process at time $t \geq s$. The uniqueness of the extension from discrete to more general initial distributions, derived in Theorem 8.3, establishes (8.50). Equation (8.51) is obvious.

We apply the Itô formula and obtain

$$
\begin{aligned}
\langle \varphi, \mathcal{Y}(t, \mathcal{X}_s)\rangle &= \langle \varphi, \mathcal{Y}_s\rangle + \Big\langle \int_s^t \{\frac{1}{2}\sum_{k,\ell=1}^{d} (\partial^2_{k\ell}\varphi)(r(u, \tilde{\mathcal{Y}}, \cdot))D_{k\ell}(\tilde{\mathcal{Y}}, \cdot, u) \\
&\quad + (\triangledown\varphi)(r(u, \tilde{\mathcal{Y}}, \cdot)) \cdot F(\cdot, \tilde{\mathcal{Y}}, u)\}\mathrm{d}u, \mathcal{Y}(s)\Big\rangle \\
&\quad + \Big\langle \int_s^t \{\triangledown\varphi(r(u, \mathcal{X}, \cdot)) \cdot \mathcal{J}(\cdot, p, \tilde{\mathcal{Y}}, u)\}w(\mathrm{d}p, \mathrm{d}u), \mathcal{Y}(s)\Big\rangle.
\end{aligned}
$$

Fubini's theorem implies immediately that we may change the order of integration in the deterministic integral with respect to du. For the stochastic integral, we employ the series representation (4.15) in the change of the order of integration and the continuity of the bilinearity $\langle \cdot, \cdot\rangle$. This establishes the first identity in (8.52). The second follows from (8.50). □

Remark 8.13. We show that the SPDEs (8.25)/(8.26) are noncoercive. In (8.26) we have

$$
\left.
\begin{aligned}
A(t, X) &:= \Big(\frac{1}{2}\sum_{k,\ell=1}^{d} \partial^2_{k\ell,r}(D_{k\ell}(X, \cdot, t)X\Big), \\
B(s, X)\mathrm{d}W &:= -\triangledown\cdot(X\int \mathcal{J}(\cdot, p, X)w(\mathrm{d}p, \mathrm{d}t)).
\end{aligned}
\right\} \tag{8.53}
$$

Employing (4.15) we can easily rewrite the stochastic term in terms of an infinite dimensional (cylindrical) Wiener process. Apart from the fact that in (8.25)/(8.26) the Wiener process is cylindrical (cf. Sect. 15.2.2, Proposition 15.29), the differential operator in (8.53) is quasilinear. More importantly, the Itô formula (8.43) shows that the SPDEs (8.26) is not coercive in the sense of definition (8.3), where we now consider the interpolation triplet[39]

$$\mathbf{H}_1 \subset \mathbf{H}_0 = \mathbf{H}_0' \subset \mathbf{H}_{-1} := \mathbf{H}_1'.$$

As a consequence of the noncoercivity of the pair, the smoothness of the solution of (8.26) is, at best, equal to the smoothness of the initial condition. (Similarly for (8.25).) □

[38] Expression (8.52) provides another proof that the empirical processes $\mathcal{Y}(\cdot)$ and $\mathcal{X}(\cdot)$ are weak solutions of (8.25) and (8.26), respectively.

[39] Cf. (11.37).

8.4 Coercive and Noncoercive SPDEs

A more general class of SPDEs, which can be coercive, is analyzed.

We now discuss a more general case which contains the noncoercive SPDEs of Sect. 8.3 as a special case and includes a class of coercive SPDEs as well. Most of what follows is, to some extent, a special case of the Kurtz and Xiong result, namely the case where the mass is conserved. Within the class of mass conserved equations, however, our method of extension by continuity appears to allow more generality than the use of exchangeability employed by Kurtz and Xiong. Accordingly, we will work with the empirical processes of (4.9) and (4.10), respectively.

The second martingale term in (4.9)/(4.10) is tagged by the noise processes $\{\beta^{\perp,i}\}$ and the coefficients $\{\sigma_i^{\perp}\}$. With respect to the latter, we introduce the following simplifying assumption for this chapter.

Hypothesis 8.2

$$\forall i \;\; \sigma_i^{\perp} = \sigma_1^{\perp} =: \sigma^{\perp}.$$

Assuming (8.35) in addition to Hypotheses 4.1 and 8.2, the following SPDE is shown to be the continuum limit of the empirical process associated to (4.10):

$$\left.\begin{aligned} d\mathcal{X} = &\left(\frac{1}{2}\sum_{k,\ell=1}^{d} \partial^2_{k\ell,r}\left(\overline{\overline{D}}_{k\ell}(\mathcal{X},\cdot,t)\mathcal{X}\right) - \nabla\cdot(\mathcal{X}F(\mathcal{X},t))\right)dt \\ &-\nabla\cdot\left(\mathcal{X}\int \mathcal{J}(\cdot,p,\mathcal{X},t)w(dp,dt)\right). \end{aligned}\right\} \tag{8.54}$$

The coefficients $\mathcal{J}$, F, and the Gaussian white noise $w(dq, dt)$ in (8.54) are the same as in (8.25)/(8.26). The only difference is the diffusion matrix. Recalling the definition of $D_{k\ell}(\mu, r, t)$ from (8.23), the diffusion matrix in (8.54) is defined by

$$\overline{\overline{D}}(\mu), r, t) = D(\mu, r, t) + (\sigma^{\perp}(r, \mu, t))^2. \tag{8.55}$$

Remark 8.14. Before showing that (8.54) has a (unique) solution for certain initial values, let us assume there is a smooth solution $X(\cdot)$ in $\mathbf{H}_0$ which is in the domain of the operator

$$\left.\begin{aligned} &\overline{\overline{A}}(t,X) := A(t,X) + \tilde{A}(t,X), \\ &\text{where} \\ &A(t,X) := \frac{1}{2}\sum_{k,\ell=1}^{d} \partial^2_{k\ell,r}(D_{k\ell}(X,\cdot,t)X), \\ &\tilde{A}(t,X) := \frac{1}{2}\sum_{k,\ell=1}^{d} \partial^2_{k\ell,r}(((\sigma^{\perp}(\cdot,X,t))^2)_{k\ell}X). \end{aligned}\right\} \tag{8.56}$$

A straightforward generalization of the Itô formula (8.43) yields

$$\left.\begin{aligned}&\|X(t)\|_0^2 = \|X(0)\|_0^2\\&+\int_0^t \{\langle 2\tilde{A}(s, X(s)), X(s)\rangle_0 - \frac{1}{2}\sum_{k,\ell=1}^{\mathrm{d}}\int_s^t \int Y^2(u,r)\mathrm{d}r\mathcal{L}_{k,\ell,1_k+1_\ell}(u)\mathrm{d}u\\&+\sum_{\ell=1}^{d}\int_s^t \int Y^2(u,r)\partial_\ell F_\ell(r,u)\mathrm{d}r\ \mathrm{d}u + \sum_{\ell=1}^{d}\int_s^t \int Y^2(u,r)\mathrm{d}\partial_\ell m_\ell(u,r))\mathrm{d}r,\end{aligned}\right\} \tag{8.57}$$

where, recalling the abbreviations from (8.34),

$$\mathrm{d}m(s) := \int \mathcal{J}(\cdot, p, s)w(\mathrm{d}p, \mathrm{d}s).$$

The trace of the unbounded operator part in the tensor quadratic variation of the martingale noise term has already disappeared, as was shown in the proof of Theorem 8.8 that it canceled against the term $\langle 2A(s, X(s)), X(s)\rangle_0$.[40] Recall that, by assumption (8.35), the functions $\mathcal{L}_{k,\ell,1_k+1_\ell}(u)$ and $\partial_\ell F_\ell(r,u)$ are bounded. Further, suppose, in addition to the previous assumptions, that $\sigma^\perp$ is positive definite. Then integration by parts implies

$$\langle 2\tilde{A}(s, X(s)), X(s)\rangle_0 \le -\gamma_1\|X(s)\|_1^2 + \gamma_2\|X(s)\|_0^2, \tag{8.58}$$

where γ_1 and γ_2 are both positive constants and $\|\cdot\|$ is the Hilbert norm on $\mathbf{H}_1$. Hence, we obtain from (8.55)

$$\left.\begin{aligned}&\|X(t)\|_0^2 \le \|X(0)\|_0^2 - \int_0^t \gamma_1\|X(s)\|_1^2\mathrm{d}s + \text{const.}\int_0^t \|X(s)\|^2\mathrm{d}s\\&+\sum_{\ell=1}^{\mathrm{d}}\int_s^t\int Y^2(u,r)\mathrm{d}\partial_\ell M_\ell(u,r))\mathrm{d}r.\end{aligned}\right\} \tag{8.59}$$

This is essentially the same bound as in (8.7), whence we have shown that (8.54) is coercive.

We now construct a solution of (8.54) as the empirical distribution of an infinite system of SODEs, extending the finite system (4.10). As in Chap. 2, we first add an empty state to $\mathbf{R}^d$ which will be again denoted $\diamond$. Recall $\hat{\mathbf{R}}^d := \mathbf{R}^d \cup \{\diamond\}$, and that the usual metric, $\rho(\cdot)$, on $\mathbf{R}^d$ is extended in (2.5) to the metric $\hat{\rho}(\cdot)$ on $\hat{\mathbf{R}}^d$. We embed finite and infinite $\mathbf{R}^d$-valued sequences into $\hat{\mathbf{R}}^d$-valued infinite sequences as follows

$$\left.\begin{aligned}&(r^1, r^2, \ldots, r^N) \in \mathbf{R}^{Nd} \longrightarrow (r^1, r^2, \ldots, r^N, \diamond, \ldots) \in (\hat{\mathbf{R}}^d)^{\mathbf{N}}, \text{ if the } \mathbf{R}^d\text{-valued sequence is finite,}\\&(r^1, r^2, \ldots, r^N, \ldots) \in (\mathbf{R}^d)^{\mathbf{N}} \longrightarrow (r^1, r^2, \ldots, r^N, \ldots) \in (\hat{\mathbf{R}}^d)^{\mathbf{N}}, \text{ if the } \mathbf{R}^d\text{-valued sequence is infinite,}\end{aligned}\right\} \tag{8.60}$$

[40] Cf. (11.66).

i.e., infinite sequences remain unchanged, and finite sequences of length N become infinite sequences by adding $\diamond$ at all coordinates $N+i, i = 1, 2, \ldots$ We denote infinite sequences by $r^{(\cdot)}$ and finite sequences of length N by $r^{(\cdot \wedge N)}$. Further, we endow $(\hat{\mathbf{R}}^d)^{\mathbf{N}}$ with the metric

$$d_\infty\left(r^{(\cdot)}, q^{(\cdot)}\right) := \sum_{i=1}^{\infty} \left(\frac{1}{2}\right) n/2\hat{\rho}(r^i, q^i). \tag{8.61}$$

We note that $((\hat{\mathbf{R}}^d)^{\mathbf{N}}, d_\infty)$ is a complete separable metric space. Since $(\mathbf{M}_f, \gamma_f)$ is also complete and separable, the same holds for the Cartesian product

$$((\hat{\mathbf{R}}^d)^{\mathbf{N}} \times \mathbf{M}_f, \hat{d}), \quad \text{where} \quad \hat{d}((r^{(\cdot)}, \mathcal{X}), (q^{(\cdot)}, \mathcal{Y})) := \sqrt{d_\infty^2(r^{(\cdot)}, q^{(\cdot)}) + \gamma_f^2(\mathcal{X} - \mathcal{Y})}. \tag{8.62}$$

Consider the following subset of $(\hat{\mathbf{R}}^d)^{\mathbf{N}} \times \mathbf{M}_f$, which consists of all finite sequences and all possible associated empirical measures:

$$\hat{\mathbf{M}}_{fin} := \left\{\left(r^{(\cdot \wedge N)}, \sum_{i=1}^{N} m_i \delta_{r^i}\right), r^i \in \mathbf{R}^d, m_i \geq 0, i = 1, \ldots, N, \quad N \geq 1\right\}. \tag{8.63}$$

Set

$$\hat{\mathbf{M}}_f := \overline{\hat{\mathbf{M}}_{fin}}, \tag{8.64}$$

where the right-hand side in (8.64) is the closure of $\hat{\mathbf{M}}_{fin}$ in $((\hat{\mathbf{R}}^d)^{\mathbf{N}} \times \mathbf{M}_f, \hat{d})$. The following properties are obvious.

Proposition 8.15. *$(\hat{\mathbf{M}}_f, \hat{d})$ is a complete separable metric space, and the set $\hat{\mathbf{M}}_{fin}$ is dense in $(\hat{\mathbf{M}}_f, \hat{d})$.* □

The following lemma provides an a priori estimate which we need for the extension by continuity.

Lemma 8.16. *Assume Hypotheses 4.1 and 8.2 in addition to the assumptions of Theorem 4.7. Let $\{r^i(\cdot)\}^N, \{q^i(\cdot)\}^M$ be two solutions of (4.10) with empirical processes $\mathcal{X}_N(\cdot)$ and $\mathcal{Y}_M(\cdot)$, respectively, where $M \leq N$ and suppose that both $\mathcal{X}_N(s)$ and $\mathcal{Y}_M(s)$ are in $L_{2,\mathcal{F}_s}(\mathbf{M}_f)$. Then for any $T > 0$ there is a finite constant c_T, depending only on the coefficients of the equation (4.10) and on the bounds on the second moments of the initial empirical distributions, such that*

$$\left.\begin{aligned} &E \sup_{s \leq t \leq T} \hat{d}^2\left(\left(r^{(\cdot \wedge N)}(t), \mathcal{X}_N(t)\right), \left(q^{(\cdot \wedge M)}(t), \mathcal{Y}_M(t)\right)\right) \\ &\leq c_T E\hat{d}^2\left(\left(r^{(\cdot \wedge N)}(s), \mathcal{X}_N(s)\right), \left(q^{(\cdot \wedge M)}(s), \mathcal{Y}_{(s)}\right)\right). \end{aligned}\right\} \tag{8.65}$$

Proof. By a slight abuse of notation we will formally add up all components in the finite sequences from 1 to ∞, bearing in mind that the infinite tails of those sequences consist of the constant value $\diamond$. Without loss of generality, we assume $s = 0$.

$$
\begin{aligned}
&E \sup_{0\le t\le T} \hat{d}^2((r^{(\cdot\wedge N)}(t), \mathcal{X}_N(t)), (q^{(\cdot\wedge M)}(t), \mathcal{Y}_M(t))) \\
&= E \sup_{0\le t\le T} \sum_{i=1}^{\infty} \frac{1}{2^i}\hat{\rho}^2(r^i(t), q^i(t)) + E \sup_{0\le t\le T} \gamma_f^2(\mathcal{X}_N(t) - \mathcal{Y}_M(t)) \\
&=: I_T + II_T.
\end{aligned}
$$

By (4.17) in Theorem 4.5 there is a finite $\tilde{c}_T$ such that

$$
\begin{aligned}
I_T &\le \sum_{i=1}^{M} \frac{1}{2^i} E \sup_{0\le t\le T} \hat{\rho}^2(r^i(t), q^i(t)) + \sum_{i=M+1}^{N} \frac{1}{2^i} \\
&\le \tilde{c}_T \sum_{i=1}^{M} \frac{1}{2^i} E\hat{\rho}^2(r^i(0), q^i(0)) + \sum_{i=M+1}^{N} \frac{1}{2^i} \\
&\quad + \sum_{i=1}^{M} \frac{1}{2^i} \int_0^T \gamma_f^2(\mathcal{X}_N(s) - \mathcal{Y}_M)(s)\mathrm{d}s \\
&= \tilde{c}_T E d_\infty^2(r^{(\cdot)}(0), q^{(\cdot)}(0)) + \sum_{i=1}^{M} \frac{1}{2^i} \int_0^T \gamma_f^2(\mathcal{X}_N(s), \mathcal{Y}_M(s))\mathrm{d}s \\
&\le \tilde{c}_T E d_\infty^2(r^{(\cdot)}(0), q^{(\cdot)}(0)) + T E \sup_{0\le t\le T} \gamma_f^2(\mathcal{X}_N(t) - \mathcal{Y}_M(t)).
\end{aligned}
$$

Hence, (8.65) follows with $c_T := \tilde{c}_T + T$. □

We are now ready to derive the extension by continuity, which is a direct generalization of Theorem 8.4.

Theorem 8.17. *Suppose $\mathcal{X}(s) \in L_{0,\mathcal{F}_s}(\mathbf{M}_f)$ in addition to Hypotheses 4.1 and 8.2. Choose an infinite system of $\mathcal{F}_s$-measurable random variables $r_s^{(\cdot)}$ and $m_{i,N} > 0$, $i = 1, 2, \ldots$, such that $\sup_N \sum_{i=1}^{N} m_{i,N} < \infty$ and such that*

$$
\mathcal{X}(s) = \lim_{N\to\infty} \sum_{i=1}^{N} m_{i,N}\delta_{r_s^i} \quad \text{in } L_{0,\mathcal{F}_s}(\mathbf{M}_f)\,.
$$

Then there is a unique empirical process $\mathcal{X}(\cdot) \in L_{loc,2,\mathcal{F}}(C((s,T];\mathbf{M}_f))$, associated with a unique system of solutions of (4.10), $r^{(\cdot)}(t) := r(\cdot,\mathcal{X},r_s)^{(\cdot)} := (r^1(\cdot,\mathcal{X},r_s^1), r^2(\cdot,\mathcal{X},r_s^2),\ldots)$, such that for all $t \geq s$

$$E \sup_{s\leq t\leq T\wedge\tau} \gamma_f^2(\mathcal{X}(t) - \mathcal{X}_N(t)) \longrightarrow 0, \text{ as } N \to \infty\,. \tag{8.66}$$

Further, let there be another $\mathcal{Y}(s) \in L_{0,\mathcal{F}_s}(\mathbf{M}_f)$ and an infinite system of $\mathcal{F}_s$-measurable random variables $q_s^{(\cdot)}$ and point masses $\tilde{m}_{i,M} > 0$ such that $\sup_M \sum_{i=1}^M \tilde{m}_{i,M} < \infty$ and $\mathcal{Y}(s) = \lim_{M\to\infty} \sum_{i=1}^\infty \tilde{m}_{i,M}\delta_{q_s^i}$ in $L_{0,\mathcal{F}_s}(\mathbf{M}_f)$. Let $\mathcal{Y}(\cdot)$ denote the associated empirical process for the system $q^{(\cdot)}(t) := q(\cdot,\mathcal{X},q_s)^{(\cdot)} := (q^1(\cdot,\mathcal{X},q_s^1), q^2(\cdot,\mathcal{X},q_s^2),\ldots)$ of solutions of (4.10) and τ be a stopping time, which is localizing for both initial conditions. The following estimate then holds:

$$\left.\begin{array}{l} E\left\{\sup_{s\leq t\leq T\wedge\tau} \hat{d}^2((r^{(\cdot)}(t),\mathcal{X}(t)),(q^{(\cdot)}(t),\mathcal{Y}(t)))1_{\{\tau>s\}}\right\} \\ \leq c^2_{T,F,\mathcal{J},\sigma^\perp,\tau} E\{\hat{d}^2((r_s^{(\cdot)},\mathcal{X}(s)),(q_s^{(\cdot)},\mathcal{Y}(s)))1_{\{\tau>s\}}\}. \end{array}\right\} \tag{8.67}$$

Moreover, if $m(\omega) := \mathcal{X}_s(\omega,\mathbf{R}^d)$ is the random mass at time s, then with probability 1 uniformly in $t \in [s,T]$

$$\mathcal{X}(t,\omega,\mathcal{X}_s(\omega)) \in \mathbf{M}_{m(\omega)}. \tag{8.68}$$

Proof. Expression (8.67) follows from (8.65) and the extension by continuity. Expression (8.68) is a consequence of the the mass conservation. □

Define the projections

$$\left.\begin{array}{l} \pi_2 : \hat{\mathbf{M}}_f \longrightarrow \mathbf{M}_f \\ \pi_2((r^{(\cdot)},\mu)) = \mu\,. \end{array}\right\} \tag{8.69}$$

The following theorem generalizes Theorem 8.5.

Theorem 8.18. *Suppose $\mathcal{X}_s \in L_{0,\mathcal{F}_s}(\mathbf{M}_f)$ and that the conditions of Theorem 4.7 hold in addition to Hypotheses 4.1 and 8.2. Further, suppose on the initial mass distribution*

$$\max_{i=1,\ldots,N} m_{i,N} \longrightarrow 0 \text{ as } N \longrightarrow \infty\,. \tag{8.70}$$

Then $\mathcal{X}(\cdot,\mathcal{X}_s)$, the projection π_2 of process obtained in Theorem 8.17, is a solution of (8.54).

Proof.

(i) Without loss of generality, $s = 0$. Let $\varphi \in C_c^3(\mathbf{R}^d;\mathbf{R})$ and $\mathcal{X}_N(t) := \sum_{i=1}^N m_{i,N}\delta_{r^i(t)}$.

Itô's formula yields

$$
\begin{aligned}
\langle \mathrm{d}\mathcal{X}_N(t), \varphi\rangle &= \mathrm{d}\langle \mathcal{X}_N(t), \varphi\rangle = \sum\nolimits_{j=1}^{N} m_{j,N} d\varphi(r^j(t)) \\
&= \sum\nolimits_{j=1}^{N} m_{j,N} (\nabla\varphi)(r^j(t)) \cdot F(r^j(t), \mathcal{X}_N, t)\mathrm{d}t \\
&\quad + \sum\nolimits_{j=1}^{N} m_{j,N} (\nabla\varphi)(r^j(t)) \cdot (\sigma^{\perp}(r^j(t), \mathcal{X}_N, t)\beta^{\perp,j}(\mathrm{d}t)) \\
&\quad + \sum_{j=1}^{N} m_{j,N} (\nabla\varphi)(r^j(t)) \cdot \int \mathcal{J}(r^j(t), p, \mathcal{X}_N, t) w(\mathrm{d}p, \mathrm{d}t) \\
&\quad + \sum\nolimits_{j=1}^{N} m_{j,N} \tfrac{1}{2} \sum\nolimits_{k,\ell=1}^{d} (\partial^2_{k\ell}\varphi)(r^j(t))(\sigma^{\perp})^2_{k\ell}(r^j(t), \mathcal{X}_N, t)\mathrm{d}t \\
&\quad + \sum_{j=1}^{N} m_{j,N} \frac{1}{2} \sum_{k,\ell=1}^{d} (\partial^2_{k\ell}\varphi)(r^j(t)) D_{k\ell}(\mathcal{X}_N, r^j(t), t)\mathrm{d}t \\
&= \langle \mathcal{X}_N(t), (\nabla\varphi)(\cdot) \cdot F(\cdot, \mathcal{X}_N, t)\mathrm{d}t\rangle \\
&\quad + \langle \mathcal{X}_N(t), (\nabla\varphi)(\cdot) \cdot \int \mathcal{J}(\cdot, p, \mathcal{X}_N, t) w(\mathrm{d}p, \mathrm{d}t)\rangle \\
&\quad + \left\langle \mathcal{X}_N(t), \tfrac{1}{2} \sum\nolimits_{k,\ell=1}^{d} (\partial^2_{k\ell}\varphi)(\cdot) D_{k\ell}(\mathcal{X}_N, \cdot, t)\right\rangle \mathrm{d}t \\
&\quad + \left\langle \mathcal{X}_N(t), \tfrac{1}{2} \sum\nolimits_{k,\ell=1}^{d} (\partial^2_{k\ell}\varphi)(\cdot)(\sigma^{\perp})^2_{k\ell}(\cdot, \mathcal{X}_N, t)\right\rangle \mathrm{d}t \\
&\quad + \sum_{j=1}^{N} m_{j,N} (\nabla\varphi)(r^j(t)) \cdot (\sigma^{\perp}(r^j(t), \mathcal{X}_N, t)\beta^{\perp,j}(\mathrm{d}t)) \\
&\quad \text{(by the definition of the duality between the measure } \mathcal{X}_N(t) \\
&\qquad \text{and the corresponding test functions)} \\
&= \left\langle \frac{1}{2} \sum_{k,\ell=1}^{d} D_{k\ell}(\mathcal{X}_N, \cdot, t)\mathcal{X}_N(t), \partial^2_{k\ell}\varphi \right\rangle \mathrm{d}t \\
&\quad + \left\langle \frac{1}{2} \sum_{k,\ell=1}^{d} (\sigma^{\perp})^2_{k\ell}(\cdot, \mathcal{X}_N, t)\mathcal{X}_N(t), \partial^2_{k\ell}\varphi \right\rangle \mathrm{d}t \\
&\quad + \sum_{k=1}^{d} \langle \mathcal{X}_N(t) F_k(\cdot, \mathcal{X}_N, t)\mathrm{d}t, \partial_k\varphi\rangle \\
&\quad + \sum_{k=1}^{d} \langle \mathcal{X}_N(t) \int \mathcal{J}_{k\cdot}(\cdot, p, \mathcal{X}_N, t) w(\mathrm{d}p, \mathrm{d}t), \partial_k\varphi\rangle \\
&\quad + \sum_{j=1}^{N} m_{j,N} (\nabla\varphi)(r^j(t)) \cdot (\sigma^{\perp}(r^j(t), \mathcal{X}_N, t)\beta^{\perp,j}(\mathrm{d}t)) \\
&\quad \text{(by rearranging terms in the duality and changing the order} \\
&\qquad \text{of summation of the brackets } \langle \cdot, \cdot \rangle),
\end{aligned}
$$

i.e.,

$$
\begin{aligned}
\langle \mathrm{d}\mathcal{X}_N(t), \varphi\rangle = & \left\langle \frac{1}{2}\sum_{k,\ell=1}^{d} \partial^2_{k\ell}(D_{k\ell}(\mathcal{X}_N, \cdot, t)\mathcal{X}_N(t))\mathrm{d}t, \varphi \right\rangle \\
& + \left\langle \frac{1}{2}\sum_{k,\ell=1}^{d} \partial^2_{k\ell}((\sigma^{\perp})^2_{k\ell}(\cdot, \mathcal{X}_N, t)\mathcal{X}_N(t))\mathrm{d}t, \varphi \right\rangle \\
& - \langle \nabla \cdot (\mathcal{X}_N(t)F(\cdot, \mathcal{X}_N, t))\mathrm{d}t, \varphi\rangle \\
& - \langle \nabla \cdot (\mathcal{X}_N(t)\int \mathcal{J}(\cdot, p, \mathcal{X}_N, t)w(\mathrm{d}p, \mathrm{d}t)), \varphi\rangle \\
& + \sum_{j=1}^{N} m_{j,N}(\nabla\varphi)(r^j(t)) \cdot (\sigma^{\perp}(r^j(t), \mathcal{X}_N, t)\beta^{\perp,j}(\mathrm{d}t)) \\
& \text{(integrating by parts in the generalized sense).}
\end{aligned}
$$

Recalling the abbreviation of $\overline{\overline{D}}_{k\ell}(\mathcal{X}_N, \cdot, t)$ from (8.55), we obtain from the last equation

$$
\left.
\begin{aligned}
\langle \mathrm{d}\mathcal{X}_N(t), \varphi\rangle = & \left\langle \frac{1}{2}\sum_{k,\ell=1}^{d} \partial^2_{k\ell}(\overline{\overline{D}}_{k\ell}(\mathcal{X}_N, \cdot, t)\mathcal{X}_N(t))\mathrm{d}t, \varphi \right\rangle \\
& - \langle \nabla \cdot (\mathcal{X}_N(t)F(\cdot, \mathcal{X}_N, t))\mathrm{d}t, \varphi\rangle \\
& - \langle \nabla \cdot (\mathcal{X}_N(t)\int \mathcal{J}(\cdot, p, \mathcal{X}_N, t)w(\mathrm{d}p, \mathrm{d}t)), \varphi\rangle \\
& + \sum_{j=1}^{N} m_{j,N}(\nabla\varphi)(r^j(t)) \cdot (\sigma^{\perp}(r^j(t), \mathcal{X}_N, t)\beta^{\perp,j}(\mathrm{d}t)) \\
& \text{(integrating by parts in the generalized sense).}
\end{aligned}
\right\}
\tag{8.71}
$$

Apparently, the last term cannot be converted into a functional of measures, independent of N, because the noise terms are tagged. If, however, in the limit this term disappears we obtain a weak form of an SPDE for the limiting empirical process (cf. (8.24)).

(ii) We estimate the variance of the tagged term from (8.71)[41]:

$$
\left.
\begin{aligned}
& E\left\{\int_0^t \sum_{i=1}^{N} m_{i,N}(\nabla\varphi)(r_i(s)) \cdot (\sigma^{\perp}(r^i(s), \mathcal{X}_N, s)\beta^{\perp,i}(\mathrm{d}s))\right\}^2 \\
& = \sum_{i=1}^{N} m^2_{i,N}\int_0^t E\frac{\partial}{\partial r_k}\varphi(r^i(s))\frac{\partial}{\partial r_\ell}\varphi(r^i(s))\sum_{m=1}^{d}\sigma^{\perp}_{km}(r^i(s), \mathcal{X}_N, s)\sigma^{\perp}_{\ell m}(r^i(s), \mathcal{X}_N, s)\mathrm{d}s \\
& \le \max_{1\le j\le N} m_{j,N}\sum_{i=1}^{N} m_{i,N}c_{T,\varphi,\sigma^{\perp}} \longrightarrow 0, \ \text{as } N \longrightarrow \infty\,.
\end{aligned}
\right\}
\tag{8.72}
$$

[41] Cf. (8.16).

(iii) By (8.66) $\gamma_f^2(\mathcal{X}(t) - \mathcal{X}_N(t)) \longrightarrow 0$, uniformly on suitably bounded intervals a.s., as $N \longrightarrow \infty$. The boundedness and Lipschitz property of $\sigma^\perp(r, \mu, t)$ imply that $(\sigma^\perp)^2(r, \mu, t)$ is Lipschitz in μ, similarly to the derivation of (8.32). Therefore, we may adjust the proof of Theorem 8.5 to show that the limiting empirical process $\mathcal{X}(\cdot)$ is a weak solution of (8.54). □

Remark 8.19. Suppose $\mathcal{X}_s \in L_{0,\mathcal{F}_s}(\mathbf{M}_f)$. Further, suppose $\tilde{\mathcal{Y}} \in L_{\text{loc},2,\mathcal{F}}(C((s,T];\mathbf{M}))$ and that the conditions of Theorems 4.5 and 8.18 hold. With the same methods (and less effort) we may show that the empirical process associated with (4.9) and the initial "smooth" distribution[42] $\mathcal{Y}_s \in$ is the solution of the following "bilinear" SPDE

$$\left.\begin{aligned} d\mathcal{Y} = &\left(\frac{1}{2}\sum_{k,\ell=1}^{d} \partial_{k\ell,r}^2 (\bar{\bar{D}}_{k\ell}(\tilde{\mathcal{Y}}, \cdot, t)\mathcal{Y}) - \nabla \cdot (\mathcal{Y}F(\tilde{\mathcal{Y}}, t))\right) dt \\ &- \nabla \cdot (\mathcal{Y} \int \mathcal{J}(\cdot, p, \tilde{\mathcal{Y}}, t) w(dp, dt)). \end{aligned}\right\} \tag{8.73}$$

The uniqueness of the solution of (8.73) follows from the bilinearity (cf. the proof of Theorem 8.8). □

In what follows, we provide sufficient conditions for "smooth" solutions of (8.54) and (8.73), generalizing the corresponding statements of Theorem 8.6. The following assumptions on $\sigma\perp$ must be added to (8.35), (8.45) and Hypotheses 4.1 and 8.2 to guarantee smoothness and uniqueness of solutions:

$$\max_{1\le k,\ell\le d} \operatorname{ess} \sup_{(t,\mu)\in[0,T]\times\mathbf{M}_1} |||\sigma_{k\ell}^\perp(\cdot, \mu, t)|||_{m+3} \le d_{3,T}, \ m \in \mathbf{N} \cup \{0\}, \tag{8.74}$$

$$\sup_{t\ge 0} \sum_{m,\ell=1}^{d} \sum_{|\mathbf{n}|\le 2} [\partial_r^{\mathbf{n}} \sigma_{\ell m}^\perp(r, \mu_1) - \partial_r^{\mathbf{n}} \sigma_{\ell m}^\perp(r, \mu_2)]^2 \le c\gamma_\varpi^2(\mu_1 - \mu_2). \tag{8.75}$$

Theorem 8.20.

(i) Suppose, in addition to the conditions of Theorem 8.18 and Remark 8.19, that (8.35) and (8.74) hold. Further, assume that

$$X_s,\ Y_s \in L_{0,\mathcal{F}_s}(\mathbf{W}_{m,p,1}) \cap L_{0,\mathcal{F}_s}(\mathbf{M}_f), \tag{8.76}$$

where $m \in \mathbf{N} \cup \{0\}$. Then the solutions of (8.54) and (8.73) have densities with respect to the Lebesgue measure $X(\cdot, X_s)$ and $Y(\cdot, Y_s) := Y(\cdot, \tilde{\mathcal{Y}}, Y_s)$, respectively, such that for any $T > s$

$$X(\cdot, X_s),\ Y(\cdot, Y_s) \in L_{p,\mathcal{F}}((s,T] \times \Omega; \mathbf{W}_{m,p,1}) \cap L_{0,\mathcal{F}}(C([s,T]; \mathbf{M}_f)). \tag{8.77}$$

[42] We assume that (8.70) holds.

(ii) If $\sigma^{\perp}(r,\mu,t)$ is positive definite for all (r,μ,t) and the previous assumptions hold with $m = 0$, $p = 2$ such that X_s , $Y_s \in L_{0,\mathcal{F}_s}(\mathbf{H}_0) \cap L_{0,\mathcal{F}_s}(\mathbf{M}_f)$, then we also have[43]

$$X(\cdot, X_s)\ ,\ Y(\cdot, s) \in L_{2,\mathcal{F}}((s,T];\mathbf{H}_1 \cap \mathbf{M}_f) \cap L_{0,\mathcal{F}}(C([s,T];\mathbf{H}_0)). \quad (8.78)$$

(iii) Suppose in addition to the previous assumptions that (8.45) and (8.75) hold. Then the solution of (8.54) is unique.

Proof. The proof of the first statement (as for Theorem 8.6) is given in Chap. 11. To prove the second statement, we "freeze" the solutions in the coefficients and treat (8.54) as a bilinear SPDE (of type (8.73)). This bilinear SPDE is coercive and satisfies the assumptions of Krylov and Rozovsky (1979), whence we obtain improved smoothness of the solution, as well as the existence of continuous sample paths with values in $\mathbf{H}_0$. In the coefficients, we may then replace the solution by its smoother version, which does not change the values of the coefficients. Finally, the uniqueness follows from the calculations in Chap. 12. □

Remark 8.21. Kurtz and Xiong define uniqueness of the SPDE as uniqueness of the solutions of the infinite version of (4.10) and its empirical processes, i.e., as a system of stochastic equations on $(\hat{\mathbf{M}}_f, \hat{d})$. If we adopt this restrictive definition of uniqueness, then by Theorem 8.17 we have uniqueness without any additional smoothness assumptions. Unfortunately, we do not know (a priori) whether all solutions of (8.54) (or of the special case (8.25)) are generated by particle methods. □

8.5 General SPDEs

1. With some modifications we may derive solutions of quasilinear SPDEs from the system of SODEs (4.10) under (4.12). In this case, we need to use a different version of the Wasserstein metric, namely a version of $\tilde{\gamma}_p(\mu,\tilde{\mu})$, which is defined as in (15.39), using the Euclidean distance $|\cdot|$ instead of $\rho(\cdot)$ where μ and $\tilde{\mu}$ are probability measures. We refer to Dorogovtsev (2004b) and the references therein.
2. The original motivation for the particle approach arose from the derivation of the stochastic Navier–Stokes equation (8.15) for the vorticity in a 2D fluid[44] which required to consider signed measure valued processes. Under additional assumptions, quasilinear SPDE (8.26) can be extended to signed measures.[45]
3. The strength of the particle approach to SPDEs is its derivation from the solutions of SODEs, which allows us to solve a large class of quasilinear SPDEs. It follows from this derivation that the driving martingale process in those SPDEs will

[43] Cf. (8.4).

[44] Cf. Kotelenez (1995a).

[45] Cf. Kurtz and Xiong (1999) as well as Kotelenez (1995a,b) for results on the particle approach to quasilinear SPDEs on the space of signed measures.

always be state dependent and contain the gradient, applied to the solution times noise. This approach, however, also exposes the limitations of the particle approach. If, e.g., we need a noise term which is state independent, like in central limit theorem phenomena,[46] then other methods must be applied. By employing the fractional step method[47] we can combine the particle method with other methods to obtain a more general class of SPDEs and their solutions. Typically, this class has the characteristics of both approaches. For a further discussion we refer the reader to Chap. 13.

8.6 Semilinear Stochastic Partial Differential Equations in Stratonovich Form

We obtain representations of the parabolic Itô SPDEs (8.25)/(8.26) in terms of Stratonovich SPDEs under the assumption that the diffusion coefficients do not depend on the solutions of the SPDEs. In particular, we show that the solutions of the stochastic noncoercive parabolic SPDEs (8.25)/(8.26), driven by Itô differentials, can be rewritten as a stochastic first-order (transport) SPDE, driven by Stratonovich differentials.[48]

Theorem 8.23. *Consider the SPDEs (8.26) and (8.54) and suppose that neither $\mathcal{J}$ nor $\sigma^{\perp}$ depend upon the measure variable. Further, in addition to the conditions of Theorem 8.20, guaranteeing unique solutions of these SPDEs, suppose that*[49]

$$\sup_{r\in\mathbf{R}^d}\sum_{i,j,k,\ell=1}^{d}\int_0^T\int(\partial_{ij}^2\mathcal{J}_{k,\ell})^2(r,q,s)\mathrm{d}q\,\mathrm{d}s<\infty$$

and

$$\sum_{k=1}^{d}(\partial_k\tilde{D})_{k\ell}(0,t)\equiv 0\ \forall\ell.$$

Let $\mathcal{X}(\cdot):=\mathcal{X}(\cdot,\mathcal{X}_0)$ be the Itô solution of the SPDE (8.54).[50] *Then $\mathcal{X}(\cdot)$ is a weak solution of the following SPDE in Stratonovich form:*

---

[46] Cf. Holley and Stroock (1978), Itô (1984), or Kotelenez (1984) and the references therein.

[47] Cf. Chap. 15.3 and Goncharuk and Kotelenez (1998).

[48] Recall that we have used the Itô differential in the SODEs, whence the associated SPDEs were obtained through Itô's formula. We refer the reader to Sects. 15.2.5 and 15.2.6 for the most important properties of Itô and Stratonovich differentials, respectively. Theorem 8.23 was obtained by Kotelenez (2007).

[49] These conditions are the conditions (15.192) and (15.195) from Sect. 15.2.6.

[50] If $\sigma^{\perp}\equiv 0$, (8.54) becomes (8.26).

$$
\left.\begin{aligned}
&\mathrm{d}\mathcal{X} = \frac{1}{2}\sum_{k,\ell=1}^{d} \partial^2_{k\ell,r}(((\sigma^{\perp}(\cdot,t))^2)_{k\ell}\mathcal{X}) - \nabla\cdot(\mathcal{X}F(\cdot,\mathcal{X},t))\mathrm{d}t\\
&-\nabla\cdot(\mathcal{X}\int \mathcal{J}(\cdot,p,t))w(\mathrm{d}p,\circ\mathrm{d}t),\\
&\mathcal{X}(0)=\mathcal{X}_0,
\end{aligned}\right\}\qquad (8.79)
$$

where, as before, "∘" denotes Stratonovich differentials.

Proof. As before, let $r(t,\mathcal{X},q)$ be the solution of (4.10), based on Theorem 4.5, Part 3.[51] Let $\varphi\in C_c^3(\mathbf{R}^d,\mathbf{R})$. In the proof of Theorem 8.18, step (i), we apply (15.210) of Sect. 15.2.6 and obtain

$$
\left.\begin{aligned}
&\sum_{j=1}^{N} m_{j,N}(\nabla\varphi)(r^j(t))\cdot\int \mathcal{J}(r^j(t),p,\mathcal{X}_N,t)w(\mathrm{d}p,\mathrm{d}t)\\
&\quad+\sum_{j=1}^{N} m_{j,N}\frac{1}{2}\sum_{k,\ell=1}^{d}(\partial^2_{k\ell}\varphi)(r^j(t))D_{k\ell}(\tilde{\mathcal{X}}_N,r^j(t),t)\mathrm{d}t\\
&=\sum_{j=1}^{N} m_{j,N}(\nabla\varphi)(r^j(t))\cdot\int \mathcal{J}(r^j(t),p,\mathcal{X}_N,t)w(\mathrm{d}p,\circ\mathrm{d}t).
\end{aligned}\right\}\qquad (8.80)
$$

Thus, we replace (8.71) by

$$
\left.\begin{aligned}
\langle\mathrm{d}\mathcal{X}_N(t),\varphi\rangle &= \left\langle\frac{1}{2}\sum_{k,\ell=1}^{d}\partial^2_{k\ell}(((\sigma^{\perp}(\cdot,t))^2)_{k\ell}\mathcal{X}_N(t))\mathrm{d}t,\varphi\right\rangle\\
&\quad-\langle\nabla\cdot(\mathcal{X}_N(t)F(\cdot,\mathcal{X}_N,t))\mathrm{d}t,\varphi\rangle\\
&\quad-\langle\nabla\cdot(\mathcal{X}_N(t)\int \mathcal{J}(\cdot,p,\mathcal{X}_N,t)w(\mathrm{d}p,\circ\mathrm{d}t)),\varphi\rangle\\
&\quad+\sum_{j=1}^{N} m_{j,N}(\nabla\varphi)(r^j(t))\cdot(\sigma^{\perp}(r^j(t),\mathcal{X}_N,t)\beta^{\perp,j}(\mathrm{d}t))\\
&\quad\text{(integrating by parts in the generalized sense).}
\end{aligned}\right\}\qquad (8.81)
$$

As the assumptions imply that the tagged noise term disappears in the limit, and the existence of the limit has been established by Theorem 8.17, we may repeat the proof of Theorem 8.18 to derive (8.79) in weak form. □

Remark 8.24.

- In the Stratonovich representation the second-order partial operator comes from the coercive contribution alone.

[51] As before in similar applications of Theorem 4.5, Part 3, the empirical measure process $\mathcal{X}(\cdot)$ is treated as an input process.

- If we restrict Theorem (8.23) to the noncoercive SPDEs (8.25)/(8.26), i.e., if $\sigma^{\perp} \equiv 0$, the proof is a simple consequence of (8.52) and (15.219). For this case, the Stratonovich representation of (8.25)/(8.26) takes the form of a first-order transport SPDE.[52] □

We state the last observation as an independent theorem.

Theorem 8.25. *Suppose that, in addition to the assumptions of Theorem 8.24, $\sigma^{\perp} \equiv 0$. Then the solution of the noncoercive parabolic Itô SPDE (8.26) is the solution of the first-order transport Stratonovich SPDE*

$$\begin{aligned} d\mathcal{X} &= -\nabla \cdot (\mathcal{X} F(\cdot, \mathcal{X}, t))dt - \nabla \cdot (\mathcal{X} \int \mathcal{J}(\cdot, p, t)) w(dp, \circ dt), \\ \mathcal{X}(0) &= \mathcal{X}_0. \end{aligned} \tag{8.82}$$

An analogous statement holds for the bilinear SPDE (8.25). □

8.7 Examples

Example 8.26. To show the relation between the symmetry assumption in (8.35) and frame-indifference of the generating kernels, let us analyze the last Example (4.53) under the additional assumption of frame-indifference.

Suppose $\Gamma = (\Gamma_{k\ell}) \in \mathcal{M}_{d\times d}$ with its one-dimensional components $\Gamma_{k\ell} \in C^{2_{b^\nu}}(\mathbf{R}^d, \mathbf{R}) \cap \mathbf{H}_2$ and suppose that Γ is frame-indifferent. Set

$$\tilde{\mathcal{J}}(r, p, \mu) := \int \Gamma(r - p - q)\varpi(q)\mu(dq) \cdot$$

Employing the notation from (8.34) with μ instead of $\tilde{\mathcal{Y}}$ we set

$$\tilde{D}_{k\ell}(r - q) := \sum_{m=1}^{d} \int \tilde{\mathcal{J}}_{km}(r, p, \mu)\tilde{\mathcal{J}}_{\ell m}(q, p, \mu)dp. \tag{8.83}$$

Then,

$$\begin{aligned} \tilde{D}_{k\ell}(r - q) &= \int\int \left[\sum_{m=1}^{d} \int \Gamma_{km}(r - p - \tilde{p})\Gamma_{\ell m}(q - p - \hat{p})dp\right] \varpi(\tilde{p})\mu(d\tilde{p})\varpi(\hat{p})\mu(d\hat{p}) \\ &= \int\int \lambda_{\perp}(|r - q - \tilde{p} + \hat{p}|^2)P^{\perp}(r - q - \tilde{p} + \hat{p}) \\ &\quad + \lambda(|r - q - \tilde{p} + \hat{p}|^2)P(r - q - \tilde{p} + \hat{p})\varpi(\tilde{p})\mu(d\tilde{p})\varpi(\hat{p})\mu(d\hat{p}) \end{aligned}$$

(employing a representation as in (5.29))

[52] The interpretation of the SPDEs (8.25)/(8.26) as stochastic transport equations was suggested to the author by J.P. Fouque (1994).

$$= \int \int \lambda_\perp(|q - r - \hat{p} + \tilde{p}|^2) P^\perp(q - r - \hat{p} + \tilde{p})$$

$$+ \lambda(|q - r - \hat{p} + \tilde{p}|^2) P(q - r - \hat{p} + \tilde{p}) \varpi(\tilde{p}) \mu(\mathrm{d}\tilde{p}) \varpi(\hat{p}) \mu(\mathrm{d}\hat{p})$$

(because all coordinates are squared in the above representation)

$$= \int \int \lambda_\perp(|q - r - \tilde{p} + \hat{p}|^2) P^\perp(q - r - \tilde{p} + \hat{p})$$

$$+ \lambda(|q - r - \tilde{p} + \hat{p}|^2) P(q - r - \tilde{p} + \hat{p}) \varpi(\tilde{p}) \mu(\mathrm{d}\tilde{p}) \varpi(\hat{p}) \mu(\mathrm{d}\hat{p})$$

(by Fubini's theorem in the double integral and change of variables $\hat{p} \leftrightarrow \tilde{p}$)

$$= \tilde{D}_{k\ell}(q - r),$$

i.e.,

$$\tilde{D}_{k\ell}(r - q) = \tilde{D}_{k\ell}(q - r). \tag{8.84}$$

□

Chapter 9
Stochastic Partial Differential Equations: Infinite Mass

9.1 Noncoercive Quasilinear SPDEs for Infinite Mass Evolution

Assuming Hypothesis 4.1, existence and uniqueness of a solution of a noncoercive quasilinear SPDE is derived in the space of σ-finite Borel measures $\mathbf{M}_{\infty,\varpi}$ (cf. (4.5)). As the main step, we construct suitable flows of SODEs which satisfy an a priori estimate.

We have not yet shown existence of a solution of (8.26) in the case of infinite mass, although we obtained, in the previous section, an extension of the solutions of the bilinear SPDE (8.25) to infinite mass and a uniqueness result even for the quasilinear SPDE (8.26). Without loss of generality, we assume $s = 0$ in what follows, and we suppress the dependence of the coefficients F and $\mathcal{J}$ on t in the formulas. Suppose

$$\mathcal{X}_0 \in L_{0,\mathcal{F}_0}(\mathbf{M}_{\varpi,\infty}). \tag{9.1}$$

Let $b > 0$ be a constant and set for a measure $\mu \in \mathbf{M}_{\infty,\varpi}$

$$\mu_b := \begin{cases} \mu, & \text{if} \quad \gamma_\varpi(\mu) < b, \\ \frac{\mu}{\gamma_\varpi(\mu)} b, & \text{if} \quad \gamma_\varpi(\mu) \geq b. \end{cases} \tag{9.2}$$

This truncation allows us to define recursively solutions of (8.25) as follows: Let $r_1(t, \mathcal{X}_{0,b}, \omega, q)$ be the unique solution of (4.9), driven by the input process $\mathcal{X}_{0,b}$. Henceforth, we will tacitly assume that the solutions of (4.9) are the space–time measurable versions, whose existence follows from Part 3) of Theorem 4.5. For simplicity of notation, we drop the "bar" over those versions. Define the empirical measure process associated with $r_1(t, \ldots)$ as follows

$$\mathcal{X}_1(t, b, \omega) := \int \delta_{r_1(t,\mathcal{X}_{0,b},\omega,q)} \mathcal{X}_{0,b}(\omega, dq).$$

$\mathcal{X}_1(t, b, \omega)$ is continuous and $\mathcal{F}_t$-adapted, whence the same holds for $\gamma_\varpi(\mathcal{X}_1(t, b, \omega))$ (cf. (8.50)/(8.51)). Moreover, by the boundedness of the coefficients of (4.9), $\varphi(t, q) := r_1(t, \mathcal{X}_{0,b}, \omega, q)$ satisfies the assumptions of Proposition 4.3. Therefore, (4.7) implies

$$E \sup_{0 \le t \le T} \gamma_{\varpi}^2(\mathcal{X}_1(t,b)) \le c_T E \gamma_{\varpi}^2(\mathcal{X}_{0,b}) \le c_T b.$$

Hence, $\mathcal{X}_1(t,b) \in L_{2,\mathcal{F}}(C([0,T];\mathbf{M}_{\infty,\varpi}))$ and we can define $r_2(t,\mathcal{X}_1(b),\omega,q)$ as the unique solution of (4.9), driven by the input process $\mathcal{X}_1(t,b)$. Suppose we have already defined $r_m(t,\mathcal{X}_{m-1}(b),\omega,q)$ as (unique) solutions of (4.9) with empirical measure processes $\mathcal{X}_{m-1}(t,b) \in L_{2,\mathcal{F}}(C([0,T];\mathbf{M}_{\infty,\varpi}))$ for $m = 1,\ldots n$. As before, the solution $r_n(t,\mathcal{X}_{n-1}(b),\omega,q$ satisfies the assumptions of Proposition 4.3 and we define the empirical measure process of $r_n(t,\mathcal{X}_{n-1}(b),\omega,q)$ by

$$\mathcal{X}_n(t,b,\omega) := \int \delta_{r_n(t,\mathcal{X}_{n-1,b},\omega,q)} \mathcal{X}_{0,b}(\omega,\mathrm{d}q). \tag{9.3}$$

Again (4.7) implies

$$E \sup_{0 \le t \le T} \gamma_{\varpi}^2(\mathcal{X}_n(t,b)) \le c_T E \gamma_{\varpi}^2(\mathcal{X}_{0,b}), \tag{9.4}$$

where the constant c_T depends only on T and the coefficients of (4.9).

Set

$$\Omega_b := \{\omega : \gamma_{\varpi}(\mathcal{X}_0(\omega)) \le b\}. \tag{9.5}$$

Clearly,

$$\Omega_{b_2} \supset \Omega_{b_1}, \quad \text{if } b_2 \ge b_1 > 0\,.$$

Proposition 9.1. *Suppose $b_2,\ b_1 > 0$. Then, a.s., for all n*

$$\mathcal{X}_n(\cdot,b_2,\omega)1_{\Omega_{b_1}\,\cap\,\Omega_{b_2}} \equiv \mathcal{X}_n(\cdot,b_1,\omega)1_{\Omega_{b_1}\,\cap\,\Omega_{b_2}} \tag{9.6}$$

Proof.
(i) The fact that $\Omega_b \in \mathcal{F}_0$ allows us to apply (4.17) with the conditional expectation with respect to the events from $\mathcal{F}_0$. Suppose $b_2 \ge b_1 > 0$. Hence,

$$\left.\begin{aligned} &E \sup_{0 \le t \le T} \rho^2(r_n(t,\mathcal{X}_{n-1}(b_2),q) - r_n(t,\mathcal{X}_{n-1}(b_1),q))1_{\Omega_{b_1}} \\ &\le c_T \int_0^T \gamma_{\varpi}^2(\mathcal{X}_{n-1}(s,b_2) - \mathcal{X}_{n-1}(s,b_1))1_{\Omega_{b_1}}\mathrm{d}s \\ &= c_T \int_0^T \gamma_{\varpi}^2(\mathcal{X}_{n-1}(s,b_2)1_{\Omega_{b_1}} - \mathcal{X}_{n-1}(s,b_1)1_{\Omega_{b_1}})1_{\Omega_{b_1}}\mathrm{d}s, \end{aligned}\right\} \tag{9.7}$$

where the last identity is obvious, as for $\omega \notin \Omega_{b_1}$ the expression $\gamma_{\varpi}^2(\mathcal{X}_{n-1}(s,b_2) - \mathcal{X}_{n-1}(s,b_1))1_{\Omega_{b_1}} = 0$.

(ii) We show by induction that the right-hand side of (9.7) equals 0. First of all, note that

$$\mathcal{X}_{0,b_2}1_{\Omega_{b_1}} = \mathcal{X}_{0,b_1}1_{\Omega_{b_1}},$$

because both sides equal $\mathcal{X}_0 1_{\Omega_{b_1}}$. If

$$\mathcal{X}_{n-1}(\cdot,b_2,\omega)1_{\Omega_{b_1}} \equiv \mathcal{X}_{n-1}(\cdot,b_1,\omega)1_{\Omega_{b_1}},$$

then (9.7) implies a.s.

$$r_n(\cdot, \mathcal{X}_{n-1}(b_2), q)1_{\Omega_{b_1}} \equiv r_n(\cdot, \mathcal{X}_{n-1}(b_1), q)1_{\Omega_{b_1}}.$$

Therefore,

$$\int \delta_{r_n(\cdot,\mathcal{X}_{n-1}(b_2),\omega,q)}\mathcal{X}_{0,b_2}(\omega, \mathrm{d}q)1_{\Omega_{b_1}}(\omega) \equiv \int \delta_{r_n(\cdot,\mathcal{X}_{n-1}(b_1),\omega,q)}\mathcal{X}_{0,b_1}(\omega, \mathrm{d}q)1_{\Omega_{b_1}}(\omega),$$

whence by (9.3)

$$\mathcal{X}_n(\cdot, b_2, \omega)1_{\Omega_{b_1}} \equiv \mathcal{X}_n(\cdot, b_1, \omega)1_{\Omega_{b_1}}.$$

□

For a second truncation, let $c > 0$ and define stopping times as follows

$$\left.\begin{aligned} \tau_n(c, b, \omega) &:= \inf\{t \geq 0 : \gamma_\varpi(\mathcal{X}_n(t, b, \omega)) \geq c\}, \\ \tau(c, b, \omega) &:= \inf_n \tau_n(c, b, \omega). \end{aligned}\right\} \tag{9.8}$$

By (9.6) we have for $b_2, b_1 > 0$ and $\forall c > 0$

$$\left.\begin{aligned} \tau_n(c, b_2, \omega)_{|\Omega_{b_1} \cap \Omega_{b_2}} &= \tau_n(c, b_1, \omega)_{|\Omega_{b_1} \cap \Omega_{b_2}}, \\ \tau(c, b_2, \omega)_{|\Omega_{b_1} \cap \Omega_{b_2}} &= \tau(c, b_1, \omega)_{|\Omega_{b_1} \cap \Omega_{b_2}}. \end{aligned}\right\} \tag{9.9}$$

Lemma 9.2.

$$P\{\omega : \lim_{c\to\infty} \tau(c, b, \omega) = \infty\} = 1 \quad \forall b > 0. \tag{9.10}$$

Proof. Suppose (9.10) is not correct and note that $\tau(c, b)$ is monotone increasing as c increases. Therefore, there must be a $T > 0$, $b > 0$, and $\delta > 0$ such that

$$P(\cap_{m\in\mathbf{N}}\{\omega : \tau(m, b, \omega) < T\}) \geq \delta.$$

Set

$$\underline{\tau}_n(c, b) := \tau_1(c, b) \wedge \tau_2(c, b) \wedge \cdots \wedge \tau_n(c, b).$$

Then

$$\underline{\tau}_n(c, b) \downarrow \tau(c, b) \ \text{a.s., as } n \longrightarrow \infty.$$

Thus,

$$\cap_{m\in\mathbf{N}}\{\omega : \tau(m, b, \omega) < T\} = \cap_{n\in\mathbf{N}} \cap_{m\in\mathbf{N}} \{\omega : \underline{\tau}_n(m, b, \omega) < T\}.$$

As

$$B_n := \cap_{m\in\mathbf{N}}\{\omega : \underline{\tau}_n(m, b, \omega) < T\} \uparrow, \ \text{as } n \longrightarrow \infty,$$

there must be an $\bar{n}$ such that

$$P(B_{\bar{n}}) \geq \frac{\delta}{2}.$$

The definition of the stopping times implies

$$B_n \subset \cap_{m\in\mathbf{N}} \left\{\omega : \max_{k=1,\ldots,n} \sup_{0\leq t\leq T} \gamma_\varpi(X_k(t, b, \omega)) \geq m\right\}.$$

Set

$$A_m(n) := \left\{ \omega : \max_{k=1,\ldots,n} \sup_{0\le t\le T} \gamma_\varpi(X_k(t,b,\omega)) \ge m \right\}.$$

Obviously, the $A_m(n)$ are monotone decreasing as m increases. Therefore, the assumption on $P(B_{\bar{n}})$ implies

$$P(\cap_{m\in\mathbf{N}} A_m(\bar{n})) \ge \frac{\delta}{2}.$$

However, by (9.4), a.s. $\forall k$,

$$\sup_{0\le t\le T} \gamma_\varpi(X_k(t,b,\omega)) < \infty.$$

Hence, we have also a.s. $\forall n$

$$\max_{k=1,\ldots,n} \sup_{0\le t\le T} \gamma_\varpi(X_k(t,b,\omega)) < \infty.$$

This implies

$$\forall n \;\; P(A_m(n)) \downarrow 0, \;\; \text{as } m \longrightarrow \infty,$$

contradicting $P(\cap_{m\in\mathbf{N}} A_m(\bar{n})) \ge \frac{\delta}{2}$. □

Abbreviate in what follows

$$\left. \tau := \tau(c,b), \quad r_i(t,q) := r_i(t,\mathcal{X}_{i-1},b,q). \right\} \tag{9.11}$$

Lemma 9.3. *For any $T > 0$ there is a finite constant $c_T := c(T,F,\mathcal{J},c,b,\gamma)$ such that $\forall t \in [0,T]$*

$$\left.\begin{aligned} E\int \sup_{0\le s\le t} \rho^4(r_n(s\wedge\tau,q) - r_m(s\wedge\tau,q))[\varpi(r_n(s\wedge\tau,q)) \\ +\varpi(r_m(s\wedge\tau,q))]\mathcal{X}_{0,b}(\mathrm{d}q) \\ \le c_T \int_0^t E\gamma_\varpi^4(\mathcal{X}_{n-1}(s\wedge\tau,b) - \mathcal{X}_{m-1}(s\wedge\tau,b))\mathrm{d}s. \end{aligned}\right\} \tag{9.12}$$

Proof.
(i) Let $\chi \in C_b^2(\mathbf{R};\mathbf{R})$ be an odd function such that there are constants $0 < c_1 \le c_2 < \infty$ and the following properties hold

$$\left.\begin{aligned} \chi(u) = u \;\; \forall u \in [-1,1], \\ \chi'(u) > 0 \;\; \forall u \in \mathbf{R}, \\ |\chi''(u)u^2| \le c_1|\chi'(u)u| \le c_2|\chi(u)| \;\; \forall u \in \mathbf{R}, \\ |||\chi||| \le 2, \\ |||\chi'||| \le 1, \end{aligned}\right\} \tag{9.13}$$

where χ' and χ'' are the first and second derivatives of χ, respectively. Then

$$\rho^2(r) \le \chi(|r|^2) \le 2\rho^2(r). \tag{9.14}$$

(ii) Abbreviate

$$f(s,q) := r_n(t \wedge \tau, q) - r_m(t \wedge \tau, q), \quad g(t,q) := r_i(t,q), \quad i \in \{n,m\}. \tag{9.15}$$

Note that $f(0,q) = 0$ and $g(0,q) = q$, whence we have $\chi(|f(0,q)|^2) = 0$ and $\varpi(g(0,q)) = \varpi(q)$. Itô's formula yields

$$\left.\begin{aligned} \chi(|f(t,q)|^2) = &\int_0^t \chi'(|f(s,q)|^2)2f(s,q)\cdot \mathrm{d}f(s,q) \\ &+ \int_0^t (\chi''(|f(s,q)|^2)2\sum_{k,\ell=1}^{\mathrm{d}} f_k(s,q)f_\ell(s,q) \\ &+ \chi'(|f(s,q)|^2)\delta_{k\ell})\mathrm{d}[f_k(s,q), f_\ell(s,q)] \end{aligned}\right\} \tag{9.16}$$

and for $\varpi_\beta := \varpi^\beta, \beta > 0$,

$$\left.\begin{aligned} \varpi_\beta(g(t,q)) = \varpi_\beta(q) &+ \int_0^t (\nabla\varpi_\beta)(g(s,q))\cdot \mathrm{d}g(s,q) \\ &+ \frac{1}{2}\sum_{k,\ell=1}^{d}\int_0^t (\partial^2_{k\ell}\varpi_\beta)(g(s,q))\mathrm{d}[g_k(s,q), g_\ell(s,q)]. \end{aligned}\right\} \tag{9.17}$$

Next, we apply Itô's formula to the product of the left-hand sides of (9.16) and (9.17):

$$\left.\begin{aligned} \chi(|f(t,q)|^2)\varpi_\beta(g(t,q)) = &\int_0^t \varpi_\beta(g(s,q))\mathrm{d}\chi(|f(s,q)|^2) \\ &+ \int_0^t \chi(|f(s,q)|^2)\mathrm{d}\varpi_\beta(g(s,q)) \\ &+ \int_0^t \mathrm{d}[\chi(|f(s,q)|^2), \varpi_\beta(g(s,q))]. \end{aligned}\right\} \tag{9.18}$$

In what follows, we first derive bounds for the deterministic differentials in (9.18) employing (4.11). By (9.16)

$$\left.\begin{aligned} \mathrm{d}\chi(|f(s,q)|^2) = &\chi'(|f(s,q)|^2)2f(s,q)\cdot \mathrm{d}f(s,q) \\ &+ (\chi''(|f(s,q)|^2)2\sum_{k,\ell=1}^{d} f_k(s,q)f_\ell(s,q) \\ &+ \chi'(|f(s,q)|^2)\delta_{k\ell})\mathrm{d}[f_k(s,q), f_\ell(s,q)]. \end{aligned}\right\} \tag{9.19}$$

First,

$$\begin{aligned} \mathrm{d}f(s,q) = &(F(r_n(s\wedge\tau,q), \mathcal{X}_{n-1}(b)) - F(r_m(s\wedge\tau,q), \mathcal{X}_{m-1}(b)))\mathrm{d}s \\ &+ \int (\mathcal{J}(r_n(s\wedge\tau,q), p, \mathcal{X}_{n-1}(b)) \\ &-\mathcal{J}(r_m(s\wedge\tau,q), p, \mathcal{X}_{m-1}(b)))w(\mathrm{d}p, \mathrm{d}s). \end{aligned}$$

Recall that by (9.8) $\varpi_\beta(\mathcal{X}_{i-1}(t\wedge\tau, b)) \le c \;\; \forall i \in \mathbf{N}$. Let $\tilde{c}$ be a positive constant, depending on c, F, $\mathcal{J}$, T, b, β and γ_ϖ, and the value of $\tilde{c}$ may be different for different steps in the following estimates. Therefore, (4.11) yields

$$
\left.\begin{aligned}
&|F(r_n(s\wedge\tau,q),\mathcal{X}_{n-1}(b))-F(r_m(s\wedge\tau,q),\mathcal{X}_{m-1}(b))|\\
&\le\tilde{c}(\rho(r_n(s\wedge\tau,q)-r_m(s\wedge\tau,q))+\gamma_\varpi(\mathcal{X}_{n-1}(s\wedge\tau,b)-\mathcal{X}_{m-1}(s\wedge\tau,b))).
\end{aligned}\right\}
\tag{9.20}
$$

Again by (4.11),

$$
\left.\begin{aligned}
&\mathrm{d}[f_k(s,q),f_\ell(s,q)]\\
&=\sum_{j=1}^d\int(\mathcal{J}_{kj}(r_n(s\wedge\tau,q),p,\mathcal{X}_{n-1}(s\wedge\tau,b))\\
&\quad-\mathcal{J}_{kj}(r_m(s\wedge\tau,q),p,\mathcal{X}_{m-1}(s\wedge\tau,b)))\\
&\quad\times(\mathcal{J}_{\ell j}(r_n(s\wedge\tau,q),p,\mathcal{X}_{n-1}(b))-\mathcal{J}_{\ell j}(r_m(s\wedge\tau,q),p,\mathcal{X}_{m-1}(b)))\mathrm{d}p\,\mathrm{d}s\\
&\le\tilde{c}(\rho^2(r_n(s\wedge\tau,q)-r_m(s\wedge\tau,q))+\gamma_\varpi^2(\mathcal{X}_{n-1}(s\wedge\tau,b)-\mathcal{X}_{m-1}(s\wedge\tau,b)))\mathrm{d}s\\
&\text{and also}\\
&\sum_{j=1}^d\int(\mathcal{J}_{kj}(r_n(s\wedge\tau,q),p,\mathcal{X}_{n-1}(s\wedge\tau,b))\\
&\quad-\mathcal{J}_{kj}(r_m(s\wedge\tau,q),p,\mathcal{X}_{m-1}(s\wedge\tau,b)))\\
&\quad\times(\mathcal{J}_{\ell j}(r_n(s\wedge\tau,q),p,\mathcal{X}_{n-1}(b))-\mathcal{J}_{\ell j}(r_m(s\wedge\tau,q),p,\mathcal{X}_{m-1}(b)))\mathrm{d}p\,\mathrm{d}s\\
&\le\tilde{c}(\rho(r_n(s\wedge\tau,q)-r_m(s\wedge\tau,q))+\gamma_\varpi(\mathcal{X}_{n-1}(s\wedge\tau,b)-\mathcal{X}_{m-1}(s\wedge\tau,b)))\mathrm{d}s,
\end{aligned}\right\}
\tag{9.21}
$$

where in the last inequality we used the boundedness of the coefficients. Employing (9.20) and (9.21), (9.13) yields the following estimates

$$
\left.\begin{aligned}
&|\chi'(|f(s,q)|^2)2f(s,q)|\rho(f(s,q))\le\tilde{c}\chi(|f(s,q)|^2),\\
&|\chi'(|f(s,q)|^2)2f(s,q)|\le\tilde{c}\\
&|(\chi''(|f(s,q)|^2)2\sum_{k,\ell=1}^d f_k(s,q)f_\ell(s,q)|\le\tilde{c}\chi(|f(s,q)|^2),\\
&|\chi'(|f(s,q)|^2)|\le 1,\\
&|\chi''(|f(s,q)|^2)2\sum_{k,\ell=1}^d f_k(s,q)f_\ell(s,q)|\mathrm{d}[f_k(s,q),f_\ell(s,q)]\\
&\le\tilde{c}\chi(|f(s,q)|^2)[\rho^2(f(s,q))+\gamma_\varpi^2(\mathcal{X}_{n-1}(s\wedge\tau,b)-\mathcal{X}_{m-1}(s\wedge\tau,b))]\mathrm{d}s,\\
&\chi'(|f(s,q)|^2))\sum_{k=1}^d\mathrm{d}[f_k(s,q),f_k(s,q)]\le\tilde{c}\{\chi(|f(s,q)|^2)\mathrm{d}s\\
&+\gamma_\varpi^2(\mathcal{X}_{n-1}(s\wedge\tau,b)-\mathcal{X}_{m-1}(s\wedge\tau,b))\mathrm{d}s\},
\end{aligned}\right\}
\tag{9.22}
$$

where the last inequality follows from

$$
\begin{aligned}
&\chi'(|f(s,q)|^2))\sum_{k=1}^{d} \mathrm{d}[f_k(s,q), f_k(s,q)] \\
&\quad \le \chi'(|f(s,q)|^2)\tilde{c}(\rho^2(r_n(s\wedge\tau,q)-r_m(s\wedge\tau,q)) \\
&\qquad +\gamma_{\varpi}^2(\mathcal{X}_{n-1}(s\wedge\tau,b)-\mathcal{X}_{m-1}(s\wedge\tau,b)))\mathrm{d}s \\
&\quad \le \tilde{c}\chi(|f(s,q)|^2)\mathrm{d}s+\tilde{c}\gamma_{\varpi}^2(\mathcal{X}_{n-1}(s\wedge\tau,b)-\mathcal{X}_{m-1}(s\wedge\tau,b))\mathrm{d}s.
\end{aligned}
$$

We note that

$$|\chi(|f(s,q)|^2)\le 2.$$

Therefore, by (9.19), (9.20), and (9.22),

$$
\left.
\begin{aligned}
\mathrm{d}\chi(|f(s,q)|^2) &\le \tilde{c}(\chi(|f(s,q)|^2)+\gamma_{\varpi}^2(\mathcal{X}_{n-1}(s\wedge\tau,b)-\mathcal{X}_{m-1}(s\wedge\tau,b)))\mathrm{d}s \\
&\quad +\chi'(|f(s,q)|^2)2f(s,q)\cdot\int(\mathcal{J}(r_n(s\wedge\tau,q),p,\mathcal{X}_{n-1}(b)) \\
&\quad -\mathcal{J}(r_m(s\wedge\tau,q),p,\mathcal{X}_{m-1}(b)))w(\mathrm{d}p,\mathrm{d}s).
\end{aligned}
\right\}
\tag{9.23}
$$

Set

$$
\begin{aligned}
\mathrm{d}m_1(s,q) := \int&(\mathcal{J}(r_n(s\wedge\tau,q),p,\mathcal{X}_{n-1}(b)) \\
&-\mathcal{J}(r_m(s\wedge\tau,q),p,\mathcal{X}_{m-1}(b)))w(\mathrm{d}p,\mathrm{d}s).
\end{aligned}
\tag{9.24}
$$

Hence,

$$
\left.
\begin{aligned}
&\int_0^t \varpi_\beta(g(s,q))\mathrm{d}\chi(|f(s,q)|^2)\le\tilde{c}\int_0^t \varpi_\beta(g(s,q))\chi(|f(s,q)|^2) \\
&\quad +\varpi_\beta(g(s,q))\gamma_{\varpi}^2(\mathcal{X}_{n-1}(s\wedge\tau,b)-\mathcal{X}_{m-1}(s\wedge\tau,b)))\mathrm{d}s \\
&\quad +\int_0^t \varpi_\beta(g(s,q))\chi'(|f(s,q)|^2)2f(s,q)\cdot\mathrm{d}m_1(s,q).
\end{aligned}
\right\}
\tag{9.25}
$$

Next, by (9.17)

$$
\left.
\begin{aligned}
\mathrm{d}\varpi_\beta(g(s,q)) &= (\nabla\varpi_\beta)(g(s,q))\cdot\mathrm{d}g(s,q) \\
&\quad +\frac{1}{2}\sum_{k,\ell=1}^{d}(\partial_{k\ell}^2\varpi_\beta)(g(s,q))\mathrm{d}[g_k(s,q),g_\ell(s,q)] \\
&= (\nabla\varpi_\beta)(r_i(s\wedge\tau,q))\cdot F(r_i(s\wedge\tau,q),\mathcal{X}_{i-1}(b))\mathrm{d}s \\
&\quad +\frac{1}{2}\sum_{k,\ell=1}^{d}(\partial_{k\ell}^2\varpi_\beta)(r_i(s\wedge\tau,q))\mathrm{d}[r_{i,k}(s\wedge\tau,q),r_{i,\ell}(s\wedge\tau,q)] \\
&\quad +(\nabla\varpi_\beta)(r_i(s\wedge\tau,q))\int\mathcal{J}(r_i(s\wedge\tau,q)-p,\mathcal{X}_{i-1}(b))w(\mathrm{d}p,\mathrm{d}s).
\end{aligned}
\right\}
\tag{9.26}
$$

By (4.11), (15.46), and the definition of τ

$$
\left.\begin{array}{r}
|\nabla \varpi_\beta)(r_i(s\wedge\tau,q))||F(r_i(s\wedge\tau,q),\mathcal{X}_{i-1}(b))| \le \tilde{c}\varpi_\beta(r_i(s\wedge\tau,q)),\\
\frac{1}{2}\sum_{k,\ell=1}^{d}(\partial^2_{k\ell}\varpi_\beta(r_i(s\wedge\tau,q))\mathrm{d}[r_{i,k}(s\wedge\tau,q),r_{i,\ell}(s\wedge\tau,q)] \le \tilde{c}\varpi_\beta(r_i(s\wedge\tau,q)).
\end{array}\right\} \tag{9.27}
$$

Abbreviating

$$
\mathrm{d}m_2(s,q) := \int \mathcal{J}(g(s,q),p,\mathcal{X}_{i-1}(b))w(\mathrm{d}p,\mathrm{d}s), \tag{9.28}
$$

we obtain

$$
\left.\begin{array}{l}
\left|\int_0^t \chi(|f(s,q)|^2)\mathrm{d}\varpi_\beta(g(s,q))\right| \le \tilde{c}\int_0^t \chi(|f(s,q)|^2)\varpi_\beta(g(s,q))\mathrm{d}s\\
\qquad\qquad + \int_0^t \chi(|f(s,q)|^2)(\nabla\varpi_\beta)(g(s,q))\cdot \mathrm{d}m_2(s,q).
\end{array}\right\} \tag{9.29}
$$

Finally,

$$
\left.\begin{array}{l}
\left[\chi(|f(t,q)|^2),\varpi_\beta(g(t,q))\right]\\
\quad = \left[\int_0^t \chi'(|f(s,q)|^2)2f(s,q)\cdot \mathrm{d}m_1(s), \int_0^t (\nabla\varpi_\beta)(g(s,q))\cdot \mathrm{d}m_2(s)\right].
\end{array}\right\} \tag{9.30}
$$

Hence,

$$
\left.\begin{array}{l}
\left[\chi(|f(t,q)|^2),\varpi_\beta(g(t,q))\right]\\
\quad = \sum_{k,\ell=1}\int_0^t \chi'(|f(s,q)|^2)2f_k(s,q)(\partial_\ell\varpi_\beta)(g(s,q))\mathrm{d}[m_{1,k}(s),m_{2,\ell}(s)].
\end{array}\right\} \tag{9.31}
$$

We apply (9.21) and the Cauchy–Schwarz inequality in addition to (4.11) and obtain

$$
\mathrm{d}[m_{1,k}(s),m_{2,\ell}(s)] \le \tilde{c}(\rho(f(s,q)) + \gamma_\varpi(\mathcal{X}_{n-1}(b)-\mathcal{X}_{m-1}(b)))\mathrm{d}s. \tag{9.32}
$$

Consequently, by (15.46) and (9.14) as well as by $|f_j(s,q)| \le |f(s,q)|$,

$$
\left.\begin{aligned}
&|[\chi(|f(t,q)|^2), \varpi_\beta(g(t,q))]| \\
&\quad \le \tilde{c}\int_0^t |\chi'(|f(s,q)|^2)||f(s,q)|\varpi_\beta(g(s,q))(\rho(f(s,q)) \\
&\qquad +\gamma_\varpi(\mathcal{X}_{n-1}(s\wedge\tau,b) - \mathcal{X}_{m-1}(s\wedge\tau,b)))\mathrm{d}s \\
&\quad \le \tilde{c}\int_0^t \chi(|f(s,q)|^2)\varpi_\beta(g(s,q))\mathrm{d}s \\
&\qquad +\tilde{c}\int_0^t \varpi_\beta(g(s,q))|\chi'(|f(s,q)|^2)||f(s,q)|\gamma_\varpi(\mathcal{X}_{n-1}(s\wedge\tau,b) \\
&\qquad -\mathcal{X}_{m-1}(s\wedge\tau,b))\mathrm{d}s \\
&\quad \le \tilde{c}\int_0^t \chi(|f(s,q)|^2)\varpi_\beta(g(s,q))ds \\
&\qquad +\tilde{c}\int_0^t \varpi_\beta(g(s,q))[|\chi'(|f(s,q)|^2)|^2|f(s,q)|^2 \\
&\qquad +\gamma_\varpi^2(\mathcal{X}_{n-1}(s\wedge\tau,b) - \mathcal{X}_{m-1}(s\wedge\tau,b))]\mathrm{d}s \\
&\text{(again employing } ab \le \tfrac{1}{2}(a^2+b^2)\text{)} \\
&\quad \le \tilde{c}\int_0^t \chi(|f(s,q)|^2)\varpi_\beta(g(s,q))\mathrm{d}s \\
&\qquad +\tilde{c}\int_0^t \varpi_\beta(g(s,q))[\chi(|f(s,q)|^2) \\
&\qquad +\gamma_\varpi^2(\mathcal{X}_{n-1}(s\wedge\tau,b) - \mathcal{X}_{m-1}(s\wedge\tau,b))]\mathrm{d}s \\
&\text{(because } |\chi'(|f(s,q)|^2)|^2|f(s,q)|^2 \le |\chi'(|f(s,q)|^2)|\chi(|f(s,q)|^2) \\
&\quad \le \tilde{c}\chi(|f(s,q)|^2)).
\end{aligned}\right\}\quad(9.33)
$$

Altogether we obtain from the previous calculations

$$
\left.\begin{aligned}
&\chi(|f(t,q)|^2)\varpi_\beta(g(t,q)) \\
&\quad \le \tilde{c}\int_0^t \big[\chi(|f(s,q)|^2)\varpi_\beta(g(s,q)) + \varpi_\beta(g(s,q))\gamma_\varpi^2(\mathcal{X}_{n-1}(s\wedge\tau,b) \\
&\quad -\mathcal{X}_{m-1}(s\wedge\tau,b))\big]\mathrm{d}s \\
&\quad +\int_0^t \varpi_\beta(g(s,q))\chi'(|f(s,q)|^2)2f(s,q)\mathrm{d}m_1(s,q) \\
&\quad +\int_0^t \chi(|f(s,q)|^2)(\nabla\varpi_\beta)(g(s,q))\cdot\mathrm{d}m_2(s,q).
\end{aligned}\right\}\quad(9.34)
$$

We next derive bounds for the quadratic variations of the martingales. Doob's inequality yields[1]

$$
\begin{aligned}
&E\sup_{0\le s\le t}\left\{\int_0^s \varpi_\beta(g(u,q))\chi'(|f(s,q)|^2)2f(s,q)\cdot\mathrm{d}m_1(u,q)\right\}^2 \\
&\qquad \le 4E\int_0^t \varpi_\beta^2(g(u,q))(\chi'(|f(s,q)|^2)2|f(s,q)|)^2\mathrm{d}[m_1(u,q)].
\end{aligned}
$$

[1] Cf. Sect. 15.2.3, Theorem 15.32.

By (9.22), (9.21), and (9.14),

$$\left.\begin{array}{l}\mathrm{d}[\chi'(|f(s,q)|^2)2f(s,q)\cdot m_1(s,q)]\\ \quad\le 4|\chi'(|f(s,q)|^2)||f(s,q)|^2\sum_{k,\ell=1}^{d}\mathrm{d}[f_k(s,q),f_\ell(s,q)]\\ \quad\le\tilde{c}\chi(|f(s,q)|^2)(\rho^2(r_n(s\wedge\tau,q)-r_m(s\wedge\tau,q))\\ \qquad+\gamma_{\varpi}^2(\mathcal{X}_{n-1}(s\wedge\tau,b)-\mathcal{X}_{m-1}(s\wedge\tau,b)))\mathrm{d}s\\ \quad\le\tilde{c}\{\chi^2(|f(s,q)|^2)+\chi(|f(s,q)|)\gamma_{\varpi}^2(\mathcal{X}_{n-1}(s\wedge\tau,b)-\mathcal{X}_{m-1}(s\wedge\tau,b))\}\mathrm{d}s\\ \quad\le\tilde{c}\{\chi^2(|f(s,q)|^2)+\gamma_{\varpi}^4(\mathcal{X}_{n-1}(s\wedge\tau,b)-\mathcal{X}_{m-1}(s\wedge\tau,b))\}\mathrm{d}s\\ \text{(employing } ab\le\frac{1}{2}(a^2+b^2)\text{ in the last inequality).}\end{array}\right\}\tag{9.35}$$

Thus,

$$\left.\begin{array}{l}E\sup_{0\le s\le t}\left|\int_0^s\varpi_\beta(g(u,q))\chi'(|f(s,q)|^2)2f(s,q)\cdot\mathrm{d}m_1(u,q)\right|^2\\ \quad\le\tilde{c}E\int_0^t\varpi_\beta^2(g(u,q))\left\{\chi^2(|f(u,q)|^2)\right.\\ \qquad\left.+\gamma_{\varpi}^4(\mathcal{X}_{n-1}(u\wedge\tau,b)-\mathcal{X}_{m-1}(u\wedge\tau,b))\right\}\mathrm{d}u.\end{array}\right\}\tag{9.36}$$

It is now convenient to square the terms in (9.18). Expressions (9.25) and (9.36) imply

$$\left.\begin{array}{l}E\sup_{0\le s\le t}\left|\int_0^s\varpi_\beta(g(u,q))\mathrm{d}\chi(|f(u,q)|^2)\right|^2\\ \quad\le\tilde{c}E\int_0^t\varpi_\beta^2(g(u,q))\{\chi^2(|f(u,q)|^2)\\ \qquad+\gamma_{\varpi}^4(\mathcal{X}_{n-1}(u\wedge\tau,b)-\mathcal{X}_{m-1}(u\wedge\tau,b))\}\mathrm{d}u.\end{array}\right\}\tag{9.37}$$

Similarly, employing (4.11) (i.e., boundedness of $\mathcal{J}$) and (15.46), Doob's inequality implies

$$\left.\begin{array}{l}E\sup_{0\le s\le t}\left|\int_0^s\chi(|f(u,q)|^2)(\nabla\varpi_\beta)(g(u,q))\cdot\mathrm{d}m_2(u,q)\right|^2\\ \quad\le\tilde{c}E\int_0^t\chi^2(|f(u,q)|^2)\varpi_\beta^2(g(u,q))\mathrm{d}u.\end{array}\right\}\tag{9.38}$$

Hence, by (9.29) and (9.38)

$$\left.\begin{array}{l}E\sup_{0\le s\le t}\left(\int_0^s\chi(|f(u,q)|^2)\mathrm{d}\varpi_\beta(g(u,q))\right)^2\\ \quad\le\tilde{c}E\int_0^t\chi^2(|f(u,q)|^2)\varpi_\beta^2(g(u,q))\mathrm{d}u.\end{array}\right\}\tag{9.39}$$

We also square both sides in (9.33) and employ the Cauchy–Schwarz inequality:

$$\left.\begin{aligned}&[\chi(|f(t,q)|^2), \varpi_\beta(g(t,q))]^2 \le \tilde{c}\int_0^t \chi^2(|f(s,q)|^2)\varpi_\beta^2(g(s,q))\mathrm{d}s\\&\quad+\tilde{c}\int_0^t \varpi_\beta^2(g(s,q))\gamma_\varpi^4(\mathcal{X}_{n-1}(s\wedge\tau,b)-\mathcal{X}_{m-1}(s\wedge\tau,b))\mathrm{d}s.\end{aligned}\right\} \tag{9.40}$$

Choosing $\beta := \frac{1}{2}$, the previous steps imply

$$\left.\begin{aligned}&E\sup_{0\le s\le t}\chi^2(|f(t,q)|^2)\varpi(g(t,q)) \le \tilde{c}E\int_0^t \Big[\varpi(g(u,q))\chi^2(|f(u,q)|^2)\\&\quad+\varpi(g(u,q))\gamma_\varpi^4(\mathcal{X}_{n-1}(u\wedge\tau,b)-\mathcal{X}_{m-1}(u\wedge\tau,b))\Big]\mathrm{d}u.\end{aligned}\right\} \tag{9.41}$$

By estimate (9.14) this implies

$$\left.\begin{aligned}&E\sup_{0\le s\le t}\rho^4(f(t,q))\varpi(g(t,q)) \le \tilde{c}E\int_0^t \Bigg[\sup_{0\le u\le s}\varpi(g(u,q))\rho^4(f(u,q)))\\&\quad+\varpi(g(s,q))\gamma_\varpi^4(\mathcal{X}_{n-1}(s\wedge\tau,b)-\mathcal{X}_{m-1}(s\wedge\tau,b))\Bigg]\mathrm{d}s.\end{aligned}\right\} \tag{9.42}$$

Gronwall's inequality implies

$$\begin{aligned}E\sup_{0\le s\le t}\rho^4(f(t,q))\varpi(g(t,q)) \le c_T\int_0^t &E\varpi(g(s,q))\gamma_\varpi^4(\mathcal{X}_{n-1}(s\wedge\tau,b)\\&-\mathcal{X}_{m-1}(s\wedge\tau,b))\mathrm{d}s.\end{aligned} \tag{9.43}$$

Adding up the inequalities of (9.43) for $g(t,q) = r_n(t\wedge\tau,q)$ and for $g(t,q) = r_m(t\wedge\tau,q)$ and integrating against $\mathcal{X}_{0,b}(\mathrm{d}q)$ yields

$$\left.\begin{aligned}&E\int\sup_{0\le s\le t}\rho^4(r_n(s\wedge\tau,q)-r_m(s\wedge\tau,q))[\varpi(r_n(s\wedge\tau,q))\\&+\varpi(r_m(s\wedge\tau,q))]\mathcal{X}_{0,b}(\mathrm{d}q) \le c_T\int_0^t E\int[\varpi(r_n(s\wedge\tau,q))\\&+\varpi(r_m(s\wedge\tau,q))]\mathcal{X}_{0,b}(\mathrm{d}q)\gamma_\varpi^4(\mathcal{X}_{n-1}(s\wedge\tau,b)-\mathcal{X}_{m-1}(s\wedge\tau,b))\mathrm{d}s.\end{aligned}\right\} \tag{9.44}$$

Taking into account that

$$\begin{aligned}&\int[\varpi(r_n(s\wedge\tau,q))+\varpi(r_m(s\wedge\tau,q))]\mathcal{X}_{0,b}(\mathrm{d}q)\\&\quad=\gamma_\varpi(\mathcal{X}_n(s\wedge\tau,b))+\gamma_\varpi(\mathcal{X}_m(s\wedge\tau,b))\le 2c,\end{aligned}$$

we obtain (9.12).

Proposition 9.4. *For any $T > 0$ there is a finite constant $c_T := c(T, F, \mathcal{J}, c, b, \gamma)$ such that*

$$\begin{aligned} &E \sup_{0 \le t \le T} \gamma_{\varpi}^4 (\mathcal{X}_n((t \wedge \tau, b) - \mathcal{X}_m(t \wedge \tau, b))) \\ &\quad \le c_T \int_0^T E \sup_{0 \le s \le t} \gamma_{\varpi}^4 (\mathcal{X}_{n-1}(s \wedge \tau, b) - \mathcal{X}_{m-1}(s \wedge \tau, b)) \mathrm{d}t. \end{aligned} \tag{9.45}$$

Proof.

$$\begin{aligned} E \sup_{0 \le t \le T} \gamma_{\varpi}^4 (\mathcal{X}_n((t \wedge \tau, b) - \mathcal{X}_m(t \wedge \tau, b)) &= E \sup_{0 \le t \le T} \sup_{\|f\|_{L,\infty} \le 1} \\ &\times \Big\{ \int \big[f(r_n(t \wedge \tau, \mathcal{X}_{n-1}(b), q)) \varpi(r_n(t \wedge \tau, \mathcal{X}_{n-1}(b), q)) \\ &\quad - f(r_m(t \wedge \tau, \mathcal{X}_{n-1}(b), q)) \varpi(r_m(t \wedge \tau, \mathcal{X}_{n-1}(b), q)) \big] \mathcal{X}_{0,b}(\mathrm{d}q) \Big\}^4, \end{aligned}$$

Further,

$$\left. \begin{aligned} &|f(r_n(t \wedge \tau, \mathcal{X}_{n-1}(b), q)) \varpi(r_n(t \wedge \tau, \mathcal{X}_{n-1}(b), q)) \\ &\quad - f(r_m(t \wedge \tau, \mathcal{X}_{m-1}(b), q)) \varpi(r_m(t \wedge \tau, \mathcal{X}_{m-1}(b), q))| \\ &\quad \le |f(r_n(t \wedge \tau, \mathcal{X}_{n-1}(b), q)) \\ &\quad - f(r_m(t \wedge \tau, \mathcal{X}_{m-1}(b), q))| \varpi(r_n(t \wedge \tau, \mathcal{X}_{n-1}(b), q)) \\ &\quad + |f(r_m(t \wedge \tau, \mathcal{X}_{m-1}(b), q))| |\varpi(r_n(t \wedge \tau, \mathcal{X}_{n-1}(b), q)) \\ &\quad - \varpi(r_m(t \wedge \tau, \mathcal{X}_{m-1}(b), q))| \\ &\le \rho(r_n(t \wedge \tau, \mathcal{X}_{n-1}(b), q) - r_m(t \wedge \tau, \mathcal{X}_{m-1}(b), q)) \varpi(r_n(t \wedge \tau, \mathcal{X}_{n-1}(b), q)) \\ &\quad + \gamma \rho(r_n(t \wedge \tau, \mathcal{X}_{n-1}(b), q) - r_m(t \wedge \tau, \mathcal{X}_{m-1}(b), q)) \\ &\quad [\varpi(r_n(t \wedge \tau, \mathcal{X}_{n-1}(b), q)) + \varpi(r_m(t \wedge \tau, \mathcal{X}_{m-1}(b), q))] \\ &\quad \text{(by the definition of } f \text{ and (15.46))} \\ &\le (\gamma + 1) \rho(r_n(t \wedge \tau, \mathcal{X}_{n-1}(b), q) - r_m(t \wedge \tau, \mathcal{X}_{m-1}(b), q)) \\ &\quad [\varpi(r_n(t \wedge \tau, \mathcal{X}_{n-1}(b), q)) + \varpi(r_m(t \wedge \tau, \mathcal{X}_{m-1}(b), q))]. \end{aligned} \right\} \tag{9.46}$$

Thus,

$$
\left.\begin{aligned}
&E \sup_{0\le t\le T} \gamma_{\varpi}^{4} \mathcal{X}_n((t\wedge\tau, b) - \mathcal{X}_m(t\wedge\tau, b)) \\
&\le (\gamma+1)^4 E \sup_{0\le t\le T} \Bigg\{ \int \rho(r_n(t\wedge\tau, \mathcal{X}_{n-1}(b), q) \\
&\qquad -r_m(t\wedge\tau, \mathcal{X}_{m-1}(b), q)) \\
&\qquad \times(\varpi(r_n(t\wedge\tau, \mathcal{X}_{n-1}(b), q)) \\
&\qquad +\varpi(r_m(t\wedge\tau, \mathcal{X}_{n-1}(b), q)))\mathcal{X}_{0,b}(\mathrm{d}q)\Bigg\}^4 \\
&\le (\gamma+1)^4 E \sup_{0\le t\le T} \int \rho^4(r_n(t\wedge\tau, \mathcal{X}_{n-1}(b), q) \\
&\qquad -r_m(t\wedge\tau, \mathcal{X}_{m-1}(b), q)) \\
&\qquad \times(\varpi(r_n(t\wedge\tau, \mathcal{X}_{n-1}(b), q)) \\
&\qquad +\varpi(r_m(t\wedge\tau, \mathcal{X}_{n-1}(b), q)))\mathcal{X}_{0,b}(\mathrm{d}q) \\
&\qquad \times\Bigg\{ \int (\varpi(r_n(t\wedge\tau, \mathcal{X}_{n-1}(b), q)) \\
&\qquad +\varpi(r_m(t\wedge\tau, \mathcal{X}_{n-1}(b), q)))\mathcal{X}_{0,b}(\mathrm{d}q)\Bigg\}^3 \\
&\qquad \text{(by Hölder's inequality)} \\
&\le \tilde{c} \int E \sup_{0\le t\le T} \rho^4(r_n(t\wedge\tau, \mathcal{X}_{n-1}(b), q) \\
&\qquad -r_m(t\wedge\tau, \mathcal{X}_{m-1}(b), q)) \\
&\qquad \times(\varpi(r_n(t\wedge\tau, \mathcal{X}_{n-1}(b), q)) \\
&\qquad +\varpi(r_m(t\wedge\tau, \mathcal{X}_{n-1}(b), q)))\mathcal{X}_{0,b}(\mathrm{d}q) \\
&\qquad \Big(\text{because} \int \varpi(r_i(t\wedge\tau, \mathcal{X}_{i-1}(b), q))\mathcal{X}_{0,b}(\mathrm{d}q) \\
&\qquad = \gamma_{\varpi}(\mathcal{X}_i(t\wedge\tau, b)) \le c, \quad i = m, n\Big).
\end{aligned}\right\}
\tag{9.47}
$$

The proof follows from (9.12). □

Corollary 9.5. *There is a unique* $\bar{\mathcal{X}}(\cdot) \in L_{loc,4,\mathcal{F}}(C((0,T];\mathbf{M}_{\infty,\varpi}))$*, and for any positive rational numbers b and c and stopping time* $\tau = \tau(c,b)$

$$
E \sup_{0\le t\le T} \gamma_{\varpi}^{4}(\mathcal{X}_n(t\wedge\tau(c,b), b)1_{\Omega_b} - \bar{\mathcal{X}}(t\wedge\tau(c,b))1_{\Omega_b}) \longrightarrow 0, \quad \text{as } n\longrightarrow\infty. \tag{9.48}
$$

Proof. The space $L_{loc,4,\mathcal{F}}(C((0,T];\mathbf{M}_{\infty,\varpi}))$ is complete. Therefore, we obtain from Proposition 9.4 a unique $\tilde{\mathcal{X}}(\cdot, c, b)$ such that

$$
E \sup_{0\le t\le T} \gamma_{\varpi}^{4}(\mathcal{X}_n(t\wedge\tau(c,b), b) - \tilde{\mathcal{X}}(t,c,b)) \longrightarrow 0, \quad \text{as } n\longrightarrow\infty. \tag{9.49}
$$

Further, the monotonicity of $\tau(c,b)$ in c implies that for $c_2 \geq c_1$ and $b > 0$

$$\mathcal{X}_n(\cdot \wedge \tau(c_2,b) \wedge \tau(c_1,b), b, \omega) \equiv \mathcal{X}_n(\cdot \wedge \tau(c_1,b), b, \omega) \quad \text{a.e.}$$

Hence,

$$\tilde{\mathcal{X}}(\cdot, c_2, b, \omega)1_{\{t \leq \tau(c_1,b)\}} \equiv \tilde{\mathcal{X}}(\cdot, c_1, b, \omega)1_{\{t \leq \tau(c_1,b)\}} \quad \text{a.e.} \tag{9.50}$$

Let $\tau(0,b) \equiv 0$ and define

$$\bar{\mathcal{X}}(t,b,\omega) := \sum_{i \in \mathbf{N}} \tilde{\mathcal{X}}(t,i,b,\omega)1_{\{\tau(i-1,b) < t \leq \tau(i,b)\}} \tag{9.51}$$

Expression (9.50) implies that

$$\begin{aligned} &\tilde{\mathcal{X}}(t \wedge \tau(c,b), i, b, \omega)1_{\{\tau(i-1,b) < t \leq \tau(i,b)\}} \\ &\quad \equiv \tilde{\mathcal{X}}(t \wedge \tau(c,b,c), b, \omega)1_{\{\tau(i-1,b) < t \leq \tau(i,b)\}} \quad \text{a.e., uniformly in } t. \end{aligned} \tag{9.52}$$

Indeed, for $i \leq c$, (9.52) follows since we consider only values of $t \in (\tau(i-1,b), \tau(i,b)]$. Therefore, we may apply (9.50) for $c = c_2$ and $i = c_1$. If, however, $i > c$, then the stopping at $\tau(c,b)$ restricts the values of t to those where $t \leq \tau(c,b)$. Hence, (9.50) applies with $c_2 = i$ and $c_1 = c$.

Thus,

$$\bar{\mathcal{X}}(t \wedge \tau(c,b), b) = \sum_{i \in \mathbf{N}} \tilde{\mathcal{X}}(t \wedge \tau(c,b), c, b)1_{\{\tau(i-1,b) < t \leq \tau(i,b)\}} \quad \text{a.e., uniformly in } t,$$

whence,

$$\bar{\mathcal{X}}(t \wedge \tau(c,b), b) \equiv \tilde{\mathcal{X}}(t \wedge \tau(c,b), c, b) \equiv \tilde{\mathcal{X}}(t,c,b) \quad \text{a.e., uniformly in } t\,. \tag{9.53}$$

Further, note that for $b_2 \geq b_1 > 0$ a.e., uniformly in t,

$$\begin{aligned} &\tilde{\mathcal{X}}(t,i,b_2)1_{\{\tau(i-1,b_2) < t \leq \tau(i,b_2)\}}1_{\Omega_{b_1}} \\ &= \tilde{\mathcal{X}}(t,i,b_2)1_{\{\tau(i-1,b_1) < t \leq \tau(i,b_1)\}}1_{\Omega_{b_1}} \quad \text{(by (9.9))} \\ &= \lim_{n\to\infty} \mathcal{X}_n(t \wedge \tau(i,b_2), b_2)1_{\{\tau(i-1,b_1) < t \leq \tau(i,b_1)\}}1_{\Omega_{b_1}} \quad \text{(by (9.49))} \\ &= \lim_{n\to\infty} \mathcal{X}_n(t \wedge \tau(i,b_1), b_1)1_{\{\tau(i-1,b_1) < t \leq \tau(i,b_1)\}}1_{\Omega_{b_1}} \quad \text{(by (9.6) and (9.9))} \\ &= \tilde{\mathcal{X}}(t,i,b_1)1_{\{\tau(i-1,b_1) < t \leq \tau(i,b_1)\}}1_{\Omega_{b_1}} \quad \text{(by (9.49))} \end{aligned}$$

i.e., a.e., uniformly in t,

$$\left.\begin{aligned} \tilde{\mathcal{X}}(t,i,b_2)1_{\{\tau(i-1,b_2) < t \leq \tau(i,b_2)\}}1_{\Omega_{b_1}} &= \tilde{\mathcal{X}}(t,i,b_1)1_{\{\tau(i-1,b_1) < t \leq \tau(i,b_1)\}}1_{\Omega_{b_1}}, \\ \bar{\mathcal{X}}(t,b_2)1_{\Omega_{b_1}} &= \bar{\mathcal{X}}(t,b_1)1_{\Omega_{b_1}}, \end{aligned}\right\} \tag{9.54}$$

where the second line follows from the first line and (9.51). We now set

$$\mathcal{X}(t) := \sum_{j\in\mathbf{N}} \bar{\mathcal{X}}(t,j)1_{\Omega_j\setminus\Omega_{j-1}}, \tag{9.55}$$

and we note that $\mathcal{X}(\cdot)\in L_{0,\mathcal{F}}(C([0,\infty);\mathbf{M}_{\infty,\varpi}))\cap_{p\geq 1} L_{\mathrm{loc},p,\mathcal{F}}(C((0,\infty);\mathbf{M}_{\infty,\varpi}))$. We verify that a.e. uniformly in t,

$$\begin{aligned}
&\mathcal{X}(t\wedge\tau(c,b))1_{\Omega_b}\\
&=\sum_{j\in\mathbf{N}}\bar{\mathcal{X}}(t\wedge\tau(c,b),j)1_{\Omega_j\setminus\Omega_{j-1}}1_{\Omega_b}\\
&=\sum_{j\in\mathbf{N}}\sum_{i\in\mathbf{N}}\tilde{\mathcal{X}}(t\wedge\tau(c,b),i,j)1_{\{\tau(i-1,j)<t\leq\tau(i,j)\}}1_{\Omega_j\setminus\Omega_{j-1}}1_{\Omega_b}\\
&=\sum_{j\in\mathbf{N}}\sum_{i\in\mathbf{N}}\tilde{\mathcal{X}}(t\wedge\tau(c,b),c,j)1_{\{\tau(i-1,j)<t\leq\tau(i,j)\}}1_{\Omega_j\setminus\Omega_{j-1}}1_{\Omega_b}\quad\text{(by (9.52))}\\
&=\sum_{j\in\mathbf{N}}\sum_{i\in\mathbf{N}}\tilde{\mathcal{X}}(t\wedge\tau(c,b),c,b)1_{\{\tau(i-1,j)<t\leq\tau(i,j)\}}1_{\Omega_j\setminus\Omega_{j-1}}1_{\Omega_b}\quad\text{(by (9.54))}\\
&=\sum_{j\in\mathbf{N}}\sum_{i\in\mathbf{N}}\tilde{\mathcal{X}}(t\wedge\tau(c,b),i,b)1_{\{\tau(i-1,j)<t\leq\tau(i,j)\}}1_{\Omega_j\setminus\Omega_{j-1}}1_{\Omega_b}\quad\text{(by (9.52))}\\
&=\bar{\mathcal{X}}(t\wedge\tau(c,b),b)1_{\Omega_b}\quad\text{(by (9.51))}\\
&=\tilde{\mathcal{X}}(t\wedge\tau(c,b),c,b)1_{\Omega_b}\quad\text{(by (9.53))}\,,
\end{aligned}$$

i.e., for any positive rational numbers b and c

$$\begin{aligned}
\mathcal{X}(t\wedge\tau(c,b))1_{\Omega_b} &= \bar{\mathcal{X}}(t\wedge\tau(c,b),b)1_{\Omega_b}\\
&=\tilde{\mathcal{X}}(t\wedge\tau(c,b),c,b)1_{\Omega_b}\quad\text{a.e., uniformly in } t.
\end{aligned}\tag{9.56}$$

(9.56) in addition to (9.49) implies (9.48). □

Let $r(\cdot,\bar{\mathcal{X}}(b),\omega,q)$ be the solution of (4.9) with input process $\bar{\mathcal{X}}(\cdot,b)$. Note that (9.49) in addition to (9.53) in the proof of Corollary 9.5 implies

$$E\sup_{0\leq t\leq T}\gamma_{\varpi}^{4}(\mathcal{X}_n(t\wedge\tau(c,b),b)-\bar{\mathcal{X}}(t\wedge\tau(c,b),b))\longrightarrow 0,\quad\text{as } n\longrightarrow\infty\,. \tag{9.57}$$

Therefore, by Theorem 4.5,

$$\left.\begin{aligned}
&\sup_q E\sup_{0\leq t\leq T}\varrho^2(r(t\wedge\tau(c,b),\bar{\mathcal{X}}(b),q)-r_n(t\wedge\tau(c,b),\mathcal{X}_{n-1}(b),q))\\
&\leq c_T\int_0^T E\gamma_{\varpi}^{2}(\bar{\mathcal{X}}(s\wedge\tau(c,b),b)-\mathcal{X}_{n-1}(s\wedge\tau(c,b),b))ds\\
&\leq c_T\sqrt{\int_0^T E\gamma_{\varpi}^{4}(\bar{\mathcal{X}}(s\wedge\tau(c,b),b)-\mathcal{X}_{n-1}(s\wedge\tau(c,b),b))ds}\longrightarrow 0,\\
&\text{as } n\longrightarrow\infty\,.
\end{aligned}\right\}\tag{9.58}$$

We set

$$\bar{\bar{\mathcal{X}}}(\cdot, b, \omega) := \int \delta_{r(\cdot,\bar{\mathcal{X}}(b),\omega,q)} \mathcal{X}_{0,b}(\mathrm{d}q). \tag{9.59}$$

By Proposition 4.3

$$\bar{\bar{\mathcal{X}}}(\cdot, b, \omega) \in L_{p,\mathcal{F}}(C([0,\infty); \mathbf{M}_{\infty,\varpi}))\ \forall p \geq 1. \tag{9.60}$$

Proposition 9.6. *For all positive rational numbers b and c*

$$E\int_0^T \gamma_\varpi^4(\bar{\bar{\mathcal{X}}}(t\wedge\tau(c,b),b) - \bar{\mathcal{X}}(t\wedge\tau(c,b),b))\mathrm{d}t = 0. \tag{9.61}$$

Proof. By (9.57),

$$\begin{aligned}
&E\int_0^T \gamma_\varpi^4(\bar{\bar{\mathcal{X}}}(t\wedge\tau(c,b),b) - \bar{\mathcal{X}}(t\wedge\tau(c,b),b))\mathrm{d}t \\
&= \lim_{n\to\infty}\int_0^T E\gamma_\varpi^4(\bar{\bar{\mathcal{X}}}(s\wedge\tau(c,b),b) - \mathcal{X}_{n-1}(s\wedge\tau(c,b),b))\mathrm{d}s \\
&= \lim_{n\to\infty} E\int_0^T \sup_{\|f\|_{L,\infty}\leq 1} \\
&\quad\times\Bigg[\int [f(r(t\wedge\tau(c,b),\bar{\mathcal{X}}(b),q))\varpi(r(t\wedge\tau,\bar{\mathcal{X}}(b),q)) \\
&\quad -f(r_n(t\wedge\tau(c,b),\mathcal{X}_{n-1}(b),q))\varpi(r_n(t\wedge\tau(c,b),\mathcal{X}_{n-1}(b),q))]\mathcal{X}_{0,b}(\mathrm{d}q)\Bigg]^4 \\
&\leq \tilde{c}\lim_{n\to\infty} E\int_0^T\int \mathcal{X}_{0,b}(\mathrm{d}q) \\
&\quad\times\varrho^4(r(t\wedge\tau(c,b),\bar{\mathcal{X}}(b),q) - r_n(t\wedge\tau(c,b),\mathcal{X}_{n-1}(b),q)) \\
&[\varpi(r(t\wedge\tau(c,b),\bar{\mathcal{X}}(b),q)) + \varpi(r_n(t\wedge\tau(c,b),\mathcal{X}_{n-1}(b),q))]
\end{aligned}$$

(as in (9.46) and (9.47) by Hölder's inequality).

By (15.46) and the boundedness of $\varpi(\cdot)$, Itô's formula in addition to (9.27) implies that $\varpi(r_n(t\wedge\tau(c,b),\mathcal{X}_{n-1}(b),\omega,q))$ is uniformly integrable with respect to the product measure $P(d\omega)\otimes\mathcal{X}_{0,b}(\mathrm{d}q)$, since $\mathcal{X}_{0,b}(\mathrm{d}q)$ is σ-finite.[2] Therefore,

$$\begin{aligned}
&\tilde{c}\lim_{n\to\infty} E\int_0^T\int \mathcal{X}_{0,b}(\mathrm{d}q)\times\varrho^4(r(t\wedge\tau(c,b),\bar{\mathcal{X}}(b),q) \\
&-r_n(t\wedge\tau(c,b),\mathcal{X}_{n-1}(b),q)) \\
&[\varpi(r(t\wedge\tau(c,b),\bar{\mathcal{X}}(b),q)) + \varpi(r_n(t\wedge\tau(c,b),\mathcal{X}_{n-1}(b),q))] \\
&\longrightarrow 0, \text{ as } n\longrightarrow\infty \text{ (by (9.50)}
\end{aligned}$$

and Lebesgue's dominated convergence theorem).

□

[2] Cf., Bauer (1968), Sect. 20, Corollary 20.5.

Since for $\mathcal{X}(0) \in L_{0,\mathcal{F}_0}(\mathbf{M}_{\varpi,\infty})$ we have $\mathcal{X}(\cdot)$, defined in (9.47), is in $L_{0,\mathcal{F}}(C([0,\infty);\mathbf{M}_{\infty,\varpi}))$, the solution of (4.9) with input process $\mathcal{X}(\cdot)$ and start q is well defined. Employing the measurable version of Part 3 of Theorem 4.5 we set

$$\left.\begin{aligned}&\sum_{j\in\mathbf{N}}\int \delta_{r(\cdot,\mathcal{X},\omega,q)}\mathcal{X}_{0,j}(\omega,dq)1_{\Omega_j\setminus\Omega_{j-1}}(\omega)\\&=\int \delta_{r(\cdot,\mathcal{X},\omega,q)}\sum_{j\in\mathbf{N}}\mathcal{X}_{0,j}(\omega,dq)1_{\Omega_j\setminus\Omega_{j-1}}(\omega)\\&=\int \delta_{r(\cdot,\mathcal{X},\omega,q)}\mathcal{X}_0(\omega,\mathrm{d}q).\end{aligned}\right\} \tag{9.62}$$

Therefore, Proposition 9.6 in addition to (9.48), (9.57), and (9.58) implies Theorem 9.7.

Theorem 9.7. *(i) Assume Hypothesis 4.1. Then the measure valued process $\mathcal{X}(\cdot)$, defined by (9.55), is a solution of the quasilinear SPDE (8.26) with initial condition $\mathcal{X}_0 \in L_{0,\mathcal{F}_0}(\mathbf{M}_{\varpi,\infty})$. $\mathcal{X}(\cdot) \in L_{0,\mathcal{F}}(C([0,\infty);\mathbf{M}_{\infty,\varpi}))$ and it has the flow representation*

$$\mathcal{X}(\cdot,\omega)=\int \delta_{r(\cdot,\mathcal{X},\omega,q)}\mathcal{X}_0(\omega,\mathrm{d}q). \tag{9.63}$$

(ii) Suppose that in addition to (4.11) the conditions of Theorem 8.11 hold and that $\mathcal{X}_0 \in L_{0,\mathcal{F}_0}(\mathbf{M}_{\varpi,\infty}) \cap L_{2,\mathcal{F}_0}(\mathbf{H}_{0,\varpi})$. Then (8.26) has a unique solution.

9.2 Noncoercive Semilinear SPDEs for Infinite Mass Evolution in Stratonovich Form

Completely analogous to Theorem 8.25 we obtain the Stratonovich representation of the solution of (8.26).

Theorem 9.8. *Under the conditions of Theorems 8.25 and 9.7 the solution of the noncoercive parabolic Itô SPDE (8.26), $\mathcal{X}(\cdot)$, driven by the (possibly infinite) initial mass distribution $\mathcal{X}(0) \in L_{0,\mathcal{F}_0}(\mathbf{M}_{\varpi,\infty})$ is the solution of the first order transport Stratonovich SPDE*

$$\begin{aligned}&\mathrm{d}\mathcal{X}=-\nabla\cdot(\mathcal{X}F(\cdot,\mathcal{X},t))\mathrm{d}t-\nabla\cdot\Big(\mathcal{X}\int\mathcal{J}(\cdot,p,t)\Big)w(\mathrm{d}p,\circ\mathrm{d}t),\\&\mathcal{X}(0)=\mathcal{X}_0,\end{aligned} \tag{9.64}$$

An analogous statements for the bilinear SPDE (8.25) is obvious.

Chapter 10
Stochastic Partial Differential Equations: Homogeneous and Isotropic Solutions

The coefficients F *and* J *of (4.9) are assumed to be time independent. Conditions are provided under which* F *and* J *allow rotational and shift invariant solutions of the bilinear SPDE (8.25). Employing Theorem 4.8, homogeneous and isotropic solutions of the bilinear SPDE (8.25) are derived (Theorem 10.2). This result is applied to show that, under the additional assumptions of Theorem 9.7, the solution of the quasilinear SPDE (8.26) is homogeneous and isotropic (Theorem 10.3).*

$\mathcal{O}(d) := \{\sigma \in \mathcal{M}_{d\times d} : \det\sigma = \pm 1\}$. For $\sigma \in \mathcal{O}(d)$ we define the rotation operator S_σ on $\mathbf{H}_{0,\varpi}$ by

$$(S_\sigma f)(r) := f_\sigma(r) := f(\sigma r).$$

Similarly, we define the shift operator U_h on $\mathbf{H}_{0,\varpi}$, where $h \in \mathbf{R}^d$, by

$$(U_h f)(r) := f_h(r) := f(r+h).$$

Moreover, we use $f_{\sigma h}(r)$ to denote the image of the composition $S_\sigma U_h$:

$$(S_\sigma U_h f)(r) := f_{\sigma h}(r) := f(\sigma(r+h)).$$

If the coefficients F and $\mathcal{J}$ are generated by homogeneous kernels, they may give rise to homogeneous solutions (cf. G_ε from (1.2)). If, in addition, $F(r,\cdot)$ and $\mathcal{J}(r,\cdot)$ depend only upon $|r|$ (as, e.g., the fundamental solution of the heat equation, $G(u,r)$) we can also expect to obtain isotropic solutions of (8.25)/(8.26).

For $h \in \mathbf{R}^d$ and $\sigma \in \mathcal{O}(d)$ $\tilde{h}$ and $\tilde{\sigma}$ will denote arbitrary elements from $\{h, 0\}$ and $\{\sigma, I_d\}$, respectively, where I_d is the identity matrix in $\mathbf{R}^d$. We define for $f \in \mathbf{H}_{0,\varpi}$

$$\left.\begin{aligned} U_h^{\sim} f &:= f_{\tilde{h}}, \\ S_\sigma^{\sim} f &:= f_{\tilde{\sigma}}. \end{aligned}\right\} \tag{10.1}$$

Set

$$Q(r, Z(t), w(\mathrm{d}\cdot,\mathrm{d}t)) := \left(F(r, Z(t))\mathrm{d}t, \int \mathcal{J}(r,p,Z(t))w(\mathrm{d}p,\mathrm{d}t)\right). \tag{10.2}$$

Z and w will be called the *coordinates* of Q or of the pair $(F,\mathcal{J})$, where $Z \in L_{\mathrm{loc},2,\mathcal{F}}(C((s,T];\mathbf{H}_{0,\varpi,+})) \subset L_{\mathrm{loc},2,\mathcal{F}}(C((s,T];\mathbf{M}_{\infty,\varpi},))$ (cf. (8.47)).

Definition 10.1. *Let* $\sigma \in \mathcal{O}(d)$ *and* $h \in \mathbf{R}^d$. *The pair* $(F, \mathcal{J})$ *is called* σ–h-admissible, *if* $\mathcal{J}$ *is spatially homogeneous,*[1] *i.e., if*

$$\mathcal{J}(r, q, \mu) = \mathcal{J}(r - q, \mu)\ \forall r, q$$

and

$$(S_\sigma U_h Q)(r, Z(t), w(\mathrm{d}\cdot, \mathrm{d}t)) = Q(r, Z_{\tilde{\sigma}\tilde{h}}(t), w_{\tilde{\sigma}\tilde{h}}(\mathrm{d}\cdot, \mathrm{d}t)), \tag{10.3}$$

where $\tilde{\sigma}$ *and* $\tilde{h}$ *may vary with the coordinates.* $(F, \mathcal{J})$ *is called* admissible, *if it is* σ–h*-admissible for all* $(\sigma, h) \in \mathcal{O}(d) \times \mathbf{R}^d$. □

$(F, \mathcal{J})$ and $(F, \tilde{\mathcal{J}})$ from Sect. 4.4 (Examples) are admissible. The coordinates in (10.3), where $\tilde{h} = h$, combined with the coordinates, where $\tilde{\sigma} = \sigma$, determine a *type* of the admissible pair $(F, \mathcal{J})$. We observe that a given pair may have several types. Since we can obviously restrict the notion of a type to any of the two components on the right-hand side of (10.2), admissibility implies that neglecting the additional coordinate $w(\mathrm{d}\cdot, \mathrm{d}t)$ of the second component of Q, F, and $\int \mathcal{J}(\cdot, p, \cdot)w(\mathrm{d}p, \mathrm{d}t)$ must have at least one common type. Note that U_h and S_σ are homomorphisms (on the group of functions with pointwise multiplication). Therefore, the types of the mutual quadratic variations $\mathrm{d}[m_k(t), m_\ell(t)]$ coincide with the types of $\mathrm{d}m(t) := \int \mathcal{J}(\cdot, p, \cdot, \cdot)w(\mathrm{d}p, \mathrm{d}t)$.

In what follows, we identify $\int_0^t w(\mathrm{d}p, \mathrm{d}s)$ with the cylindrical Brownian motion $W(t)$ from (4.28).

Theorem 10.2. *Suppose (8.35),* $m \ge 0$ *and* $Z \in L_{loc,2,\mathcal{F}}(C((s, T]; \mathbf{H}_{0,\varpi,+}))$. *Further, let* $h \in \mathbf{R}^d$ *and* $\sigma \in \mathcal{O}(d)$, *and assume* $F, \mathcal{J}$ *is* σ–h*-admissible. Let* $Y(\cdot, Z, Y_0, w)$ *be the solution of (8.25) with* $Y_0 \in L_{2,\mathcal{F}_0}(\mathbf{H}_{0,\varpi,+})$ *and* w *as in (4.9). Let* $(\tilde{\sigma}, \tilde{h})$ *be a collection of (coordinate-dependent) values so that (10.3) holds. In addition to the preceding assumptions suppose that in* $C([s, T]; \mathbf{M}_{\infty,\varpi}) \times \mathbf{H}_{0,\varpi,+} \times C([s, T]; \mathbf{H}_{w,d})$[2]

$$(S_{\tilde{\sigma}} U_{\tilde{h}} Z, S_\sigma U_h Y_0, S_{\tilde{\sigma}} U_{\tilde{h}} W) \sim (Z, Y_0, W). \tag{10.4}$$

Then, for any $t \in [s, T]$, *in* $C([s, T]; \mathbf{H}_{0,\varpi,+}) \times C([s, T]; \mathbf{M}_{\infty,\varpi}) \times \mathbf{H}_{0,\varpi,+} \times C([s, T]; \mathbf{H}_{0,\varpi,+})$

$$\left.\begin{array}{l} ((\pi_t S_\sigma U_h Y)(\cdot, Z, Y_0, W), S_{\tilde{\sigma}} U_{\tilde{h}} Z, S_\sigma U_h Y_0, S_\sigma U_{\tilde{h}} W) \\ \sim ((\pi_t Y)(\cdot, Z, Y_0, W), Z, Y_0, W). \end{array}\right\} \tag{10.5}$$

Proof.

(i) We only show the shift invariance, assuming $\sigma = I$, since the proof for $\sigma \neq I$ follows the same pattern. Again, we assume $s = 0$.

---

[1] The spatial homogeneity of $\mathcal{J}$ implies (8.33).

[2] $\mathbf{H}_{w,d}$ has been defined in (4.31).

(ii) Let $\varphi \in C_c^2(\mathbf{R}^d; \mathbf{R})$. Since differentiation and shift commute on $C_c^2(\mathbf{R}^d; \mathbf{R})$, we obtain from the preceding remarks

$$
\begin{aligned}
\langle (U_h Y)(t, Z, Y_0, W), \varphi\rangle_0 &= \langle Y(t, Z, Y_0, W), U_{-h}\varphi\rangle_0 = \langle Y_0, U_{-h}\varphi\rangle_0 \\
&+ \int_0^t \frac{1}{2}\sum_{k,\ell=1}^d \langle Y(s, Z, Y_0, W), D_{k\ell}(Z(s))U_{-h}\partial_{k\ell}^2 \varphi\rangle_0 \mathrm{d}s \\
&+ \int_0^t \langle Y(s, Z, Y_0, W), F(\cdot, Z(s)) \cdot (U_{-h} \triangledown \varphi)\rangle_0 \mathrm{d}s \\
&+ \int_0^t \langle Y(s, Z, Y_0, W), \int \mathcal{J}(\cdot - p, Z(s))w(\mathrm{d}p, \mathrm{d}s) \cdot (U_{-h} \triangledown \varphi)\rangle_0 \mathrm{d}s \\
= \langle Y_{0,h}, \varphi\rangle_0 & \\
&+ \int_0^t \frac{1}{2}\sum_{k,\ell=1}^d \langle Y_h(s, Z, Y_0, W), D_{k\ell}(Z_h^{\sim}(s))\partial_{k\ell}^2 \varphi\rangle_0 \mathrm{d}s \\
&+ \int_0^t \langle Y_h(s, Z, Y_0, W), F(\cdot, Z_h^{\sim}(s)) \cdot \triangledown\varphi\rangle_0 \mathrm{d}s \\
&+ \int_0^t \langle Y_h(s, Z, Y_0, W), \int \mathcal{J}(\cdot - p, Z_h^{\sim}(s))w_h^{\sim}(\mathrm{d}p, \mathrm{d}s) \cdot \triangledown\varphi\rangle_0.
\end{aligned}
$$

Here, $w_h^{\sim}$ is the shifted white noise.[3] However, the right-hand side of the last equation is equivalent to the statement $Y_h(\cdot) := Y_h(\cdot, Z, Y_0, W)$ is the solution of the bilinear SPDE (8.25) with input process $Z_h^{\sim}$, initial condition $Y_{0,h}$ and driving Brownian space–time white noise $w_h^{\sim}(\mathrm{d}p, \mathrm{d}s)$ ($\sim W_h^{\sim}$). By Theorem 8.6 this solution, denoted $Y(\cdot, Z_h^{\sim}, Y_{0,h}, W_h^{\sim})$, is unique. Hence, a.s.,

$$(U_h Y)(\cdot, Z, Y_0, W) = Y(\cdot, Z_h^{\sim}, Y_{0,h}, W_h^{\sim}). \tag{10.6}$$

(iii) To show that $Y(\cdot, Z_h^{\sim}, Y_{0,h}, W_h^{\sim}) \sim Y(\cdot, Z, Y_0, W)$, we apply Theorem 4.8. In a first step, we discretize $\mathbf{H}_{0,\varpi}$ as follows: Set $C_n := \{r \in \mathbf{R}^d : -n \le r_k < n$ for $k = 1, \ldots, d\}$, where $n \in \mathbf{N}$. We then partition C_n into N boxes of equal volume $(2n)^d/N$ (cf. Chap. 2), where we assume that the left end points of those boxes (or d-dimensional intervals) are excluded and the right end points belong to the box. Let us denote the midpoints by $\{\bar{r}^1, \ldots, \bar{r}^N\}$ and identify the ith box with $\bar{r}^i$, $i = 1, \ldots, N$. If $f \in \mathbf{H}_{0,\varpi}$ we set

$$(\pi_{N,n} f)(r) := \left\{ \begin{array}{ll} \frac{N}{(2n)^d} \int_{(\bar{r}^i]} f(q)\mathrm{d}q, & r \in (\bar{r}^i], i = 1, \ldots N, \\ 0, & r \notin C_n, \end{array} \right\} \tag{10.7}$$

where $(\bar{r}^i]$ is the ith box. Lebesgue's differentiation theorem and Lebesgue's dominated convergence theorem imply for any $f \in \mathbf{H}_{0,\varpi}$

$$\|f - \pi_{N,n} f\|_{0,\varpi} \to 0, \text{ as } N \to \infty, n \to \infty, \tag{10.8}$$

[3] Cf. (5.22).

and also for any $X_0 \in \mathcal{H}_{0,\varpi,0}$ and any stopping time τ which is localizing for X_0

$$E(\|X_0 - \pi_{N,n} X_0\|_{0,\varpi}^2 1_{\{\tau>0\}}) \to 0, \text{ as } N \to \infty, n \to \infty. \tag{10.9}$$

Moreover, for fixed n and N we define a map

$$\left.\begin{array}{ll} g_N : \pi_{N,n}\mathbf{H}_{0,\varpi} & \to \mathbf{R}^N \\ \pi_{N,n} f & \mapsto a(N, f) \end{array}\right\} \tag{10.10}$$

where the ith component of $a(N, f)$ is the integral in (10.7) (the "weight," if $f \geq 0$). The restriction of g_N to nonnegative values defines

$$g_N : \pi_{N,n}\mathbf{H}_{0,\varpi,+} \to \mathbf{R}_+^N.$$

We check that g_N is continuous, where $\pi_{N,n}\mathbf{H}_{0,\varpi,+}$ is endowed with $\|\cdot\|_{0,\varpi}$. Since $\pi_{N,n}$ is a projection into a closed subspace of $\mathbf{H}_{0,\varpi,+}$, the map

$$\begin{array}{rl} C([0,T];\mathbf{M}_{\infty,\varpi}) \times \mathbf{H}_{0,\varpi,+} \times C([0,T];\mathbf{H}_{w,d}) \longmapsto & C([0,T];\mathbf{M}_{\infty,\varpi}) \\ & \times \mathbf{R}_+^N \times C([0,T];\mathbf{H}_{w,d}) \\ (\eta, \xi_0, \overline{W}) \longmapsto & (\eta, g_N \pi_{N,n}\xi_0, \overline{W}) \end{array}$$

is continuous. Therefore, (10.4) implies

$$(U_h^{\sim} Z, g_N \pi_{N,n} U_h Y_0, U_h^{\sim} W) \sim (Z, g_N \pi_{N,n} Y_0, W) \tag{10.11}$$

in $C([0,T];\mathbf{M}_{\infty,\varpi}) \times \mathbf{R}_+^N \times C([0,T];\mathbf{H}_{w,d})$.

(iv) By the proof of Theorem 4.8 there is a solution map $\overline{\overline{r}}_N$ for (4.9) (on the "canonical" probability space). Using the notation which was introduced in Chap. 4 after (4.35), it follows that for any $t \in [0,T]$ the following map is measurable as indicated:

$$\left.\begin{array}{rl} \mathbf{R}_+^N \times \mathbf{R}^{dN} \times C([0,T];\mathbf{M}_{\infty,\varpi} \times \mathbf{H}_{w,d}) \longmapsto & \mathbf{R}_+^N \times C([0,T];\mathbf{R}^{dN} \\ & \times \mathbf{M}_{\infty,\varpi} \times \mathbf{H}_{w,d}) \\ (a(N), r_N, \eta, \overline{W}) \longmapsto & (a(N), (\pi_t \overline{\overline{r}}_N)(\cdot, \overline{W}, \eta, r_N), \eta, \overline{W}) \\ \mathcal{B}^N \otimes \mathcal{B}^{dN} \otimes \bar{\mathcal{F}}_{\mathbf{M}_{\varpi,\infty}\times\mathbf{H}_{w,d},T} - & \mathcal{B}^N \otimes \bar{\mathcal{F}}_{\mathbf{R}^{dN}\times\mathbf{M}_{\varpi,\infty}\times\mathbf{H}_{w,d},t} \end{array}\right\} \tag{10.12}$$

where $\overline{W}$ denotes a typical element of $C([0,T];\mathbf{H}_{w,d})$. Hence, by (10.11), for any $t \geq 0$

$$\left.\begin{array}{l} (g_N \pi_{N,n} U_h Y_0, (\pi_t \overline{\overline{r}}_N)(\cdot, U_h^{\sim} W, U_h^{\sim} Z, r_N), U_h^{\sim} Z, U_h^{\sim} W) \\ \sim (g_N \pi_{N,n} Y_0, (\pi_t \overline{\overline{r}}_N)(\cdot, W, Z, r_N), Z, W) \end{array}\right\} \tag{10.13}$$

in $\mathbf{R}_+^N \times C([0,T];\mathbf{R}^{dN} \times \mathbf{M}_{\infty,\varpi} \times \mathbf{H}_{w,d})$.

(v) Let $a(N) := (a_1, \ldots, a_N) \in \mathbf{R}_+^N$ (the "weights") and $q_N := (q^1, \ldots, q^N) \in \mathbf{R}^{dN}$ (the "positions"). Then,

$$h_N(a(N), q_N) := \sum_{i=1}^{N} a_i \delta_{q^i} \tag{10.14}$$

is continuous from $\mathbf{R}_+^N \times \mathbf{R}^{dN}$ into $(\mathbf{M}_{\infty,\varpi}, \gamma_\varpi)$. Let $\tilde{\pi}_{N,n} Y_0$ and $\tilde{\pi}_{N,n} U_h Y_0$ be the sum of point measures associated with the discretization, i.e.,

$$\tilde{\pi}_{N,n} Y_0 := \sum_{i=1}^{N} a_{\bar{r}^i}(N, Y_0) \delta_{\bar{r}^i},$$

where $a_{\bar{r}^i}(N, Y_0)$ has been defined in (10.10), and similarly for $\tilde{\pi}_{N,n} U_h Y_0$. Further, let $\mathcal{Y}(\cdot)$ denote the empirical measure process associated with $\bar{\bar{r}}_N(\cdot) = (\bar{\bar{r}}(\cdot, \bar{r}^1) \cdots \bar{\bar{r}}(\cdot, \bar{r}^N))^T$, i.e.,

$$\mathcal{Y}(t) := \sum_{i=1}^{N} a_{\bar{r}^i}(N, Y_0) \delta_{\bar{\bar{r}}(t, \bar{r}^i)}.$$

Then (10.7), (10.13), and (10.14) imply

$$\left.\begin{array}{l} (\pi_{N,n} U_h Y_0, (\pi_t \mathcal{Y})(\cdot, U_h^{\sim} W, U_h^{\sim} Z, \tilde{\pi}_{N,n} U_h Y_0), U_h^{\sim} Z, U_h^{\sim} W) \\ \sim (\pi_{N,n} Y_0, (\pi_t \mathcal{Y})(\cdot, W, Z, \tilde{\pi}_{N,n} Y_0), Z, W) \end{array}\right\} \tag{10.15}$$

in $\mathbf{H}_{0,\varpi} \times C([0,T]; \mathbf{M}_{\infty,\varpi} \times \mathbf{H}_{w,d})$. Here, we used the fact that g_N from (10.10) has a continuous inverse.

(vi) The Cauchy–Schwarz inequality implies that Y_0 (and $U_h Y_0$), restricted to C_n, is in $L_{0,\mathcal{F}_0}(\mathbf{W}_{0,1,1}) (\subset L_{0,\mathcal{F}_0}(\mathbf{M}_f) \subset L_{0,\mathcal{F}_0}(\mathbf{M}_{\infty,\varpi}))$. Therefore, for any stopping time τ which is localizing for Y_0,

$$E(\gamma_\varpi^2(\tilde{\pi}_{N,n} Y_0, 1_{C_n} Y_0) 1_{\{\tau > 0\}}) \to 0, \text{ as } N \to \infty \tag{10.16}$$

and, trivially,

$$E\|1_{C_n} Y_0 - Y_0\|_{0,\varpi}^2 \to 0 \text{ as } n \to \infty. \tag{10.17}$$

The same holds for $U_h Y_0$.

Since $\mathcal{Y}$ is the solution of the bilinear SPDE (8.25), it follows from Theorem 8.6 that $\mathcal{Y}(\cdot, \ldots, 1_{C_n} Y_0) = Y(\cdot, \ldots, 1_{C_n} Y_0) \in L_{loc,2,\mathcal{F}}(C([0,T]; \mathbf{M}_f)) \cap L_{loc,2,\mathcal{F}}(C([0,T]; \mathbf{H}_{0,\varpi,+}))$, and similarly for $1_{C_n} U_h Y_0$. Hence, (10.16) in addition to the continuous mapping theorem[4] implies the equivalence, as stated in (10.15), but with $\pi_{N,n}$ and $\tilde{\pi}_{N,n}$ replaced by 1_{C_n}. Expression (10.17) in addition to (10.6) and the continuous mapping theorem yields (10.5). □

[4] Cf. Sect. 15.2.1, Theorem 15.22.

We now extend the results of Theorem 10.2 to the solution X of the quasilinear SPDE (8.26) under the assumptions of Theorems 8.11 and 9.7.

Theorem 10.3. *Suppose the pair $(F, \mathcal{J})$ is σ–h-admissible. Further, suppose (8.35) with smoothness degree $\overline{m} > d+2$ and (8.45) in addition to (4.11). Let $h \in \mathbf{R}^d$ and $\sigma \in \mathcal{O}(d)$ and assume that $(F, \mathcal{J})$ is σ–h-admissible. Let $(\tilde{\sigma}, \tilde{h})$ be a collection of (coordinate-dependent) values so that (10.3) holds with $Z := X$. Further, suppose we have that in $\mathbf{H}_{0,\varpi,+} \times C([0,T]; \mathbf{H}_{w,d})$*

$$(X_0, S_\sigma U_h W) \sim (X_0, W). \tag{10.18}$$

Then, for any $t \geq 0$, in $C([0,T]; \mathbf{H}_{0,\varpi,+}) \times \mathbf{H}_{0,\varpi,+} \times C([0,T]; \mathbf{H}_{w,d})$

$$\left.\begin{array}{l} ((\pi_t S_\sigma U_h X)(\cdot, X, X_0, W), X_0, S_{\tilde{\sigma}} U_{\tilde{h}} W) \\ \sim ((\pi_t X)(\cdot, X, X_0, W), X_0, W). \end{array}\right\} \tag{10.19}$$

In particular, (10.19) implies (in $C([0,T]; \mathbf{H}_{0,\varpi,+})$)

$$(S_\sigma U_h X)(\cdot, X, X_0, W) \sim X(\cdot, X, X_0, W). \tag{10.20}$$

Proof.

(i) Consider the recursive sequence $\mathcal{X}_n(t,b)$, defined by (9.3). Employing (9.6), we may set

$$\mathcal{X}_n(t) := \sum_{j \in \mathbf{N}} \mathcal{X}_n(t,j) 1_{\Omega_j \setminus \Omega_{j-1}}, \tag{10.21}$$

which is analogous to (9.55) for the limiting process. It follows that for all $n \in \mathbf{N}$ $\mathcal{X}_n(\cdot) \in L_{0,\mathcal{F}}(C([0,\infty); \mathbf{M}_{\infty,\varpi})) \cap_{p \geq 1} L_{\mathrm{loc},p,\mathcal{F}}(C([0,\infty); \mathbf{M}_{\infty,\varpi}))$. Therefore, there is a unique solution of (4.9), $r(\cdot, \mathcal{X}_{n-1}, q)$, with input $\mathcal{X}_{n-1}(\cdot)$ and start q which, by Part 3 of Theorem 4.5, is measurable in all variables. Set

$$\tau_j(\omega) := \begin{cases} T, & \text{if } \omega \in \Omega_j, \\ 0, & \text{otherwise.} \end{cases}$$

Since $\Omega_j \in \mathcal{F}_0$, it follows that τ_j is a stopping time and $1_{\{\tau_j > 0\}} = 1_{\Omega_j}$. Therefore,

$$\gamma(\mathcal{X}_n(u) - \mathcal{X}_n(u,j)) 1_{\{\tau_j > 0\}} = 0 \ \forall u \in [0,T],$$

whence

$$r(u, \mathcal{X}_{n-1}, q) 1_{\Omega_j} = r(u, \mathcal{X}_{n-1}(j), q) 1_{\Omega_j} \ \forall u \in [0,T].$$

This implies

$$\mathcal{X}_n(t,\omega) 1_{\Omega_j \setminus \Omega_{j-1}} = \int \delta_{r_n(t, \mathcal{X}_{n-1}, \omega, q)} \mathcal{X}_0(dq) 1_{\Omega_j \setminus \Omega_{j-1}} = \mathcal{X}_n(t,j) 1_{\Omega_j \setminus \Omega_{j-1}}. \tag{10.22}$$

Consequently, we obtain from (9.49) and (9.55) for all $c, j \in \mathbf{N}$

$$E \sup_{0 \leq t \leq T} \gamma_\varpi^4 (\mathcal{X}_n(t \wedge \tau(c,j)) - \mathcal{X}(t \wedge \tau(c,j))) \longrightarrow 0, \quad \text{as } n \longrightarrow \infty. \tag{10.23}$$

(ii) We have

$$\mathcal{X}_{n+1}(\cdot) = \mathcal{X}_{n+1}(\cdot, \mathcal{X}_n, X_0).$$

Next, we approximate $X(\cdot, X, X_0)$ by $\mathcal{X}_{n+1}(\cdot, \mathcal{X}_n, X_0)$, as shown in the proof of Proposition 9.4 and Corollary 9.5, employing (10.23). Although this procedure uses the topology of $L_{\mathrm{loc},2,\mathcal{F}}(C((0,T]); \mathbf{M}_{\infty,\varpi}))$, the smoothness of X_0 in addition to (8.35) implies that for each n $\mathcal{X}_{n+1}(\cdot, \mathcal{X}_n, X_0) \in L_{\mathrm{loc},2,\mathcal{F}}(C((0,T]; \mathbf{H}_{0,\varpi,+}))$. Consequently, by (10.5) and (10.18),

$$\left.\begin{aligned}&((\pi_t S_\sigma U_h \mathcal{X}_{n+1})(\cdot, \mathcal{X}_n, \mathcal{X}_0, W), S_\sigma^\sim U_h^\sim \mathcal{X}_n, \mathcal{X}_0, S_\sigma^\sim U_h^\sim W))\\ &\quad\sim ((\pi_t \mathcal{X}_{n+1})(\cdot, \mathcal{X}_n, \mathcal{X}_0, W), \mathcal{X}_n, \mathcal{X}_0, W).\end{aligned}\right\} \tag{10.24}$$

Since by (10.23), $\mathcal{X}_n \to \mathcal{X}$, $S_\sigma U_h \mathcal{X}_n \to S_\sigma U_h \mathcal{X}$, $S_\sigma^\sim U_h^\sim \mathcal{X}_n \to S_\sigma^\sim U_h^\sim \mathcal{X}$ in $L_{\mathrm{loc},2,\mathcal{F}}(C((0,T]); \mathbf{M}_{\infty,\varpi}))$ and since $\mathcal{X} \in L_{\mathrm{loc},2,\mathcal{F}}(C((0,T]; \mathbf{H}_{0,\varpi,+}))$, Expression (10.24) in addition to the continuous mapping theorem implies (10.19) and (10.20). □

Chapter 11
Proof of Smoothness, Integrability, and Itô's Formula

We first derive basic estimates and define norms on the relevant state spaces. Smoothness and imbedding properties are derived. Finally, the proof of the Itô formula (8.42) is provided.

11.1 Basic Estimates and State Spaces

To describe the order of differentiability of $f : \mathbf{R}^d \to \mathbf{R}$ we recall the notation on multiindices from Sect. 4:

$$\mathbf{n} = (n_1, \dots n_d) \in (\mathbf{N} \cup \{0\})^d.$$

$|\mathbf{n}| := n_1 + \cdots + n_d$. If $\mathbf{n} = (n_1, \dots, n_d)$ and $\mathbf{m} = (m_1, \dots, m_d)$, then $\mathbf{n} \le \mathbf{m}$ iff $n_i \le m_i$ for $i = 1, \dots, d$. $\mathbf{n} < \mathbf{m}$ iff $\mathbf{n} \le \mathbf{m}$ and $|\mathbf{n}| < |\mathbf{m}|$. For arbitrary $\mathbf{n}, \mathbf{m}$ we define

$$\mathbf{n} - \mathbf{m} := (k_1, \dots, d_d),$$

where $k_i = (n_i - m_i) \vee 0$, $i = 1, \dots, d$, with "$\vee$" denoting the maximum and $\mathbf{n} = (n_1, \dots, n_d)$, $\mathbf{m} = (m_1, \dots, m_d)$. For $\mathbf{n} = (n_1, \dots, n_d) \in \mathbf{Z}^d$ we set

$$\pi_{\mathbf{n}}(r) := \begin{cases} \Pi_{i=1}^d r_i^{n_i}, & \text{for all } i = 1, \dots, d \quad n_i \ge 0 \\ 0, & \text{otherwise.} \end{cases}$$

If $x \in \mathbf{R}$, then we set $[x] := \max\{k \in \mathbf{Z} : k \le x\}$. Further, we abbreviate

$$G(u, r) := (2\pi u)^{-d/2} \exp\left(\frac{-|r|^2}{2u}\right), u > 0 \tag{11.1}$$

with the other usual assignments for the cases $u = 0$ and ($|r| = 0$ or $|r| > 0$).

Lemma 11.1. *For any $\mathbf{n} \ge \mathbf{0}$ there are finite constants $\gamma_{\mathbf{i}} := \gamma_{\mathbf{i},\mathbf{n}}$ such that*

$$\partial^{\mathbf{n}} G(u, r) = \sum_{\mathbf{0} \le 2\mathbf{i} \le \mathbf{n}} \gamma_{\mathbf{i}} \frac{\pi_{\mathbf{n}-2\mathbf{i}}(r)}{u^{|\mathbf{n}|-|\mathbf{i}|}} G(u, r), \tag{11.2}$$

where $\gamma_{\mathbf{0}} = (-1)^{|\mathbf{n}|}$.

Proof. We use induction in $|\mathbf{n}|$. Expression (11.2) is obviously correct for $|\mathbf{n}| = 0$. Assume (11.2) is correct for all $\mathbf{n}$ such that $|\mathbf{n}| \le m - 1$ and $m - 1 \ge 0$. Choose $\mathbf{n}$ with $|\mathbf{n}| = m$. Then, recalling the notation 1_k from Chap. 8 (before (8.39)),

$$\partial^{\mathbf{n}} G(u,r) = \partial_k \partial^{\mathbf{n}-1_k} G(u,r)$$

for some $k \in \{1, \ldots, d\}$

$$= \partial_k \left\{ \sum_{|\mathbf{i}|=0}^{\left[\frac{|\mathbf{n}|-1}{2}\right]} \gamma_{\mathbf{i}} \frac{\pi_{\mathbf{n}-1_k-2\mathbf{i}}(r)}{u^{|\mathbf{n}|-1-|\mathbf{i}|}} G(u,r) \right\}$$

(by induction hypothesis)

$$= \sum_{|\mathbf{i}|=0}^{\left[\frac{|\mathbf{n}|-1}{2}\right]_2} \gamma_{\mathbf{i}} \frac{1}{u^{|\mathbf{n}|-1-|\mathbf{i}|}} \left(c_{k,\mathbf{n},\mathbf{i}} \pi_{\mathbf{n}-2\cdot 1_k-2\mathbf{i}}(r) - \frac{\pi_{\mathbf{n}-1_k-2\mathbf{i}}(r) r_k}{u} \right) G(u,r)$$

$$= \sum_{|\mathbf{i}|=0}^{\left[\frac{|\mathbf{n}|-1}{2}\right]_2} \gamma_{\mathbf{i}} c_{k,\mathbf{n},\mathbf{i}} \frac{\pi_{\mathbf{n}-2(\mathbf{i}+1_k)}(r)}{u^{|\mathbf{n}|-(|\mathbf{i}|+1)}} G(u,r) - \sum_{|\mathbf{i}|=0}^{\frac{|\mathbf{n}|-1}{2}} \gamma_{\mathbf{i}} \frac{\pi_{\mathbf{n}-2\mathbf{i}}(r)}{u^{|\mathbf{n}|-|\mathbf{i}|}} G(u,r)$$

(where $c_{k,\mathbf{n},\mathbf{i}} := (n_k - 1 - 2\mathbf{i}_k) \vee 0$)

$$= \sum_{0 \le 2\mathbf{i} \le \mathbf{n}} \tilde{\gamma}_{\mathbf{i}} \frac{\pi_{\mathbf{n}-2\mathbf{i}}(r)}{u^{|\mathbf{n}|-|\mathbf{i}|}} G(u,r)$$

for some finite constants $\tilde{\gamma}_{\mathbf{i}} := \tilde{\gamma}_{\mathbf{i},\mathbf{n}}$, where $\tilde{\gamma}_{\mathbf{0}} = (-1)\gamma_{\mathbf{0}}$, as the upper limit in the summation is restricted to $\mathbf{n}$ by the definition of $\pi_{\mathbf{n}-2\mathbf{i}}(r)$. □

Lemma 11.2. *For any* $\mathbf{n} \ge \mathbf{0}$ *there are finite constants* $\gamma_{\mathbf{i}} := \gamma_{\mathbf{i},\mathbf{n}}$ *such that*

$$G(u,r)\pi_{\mathbf{n}}(r) = \sum_{0 \le 2\mathbf{i} \le \mathbf{n}} u^{|\mathbf{n}|-|\mathbf{i}|} \gamma_{\mathbf{i}} \partial^{\mathbf{n}-2\mathbf{i}} G(u,r). \tag{11.3}$$

Proof. Expression (11.3) is obvious for $|\mathbf{n}| = 0$ and $|\mathbf{n}| = 1$. Assume (11.3) is correct for all $\mathbf{n} \ge \mathbf{0}$ with $|\mathbf{n}| \le m - 1$, $m - 1 \ge 1$. Choose $\mathbf{n}$ with $|\mathbf{n}| = m$. Then, by Lemma 11.1,

$$G(u,r)\frac{\pi_{\mathbf{n}}(r)}{u^{|\mathbf{n}|}} - (-1)^{|\mathbf{n}|} \partial^{\mathbf{n}} G(u,r)$$

$$= \sum_{0 < 2\mathbf{i} \le \mathbf{n}} \frac{-(-1)^{|\mathbf{n}|} \gamma_{\mathbf{i}}}{u^{|\mathbf{n}|-|\mathbf{i}|}} \pi_{\mathbf{n}-2\mathbf{i}}(r) G(u,r)$$

$$= \sum_{\mathbf{0}<2\mathbf{i}\le\mathbf{n}} \frac{-(-1)^{|\mathbf{n}|}\gamma_i}{u^{|\mathbf{n}|-|\mathbf{i}|}} \left(\sum_{\mathbf{0}\le 2\mathbf{j}\le\mathbf{n}-2\mathbf{i}} u^{|\mathbf{n}|-2|\mathbf{i}|-|\mathbf{j}|}\tilde{\gamma}_{\mathbf{j}}\partial^{\mathbf{n}-2\mathbf{i}-2\mathbf{j}}G(u,r) \right)$$

(by induction hypothesis with finite constants $\tilde{\gamma}_{\mathbf{j}}$ and by $|\mathbf{n}-2\mathbf{i}| = |\mathbf{n}| - 2|\mathbf{i}|$ for $\mathbf{0} \le 2\mathbf{i} \le \mathbf{n}$)

$$= \sum_{\mathbf{0}<2\mathbf{k}\le\mathbf{n}} \beta_{\mathbf{k}} u^{-|\mathbf{k}|}\partial^{\mathbf{n}-2\mathbf{k}}G(u,r)$$

(with $\mathbf{k} = \mathbf{i}+\mathbf{j}$ and $\beta_{\mathbf{k}} = -(-1)^{|\mathbf{n}|} \sum_{\mathbf{i}+\mathbf{j}=k} \gamma_{\mathbf{i}}\tilde{\gamma}_{\mathbf{j}}$). Hence,

$$G(u,r)\pi_{\mathbf{n}}(r) = u^{|\mathbf{n}|}(-1)^{|\mathbf{n}|}\partial^{\mathbf{n}}G(u,r) + \sum_{\mathbf{0}<2\mathbf{k}\le\mathbf{n}} \beta_{\mathbf{k}} u^{|\mathbf{n}|-|\mathbf{k}|}\partial^{\mathbf{n}-2\mathbf{k}}G(u,r).$$

This implies (11.3). □

Expression (11.2) immediately implies

Corollary 11.3. *Let* $\mathbf{n},\mathbf{m} \ge \mathbf{0}$. *Then there are finite constants* $\gamma_{\mathbf{i}}$ *such that*

$$\left.\begin{aligned}(\partial^{\mathbf{m}}G(u,r))\pi_{\mathbf{n}}(r) &= u^{\frac{|\mathbf{n}|}{2}} \sum_{\mathbf{0}\le 2\mathbf{i}\le\mathbf{m}} \gamma_{\mathbf{i}} \frac{\pi_{\mathbf{m}-2\mathbf{i}+\mathbf{n}}(r)}{u^{|\mathbf{m}|-|\mathbf{i}|+\frac{|\mathbf{n}|}{2}}} G(u,r)\\ &= u^{\frac{|\mathbf{n}|-|\mathbf{m}|}{2}} \sum_{\mathbf{0}\le 2\mathbf{i}\le\mathbf{m}} \gamma_{\mathbf{i}} \frac{\pi_{\mathbf{m}-2\mathbf{i}+\mathbf{n}}(r)}{u^{\frac{|\mathbf{m}|}{2}-|\mathbf{i}|+\frac{|\mathbf{n}|}{2}}} G(u,r).\end{aligned}\right\} \quad (11.4)$$

□

Corollary 11.4. *Let* $\mathbf{n},\mathbf{m} \ge \mathbf{0}$. *Then there are finite constants* $\hat{\gamma}_{\tilde{\mathbf{n}}}$ *such that*

$$(\partial^{\mathbf{m}}G(u,r))\pi_{\mathbf{n}}(r) = \sum_{\mathbf{0}\le 2\tilde{\mathbf{n}}\le(\mathbf{n}+\mathbf{m})} u^{|\mathbf{n}|-|\tilde{\mathbf{n}}|}\hat{\gamma}_{\tilde{\mathbf{n}}}\partial^{\mathbf{m}+\mathbf{n}-2\tilde{\mathbf{n}}}G(u,r). \quad (11.5)$$

Proof.

$$(\partial^{\mathbf{m}}G(u,r))\pi_{\mathbf{n}}(r) = \sum_{\mathbf{0}\le 2\mathbf{i}\le\mathbf{m}} \gamma_{\mathbf{i}} \frac{\pi_{\mathbf{m}-2\mathbf{i}}(r)}{u^{|\mathbf{m}|-|\mathbf{i}|}} \pi_{\mathbf{n}}(r)G(u,r)$$

(by (11.2))

$$= \sum_{\mathbf{0}\le 2\mathbf{i}\le\mathbf{m}} \gamma_{\mathbf{i}} \frac{\pi_{\mathbf{m}+\mathbf{n}-2\mathbf{i}}(r)}{u^{|\mathbf{m}|-|\mathbf{i}|}} G(u,r).$$

Further, by (11.3),

$$\pi_{\mathbf{m}+\mathbf{n}-2\mathbf{i}}(r)G(u,r) = \sum_{\mathbf{0}\le 2\mathbf{j}\le(\mathbf{n}+\mathbf{m}-2\mathbf{i})} u^{|\mathbf{n}+\mathbf{m}-2\mathbf{i}|-|\mathbf{j}|}\tilde{\gamma}_{\mathbf{j}}\partial^{\mathbf{n}+\mathbf{m}-2\mathbf{i}-2\mathbf{j}}G(u,r).$$

Since $2\mathbf{i} \le \mathbf{m}$ we have $|\mathbf{n}+\mathbf{m}-2\mathbf{i}|-|\mathbf{j}| = |\mathbf{n}|+|\mathbf{m}|-2|\mathbf{i}|-|\mathbf{j}|$, the above calculations imply

$$(\partial^{\mathbf{m}}G(u,r))\pi_{\mathbf{n}}(r) = \sum_{\mathbf{0}\le 2\mathbf{i}\le \mathbf{m}} \sum_{\mathbf{0}\le 2\mathbf{j}\le(\mathbf{n}+\mathbf{m}-2\mathbf{i})} u^{|\mathbf{n}|-|\mathbf{i}|-|\mathbf{j}|}\gamma_{\mathbf{i}}\tilde{\gamma}_{\mathbf{j}}\partial^{\mathbf{n}+\mathbf{m}-2\mathbf{i}-2\mathbf{j}}G(u,r)$$

Setting $\hat{\gamma}_{\tilde{\mathbf{n}}} := \sum_{\mathbf{i}+\mathbf{j}=\tilde{n}} \gamma_{\mathbf{i}}\tilde{\gamma}_{\mathbf{j}}$ the last equation implies (11.5). □

Let $\beta > 1$ and note that

$$G(u,r) = \beta^{d/2}\exp\left(-\frac{|r|^2(\beta-1)}{2\beta u}\right)G(\beta u,r) \tag{11.6}$$

Obviously, for any $\mathbf{k}$

$$\sup_{r,u\ge 0}\left|\frac{\pi_{\mathbf{k}}(r)}{u^{|\mathbf{k}|/2}}\exp\left(-\frac{|r|^2(\beta-1)}{2\beta u}\right)\right| \le c_{\mathbf{k},\beta} < \infty. \tag{11.7}$$

Therefore, we obtain from (11.4)

$$|(\partial^{\mathbf{m}}G(u,r))\pi_{\mathbf{n}}(r)| \le c_{m,n,\beta}u^{(|\mathbf{n}|-|\mathbf{m}|)/2}G(\beta u,r), \tag{11.8}$$

where $c_{m,n,\beta} < \infty$.

Let $\triangle$ be the Laplacian on $(\mathbf{H}_0, \|\cdot\|_0)$, considered as a closed self-adjoint operator. In what follows let $\mu \ge 1$, and let $(\mu - \frac{1}{2}\triangle)^{-1}$ be the resolvent of $\frac{1}{2}\triangle$ at μ. Set $R_\mu := \mu(\mu - \frac{1}{2}\triangle)^{-1}$.

Next, let $\Phi \in \{1, \varpi\}$. We refer to (15.47) for a verification of the following estimate:

$$\varpi^{-1}(q)\varpi(r) = \left(\frac{1+|q|^2}{1+|r|^2}\right)^{\gamma} \le 2^{\gamma}\left(1+|r-q|^2\right)^{\gamma}.$$

If $\gamma \notin \mathbf{N}$ choose, in what follows, $\beta > 1$ the smallest number such that $\gamma\beta \in \mathbf{N}$. If $\gamma \in \mathbf{N}$ choose $\beta = 1$. Hölder's inequality allows us to compute the usual moments of the normal density as follows:

$$\left.\begin{aligned}\int G(u,r)(1+|r|^2)^{\gamma}\,\mathrm{d}r &=\le \left(\int G(u,r)(1+|r|^2)^{\gamma\beta}\mathrm{d}r\right)^{1/\beta}\\ &\le c_{1,\gamma} + c_{2,\gamma}u^{\gamma}.\end{aligned}\right\} \tag{11.9}$$

Expression (11.9) in addition to (15.47) implies, in particular, the first inequality of the following (11.10) for the case $\Phi = \varpi$. The second inequality follows from the simple observation that $|||\varpi||| \le 1$. If we choose $\Phi \equiv 1$ both bounds can be replaced by 1:

$$\left.\begin{aligned}\sup_q \int G(u,r-q)\varpi^{-1}(q)\varpi(r)\mathrm{d}r &= \sup_r \int G(u,r-q)\varpi^{-1}(r)\varpi(q)\mathrm{d}q\\ &\le 2^{\gamma}(c_{1,\gamma} + c_{2,\gamma}u^{\gamma}),\\ \sup_q \int G(u,r-q)\varpi(r)dr &= \sup_r \int G(u,r-q)\varpi(q)dq \le 1,\end{aligned}\right\} \tag{11.10}$$

where we used on the left-hand sides the symmetry in the variables q and r.

Let $f \in \mathbf{W}_{n,p,\Phi}$, $n \in \mathbf{N} \cup \{0\}$, $p \in [1, \infty)$. Setting

$$K(t,r,q) := G(t, r-q)\Phi^{-1}(q),$$

Expression (11.10) implies

$$\sup_q \int K(t,r,q)\varpi(r)\mathrm{d}r = \sup_r \int K(t,r,q)\varpi(q)\mathrm{d}q \le 2^{\gamma}(c_{1,\gamma} + c_{2,\gamma}u^{\gamma}). \tag{11.11}$$

Since $\varpi(r) = \varpi^{-1}(q)\varpi(q)\varpi(r)$, we may write

$$\int G(t, r-q)\partial_q^{\mathbf{n}} f(q)\mathrm{d}q = \int K(t,r,q)\partial_q^{\mathbf{n}} f(q)\varpi(q)\mathrm{d}q.$$

By Proposition 15.6 from Sect. 15.1.2 we obtain

$$\int \left| \int G(t, r-q)\partial_q^{\mathbf{n}} f(q)\mathrm{d}q \right|^p \Phi(r)\mathrm{d}r$$

$$\le \begin{cases} 2^{\gamma}(c_{1,\gamma} + c_{2,\gamma}t^{\gamma}) \int |\partial_q^{\mathbf{n}} f(r)|^p \varpi(r)\mathrm{d}r, & \text{if } \Phi = \varpi, \\ \int |\partial_q^{\mathbf{m}} f(r)|^p \mathrm{d}r, & \text{if } \Phi \equiv 1. \end{cases} \tag{11.12}$$

Hence, the heat semigroup extends to a semigroup of bounded operators on $\mathbf{W}_{n,p,\Phi}$. For $\Phi \equiv 1$ the heat semigroup is a contraction. Further, for $\Phi \equiv 1$ the proof of strong continuity can be found in Tanabe (1979), Sect. 3.1, p. 52, Example 3. This proof is generalized, using the following result.

Lemma 11.5. *Let $f \in \mathbf{W}_{0,p,\Phi}$. Then*

$$\int |f(h+q) - f(q)|^p \Phi(q)\mathrm{d}q \longrightarrow 0 \ \ as\ |h| \longrightarrow 0. \tag{11.13}$$

Proof. The proof for $\Phi \equiv 1$ can be found in Bauer (loc. cit., 48.1 Lemma). For $\Phi = \varpi$ we generalize Bauer's proof as follows, where we use a simple inequality and the shift invariance of the Lebesgue measure:

$$\begin{aligned}
&\int |f(h+q) - f(q)|^p \varpi(q)\mathrm{d}q \\
&= \int \left| f(h+q)\left[\varpi^{1/p}(q) - \varpi^{1/p}(h+q) + \varpi^{1/p}(h+q)\right] - \varpi^{1/p}(q)f(q)\right|^p \mathrm{d}q \\
&\le 2^{p-1}\int \left| f(h+q)\left[\varpi^{1/p}(q) - \varpi^{1/p}(h+q)\right]\right|^p \mathrm{d}q + 2^{p-1} \\
&\quad \times \int \left| f(h+q)\varpi^{1/p}(h+q) - f(q)\varpi^{1/p}(q)\right|^p \mathrm{d}q \\
&= 2^{p-1}\int \left| f(q)\left[\varpi^{1/p}(q-h) - \varpi^{1/p}(q)\right]\right|^p \mathrm{d}q + 2^{p-1} \\
&\quad \times \int \left| f(h+q)\varpi^{1/p}(h+q) - f(q)\varpi^{1/p}(q)\right|^p \mathrm{d}q
\end{aligned}$$

(by the shift invariance of the Lebesgue measure).

Note that by (15.47)

$$\varpi^{1/p}(q-h)\varpi^{-1/p}(q) \le 2^{\gamma/p}(1+|h|^2)^{\gamma/p}.$$

Hence,

$$|f(q)[\varpi^{1/p}(q-h)-\varpi^{1/p}(q)]| \le |f(q)[2^{\gamma/p}(1+|h|^2)^{\gamma/p}+1]\varpi^{1/p}(q)|.$$

Therefore,

$$2^{p-1}\int \left|f(h+q)\left[\varpi^{1/p}(q)-\varpi^{1/p}(h+q)\right]\right|^p \mathrm{d}q \longrightarrow 0, \text{ as } |h| \longrightarrow 0$$

by the uniform continuity and boundedness of ϖ and Lebesgue's dominated convergence theorem.

$$2^{p-1}\int |f(h+q)\varpi^{1/p}(h+q)-f(q)\varpi^{1/p}(q)|^p \mathrm{d}q \longrightarrow 0, \quad \text{as } |h| \longrightarrow 0$$

by the existing result for $\Phi \equiv 1$.

Altogether, this establishes (11.13) also for the case $\Phi = \varpi$. □

Abbreviating $\tilde{f}(q) := \left(\partial_q^{\mathbf{n}} f\right)(q)$, we change coordinates $\hat{q} := \frac{1}{\sqrt{t}}(r-q)$ and obtain

$$\int G(t, r-q)[\tilde{f}(q)-\tilde{f}(r)]\mathrm{d}q = \int G(1,\hat{q})[\tilde{f}(r+\sqrt{t}\hat{q})-\tilde{f}(r)]\mathrm{d}\hat{q}.$$

Thus, we have for $\Phi \in \{1, \varpi\}$

$$\begin{aligned}
&\int \left|\int G(t,r-q)[(\partial_q^{\mathbf{n}} f)(q)-(\partial_q^{\mathbf{n}} f)(r)]dq\right|^p \Phi(r)\mathrm{d}r \\
&= \int \left|\int G(1,q)[(\partial_q^{\mathbf{n}} f)(r+\sqrt{t}q)-(\partial_q^{\mathbf{n}} f)(r)]dq\right|^p \Phi(r)\mathrm{d}r \\
&\le \int\int G(1,q)|(\partial_q^{\mathbf{n}} f)(r+\sqrt{t}q)-(\partial_q^{\mathbf{n}} f)(r)|^p \mathrm{d}q \left[\int G(1,q)\mathrm{d}q\right]_{/(p-1)/p}\Phi(r)\mathrm{d}r
\end{aligned}$$

(by Hölder's inequality with $G(1,q) = G^{1/p}(1,q)G^{p-1/p}(1,q)$)

$$= \int\int G(1,q)|(\partial_q^{\mathbf{n}} f)(r+\sqrt{t}q)-(\partial_q^{\mathbf{n}} f)(r)|^p \mathrm{d}q\Phi(r)\mathrm{d}r$$

(since $\int G(1,q)\mathrm{d}q = 1$)

$$= \int G(1,q)\left[\int |(\partial_q^{\mathbf{n}} f)(r+\sqrt{t}q)-(\partial_q^{\mathbf{n}} f)(r)|^p \Phi(r)\mathrm{d}r\right]\mathrm{d}q$$

(by Fubini's theorem).

Next, we note that

$$
\begin{aligned}
&|(\partial_q^{\mathbf{n}} f)(r+\sqrt{t}q) - (\partial_q^{\mathbf{n}} f)(r)|^p \le 2^{p-1}[|f(r+\sqrt{t}q)|^p + |f(r)|^p] \\
&\text{(by Hölder's inequality, if } p > 1).
\end{aligned}
\tag{11.14}
$$

Let us first focus on the (more difficult) case $\Phi = \varpi$.

$$
\begin{aligned}
&\int 2^{p-1}|f(r+\sqrt{t}q)|^p \varpi(r)\mathrm{d}r\Big] \\
&\quad = \int 2^{p-1}|f(r)|^p \varpi(r-\sqrt{t}q))\mathrm{d}r\Big] \\
&\text{(by change of variables)} \\
&\quad = \int 2^{p-1}|f(r)|^p \varpi(r-\sqrt{t}q))\varpi^{-1}(r)\varpi(r)\mathrm{d}r\Big] \\
&\quad \le 2^{\gamma}(1+t|q|^2)^{\gamma} \int 2^{p-1}|f(r)|^p \varpi(r)\mathrm{d}r\Big] \quad \text{(by (15.47))}
\end{aligned}
$$

Hence, we obtain

$$
\begin{aligned}
G(1,q)&\Big[\int |(\partial_q^{\mathbf{n}} f)(r+\sqrt{t}q) - (\partial_q^{\mathbf{n}} f)(r)|^p \varpi(r)\mathrm{d}r \\
&\le G(1,q)2^{\gamma}(1+t|q|^2)^{\gamma}\int 2^{p-1}|f(r)|^p\varpi(r)\mathrm{d}r + K(1,q)\int 2^{p-1}|f(r)|^p\varpi(r)\mathrm{d}r, \\
&\le 2^p G(1,q)2^{\gamma}(1+t|q|^2)^{\gamma}\int |f(r)|^p\varpi(r)\mathrm{d}r,
\end{aligned}
$$

which, by (11.9), is integrable with respect to $\mathrm{d}q$.

For the case $\Phi \equiv 1$, we employ (11.14) and obtain by the shift invariance of the Lebesgue measure

$$
\begin{aligned}
G(1,q)&\Big[\int |(\partial_q^{\mathbf{n}} f)(r+\sqrt{t}q) - (\partial_q^{\mathbf{n}} f)(r)|^p \mathrm{d}r \\
&\le G(1,q)\int 2^{p-1}|f(r)|^p\mathrm{d}r + K(1,q)\int 2^{p-1}|f(r)|^p\mathrm{d}r, \\
&\le 2^p G(1,q)\int |f(r)|^p\varpi(r)\mathrm{d}r, \\
&\text{which is, of course, integrable with respect to } \mathrm{d}q.
\end{aligned}
$$

So, by (11.13) and Lebesgue's dominated convergence theorem,

$$
\int \Big| \int G(t,r-q)\Big[(\partial_q^{\mathbf{n}} f)(q) - (\partial_q^{\mathbf{n}} f)(r)\Big]\mathrm{d}q|^p \Phi(r)\mathrm{d}r \longrightarrow 0, \quad \text{as } t \downarrow 0.
\tag{11.15}
$$

The previous analysis shows that, for any $D > 0$, the heat semigroup

$$
(T_D(t)f)(r) := \int G(Dt, r-q)f(q)\mathrm{d}q
\tag{11.16}
$$

defines a strongly continuous semigroup on $\mathbf{W}_{n,p,\Phi}, n \in \mathbf{N} \cup \{0\}, p \in [1,\infty)$, $\Phi \in \{1, \varpi\}$.[1] Let $t > 0$. Then

$$\left.\begin{aligned}\left|\frac{\partial}{\partial t}G(t,r)\right| &= \left|\frac{1}{t}\left[-\frac{d}{2}G(t,r) + \frac{|r|^2}{2t}G(t,r)\right]\right| \\ &\leq \frac{c_\beta}{t}\left[G(t,r) + G(\beta t, r)\right] \\ &\text{(by (11.6) and (11.7) for } \beta > 1).\end{aligned}\right\} \tag{11.17}$$

Choose again $f \in \mathbf{W}_{n,p,\Phi}$. Then

$$\left.\begin{aligned}\left|\frac{1}{2}\triangle \int G(t,r-q)f(q)\mathrm{d}q\right| &= \left|\int \frac{\partial}{\partial t}G(t,r-q)f(q)\mathrm{d}q\right| \\ \leq \frac{c_\beta}{t}&\left[\int G(t,r-q)f(q)\mathrm{d}q + \int G(\beta t, r-q)f(q)\mathrm{d}q\right].\end{aligned}\right\} \tag{11.18}$$

Note that for any $\theta > 0$ there is a $K_\theta > 0$

$$2^\gamma\left(c_{1,\gamma} + c_{2,\gamma}t^\gamma\right) \leq K_\theta \exp(\theta t) \;\; \forall t \geq 0, \tag{11.19}$$

where the left-hand side in (11.19) is the bound in (11.10). Hence, $T(t)$ maps $\mathbf{W}_{n,p,\Phi}$ into the domain of its generator and by (11.12) in addition to (11.18) for any $D > 0$[2]

$$\left.\begin{aligned}&\|T(Dt)f\|_{n,p,\Phi} \leq K_\theta \exp(\theta Dt)]\|f\|_{n,p,\Phi} \wedge 2^\gamma\left(c_{1,\gamma} + c_{2,\gamma}t^\gamma\right)\|f\|_{n,p,\Phi}, \\ &\left\|\frac{1}{2}\triangle T(Dt)f\right\|_{n,p,\Phi} \leq \frac{c_\beta}{Dt}\left[1 + (K_\theta \exp(\theta Dt) \wedge 2^\gamma\left(c_{1,\gamma} + c_{2,\gamma}t^\gamma\right))\right]\|f\|_{n,p,\Phi}.\end{aligned}\right\} \tag{11.20}$$

This implies that the heat semigroup is analytic (or holomorphic) on $\mathbf{W}_{n,p,\Phi}$.[3] Further, (11.12) in addition to (11.15) and (11.20) implies that the heat semigroup is an analytic contraction semigroup on $\mathbf{W}_{n,p,1}$. Setting for $\alpha > 0$.

$$\Gamma(\alpha) = \int_0^\infty u^{\alpha-1}\mathrm{e}^{-u}\mathrm{d}u$$

---

[1] Strictly speaking, we need for the the heat semigroup, defined on different spaces, different symbols. However, on the space $\mathcal{S}$, the space of infinitely often differentiable and rapidly decreasing real valued functions with domain $\mathbf{R}^d$, these versions coincide, as they are defined by the same kernel $G(t,r)$. Therefore, we use the same symbol for the heat semigroup and its generator $\frac{1}{2}\triangle$ (as the closed operator) on $\mathbf{W}_{n,p,\Phi}, n \in \mathbf{N} \cup \{0\}, p \in [1,\infty), \Phi \in \{1, \varpi\}$. We refer to Sect. 15.1.3 for the definition and properties of $\mathcal{S}$.

[2] Observe that for $\Phi \equiv 1$ we may choose $\theta = 0$ and $K_\theta = 1$ in (11.20), which will be used in what follows.

[3] Cf. Davies (loc. cit., Theorem 2.39).

for the Gamma function,[4] we may for $D > 0$ and $\alpha > 0$ define the fractional powers of the resolvent of $\frac{D}{2}\triangle$ by

$$\left(\mu - \frac{D}{2}\triangle\right)^{-\alpha} = \frac{1}{\Gamma(\alpha)}\int_0^\infty u^{\alpha-1}\mathrm{e}^{-\mu u}T(Du)\mathrm{d}u, \tag{11.21}$$

where $f \in \mathbf{W}_{n,p,\Phi}$, $n \in \mathbf{N} \cup \{0\}$, $p \in [1,\infty)$, $\Phi \in \{1, \varpi\}$.[5]

Denote by $\mathcal{L}(\mathbf{B})$ the set of linear bounded operators from a Banach space $\mathbf{B}$ into itself and the operator norm $\|\cdot\|_{\mathcal{L}(\mathbf{B})}$. Choose $\tilde{\alpha} \geq 1$ the smallest number such that $(\alpha - 1)\tilde{\alpha} \in \mathbf{N}$, i.e.,

$$(\alpha - 1)\tilde{\alpha} = [\alpha - 1] + 1.$$

Let $\mu > D\theta$ for $D > 0$. Then,

$$\left.\begin{aligned}
&\left\|\int_0^\infty u^{\alpha-1}\mathrm{e}^{-\mu u}T(Du)\mathrm{d}u\right\|_{\mathcal{L}(W_{n,p,\Phi})}\\
&\quad\leq K_\theta \int_0^\infty u^{\alpha-1}\mathrm{e}^{(D\theta-\mu)u}\mathrm{d}u\\
&\quad\leq K_\theta \left\{\int_0^\infty u^{[\alpha-1]+1}\mathrm{e}^{(D\theta-\mu)u}\mathrm{d}u\right\}^{1/\tilde{\alpha}} \left\{\int_0^\infty \mathrm{e}^{(D\theta-\mu)u}\mathrm{d}u\right\}^{(\tilde{\alpha}-1)/\tilde{\alpha}}\\
&\quad\text{(by Hölder's inequality)}\\
&\quad\leq K_\theta \frac{(((\alpha-1)\tilde{\alpha})!)^{1/\tilde{\alpha}}}{(\mu - D\theta)^{\alpha-1+1/\tilde{\alpha}}}\,\frac{1}{(\mu - D\theta)^{(\tilde{\alpha}-1)/\tilde{\alpha}}} = K_\theta \frac{(((\alpha-1)\tilde{\alpha})!)^{1/\tilde{\alpha}}}{(\mu - D\theta)^{\alpha}}.
\end{aligned}\right\} \tag{11.22}$$

Hence, for $\alpha > 0$

$$\left\|(\mu - \frac{D}{2}\triangle)^{-\alpha}\right\|_{\mathcal{L}(\mathbf{W}_{n,p,\Phi})} \leq \begin{cases} K_\theta \dfrac{(((\alpha-1)\tilde{\alpha})!)^{1/\tilde{\alpha}}}{\Gamma(\alpha)(\mu - D\theta)^\alpha} \leq \dfrac{\bar{K}_\theta}{(\mu - D\theta)^\alpha}, & \text{if } \quad \Phi = \varpi \text{ and } \mu > D\theta,\\[2ex] \dfrac{(((\alpha-1)\tilde{\alpha})!)^{1/\tilde{\alpha}}}{\Gamma(\alpha)\mu^\alpha} \leq \dfrac{\bar{K}}{\mu^\alpha}, & \text{if } \quad \Phi \equiv 1 \text{ and } \mu > 0. \end{cases} \tag{11.23}$$

Set for $\alpha > 0$

$$R^\alpha_\mu := \mu^\alpha \left(\mu - \frac{1}{2}\triangle\right)^{-\alpha}, \quad R^\alpha_{\mu,D} := \mu^\alpha\left(\mu - \frac{D}{2}\triangle\right)^{-\alpha}. \tag{11.24}$$

---

[4] For $\alpha = k \in \mathbf{N}$, we have $\Gamma(k) = (k-1)!$.

[5] Cf. Pazy (1983), Chap. 2.6, (6.9).

Expression (11.23) implies

$$\|R^{\alpha}_{\mu,D} f\|_{n,p,\Phi} \le \begin{cases} \mu^{\alpha} \dfrac{\bar{K}_{\theta}}{(\mu - D\theta)^{\alpha}} \|f\|_{n,p,\varpi}, & \text{if} \quad \Phi = \varpi \text{ and } \mu > D\theta, \\ \bar{K} \|f\|_{n,p,\Phi}, & \text{if} \quad \Phi \equiv 1 \text{ and } \mu > 0. \end{cases} \tag{11.25}$$

Lemma 11.6. *Let $f \in \mathbf{W}_{0,p,\Phi}$. Then for any $\alpha > 0$*

$$\|(R^{\alpha}_{\mu,D} - I) f\|_{n,p,\Phi} \longrightarrow 0, \ \ as\ \mu \longrightarrow \infty\,. \tag{11.26}$$

Proof.

$$\left.\begin{aligned} &\left\|(R^{\alpha}_{\mu,D} - I) f\right\|^p_{n,p,\Phi} \\ &= \left\| \frac{\mu^{\alpha}}{\Gamma(\alpha)} \int_0^{\infty} \mathrm{d}u\, u^{\alpha-1} \mathrm{e}^{-\mu u} T(Du) f - f \right\|^p_{n,p,\Phi} \\ &\le \left(\frac{\mu^{\alpha}}{\Gamma(\alpha)} \int_0^{\infty} \mathrm{d}u\, u^{\alpha-1} \mathrm{e}^{-\mu u} \, \|T(Du) f - f\|_{n,p,\Phi} \right)^p \\ &= \left(\frac{1}{\Gamma(\alpha)} \int_0^{\infty} \mathrm{d}v\, v^{\alpha-1} \mathrm{e}^{-v} \left\| T(D\frac{v}{\mu}) f - f \right\|_{n,p,\Phi} \right)^p \longrightarrow 0, \text{ as } \mu \longrightarrow \infty \end{aligned}\right\} \tag{11.27}$$

by change of variables $v := \mu u$ in addition to the strong continuity of $T(Dt)$ and Lebesgue's dominated convergence theorem, since by (11.20)

$$\|T(D\frac{v}{\mu}) f - f\|_{n,p,\Phi} \le \|T(D\frac{v}{\mu}) f\|_{n,p,\Phi} + \|f\|_{n,p,\Phi} \le 2^{\gamma} (c_{1,\gamma} + c_{2,\gamma} (D\frac{v}{\mu})^{\gamma}).$$

□

Lemma 11.7. *Let $D > 0$ and* **k** *be a multi-index and*

$$\alpha \ge \frac{|\mathbf{k}|}{2}.$$

For all $\mu > D\theta$ are in the resolvent set of $\frac{D}{2}\triangle$,[6] *we can extend the operator $\partial^{\mathbf{k}}(\theta - \frac{D}{2}\triangle)^{-\alpha}$ to a bounded operator on $\mathbf{W}_{n,p,\Phi}$. Denoting this extension by the same symbol, there is a finite constant $c_{n,\mathbf{k},\alpha,p,D,\Phi}$ such that for any $\epsilon > 0$*

$$\|\partial^{\mathbf{k}}(\mu - \frac{D}{2}\triangle)^{-\alpha}\|_{\mathcal{L}(W_{n,p,\Phi})} \le c_{n,\mathbf{k},\alpha,p,D,\Phi}(\mu - D\theta + \frac{\epsilon}{2})^{-\alpha+|\mathbf{k}|} \quad \forall \mu \ge D\theta + \epsilon. \tag{11.28}$$

Proof. Let $g \in \mathbf{W}_{n,p,\Phi}$, and $f := \partial^{\mathbf{j}} g$, where $\mathbf{j} \le \mathbf{n}$ with $|\mathbf{n}| = n$. Further, let $\overline{m}$ be an even number $> n$ and $\mu > D\theta$. Then we have

[6] As before, we may choose $\theta = 0$ if $\Phi \equiv 1$.

$$
\left.
\begin{aligned}
&\left|\partial^{\mathbf{k}}(\mu - \frac{D}{2}\triangle)^{-\alpha} R^{\overline{m}/2}_{\mu,D} f(r)\right| \\
&= \left| \frac{\mu^{\overline{m}/2}}{\left(\frac{\overline{m}}{2}-1\right)!} \int_0^\infty \mathrm{d}u\, u^{\alpha+\overline{m}/2-1} \mathrm{e}^{-\mu u} \int \partial_r^{\mathbf{k}} G(Du, r-q) f(q)\mathrm{d}q \right| \\
&= \left| \sum_{\mathbf{0}\le 2\mathbf{i}\le \mathbf{k}} \gamma_{\mathbf{i}} \frac{\mu^{\overline{m}/2}}{(\frac{\overline{m}}{2}-1)!} \int_0^\infty \mathrm{d}u\, u^{\alpha+\overline{m}/2-1} \mathrm{e}^{-\mu u} \int \frac{\pi_{\mathbf{k}-2\mathbf{i}}(r-q)}{u^{|\mathbf{k}|-2|\mathbf{i}|}} G(Du, r-q) f(q)\mathrm{d}q \right| \\
&\text{(by (11.2))} \\
&= \left| \sum_{\mathbf{0}\le 2\mathbf{i}\le \mathbf{k}} \gamma_{\mathbf{i}} \frac{\mu^{\overline{m}/2}}{(\frac{\overline{m}}{2}-1)!} \int_0^\infty \mathrm{d}u\, u^{\alpha+|\mathbf{i}|-|\mathbf{k}|/2+\overline{m}/2-1} \mathrm{e}^{-\mu u} \right. \\
&\qquad \left. \times \int \frac{\pi_{\mathbf{k}-2\mathbf{i}}(r-q)}{u^{|\mathbf{k}|/2-|\mathbf{i}|}} G(Du, r-q) f(q)\mathrm{d}q \right| \\
&\le c_{\mathbf{k},D} \sum_{\mathbf{0}\le 2\mathbf{i}\le \mathbf{k}} \gamma_{\mathbf{i}} \frac{\mu^{\overline{m}/2}}{(\frac{\overline{m}}{2}-1)!} \int_0^\infty \mathrm{d}u\, u^{\alpha+|\mathbf{i}|-|\mathbf{k}|/2+\overline{m}/2-1} \mathrm{e}^{-\mu u} \int G(\beta u, r-q)|f(q)|\mathrm{d}q \\
&\text{(by (11.6) and (11.7) with } \beta = \beta(D) > 1 \text{ such that } \mu > \beta D\theta) \\
&= c_{\mathbf{k},D} \sum_{\mathbf{0}\le 2\mathbf{i}\le \mathbf{k}} \gamma_{\mathbf{i}} \frac{\mu^{\overline{m}/2}}{(\frac{\overline{m}}{2}-1)!} \beta^{-\alpha-|\mathbf{i}|+|\mathbf{k}|/2-\overline{m}/2} \\
&\qquad \times \int_0^\infty \mathrm{d}u\, u^{\alpha+|\mathbf{i}|-|\mathbf{k}|/2+\overline{m}/2-1} \mathrm{e}^{-\mu/\beta u} \int G(u, r-q)|f(q)|\mathrm{d}q \\
&= c_{\mathbf{k},D,\alpha} \sum_{\mathbf{0}\le 2\mathbf{i}\le \mathbf{k}} \gamma_{\mathbf{i}} \left(\frac{\mu}{\beta} - \frac{1}{2}\triangle\right)^{-\left(\alpha+|\mathbf{i}|-\frac{|\mathbf{k}|}{2}\right)} R^{\overline{m}/2}_{\mu/\beta} |f|(r).
\end{aligned}
\right\} \tag{11.29}
$$

Taking the pth power and integrating against $\Phi(r)\mathrm{d}r$, (11.29) implies

$$
\begin{aligned}
&\|\partial^{\mathbf{k}} \left(\mu - \frac{1}{2}\triangle\right)^{-\alpha} R^{\overline{m}/2}_{\mu} f\|_{0,p,\Phi} \\
&\le c_{\mathbf{k},D,\alpha,p} \sum_{\mathbf{0}\le 2\mathbf{i}\le \mathbf{k}} \gamma_{\mathbf{i}} \| \left(\frac{\mu}{\beta} - \frac{1}{2}\triangle\right)^{-(\alpha+|\mathbf{i}|-|\mathbf{k}|/2)} R^{\overline{m}/2}_{\mu/\beta} |f| \|_{0,p,\Phi}.
\end{aligned} \tag{11.30}
$$

By the analyticity of the semigroup, generated by $\frac{D}{2}\triangle$, the resolvent of $\frac{D}{2}\triangle$ maps $\mathbf{W}_{n,p,\Phi}$ into the domain of $\frac{D}{2}\triangle$. Then, for $\mu, \tilde{\mu} > D\theta$,

$$
\left(\tilde{\mu} - \frac{D}{2}\triangle\right)\left(\mu - \frac{D}{2}\triangle\right)^{-1} = I - \mu\left(\mu - \frac{D}{2}\triangle\right)^{-1} + \tilde{\mu}\left(\mu - \frac{D}{2}\triangle\right)^{-1}. \tag{11.31}
$$

Apparently, all operators on the right-hand side of (11.31) are bounded. The bound in the second operator is a constant c which is uniform for all $\mu > D\theta + \epsilon$, where $\epsilon > 0$. Choosing $\tilde{\mu} := \mu/\beta$, the bound of the last family of operator with $\mu > D\theta + \epsilon$ is c/β. Hence, for $\tilde{\mu} := \mu/\beta$ the operator on the left hand side is bounded by

$$1 + c\left(1 + \frac{1}{\beta}\right)$$

and positive powers on the left-hand side are bounded by the same positive powers of $1 + c(1 + 1/\beta)$. Consequently,

$$\left.\begin{aligned}&\|\mu^{-\overline{m}/2}(\mu - \frac{1}{2}\triangle)^{\overline{m}/2} R_{\mu/\beta}^{\overline{m}/2}\|_{\mathcal{L}_{n,p,\Phi}} \le c_{p,\Phi,\overline{m},\beta},\\ &\|R_{\mu/\beta}^{\overline{m}/2} R_{\mu}^{-\overline{m}/2} R_{\mu}^{\overline{m}/2} f\|_{0,p,\Phi} \le c_{p,\Phi,\overline{m},\beta}\|R_{\mu}^{\overline{m}/2} f\|_{0,p,\Phi},\end{aligned}\right\} \tag{11.32}$$

where the second inequality follows from the first one, since the resolvent operators and their powers commute. Since we chose $\beta = \beta(D)(> 1)$ in (11.29) and (11.30), we obtain that for any $\mu > D\theta$ and $D > 0$

$$\left.\begin{aligned}&\|\partial^{\mathbf{k}}\left(\mu - \frac{D}{2}\triangle\right)^{-\alpha} R_{\mu}^{\overline{m}/2} f\|_{0,p,\Phi} \le c_{\mathbf{k},\Phi,\alpha,\overline{m},p,D}\\ &\quad\times \sum_{0\le 2\mathbf{i}\le\mathbf{k}} \gamma_{\mathbf{i}}\|\left(\frac{\mu}{\beta} - \frac{1}{2}\triangle\right)^{-(\alpha+|\mathbf{i}|-|\mathbf{k}|/2)} R_{\mu}^{\overline{m}/2}|f|\|_{0,p,\Phi}\\ &\le \tilde{c}_{\mathbf{k},\Phi,\alpha,\overline{m},p,D} \sum_{0\le 2\mathbf{i}\le\mathbf{k}} \gamma_{\mathbf{i}}\|(\frac{\mu}{\beta} - \frac{1}{2}\triangle)^{-(\alpha+|\mathbf{i}|-|\mathbf{k}|/2)}|f|\|_{0,p,\Phi}\\ &\text{(by (11.23))}\\ &\le c_{\mathbf{k},\Phi,\alpha,\overline{m},p,D} \sum_{0\le 2\mathbf{i}\le\mathbf{k}} \gamma_{\mathbf{i}}\|(\frac{\mu}{\beta} - \frac{1}{2}\triangle)^{-(\alpha+|\mathbf{i}|-|\mathbf{k}|/2)}\|_{\mathcal{L}(\mathbf{W}_{0,p,\Phi})}\|f\|_{0,p,\Phi}.\end{aligned}\right\} \tag{11.33}$$

For sufficiently smooth f (e.g., $f \in \mathcal{S}$), (11.27) implies

$$\|(R_{\mu,D}^{\alpha} - I)\partial^{\mathbf{k}}\left(\mu - \frac{D}{2}\triangle\right)^{-\alpha} f\|_{n,p,\Phi}^{p} \longrightarrow 0, \quad \text{as } \mu \longrightarrow \infty .$$

Hence, we may pass to the limit on the left-hand side for (11.33) (assuming f to be smooth) and obtain

$$
\left\| \partial^{\mathbf{k}} (\mu - \frac{D}{2}\triangle)^{-\alpha} R_{\mu}^{\overline{m}/2} f \right\|_{0,p,\Phi}
$$

$$
\leq c_{\mathbf{k},\Phi,\alpha,\overline{m},p,D} \sum_{\mathbf{0}\leq 2\mathbf{i} \leq \mathbf{k}} \gamma_{\mathbf{i}} \left\| \left(\frac{\mu}{\beta} - \frac{1}{2}\triangle \right)^{-(\alpha+|\mathbf{i}|-|\mathbf{k}|/2)} \right\|_{\mathcal{L}(\mathbf{W}_{0,p,\Phi})} \|f\|_{0,p,\Phi}. \tag{11.34}
$$

Since $\mathcal{S}$ is dense in $\mathbf{W}_{0,p,\Phi}$ and $\alpha \geq \frac{|\mathbf{k}|}{2}$ by our assumption, (11.34) extends to all $f \in \mathbf{W}_{0,p,\Phi}$, whence by (11.25) $\partial^{\mathbf{k}}(\mu - \frac{D}{2}\triangle)^{-\alpha}$ is a bounded operator on $\mathbf{W}_{0,p,\Phi}$. Next, we recall the abbreviation $f = \partial^{i} g$, with $|\mathbf{i}| \leq n$. Summing up over all $|\mathbf{i}| \leq n$ in (11.34) implies the extendibility of $\alpha \geq \frac{|\mathbf{k}|}{2}$ as a bounded operator on $\mathbf{W}_{n,p,\Phi}$. The inequality (11.28) follows from (11.34), choosing

$$
\beta = \beta(D) > 1 \qquad \text{such that } (\beta - 1)D\theta \leq \tfrac{\epsilon}{2}.
$$

□

Corollary 11.8. *Let $D > 0$ and $\mathbf{k}$ be a multi-index. Then, for any $\alpha > |\mathbf{k}|/2$*

$$
\lim_{\mu\to\infty} \left\| \partial^{\mathbf{k}} \left(\mu - \frac{D}{2}\triangle \right)^{-\alpha} \right\|_{\mathcal{L}(W_{n,p,\Phi})} = 0. \tag{11.35}
$$

Proof. Expression (11.28). □

Corollary 11.9. *Let $D > 0$, $\mu > D\theta$ and $\alpha \in \mathbf{N}$. Then (suitably indexed versions of) $(\mu - \frac{D}{2}\triangle)^{-\alpha/2}$ are bounded linear operator from $\mathbf{W}_{n,p,\Phi}$ into $\mathbf{W}_{n+\alpha,p,\Phi}$ for all $n \in \mathbf{N} \cup \{0\}$ and (suitably indexed versions of) $(\mu - \frac{D}{2}\triangle)^{\alpha/2}$ are bounded linear operator from $\mathbf{W}_{n,p,\Phi}$ into $\mathbf{W}_{n-\alpha,p,\Phi}$ for all $n \geq \alpha$.* □

We develop the corresponding Hilbert scales which are better known and, for negative powers, somewhat easier to work with. Abbreviate

$$
R := R_1.
$$

Note that for the case of $\Phi \equiv 1$ and $p = 2$, the definition of the fractional powers for resolvents of analytic semigroups coincide with those defined by the method of spectral resolution.[7] In this Hilbert space setup, both positive and negative powers of R are well defined,[8] and the fractional powers R^{α} are positive and self-adjoint operators on $\mathbf{H}_0$. We now consider $\alpha \geq 0$ and define the norms

$$
|f|_{\alpha} := \|R^{\frac{\alpha}{2}} f\|_0, \tag{11.36}
$$

where $f \in \mathcal{S}$. The first observation is that for $m \in \mathbf{N}$ the norms $|f|_m$ and $\|f\|_m := \|f\|_{m,2,1}$ are equivalent. A proof of this equivalence can be constructed

[7] Cf. Tanabe (1979) Sect. 2.3.3.

[8] Cf., e.g., Pazy (loc. cit) or Tanabe (loc. cit.), Chap. 2.3.

using interpolation inequalities.[9] We denote this equivalence by $|f|_m \sim \|f\|_m$. The second observation is that $|\cdot|_\alpha$ defines Hilbert norms on $\mathcal{S}$ for all $\alpha \in \mathbf{R}$. Further, let $\mathcal{S}'$ be the strong dual of $\mathcal{S}$, i.e., the Schwarz space of tempered distributions.[10] We define

$$\mathbf{H}_\alpha := \{F \in \mathcal{S}' : |F|_\alpha < \infty\}.$$

$\mathbf{H}_\alpha$ is the completion of $\mathcal{S}$ with respect to $|\cdot|_\alpha$ in $\mathcal{S}'$. Identifying $\mathbf{H}_0$ with its strong dual, we now obtain for $\alpha \geq \beta \geq 0$ the chain of spaces

$$\mathbf{H}_\alpha \subset \mathbf{H}_\beta \subset \mathbf{H}_0 = \mathbf{H}_0' \subset \mathbf{H}_{-\beta} \subset \mathbf{H}_{-\alpha} \tag{11.37}$$

with dense continuous imbeddings.[11] As the operators R_μ^α commute with R^β we can extend (resp. restrict) the operators R_μ^α in an obvious way to $\mathbf{H}_\beta$ for all $\alpha, \beta \in \mathbf{R}$. In particular, for $\alpha/2$ the extension (or restriction) of $R_\mu^{\alpha/2}$ to $\mathbf{H}_\beta$ is a bounded linear operator from $\mathbf{H}_\beta$ into $\mathbf{H}_{\alpha+\beta}$,[12] i.e.,

$$R_\mu^{\alpha/2} \in \mathcal{L}(\mathbf{H}_\beta, \mathbf{H}_{\alpha+\beta}), \ \alpha, \beta \in \mathbf{R}. \tag{11.38}$$

Thus, for $\mu > D\theta$ and $\alpha, \tilde{\alpha}, \beta \in \mathbf{R}$

$$R_\mu^{\alpha/2}(\theta - \tfrac{D}{2}\triangle)^{\tilde{\alpha}} \in \mathcal{L}(\mathbf{H}_\beta, \mathbf{H}_{\alpha-\tilde{\alpha}+\beta}). \tag{11.39}$$

Further, by the Sobolev imbedding theorem[13] we obtain that for any $\alpha > d/2$

$$\mathbf{H}_\alpha \subset C_b(\mathbf{R}^d; \mathbf{R}). \tag{11.40}$$

Therefore, by the (square) integrability of the elements of $\mathbf{H}_\alpha$

$$\mathbf{H}_\alpha \subset C_0(\mathbf{R}^d; \mathbf{R}).$$

Recall from (15.48) the definition of distance in the total variation, $\|\mu - \nu\|_f$ for $\mu, \nu \in \mathbf{M}_f$. By the Riesz representation theorem[14]

$$\|\mu - \nu\|_f = \sup_{\{\|\|\varphi\|\| \leq 1, \varphi \in C_0(\mathbf{R}^d;\mathbf{R})\}} |<\mu - \nu, \varphi>|,$$

whence

$$(\mathbf{M}_f, \|\cdot\|_f) \subset (\mathbf{H}_{-\alpha}, |\cdot|_{-\alpha}) \quad \text{for any } \alpha > \tfrac{d}{2},$$

---

[9] Cf., e.g., Adams (1975), Chap. IV, Theorems 4.13 and 4.14.

[10] Cf. Sect. 15.1.3 (15.32).

[11] Cf. Kotelenez (1985). Observe that the main difference to (15.32) is that in (15.32) we have Hilbert–Schmidt imbeddings if the indices are sufficiently far apart. Cf. also (13.10) and (13.13).

[12] Cf. Kotelenez (1985).

[13] Cf. Triebel (1978), Sect. 2.8.1, (16).

[14] Cf. Sect. 15.1.4, (15.49).

Further, it follows that the imbedding from Proposition 15.13 that

$$(\mathbf{M}_f, \gamma_f(\cdot)) \subset (\mathbf{M}_f, \|\cdot\|_f)$$

is continuous.[15]

Altogether, we obtain the following continuous imbeddings:

$$(\mathbf{M}_f, \gamma_f(\cdot)) \subset (\mathbf{M}_f, \|\cdot\|_f) \subset (\mathbf{H}_{-\alpha}, |\cdot|_{-\alpha}) \quad \text{for any } \alpha > \tfrac{d}{2}. \tag{11.41}$$

Let $f \in \mathcal{S}$. The definition of $\|\cdot\|_{i,p,\Phi}$, the Sobolev imbedding theorem[16] and an elementary calculation imply for $i \in \mathbf{N} \cup \{0\}$, $p \geq 2$ and $\alpha > \frac{d}{2}$

$$\left.\begin{aligned} \|f\|_{i,p,\Phi}^p &\leq \|f\|_{i,p}^p \leq |||f|||_i^{p-2} \cdot \|f\|_i^2 \\ &\leq c_2^{p-2} \|f\|_{\alpha+i}^{p-2} \cdot \|f\|_i^2 \\ &\leq c_2^{p-2} \|f\|_{\alpha+i}^p, \end{aligned}\right\} \tag{11.42}$$

where c_2 is a finite positive constant. Let $m \in \mathbf{N} \cup \{0\}$ and abbreviate

$$\overline{m} := \overline{m}(m) := \min\{n \in \mathbf{N} : n \text{ even and } n > m + 2 + d\}. \tag{11.43}$$

By (11.42) for $i \in \mathbf{N} \cup \{0\}$, $p \geq 2$ and sufficiently small $\epsilon > 0$

$$\mathbf{H}_{\overline{m}-m-d/2-\epsilon} \subset \mathbf{H}_{\overline{m}-m-d/2-2+i-\epsilon} \subset \mathbf{H}_{\frac{d}{2}+\epsilon+i} \subset \mathbf{W}_{i,p,\Phi} \tag{11.44}$$

with continuous (and dense) imbeddings.

Lemma 11.10. *Let $\bar{\mathbf{n}}$ be some multi-index and assume for some other multi-index $\bar{\bar{\mathbf{n}}}$*

$$|\bar{\bar{\mathbf{n}}}| \geq |\bar{\mathbf{n}}|. \tag{11.45}$$

Further, let $\overline{m} \geq \overline{m}(0)$, where the latter is given by (11.43). Suppose $f \in \mathcal{S}$, $D > 0$, and $\mu > D\theta$, and consider the integral operator

$$\left.\begin{aligned} (\bar{K}_\mu f)(r) &:= (\bar{K}_{\mu,\overline{m},\bar{\mathbf{n}},\bar{\bar{\mathbf{n}}},D} f)(r) \\ &:= \frac{\mu^{\overline{m}/2}}{(\frac{\overline{m}}{2}-1)!} \int_0^\infty \mathrm{d}u u^{\overline{m}/2-1} \mathrm{e}^{-\mu u} \int \left[\partial_r^{\bar{\mathbf{n}}} G(Du, r-q)\right] \pi_{\bar{\bar{\mathbf{n}}}}(r-q) f(q) \mathrm{d}q. \end{aligned}\right\} \tag{11.46}$$

---

[15] For probability measures the statement also follows from Theorem 15.24 of Chap. 15.2.1. Cf. also Bauer (1968), Chap. VII, Satz 45.7.

[16] Cf. Triebel, loc. cit, Sect. 2.8.1, (16), or Adams (1975), Sect. 5.4.

Then, for sufficiently small $\epsilon > 0$ and $p \geq 1$, $\bar{K}_{\mu,\overline{m},\bar{\mathbf{n}},\bar{\bar{\mathbf{n}}}}$ has an extension to all $f \in \mathbf{H}_{-d/2-\epsilon}$ to a bounded linear operator from $\mathbf{H}_{-d/2-\epsilon}$ into $\mathbf{W}_{i,p,\Phi}$, $i = 0, 1, 2$, such that for $\mu > D\theta$

$$\left.\begin{aligned}\left\|(\bar{K}_{\mu,\overline{m},\bar{\mathbf{n}},\bar{\bar{\mathbf{n}}},D} f)\right\|_{i,p,\Phi} &\leq c_{\Phi,\overline{m},\bar{\mathbf{n}},\bar{\bar{\mathbf{n}}},D}(\mu - D\theta)^{-|\bar{\bar{\mathbf{n}}}|-|\bar{\mathbf{n}}|/2} \left\| R_{\mu}^{\overline{m}/2} f\right\|_{i,p,\Phi} \quad (i = 0, 1, 2)\\ &\leq \mu^{\frac{\overline{m}}{2}} \bar{c}_{\Phi,\overline{m},\bar{\mathbf{n}},\bar{\bar{\mathbf{n}}},D}(\mu - D\theta)^{-|\bar{\bar{\mathbf{n}}}|-|\bar{\mathbf{n}}|/2} \| f\|_{-d/2-\epsilon}.\end{aligned}\right\} \tag{11.47}$$

Proof. By (11.5)

$$(\partial^{\bar{\mathbf{n}}} G(Du, r))\pi_{\bar{\bar{\mathbf{n}}}}(r) = \sum_{\mathbf{0} \leq 2\tilde{\mathbf{n}} \leq \bar{\bar{\mathbf{n}}}+\bar{\mathbf{n}}} (Du)^{|\bar{\bar{\mathbf{n}}}|-|\tilde{\mathbf{n}}|} \gamma_{\tilde{\mathbf{n}}} \partial^{\bar{\mathbf{n}}+\bar{\bar{\mathbf{n}}}-2\tilde{\mathbf{n}}} G(Du, r), \tag{11.48}$$

we have

$$\left.\begin{aligned}&\frac{\mu^{\overline{m}/2}}{(\frac{\overline{m}}{2}-1)!}\int_0^{\infty} \mathrm{d}u\, u^{|\bar{\bar{\mathbf{n}}}|-|\tilde{\mathbf{n}}|+\overline{m}/2-1} \mathrm{e}^{-\mu u} \int \partial_r^{\bar{\mathbf{n}}+\bar{\bar{\mathbf{n}}}-2\tilde{\mathbf{n}}} G(Du, r-q) f(q) \mathrm{d}q\\ &= \left(\left(\mu - \frac{D}{2}\triangle\right)^{-(|\bar{\bar{\mathbf{n}}}|-|\tilde{\mathbf{n}}|)} \partial^{\bar{\mathbf{n}}+\bar{\bar{\mathbf{n}}}-2\tilde{\mathbf{n}}} R_{\mu,D}^{\overline{m}/2} f\right)(r),\end{aligned}\right\} \tag{11.49}$$

and $|\bar{\bar{\mathbf{n}}}| \geq |\tilde{\mathbf{n}}|$, since $|\bar{\bar{\mathbf{n}}}| \geq |\bar{\mathbf{n}}|$. By Lemma 11.7, the operator

$$\left(\mu - \frac{D}{2}\triangle\right)^{-1/2(|\bar{\mathbf{n}}+\bar{\bar{\mathbf{n}}}|-2|\tilde{\mathbf{n}}|)} \partial^{\bar{\mathbf{n}}+\bar{\bar{\mathbf{n}}}-2\tilde{\mathbf{n}}}$$

has a bounded extension on $\mathbf{W}_{m,p,\Phi}$ for all $m \geq 0$. Therefore, summing up, we obtain the first inequality in (11.47) from (11.28). Moreover, for sufficiently small $\epsilon > 0$, $R_{\mu}^{\overline{m}/2}$ is a bounded operator from $\mathbf{H}_{-d/2-\epsilon}$ into $\mathbf{H}_{d/2+2\epsilon+2}$ by (11.38). By (11.42) this implies that $R_{\mu}^{\overline{m}/2}$ also bounded operator from $\mathbf{H}_{-d/2-\epsilon}$ into $\mathbf{W}_{2,p,\Phi}$. Hence, we obtain the second inequality in (11.47). □

Lemma 11.11 was proved in Kotelenez (1995b, Lemma 4.12) for the case $\Phi \equiv 1$. Its proof can be generalized to the case $\Phi \in \{1, \varpi\}$. However, we provide a different proof, based on the preceding calculations.

Lemma 11.11. Suppose $f, g \in \mathbf{H}_{0,\Phi}$. Set for $\overline{m} \geq \overline{m}(0)$ and $D > 0$

$$g_{\mu} := |(R_{\mu,D}^{\overline{m}/2} - I)g| + |g|.$$

Then,

$$\begin{aligned}\lim_{\mu\to\infty} \frac{\mu^{\overline{m}/2}}{\left(\frac{\overline{m}}{2}-1\right)!}&\int_0^{\infty} \mathrm{d}u\, \mathrm{e}^{-\mu u} u^{\overline{m}/2-1} \int \mathrm{d}r\, \Phi(r)\\ &\times \int \mathrm{d}q\, G(u, r-q)|f(q) - f(r)| g_{\mu}(r) = 0.\end{aligned} \tag{11.50}$$

Proof.

$$\left.\begin{array}{l}\displaystyle\int\!\!\int \mathrm{d}r\,\Phi(r)\mathrm{d}q\,G(u,r-q)|f(q)-f(r)|g_\mu(r)\\[2ex]
\displaystyle\quad\le\sqrt{\int \mathrm{d}r\,\Phi(r)\left[\int \mathrm{d}q\,G(u,r-q)|f(q)-f(r)|\right]^2}\\[2ex]
\displaystyle\qquad\times\sqrt{\int \mathrm{d}r\,\Phi(r)\left[\int \mathrm{d}q\,G(u,r-q)g_\mu(r)\right]^2}\\[2ex]
\text{(by the Cauchy–Schwarz inequality, employing } G(u,q)=\sqrt{G(u,q)}\sqrt{G(u,q)}\text{)}\\[2ex]
\displaystyle\quad\le c_\Phi\sqrt{\int \mathrm{d}r\,\Phi(r)\left[\int \mathrm{d}q\,G(u,r-q)|f(q)-f(r)|\right]^2}\,\|g\|_{0,\Phi}\\[2ex]
\displaystyle\text{(since } \int \mathrm{d}q\,G(u,r-q)=1 \text{ and } R^{\overline{m}/2}_{\mu,D} \text{ is a bounded operator on } \mathbf{H}_{0,\Phi}\text{)}\\[2ex]
\quad\longrightarrow 0,\quad \text{as } u\longrightarrow\infty \text{ by (11.15).}
\end{array}\right\}\tag{11.51}$$

We next change variables $v := \mu u$, whence

$$\left.\begin{array}{l}\displaystyle\frac{\mu^{\overline{m}/2}}{\left(\frac{\overline{m}}{2}-1\right)!}\int_0^\infty \mathrm{d}u\,\mathrm{e}^{-\mu u}u^{\overline{m}/2-1}\int \mathrm{d}r\,\Phi(r)\int \mathrm{d}q\,G(u,r-q)|f(q)-f(r)|g_\mu(r)\\[2ex]
\displaystyle\quad=\int_0^\infty \mathrm{d}v\,\mathrm{e}^{-v}v^{\overline{m}/2-1}\int \mathrm{d}r\,\Phi(r)\int \mathrm{d}q\,G\left(\frac{v}{\mu},r-q\right)|f(q)-f(r)|g_\mu(r).
\end{array}\right\}\tag{11.52}$$

By (11.51) for any $v>0$ and $r,q\in\mathbf{R}^d$ the integrand on the right-hand side of (11.52) tends to 0, as $\mu\to\infty$. It remains to find an upper bound which is integrable with respect to $\mathrm{d}v\,\mathrm{e}^{-v}v^{\overline{m}/2-1}$ and independent of μ.

$$\begin{aligned}&\int \mathrm{d}r\,\Phi(r)\left[\int \mathrm{d}q\,G(u,r-q)|f(q)-f(r)|\right]^2\\&\quad\le 2\|f\|^2_{0,\Phi}+2\int \mathrm{d}r\,\Phi(r)\left[\int \mathrm{d}q\,G(u,r-q)|f(q)|\right]^2.\end{aligned}$$

As, before, by the Cauchy–Schwarz inequality and the fact that $\int \mathrm{d}q\,G(u,r-q)=1$

$$\int \mathrm{d}r\,\Phi(r)\left[\int \mathrm{d}q\,G(u,r-q)|f(q)|\right]^2\le\int \mathrm{d}r\,\Phi(r)\int \mathrm{d}q\,G(u,r-q)|f(q)|^2.$$

If $\Phi \equiv 1$ we are done. If $\Phi = \varpi$ we employ (15.47) and obtain

$$\begin{aligned}
&\int dr \int dq\, \varpi(r)\varpi^{-1}(q)\varpi(q)G(u, r-q)|f(q)|^2 \\
&\le 2^\gamma \int dq \left(\int dr\, \varpi(q)G(u, r-q)(1+|r-q|^2)^\gamma\right)|f(q)|^2 \\
&\le c_\gamma \left(1+u^\gamma\right)\|f\|_{0,\varpi}^2
\end{aligned}$$

by the integrability of the moments of the normal distribution. Thus,

$$\left.\begin{aligned}
&\int dr\, \varpi(r) \int dq\, G(\frac{v}{\mu}, r-q)|f(q) - f(r)|g_\mu(r) \\
&\le c_{\gamma,\Phi}\sqrt{\left(1+\left(\frac{v}{\mu}\right)^\gamma\right)}\|f\|_{0,\varpi}\|g\|_{0,\varpi} \\
&\le c_{\gamma,\varpi}(1+v^\gamma)\|f\|_{0,\varpi}\|g\|_{0,\varpi} \text{for } \mu \ge 1.
\end{aligned}\right\} \tag{11.53}$$

The right-hand side in the last inequality of (11.53) is integrable with respect to $dv\, e^{-v} v^{\overline{m}/2-1}$. □

Lemma 11.12. *Let $g \in C_b^m(\mathbf{R}^d; \mathbf{R}), m \in \mathbf{N}$. For $f \in \mathbf{H}_0$ let fg define point-wise multiplication. Then the multiplication operator on $\mathbf{H}_0$, defined by*

$$f \longmapsto fg$$

can be extended to a bounded operator on $\mathbf{H}_{-m}$.

Proof. If $\|\varphi\|_m \le 1$, then $\|g\varphi\|_m \le c$, where c is the norm of g in $C_b^m(\mathbf{R}^d; \mathbf{R})$. Let $f \in \mathbf{H}_0$.

$$\begin{aligned}
\|fg\|_{-m} &= \sup_{\{\|\varphi\|_m \le 1\}} |\langle fg, \varphi\rangle| \\
&= \sup_{\{\|\varphi\|_m \le 1\}} |\langle f, g\varphi\rangle| \le \sup_{\{\|\psi\|_m \le c\}} |\langle f, \psi\rangle| \le c\|f\|_{-m}.
\end{aligned}$$

Since $\mathbf{H}_0$ is densely and continuously imbedded in $\mathbf{H}_{-m}$, the preceding estimates imply the extendibility of the multiplication operator g onto $\mathbf{H}_{-m}$. □

11.2 Proof of Smoothness of (8.25) and (8.73)

(i) Assume for notational simplicity $s = 0$. Further, it follows from the estimates of the martingale part driven by space–time white noise that we may, without loss of generality, assume $\sigma^\perp(r, \mu, t) \equiv 0$. Let $\mathbf{m} \ge \mathbf{0}$ such that $|\mathbf{m}| = m$. For $0 \le \mathbf{n} \le \mathbf{m}$ we set

$$Q^{\mathbf{n}}_{\mu} := \partial^{\mathbf{n}} R^{\overline{m}/2}_{\mu}, \quad Y_{\mu,\mathbf{n}}(t) := Q^{\mathbf{n}}_{\mu} Y(t), \tag{11.54}$$

where $Y(t)$ is the solution of (8.25) and $\overline{m}$ has been defined in (11.43). Since $Y(\cdot) \in C([0,\infty); \mathbf{M}_f)$ a.s. we also have by (11.41) for any $\epsilon > 0$

$$Y(\cdot) \in C([0,\infty); \mathbf{H}_{-d/2-\epsilon}).$$

Therefore, for any $\beta \in \mathbf{R}$,

$$R^{\beta/2}_{\mu} Y(\cdot) \in C([0,\infty); \mathbf{H}_{-d/2-\epsilon+\beta}).$$

Further, let $\mathbf{n}$ be a multi-index. By (11.21), (11.24), and the equivalence of the norms $|f|_m$ and $\|f\|_m$ for $m \in \mathbf{N}$, we easily see that

$$\partial^{\mathbf{n}} R^{\overline{m}/2}_{\mu} \mathbf{H}_{-\alpha} \subset \mathbf{H}_{-\alpha+\overline{m}-|\mathbf{n}|}. \tag{11.55}$$

Set

$$\eta := \overline{m} - m - 2 - d \in (0,2],$$

and choose $\delta > 0$ small enough so that

$$\epsilon := \frac{\eta - \delta}{2} > 0.$$

Further, assume, in what follows,

$$|\mathbf{n}| \le m.$$

Then, with $\alpha = d/2 + \delta$ we obtain from (11.41), (11.44), and (11.55) that with probability 1 for $p \ge 2$

$$Y_{\mu,\mathbf{n}}(\cdot) \in C([0,\infty); \mathbf{H}_{2+d/2+\epsilon}) \subset C([0,\infty); \mathbf{W}_{2,p,\Phi}). \tag{11.56}$$

As a result, the following equation holds in $\mathbf{H}_{d/2+\epsilon} \subset \mathbf{H}_0 \cap \mathbf{W}_{0,p,\Phi} \subset \mathbf{H}_0$:

$$\begin{aligned} Y_{\mu,\mathbf{n}}(t) &= Y_{\mu,\mathbf{n}}(0) + \int_0^t \frac{1}{2} \sum_{k,\ell=1}^{d} D_{k\ell}(s) \partial^2_{k\ell} Y_{\mu,\mathbf{n}}(s) \mathrm{d}s \\ &\quad - \int_0^t \nabla \cdot Q^{\mathbf{n}}_{\mu}(Y(s)F(s)) \mathrm{d}s - \int_0^t \nabla \cdot Q^{\mathbf{n}}_{\mu}(Y(s)\mathrm{d}m(s)). \end{aligned} \tag{11.57}$$

Recall that $Y(t,\omega) \in \mathbf{M}_f$ for all ω. Expression (11.41), in addition to mass conservation, implies for $\epsilon > 0$

$$\|Y(t,\omega)\|_{-d/2-\epsilon} \le c\gamma_f(Y)(t,\omega)) = c\gamma_f(Y(0,\omega)), \tag{11.58}$$

where we may choose $\epsilon > 0$ small. By (11.55)

$$\partial^{\mathbf{n}} R_{\mu}^{\overline{m}/2} \in \mathcal{L}(\mathbf{H}_{-\alpha}, \mathbf{H}_{-\alpha+\overline{m}-|\mathbf{n}|}),$$

whence for small $\epsilon > 0$

$$\partial^{\mathbf{n}} R_{\mu}^{\overline{m}/2} \in \mathcal{L}(\mathbf{H}_{-\alpha}, \mathbf{H}_{-\alpha+d+2+2\epsilon}).$$

Choosing $\alpha = d/2 + \epsilon$, this imlies

$$\partial^{\mathbf{n}} R_{\mu}^{\overline{m}/2} \in \mathcal{L}(\mathbf{H}_{-d/2-\epsilon}, \mathbf{H}_{d/2+2+\epsilon}).$$

Hence, by (11.41) and (11.42), there is a $c_{\mu} < \infty$ such that uniformly in (t, ω), $\mathbf{0} \le \mathbf{n} \le \mathbf{m}$, $p \ge 2, i \le 2$, and any $\bar{n} \in \mathbf{N}$

$$\left.\begin{aligned} &\|Y_{\mu,\mathbf{n}}(t,\omega)\|_{i,p,\Phi}^{\bar{n}p} \le c_2^{\bar{n}(p-2)} \|Y_{\mu,\mathbf{n}}(t,\omega)\|_{2+d/2+\epsilon}^{\bar{n}p} \\ &\le c_{\mu}^{\bar{n}p} c_2^{\bar{n}(p-2)} \|Y(t,\omega)\|_{-d/2-\epsilon}^{\bar{n}p} \le c_{\mu}^{\bar{n}p} c c_2^{\bar{n}(p-2)} \gamma_f^{\bar{n}p}(Y(0,\omega)) =: c_{\mu}^{\bar{n}p} c_{\omega}^{\bar{n}p} < \infty, \end{aligned}\right\} \tag{11.59}$$

where we used (11.58) in the last inequality. Thus,

$$\begin{aligned} \max_{\mathbf{0}\le\mathbf{n}\le\mathbf{m}} \sup_{t\ge s} \|Y_{\mu,\mathbf{n}}(t,\omega)\|_{i,p,\Phi}^{\bar{n}p} &\le c_2^{(p-2)\bar{n}} \max_{\mathbf{0}\le\mathbf{n}\le\mathbf{m}} \sup_{t\ge s} \|Y_{\mu,\mathbf{n}}(t,\omega)\|_{2+d/2+\epsilon}^{\bar{n}p} \\ &\le (c_{\mu} c_{\omega})^{\bar{n}p} < \infty. \end{aligned} \tag{11.60}$$

We conclude that, with arbitrarily large probability, $\|Y_{\mu,\mathbf{n}}(t,\omega)\|_{i,p,\Phi}^{\bar{n}p}$ is integrable with respect to $\mathrm{d}t \otimes \mathrm{d}P$, $i = 0, 1, 2$. By mass conservation, this probability depends only on the initial mass distribution.

We will initially analyze the case $\bar{n} = 1$. The preceding considerations allow us to apply Itô's formula for $p \ge 2$ and $\mathbf{n} \le \mathbf{m}$:

$$\left.\begin{aligned} \|Y_{\mu,\mathbf{n}}(t)\|_{0,p,\Phi}^{p} &= \|Y_{\mu,\mathbf{n}}(0)\|_{0,p,\Phi}^{p} \\ &\quad + \frac{p}{2} \sum_{k,\ell=1}^{d} \int_0^t \langle Y_{\mu,\mathbf{n}}^{p-1}(s), \partial_{k\ell}^2 Y_{\mu,\mathbf{n}}(s)\rangle_{0,\Phi} D_{k\ell}(s)\mathrm{d}s \\ &\quad - p \int_0^t \langle Y_{\mu,\mathbf{n}}^{p-1}(s), \nabla\cdot Q_{\mu}^{\mathbf{n}}(Y(s)F(s))\rangle_{0,\Phi}\,\mathrm{d}s \\ &\quad - p \int_0^t \langle Y_{\mu,\mathbf{n}}^{p-1}(s), \nabla\cdot Q_{\mu}^{\mathbf{n}}(Y(s)\mathrm{d}m(s))\rangle_{0,\Phi} \\ &\quad + \frac{p(p-1)}{2} \int_0^t \langle Y_{\mu,\mathbf{n}}^{p-2}(s), [\nabla\cdot Q_{\mu}^{\mathbf{n}}(Y(s)\mathrm{d}m(s))]\rangle_{0,\Phi} \\ &=: \|Y_{\mu,\mathbf{n}}(0)\|_{0,p,\Phi}^{p} + \sum_{i=2}^{5} A_{i,\mathbf{n},Y}(t) \\ &= \|Y_{\mu,\mathbf{n}}(0)\|_{0,p,\Phi}^{p} + \sum_{i\in\{2,3,5\}} \int_0^t a_{i,\mathbf{n},Y}(s)\mathrm{d}s + A_{4,\mathbf{n},Y}(t), \end{aligned}\right\} \tag{11.61}$$

where the $a_{i,\mathbf{n},Y}(s)$ denote the integrands in the deterministic integrals of (11.61).

(ii) We first decompose $a_{2,\mathbf{n},Y}(t)$. Integrating by parts, we obtain

$$
\left.\begin{aligned}
p\int Y_{\mu,\mathbf{n}}^{p-1}(s,r)(\partial_\ell Y_{\mu,\mathbf{n}}(s,r))\partial_k\Phi(r)\mathrm{d}r &= \int \partial_\ell(Y_{\mu,\mathbf{n}}^{p}(s,r))\partial_k\Phi(r)\mathrm{d}r\\
&= -\int Y_{\mu,\mathbf{n}}^{p}(s,r)\partial_{k\ell}^2\Phi(r)\mathrm{d}r.
\end{aligned}\right\}
\tag{11.62}
$$

Therefore,

$$
\left.\begin{aligned}
&a_{2,\mathbf{n},Y}(s)\\
&= -\frac{p(p-1)}{2}\sum_{k,\ell=1}^{d}\int (Y_{\mu,\mathbf{n}}(s,r))^{p-2}(\partial_k Y_{\mu,\mathbf{n}}(s,r))(\partial_\ell Y_{\mu,\mathbf{n}}(s,r))\Phi(r)\mathrm{d}r\, D_{k\ell}(s)\\
&\quad +\frac{1}{2}\sum_{k,\ell=1}^{d}\int (Y_{\mu,\mathbf{n}}(s,r))^{p}\partial_{k\ell}^2\Phi(r)\mathrm{d}r\, D_{k\ell}(s)\\
&=: a_{2,\mathbf{n},1,Y}(s) + a_{2,\mathbf{n},2,Y}(s).
\end{aligned}\right\}
\tag{11.63}
$$

Next, recalling the homogeneity assumption (8.33),

$$
\begin{aligned}
&a_{5,\mathbf{n},Y}(s)\\
&= \frac{p(p-1)}{2}\sum_{k,\ell=1}^{d}\int (Y_{\mu,\mathbf{n}}(s,r))^{p-2}\frac{\mu^{\overline{m}}}{[(\frac{\overline{m}}{2}-1)!]^2}\int_0^\infty\int_0^\infty \mathrm{d}u\,\mathrm{d}v\,\mathrm{e}^{-\mu(u+v)}u^{\overline{m}/2-1}v^{\overline{m}/2-1}\\
&\quad\times\int\int \partial_{k,r}\partial_r^{\mathbf{n}}G(u,r-q)\partial_{\ell,r}\partial_r^{\mathbf{n}}G(v,r-\tilde{q})Y(s,q)Y(s,\tilde{q})\tilde{D}_{k,\ell}(s,q-\tilde{q})\Phi(r)\mathrm{d}q\,\mathrm{d}\tilde{q}\,\mathrm{d}r
\end{aligned}
$$

By Chap. 8, ((8.33) and (8.34)), $D_{k\ell}(s) := \tilde{D}_{k\ell}(0,s)$. Set

$$
\hat{D}_{k\ell}(s,r-\tilde{q}) := -\tilde{D}_{k\ell}(s,r-\tilde{q}) + D_{k\ell}(s),
\tag{11.64}
$$

where, by the symmetry assumption on $\tilde{D}_{k\ell}(s,\tilde{q}-r)$ we have

$$
\hat{D}_{k\ell}(s,r-\tilde{q}) = \hat{D}_{k\ell}(s,\tilde{q}-r).
$$

Altogether, we obtain

$$
\left.\begin{aligned}
&a_{5,\mathbf{n},Y}(s)\\
&= \frac{p(p-1)}{2}\sum_{k,\ell=1}^{d}\int (Y_{\mu,\mathbf{n}}(s,r))^{p-2}\frac{\mu^{\overline{m}}}{[(\frac{\overline{m}}{2}-1)!]^2}\int_0^\infty\int_0^\infty \mathrm{d}u\,\mathrm{d}v\,\mathrm{e}^{-\mu(u+v)}u^{\overline{m}/2-1}v^{\overline{m}/2-1}\\
&\quad\times\int\int \partial_{k,r}\partial_r^{\mathbf{n}}G(u,r-q)\partial_{\ell,r}\partial_r^{\mathbf{n}}G(v,r-\tilde{q})Y(s,\mathrm{d}q)Y(s,\mathrm{d}\tilde{q})D_{k\ell}(s)\Phi(r)\mathrm{d}r\\
&\quad -\frac{p(p-1)}{2}\sum_{k,\ell=1}^{d}\int (Y_{\mu,\mathbf{n}}(s,r))^{p-2}\frac{\mu^{\overline{m}}}{[(\frac{\overline{m}}{2}-1)!]^2}\int_0^\infty\int_0^\infty \mathrm{d}u\,\mathrm{d}v\,e^{-\mu(u+v)}u^{\overline{m}/2-1}v^{\overline{m}/2-1}\\
&\quad\times\int\int \partial_{k,r}\partial_r^{\mathbf{n}}G(u,r-q)\partial_{\ell,r}\partial_r^{\mathbf{n}}G(v,r-\tilde{q})Y(s,\mathrm{d}q)Y(s,\mathrm{d}\tilde{q})\hat{D}_{k,\ell}(s,q-\tilde{q})\Phi(r)\mathrm{d}r\\
&=: a_{5,\mathbf{n},1,Y}(s) + a_{5,\mathbf{n},2,Y}(s).
\end{aligned}\right\}
\tag{11.65}
$$

Apparently,

$$a_{2,\mathbf{n},1,Y}(s) + a_{5,\mathbf{n},1,Y}(s) \equiv 0. \tag{11.66}$$

(iii) $Y(s) \in \mathbf{H}_{-d/2-\epsilon}\ \forall \epsilon > 0$. The highest derivative involved in $a_{5,\mathbf{n},2,Y}(s)$ is of the order of $n+1$, and after one integration by parts with respect to a scalar coordinate the order can increase to $n+2$. Applied to $Y(s)$ (in the generalized sense) this would map $Y(s)$ into $\mathbf{H}_{-d/2-\epsilon-n-2} \subset \mathbf{H}_{-\overline{m}+d/2+\epsilon}$ for all small $\epsilon > 0$, since $\overline{m} > d+m+2 \geq d+n+2$.

Recall from (8.40) the abbreviation for $\mathcal{L}_{k\ell,\mathbf{n}}(s)$, where $\mathbf{n} \geq \mathbf{0}$ with $|\mathbf{n}| \leq m+1$. We have

$$\mathcal{L}_{k\ell,\mathbf{n}}(s) := \partial^{\mathbf{n}} \tilde{D}_{k\ell}(s,r)|_{r=0} = -\partial^{\mathbf{n}} \hat{D}_{k\ell}(s,r)|_{r=0}. \tag{11.67}$$

We expand $\hat{D}_{k\ell}(s, q-\tilde{q}) = \hat{D}_{k\ell}(s, q-r+r-\tilde{q})$ in the variable $r-\tilde{q}$, using Taylor's formula. Observe that, by symmetry assumption,

$$\mathcal{L}_{k\ell,\mathbf{n}}(s) = (-1)^{|\mathbf{n}|}\mathcal{L}_{k\ell,\mathbf{n}}(s),$$

whence $\mathcal{L}_{k\ell,\mathbf{n}}(s) \equiv 0$, if $|\mathbf{n}|$ is odd. Further, we recall that $\hat{D}_{k\ell}(s,0) = 0$. Therefore, the degree of the polynomials $\pi_{\tilde{\mathbf{n}}_j}(q-r)\pi_{\mathbf{n}_i}(r-\tilde{q})$, $|\tilde{\mathbf{n}}_j + \mathbf{n}_i|$, in the following expansion (11.68) is an even number ≥ 2.

$$\left.\begin{aligned}
&\hat{D}_{k\ell}\,(s, q-r+r-\tilde{q})\\
&= \sum_{i=0}^{m+1} \sum_{|\mathbf{n}_i|=i} \left(\partial_{q-r}^{\mathbf{n}_i} \frac{\hat{D}_{k\ell}\,(s,q-r)}{i!} \right) \pi_{\mathbf{n}_i}\,(r-\tilde{q})\\
&\quad + \sum_{|\mathbf{n}_{m+2}|=m+2} \hat{\theta}_{k\ell,\mathbf{n}_{m+2}}(s,q,r,\tilde{q})\pi_{\mathbf{n}_{m+2}}\,(r-\tilde{q})\\
&= -\sum_{i=0}^{m+1} \sum_{|\mathbf{n}_i|=i} \sum_{j=0}^{m+1-i} \sum_{|\tilde{\mathbf{n}}_j|=j} 1_{\{|\tilde{\mathbf{n}}_j|+|\mathbf{n}_i|\geq 2\}} \frac{1}{i!j!} \mathcal{L}_{k\ell,\mathbf{n}_i+\tilde{\mathbf{n}}_j}(s)\pi_{\tilde{\mathbf{n}}_j}(q-r)\pi_{\mathbf{n}_i}\,(r-\tilde{q})\\
&\quad + \sum_{|\mathbf{n}_{m+2}|=m+2} \hat{\theta}_{k\ell,\mathbf{n}_{m+2}}(s,q,r,\tilde{q})\pi_{\mathbf{n}_{m+2}}(r-\tilde{q})\\
&\quad + \sum_{i=0}^{m+1} \sum_{|\mathbf{n}_i|=i} \sum_{|\mathbf{n}_{m+2}-\mathbf{n}_i|=m+2-i} \tilde{\theta}_{k\ell,\mathbf{n}_{m+2}-\mathbf{n}_i}(s,q,r)\pi_{\mathbf{n}_{m+2}-\mathbf{n}_i}(q-r)\pi_{\mathbf{n}_i}\,(r-\tilde{q})\,.\\
&=: I_{k\ell}(s,q,\tilde{q},r) + II_{s,k\ell}(q,\tilde{q},r) + III_{s,k\ell}(q,\tilde{q},r),
\end{aligned}\right\} \tag{11.68}$$

where by (8.35) the remainder terms $\hat{\theta}$ and $\tilde{\theta}$ have derivatives which are bounded uniformly in all parameters. By assumption (8.35), for any $T > 0$

$$\left.\begin{aligned}
&\sum_{0\leq|\mathbf{n}|\leq(m+1)} \operatorname*{ess\,sup}_{\omega} \sup_{0\leq s\leq T,q,r,\tilde{q}} [|\mathcal{L}_{k\ell,\mathbf{n}}(s,\omega)| + |||\hat{\theta}_{k\ell,\mathbf{n}}(s,\cdot,r,\tilde{q},\omega)|||_1\\
&\quad +|||\tilde{\theta}_{k\ell,\mathbf{n}}(s,q,\cdot,\omega)|||_1] < \infty.
\end{aligned}\right\} \tag{11.69}$$

We replace $\hat{D}_{k,\ell}(s,q-p)$ in $a_{5,\mathbf{n},2,Y}(s)$ with terms from the expansion (11.68). If a term from that expansion does not depend on $\tilde{q}$ (or on q), we can integrate by parts to simplify the resulting expression. Let us assume that, without loss of generality, the term does not depend on $\tilde{q}$. Note that for any $Z \in \mathbf{M}_f$

$$\frac{\mu^{\frac{\overline{m}}{2}}}{\left[(\frac{\overline{m}}{2}-1)!\right]}\int_0^\infty \mathrm{d}v\, \mathrm{e}^{-\mu v} v^{\overline{m}/2-1}\int \partial_{\ell,r}\partial_r^{\mathbf{n}} G(v,r-\tilde{q})Z(\mathrm{d}\tilde{q}) = \left(\partial_{\ell,r}\partial^{\mathbf{n}} R_\mu^{\overline{m}/2} Z\right)(r).$$

Thus,[17]

$$\left(\left(\partial^{\mathbf{n}} R_\mu^{\overline{m}/2} Z\right)(r)\right)^{p-2}\frac{\mu^{\overline{m}/2}}{\left[\left(\frac{\overline{m}}{2}-1\right)!\right]}\int_0^\infty \mathrm{d}v\, \mathrm{e}^{-\mu v} v^{\overline{m}/2-1}\int \partial_{\ell,r}\partial_r^{\mathbf{n}} G(v,r-\tilde{q})Z(\mathrm{d}\tilde{q})$$

$$= \frac{1}{p-1}\partial_{\ell,r}((\partial^{\mathbf{n}} R_\mu^{\overline{m}/2} Z)(r))^{p-1}. \tag{11.70}$$

Step 1: $|\mathbf{n}| = 0$. Let us first analyze $I_{k\ell}(s,q,\tilde{q},r)$. Replacing $\hat{D}_{k,\ell}(s,q-p)$ with a term from $I_{k\ell}(q,\tilde{q},r)$ in $a_{5,0,2,Y}(s)$ yields

$$\left.\begin{aligned}
&-\frac{p(p-1)}{2}\sum_{k,\ell=1}^{d}\int (Y_{\mu,\mathbf{n}}(s,r))^{p-2}\frac{\mu^{\overline{m}}}{[(\frac{\overline{m}}{2}-1)!]^2}\\
&\times\int_0^\infty\int_0^\infty \mathrm{d}u\,\mathrm{d}v\,\mathrm{e}^{-\mu(u+v)}u^{(\overline{m}/2)-1}v^{(\overline{m}/2)-1}\\
&\times\int\int(\partial_{k,r}G(u,r-q))(\partial_{\ell,r}G(v,r-\tilde{q}))\\
&\times Y(s,\mathrm{d}q)Y(s,\mathrm{d}\tilde{q})\pi_{\tilde{\mathbf{n}}_j}(q-r)\pi_{\mathbf{n}_i}(r-\tilde{q})\Phi(r)\mathrm{d}r\\
&\times 1_{\{|\tilde{\mathbf{n}}_j|+|\mathbf{n}_i|\ge 2\}}\frac{1}{i!j!}\mathcal{L}_{k\ell,\mathbf{n}_i+\tilde{\mathbf{n}}_j}(s),
\end{aligned}\right\} \tag{11.71}$$

where

$$\tilde{\mathbf{n}}_j := \mathbf{n}_{m+2}-\mathbf{n}_i.$$

Let us first consider the case $|\mathbf{n}_i| = 0$ and employ (11.70) with $Z := Y(s)$. Integrating by parts in $a_{5,0,2,Y}(s)$ with $\pi_{\tilde{\mathbf{n}}_j}(q-r)\frac{1}{j!}\mathcal{L}_{k\ell,\tilde{\mathbf{n}}_j}(s)$ replacing $\hat{D}_{k,\ell}(s,q-p)$, we obtain

[17] We provide the simple formula (11.70) in a more abstract formulation which allows us to employ it later to smoothed versions of $Y(s)$ (cf. (11.92)).

$$
\left.
\begin{aligned}
&-\frac{p(p-1)}{2}\sum_{k,\ell=1}^{d}\int (Y_{\mu,0}(s,r))^{p-2}\frac{\mu^{\overline{m}}}{[(\frac{\overline{m}}{2}-1)!]^2}\\
&\times\int_0^\infty\int_0^\infty \mathrm{d}u\,\mathrm{d}v\,\mathrm{e}^{-\mu(u+v)}u^{(\overline{m}/2)-1}v^{(\overline{m}/2)-1}\\
&\times\int\int(\partial_{k,r}G(u,r-q))(\partial_{\ell,r}G(v,r-\tilde{q}))Y(s,\mathrm{d}q)Y(s,\mathrm{d}\tilde{q})\pi_{\tilde{\mathbf{n}}_j}(q-r)\Phi(r)\mathrm{d}r\\
&=-\frac{p(p-1)}{2}\sum_{k,\ell=1}^{d}\int (Y_{\mu,0}(s,r))^{p-2}\frac{\mu^{\overline{m}/2}}{(\frac{\overline{m}}{2}-1)!}\int_0^\infty \mathrm{d}u\,\mathrm{e}^{-\mu u}u^{(\overline{m}/2)-1}\\
&\quad\times\int(\partial_{k,r}G(u,r-q))Y(s,\mathrm{d}q)\pi_{\tilde{\mathbf{n}}_j}(q-r)(\partial_{\ell,r}Y_{\mu,0})(s,r)\Phi(r)\mathrm{d}r\\
&=\frac{p}{2}\sum_{k,\ell=1}^{d}\int (Y_{\mu,0}(s,r))^{p-1}\frac{\mu^{\overline{m}/2}}{(\frac{\overline{m}}{2}-1)!}\int_0^\infty \mathrm{d}u\,\mathrm{e}^{-\mu u}u^{(\overline{m}/2)-1}\int(\partial^2_{k\ell,r}G(u,r-q))\\
&\quad\times Y(s,\mathrm{d}q)\pi_{\tilde{\mathbf{n}}_j}(q-r)\Phi(r)\mathrm{d}r\\
&\quad+\frac{p}{2}\sum_{k,\ell=1}^{d}\int (Y_{\mu,0}(s,r))^{p-1}\frac{\mu^{\overline{m}/2}}{(\frac{\overline{m}}{2}-1)!}\int_0^\infty \mathrm{d}u\,\mathrm{e}^{-\mu u}u^{(\overline{m}/2)-1}\int(\partial_{k,r}G(u,r-q))\\
&\quad\times Y(s,\mathrm{d}q)(\partial_{\ell,r}\pi_{\tilde{\mathbf{n}}_j}(q-r))\Phi(r)\mathrm{d}r\\
&\quad+\frac{p}{2}\sum_{k,\ell=1}^{d}\int (Y_{\mu,0}(s,r))^{p-1}\frac{\mu^{\overline{m}/2}}{(\frac{\overline{m}}{2}-1)!}\int_0^\infty \mathrm{d}u\mathrm{e}^{-\mu u}u^{\overline{m}/2-1}\int(\partial_{k,r}G(u,r-q))\\
&\quad\times Y(s,\mathrm{d}q)\pi_{\tilde{\mathbf{n}}_j}(q-r)\partial_{\ell,r}\Phi(r)\mathrm{d}r.
\end{aligned}
\right\}
\tag{11.72}
$$

The first term on the right-hand side of (11.72) can be estimated, employing Lemma 11.10. We set

$$\bar{\mathbf{n}} := 1_k + 1_\ell,\ \ \bar{\bar{\mathbf{n}}} := \tilde{\mathbf{n}}_j.$$

We note that $|\bar{\bar{\mathbf{n}}}| = |\tilde{\mathbf{n}}_j| \geq |\bar{\mathbf{n}}|$, since $|\mathbf{n}_i| = 0$ and $|\mathbf{n}_i| + |\tilde{\mathbf{n}}_j| \geq 2$ (recalling that the sum has to be even and ≥ 2). Employing Hölder's inequality to separate $(Y_{\mu,0}(s,r))^{p-1}$ from the integral operator in addition to (11.47), we obtain that the first term is estimated above by

$$\left.\begin{aligned}&\frac{p}{2}\sum_{k,\ell=1}^{d}\Big|\int (Y_{\mu,0}(s,r))^{p-1}\frac{\mu^{\overline{m}/2}}{(\frac{\overline{m}}{2}-1)!}\int_0^\infty \mathrm{d}u\,\mathrm{e}^{-\mu u}u^{(\overline{m}/2)-1}\int(\partial^2_{k\ell,r}G(u,r-q))\\&\times Y(s,\mathrm{d}q)\pi_{\tilde{\mathbf{n}}_j}(q-r)\Phi(r)\mathrm{d}r\Big|\\&\le \|Y^p_{\mu,0}(s)\|^{(p-1)/p}_{0,p,\Phi}c_{\Phi,\overline{m},k,\ell,\tilde{\mathbf{n}}_j,D}\|Y_{\mu,0}(s)\|_{0,p,\Phi}\\&\qquad\text{(assuming } \mu\ge D\theta+1)\\&=c_{\Phi,\overline{m},k,\ell,\tilde{\mathbf{n}}_j,D}\|Y_{\mu,0}(s)\|^p_{0,p,\Phi}.\end{aligned}\right\}\tag{11.73}$$

The second and third terms are estimated in the same way, where for the third term we also use (15.46).

Consider the case when $|\mathbf{n}_i|\ge 1$ and $|\tilde{\mathbf{n}}_j|\ge 1$ in $I_{k\ell}(s,q,\tilde{q},r)$. (The case $|\tilde{\mathbf{n}}_j|=0$ obviously leads to the same estimate as the previous case.) We now do not integrate by parts, but otherwise we argue as in the previous step, applying Lemma 11.10 to both factors containing $Y(s,\mathrm{d}q)$ and $Y(s,\mathrm{d}\tilde{q})$ in addition to Hölder's inequality. Thus, we obtain a similar bound as in (11.73):

$$\left.\begin{aligned}&\frac{p(p-1)}{2}\sum_{k,\ell=1}^{d}\Big|\int (Y_{\mu,0}(s,r))^{p-2}\frac{\mu^{\overline{m}}}{[(\frac{\overline{m}}{2}-1)!]^2}\\&\times\int_0^\infty\int_0^\infty \mathrm{d}u\,\mathrm{d}v\,\mathrm{e}^{-\mu(u+v)}u^{(\overline{m}/2)-1}v^{(\overline{m}/2)-1}\\&\times\int\int(\partial_{k,r}G(u,r-q))(\partial_{\ell,r}G(v,r-\tilde{q}))Y(s,\mathrm{d}q)\\&\times Y(s,\mathrm{d}\tilde{q})\pi_{\tilde{\mathbf{n}}_j}(q-r)\pi_{\mathbf{n}_i}(r-\tilde{q})\Phi(r)\mathrm{d}r\Big|\le c_{\Phi,\overline{m},k,\ell,\tilde{\mathbf{n}}_j,D}\|Y_{\mu,0}(s)\|^p_{0,p,\Phi}.\end{aligned}\right\}\tag{11.74}$$

We now estimate $III_{s,k\ell}(q,\tilde{q},r)$. The estimate of $II_{k\ell}(s,q,\tilde{q},r)$ is easier. It suffices to restrict ourselves to the case $|\mathbf{n}_i|=0$ and follow the pattern in the estimate of $I_{k\ell}(q,\tilde{q},r)$. The other cases can be handled similarly. We obtain

$$
\left\{
\begin{aligned}
&\left|\frac{p(p-1)}{2}\sum_{k,\ell=1}^{d}\int (Y_{\mu,0}(s,r))^{p-2}\frac{\mu^{\overline{m}}}{[(\frac{\overline{m}}{2}-1)!]^2}\right. \\
&\int_0^\infty\int_0^\infty \mathrm{d}u\,\mathrm{d}v\,\mathrm{e}^{-\mu(u+v)}u^{(\overline{m}/2)-1}v^{(\overline{m}/2)-1} \\
&\quad\times\int\int(\partial_{k,r}G(u,r-q))(\partial_{\ell,r}G(v,r-\tilde{q}))Y(s,\mathrm{d}q) \\
&\quad\left.\times Y(s,\mathrm{d}\tilde{q})\tilde{\theta}_{k\ell,\mathbf{n}_{m+2}}(s,q,r)\pi_{\mathbf{n}_{m+2}(q-r)}\Phi(r)\mathrm{d}r\right| \\
&\le\left|\frac{p}{2}\sum_{k,\ell=1}^{d}\int (Y_{\mu,0}(s,r))^{p-1}\frac{\mu^{\overline{m}/2}}{[(\frac{\overline{m}}{2}-1)!]}\int_0^\infty \mathrm{d}u\,\mathrm{e}^{-\mu u}u^{(\overline{m}/2)-1}\right. \\
&\left.\times\int(\partial^2_{k\ell,r}G(u,r-q))Y(s,\mathrm{d}q)\tilde{\theta}_{k\ell,\mathbf{n}_{m+2}}(s,q,r)\pi_{\mathbf{n}_{m+2}}(q-r)\Phi(r)\mathrm{d}r\right| \\
&+\left|\frac{p}{2}\sum_{k,\ell=1}^{d}\int (Y_{\mu,0}(s,r))^{p-1}\frac{\mu^{\overline{m}/2}}{[(\frac{\overline{m}}{2}-1)!]}\int_0^\infty \mathrm{d}u\,\mathrm{e}^{-\mu u}u^{(\overline{m}/2)-1}\right. \\
&\left.\times\int(\partial_{k,r}G(u,r-q))Y(s,\mathrm{d}q)(\partial_{\ell,r}\tilde{\theta}_{k\ell,\mathbf{n}_{m+2}}(s,q,r))\pi_{\mathbf{n}_{m+2}}(q-r)\Phi(r)\mathrm{d}r\right| \\
&+\left|\frac{p}{2}\sum_{k,\ell=1}^{d}\int (Y_{\mu,0}(s,r))^{p-1}\frac{\mu^{\overline{m}/2}}{[(\frac{\overline{m}}{2}-1)!]}\int_0^\infty \mathrm{d}u\,\mathrm{e}^{-\mu u}u^{(\overline{m}/2)-1}\right. \\
&\left.\times\int(\partial_{k,r}G(u,r-q))Y(s,\mathrm{d}q)\tilde{\theta}_{k\ell,\mathbf{n}_{m+2}}(s,q,r)(\partial_{\ell,r}\pi_{\mathbf{n}_{m+2}}(q-r))\Phi(r)\mathrm{d}r\right| \\
&+\left|\frac{p}{2}\sum_{k,\ell=1}^{d}\int (Y_{\mu,0}(s,r))^{p-1}\frac{\mu^{\overline{m}/2}}{[(\frac{\overline{m}}{2}-1)!]}\int_0^\infty \mathrm{d}u\,\mathrm{e}^{-\mu u}u^{(\overline{m}/2)-1}\right. \\
&\left.\times\int(\partial_{k,r}G(u,r-q))Y(s,\mathrm{d}q)\tilde{\theta}_{k\ell,\mathbf{n}_{m+2}}(s,q,r)\pi_{\mathbf{n}_{m+2}}(q-r)\partial_{\ell,r}\Phi(r)\mathrm{d}r\right|.
\end{aligned}
\right\}
\tag{11.75}
$$

By the boundedness of $\partial_{\ell,r}\tilde{\theta}_{k\ell,\mathbf{n}_{m+2}}(s,q,r)$ and the fact that $\partial_{\ell,r}\pi_{\mathbf{n}_{m+2}}(q-r)$ has degree ≥ 1 in addition to (15.46), it is sufficient to estimate the first term on the right-hand side of (11.75). By (11.8), we obtain

$$
\left.\begin{array}{l}
\left|\frac{p}{2}\sum_{k,\ell=1}^{d}\int (Y_{\mu,0}(s,r))^{p-1}\frac{\mu^{\overline{m}/2}}{[(\frac{\overline{m}}{2}-1)!]}\int_0^\infty \mathrm{d}u\,\mathrm{e}^{-\mu u}u^{(\overline{m}/2)-1}\right.\\
\quad\left.\times\int(\partial^2_{k\ell,r}G(u,r-q))Y(s,\mathrm{d}q)\tilde{\theta}_{k\ell,\mathbf{n}_{m+2}}(s,q,r)\pi_{\mathbf{n}_{m+2}}(q-r)\Phi(r)\mathrm{d}r\right|\\
\le c_{m,\beta}\left|\frac{p}{2}\sum_{k,\ell=1}^{d}\right.\\
\quad\times\int|Y_{\mu,0}(s,r)|^{p-1}\frac{\mu^{\overline{m}/2}}{[(\frac{\overline{m}}{2}-1)!]}\int_0^\infty \mathrm{d}u\,\mathrm{e}^{-\mu u}u^{(\overline{m}+m/2)-1}\int G(\beta u,r-q))\\
\quad\left.\times Y(s,\mathrm{d}q)|\tilde{\theta}_{k\ell,\mathbf{n}_{m+2}}(s,q,r)|\Phi(r)\mathrm{d}r\right|\\
\le c_{m,\beta}[\sup_{q,r.s}|\tilde{\theta}_{k\ell,\mathbf{n}_{2m+2}}(s,q,r)|]\left|\frac{p}{2}\sum_{k,\ell=1}^{d}\right.\\
\quad\times\int|Y_{\mu,0}(s,r)|^{p-1}\frac{\mu^{\overline{m}/2}}{[(\frac{\overline{m}}{2}-1)!]}\int_0^\infty \mathrm{d}u\,\mathrm{e}^{-\mu u}u^{(\overline{m}+m/2)-1}\int G(\beta u,r-q))\\
\quad\left.\times Y(s,\mathrm{d}q)\Phi(r)\mathrm{d}r\right|\\
=\bar{c}_{m,\beta}|\frac{p}{2}\sum_{k,\ell=1}^{d}\int|Y_{\mu,0}(s,r)|^{p-1}\mu^{\overline{m}/2}(\mu-\frac{\beta}{2}\triangle)^{-(\overline{m}/2)-(m-|\mathbf{n}|/2)}(Y(s))(r)\Phi(r)\mathrm{d}r|\\
\text{(by (11.21) and (11.24))}\\
\quad\le\bar{c}_{m,\beta,\Phi}\frac{p}{2}\sum_{k,\ell=1}^{d}\|Y_{\mu,0}(s)\|^{p-1}_{0,p,\Phi}\|\mu^{\overline{m}/2}(\mu-\frac{\beta}{2}\triangle)^{-\overline{m}/2}Y(s))\|_{0,p,\Phi}\\
\text{(by Hölder's inequality and since } m\ge 0)\\
\quad\le\tilde{c}_{m,\beta,\Phi}\frac{p}{2}\sum_{k,\ell=1}^{d}\|Y_{\mu,0}(s)\|^{p-1}_{0,p,\Phi}\|R_\mu^{\overline{m}/2}Y(s)\|_{0,p,\Phi}\\
\text{(by (11.24), (11.25), and (11.32))}\\
\quad\le\tilde{c}_{m,\beta,\Phi}\frac{p}{2}\sum_{k,\ell=1}^{d}(\|Y_{\mu,0}(s)\|^{p}_{0,p,\Phi}+\|R_\mu^{\overline{m}/2}Y(s)\|^{p}_{0,p,\Phi}),
\end{array}\right\}
\tag{11.76}
$$

where the last step follows from the definition of the norm $\|\cdot\|_{m,p,\Phi}$ and the following simple inequality[18]

$$ab \le \frac{a^p}{p} + \frac{b^{\bar p}}{\bar p} \quad \text{for nonnegative numbers } a \text{ and } b, \text{ and } p > 1 \text{ and } \bar p := \tfrac{p}{p-1}.$$

Recalling that the functions $\frac{1}{i!j!}\mathcal{L}_{k\ell,\mathbf{n}_i+\tilde{\mathbf{n}}_j}(s)$ are uniformly bounded in s and ω, we obtain

$$a_{5,0,2,Y}(s) \le \tilde c_{\bar m,\Phi}(\|Y_{\mu,0}(s)\|^p_{0,p,\Phi} + \|R_\mu^{\overline{m}/2}Y(s)\|^p_{0,p,\Phi}). \tag{11.77}$$

(iv) Now we estimate $a_{3,0,Y}(s)$. Taylor's expansion yields

$$F_k(s,q) = F_k(s,r) + \sum_{|\mathbf{j}|=1}^{|\mathbf{m}|} \frac{1}{|\mathbf{j}|!}(\partial^{\mathbf{j}} F_k(s,r))\pi_{\mathbf{j}}(q-r) + \sum_{|\mathbf{m}|=m} \bar\theta_{k,\mathbf{m}+1}(s,q,r)\pi_{\mathbf{m}+1}(q-r), \tag{11.78}$$

where by assumption (8.35), for any $T > 0$

$$\operatorname*{ess\,sup}_{\omega} \sup_{0\le s\le T, q,} |||\hat\theta_{k\ell,\mathbf{m}+1}(s,q,\cdot,\omega)|||] < \infty. \tag{11.79}$$

Then,

$$\left.\begin{aligned}
&-p\sum_{k=1}^d \int (Y_{\mu,0}(s,r))^{p-1}\frac{\mu^{\overline{m}/2}}{(\frac{\overline{m}}{2}-1)!}\\
&\quad\times\int_0^\infty du\, e^{-\mu u}u^{(\overline{m}/2)-1}\int (\partial_{k,r}G(u,r-q))F_k(s,q)\\
&\quad\times Y(s,dq)\Phi(r)dr\\
&= -p\sum_{k=1}^d \int (Y_{\mu,0}(s,r))^{p-1}\frac{\mu^{\overline{m}/2}}{(\frac{\overline{m}}{2}-1)!}\int_0^\infty du\, e^{-\mu u}u^{(\overline{m}/2)-1}\int (\partial_{k,r}G(u,r-q))\\
&\quad\times Y(s,dq)F_k(s,r)\Phi(r)dr\\
&\quad -p\sum_{k=1}^d \int (Y_{\mu,0}(s,r))^{p-1}\frac{\mu^{\overline{m}/2}}{(\frac{\overline{m}}{2}-1)!}\int_0^\infty du\, e^{-\mu u}u^{(\overline{m}/2)^{-1}}\\
&\quad\times\int (\partial_{k,r}G(u,r-q))\left[\sum_{|\mathbf{j}|=1}^{|\mathbf{m}|}\frac{1}{|\mathbf{j}|!}\pi_{\mathbf{j}}(q-r)\right]Y(s,dq)(\partial^{\mathbf{j}}F_k(s,r))\Phi(r)dr\\
&\quad -p\sum_{k=1}^d \int (Y_{\mu,0}(s,r))^{p-1}\frac{\mu^{\overline{m}/2}}{(\frac{\overline{m}}{2}-1)!}\int_0^\infty du\, e^{-\mu u}u^{(\overline{m}/2)-1}\\
&\quad\times\int (\partial_{k,r}G(u,r-q))\sum_{|\mathbf{m}|=m}\bar\theta_{k,\mathbf{m}+1}(s,q,r)\pi_{\mathbf{m}+1}(q-r)Y(s,dq)\Phi(r)dr\\
&=: \sum_{k=1}^d [I_k(s) + II_k(s) + III_k(s)].
\end{aligned}\right\} \tag{11.80}$$

[18] Cf. (15.2) in Sect. 15.1.2 for a proof.

We estimate $I_k(s)$ for $\mathbf{n} \le \mathbf{m}$ as in (11.72), integrating by parts

$$
\left.\begin{aligned}
&-p\sum_{k=1}^{d}\int (Y_{\mu,0}(s,r))^{p-1}\frac{\mu^{\overline{m}/2}}{(\frac{\overline{m}}{2}-1)!}\int_0^\infty \mathrm{d}u\, \mathrm{e}^{-\mu u}u^{\overline{m}-1}\\
&\quad\times\int(\partial_{k,r}G(u,r-q))Y(s,\mathrm{d}q)F_k(s,r)\Phi(r)\mathrm{d}r\\
&=-\sum_{k=1}^{d}\int(\partial_{k,r}((Y_{\mu,0}(s,r))^p))F_k(s,r)\Phi(r)\mathrm{d}r\\
&=\sum_{k=1}^{d}\int(Y_{\mu,0}(s,r))^p(\partial_{k,r}F_k(s,r))\Phi(r)\mathrm{d}r\\
&\quad+\sum_{k=1}^{d}\int(Y_{\mu,0}(s,r))^pF_k(s,r)\partial_{k,r}\Phi(r)\mathrm{d}r.
\end{aligned}\right\}\tag{11.81}
$$

Applying the assumptions on F in addition to (15.46) yields

$$
|p\sum_{k=1}^{d}|I_k(s)|\le c_{F,\Phi}\|Y_{\mu,0}(s)\|_{0,p,\Phi}^p. \tag{11.82}
$$

Repeating the arguments in the estimate of $a_{5,0,2,Y}(s)$ with respect to $II_k(t)$ and $III_k(t)$, we obtain altogether

$$
|a_{3,0,Y}(s)|\le c_{\overline{m},p,\Phi}(\|Y_{\mu,0}(s)\|_{0,p,\Phi}^p+\|R_\mu^{\overline{m}/2}Y(s)\|_{0,p,\Phi}^p). \tag{11.83}
$$

(vi) The previous steps imply

$$
\left.\begin{aligned}
&|a_{2,0,Y}(s)+a_{5,0,Y}(s)|+|a_{3,0,Y}(s)|\\
&\le\hat{c}_{p,\Phi,\overline{m}}(\|Y_{\mu,0}(s)\|_{0,p,\Phi}^p+\|R_\mu^{\overline{m}/2}|Y|(s)\|_{0,p,\Phi}^p)\\
&=\hat{c}_{p,\Phi}(\|Y_{\mu,0}(s)\|_{n,p,\Phi}^p+\|R_\mu^{\overline{m}/2}Y(s)\|_{0,p,\Phi}^p).
\end{aligned}\right\}\tag{11.84}
$$

Integrating both sides of (11.84) against ds from 0 to t we obtain from (11.61)

$$
\left.\begin{aligned}
&\|Y_{\mu,0}(t)\|_{0,p,\Phi}^p\le\|Y_{\mu,0}(0)\|_{0,p,\Phi}^p+\hat{c}_{p,\Phi,\overline{m}}\int_0^t(\|Y_{\mu,0}(s)\|_{0,p,\Phi}^p\\
&+\|R_\mu^{\overline{m}/2}Y(s)\|_{0,p,\Phi}^p)\mathrm{d}s-p\int_0^t\langle Y_{\mu,0}^{p-1}(s),\nabla\cdot R_\mu^{\overline{m}/2}(Y(s)\mathrm{d}M(s))\rangle_{0,\Phi}.
\end{aligned}\right\}\tag{11.85}
$$

We take the mathematical expectation on both sides and may apply the Gronwall lemma in addition to (11.25) (for the norms of the initial condition) and obtain

$$\sup_{0\le t\le T} E\|R_\mu^{\overline{m}/2}Y(t)\|_{0,p,\Phi}^p \le \hat{c}_{\overline{m},p,T,\Phi}E\|Y(0)\|_{0,p,\Phi}^p. \tag{11.86}$$

As $\mu \to \infty$, Fatou's lemma implies the integrability part of (8.38) for $|\mathbf{n}| = 0$.

Step 2: $|\mathbf{n}| > 0$ – Estimates with Spatially Smooth Processes.

The problem for $|\mathbf{n}| > 0$ is that a generalization of Lemma 11.10 to integral operators from (11.46) to

$$(\bar{K}_\mu f)(r) := (\bar{K}_{\mu,\overline{m},\bar{\mathbf{n}},\bar{\bar{\mathbf{n}}},D} f)(r)$$
$$:= \frac{\mu^{\overline{m}/2}}{(\frac{\overline{m}}{2}-1)!}\int_0^\infty \mathrm{d}u\, u^{\overline{m}/2-1}\mathrm{e}^{-\mu u}\int [\partial_r^{\bar{\mathbf{n}}}\partial^{\mathbf{n}}G(Du,r-q)]\pi_{\bar{\bar{\mathbf{n}}}}(r-q)f(q)\mathrm{d}q$$

does not yield good upper bounds, which could be used in estimating $a_{5,n,2,Y}(s)$ and $a_{3,\mathbf{n},Y}(s)$. On the other hand, if $Y(\cdot)$ were already smooth, we could proceed differently. Let us, therefore, replace $Y(\cdot)$ in $a_{5,n,2,Y}(s)$ and $a_{3,\mathbf{n},Y}(s)$ by

$$Y_\lambda(\cdot) := R_\lambda^{\overline{m}/2}Y(\cdot) \text{ and } Y_{\mu,\mathbf{n},\lambda}(s) := \partial^{\mathbf{n}}R_\mu^{\overline{m}/2}R_\lambda^{\overline{m}/2}Y(s) = Q_\mu^{\mathbf{n}}Y_\lambda(s), \tag{11.87}$$

where $\lambda > D\theta + 1$ (cf. 11.54). Then, from (11.65)

$$\left.\begin{aligned}
&a_{5,\mathbf{n},2,Y_\lambda}(s)\\
&= -\frac{p(p-1)}{2}\sum_{k,\ell=1}^d\int (Y_{\mu,\mathbf{n},\lambda}(s,r))^{p-2}\frac{\mu^{\overline{m}}}{[(\frac{\overline{m}}{2}-1)!]^2}\\
&\times\int_0^\infty\int_0^\infty \mathrm{d}u\,\mathrm{d}v\,\mathrm{e}^{-\mu(u+v)}u^{\overline{m}/2-1}v^{\overline{m}/2-1}\\
&\times\int\int \mathrm{d}q\,\mathrm{d}\tilde{q}\,\partial_{k,r}\partial_r^{\mathbf{n}}G(u,r-q)\partial_{\ell,r}\partial_r^{\mathbf{n}}G(v,r-\tilde{q})Y_\lambda(s,q)\\
&\times Y_\lambda(s,\tilde{q})\hat{D}_{k,\ell}(s,q-\tilde{q})\Phi(r)\mathrm{d}r.
\end{aligned}\right\} \tag{11.88}$$

As in Step 1 we first replace $\hat{D}_{k,\ell}(s,q-p)$ with a term from $I_{k\ell}(q,\tilde{q},r)$ in $a_{5,\mathbf{n},2,Y_\lambda}(s)$:

$$\left.\begin{aligned}
&-\frac{p(p-1)}{2}\sum_{k,\ell=1}^d\int (Y_{\mu,\mathbf{n},\lambda}(s,r))^{p-2}\frac{\mu^{\overline{m}}}{[(\frac{\overline{m}}{2}-1)!]^2}\\
&\times\int_0^\infty\int_0^\infty \mathrm{d}u\,\mathrm{d}v\,\mathrm{e}^{-\mu(u+v)}u^{\overline{m}/2-1}v^{\overline{m}/2-1}\\
&\times\int\int \mathrm{d}q\,\mathrm{d}\tilde{q}(\partial_{k,r}\partial_r^{\mathbf{n}}G(u,r-q))(\partial_{\ell,r}\partial_r^{\mathbf{n}}G(v,r-\tilde{q}))Y_\lambda(s,q)\\
&\times Y_\lambda(s,\tilde{q})\pi_{\tilde{\mathbf{n}}_j}(q-r)\pi_{\mathbf{n}_i}(r-\tilde{q})\Phi(r)\mathrm{d}r\\
&\times 1_{\{|\tilde{\mathbf{n}}_j|+|\mathbf{n}_i|\ge 2\}}\frac{1}{i!j!}\mathcal{L}_{k\ell,\mathbf{n}_i+\tilde{\mathbf{n}}_j}(s).
\end{aligned}\right\} \tag{11.89}$$

Again, we consider first the case $|\mathbf{n}_i| = 0$ and employ (11.70) with $Z := Y_\lambda(s)$. Integrating by parts in $a_{5,\mathbf{n},2,Y_\lambda}(s)$ with $\pi_{\tilde{\mathbf{n}}_j}(q-r)\frac{1}{j!}\mathcal{L}_{k\ell,\tilde{\mathbf{n}}_j}(s)$, replacing $\hat{D}_{k,\ell}(s,q-p)$, we obtain in place of (11.72)

$$
\left.
\begin{aligned}
&-\frac{p(p-1)}{2}\sum_{k,\ell=1}^{d}\int (Y_{\mu,\mathbf{n},\lambda}(s,r))^{p-2}\frac{\mu^{\overline{m}}}{[(\frac{\overline{m}}{2}-1)!]^2}\\
&\times\int_0^\infty\int_0^\infty \mathrm{d}u\,\mathrm{d}v\,\mathrm{e}^{-\mu(u+v)}u^{\overline{m}/2-1}v^{\overline{m}/2-1}\\
&\times\int\int \mathrm{d}q\,\mathrm{d}\tilde{q}(\partial_{k,r}\partial_r^{\mathbf{n}}G(u,r-q))(\partial_{\ell,r}\partial_r^{\mathbf{n}}G(v,r-\tilde{q}))Y_\lambda(s,q)\\
&\times Y_\lambda(s,\tilde{q})\pi_{\tilde{\mathbf{n}}_j}(q-r)\Phi(r)\mathrm{d}r\\
&=-\frac{p(p-1)}{2}\sum_{k,\ell=1}^{d}\int (Y_{\mu,\mathbf{n},\lambda}(s,r))^{p-2}\frac{\mu^{\overline{m}/2}}{(\frac{\overline{m}}{2}-1)!}\int_0^\infty \mathrm{d}u\,\mathrm{e}^{-\mu u}u^{\overline{m}/2-1}\\
&\times\int(\partial_{k,r}\partial_r^{\mathbf{n}}G(u,r-q))Y_\lambda(s,q)\pi_{\tilde{\mathbf{n}}_j}(q-r)(\partial_{\ell,r}Y_{\mu,\mathbf{n},\lambda})(s,r)\Phi(r)\mathrm{d}r\\
&=\frac{p}{2}\sum_{k,\ell=1}^{d}\int (Y_{\mu,\mathbf{n},\lambda}(s,r))^{p-1}\frac{\mu^{\overline{m}/2}}{(\frac{\overline{m}}{2}-1)!}\int_0^\infty \mathrm{d}u\,\mathrm{e}^{-\mu u}u^{\overline{m}/2-1}\\
&\times\int \mathrm{d}q(\partial^2_{k\ell,r}\partial_r^{\mathbf{n}}G(u,r-q))Y_\lambda(s,q)\pi_{\tilde{\mathbf{n}}_j}(q-r)\Phi(r)\mathrm{d}r\\
&+\frac{p}{2}\sum_{k,\ell=1}^{d}\int (Y_{\mu,\mathbf{n},\lambda}(s,r))^{p-1}\frac{\mu^{\overline{m}/2}}{(\frac{\overline{m}}{2}-1)!}\int_0^\infty \mathrm{d}u\,\mathrm{e}^{-\mu u}u^{\overline{m}/2-1}\\
&\times\int \mathrm{d}q(\partial_{k,r}\partial_r^{\mathbf{n}}G(u,r-q))Y_\lambda(s,q)(\partial_{\ell,r}\pi_{\tilde{\mathbf{n}}_j}(q-r))\Phi(r)\mathrm{d}r\\
&+\frac{p}{2}\sum_{k,\ell=1}^{d}\int (Y_{\mu,\mathbf{n},\lambda}(s,r))^{p-1}\frac{\mu^{\overline{m}/2}}{(\frac{\overline{m}}{2}-1)!}\int_0^\infty \mathrm{d}u\,\mathrm{e}^{-\mu u}u^{\overline{m}/2-1}\\
&\times\int \mathrm{d}q(\partial_{k,r}\partial_r^{\mathbf{n}}G(u,r-q))Y_\lambda(s,q)\pi_{\tilde{\mathbf{n}}_j}(q-r)\partial_{\ell,r}\Phi(r)\mathrm{d}r.
\end{aligned}
\right\}
\tag{11.90}
$$

Let us now estimate the first term on the right-hand side of (11.90). By the homogeneity of $G(u,r-q)$ we have

$$
\partial^2_{k\ell,r}\partial_r^{\mathbf{n}}G(u,r-q) = (-1)^{2+|\mathbf{n}|}\partial^2_{k\ell,q}\partial_q^{\mathbf{n}}G(u,r-q),
$$

whence, integrating by parts

$$
\begin{aligned}
&\int \mathrm{d}q (\partial^2_{k\ell,r}\partial^{\mathbf{n}}_r G(u, r-q)) Y_\lambda(s,q)\pi_{\tilde{\mathbf{n}}_j}(q-r)\\
&= \int \mathrm{d}q\, G(u, r-q)\partial^2_{k\ell,q}\partial^{\mathbf{n}}_q (Y_\lambda(s,q)\pi_{\tilde{\mathbf{n}}_j}(q-r))\\
&= \int \mathrm{d}q\, G(u, r-q) \sum_{\{\tilde{\mathbf{n}}\le \mathbf{n}+1_k+1_\ell\}} \gamma_{\tilde{\mathbf{n}}}(\partial^{\tilde{\mathbf{n}}}_q Y_\lambda(s,q))(\partial_q^{\mathbf{n}+1_k+1_\ell-\tilde{\mathbf{n}}}\pi_{\tilde{\mathbf{n}}_j}(q-r))
\end{aligned}
$$

$$
\begin{aligned}
&= \int \mathrm{d}q\, G(u, r-q) \sum_{\{\tilde{\mathbf{n}}\le \mathbf{n}+1_k+1_\ell, |\tilde{\mathbf{n}}|\le|\mathbf{n}|\}} \gamma_{\tilde{\mathbf{n}}}(\partial^{\tilde{\mathbf{n}}}_q Y_\lambda(s,q))(\partial_q^{\mathbf{n}+1_k+1_\ell-\tilde{\mathbf{n}}}\pi_{\tilde{\mathbf{n}}_j}(q-r))\\
&+ \int \mathrm{d}q\, G(u, r-q) \sum_{\{\tilde{\mathbf{n}}\le \mathbf{n}+1_k+1_\ell, |\tilde{\mathbf{n}}|>|\mathbf{n}|\}} \gamma_{\tilde{\mathbf{n}}}(\partial^{\tilde{\mathbf{n}}}_q Y_\lambda(s,q))(\partial_q^{\mathbf{n}+1_k+1_\ell-\tilde{\mathbf{n}}}\pi_{\tilde{\mathbf{n}}_j}(q-r))\\
&=: I + II
\end{aligned}
$$

In I the order of derivative at Y_λ is $\le |\mathbf{n}|$ which need not be changed. In II we simplify the notation in the terms to

$$
(\partial^{\tilde{\mathbf{n}}}_q Y_\lambda(s,q))G(u, r-q)\partial^{\mathbf{i}}_q \pi_{\tilde{\mathbf{n}}_j}(q-r),
$$

where $\mathbf{i} := \mathbf{n} + 1_k + 1_\ell - \tilde{\mathbf{n}}$. In this case $|\tilde{\mathbf{n}}|$ equals either $|\mathbf{n}| + 1$ or $|\mathbf{n}| + 2$. In the first case, $|i| = 1$, in the latter case $|i| = 0$. We again integrate by parts once or twice against the product $G(u, r-q)\partial^{\mathbf{i}}_q \pi_{\tilde{\mathbf{n}}_j}(q-r)$ to reduce the order to $|\mathbf{n}|$. The resulting expression in the product $G(u, r-q)\partial^{\mathbf{i}}_q \pi_{\tilde{\mathbf{n}}_j}(q-r)$ is a linear combination of terms of the following kind

$$
(\partial^{\tilde{\mathbf{j}}}_q G(u, r-q))(\partial^{\hat{\mathbf{j}}}_q \partial^{\mathbf{i}}_q \pi_{\tilde{\mathbf{n}}_j}(q-r)), \quad |\tilde{\mathbf{j}}| = 2 - |\mathbf{i}| - |\hat{\mathbf{j}}|.
$$

Hence, the order in the derivative at $G(u, r-q)$ is the same as the degree of the resulting polynomial $(\partial^{\hat{\mathbf{j}}}_q \partial^{\mathbf{i}}_q \pi_{\tilde{\mathbf{n}}_j}(q-r))$. Thus,

$$
\left.
\begin{aligned}
&\int \mathrm{d}q (\partial^2_{k\ell,r}\partial^{\mathbf{n}}_r G(u, r-q)) Y_\lambda(s,q)\pi_{\tilde{\mathbf{n}}_j}(q-r)\\
&= \int \mathrm{d}q \sum_{\{|\tilde{\mathbf{j}}|\le 2-|\mathbf{i}|-|\hat{\mathbf{j}}|\}} \sum_{\{\tilde{\mathbf{n}}+\mathbf{i}=\mathbf{n}+1_k+1_\ell, |\tilde{\mathbf{n}}|\le|\mathbf{n}|\}} \gamma_{\tilde{\mathbf{n}},\tilde{\mathbf{j}},\hat{\mathbf{j}}}\\
&\times (\partial^{\tilde{\mathbf{j}}}_q G(u, r-q))(\partial^{\hat{\mathbf{j}}}_q \partial^{\mathbf{i}}_q \pi_{\tilde{\mathbf{n}}_j}(q-r))(\partial^{\tilde{\mathbf{n}}}_q Y_\lambda(s,q)).
\end{aligned}
\right\} \tag{11.91}
$$

Employing Lemma 11.10,

$$
\left.
\begin{aligned}
&\left\{\left|\int_0^\infty \mathrm{d}u\, \mathrm{e}^{-\mu u} u^{\overline{m}/2-1} \int \mathrm{d}q (\partial^2_{k\ell,r}\partial^{\mathbf{n}}_r G(u, r-q)) Y_\lambda(s,q)\pi_{\tilde{\mathbf{n}}_j}(q-r)\right|^p \Phi(r)\mathrm{d}r\right\}^{1/2}\\
&\le c_{\beta,\mathbf{n},\Phi,\overline{m}} \sum_{\{|\tilde{\mathbf{n}}|\le|\mathbf{n}|\}} \left\| R_\mu^{\overline{m}/2}\partial^{\tilde{\mathbf{n}}}_q Y_\lambda(s)\right\|_{0,p,\Phi}.
\end{aligned}
\right\} \tag{11.92}
$$

As a result of (11.91) and (11.92) we can estimate the first term on the right-hand side of (11.90) by Hölder's inequality and obtain

$$
\left.\begin{array}{l}
\left|\frac{p}{2}\sum_{k,\ell=1}^{d}\int (Y_{\mu,\mathbf{n},\lambda}(s,r))^{p-1}\frac{\mu^{\overline{m}/2}}{(\frac{\overline{m}}{2}-1)!}\int_0^\infty \mathrm{d}u\, \mathrm{e}^{-\mu u}u^{\overline{m}/2-1}\right.\\
\left.\times\int \mathrm{d}q(\partial^2_{k\ell,r}\partial^{\mathbf{n}}_r G(u,r-q))Y_\lambda(s,q)\pi_{\tilde{\mathbf{n}}_j}(q-r)\Phi(r)\mathrm{d}r\right|\\
\le c_{\beta,\mathbf{n},d}\frac{p}{2}\left\|Y_{\mu,\mathbf{n},\lambda}(s)\right\|_{0,p,\Phi}^{p-1/p}\sum_{\{|\tilde{\mathbf{n}}|\le|\mathbf{n}|\}}\left\|R_\mu^{\overline{m}/2}\partial_q^{\tilde{\mathbf{n}}}Y_\lambda(s)\right\|_{0,p,\Phi}=c_{\beta,\mathbf{n},d}\frac{p}{2}\\
\times\left\|Y_{\mu,\mathbf{n},\lambda}(s)\right\|_{0,p,\Phi}^{p-1/p}\left\|R_\mu^{\overline{m}/2}Y_\lambda(s)\right\|_{n,p,\Phi}\\
\le c_{\beta,\mathbf{n},d}\frac{p}{2}\left\|R_\mu^{\overline{m}/2}Y_\lambda(s)\right\|_{n,p,\Phi}^{p}.
\end{array}\right\}\quad (11.93)
$$

We obtain similar bounds by the same arguments (in addition to (15.46)) for the second and third terms on the right-hand side of (11.90). For $|\mathbf{n}_i| > 0$ we do not integrate by parts and proceed for the $\mathrm{d}q$- and $\mathrm{d}\tilde{q}$ integrals as in the arguments leading to (11.91). A simple adjustment of the arguments leading from (11.91) to (11.93) yields the same bounds as in (11.93).

The term $III_{s,k\ell}(q,\tilde{q},r)$ can be directly estimated as in Step 1, since for those estimates the derivatives at $G(u,r-q)$ are compensated by the correspondingly high order of the polynomials $\pi_{\mathbf{n}_{m+2}}(q-r)$. The bounds for $a_{3,\mathbf{n},Y}(s)$ can be obtained in the same way. Finally, we obtain, as in (11.66)

$$a_{2,\mathbf{n},1,Y_\lambda}(s)+a_{5,\mathbf{n},1,Y_\lambda(s)}\equiv 0. \quad (11.94)$$

Employing (11.94), we obtain altogether the estimate, corresponding to (11.84),

$$
\begin{aligned}
&(|a_{2,\mathbf{n},Y_\lambda}(s)+a_{5,\mathbf{n},Y_\lambda}(s)|+|a_{3,\mathbf{n},Y_\lambda}(s)|)^{\bar{n}}\le \hat{c}_{p,\Phi,\overline{m},\bar{n}}(\|R_\mu^{\overline{m}/2}Y_\lambda(s)\|_{n,p,\Phi}^{\bar{n}p}\\
&+\|R_\mu^{\overline{m}/2}Y_\lambda(s)\|_{0,p,\Phi}^{\bar{n}p})\quad \forall \bar{n}\in\mathbf{N},
\end{aligned}\quad (11.95)
$$

where the estimate for $\bar{n} > 1$ is obtained by taking the $\bar{n}$th power on both sides of the inequality for the case $\bar{n}=1$. Note that, by (11.25),

$$\|R_\mu^{\overline{m}/2}Y_\lambda(s)\|_{0,p,\Phi}^p=\|R_\mu^{\frac{\overline{m}}{2}}R_\lambda^{\overline{m}/2}Y(s)\|_{0,p,\Phi}^p\le c_\Phi\|R_\mu^{\overline{m}/2}Y(s)\|_{0,p,\Phi}^p.$$

Taking mathematical expectations, this last inequality in addition to (11.86) implies $\forall\bar{n}\in\mathbf{N}$:

$$
\left.\begin{array}{rl}
E[|a_{2,\mathbf{n},Y_\lambda}(s)+a_{5,\mathbf{n},Y_\lambda}(s)|+|a_{3,\mathbf{n},Y_\lambda}(s)|]^{\bar{n}} & \le \hat{c}_{p,\Phi,\overline{m},\bar{n}}(E\|R_\mu^{\frac{\overline{m}}{2}}Y_\lambda(s)\|_{n,p,\Phi}^{\bar{n}p} \\
& \quad +c_T E\|Y(0)\|_{0,p,\Phi}^{\bar{n}p}) \\
& \le \tilde{c}_{p,\Phi,\overline{m},\bar{n}}(E\|R_\mu^{\frac{\overline{m}}{2}}Y(s)\|_{n,p,\Phi}^{\bar{n}p} \\
& \quad +c_T E\|Y(0)\|_{0,p,\Phi}^{\bar{n}p}) \\
& (\text{since} R_\mu^{\frac{\overline{m}}{2}}Y_\lambda(s)=R_\lambda^{\frac{\overline{m}}{2}}R_\mu^{\frac{\overline{m}}{2}}Y(s).
\end{array}\right\} \tag{11.96}
$$

Step 3: $|\mathbf{n}|\ge 0$ – **Extension of the Estimates.**

We need to show that $E\sum\limits_{i\in\{2,3,5\}}\int_0^t(a_{i,\mathbf{n},Y}(s)-a_{i,\mathbf{n},Y_\lambda}(s))\mathrm{d}s\longrightarrow 0$, as $\lambda\longrightarrow\infty$ and apply the Gronwall Lemma. The first observation is that, by (11.66) and (11.94),

$$
\sum_{i\in\{2,5\}}(a_{i,\mathbf{n},Y}(s)-a_{i,\mathbf{n},Y_\lambda}(s))=a_{5,\mathbf{n},2,Y}(s)-a_{5,\mathbf{n},2,Y_\lambda}(s). \tag{11.97}
$$

Recalling (11.65), we set

$$
\left.\begin{array}{l}
\Psi_2(g,f,f,s,k,\ell) \\
:=\displaystyle\int g(s,r)\frac{\mu^{\overline{m}}}{[(\frac{\overline{m}}{2}-1)!]^2}\int_0^\infty\int_0^\infty \mathrm{d}u\,\mathrm{d}v\,\mathrm{e}^{-\mu(u+v)}u^{\overline{m}/2-1}v^{\overline{m}/2-1} \\
\times\displaystyle\int\int \mathrm{d}q\mathrm{d}\tilde{q}\partial_{k,r}\partial_r^{\mathbf{n}}G(u,r-q)\partial_{\ell,r}\partial_r^{\mathbf{n}}G(v,r-\tilde{q}) \\
\times f(s,q)f(s,\tilde{q})\hat{D}_{k,\ell}(s,q-\tilde{q})\Phi(r)\mathrm{d}r,
\end{array}\right\} \tag{11.98}
$$

where g and f are suitably chosen integrable functions. Observe in what follows that, based on Step 1, we may assume that $Y(s)\in\mathbf{W}_{0,p,\Phi}$. Then,

$$
\left.\begin{array}{l}
\Psi_2(Y_{\mu,\mathbf{n}}^{p-2},Y,Y,s,k,\ell)-\Psi_2(Y_{\mu,\mathbf{n},\lambda}^{p-2},Y_\lambda,Y_\lambda,s,k,\ell) \\
=\Psi_2(Y_{\mu,\mathbf{n}}^{p-2}-Y_{\mu,\mathbf{n},\lambda}^{p-2},Y,Y,s,k,\ell)+\Psi_2(Y_{\mu,\mathbf{n},\lambda}^{p-2}, \\
\quad\times Y-Y_\lambda,Y,s,k,\ell)|+\Psi_2(Y_{\mu,\mathbf{n},\lambda}^{p-2},Y_\lambda,Y-Y_\lambda,s,k,\ell).
\end{array}\right\} \tag{11.99}
$$

We first estimate $\Psi_2(Y^{p-2}_{\mu,\mathbf{n}} - Y^{p-2}_{\mu,\mathbf{n},\lambda}, Y, Y, s, k, \ell)$, assuming $p \geq 4$, as for $p = 2$ the term equals 0 and for $p = 3$ the term can be estimated as in the case $p \geq 4$ in what follows

$$\left.\begin{aligned}
&\sup_{0\leq s\leq T} \|Y^{p-2}_{\mu,\mathbf{n}} - Y^{p-2}_{\mu,\mathbf{n},\lambda}\|_{0,\frac{p}{p-2},\Phi} \leq \|(R^{\overline{m}/2}_{\lambda} - I)Y_{\mu,\mathbf{n}}|_{0,\frac{p}{(p-2)},\Phi} \\
&\times \Big|\sum_{i=0}^{p-3}(Y^{i}_{\mu,\mathbf{n}} + Y^{p-3-i}_{\mu,\mathbf{n},\lambda})\|_{0,(p/p-2),\Phi} \\
&\leq c(\omega, \Phi, p)\|(R^{\overline{m}/2}_{\lambda} - I)Y_{\mu,\mathbf{n}}\Big|_{0,(p/p-2),\Phi} \text{ (by (11.60))} \\
&\longrightarrow 0, \text{ as } \lambda \longrightarrow \infty \text{ by (11.26) for any } T > 0.
\end{aligned}\right\} \quad (11.100)$$

Similarly for $\Psi_2(Y^{p-2}_{\mu,\mathbf{n},\lambda}, Y - Y_\lambda, Y, s, k, \ell) = \Psi_2(Y^{p-2}_{\mu,\mathbf{n},\lambda}, Y_\lambda, Y - Y_\lambda, s, k, \ell)$. Applying Hölder's inequality (cf. (11.60) for the integrability condition), it follows that for any $T > 0$

$$\sup_{0\leq s\leq T} E|a_{5,\mathbf{n},2,Y}(s) - a_{5,\mathbf{n},2,Y_\lambda}(s)|^{\bar{n}} \longrightarrow 0, \text{ as } \lambda \longrightarrow \infty \ \forall \bar{n} \in \mathbf{N}\,. \quad (11.101)$$

By similar arguments we can show that

$$\sup_{0\leq s\leq T} E|a_{3\mathbf{n},Y}(s) - a_{3,\mathbf{n},Y_\lambda}(s)|^{\bar{n}} \longrightarrow 0, \text{ as } \lambda \longrightarrow \infty \ \forall \bar{n} \in \mathbf{N}\,. \quad (11.102)$$

Altogether, for any $T > 0$

$$\Psi^{\bar{n}}_{\lambda}(T) := \sup_{0\leq s\leq T} E\Big|\sum_{i\in\{2,5\}}(a_{i,\mathbf{n},Y}(s) - a_{i,\mathbf{n},Y_\lambda}(s))\Big|^{\bar{n}} \longrightarrow 0, \text{ as } \lambda \longrightarrow \infty \ \forall \bar{n} \in \mathbf{N}\,. \quad (11.103)$$

(11.103) in addition to (11.95) implies

$$\left.\begin{aligned}
&E[|a_{2,\mathbf{n},Y}(s) + a_{5,\mathbf{n},Y}(s)| + |a_{3,\mathbf{n},Y}(s)|]^{\bar{n}} \\
&\leq \tilde{c}_{p,\Phi,\overline{m}}\Big(E\big\|R^{\overline{m}/2}_{\mu}Y(s)\big\|^{\bar{n}p}_{n,p,\Phi} + c_T E|Y(0)\big\|^{\bar{n}p}_{0,p,\Phi}\Big) + \Psi^{\bar{n}}_{\lambda}(T), \\
&\text{where } \Psi_\lambda(T) \longrightarrow 0, \text{ as } \lambda \longrightarrow \infty \ \forall \bar{n} \in \mathbf{N}.
\end{aligned}\right\} \quad (11.104)$$

Hence, employing the previous calculations for $\bar{n} = 1$, we obtain the following bound for the left-hand side of (11.61)

$$
\left.\begin{aligned}
E\|Y_{\mu,\mathbf{n}}(t)\|_{0,p,\Phi}^{p} \le E\|Y_{\mu,\mathbf{n}}(0)\|_{0,p,\Phi}^{p} + c_T E\|Y(0)\|_{0,p,\Phi}^{p} + T\Psi_{\lambda}(T)\\
+\int_0^t \tilde{c}_{p,\Phi,\overline{m}}(\|R_{\mu}^{\overline{m}/2}Y(s)\|_{n,p,\Phi}^{p}\mathrm{d}s,
\end{aligned}\right\} \tag{11.105}
$$

where $|\mathbf{n}| \le m$. Note that

$$
\|Y_{\mu,\tilde{\mathbf{n}}}(t)\|_{0,p,\Phi}^{p} \le \|R_{\mu}^{\overline{m}/2}Y(s)\|_{|\tilde{\mathbf{n}}|,p,\Phi}^{p} \le \|R_{\mu}^{\overline{m}/2}Y(s)\|_{n,p,\Phi}^{p} \quad \forall \tilde{\mathbf{n}} \le \mathbf{n},
$$

$$
\sum_{\tilde{\mathbf{n}} \le \mathbf{n}} \|Y_{\mu,\tilde{\mathbf{n}}}(t)\|_{0,p,\Phi}^{p} = \|R_{\mu}^{\overline{m}/2}Y(s)\|_{n,p,\Phi}^{p}.
$$

As a result, we obtain an "a priori" estimate for the left-hand side of (11.61)

$$
\begin{aligned}
E\|R_{\mu}^{\overline{m}/2}Y(t)\|_{n,p,\Phi}^{p} \le (c_T+1)(E\|Y_{\mu,\mathbf{n}}(0)\|_{0,p,\Phi}^{p} + E\|Y(0)\|_{0,p,\Phi}^{p}) + T\Psi_{\lambda}(T)\\
+\int_0^t \bar{c}_{p,\Phi,\overline{m}}(\|R_{\mu}^{\overline{m}/2}Y(s)\|_{n,p,\Phi}^{p}\mathrm{d}s.
\end{aligned} \tag{11.106}
$$

Employing Gronwall's inequality (cf. Sect. 15.1.2, Proposition 15.5), (11.106) and a simple calculation imply

$$
\begin{aligned}
\sup_{0\le t\le T} E\|R_{\mu}^{\overline{m}/2}Y(t)\|_{n,p,\Phi}^{p} \le c_{p,\Phi,\overline{m},T}(E\|Y_{\mu,\mathbf{n}}(0)\|_{0,p,\Phi}^{p}\\
+E\|Y(0)\|_{0,p,\Phi}^{p} + \Psi_{\lambda}(T)) \ \forall \lambda \ge D\theta + 1.
\end{aligned} \tag{11.107}
$$

Since $\Psi_{\lambda}(T) \longrightarrow 0$, as $\lambda \longrightarrow \infty$ we obtain the generalization of (11.86) to $|\mathbf{n}| \ge 0$

$$
\sup_{0\le t\le T} E\|R_{\mu}^{\overline{m}/2}Y(t)\|_{n,p,\Phi}^{p} \le c_{p,\Phi,\overline{m},T} E\|Y(0)\|_{n,p,\Phi}^{p}. \tag{11.108}
$$

Hence, again applying Fatou's lemma, as $\mu \to \infty$, implies (8.39) for $0 \le |\mathbf{n}| \le m$. The statement (8.38) is a simple consequence of the fact that $\sup_{0\le t\le T} E\|f(t)\|_{m,p,\Phi}^{p}$ is stronger than $\int_0^T E\|f(t)\|_{m,p,\Phi}^{p}\mathrm{d}t$ for suitable $f(\cdot)$ and the $\mathrm{d}P \otimes \mathrm{d}t$ measurability of $R_{\mu}^{\overline{m}/2}Y(\cdot)$ as an $\mathbf{W}_{m,p,\Phi}$-valued process. This finishes the proof of the integrability part of (8.38) for the case $\bar{n} = 1$.

Step 4: Case $\bar{n} \ge 2$.

Employing (11.61), the Itô formula yields

$$
\left.
\begin{aligned}
\|Y_{\mu,\mathbf{n}}(t)\|_{0,p,\Phi}^{\bar{n}p} &= \|Y_{\mu,\mathbf{n}}(0)\|_{0,p,\Phi}^{\bar{n}p} \\
&\quad + \frac{\bar{n}p}{2}\sum_{k,\ell=1}^{d}\int_0^t \|Y_{\mu,\mathbf{n}}(s)\|_{0,p,\Phi}^{(\bar{n}-1)p}\langle Y_{\mu,\mathbf{n}}^{p-1}(s), \partial_{k\ell}^2 Y_{\mu,\mathbf{n}}(s)\rangle_{0,\Phi} D_{k\ell}(s)\mathrm{d}s \\
&\quad -\bar{n}p\int_0^t \|Y_{\mu,\mathbf{n}}(s)\|_{0,p,\Phi}^{(\bar{n}-1)p}\langle Y_{\mu,\mathbf{n}}^{p-1}(s), \nabla\cdot Q_\mu^{\mathbf{n}}(Y(s)F(s))\rangle_{0,\Phi}\mathrm{d}s \\
&\quad -\bar{n}p\int_0^t \|Y_{\mu,\mathbf{n}}(s)\|_{0,p,\Phi}^{(\bar{n}-1)p}\langle Y_{\mu,\mathbf{n}}^{p-1}(s), \nabla\cdot Q_\mu^{\mathbf{n}}(Y(s)\mathrm{d}m(s))\rangle_{0,\Phi} \\
&\quad +\frac{\bar{n}p(p-1)}{2}\int_0^t \|Y_{\mu,\mathbf{n}}(s)\|_{0,p,\Phi}^{(\bar{n}-1)p}\langle Y_{\mu,\mathbf{n}}^{p-2}(s), [\nabla\cdot Q_\mu^{\mathbf{n}}(Y(s)\mathrm{d}m(s))]\rangle_{0,\Phi} \\
&\quad +\frac{\bar{n}(\bar{n}-1)p^2}{2}\int_0^t \|Y_{\mu,\mathbf{n}}(s)\|_{0,p,\Phi}^{(\bar{n}-2)p}[\langle Y_{\mu,\mathbf{n}}^{p-1}(s), \nabla\cdot Q_\mu^{\mathbf{n}}(Y(s)\mathrm{d}m(s))\rangle_{0,\Phi}] \\
&=: \sum_{i=1}^{6} A_{i,\bar{n},\mathbf{n},Y}(t) \\
&= \|Y_{\mu,\mathbf{n}}(0)\|_{0,p,\Phi}^{\bar{n}p} + \sum_{i\in\{2,3,5,6\}}\int_0^t \|Y_{\mu,\mathbf{n}}(s)\|_{0,p,\Phi}^{(\bar{n}-1)p} a_{i,\mathbf{n},Y}(s)\mathrm{d}s \\
&\quad + A_{4,\bar{n},\mathbf{n},Y}(t) + A_{6,\bar{n},\mathbf{n},Y}(t),
\end{aligned}
\right\}
\tag{11.109}
$$

where we used the notation for the integrands in the deterministic integrals of (11.61). By Hölder's inequality in addition to (11.104)

$$
\begin{aligned}
&E\|Y_{\mu,\mathbf{n}}(s)\|_{0,p,\Phi}^{(\bar{n}-1)p}\Big|\sum_{i\in\{2,3,5\}} a_{i,\mathbf{n},Y}(s)\Big| \\
&\le \Big\{E\|Y_{\mu,\mathbf{n}}(s)\|_{0,p,\Phi}^{\bar{n}p}\Big\}^{(\bar{n}-1)/\bar{n}}\Big\{E\Big|\sum_{i\in\{2,3,5\}} a_{i,\mathbf{n},Y}(s)\Big|^{\bar{n}}\Big\}^{1/\bar{n}} \\
&\le \Big\{E\|Y_{\mu,\mathbf{n}}(s)\|_{0,p,\Phi}^{\bar{n}p}\Big\}^{(\bar{n}-1)/\bar{n}}\Big\{\tilde{c}_{p,\Phi,\overline{m}}\Big(E\|R_\mu^{\overline{m}/2}Y(s)\|_{n,p,\Phi}^{\bar{n}p} + c_T E|Y(0)\|_{0,p,\Phi}^{\bar{n}p}\Big) \\
&\quad + \Psi_\lambda^{\bar{n}}(T)\Big\}^{1/\bar{n}}.
\end{aligned}
$$

Employing inequality (15.2) to the last term with $\eta := \bar{n}/(\bar{n}-1)$ and $\bar{\eta} := \bar{n}$ (instead of p and $\bar{p}$) and recalling $\|R_\mu^{\overline{m}/2}Y(s)\|_{n,p,\Phi} = \|Y_{\mu,\mathbf{n}}(s)\|_{0,p,\Phi}$, we obtain

$$\left.\begin{aligned}
&E\|Y_{\mu,\mathbf{n}}(s)\|_{0,p,\Phi}^{(\bar{n}-1)p}\left|\sum_{i\in\{2,3,5\}} a_{i,\mathbf{n},Y}(s)\right| \\
&\quad\le c_{p,\Phi,\overline{m},\bar{n}}\left(E\|Y_{\mu,\mathbf{n}}(s)\|_{0,p,\Phi}^{\bar{n}p}+E\|R_\mu^{\overline{m}/2}Y(s)\|_{n,p,\Phi}^{\bar{n}p}\right)+c_{\bar{n},T}E\|Y(0)\|_{0,p,\Phi}^{\bar{n}p} \\
&\qquad+\tfrac{1}{\bar{n}}\Psi_\lambda^{\bar{n}}(T) \\
&\quad\le \bar{c}_{p,\Phi,\overline{m},\bar{n}}E\|R_\mu^{\overline{m}/2}Y(s)\|_{n,p,\Phi}^{\bar{n}p}+c_{\bar{n},T}E\|Y(0)\|_{0,p,\Phi}^{\bar{n}p}+\tfrac{1}{\bar{n}}\Psi_\lambda^{\bar{n}}(T).
\end{aligned}\right\}\tag{11.110}$$

Setting

$$\bar{a}_{6,\mathbf{n}.Y}(ds):=\left[\langle Y_{\mu,\mathbf{n}}^{p-1}(s),\nabla\cdot Q_\mu^{\mathbf{n}}(Y(s)\mathrm{d}m(s))\rangle_{0,\Phi}\right],$$

we have

$$\left.\begin{aligned}
&\bar{a}_{6,\mathbf{n},Y}(ds) \\
&=\sum_{k,\ell=1}^d\int\int \mathrm{d}r\,\mathrm{d}\tilde{r}\,\Phi(r)\Phi(\tilde{r})(Y_{\mu,\mathbf{n}}(s,r))^{p-1} \\
&\quad\times(Y_{\mu,\mathbf{n}}(s,\tilde{r}))^{p-1}\frac{\mu^{\overline{m}}}{[(\frac{\overline{m}}{2}-1)!]^2}\int_0^\infty\int_0^\infty \mathrm{d}u\,\mathrm{d}v\,\mathrm{e}^{-\mu(u+v)}u^{(\overline{m}/2)-1}v^{(\overline{m}/2)-1} \\
&\quad\times\int\int\partial_{k,r}\partial_r^{\mathbf{n}}G(u,r-q)\partial_{\ell,\tilde{r}}\partial_{\tilde{r}}^{\mathbf{n}}G(v,r-\tilde{q})Y(s,q) \\
&\quad\times Y(s,\tilde{q})(\tilde{D}_{k,\ell}(s,0)-\hat{D}_{k,\ell}(s,q-\tilde{q}))\mathrm{d}q\,\mathrm{d}\tilde{q}\,\mathrm{d}s \\
&=\sum_{k,\ell=1}^d\int\int \mathrm{d}r\,\mathrm{d}\tilde{r}\,\Phi(r)\Phi(\tilde{r})(Y_{\mu,\mathbf{n}}(s,r))^{p-1} \\
&\quad\times(Y_{\mu,\mathbf{n}}(s,\tilde{r}))^{p-1}\frac{\mu^{\overline{m}}}{[(\frac{\overline{m}}{2}-1)!]^2}\int_0^\infty\int_0^\infty \mathrm{d}u\,\mathrm{d}v\,\mathrm{e}^{-\mu(u+v)}u^{(\overline{m}/2)-1}v^{(\overline{m}/2)-1} \\
&\quad\times\int\int\partial_{k,r}\partial_r^{\mathbf{n}}G(u,r-q)\partial_{\ell,\tilde{r}}\partial_{\tilde{r}}^{\mathbf{n}}G(v,r-\tilde{q})Y(s,q)Y(s,\tilde{q})\tilde{D}_{k,\ell}(s,0)\mathrm{d}q\,\mathrm{d}\tilde{q}\mathrm{d}s \\
&\quad-\sum_{k,\ell=1}^d\int\int \mathrm{d}r\,\mathrm{d}\tilde{r}\,\Phi(r)\Phi(\tilde{r})(Y_{\mu,\mathbf{n}}(s,r))^{p-1} \\
&\quad\times(Y_{\mu,\mathbf{n}}(s,\tilde{r}))^{p-1}\frac{\mu^{\overline{m}}}{[(\frac{\overline{m}}{2}-1)!]^2}\int_0^\infty\int_0^\infty \mathrm{d}u\,\mathrm{d}v\,\mathrm{e}^{-\mu(u+v)}u^{(\overline{m}/2)-1}v^{(\overline{m}/2)-1} \\
&\quad\times\int\int\partial_{k,r}\partial_r^{\mathbf{n}}G(u,r-q)\partial_{\ell,\tilde{r}}\partial_{\tilde{r}}^{\mathbf{n}}G(v,r-\tilde{q})Y(s,q)Y(s,\tilde{q})\hat{D}_{k,\ell}(s,q-\tilde{q})\mathrm{d}q\,\mathrm{d}\tilde{q}\,\mathrm{d}s \\
&=:a_{6,\mathbf{n},1,Y}(s)\mathrm{d}s+a_{6,\mathbf{n},2,Y}(s)\mathrm{d}s=:a_{6,\mathbf{n},Y}(s)\mathrm{d}s.
\end{aligned}\right\}\tag{11.111}$$

Then,

$$A_{6,\bar{n},\mathbf{n},Y}(t) = \frac{\bar{n}(\bar{n}-1)p^2}{2}\int_0^t \|Y_{\mu,\mathbf{n}}(s)\|_{0,p,\Phi}^{(\bar{n}-2)p} a_{6,\mathbf{n},Y}(s)(\mathrm{d}s).$$

We have

$$\left.\begin{aligned}
&|a_{6,\mathbf{n},1,Y}(s)\mathrm{d}s| \\
&= \Bigg| \sum_{k,\ell=1}^{d} \int\int \mathrm{d}r\,\mathrm{d}\tilde{r}\,\Phi(r)\Phi(\tilde{r})(Y_{\mu,\mathbf{n}}(s,r))^{p-1} \\
&\quad \times (Y_{\mu,\mathbf{n}}(s,\tilde{r}))^{p-1}\frac{\mu^{\overline{m}}}{[(\frac{\overline{m}}{2}-1)!]^2}\int_0^\infty\int_0^\infty \mathrm{d}u\,\mathrm{d}v\,\mathrm{e}^{-\mu(u+v)}u^{(\overline{m}/2)-1}v^{(\overline{m}/2)-1} \\
&\quad \times \int\int \partial_{k,r}\partial_r^{\mathbf{n}}G(u,r-q)\partial_{\ell,\tilde{r}}\partial_{\tilde{r}}^{\mathbf{n}}G(v,r-\tilde{q})Y(s,q)Y(s,\tilde{q})\tilde{D}_{k,\ell}(s,0)\mathrm{d}q\,\mathrm{d}\tilde{q}\Bigg| \\
&\le c_D\left(\sum_{k=1}^{d}\int \mathrm{d}r\;\Phi(r)(Y_{\mu,\mathbf{n}}(s,r))^{p-1}(\partial_k Y_{\mu,\mathbf{n}}(s,r))\right)^2 \\
&\qquad \text{(by the uniform boundedness of } \tilde{D}_{k,\ell}(s,0) \text{ in all arguments)} \\
&= c_D\left(\frac{1}{p}\sum_{k=1}^{d}\int \mathrm{d}r\;\Phi(r)\partial_k((Y_{\mu,\mathbf{n}}(s,r))^p)\right)^2 \\
&\le \frac{\bar{c}_{D,\Phi,d}}{p^2}\left(\int \mathrm{d}r\;\Phi(r)(Y_{\mu,\mathbf{n}}(s,r))^p\right)^2 \\
&\qquad \text{(integrating by parts, if } \Phi(r)=\varpi(r) \text{ and employing (15.46))} \\
&= \frac{\bar{c}_{D,\Phi,d}}{p^2}\|Y_{\mu,\mathbf{n}}(s)\|_{0,p,\Phi}^{2p} \le \frac{\bar{c}_{D,\Phi,d}}{p^2}\|R_\mu^{\overline{m}/2}Y_\lambda(s)\|_{n,p,\Phi}^{2p}.
\end{aligned}\right\} \tag{11.112}$$

Next, we employ the representation (11.68) to find bounds for $a_{6,\mathbf{n},2,Y}(s)$. We can argue as in the estimate of $a_{5,\mathbf{n},2,Y}(s)$. By a simple repeat of those arguments, we first establish the analogue of (11.95)

$$\left.\begin{aligned}
&|a_{6,\mathbf{n},2,Y_\lambda}(s)ds|^{\bar{n}/2} \\
&\le c_{p,\Phi,\overline{m},\bar{n}}\left(\|R_\mu^{(\overline{m}/2)}Y_\lambda(s)\|_{n,p,\Phi}^{\bar{n}p} + \|R_\mu^{(\overline{m}/2)}Y_\lambda(s)\|_{0,p,\Phi}^{\bar{n}p}\right) \quad \forall \bar{n}\in\mathbf{N}
\end{aligned}\right\} \tag{11.113}$$

as well as the analogue of (11.96)

$$\left.\begin{aligned}
E|a_{6,\mathbf{n},2,Y_\lambda}(s)\mathrm{d}s|^{\frac{\bar{n}}{2}} &\le \hat{c}_{p,\Phi,\overline{m},\bar{n}}\left(E\|R_\mu^{\overline{m}/2}Y_\lambda(s)\|_{n,p,\Phi}^{\bar{n}2p} + c_T E\|Y(0)\|_{0,p,\Phi}^{\bar{n}p}\right) \\
&\le \tilde{c}_{p,\Phi,\overline{m},\bar{n}}\left(E\|R_\mu^{\overline{m}/2}Y(s)\|_{n,p,\Phi}^{\bar{n}p} + c_T E\|Y(0)\|_{0,p,\Phi}^{\bar{n}p}\right).
\end{aligned}\right\} \tag{11.114}$$

Further, the analogue of (11.101) holds:

$$\tilde{\Psi}_{\lambda}^{\bar{n}}(T) := \sup_{0\le s\le T} E|a_{6,\mathbf{n},2,Y}(s) - a_{6,\mathbf{n},2,Y_\lambda}(s)|^{\bar{n}/2} \longrightarrow 0, \text{ as } \lambda \longrightarrow \infty \ \forall \bar{n} \in \mathbf{N}\,. \tag{11.115}$$

Expressions (11.112) and (11.113) in addition to (11.115) imply the analogue of (11.104)

$$\left.\begin{aligned} &E|a_{6,\mathbf{n},Y}(s)^{\bar{n}/2} \\ &\le c_{p,\Phi,\overline{m},\bar{n}}\left(E\|R_\mu^{\overline{m}/2}Y(s)\|_{n,p,\Phi}^{\bar{n}p} + c_T E|Y(0)\|_{0,p,\Phi}^{\bar{n}p}\right) + \tilde{\Psi}_{\lambda}^{\bar{n}}(T), \\ &\text{where } \tilde{\Psi}_{\lambda}(T) \longrightarrow 0, \text{ as } \lambda \longrightarrow \infty \ \forall \bar{n} \in \mathbf{N}. \end{aligned}\right\} \tag{11.116}$$

By Hölder's inequality, in addition to (15.2), we obtain the analogue of (11.110)

$$\left.\begin{aligned} &E\|Y_{\mu,\mathbf{n}}(s)\|_{0,p,\Phi}^{(\bar{n}-2)p}|a_{6,\mathbf{n},Y}(s)| \\ &\le \left\{E\|Y_{\mu,\mathbf{n}}(s)\|_{0,p,\Phi}^{\bar{n}p}\right\}^{(\bar{n}-2)/\bar{n}}\left\{\tilde{c}_{p,\Phi,\overline{m}}\left(E\|R_\mu^{\overline{m}/2}Y(s)\|_{n,p,\Phi}^{\bar{n}p} + c_T E\|Y(0)\|_{0,p,\Phi}^{\bar{n}p}\right)\right. \\ &\qquad\qquad\qquad\qquad \left. + \tilde{\Psi}_{\lambda}^{\bar{n}}(T)\right\}^{2/\bar{n}} \\ &\le c_{p,\Phi,\overline{m},\bar{n}}\left(E\|Y_{\mu,\mathbf{n}}(s)\|_{0,p,\Phi}^{\bar{n}p} + E\|R_\mu^{\overline{m}/2}Y(s)\|_{n,p,\Phi}^{\bar{n}p}\right) + c_{\bar{n},T}E\|Y(0)\|_{0,p,\Phi}^{\bar{n}p} \\ &\quad + \tfrac{2}{\bar{n}}\tilde{\Psi}_{\lambda}^{\bar{n}}(T) \\ &\le \bar{c}_{p,\Phi,\overline{m},\bar{n}}E\|R_\mu^{\overline{m}/2}Y(s)\|_{n,p,\Phi}^{\bar{n}p} + c_{\bar{n},T}E\|Y(0)\|_{0,p,\Phi}^{\bar{n}p} + \tfrac{2}{\bar{n}}\tilde{\Psi}_{\lambda}^{\bar{n}}(T). \end{aligned}\right\} \tag{11.117}$$

Altogether, we obtain from (11.109), (11.110), (11.117), and $\sum \tilde{\mathbf{n}} \le \mathbf{n}\|Y_{\mu,\tilde{\mathbf{n}}}(s)\|_{0,p,\Phi} = \|R_\mu^{\overline{m}/2}Y(s)\|_{n,p,\Phi}$

$$\left.\begin{aligned} E\|R_\mu^{\overline{m}/2}Y(t)\|_{n,p,\Phi}^{\bar{n}p} &\le E\|R_\mu^{\overline{m}/2}Y(0)\|_{n,p,\Phi}^{\bar{n}p} \\ &\quad + \bar{c}_{p,\Phi,\overline{m},\bar{n},T}\left[\int_0^t E\|R_\mu^{\overline{m}/2}Y(s)\|_{n,p,\Phi}^{\bar{n}p}\mathrm{d}s + E\|Y(0)\|_{0,p,\Phi}^{\bar{n}p}\right] \\ &\quad + \bar{c}_{p,\Phi,\overline{m},\bar{n},T}\left[\Psi_{\lambda}^{\bar{n}}(T) + \tilde{\Psi}_{\lambda}^{\bar{n}}(T)\right]. \end{aligned}\right\} \tag{11.118}$$

Employing again Gronwall's inequality

$$\begin{aligned} &E\sup_{0\le t\le T}\|R_\mu^{\overline{m}/2}Y(t)\|_{n,p,\Phi}^{\bar{n}p} \le c_{p,\Phi,\overline{m},T,\bar{n}}(E\|R_\mu^{\overline{m}/2}Y(0)\|_{n,p,\Phi}^{\bar{n}p} \\ &+\Psi_{\lambda}^{\bar{n}}(T) + \tilde{\Psi}_{\lambda}^{\bar{n}}(T)) \ \forall \lambda \ge D\theta + 1. \end{aligned} \tag{11.119}$$

Since $\Psi_\lambda(T) + \tilde{\Psi}_\lambda(T) \longrightarrow 0$, as $\lambda \longrightarrow \infty$ we obtain the generalization of (11.108) to $\bar{n} \ge 2$:

$$\sup_{0\le t\le T} E\|R_\mu^{\overline{m}/2}Y(t)\|_{n,p,\Phi}^{\bar{n}p} \le c_{p,\Phi,\overline{m},T}E\|Y(0)\|_{n,p,\Phi}^{\bar{n}p}. \tag{11.120}$$

Hence, again applying Fatou's lemma, as $\mu \to \infty$, implies the integrability part of (8.38) for $0 \le |\mathbf{n}| \le m$ and $\bar{n} \ge 1$.

Step 5: Continuous Sample Paths
The solution is almost trivial. By the previous step $Y(\cdot)$ are in $L_{\bar{n}p,\mathcal{F}}((s,T] \times \Omega; \mathbf{W}_{m,p,\Phi})$. The first and second derivatives are bounded operators from $\mathbf{W}_{m,2,\Phi}$ into $\mathbf{W}_{m-1,2,\Phi}$ and $\mathbf{W}_{m-2,2,\Phi}$, respectively. The coefficients have bounded derivatives by (8.35). Hence, the stochastic Itô integral for the strong solution is a continuous square integrable martingale with values in $\mathbf{W}_{m-1,2,\Phi} \subset \mathbf{W}_{m-2,2,\Phi}$.[19] The deterministic (Bochner) integrals are obviously continuous with values in $\mathbf{W}_{m-1,2,\Phi}$ and $\mathbf{W}_{m-2,2,\Phi}$, respectively. □

11.3 Proof of the Itô formula (8.42)

We choose $\overline{m} = \overline{m}(0)$ in (11.43). Assume, without loss of generality, $s = 0$ and set for the solution of (8.25) $Y(t) := Y(t, Y(0))$.

(i) *Smooth Initial Conditions*

For $p = 2$ and smoothness coefficient $\overline{m}$, Theorem 8.6 and the definition of $\overline{m}$ imply $Y(t)$, $\partial_k Y(t)$, $\partial^2_{k\ell} Y(t)$ are in $\mathbf{W}_{0,2,1} \subset \mathbf{W}_{0,2,\Phi}$ for $k, \ell = 1, \ldots, d$. Further, since we can always choose a decomposition $\{B_n\}_{n \in \mathbf{n}}$ of Ω with $B_n \in \mathcal{F}_0$ such that $\operatorname{ess\,sup}_{\omega} \|Y_0\|_{\overline{m}} 1_{B_n} < \infty$ for $n \ge 2$ and $P(B_1) = 0$ we may, without loss of generality, assume

$$\operatorname*{ess\,sup}_{\omega} \|Y_0(\omega)\|_2 \le c < \infty. \tag{11.121}$$

By Theorem 8.6, we may stop $\|Y(t)\|_2$ at the first exit time $\tau := \tau_N$ for the ball with center 0 and radius $N > c$ (where $\tau_N(\omega) = \infty$, if $\sup_{0 \le t < \infty} \|Y(t)\|_2 < \infty$). Hence, $\|Y(t \wedge \tau_N)\|_2 \le N$ a.s. for all $t \ge 0$. By the smoothness result, we obtain the analogue of (11.86) with $n = 0$, $\mu = \infty$, and $t \wedge \tau$ instead of t. We now proceed as in the proof of Theorem 8.6, and let us also use similar abbreviations as in that proof (with $t \mapsto t \wedge \tau$).

(ii) Integration by parts yields

$$\left.\begin{aligned} A_2(t) &= -\frac{1}{2} \sum_{k,\ell=1}^{d} \int_0^{t\wedge\tau} \int (\partial_k Y(s,r))(\partial_\ell Y(s,r)) \Phi(r) dr\, D_{k\ell}(s) \mathrm{d}s \\ &\quad + \frac{1}{2} \sum_{k,\ell=1}^{d} \int_0^{t\wedge\tau} \int (Y(s,r))^2 \partial^2_{k\ell} \Phi(r) \mathrm{d}r\, \mathrm{d}s \end{aligned}\right\} \tag{11.122}$$

[19] Cf. Metivier and Pellaumail (1980). Cf. also our sketch of the Itô integral in Sects. 15.2.5 and 15.2.3.

$$=: A_{2,1}(t) + A_{2,2}(t),$$

where the second term comes from

$$2\int Y(s,r)(\partial_\ell Y(s,r))\partial_k \Phi(r)\mathrm{d}r$$
$$= \int \partial_\ell(Y^2(s,r))\partial_k \Phi(r)\mathrm{d}r$$

and integration by parts. Next, note that

$$\mathrm{d}[\partial_k m_k(s,r), \partial_\ell m_\ell(s,r)] = \mathcal{L}_{k\ell,1_k+1_\ell}(s)\mathrm{d}s \tag{11.123}$$

and

$$-\mathcal{L}_{k\ell,0}(s) = D_{k\ell}(s)$$

Recalling, that $\mathcal{L}_{k\ell,1_i}(s) = 0$ by the symmetry assumption in (8.33), we obtain,

$$\left.\begin{aligned} &[\partial_k(Y(s,r)\mathrm{d}m_k(s,r)), \partial_\ell(Y(s,r)\mathrm{d}m_\ell(s,r))] \\ &= (\partial_k Y\ (s,r))(\partial_\ell Y(s,r))D_{k\ell}(s)\mathrm{d}s \\ &+ (Y^2(s,r))\mathcal{L}_{k\ell,1_k+1_\ell}(s)\mathrm{d}s \\ &=: \sum_{i=1}^{2} B_{k\ell,i}(s,r)\mathrm{d}s. \end{aligned}\right\} \tag{11.124}$$

The decomposition of the left-hand side of (11.95) into $\sum_{i=1}^{2} B_{k\ell,i}(s,r)\mathrm{d}s$ induces a decomposition

$$A_5(t) = \sum_{i=1}^{2} A_{5,i}(t).$$

Clearly,

$$A_{2,1}(t) + A_{5,1}(t) \equiv 0. \tag{11.125}$$

The transformation of the drift and the stochastic terms is easier. □

- *Initial Conditions* $\in L_{2,\mathcal{F}_0}(\mathbf{W}_{0,2,1})$.

Let $Y_{0,m} \in L_{2,\mathcal{F}_0}(\mathbf{W}_{2,2,\Phi})$ and assume

$$E\|Y_{0,m} - Y_0\|_{0,\Phi}^2 \longrightarrow 0, \quad \text{as } m \longrightarrow \infty\,. \tag{11.126}$$

Truncating at a level $N > 0$,

$$Y_{0,N} := Y_0 1_{\{\|Y_0\|_{0,\Phi} \le N\}}, \quad Y_{0,m,N} := Y_{0,m} 1_{\{\|Y_{0,m}\|_{0,\Phi} \le N\}} \tag{11.127}$$

implies that for $p = 1, 2$ and $\forall N > 0$:

$$\left.\begin{aligned}
&\sup_{0\le t\le T} E\|Y(t, Y_{0,N}) - Y(t, Y_{0,m,N})\|_{0,\Phi}^{2p} \le c_p E\|Y(t, Y_{0,N}) - Y(t, Y_{0,m,N})\|_{0,\Phi}^{2p} \\
&\qquad\qquad\qquad\qquad \longrightarrow 0 \ , \text{as } m \longrightarrow \infty \ , \\
&\sup_{0\le t\le T} E\|Y(t, Y_{0,N})\|_{0,\Phi}^{2p} \le c_p E\|Y(t, Y_{0,N})\|_{0,\Phi}^{2p}, \\
&\sup_{0\le t\le T} E\|Y(t, Y_{0,N})\|_{0,\Phi}^{2p} \le c_p E\|Y(t, Y_{0,N})\|_{0,\Phi}^{2p},
\end{aligned}\right\} \tag{11.128}$$

where for the first inequality we used the bilinearity of (8.25). For $\Phi = \varpi$, (11.128) follows from Theorem 8.6, since by the Cauchy–Schwarz inequality

$$\|f\|_{0,\varpi}^4 \le c_\varpi \|f\|_{0,4,\varpi}^4 .$$

For $\Phi \equiv 1$, (11.128) follows from Kotelenez (1995b), Theorem 4.11. Note that for the truncated initial conditions the convergence assumption (11.126) also holds for the fourth moment of the norm of the difference of the initial conditions (by Lebesgue's dominated convergence theorem). Since the Itô formula is a statement for almost all ω we may in what follows drop the subindex N and assume, without loss of generality, (11.128). Abbreviate

$$Y_m(t) := Y_m(t, Y_{0,m}), \quad Y(t) := Y(t, Y_0).$$

We know already from the first part (with smooth initial conditions) that the Itô formula holds for $\|Y_m(t)\|_{0,\Phi}^2$. Denote the right-hand side of (8.42) with process $Y_m(\cdot)$ by $R_m(t)$. If we replace on the right-hand side of (8.42) all Y_m terms by $Y(t)$ the right-hand side of (8.42) is a continuous semimartingale and will be denoted by $R(t)$. We need to show that for all $t \ge 0$,

$$\|Y(t)\|_{0,\Phi}^2 = R(t) \ \text{ with probability } 1,$$

which is precisely the Itô formula for $\|Y(t)\|_{0,\Phi}^2$. We easily see that the deterministic integrals in $R_m(\cdot)$ converge to the deterministic integrals of $R(\cdot)$ (in the topology of $L_{2,\mathcal{F}}([s, T]; \mathbf{W}_{0,2,\Phi})$). Therefore, we will provide details only for the more difficult stochastic integrals. Abbreviate

$$I_{m,\ell}(t) := \int\int_0^t Y_m^2(s, r)\partial_\ell m_\ell(r, \mathrm{d}s)\Phi(r)\mathrm{d}r$$

and

$$I_\ell(t) := \int\int_0^t Y^2(s, r)\partial_\ell m_\ell(r, \mathrm{d}s)\Phi(r)\mathrm{d}r.$$

The mutual quadratic variation

$$\mathrm{d}[\partial_\ell m_\ell(r, \mathrm{d}s), \partial_\ell m_\ell(\tilde{r}, \mathrm{d}s)] = -(\partial_{\ell,r\ell,r}\tilde{D})(s, r - \tilde{r})\mathrm{d}s. \tag{11.129}$$

Thus, by the boundedness of $|(\partial_{\ell,r\ell,r}\tilde{D})(s, r - \tilde{r})|$ in all variables (cf. (8.35)), we obtain

$$E|I_\ell(t) - I_{m,\ell}(t)|^2 \leq c_{\tilde{D}} E\left(\int_0^t \left(\int (Y^2(s,r) - Y_m^2(s,r))\Phi(r)\mathrm{d}r\right)^2\right). \quad (11.130)$$

A simple calculation shows that the right-hand side of (11.130) can be estimated by

$$c_{\tilde{D},\Phi}\left(\int_0^t (E\|Y(s) - Y_m(s)\|_{0,\Phi}^4)^{1/2}(E\|Y(s) + Y_m(s)\|_{0,\Phi}^4\right)^{1/2}.$$

Hence, by (11.128) in addition to (11.130) and the convergence assumption (11.126) (also holding after truncation for the fourth moment),

$$E|I_\ell(t) - I_{m,\ell}(t)|^2 \leq \bar{c}_{\tilde{D},\Phi}(E\|Y(0) - Y_m(0)\|_{0,\Phi}^4)^{1/2} \longrightarrow 0 \text{ , as } m \longrightarrow \infty . \quad (11.131)$$

Chapter 12
Proof of Uniqueness

Employing Itô's formula, the basic estimate (8.48) is proved, which ensures strong uniqueness for quasilinear SPDEs under smoothness assumptions on the coefficients and initial conditions.

We may without loss of generality assume $s = 0$ and that the coefficients are independent of t. Further, as in the smoothness proof of Chap. 11, the proof of (8.48) generalizes immediately to the coercive case, assuming in addition (8.74) and (8.75). Therefore, it suffices to provide the proof of (8.48) assuming $\sigma^{\perp}(r, \mu, t) \equiv 0$. Hence, the two solutions from Lemma 8.10 can be written as

$$\left.\begin{aligned} Y(t) &= Y_0 + \frac{1}{2}\int_0^t \sum_{k,\ell=1}^{d} \partial_{k,\ell}^2 (D_{k,\ell}(Z(s))Y(s))\mathrm{d}s - \int_0^t \nabla\big(Y(s)F(\cdot, Z(s))\big)\mathrm{d}s \\ &\quad - \int_0^t \nabla(Y(s) \int \mathcal{J}(\cdot, p, Z(s))w(\mathrm{d}p, \mathrm{d}s)), \\ \tilde{Y}(t) &= \tilde{Y}_0 + \frac{1}{2}\int_0^t \sum_{k,\ell=1}^{d} \partial_{k,\ell}^2 \big(D_{k,\ell}(\tilde{Z}(s))\tilde{Y}(s)\big)\mathrm{d}s - \int_0^t \nabla\big(\tilde{Y}(s)F(\cdot, \tilde{Z}(s))\big)\mathrm{d}s \\ &\quad - \int_0^t \nabla\big(\tilde{Y}(s) \int \mathcal{J}(\cdot, p, \tilde{Z}(s))w(\mathrm{d}p, \mathrm{d}s)\big). \end{aligned}\right\} \tag{12.1}$$

Let c denote some finite constant whose values may change throughout the steps of a series of estimates. ∂_ℓ denotes the partial derivative with respect to the ℓ's spatial coordinate. If there are two spatial variables, we will write $\partial_{\ell,r}$ to indicate that the partial derivative is to be taken with respect to the variable r, etc.

Note that our assumptions imply by Theorem 8.6 that both Y and $\tilde{Y}$ are in $L_{0,\mathcal{F}}(C([0,\infty); \mathbf{W}_{2,2,\varpi}))$. Therefore, by Itô's formula (8.42)

$$
\left.
\begin{aligned}
&\left\|Y(t)-\tilde{Y}(t)\right\|_{0,\varpi}^{2}\\
&=\int_0^t\left\langle\sum_{k,\ell=1}^{d}\partial_{k,\ell}^{2}\left(D_{k,\ell}\big(Z(s)\big)Y(s)-D_{k,\ell}\big(\tilde{Z}(s)\big)\tilde{Y}(s)\right),Y(s)-\tilde{Y}(s)\right\rangle_{0,\varpi}\mathrm{d}s\\
&\quad+\sum_{m=1}^{d}\int_0^t\int\int\left(\nabla\bullet\left(Y(r,s)\mathcal{J}_{\cdot,m}(r,p,Z(s))-\tilde{Y}(r,s)\mathcal{J}_{\cdot,m}\left(r,p,\tilde{Z}(s)\right)\right)\right)^{2}\mathrm{d}p\varpi(r)\mathrm{d}r\mathrm{d}s\\
&\quad-2\int_0^t\left\langle\nabla\bullet\left(Y(s)F(\cdot,Z(s))\right)-\nabla\bullet\left(\tilde{Y}(s)F\left(\cdot,\tilde{Z}(s)\right)\right),Y(s)-\tilde{Y}(s)\right\rangle_{0,\varpi}\mathrm{d}s\\
&\quad+\left\|Y(0)-\tilde{Y}(0)\right\|_{0,\varpi}^{2}+\text{martingale}\\
&=:\int_0^t(I(s)+II(s)+III(s))\mathrm{d}s+\left\|Y(0)-\tilde{Y}(0)\right\|_{0,\varpi}^{2}+\text{martingale}\,.
\end{aligned}
\right\}
\tag{12.2}
$$

The homogeneity assumption (8.33) implies that the diffusion coefficient is not dependent on the spatial variable. Therefore, integrating by parts,

$$
\left.
\begin{aligned}
I(s)&=\sum_{k,\ell=1}^{d}-\left\langle D_{k,\ell}(Z(s))\partial_k Y(s)-D_{k,\ell}\big(\tilde{Z}(s)\big)\partial_k\tilde{Y}(s),\partial_\ell Y(s)-\partial_\ell\tilde{Y}(s)\right\rangle_{\varpi}\\
&\qquad-\left\langle D_{k,\ell}(Z(s))\partial_k Y(s)-D_{k,\ell}\big(\tilde{Z}(s)\big)\partial_k\tilde{Y}(s),\big(Y(s)-\tilde{Y}(s)\big)\partial_\ell\varpi\right\rangle_{0}\\
&=:\sum_{k,\ell=1}^{d}I_{1,k\ell}(s)+I_{2,k\ell}(s),
\end{aligned}
\right\}
\tag{12.3}
$$

$$
\left.
\begin{aligned}
I_{1,k\ell}(s)&=-\langle D_{k,\ell}(Z(s))\partial_k Y(s),\partial_\ell Y(s),\rangle_{0,\varpi}-\left\langle D_{k,\ell}\big(\tilde{Z}(s)\big)\partial_k\tilde{Y}(s),\partial_\ell\tilde{Y}(s),\right\rangle_{0,\varpi}\\
&\quad+\left\langle D_{k,\ell}(Z(s))\partial_k Y(s),\partial_\ell\tilde{Y}(s)\right\rangle_{0,\varpi}+\left\langle D_{k,\ell}\big(\tilde{Z}(s)\big)\partial_k\tilde{Y}(s),\partial_\ell Y(s)\right\rangle_{0,\varpi}\\
&=:I_{1,1,k\ell}(s)+I_{1,2,k\ell}(s)+I_{1,3,k\ell}(s)+I_{1,4,k\ell}(s),
\end{aligned}
\right\}
\tag{12.4}
$$

$$
\left.
\begin{aligned}
I_{2,k\ell}(s)&=-\left\langle D_{k,\ell}(Z(s))\partial_k\big(Y(s)-\tilde{Y}(s)\big),\big(Y(s)-\tilde{Y}(s)\big)\partial_\ell\varpi\right\rangle_{0}\\
&\quad-\left\langle\big(D_{k,\ell}(Z(s))-D_{k,\ell}\big(\tilde{Z}(s)\big)\big)\partial_k\tilde{Y}(s),\big(Y(s)-\tilde{Y}(s)\big)\partial_\ell\varpi\right\rangle_{0}\\
&=:I_{2,1,k\ell}(s)+I_{2,2,k\ell}(s).
\end{aligned}
\right\}
\tag{12.5}
$$

In what follows we will repeatedly use the simple formula

$$
\partial_k\big(Y(s)-\tilde{Y}(s)\big)\big(Y(s)-\tilde{Y}(s)\big)=\frac{1}{2}\partial_k\big(Y(s)-\tilde{Y}(s)\big)^{2}.
\tag{12.6}
$$

Hence, by the homogeneity assumption,

$$\left.\begin{aligned}
I_{2,1,k\ell}(s) &= -D_{k,\ell}(Z(s))\left\langle \partial_k\big(Y(s)-\tilde{Y}(s)\big), \big(Y(s)-\tilde{Y}(s)\big)\partial_\ell \varpi\right\rangle_0 \\
&= -D_{k,\ell}(Z(s))\left\langle \partial_k\big(Y(s)-\tilde{Y}(s)\big)\big(Y(s)-\tilde{Y}(s)\big), \partial_\ell \varpi\right\rangle_0 \\
&= -D_{k,\ell}(Z(s))\left\langle \frac{1}{2}\partial_k\big(\big(Y(s)-\tilde{Y}(s)\big)^2\big), \partial_\ell \varpi\right\rangle_0 \\
&= \frac{1}{2}D_{k,\ell}(Z(s))\left\langle \big(Y(s)-\tilde{Y}(s)\big)^2, \partial^2_{k,\ell} \varpi\right\rangle_0,
\end{aligned}\right\} \tag{12.7}$$

using integration by parts. By (15.46)

$$\|\partial_k \varpi(r)\| + \left|\partial^2_{k,\ell}\varpi(r)\right| \le c\varpi(r).$$

It follows

$$\begin{aligned}
I_{2,1,k\ell}(s) &\le \frac{c}{2}\left|D_{k,\ell}(Z(s))\right|\left\langle \big(Y(s)-\tilde{Y}(s)\big)^2, \varpi\right\rangle_0 \\
&= \frac{c}{2}\left|D_{k,\ell}(Z(s))\right| \cdot \|Y(s)-\tilde{Y}(s)\|^2_{0,\varpi}.
\end{aligned} \tag{12.8}$$

Next, suppressing temporarily the dependence on s in the notation

$$\begin{aligned}
&\left|D_{k,\ell}(Z) - D_{k,\ell}(\tilde{Z})\right| \\
&= \left|\sum_{m=1}^{d}\int \big(\mathcal{J}_{k,m}(r,p,Z)\mathcal{J}_{\ell,m}(r,p,Z) - \mathcal{J}_{k,m}(r,p,\tilde{Z})\mathcal{J}_{\ell,m}(r,p,\tilde{Z})\big)\mathrm{d}p\right| \\
&= \Bigg|\sum_{m=1}^{d}\int \big(\mathcal{J}_{k,m}(r,p,Z) - \mathcal{J}_{k,m}(r,p,\tilde{Z})\big)\mathcal{J}_{\ell,m}(r,p,Z)\,\mathrm{d}p \\
&\quad + \int \Big(\mathcal{J}_{k,m}(r,p,\tilde{Z})\big(\mathcal{J}_{\ell,m}(r,p,Z) - \mathcal{J}_{\ell,m}(r,p,\tilde{Z})\big)\mathrm{d}p\Bigg| \\
&\le \sum_{m=1}^{d} c\Big(\int \big(\mathcal{J}_{k,m}(r,p,Z) - \mathcal{J}_{k,m}(r,p,\tilde{Z})\big)^2\mathrm{d}p\Big)^{\frac{1}{2}} \\
&\quad + c\Big(\int \big(\mathcal{J}_{\ell,m}(r,p,Z) - \mathcal{J}_{\ell,m}(r,p,\tilde{Z})\big)^2\mathrm{d}p\Big)^{\frac{1}{2}}
\end{aligned}$$

(by Hölder's inequality and the boundedness assumption on the integral of the second moment of $\mathcal{J}$)

$$\le c\|Z - \tilde{Z}\|_{0,\varpi}$$

(by Hypothesis 4.1 and (8.46)),

i.e.,

$$\left|D_{k,\ell}(Z(s)) - D_{k,\ell}(\tilde{Z}(s))\right| \le c\left\|Z(s) - \tilde{Z}(s)\right\|_{0,\varpi}. \tag{12.9}$$

Hence,

$$
\begin{aligned}
&|I_{2,2,k\ell}(s)| \\
&\quad \le \big|D_{k,\ell}(Z(s)) - D_{k,\ell}(\tilde{Z}(s))\big| \Big\langle \big|\partial_k \tilde{Y}(s)\big|, \big|Y(s) - \tilde{Y}(s)\big| \big|\partial_\ell \varpi\big| \Big\rangle_0 \\
&\quad \le c \big|D_{k,\ell}(Z(s)) - D_{k,\ell}(\tilde{Z}(s))\big| \Big\langle \big|\partial_k \tilde{Y}(s)\big|, \big|Y(s) - \tilde{Y}(s)\big| \Big\rangle_{0,\varpi} \\
&\quad \text{(by (15.46))} \\
&\quad \le c \big\| Z(s) - \tilde{Z}(s)\big\|_{\varpi} \big\|\partial_k \tilde{Y}(s)\big\|_{\varpi} \big\|Y(s) - \tilde{Y}(s)\big\|_{0,\varpi} \\
&\quad \text{(by the Cauchy–Schwarz inequality)} \\
&\quad = c \big\|\partial_k \tilde{Y}(s)\big\|_{0,\varpi} \big\| Z(s) - \tilde{Z}(s)\big\|_{0,\varpi} \big\|Y(s) - \tilde{Y}(s)\big\|_{0,\varpi},
\end{aligned}
$$

i.e., $\forall k, \ell$

$$
|I_{2,2,k\ell}(s)| \le c \|\partial_k \tilde{Y}(s)\|_{0,\varpi} \|Z(s) - \tilde{Z}(s)\|_{0,\varpi} \|Y(s) - \tilde{Y}(s)\|_{0,\varpi}. \tag{12.10}
$$

Expression (12.8) and (12.10) together yield

$$
\begin{aligned}
|I_{2,k\ell}(s)| \le c\Big[&\big\|\partial_k \tilde{Y}(s)\big\|_{0,\varpi} \big\| Z(s) - \tilde{Z}(s)\big\|_{0,\varpi} \big\|Y(s) - \tilde{Y}(s)\big\|_{0,\varpi} \\
&+ \big\|Y(s) - \tilde{Y}(s)\big\|_{\varpi}^2 \Big],
\end{aligned} \tag{12.11}
$$

$$
\left.
\begin{aligned}
&II(s) \\
&\quad = \sum_{k,\ell=1}^d \Big\{ \sum_{m=1}^d \int\int (\partial_k(Y(r,s)\mathcal{J}_{k,m}(r,p,Z(s)))) \\
&\qquad\qquad \times (\partial_\ell(Y(r,s)\mathcal{J}_{\ell,m}(r,p,Z(s))))\mathrm{d}p\, \varpi(r)\mathrm{d}r \\
&\qquad + \int\int \big(\partial_k\big(\tilde{Y}(r,s)\mathcal{J}_{k,m}\big(r,p,\tilde{Z}(s)\big)\big)\big)\big(\partial_\ell\big(\tilde{Y}(r,s)\mathcal{J}_{\ell,m}(r,p,\tilde{Z}(s))\big)\big)\mathrm{d}p\, \varpi(r)\mathrm{d}r \\
&\qquad - \int\int (\partial_k(Y(r,s)\mathcal{J}_{k,m}(r,p,Z(s))))\big(\partial_\ell\big(\tilde{Y}(r,s)\mathcal{J}_{\ell,m}(r,p,\tilde{Z}(s))\big)\big)\mathrm{d}p\, \varpi(r)\mathrm{d}r \\
&\qquad - \int\int \big(\partial_k\big(\tilde{Y}(r,s)\mathcal{J}_{k,m}\big(r,p,\tilde{Z}(s)\big)\big)\big)(\partial_\ell(Y(r,s)\mathcal{J}_{\ell,m}(r,p,Z(s)))) \\
&\qquad\qquad \times \mathrm{d}p\, \varpi(r)\mathrm{d}r \Big\} \\
&\quad =: \sum_{m=1}^d II_{1,m}(s) + II_{2,m}(s) + II_{3,m}(s) + II_{4,m}(s).
\end{aligned}
\right\} \tag{12.12}
$$

By symmetry,

$$
II_{3,m}(s) + II_{4,m}(s) = 2II_{3,m}(s) \;\; \forall m. \tag{12.13}
$$

Further, for $\bar{Z} \in \{Y, \tilde{Y}\}$, integrating by parts and employing (12.6),

$$
\int \partial_k \bar{Z}(s,r)\bar{Z}(s,r)\varpi(r)\mathrm{d}r = \frac{1}{2}\int \partial_k\big(\bar{Z}^2(s,r)\big)\varpi(r)\mathrm{d}r = -\frac{1}{2}\int \bar{Z}^2(s,r)\partial_k\varpi(r)\mathrm{d}r. \tag{12.14}
$$

Set for $\hat{Z} \in \{Z, \tilde{Z}\}$

$$\vec{C}_{\hat{Z},m} := \mathcal{J}_{\cdot,m}(r,p,\hat{Z}), \quad D_{\hat{Z},m} := div_r \vec{C}_{\hat{Z},m} (= \sum_{\ell=1}^{d} \partial_{\ell,r} \mathcal{J}_{\ell,m}(r,p,\hat{Z})).$$

Note that

$$\left.\begin{aligned}
&\sum_{k,\ell=1}^{d} \int\int (\partial_k Y(r,s)) \mathcal{J}_{k,m}(r,p,Z(s)) Y(r,s)(\partial_{\ell,r}\mathcal{J}_{\ell,m}(r,p,Z(s)))\mathrm{d}p\,\varpi(r)\mathrm{d}r \\
&= \sum_{k,\ell=1}^{d} \int (\partial_k Y(r,s)) Y(r,s) \int \mathcal{J}_{k,m}(r,p,Z(s))(\partial_{\ell,r}\mathcal{J}_{\ell,m}(r,p,Z(s)))\mathrm{d}p\,\varpi(r)\mathrm{d}r \\
&= -\sum_{k,\ell=1}^{d} \frac{1}{2} \int Y^2(r,s) \int \mathcal{J}_{k,m}(r,p,Z(s))(\partial_{\ell,r}\mathcal{J}_{\ell,m}(r,p,Z(s)))\mathrm{d}p \partial_k\,\varpi(r)\mathrm{d}r \\
&\qquad = -\frac{1}{2} \int Y^2(r,s) \Big(\int \vec{C}_{Z,m} D_{Z,m} \mathrm{d}p \Big) \cdot (\nabla \varpi)(r)\mathrm{d}r \\
&\qquad =: II_{1,3,m}(s)
\end{aligned}\right\} \quad (12.15)$$

and, similarly,

$$\left.\begin{aligned}
&\sum_{k,\ell=1}^{d} \int\int (\partial_\ell Y(r,s))(\partial_{k,r}\mathcal{J}_{k,m}(r,p,Z(s))) Y(r,s) \mathcal{J}_{\ell,m}(r,p,Z(s))\mathrm{d}p\,\varpi(r)\mathrm{d}r \\
&= -\sum_{k,\ell=1}^{d} \frac{1}{2} \int Y^2(r,s) \int (\partial_{k,r}\mathcal{J}_{\ell,m}(r,p,Z(s))\mathcal{J}_{\ell,m}(r,p,Z(s))\mathrm{d}p\,\partial_\ell \varpi(r)\mathrm{d}r \\
&\qquad =: II_{1,4,m}(s) = II_{1,3,m}(s).
\end{aligned}\right\} \quad (12.16)$$

Therefore,

$$\left.\begin{aligned}
II_{1,m}(s) &= \int\int \big(\nabla Y(r,s) \cdot \vec{C}_{Z,m}\big)^2 \mathrm{d}p\varpi(r)\mathrm{d}r \\
&\quad + \int\int Y^2(r,s) D^2_{Z,m} \mathrm{d}p\varpi(r)\mathrm{d}r \\
&\quad + 2II_{1,3,m}(s) \\
&=: II_{1,1,m}(s) + II_{1,2,m}(s) + 2II_{1,3,m}(s).
\end{aligned}\right\} \quad (12.17)$$

In the same way, we obtain

$$\left.\begin{aligned}
II_{2,m}(s) &= \int\int \Big(\nabla \tilde{Y}(r,s) \cdot \vec{C}_{\tilde{Z},m}\Big)^2 \mathrm{d}p\varpi(r)dr \\
&\quad + \int\int \tilde{Y}^2(r,s) D^2_{\tilde{Z},m} \mathrm{d}p\varpi(r)rmdr \\
&- \int \tilde{Y}^2(r,s) \Big(\int \vec{C}_{\tilde{Z},m} D_{\tilde{Z},m} \mathrm{d}p \Big) \cdot (\nabla \varpi)(r)\mathrm{d}r \\
&=: II_{2,1,m}(s) + II_{2,2,k\ell,m}(s) + 2II_{2,3,m}(s).
\end{aligned}\right\} \quad (12.18)$$

For the following elementary algebraic transformations it is safe to abbreviate

$$\int\int(\ ) := \int\left(\int(\ )\mathrm{d}p\right)\varpi(r)\mathrm{d}r.$$

Thus,

$$\left.\begin{aligned} II_{3,m}(s) = &-\int\int(\nabla Y(r,s)\cdot\vec{C}_{Z,m})(\nabla\tilde{Y}(r,s)\cdot\vec{C}_{\tilde{Z},m})\mathrm{d}p \\ &-\int\int Y(s,r)\tilde{Y}(r,s)D_{Z,m}D_{\tilde{Z},m}\,\mathrm{d}p\,\varpi(r)\mathrm{d}r \\ &-\int\int(\nabla Y(r,s)\cdot\vec{C}_{Z,m})\tilde{Y}(r,s)D_{\tilde{Z},m} \\ &-\int\int Y(s,r)D_{Z,m}(\nabla\tilde{Y}(r,s)\cdot\vec{C}_{\tilde{Z},m}) \\ =:&\ II_{3,1,m}(s)+II_{3,2,m}(s)+II_{3,3,m}(s)+II_{3,4,m}(s). \end{aligned}\right\} \tag{12.19}$$

A similar representation holds, of course, for $II_{4,m}(s)$ which by (12.13) equals $II_{3,m}(s)$. Set

$$\left.\begin{aligned} A &:= \int\int(Y-\tilde{Y})(\nabla Y+\nabla\tilde{Y})\cdot(\vec{C}_{Z,m}-\vec{C}_{\tilde{Z},m})(D_{Z,m}+D_{\tilde{Z},m}) \\ &= \int\int\left[(Y\nabla Y-\tilde{Y}\nabla\tilde{Y})+(Y\nabla\tilde{Y}-\tilde{Y}\nabla Y)\right]\cdot\left[(\vec{C}_{Z,m}D_{Z,m}-\vec{C}_{\tilde{Z},m}D_{\tilde{Z},m})\right. \\ &\qquad\qquad \left.+(\vec{C}_{Z,m}D_{\tilde{Z},m}-\vec{C}_{\tilde{Z},m}D_{Z,m})\right] \\ &= \int\int(Y\nabla Y-\tilde{Y}\nabla\tilde{Y})\cdot(\vec{C}_{Z,m}D_{\tilde{Z},m}-\vec{C}_{\tilde{Z},m}D_{Z,m}) \\ &\quad+\int\int(Y\nabla\tilde{Y}-\tilde{Y}\nabla Y)\cdot(\vec{C}_{Z,m}D_{Z,m}-\vec{C}_{\tilde{Z},m}D_{\tilde{Z},m}) \\ &\quad+\int\int Y\nabla Y\bullet\vec{C}_{Z,m}D_{Z,m}+\int\int\tilde{Y}\nabla\tilde{Y}\bullet\vec{C}_{\tilde{Z},m}D_{\tilde{Z},m} \\ &\quad-\int\int Y\nabla Y\bullet\vec{C}_{\tilde{Z},m}D_{\tilde{Z},m}-\int\int\tilde{Y}\nabla\tilde{Y}\bullet\vec{C}_{Z,m}D_{Z,m} \\ &\quad+\int\int Y\nabla\tilde{Y}\bullet\vec{C}_{Z,m}D_{\tilde{Z},m}+\int\int\tilde{Y}\nabla Y\bullet\vec{C}_{\tilde{Z},m}D_{Z,m} \\ &\quad-\int\int Y\nabla\tilde{Y}\bullet\vec{C}_{\tilde{Z},m}D_{Z,m}-\int\int\tilde{Y}\nabla Y\bullet\vec{C}_{Z,m}D_{\tilde{Z},m} \\ &\qquad =: \sum_{i=1}^{10}A_i. \end{aligned}\right\} \tag{12.20}$$

Note

$$\left.\begin{aligned} A_3 &= II_{1,3,m}(s) = II_{1,4,m}(s), \\ A_4 &= II_{2,3,m}(s) = II_{2,4,m}(s), \\ A_9 &= II_{3,4,m}(s) = II_{4,4,m}(s), \\ A_{10} &= II_{3,3,m}(s) = II_{4,3,m}(s). \end{aligned}\right\} \tag{12.21}$$

$$\left.\begin{aligned} A_5 + A_7 &= \int\int Y D_{\tilde{Z},m}[\nabla\tilde{Y} \bullet \vec{C}_{Z,m} - \nabla Y \bullet \vec{C}_{\tilde{Z},m}],\\ A_6 + A_8 &= \int\int \tilde{Y} D_{Z,m}[\nabla Y \bullet \vec{C}_{\tilde{Z},m} - \nabla\tilde{Y} \bullet \vec{C}_{Z,m}]\\ &= -\int\int \tilde{Y} D_{Z,m}[\nabla\tilde{Y} \bullet \vec{C}_{Z,m} - \nabla Y \cdot \vec{C}_{\tilde{Z},m}]. \end{aligned}\right\} \tag{12.22}$$

It follows

$$\sum_{i=5}^{8} A_i = \int\int (Y D_{\tilde{Z},m} - \tilde{Y} D_{Z,m})[\nabla\tilde{Y} \bullet \vec{C}_{Z,m} - \nabla Y \bullet \vec{C}_{\tilde{Z},m}] \tag{12.23}$$

and

$$A_3 + A_4 + A_9 + A_{10} = A - (A_1 + A_2) - \sum_{i=5}^{8} A_i. \tag{12.24}$$

Hence,

$$\left.\begin{aligned} &\sum_{k,\ell=1}^{d}\sum_{m=1}^{d}(II_{1,3,k\ell,m}(s) + II_{1,4,k\ell,m}(s) + II_{2,3,k\ell,m}(s) + II_{2,4,k\ell,m}(s))\\ &\quad + \sum_{k,\ell=1}^{d}\sum_{m=1}^{d}(II_{3,4,k\ell,m}(s) + II_{4,4,k\ell,m}(s) + II_{3,3,k\ell,m}(s) + II_{4,3,k\ell,m}(s))\\ &= 2\sum_{m=1}^{d}\int\int \big(Y(s,r) - \tilde{Y}(s,r)\big)\big(\nabla Y(s,r)\\ &\quad +\nabla\tilde{Y}(s,r)\big)\cdot(\vec{C}_{Z(s),m} - \vec{C}_{\tilde{Z}(s),m})(D_{Z(s),m} + D_{\tilde{Z}(s),m})\\ &\quad -2\sum_{m=1}^{d}\int\int \big(Y(s,r)\nabla Y(s,r)\\ &\quad -\tilde{Y}(s,r)\nabla\tilde{Y}(s,r)\big)\cdot(\vec{C}_{Z(s),m} D_{\tilde{Z}(s),m} - \vec{C}_{\tilde{Z}(s),m} D_{Z(s),m})\\ &\quad -2\sum_{m=1}^{d}\int\int \big(Y(s,r)\nabla\tilde{Y}(s,r)\\ &\quad -\tilde{Y}(s,r)\nabla Y(s,r)\big)\cdot(\vec{C}_{Z(s),m} D_{Z(s),m} - \vec{C}_{\tilde{Z}(s),m} D_{\tilde{Z}(s),m})\\ &\quad -2\sum_{m=1}^{d}\int\int \big(Y(s,r) D_{\tilde{Z}(s),m}\\ &\quad -\tilde{Y}(s,r) D_{Z(s),m}\big)\big(\nabla\tilde{Y}(s,r) \bullet \vec{C}_{Z(s),m} - \nabla Y(s,r) \bullet \vec{C}_{\tilde{Z}(s),m}\big). \end{aligned}\right\} \tag{12.25}$$

Next (cf. (12.17), (12.18) and (12.4)),

$$
\left.\begin{aligned}
\sum_{m=1}^{d} II_{1,1,m}(s) &= \sum_{m=1}^{d}\sum_{k,\ell=1}^{d}\int\int \partial_k Y(r,s), \partial_\ell Y(r,s)\\
&\qquad \int \mathcal{J}_{km}(r,p,Z)\mathcal{J}_{\ell m}(r,p,Z)\mathrm{d}p\, \varpi(r)\mathrm{d}r\\
&= \sum_{k,\ell=1}^{d} \langle \partial_k Y(s), \partial_\ell Y(s)\rangle_{0,\varpi} D_{k\ell}(Z(s)) = -I_{1,1}(s)\\
&\qquad\text{(and similarly)}\\
\sum_{m=1}^{d}(II_{2,1,m}(s) &= \sum_{k,\ell=1}^{d} \big\langle \partial_k \tilde{Y}(s), \partial_\ell \tilde{Y}(s)\big\rangle_{0,\varpi} D_{k\ell}\big(\tilde{Z}(s)\big) = -I_{1,2}(s).
\end{aligned}\right\} \quad (12.26)
$$

Thus,

$$
I_{1,1}(s) + I_{1,2}(s) + \sum_{m=1}^{d}(II_{1,1,m}(s) + II_{2,1,m}(s)) = 0. \quad (12.27)
$$

Further (cf. (12.17) – (12.19)),

$$
\left.\begin{aligned}
&\sum_{m=1}^{d}(II_{1,2,m}(s) + II_{2,2,m}(s) + II_{3,2,m}(s) + II_{4,2,m}(s))\\
&\quad = \sum_{m=1}^{d}\int\int Y^2(r,s) D^2_{Z,m}\, \mathrm{d}p\, \varpi(r)\mathrm{d}r\\
&\qquad + \sum_{m=1}^{d}\int\int \tilde{Y}^2(r,s) D^2_{\tilde{Z},m}\, \mathrm{d}p\, \varpi(r)\mathrm{d}r\\
&\qquad -2\sum_{m=1}^{d} Y(s,r)\tilde{Y}(r,s) D_{Z,m} D_{\tilde{Z},m}\, \mathrm{d}p\, \varpi(r)\mathrm{d}r\\
&\quad = \sum_{m=1}^{d}\int\int \big(Y(r,s)D_{Z(s),m} - \tilde{Y}(s,r)D_{\tilde{Z}(s),m}\big)^2 \mathrm{d}p\, \varpi(r)\mathrm{d}r,
\end{aligned}\right\} \quad (12.28)
$$

$$
\left.
\begin{aligned}
& I_{1,3}(s) + I_{1,4}(s) + \sum_{m=1}^{d} [II_{3,1,m}(s) + II_{4,1,m}(s)] \\
& = \sum_{k,\ell=1}^{d} \left[\langle \partial_k Y(s), \partial_\ell \tilde{Y}(s) \rangle_{0,\varpi} D_{k\ell}(Z(s)) + \langle \partial_k Y(s), \partial_\ell \tilde{Y}(s) \rangle_{\varpi} D_{k\ell}(\tilde{Z}(s)) \right] \\
& -2 \sum_{m=1}^{d} \int \int (\nabla Y(s,r) \cdot \vec{C}_{Z(s),m}) (\nabla \tilde{Y}(s,r) \cdot \vec{C}_{\tilde{Z}(s),m}) \, \mathrm{d}p \, \varpi(r) \mathrm{d}r \\
& = \sum_{m=1}^{d} \int \int \mathrm{d}p \, \varpi(r) \mathrm{d}r \\
& \quad \times \{ (\nabla Y(s,r) \cdot \vec{C}_{Z(s),m}) (\nabla \tilde{Y}(s,r) \cdot \vec{C}_{Z(s),m}) \\
& \quad + (\nabla Y(s,r) \cdot \vec{C}_{\tilde{Z}(s),m}) (\nabla \tilde{Y}(s,r) \cdot \vec{C}_{\tilde{Z}(s),m}) \\
& \quad - 2 (\nabla Y(s,r) \cdot \vec{C}_{Z(s),m}) (\nabla \tilde{Y}(s,r) \cdot \vec{C}_{\tilde{Z}(s),m}) \} \\
& = \sum_{m=1}^{d} \int \int \mathrm{d}p \, \varpi(r) \mathrm{d}r \\
& \quad \times \{ (\nabla Y(s,r) \cdot \vec{C}_{Z(s),m}) (\nabla \tilde{Y}(s,r) \cdot [\vec{C}_{Z(s),m} - \vec{C}_{\tilde{Z}(s),m}]) \\
& \quad + (\nabla \tilde{Y}(s,r) \cdot \vec{C}_{\tilde{Z}(s),m}) (\nabla Y(s,r) \cdot [\vec{C}_{\tilde{Z}(s),m} - \vec{C}_{Z(s),m}]) \} \\
& = \sum_{m=1}^{d} \int \int \mathrm{d}p \, \varpi(r) \mathrm{d}r \\
& \quad \times \{ [(\nabla Y(s,r) \cdot \vec{C}_{Z(s),m}) \nabla \tilde{Y}(s,r) \\
& \quad - (\nabla \tilde{Y}(s,r) \cdot \vec{C}_{\tilde{Z}(s),m}) \nabla Y(s,r)] \cdot [\vec{C}_{Z(s),m} - \vec{C}_{\tilde{Z}(s),m}] \}.
\end{aligned}
\right\}
\tag{12.29}
$$

Note that

$$
\left.
\begin{aligned}
& [(\nabla Y(s,r) \cdot \vec{C}_{Z(s),m}) \nabla \tilde{Y}(s,r) - (\nabla \tilde{Y}(s,r) \cdot \vec{C}_{\tilde{Z}(s),m}) \nabla Y(s,r)] \\
& \qquad = ([\nabla Y(s,r) - \nabla \tilde{Y}(s,r)] \cdot \vec{C}_{Z(s),m}) \nabla \tilde{Y}(s,r) \\
& \qquad + (\nabla \tilde{Y}(s,r) \cdot [\vec{C}_{Z(s),m} - \vec{C}_{\tilde{Z}(s),m}]) \nabla \tilde{Y}(s,r) \\
& \qquad + (\nabla \tilde{Y}(s,r) \cdot \vec{C}_{\tilde{Z}(s),m}) [\nabla \tilde{Y}(s,r) - \nabla Y(s,r)].
\end{aligned}
\right\}
\tag{12.30}
$$

Hence,

$$
\begin{aligned}
&I_{1,3}(s)+I_{1,4}(s)+\sum_{m=1}^{d}[II_{3,1,m}(s)+II_{4,1,m}(s)]\\
&=\sum_{m=1}^{d}\int\int \mathrm{d}p\ \varpi(r)\mathrm{d}r\\
&\quad\times\{([\nabla Y(s,r)-\nabla\tilde{Y}(s,r)]\cdot\vec{C}_{Z(s),m})\nabla\tilde{Y}(s,r)\\
&\quad+(\nabla\tilde{Y}(s,r)\cdot[\vec{C}_{Z(s),m}-\vec{C}_{\tilde{Z}(s),m}])\nabla\tilde{Y}(s,r)\\
&\quad+(\nabla\tilde{Y}(s,r)\cdot\vec{C}_{\tilde{Z}(s),m})[\nabla\tilde{Y}(s,r)-\nabla Y(s,r)])\\
&\quad\cdot(\vec{C}_{Z(s),m}-\vec{C}_{\tilde{Z}(s),m})\}\\
&:=\hat{I}_1(s)+\hat{I}_2(s)+\hat{I}_3(s).
\end{aligned}
$$

Integration by parts yields

$$
\begin{aligned}
&\hat{I}_1(s)\\
&=\sum_{m,k,\ell=1}^{d}\int\int \mathrm{d}p\ \varpi(r)\mathrm{d}r\partial_{k,r}(Y(s,r)\\
&\quad-\tilde{Y}(s,r))\mathcal{J}_{km}(r,p,Z(s))\big(\partial_{\ell,r}\tilde{Y}(s,r)\big)\big(\mathcal{J}_{\ell m}(r,p,Z(s))-\mathcal{J}_{\ell m}\big(r,p,\tilde{Z}(s)\big)\big)\\
&=-\sum_{m,k,\ell=1}^{d}\int\int \mathrm{d}p\ \varpi(r)\mathrm{d}r\big(Y(s,r)\\
&\quad-\tilde{Y}(s,r)\big)(\partial_{k,r}\mathcal{J}_{km}(r,p,Z(s)))\big(\partial_{\ell,r}\tilde{Y}(s,r)\big)\big(\mathcal{J}_{\ell m}(r,p,Z(s))\\
&\quad-\mathcal{J}_{\ell m}\big(r,p,\tilde{Z}(s)\big)\big)-\sum_{m,k,\ell=1}^{d}\int\int \mathrm{d}p\ \varpi(r)\mathrm{d}r\big(Y(s,r)\\
&\quad-\tilde{Y}(s,r)\big)\mathcal{J}_{km}(r,p,Z(s))\big(\partial^2_{k,\ell,r}\tilde{Y}(s,r)\big)\big(\mathcal{J}_{\ell m}(r,p,Z(s))\\
&\quad-\mathcal{J}_{\ell m}\big(r,p,\tilde{Z}(s)\big)\big)-\sum_{m,k,\ell=1}^{d}\int\int \mathrm{d}p\ \varpi(r)\mathrm{d}r\big(Y(s,r)\\
&\quad-\tilde{Y}(s,r)\big)\mathcal{J}_{km}(r,p,Z(s))\big(\partial_{\ell,r}\tilde{Y}(s,r)\big)\big(\partial_{k,r}\big(\mathcal{J}_{\ell m}(r,p,Z(s))\\
&\quad-\mathcal{J}_{\ell m}\big(r,p,\tilde{Z}(s)\big)\big)\big)-\sum_{m,k,\ell=1}^{d}\int\int \mathrm{d}p\partial_k\varpi(r)\mathrm{d}r\big(Y(s,r)\\
&\quad-\tilde{Y}(s,r)\big)\mathcal{J}_{km}(r,p,Z(s))\big(\partial_{\ell,r}\tilde{Y}(s,r)\big)\big(\mathcal{J}_{\ell m}(r,p,Z(s))\\
&\quad-\mathcal{J}_{\ell m}\big(r,p,\tilde{Z}(s)\big)\big).
\end{aligned}
$$

Therefore, by (15.46) in addition to our assumptions and the Cauchy–Schwarz inequality,

$$
\big|\hat{I}_1(s)\big|\le c\max_{|\mathbf{m}|\le 2}\big\|\partial^{\mathbf{m}}\tilde{Y}(s)\big\|_{0,\varpi}\big\|Y(s)-\tilde{Y}(s)\big\|_{0,\varpi}\big\|Z(s)-\tilde{Z}(s)\big\|_{0,\varpi}.
$$

A similar calculation yields the same estimate for $\|\hat{I}_3(s)\|$. For $\|\hat{I}_2(s)\|$ we obtain a bound

$$c \max_{|\mathbf{m}|\le 2} \|\partial^{\mathbf{m}}\tilde{Y}(s)\|_{0,\varpi}\|Z(s)-\tilde{Z}(s)\|^2_{0,\varpi}$$

without integration by parts. Hence,

$$\left.\begin{aligned}&\Big|I_{1,3}(s)+I_{1,4}(s)+\sum_{m=1}^{d}[II_{3,1,m}(s)+II_{4,1,m}(s)]\Big|\\&\le c\max_{|\mathbf{m}|\le 2}\|\partial^{\mathbf{m}}\tilde{Y}(s)\|_{0,\varpi}[\|Y(s)-\tilde{Y}(s)\|_{0,\varpi}\|Z(S)\\&\qquad-\tilde{Z}(s)\|_{0,\varpi}+\|Z(s)-\tilde{Z}(s)\|^2_{0,\varpi}].\end{aligned}\right\} \tag{12.31}$$

Next, we derive suitable bounds for the terms in (12.25) and (12.28). We start with (12.25):

$$\left.\begin{aligned}&\Bigg|\sum_{k,\ell=1}^{d}\sum_{m=1}^{d}\int[\mathcal{J}_{k,m}(r,p,Z(s))-\mathcal{J}_{k,m}(r,p,\tilde{Z}(s))]\\&\qquad[\partial_{\ell,r}\mathcal{J}_{\ell,m}(r,p,Z(s))+\partial_{\ell,r}\mathcal{J}_{\ell,m}(r,p,\tilde{Z}(s))]\mathrm{d}p\Bigg|\\&\le\sum_{k,\ell=1}^{d}\sum_{m=1}^{d}\left\{\int[\mathcal{J}_{k,m}(r,p,Z(s))-\mathcal{J}_{k,m}(r,p,\tilde{Z}(s))]^2\mathrm{d}p\right\}^{1/2}\\&\quad\times\left\{\int[\partial_{\ell,r}\mathcal{J}_{\ell,m}(r,p,Z(s))+\partial_{\ell,r}\mathcal{J}_{\ell,m}(r,p,\tilde{Z}(s))]^2\mathrm{d}p\right\}^{1/2}.\end{aligned}\right\} \tag{12.32}$$

By the Cauchy–Schwarz inequality, we have

$$\left.\begin{aligned}&\left|\int(\partial_{\ell,r}\mathcal{J}_{\ell,m}(r,p,Z(s))\partial_{\ell,r}\mathcal{J}_{\ell,m}(r,p,\tilde{Z}(s))\mathrm{d}p\right|^2\\&\le\int\partial_{\ell,r}\mathcal{J}^2_{\ell,m}(r,p,Z(s))\mathrm{d}p\int\partial_{\ell,r}\mathcal{J}^2_{\ell,m}(r,p,\tilde{Z}(s))\mathrm{d}p.\end{aligned}\right\}$$

Further, by assumption (8.35),

$$\left.\left|\int\partial_{\ell,r}\mathcal{J}^2_{\ell,m}(r,p,Z(s))\mathrm{d}p\right|=|-\partial_r\partial_q D_{k\ell}(s,Z,r-q)_{|q=r}|\le c\right\}. \tag{12.33}$$

Therefore,

$$\left\{\int[\partial_{\ell,r}\mathcal{J}_{\ell,m}(r,p,Z(s))+\partial_{\ell,r}\mathcal{J}_{\ell,m}(r,p,\tilde{Z}(s))]^2\mathrm{d}p\right\}^{1/2}$$

$$\le c\ \forall\ell,m\ \text{ (independent of } Z \text{ and } \tilde{Z}\text{).} \tag{12.34}$$

Consequently, by (12.32) and (12.34) in addition to Hypothesis 4.1 and (8.46),

$$\left.\begin{array}{l}\left|\sum_{k,\ell=1}^{d}\sum_{m=1}^{d}\int[\mathcal{J}_{k,m}(r,p,Z(s))-\mathcal{J}_{k,m}(r,p,\tilde{Z}(s))]\right.\\ \left.[\partial_{\ell,r}\mathcal{J}_{\ell,m}(r,p,Z(s))+\partial_{\ell,r}\mathcal{J}_{\ell,m}(r,p,\tilde{Z}(s))]\mathrm{d}p\right|\\ \qquad\leq c\|Z(s)-\tilde{Z}(s)\|_{0,\varpi}.\end{array}\right\}\tag{12.35}$$

Therefore,

$$\left.\begin{array}{l}\left|\sum_{m=1}^{d}\int\int(Y(s,r)-\tilde{Y}(s,r))(\nabla Y(s,r)\right.\\ \quad\left.+\nabla\tilde{Y}(s,r))\cdot(\vec{C}_{Z(s),m}-\vec{C}_{\tilde{Z}(s),m})(D_{Z(s),m}+D_{(s\tilde{)},m)})\right|\\ \leq\int|Y(s,r)-\tilde{Y}(s,r)|\max_{k=1,..,d}(|\partial Y(s,r)|\\ +|\partial\tilde{Y}(s,r)|)\varpi(r)\mathrm{d}r\; c\|Z(s)-\tilde{Z}(s)\|_{\varpi}\\ \leq c\max_{k=1,..,d}[\|\partial Y(s)\|_{0,\varpi}\\ +\|\partial\tilde{Y}(s)\|_{0,\varpi}]\|Y(s)-\tilde{Y}(s)\|_{0,\varpi}\|Z(s)-\tilde{Z}(s)\|_{0,\varpi}\\ \qquad\text{(by the Cauchy–Schwarz inequality).}\end{array}\right\}\tag{12.36}$$

Let $\partial^{\mathbf{m}}$ and $\partial^{\mathbf{n}}$ be two partial differential operators of orders $m := |\mathbf{m}|$ and $n := |\mathbf{n}|$, respectively. Note that the integrals in formula (12.37) are well defined for $|\mathbf{m}|, |\mathbf{n}| \leq 2$ by the assumptions in Theorem 8.6. Then,

$$\left.\begin{array}{l}\int(\partial^{\mathbf{m}}\mathcal{J}_{k,m}(r,p,Z(s))\partial^{\mathbf{n}}\mathcal{J}_{\ell,m}(r,p,\tilde{Z}(s))\\ \quad-\partial^{\mathbf{m}}\mathcal{J}_{k,m}(r,p,\tilde{Z}(s))\partial^{\mathbf{n}}\mathcal{J}_{\ell,m}(r,p,Z(s)))\mathrm{d}p\\ \quad\leq\left|\int(\partial^{\mathbf{m}}\mathcal{J}_{k,m}(r,p,Z(s))-\partial^{\mathbf{m}}\mathcal{J}_{k,m}(r,p,\tilde{Z}(s)))\partial^{\mathbf{n}}\mathcal{J}_{\ell,m}(r,p,\tilde{Z}(s))\mathrm{d}p\right|\\ \qquad+\left|\int(\partial^{\mathbf{n}}\mathcal{J}_{\ell,m}(r,p,Z(s))-\partial^{\mathbf{n}}\mathcal{J}_{\ell,m}(r,p,\tilde{Z}(s)))\partial^{\mathbf{m}}\mathcal{J}_{k,m}(r,p,\tilde{Z}(s))\mathrm{d}p\right|\\ \quad\leq\left\{\int(\partial^{\mathbf{m}}\mathcal{J}_{k,m}(r,p,Z(s))-\partial^{\mathbf{m}}\mathcal{J}_{k,m}(r,p,\tilde{Z}(s)))^{2}\mathrm{d}p\right\}^{1/2}\\ \quad\left\{\int(\partial^{\mathbf{n}}\mathcal{J}_{\ell,m}(r,p,\tilde{Z}(s)))^{2}\mathrm{d}p\right\}^{1/2}\\ \qquad+\left\{\int(\partial^{\mathbf{n}}\mathcal{J}_{\ell,m}(r,p,Z(s))-\partial^{\mathbf{n}}\mathcal{J}_{\ell,m}(r,p,\tilde{Z}(s)))^{2}\mathrm{d}p\right\}^{1/2}\\ \qquad\times\left\{\int(\partial^{\mathbf{m}}\mathcal{J}_{k,m}(r,p,\tilde{Z}(s)))^{2}\mathrm{d}p\right\}^{1/2}.\end{array}\right\}\tag{12.37}$$

Hence, by our assumptions for $|\mathbf{m}|, |\mathbf{n}| \leq 2$,[242]

$$\left.\begin{aligned}&\int (\partial^{\mathbf{m}} \mathcal{J}_{k,m}(r,p,Z(s))\partial^{\mathbf{n}} \mathcal{J}_{\ell,m}(r,p,\tilde{Z}(s))\\&-\partial^{\mathbf{m}} \mathcal{J}_{k,m}(r,p,\tilde{Z}(s))\partial^{\mathbf{n}} \mathcal{J}_{\ell,m}(r,p,Z(s)))\mathrm{d}p\\&\leq c\|Z(s)) - \tilde{Z}(s)\|_{0,\varpi}.\end{aligned}\right\} \quad (12.38)$$

Further,

$$\left.\begin{aligned}&\sum_{m=1}^{d}\Big|\int\int \mathrm{d}p\,\varpi(r)\mathrm{d}r(Y(s,r)\nabla Y(s,r)\\&\quad -\tilde{Y}(s,r)\nabla\tilde{Y}(s,r))\cdot[\vec{C}_{Z(s),m}D_{\tilde{Z}(s),m} - \vec{C}_{\tilde{Z}(s),m}D_{Z(s),m}]\Big|\\&\leq \sum_{m=1}^{d}\Big|\int\int \mathrm{d}p\,\varpi(r)\mathrm{d}r(Y(s,r)\\&\quad -\tilde{Y}(s,r))\nabla Y(s,r)\cdot[\vec{C}_{Z(s),m}D_{\tilde{Z}(s),m} - \vec{C}_{\tilde{Z}(s),m}D_{Z(s),m}]\Big|\\&\quad +\sum_{m=1}^{d}\Big|\int\int \mathrm{d}p\,\varpi(r)\mathrm{d}r\tilde{Y}(s,r)(\nabla(Y(s,r)\\&\quad -\tilde{Y}(s,r)))\cdot[\vec{C}_{Z(s),m}D_{\tilde{Z}(s),m} - \vec{C}_{\tilde{Z}(s),m}D_{Z(s),m}]\Big|\\&=: A + B.\end{aligned}\right\} \quad (12.39)$$

By (12.38)

$$\left.\begin{aligned}A &\leq c\|Z(s) - \tilde{Z}(s)\|_{0,\varpi}\sum_{k,\ell=1}^{d}\sum_{m=1}^{d}\Big|\int |Y(s,r) - \tilde{Y}(s,r)|\cdot|\partial_k Y(s,r)|\varpi(r)\mathrm{d}r\\&\leq c\|Z(s) - \tilde{Z}(s)\|_{0,\varpi}\|Y(s) - \tilde{Y}(s)\|_{0,\varpi}\max_{k=1,..,d}\|\partial_k Y(s)\|_{0,\varpi}\\&\qquad \text{(by the Cauchy–Schwarz inequality).}\end{aligned}\right\} \quad (12.40)$$

[242] Cf. (8.45).

In B, we integrate by parts to move the partial differentiation from $(Y(s,r) - \tilde{Y}(s,r))$ to the other factors:

$$\left.\begin{array}{l} B \le \sum_{m=1}^{d} \\ \left| \int\int \mathrm{d}p\, \varpi(r)\mathrm{d}r(Y(s,r) \right. \\ \quad \left. -\tilde{Y}(s,r))\nabla\tilde{Y}(s,r)\cdot[\vec{C}_{Z(s),m}D_{\tilde{Z}(s),m} - \vec{C}_{\tilde{Z}(s),m}D_{Z(s),m}] \right| \\ \quad + \left| \int\int \mathrm{d}p\, \varpi(r)\mathrm{d}r(Y(s,r) \right. \\ \quad \left. -\tilde{Y}(s,r))\tilde{Y}(s,r)[\mathrm{div}\vec{C}_{Z(s),m}D_{\tilde{Z}(s),m} - \mathrm{div}\vec{C}_{\tilde{Z}(s),m}D_{Z(s),m}] \right| \\ \quad + \left| \int\int \mathrm{d}p\, \varpi(r)\mathrm{d}r(Y(s,r) \right. \\ \quad \left. -\tilde{Y}(s,r))\tilde{Y}(s,r)[\vec{C}_{Z(s),m}\cdot\nabla D_{\tilde{Z}(s),m} - \vec{C}_{\tilde{Z}(s),m}\cdot\nabla D_{Z(s),m}] \right| \\ \quad + \left| \int\int \mathrm{d}p(\nabla\varpi(r))\mathrm{d}r(Y(s,r) \right. \\ \quad \left. -\tilde{Y}(s,r))\tilde{Y}(s,r)\cdot[\vec{C}_{Z(s),m}D_{\tilde{Z}(s),m} - \vec{C}_{\tilde{Z}(s),m}D_{Z(s),m}] \right| \\ =: B_1 + B_2 + B_3 + B_4. \end{array}\right\} \quad (12.41)$$

By (12.38)

$$\left.\begin{array}{l} B_1 \le c\|Z(s) - \tilde{Z}(s)\|_{0,\varpi}\|Y(s) - \tilde{Y}(s)\|_{0,\varpi}\max_{k=1,..,d}\|\partial_k\tilde{Y}(s)\|_{0,\varpi}, \\ B_i \le c\|Z(s) - \tilde{Z}(s)\|_{0,\varpi}\|Y(s) - \tilde{Y}(s)\|_{0,\varpi}\|\tilde{Y}(s)\|_{0,\varpi}, \quad i = 2,3,4. \end{array}\right\} \quad (12.42)$$

Hence,

$$\left.\begin{array}{l} \left| \sum_{m=1}^{d} \int\int (Y(s,r)\nabla Y(s,r) \right. \\ \qquad \left. -\tilde{Y}(s,r)\nabla\tilde{Y}(s,r))\cdot(\vec{C}_{Z(s),m}D_{\tilde{Z}(s),m} - \vec{C}_{\tilde{Z}(s),m}D_{Z(s),m}) \right| \\ \quad \le c \max_{k=1,..,d}[\|\partial_k Y(s)\|_{0,\varpi} \\ \quad +\|\partial_k\tilde{Y}(s)\|_{0,\varpi}]\|Z(s) - \tilde{Z}(s)\|_{0,\varpi}\|Y(s) - \tilde{Y}(s)\|_{0,\varpi}. \end{array}\right\} \quad (12.43)$$

Obviously, we obtain as in (12.43)

$$\left.\begin{array}{l} \left| \sum_{m=1}^{d} \int\int (Y(s,r)\nabla\tilde{Y}(s,r) \right. \\ \qquad \left. -\tilde{Y}(s,r)\nabla Y(s,r))\cdot(\vec{C}_{Z(s),m}D_{Z(s),m} - \vec{C}_{\tilde{Z}(s),m}D_{\tilde{Z}(s),m}) \right| \\ \quad \le c \max_{k=1,..,d}[\|\partial_k Y(s)\|_{0,\varpi} \\ \quad +\|\partial_k\tilde{Y}(s)\|_{0,\varpi}]\|Z(s) - \tilde{Z}(s)\|_{0,\varpi}\|Y(s) - \tilde{Y}(s)\|_{0,\varpi}. \end{array}\right\} \quad (12.44)$$

The integrand in the last integral of (12.25) can be rewritten as follows

$$(Y(s,r)D_{\tilde{Z}(s),m} - \tilde{Y}(s,r)D_{Z(s),m})(\nabla\tilde{Y}(s,r)\cdot\vec{C}_{Z(s),m} - \nabla Y(s,r)\bullet\vec{C}_{\tilde{Z}(s),m})$$
$$= [(Y(s,r)-\tilde{Y}(s,r))D_{\tilde{Z}(s),m} + \tilde{Y}(s,r)(D_{\tilde{Z}(s),m} - D_{Z(s),m})]$$
$$\times[\nabla(\tilde{Y}(s,r)-Y(s,r))\bullet\vec{C}_{Z(s),m} + \nabla Y(s,r)\bullet(\vec{C}_{Z(s),m} - \vec{C}_{\tilde{Z}(s),m})].$$

By (12.6) we simplify:

$$\int\int(Y(s,r)-\tilde{Y}(s,r))D_{\tilde{Z}(s),m}[\nabla(\tilde{Y}(s,r)-Y(s,r))]\bullet\vec{C}_{Z(s),m}\mathrm{d}p\,\varpi(r)\mathrm{d}r$$
$$= -\frac{1}{2}\int\int(Y(s,r)-\tilde{Y}(s,r))^2 D_{Z(s),m}D_{\tilde{Z}(s),m}\mathrm{d}p\,\varpi(r)\mathrm{d}r$$
$$-\frac{1}{2}\int\int(Y(s,r)-\tilde{Y}(s,r))^2\vec{C}_{Z(s),m}\bullet(\nabla D_{\tilde{Z}(s),m})\mathrm{d}p\,\varpi(r)\mathrm{d}r$$
$$-\frac{1}{2}\int\int(Y(s,r)-\tilde{Y}(s,r))^2\vec{C}_{Z(s),m}\bullet(\nabla\varpi(r))D_{\tilde{Z}(s),m})\mathrm{d}p\,\mathrm{d}r$$

(integrating by parts).

Hence,

$$\left.\begin{array}{l}
\int\int(Y(s,r)-\tilde{Y}(s,r))D_{\tilde{Z}(s),m}[\nabla(\tilde{Y}(s,r)-Y(s,r))]\bullet\vec{C}_{Z(s),m}\mathrm{d}p\,\varpi(r)\mathrm{d}r\\
\quad\le\frac{1}{2}\int\int(Y(s,r)-\tilde{Y}(s,r))^2|\vec{C}_{Z(s),m}\bullet\nabla D_{\tilde{Z}(s),m}|\varpi(r)\mathrm{d}p\,\mathrm{d}r\\
\quad+\frac{1}{2}\int\int(Y(s,r)-\tilde{Y}(s,r))^2|D_{Z(s),m}D_{\tilde{Z}(s),m}|\mathrm{d}p\,\varpi(r)\mathrm{d}r\\
\quad+\frac{1}{2}\int\int(Y(s,r)-\tilde{Y}(s,r))^2|\vec{C}_{Z(s),m}\bullet\nabla\varpi(r)|D_{\tilde{Z}(s),m}\mathrm{d}p\,\mathrm{d}r\\
\qquad\qquad\le c\|Y(s)-\tilde{Y}(s)\|^2_{0,\varpi},
\end{array}\right\}\tag{12.45}$$

employing (15.46) in addition to the boundedness assumptions on the moments of the derivatives of $\mathcal{J}$. The other terms in the last integral of (12.25) can be rewritten as in the derivation of (12.43). Therefore,

$$\left.\begin{array}{l}
\left|\sum_{m=1}^{d}\int\int(Y(s,r)D_{\tilde{Z}(s),m} - \tilde{Y}(s,r)D_{Z(s),m})\right.\\
\left.(\nabla\tilde{Y}(s,r)\bullet\vec{C}_{Z(s),m} - \nabla Y(s,r)\bullet\vec{C}_{\tilde{Z}(s),m})\right|\\
\le c\max_{k=1,..,d}[\|\partial_k Y(s)\|_{0,\varpi} + \|\partial_k\tilde{Y}(s)\|_{0,\varpi}]\|Z(s)-\tilde{Z}(s)\|_{0,\varpi}\|Y(s)-\tilde{Y}(s)\|_{0,\varpi}.
\end{array}\right\}\tag{12.46}$$

Altogether, we can bound the right-hand side of (12.25) as follows

$$\left.\begin{aligned}&\Bigg|\sum_{m=1}^{d}(II_{1,3,m}(s)+II_{1,4,m}(s)+II_{2,3,m}(s)+II_{2,4,m}(s))\\&+\sum_{m=1}^{d}(II_{3,4,m}(s)+II_{4,4,m}(s)+II_{3,3,m}(s)+II_{4,3,m}(s))\Bigg|\\&\qquad\le c\max_{k=1,..,d}[\|\partial_k Y(s)\|_{0,\varpi}+\|\partial_k\tilde{Y}(s)\|_{0,\varpi}]\\&\qquad\|Z(s)-\tilde{Z}(s)\|_{0,\varpi}\|Y(s)-\tilde{Y}(s)\|_{0,\varpi}.\end{aligned}\right\}\tag{12.47}$$

obtaining a bound for the right-hand side of (12.28) is easier. Following the preceding steps, we obtain

$$\left.\begin{aligned}&\sum_{m=1}^{d}\int\int\left(Y(r,s)D_{Z(s),m}-\tilde{Y}(s,r)D_{\tilde{Z}(s),m}\right)^2\mathrm{d}p\,\varpi(r)\mathrm{d}r\\&\quad\le 2\sum_{m=1}^{d}\int\int\left(Y(r,s)-\tilde{Y}(s,r)\right)^2D^2_{Z(s),m}\mathrm{d}p\,\varpi(r)\mathrm{d}r\\&\quad\quad+2\sum_{m=1}^{d}\int\int\tilde{Y}^2(s,r)(D_{Z(s),m}-D_{\tilde{Z}(s),m})^2\mathrm{d}p\,\varpi(r)\mathrm{d}r\\&\quad\le c\|\tilde{Y}(s)\|^2_{0,\varpi}[\|Z(s)-\tilde{Z}(s)\|^2_{0,\varpi}+\|Y(s)-\tilde{Y}(s)\|^2_{0,\varpi}].\end{aligned}\right\}\tag{12.48}$$

Using $|ab|\le\frac{1}{2}(a^2+b^2)$ for the upper bounds, we obtain the final bound for the deterministic (diffusion) integrals in (12.2):

$$\left.\begin{aligned}&\int_0^t<\sum_{k,\ell=1}^{d}\partial^2_{k,\ell}(D_{k,\ell}(Z(s))Y(s)\\&\quad-D_{k,\ell}(\tilde{Z}(s))\tilde{Y}(s)),\,Y(s)-\tilde{Y}(s)>_{0,\varpi}\,\mathrm{d}s\\&\quad+\sum_{m=1}^{d}\int_0^t\int\int(\nabla\bullet(Y(r,s)\mathcal{J}_{\cdot,m}(r,p,Z(s))\\&\quad-\tilde{Y}(r,s)\mathcal{J}_{\cdot,m}(r,p,\tilde{Z}(s))))^2\mathrm{d}p\,\varpi(r)\mathrm{d}r\,\mathrm{d}s\\&\quad\le c\int_0^t\max_{|\mathbf{m}|\le 2}\left[\|\partial^{\mathbf{m}}Y(s)\|_{0,\varpi}+\|\partial^{\mathbf{m}}\tilde{Y}(s)\|_{0,\varpi}+1\right]\\&\quad\left[\|Y(s)-\tilde{Y}(s)\|^2_{0,\varpi}+\|Z(s)-\tilde{Z}(s)\|^2_{0,\varpi}\right]\mathrm{d}s.\end{aligned}\right\}\tag{12.49}$$

It remains to estimate $III(s)$ in (12.2). We obtain

$$\left.\begin{aligned}
&III(s)\\
&= \int [(\nabla Y(s,r) \bullet F(r, Z(s))\\
&\quad -\nabla \tilde{Y}(s,r) \bullet F(r, \tilde{Z}(s)))][Y(s,r) - \tilde{Y}(x,r)]\varpi(r)\mathrm{d}r\\
&\quad + \int [Y(s,r)\mathrm{div}\, F(r, Z(s))\\
&\quad -\tilde{Y}(s,r)\mathrm{div}\, F(r, \tilde{Z}(s))][Y(s,r) - \tilde{Y}(x,r)]\varpi(r)\mathrm{d}r\\
&= \int [(\nabla(Y(s,r) - \tilde{Y}(x,r))) \bullet F(r, Z(s))\\
&\quad +(\nabla \tilde{Y}(s,r)) \bullet (F(r, Z(s)) - F(r, \tilde{Z}(s)))][Y(s,r) - \tilde{Y}(x,r)]\varpi(r)\mathrm{d}r\\
&\quad + \int [(Y(s,r) - \tilde{Y}(x,r))\mathrm{div}\, F(r, Z(s)\\
&\quad +\tilde{Y}(s,r)(\mathrm{div}\, F(r, Z(s) - \mathrm{div}\, F(r, \tilde{Z}(s))][Y(s,r) - \tilde{Y}(x,r)]\varpi(r)\mathrm{d}r.
\end{aligned}\right\} \tag{12.50}$$

By the same simple integration by parts and (12.6), we obtain from (12.49) and the assumptions of Lemma 8.10 (cf. (8.45))

$$\left.\begin{aligned}
&|\langle \nabla \bullet (Y(s)F(\cdot, Z(s))) - \nabla \bullet (\tilde{Y}(s)F(\cdot, \tilde{Z}(s))), Y(s) - \tilde{Y}(s)\rangle_{0,\varpi}|\\
&\le c \max_{k=1,..,d}[\|\partial_k \tilde{Y}(s)\|_{0,\varpi}][\|Y(s) - \tilde{Y}(s)\|^2_{0,\varpi} + \|Z(s) - \tilde{Z}(s)\|^2_{0,\varpi}].
\end{aligned}\right\} \tag{12.51}$$

Altogether, (12.49) and (12.51) imply the estimate

$$\left.\begin{aligned}
&\|Y(t) - \tilde{Y}(t)\|^2_{0,\varpi}\\
&\quad \le c \int_0^t \max_{|\mathbf{m}|\le 2}[\|\partial^{\mathbf{m}} Y(s)\|_{0,\varpi} + \|\partial^{\mathbf{m}} \tilde{Y}(s)\|_{0,\varpi} + 1]\\
&\qquad \times [\|Y(s) - \tilde{Y}(s)\|^2_{0,\varpi} + \|Z(s)) - \tilde{Z}(s)\|^2_{0,\varpi}]\mathrm{d}s\\
&\qquad +\|Y(0) - \tilde{Y}(0)\|^2_{0,\varpi} + \text{martingale}.
\end{aligned}\right\} \tag{12.52}$$

We stop

$$\tau_N := \inf\{t \ge 0 : \max_{|\mathbf{m}|\le 2}[\|\partial^{\mathbf{m}} Y(s)\|_{0,\varpi} + \|\partial^{\mathbf{m}} \tilde{Y}(s)\|_{0,\varpi} + 1] \ge N\}. \tag{12.53}$$

Now Gronwall's inequality implies (8.48).

Remark 12.1. Expression (12.3) can be estimated without the associated quadratic variation term:

$$\left.\begin{aligned}
&\left|\sum_{k,\ell=1}^{d} \langle D_{k,\ell}(Z(s))\partial_k Y(s) - D_{k,\ell}(\tilde{Z}(s))\partial_k \tilde{Y}(s), \partial_\ell Y(s) - \partial_\ell \tilde{Y}(s)\rangle_{\varpi}\right| \\
&\quad \le \left|\sum_{k,\ell=1}^{d} \langle [D_{k,\ell}(Z(s)) - D_{k,\ell}(\tilde{Z}(s))]\partial_k Y(s), \partial_\ell Y(s) - \partial_\ell \tilde{Y}(s)\rangle_{\varpi}\right| \\
&+ \left|\sum_{k,\ell=1}^{d} -\langle D_{k,\ell}(\tilde{Z}(s))\partial_k [Y(s)\tilde{Y}(s)], \partial_\ell Y(s) - \partial_\ell \tilde{Y}(s)\rangle_{\varpi}\right| \\
&\quad \le c \max_{|\mathbf{m}|\le 2} \|\partial^{\mathbf{m}} Y(s)\|_{0,\varpi}[\|Z(s) - \tilde{Z}(s)\|_{0,\varpi}^2 + \|Y(s) - \tilde{Y}(s)\|_{0,\varpi}^2],
\end{aligned}\right\} \tag{12.54}$$

where we used the previous calculations in the last step. In this formal manipulation we may replace $D_{k,\ell}(Z(s))$ by $\sigma^{\perp}(Z(s))$, and we obtain under the assumptions of Theorem 8.20 also strong uniqueness in the coercive case. □

Chapter 13
Comments on Other Approaches to SPDEs

SPDEs are classified by formal criteria such as linear, semilinear, quasilinear, and nonlinear. Each class is divided into subclasses, depending on whether the driving noise is a regular Brownian motion or cylindrical Brownian motion.

The comments in this section are not meant to be a complete overview of the field of SPDEs in its present state; although, in view of the absence of such an overview, this might be a desirable goal. Instead, we will review a small "random" sample of results. This may help the reader to learn about some other approaches to SPDEs. The approaches were motivated either by some kind of physical model or by the desire to generalize deterministic PDEs to SPDEs, adding a stochastic term to the right-hand side of the PDE. The latter approach is formally similar to the transition from an ordinary differential equation (ODE) to a stochastic ordinary differential equation (SODE). Although the present book is motivated by modeling problems, it is somewhat easier to classify SPDEs first in a formal manner by considering a time-dependent PDE and adding a stochastic term. In fact, all the models we will be considering may be formally represented in this way. The reader may also wish to consider the book, edited by Carmona and Rozovski (1999), which contains six papers, written by leading experts in their fields. Four of them are on models and two papers cover different mathematical approaches.

13.1 Classification

Let $\mathcal{O} \subset \mathbf{R}^d$ be some domain and $X(t, r)$ be some $\mathbf{R}^m$-valued (random) field, where $r \in B$ and $t \geq 0$. For most of what follows we assume that $B = \mathbf{R}^d$, and we will point out whenever this assumption does not hold. $\hat{A}(t, X, r, \nabla X, \nabla\nabla^T X, \omega)$ is some $\mathbf{R}^m$-valued functional. This functional describes the deterministic rate of change in our SPDE. For the stochastic rate of change we initially assume that $W(t, r)$ is a Brownian motion[1] in t for fixed r and measurable in all arguments. Later

[1] An equivalent term for $W(\cdot, \cdot)$ is *Wiener process*.

we will somewhat modify this description to a more rigorous infinite dimensional setup. Further, let $\hat{B}(t, X, r, \nabla X, \omega)$ be a suitable linear operator valued functional such that $\hat{B}(t, X, r, \nabla X, \omega)[W(t + \mathrm{d}t, \cdot) - W(t, \cdot)] \in \mathbf{R}^m$. As before, we write

$$W(\mathrm{d}t, \cdot) := W(t + \mathrm{d}t, \cdot) - W(t, \cdot).$$

Using this notation, the SPDEs we are mainly interested in are of second-order parabolic type. We make some additional comments and give references to some SPDEs of hyperbolic type as well. Second-order parabolic SPDEs are usually cast in the form of the following initial value problem:

$$\left.\begin{aligned} \mathrm{d}X &= \hat{A}(t, X, \cdot, \nabla X, \nabla\nabla^{\mathrm{T}} X, \cdot)\mathrm{d}t + \hat{B}(t, X, \cdot, \nabla X, \cdot)W(\mathrm{d}t, \cdot), \\ X(0) &= X_0. \end{aligned}\right\} \tag{13.1}$$

We assume that all quantities are adapted to a stochastic filtration $(\Omega, \mathcal{F}, \mathcal{F}_t, P)$ and that the stochastic differential is to be taken in the sense of Itô. In addition, we need to impose boundary conditions, if $\mathcal{O} \neq \mathbf{R}^d$. The generalization of this scheme to higher order PDE operators is simple and will not be discussed here. With the exception of some recent results in stochastic fluid mechanics, all functionals will be real valued, i.e., $m = 1$. In the following classification we assume $m = 1$, since the vector valued case can be classified by a straightforward generalization. Let us first assume $\hat{B} \equiv 0$. (13.1) then becomes the deterministic PDE

$$\left.\begin{aligned} \tfrac{\partial}{\partial t} Y &= \hat{A}(t, Y, \cdot, \nabla Y, \nabla\nabla^T Y, \cdot), \\ Y(0) &= Y_0. \end{aligned}\right\} \tag{13.2}$$

Let us now assume that $\hat{A} \equiv 0$. Expression (13.1) then becomes the purely stochastic equation

$$\left.\begin{aligned} \mathrm{d}Z &= \hat{B}(t, Z, \cdot, \nabla Z, \cdot)W(\mathrm{d}t, \cdot), \\ Z(0) &= Z_0. \end{aligned}\right\} \tag{13.3}$$

Note that we wrote Y instead of X in (13.2) and Z instead of X in (13.3). This change of notation is convenient, as there are many cases where one can decompose (13.1) into a deterministic equation (13.2) and a purely stochastic equation (13.3). Assuming that both equations are uniquely solvable in a suitably chosen state space, the method of fractional steps[2] may be used to obtain a solution of (13.1). In what follows we use the decomposition of (13.1) into (13.2) and (13.3) to classify the SPDEs given by (13.1).

If $\hat{A}$ is a nonlinear functional in the variables $\partial^2_{k\ell} X$, then (13.2) is called *nonlinear*. Next, suppose $\hat{A}$ can be written as

$$\left.\hat{A}(t, Y, \cdot, \nabla Y, \nabla\nabla^{\mathrm{T}} Y, \cdot) = \sum_{k,\ell=1}^{d} \partial^2_{k\ell}[D_{k,\ell}(t, Y, \cdot, \cdot)Y] + \hat{B}_1(t, Y, r, \nabla X, \omega),\right\} \tag{13.4}$$

[2] Cf. Goncharuk and Kotelenez (1998) and Chap. 15.3 of this book.

where $\hat{B}_1(t, Y, r, \nabla Y, \omega)$ is a real valued functional, similar to $\hat{B}(t, Y, r, \nabla Y, \omega)$ and $D_{k,\ell}(t, Y, r, \omega)$ is a symmetric and nonnegative definite matrix. In this case, (13.2) is called *quasilinear*.

If the diffusion matrix $D_{k,\ell}(t, Y, r, \omega)$ in (13.4) does not depend on Y and $\hat{B}_1(t, Y, r, \nabla Y, \omega)$ is nonlinear in the first-order derivatives, we call (13.2) *semilinear of order 1*. If $\hat{B}_1(t, Y, r, \nabla Y, \omega)$ does not depend on ∇Y but it is nonlinear in Y we call (13.2) *semilinear of order 0*. If $\hat{B}_1(t, Y, r, \nabla X, \omega)$ is linear in ∇Y and Y (13.2) will be called *linear*.

The same classification can be utilized with respect to the functional $\hat{B}(t, X, r, \nabla X, \omega)$, namely into *nonlinear*, *quasilinear*, and *semilinear of order 0*. However, the assumptions on $W(\cdot, \cdot)$ generate additional subclasses. To differentiate between two types of $W(t, \cdot)$, as has been done for most classes of SPDEs, we consider suitable spaces of functions or distributions in the space variable r, the most common of which is $L_2(\mathcal{O}, dr)$, i.e., the space of real-valued functions, defined on $\mathcal{O}$, which are square integrable with respect to the Lebesgue measure dr. The latter space is a real separable Hilbert space, and most of the other function and distribution spaces may be defined as real valued Hilbert spaces, where the step to L_p-spaces is not difficult. Therefore, we change the previous space–time setup to an infinite-dimensional one and, to keep things simple, we assume for most of what follows that the state space is a real separable Hilbert space $(\mathbf{H}, \langle\cdot,\cdot\rangle_{\mathbf{H}}$ with scalar product $\langle\cdot,\cdot\rangle_{\mathbf{H}}$. We call an $\mathbf{H}$-valued Brownian motion (Wiener process) $W(\cdot)$ *regular* if the covariance operator Q_W of $W(1)$ is nuclear, which is often also called "*trace-class*". Q_W is by definition nuclear on $\mathbf{H}$ if, for a given complete orthonormal system of vectors $\{\phi_n\}$,

$$\sum_n \|Q_W\phi_n\|_{\mathbf{H}} < \infty. \tag{13.5}$$

If Q_W is not nuclear, we call $W(\cdot)$ *cylindrical*. $W(\cdot)$ is called a *standard cylindrical Brownian motion* if $Q_W = I$, where I is the identity operator on $\mathbf{H}$.[3] All we need to do now is to combine the various cases from (13.2) and (13.3), writing the class from (13.2) in the first coordinate, the class from (13.3) in the second coordinate and the type of Wiener process or Brownian motion in the third coordinate. Further, if $\hat{B}_1(t, Y, r, \nabla Y, \omega) \equiv I$, where I is the identity operator on $\mathbf{H}$, we will just write I instead of $\hat{B}$. Thus, in our somewhat simplified setting, we obtain the following family of classes:

$$\begin{aligned}\{(\hat{A}, \hat{B}, W) :\ & \hat{A} \in \{NL, QL, SL_1, SL_0, L\},\\ & \hat{B} \in \{NL, QL, SL_0, L, I\},\ W \in \{C, R\}\},\end{aligned} \tag{13.6}$$

where we used the abbreviations,

$NL :=$ nonlinear, $QL :=$ quasilinear, $SL_1 :=$ semilinear of order 1, $SL_0 :=$ semilinear of order 0, $L :=$ linear, I is the identity operator, $R :=$ regular, $C :=$ cylindrical.

In addition, we will quote some work on stochastic wave equations.

[3] Cf. Sect. 15.2.2.

Not all classes of SPDEs in the above classification may be made well posed in full generality, even for "nice" functionals $\hat{A}$ and $\hat{B}$. The first obstacle to overcome is to give a meaning to (13.1) if $W(t)$ is cylindrical. To see which problems may arise as a result of the non regularity of $W(\cdot)$, let us first discuss the simplest SPDE, namely a linear one.

13.1.1 Linear SPDEs

In the above classification, linear SPDEs are described by a linear partial second-order differential operator $A(t,\omega)$, $\hat{B} \equiv I_{\mathbf{H}}$, where $I_{\mathbf{H}}$ is the identity on $\mathbf{H}$, and by a regular or cylindrical Brownian motion $W(t)$. Assuming in addition that $A(t,\omega) \equiv A$, independent of t and ω, we obtain (infinite dimensional) *Langevin equation*:

$$\left.\begin{aligned} \mathrm{d}X &= AX\mathrm{d}t + \mathrm{d}W, \\ X(0) &= X_0. \end{aligned}\right\} \tag{13.7}$$

Suppose that A is a closed operator (closing, if necessary, with respect to some boundary conditions, etc.) and that A generates a strongly continuous semigroup $U(t)$. The *mild solution* of (13.7), if it exists, is by definition the solution of the following integral equation:

$$X(t) = U(t)X_0 + \int_0^t U(t-s)\mathrm{d}W \tag{13.8}$$

The right-hand side of (13.8) is often called the *variation-of-constants* formula, and the solution of (13.2), if it exists, is often called an *infinite dimensional Orstein–Uhlenbeck process*. We may assume that X_0 is a square integrable $\mathbf{H}$-valued $\mathcal{F}_0$-measurable random variable. The mild solution of (13.7) exists exactly when the stochastic convolution integral on the right-hand side of (13.8) is a well-defined $\mathbf{H}$-valued process.

It is important to note that in (13.8) $U(t-s)$ becomes *anticipating* if A depends on ω and t and is adapted to $\mathcal{F}_t$. Therefore, the first restrictive assumption in the semigroup approach is the assumption that A or $A(t)$ does not depend on ω. However, if the pair is coercive[4] we can still solve infinite dimensional Langevin equations, using the variational formulation. In contrast, the semigroup approach allows to solve (13.8) also for noncoercive cases, like hyperbolic equations, etc.[5] We will continue our discussion of (13.7) under the previous assumption that A does not depend on ω.

Recall that Q_W is the covariance operator of $W(1)$ and denote by "Tr" the trace of a linear operator on $\mathbf{H}$, if it exists. A necessary and sufficient condition for the existence of a mild solution of (13.6) on some interval $[0,T]$ is the assumption that[6]

---

[4] Cf. Chap. 8.

[5] Cf. Kotelenez (1985), Sect. 3, and Da Prato and Zabczyk (1992).

[6] Cf. Da Prato and Zabczyk (1992), Sect. 5.1.2.

$$\int_0^T Tr(U(s)Q_W U^*(s))ds < \infty. \quad (13.9)$$

The nomenclature already indicates the way equations (13.7) and (13.8) are generalizations of corresponding linear equations in $\mathbf{R}^d$. Naturally, the problem is easier if we assume that Q_W is already nuclear, i.e., that the **H**-valued Wiener process is *regular*. This assumption implies (13.9).

Under the assumption that $W(\cdot)$ is regular equations of type (13.7) and (13.8) are generalizations of corresponding linear equations in $\mathbf{R}^d$. Expression (13.7)/(13.8) and with possible time-dependent families of unbounded operators $A(t)$ and two-parameter semigroups $U(t, s)$ and additional linear deterministic terms were considered by Curtain and Falb (1971), Chow (1976), Zabczyk (1977) and others (cf. also Curtain and Pritchard (1978) for more references and applications in linear systems theory). The problem of continuous versions for (13.8) under the assumption of a regular Wiener process (and its generalization to continuous locally square integrable Hilbert space valued martingales) was solved by the author[7] under the assumption that the (two-parameter) semigroup $U(t, s)$ is of contraction type (cf. also for similar results Ichikawa (1982), Kotelenez (1984), and Tubaro (1984)). Continuous versions for the general case of (13.8) were obtained by Da Prato et al. (1987). The reader may also consider Walsh (1981) for space–time regular versions of equations of type (13.8) for a standard cylindrical Brownian motion $W(\cdot)$. To our knowledge, Dawson (1972) was the first to obtain space–time regular solutions of a generalization of (13.7)/(13.8) to the following case: $\hat{A}$ linear and independent of t and ω, $\hat{B}$ semilinar of order 0 and $W(\cdot)$ a standard cylindrical Brownian motion.

Note that the condition (13.9) appears naturally if we formally compute the mean square of the norm of $\int_0^t U(t - s)dW$ in **H**. For finite mean square we may also solve (13.8) for some cylindrical Wiener processes, provided the semigroup $T(t)$ is sufficiently smoothing so that Expression (13.9) holds.[8] Expression (13.9) holds if and only if $U(s)Q_W U^*(s)$ is nuclear, i.e., if it has finite trace norm. Further, the mean square of (13.8) is finite if the trace norm $U(s)Q_W U^*(s)$ is integrable with respect to $dt\, dP$ on $[0, T] \times \Omega$. Da Prato (1982) obtains spatial regularity results for (13.8) and Kotelenez (1987) generalizes some results of Dawson (1972).

Obviously, the requirement of $U(s)Q_W U^*(s)$ being nuclear with integrable trace norm restricts the choice of admissible covariance operators Q_W. We explain this by two examples, following Kotelenez (1985), Sect 3.

(i) Suppose A is self-adjoint and has a discrete spectrum. Then there is $\beta \geq 0$ such that $\beta - A$ has a strictly positive spectrum of eigenvalues $0 < \lambda_1 < \lambda_2 < \cdots$, where $\lambda_n \to \infty$. Here $\beta - A := \beta I_{\mathbf{H}} - A$, where $I_{\mathbf{H}}$ is the identity operator on **H**. Further, there is a complete orthonormal system (CONS) of eigenvectors $\{\Phi_n\}$ in $\mathbf{H}_0$. For $\gamma \geq 0$ set

$$\tilde{\mathcal{H}}_\gamma := \{f \in \mathbf{H}_0 : |(\beta - A)^{\gamma/2} f|_0 < \infty,$$

[7] Cf. Kotelenez (1982).

[8] Cf. again Da Prato and Zabczyk (loc. cit.) and references therein.

where $(\beta - A)^{\gamma/2}$ is the fractional power of $\beta - A$.[9] Endowed $\tilde{\mathcal{H}}_\gamma$ with the scalar product

$$\langle , g\rangle_\gamma := \langle(\beta - A)^{\gamma/2} f, (\beta - A)^{\gamma/2} g\rangle_0.$$

$\tilde{\mathcal{H}}_\gamma$ becomes a real separable Hilbert space. Set

$$\Phi := \cap_{\gamma \geq \mathbf{0}} \tilde{\mathcal{H}}_\gamma .$$

Denote by Φ' the strong dual of Φ. For $\gamma \geq 0$, set

$$\tilde{\mathcal{H}}_{-\gamma} := \{F \in \Phi' : \sup_{|f|_\gamma = 1} |\langle F, f\rangle| < \infty,$$

where $\langle \cdot, \cdot\rangle$ denotes the extension of $\langle \cdot, \cdot\rangle_0$ to a duality between Φ' and Φ. Finally, we identify $\mathbf{H}_0$ with its strong dual $\mathbf{H}_0'$ and we obtain the scale of Hilbert spaces

$$\Phi \subset \tilde{\mathcal{H}}_\gamma \subset \mathbf{H}_0 \cong \mathbf{H}_0' \subset \tilde{\mathcal{H}}_{-\gamma} \subset \Phi' \tag{13.10}$$

with dense continuous imbeddings. Let $\{\phi_{\gamma,n}\}$ be a CONS in $\tilde{\mathcal{H}}_\gamma$, where now $\gamma \in \mathbf{R}$. The assumptions on the spectrum imply that there is a $\beta < \gamma$ such that

$$\sum_n |\phi_{\gamma,n}|_\beta^2 < \infty,$$

i.e., the imbedding $\tilde{\mathcal{H}}_\gamma \subset \tilde{\mathcal{H}}_\beta$ is Hilbert Schmidt. Since the semigroup $U(t)$ commutes with the fractional powers $(\beta - A)^\gamma /2$ it can be extended (restricted) to $\tilde{\mathcal{H}}_\gamma$ preserving the operator norm. All we need to do now is to assume that Q_W can be defined as a bounded operator from $\mathbf{H}_0$ into $\tilde{\mathcal{H}}_\gamma$ for some positive γ. We then choose a positive $\beta > \gamma$ such that the imbedding $\tilde{\mathcal{H}}_{-\gamma} \subset \tilde{\mathcal{H}}_{-\beta}$ is Hilbert Schmidt. On $\tilde{\mathcal{H}}_{-\beta}$, $W(\cdot)$ is regular. We now extend (13.7) onto $\tilde{\mathcal{H}}_{-\beta}$ and solve it via (13.8) on $\tilde{\mathcal{H}}_{-\beta}$.

A typical example for A is the Laplacian on a bounded domain $\mathcal{O} \subset \mathbf{R^d}$ such that its closure with respect to boundary conditions is a self-adjoint operator.

(ii) Let $A = \frac{1}{2}\triangle$ and $\mathcal{O} = \mathbf{R}^d$. We consider the Laplacian $\triangle$ to be a closed operator on $\mathbf{H} := \mathbf{H}_0 = L_2(\mathbf{R}^d, dr)$. Setting

$$G(t, r) := (2\pi t)^{-d/2} \exp\left[-\tfrac{|r|^2}{2t}\right], \tag{13.11}$$

the semigroup $U(t)$ is given by

$$(U(t)f)(r) = \int G(t, r - q) f(q) dq, \tag{13.12}$$

where $f \in \mathbf{H}_0$. In difference from the previous example, A and $U(t)$ do not have a discrete spectrum.[10] Therefore, a direct copy of the scale (13.10) does

[9] Cf., e.g., Tanabe (1979), Sect. 2.3.

[10] The spectrum of $\frac{1}{2}\triangle$ on $\mathbf{R}^d$ is the set $(-\infty, 0]$. Cf., e.g., Akhiezer and Glasman (1950), Chap. VI.

not yield Hilbert-Schmidt imbeddings between Hilbert spaces on the scale. The remedy is found by using $|r|^2 - \triangle$ instead of $\beta - \triangle$, where $|r|^2$ is considered a multiplication operator. This operator has a discrete spectrum, and the CONS of eigenfunctions are the well-known normalized Hermite functions. Thus, we define for $\gamma \in \mathbf{R}$ the fractional powers $(|r|^2 - \triangle)^{\gamma/2}$. We now choose $\gamma \geq 0$, denote corresponding Hilbert spaces by $\mathcal{H}_\gamma$ and obtain the Schwarz scale of distribution spaces[11]

$$\mathcal{S} \subset \mathcal{H}_\gamma \subset \mathbf{H}_0 \cong \mathbf{H}_0' \subset \mathcal{H}_{-\gamma} \subset \mathcal{S}'. \tag{13.13}$$

The imbedding $\subset \mathcal{H}_\gamma \subset \mathcal{H}_\beta$ is Hilbert Schmidt if, and only if, $\gamma > \beta + d$. Although $U(t)$ does not commute with $(|r|^2 - \triangle)^{\gamma/2}$ it may still be extended and restricted to a strongly continuous semigroup on the corresponding spaces $\mathcal{H}_\gamma$, which is enough to consider (13.7) on a corresponding Hilbert distribution space and solve it via (13.8).

We have seen in both examples that linear SPDEs may be solvable for a large class of cylindrical Brownian motions by simply redefining them on a suitable Hilbert space of distributions. Langevin equations of type (13.7) arise naturally in central limit phenomena in scaling limits for the mass distribution of various particle systems. We refer the reader to Itô (1983, 1984) and to the papers by Martin-Löf (1976), Holley and Stroock (1978), Gorostiza (1983), Kotelenez (1986, 1988), Bojdecki and Gorostiza (1986), Gorostiza and León (1990), and Dawson and Gorostiza (1990). There are many papers which deal with properties of infinite dimensional Ornstein–Uhlenbeck processes. Cf., e.g., Curtain (1981), Schmuland (1987), Bojdecki and Gorostiza (1991), Iscoe and McDonald (1989), and the references therein.

13.1.2 Bilinear SPDEs

Let $B(X, \mathrm{d}W)$ be a bilinear functional on $\mathbf{H} \times \mathbf{H}$. The prototype of a bilinear SPDE can be described as follows:

$$\left.\begin{aligned} \mathrm{d}X &= AX\,\mathrm{d}t + B(X, \mathrm{d}W), \\ X(0) &= X_0, \end{aligned}\right\} \tag{13.14}$$

Sometimes SPDEs of type (13.14) are also called linear in the literature. We find such a term confusing. We have already seen that for many self-adjoint operators A the linear SPDE may be redefined on a suitable distribution space where the original cylindrical Brownian motion becomes a regular one. In some cases of bilinear SPDEs, one may succeed in generalizing the variations-of-constants, considering the two-parameter random semigroup (also called random evolution operator) generated by the bilinear SPDE (cf. Curtain and Kotelenez (1987)). However, in

[11] Cf. (15.32) in Sect. 15.1.3. The completeness of the normalized Hermite functions is derived in Proposition 15.8 of Sect. 15.1.3.

many cases such a procedure may not be possible. Note that the bilinear SPDE with $B(X, \mathrm{d}W) = \nabla X \cdot \mathrm{d}W$ is a special case of our equation (8.26) for $\mathcal{J}$, $D_{k\ell}$ independent of X and $F \equiv 0$, where the noise term can be cylindrical. For more examples of (13.14) with regular $W(\cdot)$, cf. Da Prato et al. (1982), Balakrishnan (1983), Üstünel (1985), and the references therein.

The difference between the linear SPDE (13.7) and the bilinear SPDE (13.14) is best explained by the following special case of (13.14):

$$\left.\begin{aligned} \mathrm{d}X &= AX\mathrm{d}t + \sigma(X)\mathrm{d}W), \\ X(0) &= X_0, \end{aligned}\right\} \tag{13.15}$$

where σ is some nice function of $x \in \mathbf{R}$ and $\sigma(X)\mathrm{d}W$ is understood as a pointwise multiplication. We assume that $W(t)$ is a standard cylindrical Brownian motion on $\mathbf{H}_0$. Recall that, by our Fourier expansion (4.28), $W(\cdot)$ may be represented by our standard space–time white noise $w(\mathrm{d}r, \mathrm{d}t)$.[12] Further, to make things somewhat easier suppose that the semigroup generated by A, $U(t)$ has a kernel $G(t, r)$ (the *fundamental solution*, associated with the Cauchy problem for A, I.e., $U(t)$ is cast in the form of the following integral operator

$$(U(t)f)(r) := \int G(t, r-q)f(q)\mathrm{d}q,$$

where $f \in \mathbf{H}_0$. In this case the mild solution of (13.15), if it exists, by definition, is the solution of

$$X(t) = \int G(t, r-q)X_0(q)\mathrm{d}q + \int_0^t \int G(t-s, r-q)\sigma(X(t,q)w(dq, ds). \tag{13.16}$$

To see whether (13.16) can be well posed for nontrivial multiplication operators $\sigma(\cdot)$, we take a random field $Y(t, q)$, measurable, adapted, etc. We compute the variance of the stochastic integral and obtain as a condition to solve the problem on $\mathbf{H}_0 := L_2(\mathbf{R}^{\mathrm{d}}, \mathrm{d}r)$ that the variance be finite, i.e.,

$$\begin{aligned} &E\int \left[\int_0^t \int G(t-s, r-q)\sigma(Y(t,q)w(\mathrm{d}q, \mathrm{d}s)\right]^2 \mathrm{d}r \\ &= \int\int_0^t \int G^2(t-s, r-q)\sigma^2(Y(s,q))\mathrm{d}q\,\mathrm{d}s\,\mathrm{d}r < \infty. \end{aligned} \tag{13.17}$$

For the important special case

$$A = D\Delta, \text{ and } G(t, r) := (4\pi Dt)^{-d/2}\exp\left[-\tfrac{|r|^2}{2Dt}\right]. \tag{13.18}$$

It is hard to guarantee a priori the condition (13.17). Take, e.g., $\sigma(\cdot) \equiv 1$. We are then dealing with a special case of the linear SPDE (13.7) and the Integral on the right-hand side of (13.17) equals ∞. Note that this problem would not have appeared

[12] Cf. also (15.69) in Sect. 15.2.2.

if $d = 1$ and the SPDE would have been restricted to a bounded interval with Neumann or Dirichlet boundary conditions. Walsh (1986) shows that for $d = 1$ and G as in (13.15) the convolution integral

$$\int G(t-s, r-q)w(\mathrm{d}q, \mathrm{d}s)$$

is function valued. Following Kotelenez (1992a),[13] we choose the weight function $\varpi(r) := (1+|r|^2)^{-\gamma}$ with $\gamma > d/2$. This implies $\int \varpi(r)\mathrm{d}r < \infty$. Instead of working with $\mathbf{H}_0 = L_2(\mathbf{R}, \mathrm{d}r)$ we now analyze (13.16) on the weighted L_2 space $\mathbf{H}_{0,\varpi} = L_2(\mathbf{R}, \varpi(r)dr)$. Young's inequality[14] implies that for $d=1$ and G as in (13.18) the condition (13.17) holds, i.e., at least for bounded $\sigma(\cdot)$ the problem (13.16) may be well posed. The generalization to unbounded $\sigma(\cdot)$ and to $d > 1$ and higher order partial differential operators and pseudodifferential operators has been treated by Kotelenez (1992a). Finally, for the Laplacian on a bounded domain in $\mathbf{R}^d$, closed with respect to Neumann boundary conditions, and $d > 1$ the linear convolution is necessarily distribution valued. In this problem G from (13.18) must be replaced by the corresponding fundamental solution.[15]

Finally, Nualart and Zakai (1989) show that (13.15) for A is in (13.18) and $\sigma(x) \equiv x$ has a solution only in the space of generalized Brownian functionals if $\mathbf{H}_0 := L_2\mathbf{R}^d, \mathrm{d}r)$, where $d \geq 2$ and $W(t)$ is the standard cylindrical Brownian motion on $\mathbf{H}_0$. In other words, we cannot construct some state space (like $\mathcal{S}'$) and obtain solutions of (13.14) as ordinary random variables, defined on $\mathcal{S}'$).

13.1.3 Semilinear SPDEs

Semilinear SPDEs may be written as

$$\left.\begin{aligned} \mathrm{d}X = AX + B_1(\cdot, X, \nabla X)\mathrm{d}t + B(\cdot, X, \nabla X)\mathrm{d}W, \\ X(0) = X_0, \end{aligned}\right\} \quad (13.19)$$

The type (SL_0, SL_0, R) is probably the simplest and has been studied early on (cf. Chojnowska-Michalik (1976)). The difficulty increases if we consider (SL_1, SL_0, R) or (SL_1, SL_1, R). Both types have also been investigated over many years, and we wish to mention first of all the variational approach to those SPDEs in the thesis of Pardoux (1975) as well as Krylov and Rozovsky (1979), Gyöngy (1982), and the references therein. Chow and Jiang (1994) obtain space–time smooth solutions. For the semigroup approach see the paper by Da Prato (1982) and the book by Da Prato and Zabczyk (1992), which contains a rather complete description of those equations as well as many interesting references.

[13] Cf. also our Chaps. 8, 9, and Sect. 15.1.4 for the use of $\mathbf{H}_{0,\varpi}$.

[14] Cf. Theorem 15.7 in Sect. 15.1.2.

[15] Cf. Walsh (1986), Sect. 3.

As previously mentioned in the comments on linear SPDEs, Dawson (1972) obtains existence, uniqueness and smoothness for a class of SPDEs of type (SL_0, SL_0, C), using a semigroup approach. The Brownian motion in Dawson's paper is standard cylindrical, and his results hold for $d = 1$ if A is a second-order elliptic partial differential operator. Dawson also considers higher dimensions and, accordingly, A is an elliptic partial differential operator of higher order. Marcus (1974) studies stationary solutions of the mild solution of an SPDE of type (SL_0, SL_0, C). Generalization of Dawson's result have been obtained by Funaki (1983) for $d = 1$, Kotelenez (1987, 1992a,b) in a more general setting and by others. Kunita (1990) obtains solutions of SPDEs of type (SL_1, SL_1, C) and first-order SPDE for regular and certain cylindrical Brownian motions which are equivalent to our cylindrical Brownian motion $W(t) := \int_0^t \int \mathcal{J}(\cdot, q, s)w(dq, ds)$.[16] Kunita's approach is based on the method characteristics. First-order SPDEs are also analyzed by Gikhman and Mestechkina (1983), employing the method of characteristics. Krylov (1999) provides an analytic approach to solve a large class of semilinear SPDEs on L_p-spaces with $p \geq 2$. For more work in this direction, cf. Kim (2005).

We mention here Dawson's (1975) paper in which he introduces the Dawson–Watanabe SPDE (to be commented on later). This SPDE describes a scaled system of branching Brownian motions. The driving term is standard cylindrical, $A = \triangle$, and Dawson shows the existence of the process in higher dimensions as well, although the corresponding SPDE seems to be ill posed. For $d = 1$, however, one may solve the SPDE and obtain the Dawson–Watanabe process as a solution (cf. Konno and Shiga (1988) and Reimers (1989)).

For $W(\cdot)$ being standard cylindrical, type (SL_1, I, C) has a solution if $d = 1$, which can be shown by semigroup methods as well as variational methods (cf. Da Prato and Zabczyk (1992)). Note that the SPDEs driven by a Brownian motion and with deterministic coefficients are Markov processes in infinite dimensions with well-defined transition probabilities. Da Prato and Zabczyk (1991) analyze a semilinear SPDE of the type (SL_1, I, R) or (SL_1, I, C) so that our condition (13.9) holds. The semigroup, defined through the transition probabilities is called the *Kolmogorov* semigroup. This semigroup, $\mathbf{P}(t)$ is defined on $C_b(\mathbf{H}; \mathbf{R})$, the space of bounded continuous real-valued functions on $\mathbf{H}$. Let $f \in C_b(\mathbf{H}; \mathbf{R})$ and δ and δ^2 denote the first- and second-Fréchet derivatives with respect to the space variable $X \in \mathbf{H}$. Then $\mathbf{P}(t)$ satisfies the following equation:

$$\left.\begin{aligned} \frac{\partial}{\partial t}(\mathbf{P}(t)f)(X) = \frac{1}{2}\mathrm{Tr}(Q_W \delta^2(\mathbf{P}(t)f)(X) + \langle AX + B_1(X), \delta(\mathbf{P}(t)f)(X)\rangle_{\mathbf{H}}, \\ (\mathbf{P}(0)f)(X) = f(X). \end{aligned}\right\} \tag{13.20}$$

Dropping initially the semilinear part, the SPDE becomes a linear SPDE. Its Kolmogorov semigroup, $\mathbf{P}_2(t)$, is shown to have certain smoothing properties (similar to analytic semigroups), and the Kolmogorov semigroup for the semilinear

[16] Cf. (7.4) and (7.13) in Chap. 7.

SPDE may be obtained by variation of constants as the mild solution of a linear deterministic evolution equation:

$$(\mathbf{P}(t)f)(\cdot) = (\mathbf{P}_2(t)f)(\cdot) + \int_0^t \mathbf{P}_2(t-s)(\langle B_1(\cdot), \delta(\mathbf{P}(t)f)(\cdot)\rangle_{\mathbf{H}} ds. \tag{13.21}$$

For more details and results, cf. Da Prato (2004).

For $d = 2$, Albeverio and Röckner (1991) employ Dirichlet form methods to weak solutions to an SPDE related to an SPDE in Euclidean field theory. Mikulevicius and Rozovsky (1999) analyze a class of SPDEs driven by cylindrical Brownian motions in the strong dual Φ' of a nuclear space Φ (cf. our previous (13.10)). A large portion of the book by Kallianpur and Xiong (1995) treats SPDEs in a similar framework.

Finally, let us also mention some qualitative results. Wang et al. (2005) consider two families of semilinear PDE's. The first one is defined on a domain with (periodically placed) holes, and the second one is the weak (homogenization) limit of the first one, as the size of the holes tends to 0. Invariant manifolds for semilinear SPDEs with regular Brownian motion have been obtained by Duan et al. (2003, 2004).

13.1.4 Quasilinear SPDEs

Quasilinear SPDEs can be written as

$$\left.\begin{aligned} dX &= A(t,X)X + B_1(\cdot, X, \nabla X)dt + B(\cdot, X, \nabla X)dW, \\ X(0) &= X_0, \end{aligned}\right\} \tag{13.22}$$

where for fixed $f \in \mathbf{H}$ $A(t,f)$ is an unbounded operator, e.g., an elliptic differential operator. Daletskii and Goncharuk (1994) employ analytical methods in the analysis of a special case of (13.22) with regular Brownian motion. Further, let us mention some results, obtained by a particle approach: Dawson and Vaillancourt (1995), Dawson et al. (2000), Kotelenez (1995a–c 1996, 1999, 2000), Kurtz and Protter (1996), Kurtz and Xiong (1999, 2001, 2004), Dorogovstev (2004b), and Dorogovtsev and Kotelenez (2006). Goncharuk and Kotelenez (1998) employ fractional steps in addition to particle and "traditional methods" to derive quasilinear SPDEs with creation and annihilation.

13.1.5 Nonlinear SPDEs

For the type (*NL,NL,R*) we refer the reader to a paper by Lions and Souganidis (2000), who employ a viscosity solution approach. The noise in that setup consists of finitely many i.i.d. $\mathbf{R}^d$-valued Brownian motions. A direct generalization to the case of an infinite dimensional regular Brownian motion should not be too difficult.

13.1.6 Stochastic Wave Equations

Results and more references may be found in Carmona and Nualart (1988a,b), Marcus and Mizel (1991), Peszat and Zabczyk (2000), Dalang (1999), Dalang and Walsh (2002), Chow (2002), Dalang and Mueller (2003), Dalang and Nualart (2004), Dalang and Sanz-Sol (2005), Dalang et al.(2006).

13.2 Models

Applications of SPDEs within a number of different areas are reviewed.

Historically, the development of SPDEs was motivated by two main models:

- The Kushner[17] and Zakai equations in nonlinear filtering (cf. Kushner (1967) and Zakai (1969)).
- The Dawson–Watanabe equation for the mass distribution of branching Brownian motions (cf. Dawson (1975).

Other models and equations followed. We will start with the nonlinear filtering equation.

13.2.1 Nonlinear Filtering

Let $r(t)$ be a Markov diffusion process in $\mathbf{R}^d$ which is described by the following SODE:

$$r(t) = r_0 + \int_0^t a(r(s), s)\mathrm{d}s + \int_0^t b(r(s), s)\mathrm{d}\tilde{w}(\mathrm{d}s).$$

Suppose we observe the $\mathbf{R}^m$-valued process

$$q(t) := \int_0^t h(r(s))\mathrm{d}s + w(t).$$

$w(\cdot)$ and $\tilde{w}(\cdot)$ are assumed to be independent standard Brownian motions with values in $\mathbf{R}^d$ and $\mathbf{R}^m$, respectively. Under these conditions Kushner (1967) obtains a semilinear SPDE for the normalized conditional density of $r(t)$ based on the observations of $\sigma\{q(s), s \le t\}$, where the linear second-order operator A is the formal adjoint of the generator of $r(\cdot)$. Zakai (1969) obtains a bilinear SPDE for the unnormalized conditional density:

$$\left.\begin{array}{c} \mathrm{d}X(t) = AY(t)\mathrm{d}t + Q^{-1/2}h(\cdot)X \cdot q(\mathrm{d}t), \\ X(0) = X_0, \end{array}\right\} \tag{13.23}$$

where Q is the covariance operator of $w(\cdot)$ and X_0 is the density of the random variable r_0.

[17] The Kushner equation is also known as the "Kushner–Stratonovich" equation.

Besides the original papers by Kushner and Zakai we refer for more general nonlinear filtering equations to Pardoux (1979), Krylov and Rozovsky (1979), Da Prato (1982), Da Prato and Zabczyk (1992), Bhatt et al. (1995), and the references therein. Similar to the approach taken in this book, Kurtz and Xiong (2001) employ a particle method approach to the solution of the Zakai equation, and Crisan (2006) provides numerical solutions, based on particle methods.

13.2.2 SPDEs for Mass Distributions

A number of papers have been devoted to stochastic reaction-diffusion phenomena. The SPDEs in many papers are semilinear of type (SL_0, SL_0, R), (SL_0, SL_0, C) as well as (SL_0, I, R) and (SL_0, I, C). However, the last two cases, if solvable, do not have nonnegative solutions and, therefore, cannot describe the distribution of some matter (cf. Kotelenez (1995b) and Goncharuk and Kotelenez (1998) for an alternative proof of the positivity of solutions). In contrast, a particle approach automatically has the positivity property. In addition, in a particle approach we describe the microscopic dynamics first before simplifying the time evolution of the mass distribution by passing to SPDEs and PDEs. There are at least two directions in the particle approach to SPDEs:

1. Branching Brownian motions and associated superprocesses
2. Brownian motions with mean-field interaction and associated McKean–Vlasov equations

1. We will first discuss superprocesses and the Dawson–Watanabe equation. Dawson (1975) analyzes an infinite system of i.i.d. diffusion approximations to a system of branching particles. For a set of atomic measures, he obtains the stochastic evolution equation on $\mathbf{M}_f$

$$\left.\begin{aligned} dX(t) &= \alpha X(t)dt + \sqrt{X(t)}W(dt),\\ X(0) &= X_0. \end{aligned}\right\} \tag{13.24}$$

Here X_0 is supported by a countable set of points $\{a_i\}$ and $\alpha \neq 0$ is the difference between birth and death rates. The square root is taken with respect to the mass at the point of support, $W(t, a_i)$ are i.i.d. standard Brownian motions and $\sqrt{X(t, a_i)}$ is a factor at $W(dt, a_i)$, whence $\sqrt{X(t, \cdot)}$ acts as a multiplication operator on $W(dt, \cdot)$. The i.i.d assumption implies that $W(t, \cdot)$ can be extended to a standard cylindrical Brownian motion (on $\mathbf{H}_0 = L_2(\mathbf{R}^d, dr)$). Through extension by continuity, Dawson extends the process from (13.24) to starts $X_0 \in \mathbf{M}_f$. The process (13.24) is a Markov process. Dawson then considers a pure diffusion, governed by the Laplacian $\triangle$, which itself is a Markov process. The dynamics of both phenomena are linked through the Trotter product of their respective Markov semigroups, resulting in an $\mathbf{M}_f$-valued Markov process with generator

$$(\mathcal{A}f) = \langle(\triangle + \alpha)\delta f(X), X\rangle + \tfrac{1}{2}\mathrm{Tr}(\delta^2 f)(X). \tag{13.25}$$

Formally, this diffusion may be written as a solution to the following SPDE:

$$\left.\begin{array}{c} \mathrm{d}X(t) = (\triangle + \alpha)X(t)\mathrm{d}t + \sqrt{X(t)}W(dt), \\ X(0) = X_0. \end{array}\right\} \tag{13.26}$$

There are two difficulties in the interpretation of (13.26). The first and most important problem is that $\sqrt{X(t)}$ is not defined for general measures. The second difficulty is that multiplication between measures and distributions is not defined. Recalling our analysis of the one-dimensional case in (13.15)–(13.18) we conclude that (13.26) may be well posed for $d = 1$, and the only problem in (13.26) is posed by the square root, which is not Lipschitz. As mentioned earlier, Konno and Shiga (1988) and Reimers (1989) showed existence of a solution of (13.26) for $d = 1$. Consequently, the research activities on (13.26) "branched" into two directions: For $d > 1$ many papers have been written both on the generalization of (13.26) and on some qualitative properties. For $d = 1$, it became an active research area in SPDEs.

In our earlier discussion of bilinear SPDEs we have seen that for $d > 1$ we cannot interpret (13.26) as an SPDE, and most of the work in that direction goes beyond the scope of this book. We just mention Perkins (1992, 1995) who includes interaction into the dynamics of branching Brownian particles. Further, Dawson and Hochberg (1979) and Iscoe (1988) analyze the support of measure valued branching Brownian motion. Cf. also Dawson (1993) for a detailed analysis, many properties and more references.

Blount (1996) obtains for $d = 1$ a semilinear version of (13.25). For SPDEs of type (13.25), $d = 1$, cf. Mueller and Perkins (1992), Donati-Martin and Pardoux (1993), Shiga (1993), Mueller and Sowers (1995), Mueller and Tribe (1995), Mueller (2000), Dawson et al. (2003), and the references therein.

2. The second direction in particle methods is the central theme of this book. Apart from the author's contributions (cf. Kotelenez (1995–2000)), we need to mention Vaillancourt (1988) and Dawson and Vaillancourt (1995) for a somewhat different approach. For generalizations of both the author's and other results, cf. Wang (1995), Kurtz and Protter (1996), Goncharuk and Kotelenez (1998), Kurtz and Xiong (1999), Dawson et al. (2000), Kurtz and Xiong (2001), Dorogovtsev (2004b), and Dorogovtsev and Kotelenez (2006).

13.2.3 Fluctuation Limits for Particles

We discussed earlier the infinite dimensional Langevin equations, which are obtained in central limit theorem phenomena for the mass concentration of particle systems. Kurtz and Xiong (2004) obtain an interesting generalization of this classical central limit theorem as follows: Consider the empirical process $\mathcal{X}_N(\cdot)$ associated with our system of SODEs (4.2) and the limit $\mathcal{X}(\cdot)$ which is a solution of the quasilinear SPDE (8.54) in Chap. 8. Under some assumptions $\frac{1}{N}(\mathcal{X}_N - \mathcal{X})(\cdot)$ tends to the solution of a bilinear SPDE, whose coefficients depend on $\mathcal{X}(t)$. Giacomin et al. (1999) obtain a semilinear Cahn–Allen type SPDE with additive white noise for the fluctuations in a Glauber spin flip system. The space dimension is $d = 1$ for a rigorous result, and the equation is conjectured to be the limit also for $d = 2$. Solvability for $d = 2$ is not shown.

13.2.4 SPDEs in Genetics

Fleming and Viot (1979) propose a derivation of the frequency distribution of infinitely many types in a population, following Dawson's approach to measure valued diffusion. Donelly and Kurtz (1999) obtain a general particle representation which includes both some Fleming–Viot processes and the Dawson–Watanabe process.

If the domain of types is the bounded interval $[0, L]$ in the Fleming–Viot model, then this frequency distribution may be represented by the solution of the following SPDE:

$$\left.\begin{aligned} dX(t) &= (\Delta + \alpha)X(t) + g(X(t))dt + \sqrt{X(t)(1 - X(t)}W(dt), \\ X(0, r) &= X_0(r) \in [0, 1], \end{aligned}\right\} \qquad (13.27)$$

where the Laplacian is closed with respect to homogeneous Neumann boundary conditions. The term $\sqrt{X(t)(1 - X(t)}W(dt)$ is called the *random genetic drift*. For $W(\cdot)$ regular (13.27) was solved by Viot (1975) (even in higher dimensions). For $W(t)$ being standard cylindrical in $L_2([0, L]; dr)$, this equation is of the type we discussed in Sect. 13.2.2.

13.2.5 SPDEs in Neuroscience

Walsh (1981) represents a neuron by the one-dimensional interval $[0, L]$. He obtains an equation for an electrical potential, $X(t, r)$, by perturbing the (linear) cable equation by impulses of a current. These impulses are modeled by space–time white noise times a nonlinear function of the potential:

$$\left.\begin{aligned} dX(t, r) &= (\Delta - 1)X(t, r) + g(X(t, r), t)w(dr, dt), \\ \frac{\partial}{\partial r}X(0, t) &= \frac{\partial}{\partial r}X(L, t) = 0, \\ X(0, r) &= X_0(r). \end{aligned}\right\} \qquad (13.28)$$

Under a Lipschitz assumption on g, Walsh obtains existence and uniqueness for the mild solution of (13.28) as well as space–time smoothness and a version of a multiparameter Markov property. It was in part this work which motivated Walsh's general approach to semilinear SPDEs, driven by space–time white noise and its generalization, employing variation of constants (cf. Walsh, 1986).

Kallianpur and Wolpert (1984) analyze an Ornstein–Uhlenbeck approximation to a similar stochastic model, where the neuron can be represented by a compact d-dimensional manifold and the driving noise is a generalized Poisson process.

13.2.6 SPDEs in Euclidean Field Theory

These are semilinear SPDEs driven by space–time white noise, i.e., they are of type (SL_1, I, C). As we have already seen in our comments, these equations can be solvable in space dimension $d = 1$ by direct SPDEs methods. The nonlinear term is usually a polynomial $g(X)$ and in higher dimension the powers $X^m(r)$ have to be replaced by the *Wick* powers: $X^m(r)$:[18] The resulting quantization equation is written as

$$\left.\begin{aligned} dX(t) &= (\triangle X(t) + :g(X(t)):)dt + W(dt),\\ X(0) &= X_0,\\ &\text{and possible self-adjoint boundary conditions.} \end{aligned}\right\} \tag{13.29}$$

For existence, uniqueness, and properties in space dimension $d = 2$ we refer to papers of Albeverio and Röckner (1991), Albeverio et al. (2001), Da Prato and Debussche (2003), and the references therein. A version of (13.29) with a regular $\mathbf{H}_0$-valued Brownian motion and with an ordinary polynomial has been analyzed in detail by Doering (1987). This paper also contains a number of interesting comments on the problem of stochastic quantization and Euclidean field theory.

13.2.7 SPDEs in Fluid Mechanics

Monin and Yaglom (1965) treat the statistical approach to turbulence. The velocity field of a fluid is described by a space–time vector-valued random field. If the random field is stationary, homogeneous, and isotropic, the correlation matrix has a particularly simple structure which can be compared with empirical data (cf. Monin and Yaglom (loc. cit. Chap. 7). Observe that the velocity field is governed by the Navier–Stokes equations. Solutions of a forced version of these equations with random initial conditions may yield stationary, homogeneous and isotropic random fields as solutions. This begs the question whether the solutions of a suitably forced version of the Navier–Stokes are consistent with theoretical considerations, based on the theory of Kolmogorov and Obukhov and its generalizations and whether their correlation matrices are structurally similar to the empirical data.

If the fluid is incompressible and the forcing is a Gaussian random field which is white in time, we obtain stochastic Navier–Stokes equations. We recommend the review paper by Kupiainen (2004) and the references therein as a starting point for more information on turbulence, homogeneous, and isotropic solutions and (in 2D) on ergodicity results of stochastically forced Navier–Stokes equations. Let $U(t, r)$ be the velocity of the fluid. The stochastic Navier–Stokes equations, mentioned above, are cast in the form

---

[18] Cf. Da Prato and Zabczyk (1992), Introduction, 06, for a definition.

$$\left.\begin{array}{c} dU(t) = (\nu\triangle U(t) - (U\cdot\nabla)U - \nabla p)dt + W(dt), \\ \nabla\cdot U = 0, \\ U(0) = U_0, \\ \text{and possible boundary conditions.} \end{array}\right\} \tag{13.30}$$

Here, p is the pressure of the fluid and ν the kinematic viscosity. A first step in the derivation of stationary, homogeneous, and isotropic random field solutions of (13.30) is to show existence of a stationary solution of (13.30) and, if possible, ergodicity.

Stochastic Navier–Stokes equations of type (13.30) were considered by Chow (1978) for regular (vector-valued) Brownian motion. Vishik and Fursikov (1980) show existence and uniqueness (for $d = 2$) and derive a number of properties. Flandoli and Maslowski (1995) obtain an ergodicity result for a 2D version of (13.30). For more results in the 2D case, cf. Mattingly and Sinai (2001) as well as Da Prato and Debussche (2002) and the references therein. In the paper by Da Prato and Debussche the Brownian motion is standard cylindrical. Da Prato and Debussche (2003) show existence of a solution and ergodicity for $d = 3$, where the Brownian motion is regular. Mikulevicius and Rozovsky (2004) consider the fluid dynamics of a fluid particle as a stochastic flow. The time derivative of the flow is governed by the velocity field of a more general version of (13.30) and by a stochastic Stratonovich differential. The last term represents the fast oscillating component of the flow and shall model turbulence. Further, existence and uniqueness results with regard to the stochastic Navier–Stokes equations with state-dependent noise are obtained by Mikulevicius and Rozovsky (2005).

For convergence of the solutions of the Navier-Stokes equations toward random attractors, we refer the reader to Schmalfuss (1991), Crauel et al.(1997), Brzeźniak and Li (2006), and the references therein.

Kotelenez (1995a) analyzes a stochastic Navier–Stokes equation for the vorticity of a 2D fluid,[19] employing a particle approach. Apart from the fact that, in the approximation, the (positive) point masses are replaced positive and negative intensities (for the angular velocities), this equation is a semi–linear version of our SPDE (8.26). Sritharan and Sundar (2006) consider a 2D Navier–Stokes equations perturbed by multiplicative noise and establish a large deviation principle for the small noise limit. Neate and Truman (2006) employ the d-dimensional stochastic Burgers equation in the study of turbulence and intermittence (cf. also the references therein).

Baxendale and Rozovsky (1993) consider an SPDE to study the long-time asymptotics of a magnetic field in a turbulent δ-correlated flow of an ideal incompressible fluid.

[19] Cf. Chap. 8, (8.15).

13.2.8 SPDEs in Surface Physics/Chemistry

In many formulations, SPDEs arising from the study of interfaces between two phases (e.g., vapor–liquid interface) are quasilinear, where the quasilinearity results from the motion by mean curvature of the interface. Such an SPDE for a sharp interface has been analyzed by Yip (1998), where the stochastic perturbation is given by Kunitas Gaussian martingales increments $dM(t,r)$. This is equivalent to our choice of $\int G(r,q,t)w(dq,dt)$, as shown in Chap. 7. Yip obtains existence of a solution using fractional steps. Souganidis and Yip (2004) show that the solution of a similar SPDE can be unique, where the deterministic motion may have several solutions.

Funaki and Spohn (1997) derive a sharp interface model from an interacting and diffusing particle model (on a lattice) as a macroscopic limit, employing a hydrodynamic scaling. Replacing the i.i.d. Brownian motions (indexed by the lattice sites) in the paper by Funaki and Spohn by correlated Brownian motions should result in a sharp stochastic interface, which itself should be the solution of a quasilinear SPDE, similar to the SPDE obtained by Yip (loc. Cit.).

Blömker (2000) analyzes surface growth and phase separation, whose macroscopic regime is often described by the semilinear fourth-order parabolic Cahn–Hilliard equation.

13.2.9 SPDEs for Strings

For SPDEs describing the motion of strings we refer to Funaki (1983), Faris and Jona-Lasinio (1982), Mueller and Tribe (2002), and the references therein.

13.3 Books on SPDEs

To our knowledge Rozovsky's book in 1983 is first book on SPDEs. It treats bilinear SPDEs driven by regular Brownian motion with emphasis on filtering. Metivier (1988) obtains weak solutions for a class of semilinear SPDEs. Da Prato and Zabczyk (1992) develop linear, bilinear, and semilinear SPDEs using a semigroup approach. Both regular and cylindrical Brownian motions appear as stochastic driving terms. It also contains a chapter on invariant measures and large deviations as well appendixes on linear deterministic systems and control theory. The problem of invariant measures and ergodicity is treated in detail by Da Prato and Zabczyk (1996). Kallianpur and Xiong (1995) analyze linear, bilinear and some semilinear SPDEs driven by regular and cylindrical Brownian motions and quasilinear SPDEs driven by Poisson random measures. This book also provides a chapter on large deviations. Greksch and Tudor (1995) analyze semilinear SPDEs, employing both a semigroup approach and variational methods. Flandoli (1995) obtains regularity results for stochastic flows associated with the solutions of bilinear SPDEs. The book by Holden et al. (1996) deviates from all the previously mentioned monographs in that the state space is a space of generalized random variables (cf. the paper by Nualart and Zakai (1989)) and multiplication is replaced by the Wick product (cf. (13.29) above and the references quoted). Peszat and Zabczyk (2006) employ a semigroup approach to SPDEs, driven by Lévy processes with applications to models in physics and mathematical finance. Knoche–Prevot and Röckner (2006) provide an introduction to the variational approach.

Finally, we must mention Walsh (1986) and Dawson (1993). Although those St. Fleur notes have not appeared as separate books, they are as comprehensive and detailed as any good monograph. Walsh employs a semigroup approach, but uses a space–time formulation for the stochastic terms instead of infinite dimensional Brownian motions or martingales. We have adopted the integration with respect to space–time white noise from Walsh. Dawson's St. Fleur notes are, strictly speaking, not on SPDEs. However, the particle approach to SPDEs has been strongly influenced by Dawson's work.

Part IV
Macroscopic: Deterministic Partial Differential Equations

Chapter 14
Partial Differential Equations as a Macroscopic Limit

14.1 Limiting Equations and Hypotheses

We define the limiting equations and state additional hypotheses.

In this section we add the subscript ε to the coefficients of the SODEs and SPDEs which are driven by correlated Brownian noise. Further, we restrict the measures to $\mathbf{M}_1$, since we want to use results proved by Oelschläger (1984) and Gärtner (1988) on the derivation of the macroscopic McKean–Vlassov equation as the limit of interacting and diffusion particles driven by independent Brownian motions. To this end, choose a countable set of i.i.d. $\mathbf{R}^d$-valued standard Brownian motions $\{\beta^i\}_{j\in\mathbf{N}}$. Let $F_0(r,\mu,t)$ and $\mathcal{J}_0(r,\mu,t)$ be $\mathbf{R}^d$-valued and $\mathcal{M}_{d\times d}$ valued functions, respectively, jointly measurable in all arguments. Further, suppose $(r_\ell,\mu_\ell,t)\in\mathbf{R}^d\times\mathbf{M}\times\mathbf{R}, \ell=1,2$, Let $c_F, c_{\mathcal{J}}\in(0,\infty)$ and assume global Lipschitz and boundedness conditions

$$\left.\begin{array}{l}
|F_0(r_1,\mu_1,t)-F_0(r_2,\mu_2,t)|\le c_F\,\{\rho(r_1-r_2)+\gamma(\mu_1-\mu_2)\}\\
\quad\displaystyle\sum_{k,\ell=1}^{d}(\mathcal{J}_{0,k\ell}(r_1,\mu_1,t)-\mathcal{J}_{0,k\ell}(r_2,\mu_2,t))^2\\
\quad\le c_{\mathcal{J}}^2\left\{\rho^2(r_1-r_2)+\gamma^2(\mu_1-\mu_2)\right\},\\
\quad|F_0(r,\mu,t)|^2+\displaystyle\sum_{k,\ell=1}^{d}\mathcal{J}_{0,k\ell}^2(r,\mu,t)\le c_{F,\mathcal{J}}.
\end{array}\right\}\tag{14.1}$$

Consider stochastic ordinary differential equations (SODEs) for the displacement of r^i of the following type

$$\left.\begin{array}{l}
\mathrm{d}r_{0,N}^i(t)=F_0(r_{0,N}^i(t),X_N(t),t)\mathrm{d}t+\mathcal{J}_0(r_{0,N}^i(t),X_N(t),t)\mathrm{d}\beta_i(t),\\
r^i(s)=q^i, i=1,\ldots,N,\quad X_{0,N}(t):=\displaystyle\sum_{i=1}^{N}m_i\delta_{r_{0,N}^i(t)}.
\end{array}\right\}\tag{14.2}$$

The subscript 0 indicates that there is no correlation between the Brownian noises for each particles. The Lipschitz conditions are those of Oelschläger. Gärtner allows for local Lipschitz conditions. Both authors restrict the analysis to time-independent coefficients. However, assuming conditions, uniformly in t, we may assume that their results can be formulated also for the nonautonomous case. The two-particle diffusion matrix of the noise is given by

$$\tilde{D}_0(\mu, r^i, r^j, t) := \begin{cases} \mathcal{J}_0(r^i, \mu, t)\mathcal{J}_0^{\mathrm{T}}(r^j, \mu, t), & \text{if} \quad i = j, \\ \mathbf{0}, & \text{if} \quad i \neq j, \end{cases} \tag{14.3}$$

where A^{T} is the transpose of a matrix A. Set

$$D_0(\mu, r, t) := \tilde{D}_0(\mu, r, r, t). \tag{14.4}$$

Under the above assumptions

$$X_{0,N}(\cdot) \Longrightarrow X_{0,\infty}(\cdot) \quad \text{in } C([0,\infty); \mathbf{M}_1), \text{ as } n \longrightarrow \infty, \tag{14.5}$$

where $X_{0,\infty}(\cdot)$ is the unique solution of the macroscopic McKean–Vlassov equation (or "nonlinear diffusion equation"):[1]

$$\left.\begin{aligned} \frac{\partial}{\partial t} X_{0,\infty} = \frac{1}{2} \sum_{k,\ell=1}^{d} \partial_{k\ell}^2 (D_{0,k\ell}(X_{0,\infty}, \cdot, t) X_{0,\infty}) - \nabla \cdot (X_{0,\infty} F_0(\cdot, X_{0,\infty}, t)), \\ X_{0,\infty}(0) = \mu. \end{aligned}\right\} \tag{14.6}$$

Along with (14.6) we define infinitely many SODEs, whose empirical distribution is the solution of (14.6):

$$\left.\begin{aligned} \mathrm{d}r_{0,\infty}^i &= F_0(r_{0,\infty}^i, X_{0,\infty}, t)\mathrm{d}t + \mathcal{J}_0(r_{0,\infty}^i, X_{0,\infty}, t)\mathrm{d}\beta^i, \\ r_{0,\infty}^i(0) &= q^i, \quad i = 1, 2, \ldots, X_{0,\infty}(t) = \lim_{N\to\infty} \sum_{i=1}^{N} \frac{1}{N} \delta_{r_{0,\infty}^i}(t). \end{aligned}\right\} \tag{14.7}$$

Existence and uniqueness for (14.7) follow under the assumptions (14.1).[2]

Let $\sigma_\varepsilon(r, \mu, t)$ be the nonnegative square root of $D_\varepsilon(\mu, r, t)$. Since the entries of $D_\varepsilon(\mu, r, t)$ are bounded, uniformly in (r, μ, t), the same boundedness holds for $\sigma_\varepsilon(r, \mu, t)$.[3]

Hypothesis 14.1

(i) Suppose that $\sigma_i^{\perp} \equiv 0$ $\forall i$ in (4.10).
(ii) Suppose that for each $r \in \mathbf{R}^d$, $t \geq 0$ and $\mu \in \mathbf{M}_1$, $\sigma_\varepsilon(r, \mu, t)$ is invertible.

---

[1] Cf. Oelschläger (1984). Gärtner (loc. cit.) obtains the same result under local Lipschitz and linear growth conditions, making an additional assumption on the initial condition.

[2] Cf. Oelschläger (1984), Gärtner (loc. cit.), and Kurtz and Protter (1996). Cf. also Remark 14.1.

[3] In Sect. 4.3, we provide a class of coefficients which satisfy Hypothesis 14.1.

(iii) Further, suppose that for any compact subset $\mathcal{K}$ of $\mathbf{R}^d$, any compact subset $\mathcal{C} \subset \mathbf{M}_1$, any $T > 0$ and any $\delta > 0$ the following three relations hold:

$$\lim_{\varepsilon \downarrow 0} \sup_{r \in \mathcal{K}} \sup_{0 \le t \le T} \sup_{\mu \in \mathcal{C}} (|F_\varepsilon(r,\mu,t) - F_0(r,\mu,t)| + |\sigma_\varepsilon(r,\mu,t) - \mathcal{J}_0(r,\mu,t)|) = 0, \tag{14.8}$$

where $|\cdot|$ denotes the Euclidean norms in $\mathbf{R}^d$ and $\mathcal{M}_{d\times d}$, respectively.

(iv)

$$\sup_{1 \ge \varepsilon > 0} \sup_{r \in \mathcal{K}, 0 \le t \le T, \mu \in \mathcal{C}} |\sigma_\varepsilon^{-1}(r,\mu,t)| < \infty, \tag{14.9}$$

(v)

$$\lim_{\varepsilon \downarrow 0} \sup_{|r-q|>\delta} \sup_{0 \le t \le T, \mu \in \mathcal{C}} |\tilde{D}_\varepsilon(\mu, r, q, t)| = 0,\,^{4} \tag{14.10}$$

(vi)

$$\lim_{\eta \downarrow 0} \sup_{r \in \mathcal{K}} \sup_{\mu \in \mathcal{C}} \sup_{0 \le s,t \le T, |t-s| \le \eta} |F_0(r,\mu,t) - F_0(r,\mu,s)| + |\mathcal{J}_0(r,\mu,t) - \mathcal{J}_0(r,\mu,s)| = 0, \tag{14.11}$$

i.e., $F_0(r,\mu,t)$ and $\mathcal{J}_0(r,\mu,t)$ are continuous in t, uniformly in (r,μ) from compact sets $\mathcal{K} \times \mathcal{C}$.

(vii) Given the solution $X_{0,\infty}(t)$ of (14.7), suppose that $D_{0,k\ell}(X_{0,\infty}(\cdot),\cdot,\cdot)$ and $F_{0,k}(\cdot, X_{0,\infty}(\cdot),\cdot)$ as functions of (r,t) are both twice continuously differentiable, where $k,\ell = 1,\ldots,d$. □

Remark 14.1.

(i) In what follows we assume the Oelschläger–Gärtner result to hold also for nonautonomous coefficients, since one can obviously generalize their results, assuming that their hypotheses hold uniformly in t.

(ii) A closed k-dimensional submanifold Λ of $\mathbf{R}^n$ will be called *nonattainable* by an $\mathbf{R}^n$-valued diffusion, $(r_1(\cdot,q_1),\ldots,r_n(\cdot,q_n))$ with initial values $(q_1,\ldots,q_n)$ if

$$P\{(q_1,\ldots,q_n) \in \Lambda\} = 0 \text{ implies } P\{\cup_{t>0}(r_1(\cdot,q_1),\ldots,r_n(\cdot,q_n)) \in \Lambda)\} = 0. \tag{14.12}$$

□

Friedman (1976), Sect. 11, obtains general conditions for nonattainability of closed k-dimensional submanifold Λ of $\mathbf{R}^n$ for linear $\mathbf{R}^n$-valued diffusions, under the assumption that the coefficients are autonomous and twice continuously differentiable.

In Lemma 14.8, we extend Friedman's result to the nonautonomous case, defined by the first m diffusions of the system (14.7). The standard device of adding the time t as another dimension converts the nonautonomous system into an autonomous (but degenerate) system.

---

[4] Cf. the definition of $\tilde{D}_\varepsilon(\mu, r, q, t)$ before (8.23).

The differentiability assumption (iv) on the coefficients as functions of (r, t) in Hypothesis 14.1 depends on the existence of weak derivatives of the solution of the McKean–Vlasov equation (14.7) with respect to t. For an example where the assumption holds, cf. (14.3) (v).

Finally, let $X_{\varepsilon,\infty}(0)$ be the continuum limit of $X_{\varepsilon,N}(0) := \sum_{i=1}^{N} m_i \delta_{q_\varepsilon^i}$ and $Z_{0,\infty}(Y, t, X_{\varepsilon,\infty}(0))$ be the solution of the random media PDE:

$$\left.\begin{aligned} \frac{\partial}{\partial t} Z_{0,\infty}(Y) = \frac{1}{2} \sum_{k,\ell=1}^{d} \partial^2_{k\ell}(D_{0,k\ell}(Y, \cdot, t) Z_{0,\infty}(Y)) - \nabla \cdot (Z_{0,\infty}(Y) F_0(\cdot, Z_{0,\infty}(Y), t)),\\ Z_{0,\infty}(0) = \mu, \end{aligned}\right\} \tag{14.13}$$

where $Y(\cdot) \in \mathcal{M}_{loc,2,0,\infty}$.

14.2 The Macroscopic Limit for $d \geq 2$

Assuming $d \geq 2$ we prove that, under the hypotheses from 14.1 and Chap. 4, the solutions of the SPDE (8.26) converge weakly toward the solution of a corresponding quasilinear macroscopic PDE, as the correlation length tends to 0.

For $m \in \mathbf{N}$, set

$$\Lambda_m := \left\{(p^1, \ldots, p^m) \in \mathbf{R}^{d\cdot m} : \exists i \neq j,\ i, j \in \{1, \ldots, m\}, \quad \text{with } p^i = p^j\right\}.$$

Infinite sequences in $\mathbf{R}^d$ will be denoted either $(r^1, r^2, \ldots)$ or $r^{(\cdot)}$. The corresponding state space, $(\mathbf{R}^d)^\infty$, will be endowed with the metric

$$d_\infty(r^{(\cdot)}, q^{(\cdot)}) := \sum_{k=1}^{\infty} 2^{-k} \rho(r^k - q^k).$$

Theorem 14.2.[5] *Suppose Hypotheses 4.1 and 14.1. Further, suppose that $d \geq 2$ and that $\{q_\varepsilon^1, q_\varepsilon^2, \ldots\}$ is a sequence of exchangeable*[6] *initial conditions in (4.10) and (14.2), respectively, such that for all $m \in \mathbf{N}$ and $\varepsilon \geq 0$,*

$$P\left\{(q_\varepsilon^1, \ldots, q_\varepsilon^m) \in \Lambda_m\right\} = 0,$$

where $q_\varepsilon^1, \ldots, q_\varepsilon^m$ are the initial conditions in (4.10) for $\varepsilon > 0$ and in (14.2) for $\varepsilon = 0$, respectively. Finally, suppose

[5] This theorem was obtained by Kotelenez and Kurtz (2006), and, in what follows, we adopt the proof from that paper.

[6] By definition exchangeability means that $\{q_\varepsilon^1, q_\varepsilon^2, \ldots\} \sim \{q_\varepsilon^{\pi(1)}, q_\varepsilon^{\pi(1)}, \ldots\}$ whenever π is a finite permutation of $\mathbf{N}$. Cf. Aldous (1985).

$$(X_\varepsilon(0), q_\varepsilon^1, q_\varepsilon^2, \ldots) \Rightarrow (X_{0,\infty}(0), q_0^1, q_0^2, \ldots) \ \ in\ \mathbf{M}_1 \times (\mathbf{R}^d)^\infty, \ \ as\ \varepsilon \downarrow 0\,,$$

where $X_\varepsilon(0)$ *are the initial values for (8.26) and* $X_{0,\infty}(0) = \lim\limits_{N\to\infty} \frac{1}{N}\sum\limits_{j=1}^{N} \delta_{q_0^j}$ *is the initial condition of (14.6). Then*

$$X_\varepsilon \Rightarrow X_{0,\infty}\ in\ C([0,\infty); \mathbf{M}_1), \ \ as\ \varepsilon \downarrow 0 \tag{14.14}$$

and

$$Z_\varepsilon(Y) \Rightarrow Z_{0,\infty}(Y)\ in\ C([0,\infty); \mathbf{M}_1), \ \ as\ \varepsilon \downarrow 0\,, \tag{14.15}$$

where $Z_\varepsilon(Y)$ *is the solution of the SPDE (8.25) with coefficients* F_ε *and* D_ε *and* $Y(\cdot)$ *as in (14.13).*

The proof of Theorem 14.2 will be the consequence of a series of lemmas, which we will derive first. □

Lemma 14.3. *Let* $(q_\varepsilon^1, \ldots, q_\varepsilon^m, \ldots)$ *be a finite or infinite subsequence of initial conditions for (4.10) at time* 0 *and* $X_\varepsilon(0)$ *an initial condition for (8.26) such that* $(X_\varepsilon(0), q_\varepsilon^1, q_\varepsilon^2, \ldots)$ *satisfy the assumptions of Theorem 14.2.*

Then $(X_\varepsilon(\cdot), r_\varepsilon(\cdot, X_\varepsilon, q_\varepsilon^1), \ldots, r_\varepsilon(\cdot, X_\varepsilon, q_\varepsilon^m, \ldots))$ *is relatively compact in* $C([0,\infty); \mathbf{M}_1 \times (\mathbf{R}^d)^\infty)$.

Proof.

(i) It sufficient to restrict the sequences to the first m coordinates in the formulation of the Lemma.

(ii) For $s \leq t \leq T$,

$$E(\gamma^2(X_\varepsilon(t), X_\varepsilon(s))/\mathcal{F}_s)$$
$$= E \sup_{\|f\|_{L,\infty}\leq 1} \left(\int (f(r(t, X_\varepsilon, s, q)) - f(q)) X_\varepsilon(s, dq) \right)^2 / \mathcal{F}_s)$$

(by (15.22) and (8.50))

$$\leq E(\int \rho(r_\varepsilon(t,q), q) X_\varepsilon(s, \mathrm{d}q))^2/\mathcal{F}_s)$$

(since the f, used in the norm γ, are Lipschitz with Lipschitz constant 1)

$$\leq E(\int \rho^2(r_\varepsilon(t,q), q) X_\varepsilon(s, \mathrm{d}q)/\mathcal{F}_s)$$

(by the Cauchy–Schwarz inequality and $X_\varepsilon(s, \mathbf{R}^d) = 1$)

$$= \int E(\rho^2(r_\varepsilon(t,q), q)/\mathcal{F}_s) X_\varepsilon(s, \mathrm{d}q)$$
$$\leq c_T(t-s) X_\varepsilon(0, \mathbf{R}^d) = c_T(t-s)$$

independent of ε by the boundedness of F_ε and $\mathcal{J}_\varepsilon$ (cf. (4.11)) and by $X_\varepsilon(s, \mathbf{R}^d) = X_\varepsilon(0, \mathbf{R}^d) = 1$ a.s. Similarly, for the m-particle process with a constant $c_{T,m}$.

(iii) To prove relative compactness of the marginals we must show tightness. Indeed, for the m-particle process this is obvious and its tightness follows from (4.11). For the $(\mathbf{M}_1, \gamma)$-valued process this follows from the proof of a similar statement, when $\mathbf{M}_1$ is endowed with the Prohorov metric and the equivalence of γ and the Prohorov metric.[7]
To show tightness of $X_\varepsilon(t)$ let $\delta > 0$ be given. Choose a closed ball S_L in $\mathbf{R}^d$ with center 0 and radius $L = L(T)$ such that for all $\varepsilon > 0$
(α) $P\{X_\varepsilon(0, S_L^c) > \frac{\delta}{2}\} \le \frac{\delta}{2}$,

$$(\beta)\ E \sup_{q\in\mathbf{R}^d,\varepsilon>0} L^{-2}\Bigg\{\Bigg|\int_0^T F_\varepsilon(r_{\bar\varepsilon}(s,q), X_{\bar\varepsilon}(s), s)\mathrm{d}s\Bigg|^2 + \sum_{k,\ell=1}^{d}\int_0^T\int \mathcal{J}^2_{\varepsilon,k\ell}(r_\varepsilon(s,q), p, X_\varepsilon(s), s)\mathrm{d}p\,\mathrm{d}s\Bigg\} \le \frac{\delta}{2}.$$

Cf. (4.11)–here we used the abbreviation $A^c := \mathbf{R}^d\backslash A$ for $A \in \mathcal{B}^d$. Then,

$$\begin{aligned}
&P\left\{\int 1_{S^c_{2L}}(r_\varepsilon(t,q))X_\varepsilon(0,\mathrm{d}q) > \delta\right\}\\
&\le P\left\{\int (1_{S^c_L}(r_\varepsilon(t,q)-q) + 1_{S^c_L}(q))X_\varepsilon(0,\mathrm{d}q) > \delta\right\}\\
&\le E\left\{\int\left(E\frac{|r_\varepsilon(t,q)-q|^2}{L^2}\Big|\mathcal{F}_0\right)X_\varepsilon(0,\mathrm{d}q)\right\} + \frac{\delta}{2}\\
&< \frac{\delta}{2} + \frac{\delta}{2} = \delta.
\end{aligned}$$

Employing Theorems 15.26 and 15.27 of Sect. 15.2.1 completes the proof. □

In what follows let $(X_\varepsilon, r_\varepsilon^1, \ldots, r_\varepsilon^{N_\varepsilon}) := (X_{\varepsilon,N_\varepsilon}, r_\varepsilon(\cdot, X_{\varepsilon,N_\varepsilon}, q^1), \ldots, r_\varepsilon(\cdot, X_{\varepsilon,N_\varepsilon}, q^{N_\varepsilon}))$ be a convergent subsequence in $C_{\mathbf{M}_1\times(\mathbf{R}^d)^\infty}[0,\infty)$, where $\varepsilon \downarrow 0$. We allow $N_\varepsilon = \infty$. If $N_\varepsilon < \infty$, we require $N_\varepsilon \to \infty$. Set

$$w_\varepsilon^i(t) := \int_0^t\int \sigma_\varepsilon^{-1}(r_\varepsilon^i(s), X_\varepsilon(s), s)\mathcal{J}_\varepsilon(r_\varepsilon^i(s), q, X_\varepsilon(s), s)w(\mathrm{d}q, \mathrm{d}s), \tag{14.16}$$

where $X_\varepsilon := X_{\varepsilon,N_\varepsilon}$ is the empirical process for $(r_\varepsilon^1, \ldots, r_\varepsilon^{N_\varepsilon})$. By Levy's theorem[8] each w_ε^i is an $\mathbf{R}^d$-valued Brownian motion. Abbreviating

$$c_\varepsilon(r, q, \mu, t) := \sigma_\varepsilon^{-1}(r, \mu, t)\tilde{D}_\varepsilon(\mu, r, q, t)\sigma_\varepsilon^{-1}(q, \mu, t),$$

we obtain

$$[[w_\varepsilon^i, w_\varepsilon^j]](t) = \int_0^t c_\varepsilon(r_\varepsilon^i(s), r_\varepsilon^j(s), X_\varepsilon(s), s)\mathrm{d}s, \tag{14.17}$$

where the left-hand side of (14.17) is the tensor quadratic variation.[9] Moreover, hypotheses (14.9) and (14.10) imply for any $\delta > 0$, any compact subset $\mathcal{C} \subset \mathbf{M}_1$ and any T

[7] Cf. Sect. 15.2.1, Theorems 15.23 and 15.24.

[8] Cf. Theorem 15.37, Sect. 15.2.3.

[9] Cf. Sect. 15.2.3.

$$\lim_{\varepsilon \downarrow 0} \sup_{|r-q|>\delta} \sup_{\mu \in \mathcal{C}, 0 \leq t \leq T} |c_\varepsilon(r, q, \mu, t)| = 0. \tag{14.18}$$

Define for $1 > \delta > 0$ and $m < N_\varepsilon$ stopping times

$$\tau_\varepsilon^m(\delta) := \min_{1 \leq i \neq j \leq m} \inf\{t \geq 0 : \quad |r_\varepsilon^i(t) - r_\varepsilon^j(t)| \leq \delta\}. \tag{14.19}$$

Let $\mathcal{G}$ be the σ-algebra, generated by $w(dr, dt)$ and by all events from $\mathcal{F}_0$, i.e., we set

$$\mathcal{G} := \sigma \left\{ \int_{\mathbf{R}^d} \int_0^\infty 1_B(r, t) w(dr, dt), \, B \in \mathcal{B}^{d+1}, A \in \mathcal{F}_0 \right\},$$

where $\mathcal{B}^{d+1}$ is the Borel σ-algebra in $\mathbf{R}^{d+1}$. Further, let $\{\beta^i\}_{i \in \mathbf{N}}$ be i.i.d. $\mathbf{R}^d$-valued Brownian motions, defined on $(\Omega, \mathcal{F}, \mathcal{F}_t, P)$ and independent of $\mathcal{G}$. Set

$$\left(\beta_{\varepsilon,\delta,m}^1, \ldots, \beta_{\varepsilon,\delta,m}^{N_\varepsilon}\right)(t) := (w_\varepsilon^1, \ldots, w_\varepsilon^{N_\varepsilon})(t \wedge \tau_\varepsilon^m(\delta)) + \int_0^t 1_{\{u \geq \tau_\varepsilon^m(\delta)\}} (d\beta^1, \ldots, d\beta^{N_\varepsilon})(u). \tag{14.20}$$

Clearly, $(\beta_{\varepsilon,\delta,m}^1, \ldots, \beta_{\varepsilon,\delta,m}^{N_\varepsilon})$ is a continuous square integrable $\mathbf{R}^{d \cdot N_\varepsilon}$-valued martingale with tensor quadratic variation

$$\left|\left[\beta_{\varepsilon,\delta,m}^1, \ldots, \beta_{\varepsilon,\delta,m}^{N_\varepsilon}\right]\right|(t) = \left|\left[w_\varepsilon^1, \ldots, w_\varepsilon^{N_\varepsilon}\right]\right|(t \wedge \tau_\varepsilon^m(\delta)) + \int_0^t 1_{\{u \geq \tau_\varepsilon^m(\delta)\}} du \, I_{d \cdot N_\varepsilon}. \tag{14.21}$$

Here I_k denotes the identity matrix on $\mathbf{R}^k$and for $N_\varepsilon = \infty$ $I_{d \cdot \infty}$ is the identity operator in $(\mathbf{R}^d)^\infty$.

Lemma 14.4. *For each $\delta > 0$ and $m < N_\varepsilon$*

$$(\beta_{\varepsilon,\delta,m}^1, \ldots, \beta_{\varepsilon,\delta,m}^m) \Rightarrow (\beta^1, \ldots, \beta^m) \text{ in } C([0, \infty); \mathbf{R}^{m \cdot d}), \quad \text{as } \varepsilon \downarrow 0. \tag{14.22}$$

Proof. The family of marginals $\{X_\varepsilon(t), \varepsilon > 0\}$, $t \geq 0$, is relatively compact by Lemma 14.3. So, by the compact containment condition[10] in addition to Lebesgue's dominated convergence theorem

$$\left|\left[\beta_{\varepsilon,\delta,m}^1, \ldots, \beta_{\varepsilon,\delta,m}^m\right]\right|(t) \to t I_{d \cdot m}, \text{ as } \varepsilon \downarrow 0 \text{ in probability}.$$

Hence, (14.22) follows from the martingale central limit theorem.[11] □

Consider the following system of SODEs

$$\left.\begin{aligned} dr_{\varepsilon,\delta,m,N_\varepsilon}^i &= F_\varepsilon(r_{\varepsilon,\delta,m,N_\varepsilon}^i, X_\varepsilon)dt + \sigma_\varepsilon(r_{\varepsilon,\delta,m,N_\varepsilon}^i, X_\varepsilon)d\beta_{\varepsilon,\delta,m}^i \\ r_{\varepsilon,\delta,m,N_\varepsilon}^i(0) &= q_\varepsilon^i, i = 1, \ldots, N_\varepsilon, \quad X_{\varepsilon,\delta,m,N_\varepsilon}(t) := \sum_{i=1}^{N_\varepsilon} \frac{1}{N_\varepsilon} \delta_{r_{\varepsilon,\delta,m,N_\varepsilon}^i}(t), \end{aligned}\right\} \tag{14.23}$$

[10] Cf. Remark 7.3 in Sect. 3 of Ethier and Kurtz (1986) and (14.24).

[11] Cf. Theorem 15.39, Sect. 15.2.3.

where for $N_\varepsilon = \infty$ here and in what follows $i = 1, \ldots, N_\varepsilon$ shall mean $i \in \mathbf{N}$. Repeating the proof of Lemma 14.3, we show that $(X_{\varepsilon,\delta,m,N_\varepsilon}(\cdot), r^1_{\varepsilon,\delta,m,N_\varepsilon}(\cdot), \ldots)$ is relatively compact in $C([0,\infty); \mathbf{M}_1 \times \mathbf{R}^\infty)$ and, therefore, we may, without loss of generality, assume that $(X_{\varepsilon,\delta,m,N_\varepsilon}(\cdot), r^1_{\varepsilon,\delta,m,N_\varepsilon}(\cdot), \ldots)$ converges weakly. Let $(r^1_{0,\infty}, \ldots)$ be the solution of (14.7).

Lemma 14.5. *For every $\delta > 0$, $m < N_\varepsilon$*

$$(r^1_{\varepsilon,\delta,m,N_\varepsilon}, \ldots, r^{N_\varepsilon}_{\varepsilon,\delta,m,N_\varepsilon}) \Rightarrow (r^1_{0,\infty}, \ldots) \quad \textit{in } C([0,\infty) : (\mathbf{R}^d)^\infty), \textit{ as } \varepsilon \downarrow 0 .$$

Proof. Recall that $(q^1_\varepsilon, \ldots, q^{N_\varepsilon}_\varepsilon) \Rightarrow (q^1_0, \ldots)$ in $(\mathbf{R}^d)^\infty$, as $\varepsilon \downarrow 0$. Using the boundedness of the coefficients in (14.23) we may repeat the proof of Lemma 14.3 and conclude that the family $X_{\varepsilon,\delta,m,N_\varepsilon}(\cdot)$ is relatively compact in $C([0,\infty); \mathbf{M}_1)$, as a function of ε, δ, and m. Hence, the aforementioned compact containment condition of Ethier and Kurtz (loc. cit.) now reads: For every $L \in \mathbf{N}$ there is a compact subset $\mathcal{C}_L$ of $\mathbf{M}_1$ such that

$$\begin{aligned} \inf_{\{\varepsilon>0,\delta>0,m<N_\varepsilon\}} & P\{X_\varepsilon(t) \in \mathcal{C}_L \, , \, X_{\varepsilon,\delta,m,N_\varepsilon}(t) \in \mathcal{C}_L \\ & \text{and } X_{0,\infty}(t) \in \mathcal{C}_L \;\; \text{for } 0 \le t \le T \} \ge 1 - \frac{1}{L}. \end{aligned} \tag{14.24}$$

This implies the existence of an increasing sequence of compact subsets $\mathcal{C}_L$ of $\mathbf{M}_1$, $L \in \mathbf{N}$, such that $\forall \varepsilon > 0$, $\delta > 0$, $m \in \mathbf{N}$ with probability 1

$$\cup_{0\le t\le T}(\{X_\varepsilon(t,\omega)\} \cup \{X_{\varepsilon,\delta,m,N_\varepsilon}(t,\omega)\} \cup \{X_{0,\infty}(t,\omega)) \subset \cup_{L\in\mathbf{N}}\mathcal{C}_L. \tag{14.25}$$

For each $N \in \mathbf{N}$ define maps Ψ_N from $((\mathbf{R}^d)^\infty, d_\infty)$ into $(\mathbf{M}_1, \gamma_f)$ by

$$\Psi_N\left(r^{(\cdot)}\right) = \frac{1}{N}\sum_{i=1}^N \delta_{r^i},$$

where here and in what follows $(r^1, \ldots, r^N)$ are the first N coordinates of $r^{(\cdot)}$. The definition of the norm γ implies that for two sequences $r^{(\cdot)}$ and $q^{(\cdot)}$

$$\gamma\left(\Psi_N\left(r^{(\cdot)}\right) - \Psi_N\left(q^{(\cdot)}\right) \le \frac{1}{N}\sum_{i=1}^N \varrho(r^i - q^i). \tag{14.26}$$

So the maps Ψ_N are continuous. Further, set

$$\begin{aligned} D_\infty &:= \{r^{(\cdot)} \in (\mathbf{R}^d)^\infty : \text{such that } \lim_{N\to\infty} \Psi_N(r^{(\cdot)}) \text{ exists}\}, \\ \Psi_\infty(r^{(\cdot)}) &:= \begin{cases} \lim\limits_{N\to\infty} \Psi_N(r^{(\cdot)}), & \text{if} \quad r^{(\cdot)} \in D_\infty, \\ \delta_0, & \text{if} \quad r^{(\cdot)} \notin D_\infty. \end{cases} \end{aligned}$$

This definition implies that D_∞ is a Borel set in $((\mathbf{R}^d)^\infty, d_\infty)$ and that Ψ_∞ is a measurable map from $((\mathbf{R}^d)^\infty, d_\infty)$ into $(\mathbf{M}_1, \gamma_f)$. Further, by the previously mentioned density of measures of the form $\frac{1}{N}\sum\limits_{i=1}^N \delta_{r^i}$ in $(\mathbf{M}_1, \gamma)$

$$\Psi_\infty(D_\infty) = \mathbf{M}_1.$$

Finally, we set

$$D_{\infty,L} := \big\{r^{(\cdot)} \in D_\infty : \Psi_\infty(r^{(\cdot)}) \in \mathcal{C}_L\big\}.$$

Define stopping times

$$\tau_\varepsilon(\delta, m, L) := \inf\big\{t \geq 0 : X_{\varepsilon,\delta,m,N_\varepsilon}(t) \notin \mathcal{C}_L\big\}. \tag{14.27}$$

By (14.25), $\forall \varepsilon > 0, \delta > 0, m \in \mathbf{N}$ with probability 1:

$$\tau_\varepsilon(\delta, m, L) \uparrow \infty, \text{as } L \longrightarrow \infty. \tag{14.28}$$

If $X_\varepsilon(0) \in \mathcal{C}_L$ with probability 1 for some L then with probability 1

$$\big(r^1_{\varepsilon,\delta,m,N_\varepsilon}(\cdot \wedge \tau_\varepsilon(\delta, m, L)), \ldots, r^m_{\varepsilon,\delta,m,N_\varepsilon}(\cdot \wedge \tau_\varepsilon(\delta, m, L)), \ldots\big) \in D_{\infty,L}. \tag{14.29}$$

Let $\hat{C} \in \mathbf{R}^d$ and $\hat{D} \in \mathcal{M}_{d\times d}$ arbitrary but fixed elements. Define for $\varepsilon \geq 0$ maps $\hat{F}_{\varepsilon,i,L}$ and $\hat{\sigma}_{\varepsilon,i,L}$ from $D_{\infty,L} \times [0,\infty)$ into $\mathbf{R}^d$ and $\mathcal{M}_{d\times d}$, respectively:

$$\left.\begin{aligned}
&\text{For } \varepsilon > 0, \text{ we set}\\
&\hat{F}_{\varepsilon,i,\delta,m,L}(r^{(\cdot)}, t) := \begin{cases} F_\varepsilon(r^i, \Psi_{N_\varepsilon}(r^{(\cdot)}), t), & \text{if} \quad r^{(\cdot)} \in D_{\infty,L},\\ \hat{C}, & \text{if} \quad r^{(\cdot)} \notin D_{\infty,L},\end{cases}\\
&\hat{\sigma}_{\varepsilon,i,\delta,m,L}(r^{(\cdot)}, t) := \begin{cases} \sigma_\varepsilon(r^i, \Psi_{N_\varepsilon}(r^{(\cdot)}), t), & \text{if} \quad r^{(\cdot)} \in D_{\infty,L},\\ \hat{D}, & \text{if} \quad r^{(\cdot)} \notin D_{\infty,L},\end{cases}
\end{aligned}\right\} \tag{14.30}$$

and for $\varepsilon = 0$ we make corresponding definitions, employing in the above definition Ψ_∞.

Stopping the solutions of (14.23) and (14.7) at $\tau_\varepsilon(\delta, m, L)$, we see by (14.27) and (14.29) that these stopped solutions solve the following systems of SODEs in $(\mathbf{R}^d)^\infty$:

$$\left.\begin{aligned}
\mathrm{d}\hat{r}^i_{\varepsilon,\delta,m,N_\varepsilon} &= \hat{F}_{\varepsilon,i,\delta,m,L}\big(\hat{r}^{(\cdot)}_{\varepsilon,\delta,m,N_\varepsilon}, t\big)1_{D_{\infty,L}}\big(\hat{r}^{(\cdot)}_{\varepsilon,\delta,m,N_\varepsilon}\big)\mathrm{d}t\\
&\quad + \hat{\sigma}_{\varepsilon,i,\delta,m,L}\big(\hat{r}^{(\cdot)}_{\varepsilon,\delta,m,N_\varepsilon}, t\big)1_{D_{\infty,L}}\big(\hat{r}^{(\cdot)}_{\varepsilon,\delta,m,N_\varepsilon}\big)\mathrm{d}\beta^i_{\varepsilon,\delta,m},\\
r^i_{\varepsilon,\delta}(0) &= q^i_\varepsilon, i = 1, \ldots, N_\varepsilon
\end{aligned}\right\} \tag{14.31}$$

and

$$\left.\begin{aligned}
\mathrm{d}\hat{r}^i_{0,\infty} = \hat{F}_{0,i,L}\big(\hat{r}^{(\cdot)}_{0,\infty}, t\big)1_{D_{\infty,L}}\big(\hat{r}^{(\cdot)}_{0,\infty}\big)\mathrm{d}t + \hat{\sigma}_{0,i,L}\big(\hat{r}^{(\cdot)}_{0,\infty}, t\big)1_{D_{\infty,L}}\big(\hat{r}^{(\cdot)}_{0,\infty}\big)\mathrm{d}\beta^i,\\
r^i_{\varepsilon,\delta}(0) = q^i_\varepsilon, i = 1, \ldots, N_\varepsilon,
\end{aligned}\right\} \tag{14.32}$$

respectively. Conversely, up to time $\tau_\varepsilon(\delta, m, L)$ every solution of (14.31) and (14.32) is a solution of (14.23) and (14.7), respectively. Note that

$$\begin{aligned}\left|F_\varepsilon(r^i, \Psi_{N_\varepsilon}(r^{(\cdot)}), t) - F_0(r^i, \Psi_\infty(r^{(\cdot)}), t)\right| &\le \left|F_\varepsilon(r^i, \Psi_{N_\varepsilon}(r^{(\cdot)}), t)\right.\\ -F_0(r^i, \Psi_{N_\varepsilon}(r^{(\cdot)}), t)\big| &+ \left|F_0(r^i, \Psi_{N_\varepsilon}(r^{(\cdot)}), t) - F_0(r^i, \Psi_\infty(r^{(\cdot)}), t)\right|.\end{aligned}$$

The last term in the above inequality tends to 0 by the Lipschitz assumption on F_0, as $\varepsilon \downarrow 0$, if $r^{(\cdot)} \in D_\infty$. The first terms tends to 0 by (14.8), as $\varepsilon \downarrow 0$. Similarly, we can estimate the distance between the diffusion coefficients. Let $\mathcal{K}$ be a compact subset of $\mathbf{R}^d$. Our assumption (14.8) now implies that for all $i \le m$, $m \in \mathbf{N}$, $\delta > 0$, and $L \in \mathbf{N}$:

$$\begin{aligned}\limsup_{\varepsilon\downarrow 0} \sup_{r^i \in \mathcal{K}} \sup_{0\le t\le T} \sup_{r^{(\cdot)}} \Big(\big|\hat{F}_{\varepsilon,i,\delta,m,L}(r^{(\cdot)}, t) - \hat{F}_{0,i,\delta,m,L}(r^{(\cdot)}, t)\big|\\ +\big|\hat{\sigma}_{\varepsilon,i,\delta,m,L}(r^{(\cdot)}, t) - \hat{\sigma}_{0,i,\delta,m,L}(r^{(\cdot)}, t)\big|\Big) = 0\,.\end{aligned} \tag{14.33}$$

Consequently, Condition C.3 of Sect. 9 of Kurtz and Protter (loc. cit.) holds for (14.31) and (14.32). Finally, the Brownian motions in (14.23) and (14.6) are identically distributed (they are all standard). This implies that the sequence of integrators in (14.23) is uniformly tight (cf. Kurtz and Protter, loc. cit., Sect. 6). Hence, employing Theorem 9.4 of Kurtz and Protter (loc. cit.) in addition to (14.28) and the uniqueness of the solution of (14.7), we obtain that the solutions of (14.23) converge to the solutions of (14.7) for all $m < N_\varepsilon$ and all $\delta > 0$.[12] □

Let S be a complete, separable metric space. A family of S-valued random variables $\{\xi_1, \ldots, \xi_m\}$ is exchangeable if for every permutation $(\sigma_1, \ldots, \sigma_m)$ of $(1, \ldots, m)$, $\{\xi_{\sigma_1}, \ldots, \xi_{\sigma_m}\}$ has the same distribution as $\{\xi_1, \ldots, \xi_m\}$. A sequence $\xi_1, \xi_2, \ldots$ is exchangeable if every finite subfamily $\xi_1, \ldots, \xi_m$ is exchangeable. Let $P(S)$ be the set of probability measures on (the Borel sets of) S.

Lemma 14.6. *For $n = 1, 2, \ldots$, let $\{\xi_1^n, \ldots, \xi_{N_n}^n\}$ be exchangeable, S-valued random variables. (We allow $N_n = \infty$.) Let Ξ^n be the corresponding empirical measure,*

$$\Xi^n = \frac{1}{N_n}\sum_{i=1}^{N_n} \delta_{\xi_1^n},$$

where if $N_n = \infty$, we mean

$$\Xi^n = \lim_{m\to\infty} \frac{1}{m}\sum_{i=1}^{m} \delta_{\xi_i^n}.$$

We will refer to the empirical process of an exchangeable system as the DeFinitti measure for the system. By the above convention this implies its continuum limit, if the system is infinite.

---

[12] Cf. also the end of Sect. 15.2.5.

Assume that $N_n \to \infty$ and that for each $m = 1, 2, \ldots, \{\xi_1^n, \ldots, \xi_m^n\} \Rightarrow \{\xi_1, \ldots, \xi_m\}$ in S^m. Then $\{\xi_i\}$ is exchangeable and setting $\xi_i^n = s_0 \in S$ for $i > N_n$,

$$\{\Xi^n, \xi_1^n, \xi_2^n, \ldots\} \Rightarrow \{\Xi, \xi_1, \xi_2, \ldots\} \text{ in } \mathcal{P}(S) \times S^\infty, \tag{14.34}$$

where Ξ is the DeFinetti measure for $\{\xi_i\}$. If for each m, $\{\xi_1^n, \ldots, \xi_m^n\} \to \{\xi_1, \ldots, \xi_m\}$ in probability in S^m, then

$$\Xi^n \to \Xi \text{ in probability in } \mathcal{P}(S). \tag{14.35}$$

Proof. The exchangeability of $\{\xi_i\}$ follows immediately from the exchangeability of $\{\xi_i^{N_n}\}$. Assuming $m + k \leq N_n$, exchangeability implies

$$\begin{aligned}
&E\left[f(\xi_1^n, \ldots, \xi_{m+k}^n)\right] \\
&= E\left[\frac{1}{(N_n - m)\cdots(N_n - m - k + 1)}\right. \\
&\qquad \left.\sum_{\{i_1,\ldots,i_k\}\subset\{m+1,\ldots,N_n\}} f\left(\xi_1^n, \ldots, \xi_m^n, \xi_{i_1}^n, \ldots, \xi_{i_k}^n\right)\right] \\
&= E\left[\int_{S^k} f\left(\xi_1^n, \ldots, \xi_m^n, s_1, \ldots, s_k\right) \Xi^n(\mathrm{d}s_1)\cdots\Xi^n(s_k)\right] + O\left(\frac{1}{N_n}\right)
\end{aligned}$$

and hence if $f \in \bar{C}(S^{m+k})$, the bounded continuous real-valued functions on S^{m+k},

$$\begin{aligned}
&\lim_{n\to\infty} E\left[\int_{S^k} f\left(\xi_1^n, \ldots, \xi_m^n, s_1, \ldots, s_k\right) \Xi^n(\mathrm{d}s_1)\cdots\Xi^n(s_k)\right] \\
&= E\left[f(\xi_1, \ldots, \xi_{m+k})\right] \\
&= E\left[\int_{S^k} f(\xi_1, \ldots, \xi_m, s_1, \ldots, s_k)\, \Xi(\mathrm{d}s_1)\cdot\Xi(\mathrm{d}s_k)\right],
\end{aligned}$$

where the second equality follows by exchangeability. Since the space of functions on $\mathcal{P}(S) \times S^\infty$ of the form

$$F(\mu, x_1, \ldots, x_m) = \int_{S^k} f(x_1, \ldots, x_m, s_1, \ldots, s_k)\mu(\mathrm{d}s_1)\cdots\mu(\mathrm{d}s_k)$$

form a convergence determining class, the first part of the lemma follows.

If for each m, $\{\xi_1^n, \ldots, \xi_m^n\} \to \{\xi_1, \ldots, \xi_m\}$ in probability, then

$$\Xi_n^{(m)} \equiv \frac{1}{m}\sum_{i=1}^m \delta_{\xi_i^n} \to \frac{1}{m}\sum_{i=1}^m \delta_{\xi_i}$$

in probability in $\mathcal{P}(S)$, and the convergence of Ξ_n to Ξ follows by approximation, that is, by exchangeability, for each $\varepsilon > 0$ and $\varphi \in \bar{C}(S)$,

$$\lim_{m\to\infty} \sup_n P\left\{\left|\langle\varphi, \Xi_n^{(m)}\rangle - \langle\varphi, \Xi_n\rangle\right| > \varepsilon\right\} = 0.$$

□

Corollary 14.7. *Assuming the above hypotheses we obtain that for all $m \in \mathbf{N}$ and $\delta > 0$*

$$X_{\varepsilon,\delta,m,N_\varepsilon} \Rightarrow X_{0,\infty}, \quad as\ \varepsilon \downarrow 0\,. \tag{14.36}$$

Proof.

(i) Recall that, by assumption, $\{q_\varepsilon^i\}_{i\in\mathbf{N}}$ is exchangeable and $P\{q_\varepsilon^i = q_\varepsilon^j\} = 0$ for $i \neq j$ and $\varepsilon \geq 0$. The assumptions on F_ε and $\mathcal{J}_\varepsilon$ as well as the usual properties of the conditional expectation and of $\{\beta^i\}_{i\in\mathbf{N}}$ imply that $\{r^1_{\varepsilon,\delta,m,N_\varepsilon}(\cdot),\ldots,r^{N_\varepsilon}_{\varepsilon,\delta,m,N_\varepsilon}(\cdot)\}$ is exchangeable. Denote by $\Xi_{\varepsilon,\delta,m,N_\varepsilon}$ the empirical process for $\{r^1_{\varepsilon,\delta,m,N_\varepsilon}(\cdot),\ldots,r^{N_\varepsilon}_{\varepsilon,\delta,m,N_\varepsilon}(\cdot)\}$, which is an element of $\mathbf{M}_1(C([0,\infty);\mathbf{R}^d))$, i.e., of the space of probability measures on $C([0,\infty);\mathbf{R}^d)$. The existence for $N_\varepsilon = \infty$ follows from DeFinetti's theorem (cf. Dawson (1993), Sect. 11.2). Let $\Xi_{0,\infty}$ be the DeFinetti measure for (14.7). By our Lemma 14.6,

$$\Xi_{\varepsilon,\delta,m,N_\varepsilon} \Longrightarrow \Xi_{0,\infty}, \quad \text{as } \varepsilon \downarrow 0\,. \tag{14.37}$$

Since $X_{\varepsilon,\delta,m,N_\varepsilon}(\cdot)$ converges weakly, (14.36) follows. □

Recall the definition

$$\Lambda_m := \{(p^1,\ldots,p^m) \in \mathbf{R}^{d\cdot m} : \exists i \neq j,\ i,j \in \{1,\ldots,m\}, \quad \text{with } p^i = p^j\}.$$

Lemma 14.8. *Λ_m is nonattainable by the $\mathbf{R}^{d\cdot m}$-valued diffusion, $(r^1_{0,\infty}(\cdot,q^1),\ldots,r^m_{0,\infty}(\cdot,q^m))$ with initial values $(q^1,\ldots,q^m)$, which is defined by the first m d-dimensional equations of the system (14.7).*

Proof. Recall that we assumed $d \geq 2$. It is obviously sufficient to prove the Lemma for $m = 2$, and, in what follows, we use a notation similar to Friedman (loc. cit.). We convert the nonautonomous equation into an autonomous one by adding time t as an additional dimension. Abbreviate

$$F(r,t) := F_0(r, X_{0,\infty}(t), t), \quad \sigma(r,t) := \mathcal{J}_0(r, X_{0,\infty}(t), t)$$

and set

$$x := (r^1, r^2, t)^T \in \mathbf{R}^{2d+1},$$

where, as before, "A^{T}" denotes the transpose of a matrix A. Further, set

$$\hat{F}(x) := \begin{pmatrix} 1 \\ F(r^1,t) \\ F(r^2,t) \end{pmatrix}, \quad \hat{\sigma}(x) = \begin{pmatrix} 0 & 0 & 0 \\ 0 & \sigma(r^1,t) & 0 \\ 0 & 0 & \sigma(r^2,t) \end{pmatrix}, \tag{14.38}$$

where σ is the appropriate $d \times d$ matrix and $\hat{\sigma}(x)$ is a block-diagonal matrix with entries in the first row and the first column all equal to 0. Let $\beta(t)$ be a one-dimensional standard Brownian motion, independent of the d-dimensional standard Brownian motions $\beta^1(t)$ and $\beta^2(t)$ and set

$$\hat{\beta}(t) := (\beta(t), \beta^1(t), \beta^2(t))^{\mathrm{T}}.$$

Then the stochastic SODEs for the two solutions of (14.7) can be written as an SODE in $\mathbf{R}^{2d+1}$:

$$\left.\begin{aligned} \mathrm{d}x &= \hat{F}(x)\mathrm{d}t + \hat{\sigma}(x)\mathrm{d}\hat{\beta}, \\ x(0) &= (0, q^1, q^2)^{\mathrm{T}}. \end{aligned}\right\} \tag{14.39}$$

Imbedded into $\mathbf{R}^{2d+1}$, Λ_2 becomes the d-dimensional submanifold:

$$\hat{\Lambda}_2 = \{(t, r^1, r^2) : r^1 = r^2\}. \tag{14.40}$$

Normal vectors to $\hat{\Lambda}_2$ are given by

$$N^i := \frac{1}{\sqrt{2}}(0, \ldots, 0, 1, 0, \ldots, 0, -1, \ldots, 0)^{\mathrm{T}},$$

where the 1 is at the $(i+1)$th coordinate and the -1 at the $(i+d+1)$th coordinate and all other coordinates are 0, $i = 1, \ldots, d$. The linear subspace, spanned by $\{N^i, \mathrm{i} = 1, \ldots, d\}$, is d-dimensional and will be denoted $\hat{\Lambda}_2^{\perp}$. Further, let "•" denote here the scalar product in $\mathbf{R}^{2d+1}$.Note that the diffusion matrix

$$a(x) := \hat{\sigma}(x)\hat{\sigma}^T(x)$$

has a block-diagonal structure, similar to $\hat{\sigma}(x)$. Indeed, for $k \in \{2, \ldots, d+1\}$ and $\ell \in \{d+2, \ldots, 2d+1\}$,

$$\left.\begin{aligned} a_{\ell k}(x) &= a_{k\ell}(x) \quad \text{(by symmetry)} \\ &= \sum_{j=1}^{2d+1} \hat{\sigma}_{kj}(x)\hat{\sigma}_{\ell j}(x) \\ &= \sum_{j=1}^{d+1} \hat{\sigma}_{kj}(x)\hat{\sigma}_{\ell j}(x) + \sum_{j=d+2}^{2d+1} \hat{\sigma}_{kj}(x)\hat{\sigma}_{\ell j}(x) \\ &= 0 + 0, \end{aligned}\right\} \tag{$*$}$$

since $\hat{\sigma}_{kj}(x) = 0$ for $k \leq d+1$ and $j \geq d+2$ and $\hat{\sigma}_{\ell j}(x) = 0$ for $\ell \geq d+2$ and $j \leq d+1$. Further, since

$$\hat{\sigma}_{1j}(x) = \hat{\sigma}_{i1}(x) = 0 \;\; \forall i, j,$$

we have

$$a_{k\ell}(x) = 0 \quad \text{if } k = 0 \text{ or } \ell = 0. \tag{$**$}$$

So

$$(a(x)N^i) \cdot N^j = \sqrt{2}\sum_{k=1}^{d} \sigma_{jk}(r^1, t)\sigma_{ik}(r^1, t), \quad \text{if } x = (t, r^1, r^1)^{\mathrm{T}} \in \hat{\Lambda}_2 \,. \tag{14.41}$$

Since $\sigma_{ji}(r, t)$ is invertible for all (r, t), we obtain that for $x \in \hat{\Lambda}_2$ the rank of $a(x)$, restricted to $\hat{\Lambda}_2^{\perp}$, equals d. The statement of the Lemma now follows from Friedman (loc. cit., Chap. 11, (1.8) and Theorem 4.2). □

Lemma 14.9. *Assuming the above hypotheses we obtain that for all $t \geq 0$ there is a sequence $\delta_m := \delta_m(t) \downarrow 0$ as $m \to \infty$ such that*

$$\lim_{m\to\infty} \overline{\lim}_{\varepsilon\downarrow 0} P\{\tau_\varepsilon^m(\delta_m) \leq t\} = 0. \tag{14.42}$$

Proof.

For $\delta \geq 0$, set

$$\Lambda_m(\delta) := \left\{(p^1,\ldots,p^m) \in \mathbf{R}^{d\cdot m} : \exists i \neq j,\ i,j \in \{1,\ldots,m\},\quad \text{with } |p^i - p^j| \leq \delta\right\} \tag{14.43}$$

and note that $\Lambda_m(0) = \Lambda_m$, where Λ_m was defined before Theorem 14.2. Define maps $\tau^m(\delta)$ from $C_{\mathbf{R}^{dm}}[0,\infty)$ into $[0,\infty)$ through

$$\tau^m(\delta)((r^1(\cdot),\ldots,r^m(\cdot))) := \inf\{t : (r^1(t),\ldots,r^m(t)) \in \Lambda_m(\delta)\}.$$

Observe that

$$\begin{aligned}&\left\{(r^1(\cdot),\ldots,r^m(\cdot)) : \tau^m(\delta)((r^1(\cdot),\ldots,r^m(\cdot))) \leq t\right\}\\ &\quad = \left\{(r^1(\cdot),\ldots,r^m(\cdot)) : \exists s \in [0,t], i \neq j \text{ such that } |r^i(s) - r^j(s)| \leq \delta\right\},\end{aligned}$$

and we verify that the latter set is closed in $C([0,\infty);\mathbf{R}^{dm})$, which is endowed with the metric of uniform convergence on compact intervals $[0,T]$. Setting

$$\tau_0^m(\delta) := \min_{1\leq i\neq j\leq m} \inf\left\{t : \left|r^i_{0,\infty}(t) - r^j_{0,\infty}(t)\right| \leq \delta\right\},$$

we have

$$\begin{aligned}&\tau^m(\delta)((r^1_{0,\infty}(\cdot,\omega),\ldots,r^m_{0,\infty}(\cdot,\omega)))\\ &\quad = \tau_0^m(\delta,\omega) \text{ and for } \delta > 0 \quad \tau^m(\delta)((r^1_\varepsilon(\cdot,,\omega),\ldots,r^m_\varepsilon(\cdot,\omega))) = \tau_\varepsilon^m(\delta,\omega)\,.\end{aligned}$$

Hence, by a standard theorem in weak convergence,[13] *for all $\delta > 0$ and $m < N_\varepsilon$*

$$\overline{\lim}_{\varepsilon\downarrow 0} P\left\{\tau_\varepsilon^m(\delta) \leq t\right\} \leq P\left\{\tau_0^m(\delta) \leq t\right\}. \tag{14.44}$$

By Lemma 14.8 for all m

$$\lim_{\delta\downarrow 0} P\left\{\tau_0^m(\delta) \leq t\right\} = 0 \ \forall t > 0. \tag{14.45}$$

Hence there is a sequence $\delta_m = \delta_m(t) \downarrow 0$, as $m \to \infty$ such that

$$\lim_{m\to\infty} P\left\{\tau_0^m(\delta_m) \leq t\right\} = 0. \tag{14.46}$$

Expressions (14.44) and (14.46) together imply (14.42). □

[13] Cf. Theorem 15.24, Sect. 15.2.1.

Note that w.p. 1

$$r^i_{\varepsilon,\delta,N_\varepsilon}(t) \equiv r^i_\varepsilon(t) \text{ on } \left[0, \tau^m_\varepsilon(\delta)\right], i = 1, \ldots, m \le N_\varepsilon, \tag{14.47}$$

where $\{r^i_\varepsilon(t)\}_{i=1,\ldots,N_\varepsilon}$ is the solution of (4.10) with $\sigma_i^\perp \equiv 0\ \forall i$.

Proof of Theorem 14.2

Employing (14.42), Lemma 14.5, and Corollary 14.7, we obtain

$$\left.\begin{array}{l}(X_{\varepsilon,N_\varepsilon}, r_\varepsilon(\cdot, X_{\varepsilon,N_\varepsilon}, q^1_\varepsilon), \ldots, r_\varepsilon(\cdot, X_{\varepsilon,N_\varepsilon}, q^{N_\varepsilon}_\varepsilon)) \Rightarrow (X_{0,\infty}, r_{0,\infty}(\cdot, X_{0,\infty}, q^1), \ldots) \\ \quad \text{in } C([0,\infty); \mathbf{M} \times (\mathbf{R}^d)^\infty), \text{ as } \varepsilon \to 0\,. \end{array}\right\} \tag{14.48}$$

Since the solution $X_{0,\infty}$ of (14.6) is unique (cf. Remark 14.1), (14.14) follows from Lemma 14.3.

The proof of (14.15) is easier. □

14.3 Examples

We provide examples of kernels, satisfying the additional hypotheses from Sect. 5.1.

(i) $F_\varepsilon(r,\mu) \equiv F_0(r,\mu)$ for $\varepsilon \ge 0$. Let $\alpha > 0$. Set
$\tilde{\Gamma}_\alpha(r) := \left(\frac{1}{(2\pi\alpha)}\right)^{d/4} \exp\left(-\frac{|r|^2}{4\alpha}\right)$ and $\Gamma_\alpha(r)$ the diagonal $d \times d$ matrix, whose entries on the main diagonal are all $\tilde{\Gamma}_\varepsilon(r)$. Set

$$\mathcal{J}_\varepsilon(r,p,\mu) := \int \Gamma_\varepsilon(r-p)\Gamma_2(p-q)\mu(\mathrm{d}q).$$

Then,

$$\tilde{D}_\varepsilon(\mu,r,q) = \int\int\int \Gamma_\varepsilon(r-p)\Gamma_\varepsilon(q-p)\Gamma_2(p-\tilde{q})\Gamma_2(p-\hat{q})\mu(\mathrm{d}\tilde{q})\mu(\mathrm{d}\hat{q})\mathrm{d}p.$$

Since we are dealing with diagonal matrices with identical entries we may, in what follows, assume $d = 1$. Then,

$$\sigma^2_\varepsilon(r,\mu) = D_\varepsilon(\mu,r) = \int\int\int \Gamma^2_\varepsilon(r-p)\Gamma_2(p-\tilde{q})\Gamma_2(p-\hat{q})\mu(\mathrm{d}\tilde{q})\mu(\mathrm{d}\hat{q})dp$$

and, since $\int \Gamma^2_\varepsilon(r-p)\mathrm{d}p = 1$,

$$\begin{aligned}\sigma^2_0(r,\mu) &= \int\int \Gamma_2(r-\tilde{q})\mu(\mathrm{d}\tilde{q})\Gamma_2(r-\hat{q})\mu(\mathrm{d}\hat{q}) \\ &= \int\int\int \Gamma^2_\varepsilon(r-p)\Gamma_2(r-\tilde{q})\mu(\mathrm{d}\tilde{q})\Gamma_2(r-\hat{q})\mu(\mathrm{d}\hat{q})\mathrm{d}p.\end{aligned}$$

We employ the Chapman–Kolmogorov equation to "expand" $\Gamma_2(q-\bar{q}) = c_1 \int \Gamma_1(q-\bar{p})\Gamma_1(\bar{p}-\bar{q})\mathrm{d}\bar{p})$ with $\sup_q \Gamma_1(q) \le c_2 < \infty$. Hence,

$$\begin{aligned}
&\left| \int\int \Gamma_2(p-\tilde{q})\mu(\mathrm{d}\tilde{q})\Gamma_2(p-\hat{q})\mu(\mathrm{d}\hat{q}) - \int\int \Gamma_2(r-\tilde{q})\mu(\mathrm{d}\tilde{q})\Gamma_2(r-\hat{q})\mu(\mathrm{d}\hat{q}) \right| \\
&\quad = c_1^2 \Big| \int\int \left[\Gamma_1(p-\tilde{p})\Gamma_1(p-\hat{p}) - \Gamma_1(r-\tilde{p})\Gamma_1(r-\hat{p})\right] \\
&\qquad \left[\int\int \Gamma_1(\tilde{p}-\tilde{q})\mu(\mathrm{d}\tilde{q})\Gamma_1(\hat{p}-\hat{q})\mu(\mathrm{d}\hat{q})\right] \mathrm{d}\tilde{p}\,\mathrm{d}\hat{p} \Big| \\
&\quad \le c_1^2 c_2^2 \int\int |\Gamma_1(p-\tilde{p})\Gamma_1(p-\hat{p}) - \Gamma_1(r-\tilde{p})\Gamma_1(r-\hat{p})| \mathrm{d}\tilde{p}\,\mathrm{d}\hat{p},
\end{aligned}$$

where we used $\mu(\mathbf{R}^d) = 1$. So,

$$\begin{aligned}
&|\sigma_\varepsilon^2(r,\mu) - \sigma_0^2(r,\mu)| \\
&\quad \le c \int \Gamma_\varepsilon^2(r-p)|\Gamma_1(p-\tilde{p})\Gamma_1(p-\hat{p}) - \Gamma_1(r-\tilde{p})\Gamma_1(r-\hat{p})| \mathrm{d}\tilde{p}\,\mathrm{d}\hat{p}\,\mathrm{d}p \\
&\quad = c \int \Gamma_\varepsilon^2(p) f(r,p)\mathrm{d}p
\end{aligned}$$

by change of variables with $f(r,p) := \int\int |\Gamma_1(r+p-\tilde{p})\Gamma_1(r+p-\hat{p}) - \Gamma_1(r-\tilde{p})\Gamma_1(r-\hat{p})| \mathrm{d}\tilde{p}\,\mathrm{d}\hat{p}$. f is continuous in $\mathbf{R}^{2d}$ with $f(r,0) = 0\ \forall r$. Since $\Gamma_\varepsilon^2(p) \longrightarrow \delta_0$ as $\varepsilon \downarrow 0$, we have for a compact set $K \subset \mathbf{R}^d$

$$\sup_{r\in K} \int \Gamma_\varepsilon^2(p) f(r,p)\mathrm{d}p \to 0, \text{ as } \varepsilon \downarrow 0.$$

This implies (14.8).

(ii) Let $\mu \in \mathcal{C}$, where $\mathcal{C}$ is a compact subset of $\mathbf{M}_1$. By Prohorov's theorem for any $\delta > 0$ there is an $L > 0$ such that $\inf_{\mu\in\mathcal{C}} \mu(S_L) \ge 1-\delta$. Let $S_L := \{p \in \mathbf{R}^d; |p| \le L\}$. Set

$$c_\delta := \inf_{p\in S_L, \tilde{q}\in S_L} \Gamma_2(p-q).$$

Obviously, $c_L > 0$. So,

$$\sigma_\varepsilon^2(r,\mu) = \int \Gamma_\varepsilon^2(r-p)\left[\int \Gamma_2(p-\tilde{q})\mu(\mathrm{d}\tilde{q})\right]^2 \mathrm{d}p$$

$$\begin{aligned}
\sigma_\varepsilon^2(r,\mu) &= \int \Gamma_\varepsilon^2(r-p)\left[\int \Gamma_2(p-\tilde{q})\mu(\mathrm{d}\tilde{q})\right]^2 \mathrm{d}p \\
&\ge \int_{S_L} \Gamma_\varepsilon^2(r-p)\left[\int_{S_L} \Gamma_2(p-\tilde{q})\mu(\mathrm{d}\tilde{q})\right]^2 \mathrm{d}p \\
&\ge \int_{S_L} \Gamma_\varepsilon^2(r-p) c_\delta^2 [1-\delta]^2 \,\mathrm{d}p.
\end{aligned}$$

Since $\{r\}$ is bounded in the formulation of assumption in (14.9), we may assume $|r| \le \frac{L}{2}$. Changing variables in $\int_{S_L} \Gamma_\varepsilon^2(r-p)\mathrm{d}p$ we obtain that for $|r| \le \frac{L}{2}$ and $0 < \varepsilon \le 1$,

$$\int_{S_L} \Gamma_\varepsilon^2(r-p)\mathrm{d}p \ge \int_{\{|p| \le \frac{L}{2}\}} \Gamma_1^2(p)\mathrm{d}p \ge c_L > 0.$$

Altogether we obtain (14.9).

(iii) Set

$$c_2 := \sup_q \Gamma_2(q),$$

which is obviously finite. Since $\mu(\mathbf{R}^d) = 1$, we obtain

$$\begin{aligned} D_\varepsilon(r,q,\mu) &= \int \Gamma_\varepsilon(r-p)\Gamma_\varepsilon(q-p)\left[\int \Gamma_2(p-\tilde{q})\mu(\mathrm{d}\tilde{q})\right]^2 \mathrm{d}p \\ &\le c_2^2 \int \Gamma_\varepsilon(r-p)\Gamma_\varepsilon(q-p)\mathrm{d}p = c_2^2 \exp\left(-\frac{|r-q|^2}{8\varepsilon}\right), \end{aligned}$$

whence we obtain (14.10).

(iv) We are checking the Lipschitz condition (4.11) for our example, assuming again without loss of generality $d = 1$.

$$\begin{aligned} &\mathcal{J}_\varepsilon(r,p,\mu) := \int \Gamma_\varepsilon(r-p)\Gamma_2(p-q)\mu(\mathrm{d}q). \\ &\int (\mathcal{J}_\varepsilon(r_1,p,\mu_1) - \mathcal{J}_\varepsilon(r_2,p,\mu_1))^2 \mathrm{d}p \\ &= \int \left\{[\Gamma_\varepsilon(r_1-p) - \Gamma_\varepsilon(r_2-p)] \int \Gamma_2(p-q)\mu(\mathrm{d}q)\right\}^2 \mathrm{d}p \\ &\le c_2^2 \int [\Gamma_\varepsilon(r_1-p) - \Gamma_\varepsilon(r_2-p)]^2 \mathrm{d}p = 2c_2^2\left[1 - \exp\left(-\frac{|r_1-r_2|^2}{8\varepsilon}\right)\right] \\ &\le 2c_2^2\left[\frac{|r_1-r_2|^2}{8\varepsilon} \wedge 1 \le c_\varepsilon \rho^2(r_1-r_2).\right. \end{aligned}$$

$\Gamma_2(p - \cdot)$ is a bounded function of q with bounded derivatives, where the bound can be taken uniform in p. Therefore,

$$|\Gamma_2(p-q_1) - \Gamma_2(p-q_2)| \le c_2\rho(q_1-q_2).$$

Hence,

$$\begin{aligned} &\int (\mathcal{J}_\varepsilon(r,p,\mu_1) - \mathcal{J}_\varepsilon(r,p,\mu_2))^2 \mathrm{d}p \\ &= \int \left\{\Gamma_\varepsilon(r-p) \int \Gamma_2(p-q)[\mu_1(\mathrm{d}q) - \mu_2(\mathrm{d}q)]\right\}^2 \mathrm{d}p \\ &\le c^2\gamma^2(\mu_1-\mu_2) \int \Gamma_\varepsilon^2(r-p)\mathrm{d}p = c^2\gamma^2(\mu_1-\mu_2). \end{aligned}$$

(v) Assume, without loss of generality, $F \equiv 0$. We have

$$D_0(r,t) := \int\int \Gamma_2(r-\tilde{q})X_{0,\infty}(t,\mathrm{d}\tilde{q})\Gamma_2(r-\hat{q})X_{0,\infty}(t,\mathrm{d}\hat{q}).$$

Obviously, $D_0(r,t)$ is infinitely often differentiable with respect to r with derivative continuous in (r,t). Note that the "test functions" $\Gamma_2(r-q)$ are in the Schwarz space of infinitely often differentiable functions, with all derivatives rapidly decreasing. Taking all derivatives in the integrals in the distributional sense,

$$\frac{\partial}{\partial t}D_0(r,t) = 2\int\int \Gamma_2(r-\tilde{q})\frac{\partial}{\partial t}X_{0,\infty}(t,\mathrm{d}\tilde{q})\Gamma_2(r-\hat{q})X_{0,\infty}(t,\mathrm{d}\hat{q}).$$

Further, by (14.6), we may replace the partial with respect to t by the (quasilinear) second partial with respect to the space variable, i.e.,

$$\frac{\partial}{\partial t}D_0(r,t) = \int\int \Gamma_2(r-\tilde{q})\partial_{\tilde{q}\tilde{q}}[D_0(\tilde{q},t)X_{0,\infty}(t,\mathrm{d}\tilde{q})]\Gamma_2(r-\hat{q})X_{0,\infty}(t,\mathrm{d}\hat{q}).$$

By definition ("integrating by parts") the right-hand side can be rewritten, taking the second derivative of the "test function" $\Gamma_2(r-\tilde{q})$, with respect to $\tilde{q}$. By the homogeneity of $\Gamma_2(r-\tilde{q})$, this second derivative equals the second derivative with respect to r. Hence,

$$\frac{\partial}{\partial t}D_0(r,t) = \int\int (\partial_{rr}\Gamma_2)(r-\tilde{q})D_0(\tilde{q},t)X_{0,\infty}(t,\mathrm{d}\tilde{q})\Gamma_2(r-\hat{q})X_{0,\infty}(t,\mathrm{d}\hat{q}).$$

We verify that $\frac{\partial}{\partial t}D_0(r,t)$ is continuously differentiable in all variables. Hence, we may compute the second derivative with respect to t, obtaining two terms. One of them contains $\frac{\partial}{\partial t}D_0(r,t)$ as a factor in the integral which we may replace by the right-hand side of the last equation. The other term contains $\frac{\partial}{\partial t}X_{0,\infty}(t,\mathrm{d}\tilde{q})$ as a factor in the integral. As before, we replace this term by $\partial_{\tilde{q}\tilde{q}}[D_0(\tilde{q},t)X_{0,\infty}(t,\mathrm{d}\tilde{q})]$ and integrate by parts.
Altogether, we obtain that $D_0(r,t)$ is twice continuously differentiable with respect to all variables. The statement for $d \times d$ matrix-valued coefficients follows from the one-dimensional case.

14.4 A Remark on $d = 1$

Based on a result by Dorogovtsev (2005) we expect in the one-dimensional case the solution of (8.26) not to converge to the solution of a macroscopic PDE but to the solution of an SPDE driven by one standard Brownian motion $\beta(t)$ instead of the space–time white noise $w(\mathrm{d}q,\mathrm{d}t)$. A rigorous derivation of this SPDE is planned for a future research project.

14.5 Convergence of Stochastic Transport Equations to Macroscopic Parabolic Equations

We obtain that a sequence of first-order stochastic transport equations, driven by Stratonovich differentials converges to the solution of a semilinear parabolic PDE, as the correlation length tend to 0.

We reformulate Theorem 14.2, employing the representation of the solutions of (8.26) via a first-order stochastic transport equation, driven by Stratonovich differentials.

Theorem 14.3. *Suppose the conditions of Theorem 14.2 with diffusion kernel $\mathcal{J}_\varepsilon(r, q, u)$ being independent of the measure variables. Consider the sequence of solutions of*

$$\begin{aligned} dX_\varepsilon &= -\nabla \cdot (X_\varepsilon F(\cdot, X_\varepsilon, t))dt - \nabla \cdot \left(X_\varepsilon \int \mathcal{J}_\varepsilon(\cdot, p, t)\right) w(dp, \circ dt), \\ X_\varepsilon(0) &= X_{\varepsilon,0}. \end{aligned} \tag{14.49}$$

Then

$$X_\varepsilon \Rightarrow X_{0,\infty} \text{ in } C([0,\infty); \mathbf{M}_1), \quad \text{as } \varepsilon \downarrow 0, \tag{14.50}$$

where $X_{0,\infty}(\cdot)$ is the solution of the semilinear McKean–Vlasov equation

$$\left.\begin{aligned} \frac{\partial}{\partial t} X_{0,\infty} = \frac{1}{2} \sum_{k,\ell=1}^{d} \partial^2_{k\ell}(D_{0,k\ell}(\cdot, t) X_{0,\infty}) - \nabla \cdot (X_{0,\infty} F_0(\cdot, X_{0,\infty}, t)), \\ X_{0,\infty}(0) = \mu. \end{aligned}\right\} \tag{14.51}$$

Proof. Theorem 14.2 in addition to (8.81). □

Part V
General Appendix

Chapter 15
Appendix

15.1 Analysis

15.1.1 Metric Spaces: Extension by Continuity, Contraction Mappings, and Uniform Boundedness

The proof of the following statement may be found in Dunford and Schwartz ((1958), Sect. I.6., Theorem 17).

Proposition 15.1. *Principle of Extension by Continuity*
Let $\mathbf{B}_1$ *and* $\mathbf{B}_2$ *be metric spaces and let* $\mathbf{B}_2$ *be complete. If*

$$f : \mathbf{A} \mapsto \mathbf{B}_2$$

is uniformly continuous on the dense subset $\mathbf{A}$ *of* $\mathbf{B}_1$*, then* f *has a unique continuous extension*

$$\bar{f} : \mathbf{B}_1 \mapsto \mathbf{B}_2.$$

This unique extension is uniformly continuous on $\mathbf{B}_1$. □

The following theorem is also known as "*Banach's fixed point theorem.*" For the proof we refer to Kantorovich and Akilov (1977), Chap. XVI.1.

Theorem 15.2. *Contraction Mapping Principle*

Let $(\mathbf{B}, d_{\mathbf{B}})$ *be a complete metric space and* B *a closed subset of* $\mathbf{B}$*. Suppose there is a mapping*

$$\Psi : B \to B$$

such that there is a $\delta \in (0, 1)$ *and the following holds:*

$$d_{\mathbf{B}}(\Psi(f), \Psi(g)) \leq \delta d_{\mathbf{B}}(f, g) \ \ \forall f, g \in \mathbf{B}.$$

Then there is a unique $f^* \in B$ such that[1]

$$\Psi(f^*) = f^*. \qquad \square$$

The following uniform boundedness principle and the preceding definitions are found in Dunford and Schwartz (1958), Sect. II, Theorem 11, and Definitions 7 and 10 and Yosida (1968), Theorem 1 and Corollary 1 (Resonance Theorem). First, we require some terminology

Let $(\mathbf{B}, d_{\mathbf{B}})$ be a metric space. The metric $d_{\mathbf{B}}(f, g)$ is called a "*Fréchet metric*" (or "*quasi-norm*") if

$$d_{\mathbf{B}}(f, g) = d_{\mathbf{B}}(f - g, 0) \;\; \forall f, g \in \mathbf{M}.$$

- If the metric space $\mathbf{B}$, endowed with a Fréchet metric, $d_{\mathbf{B}}$ is also linear topological space[2] it is called a "*quasi-normed linear space.*"
- If a quasi-normed space $\mathbf{B}$ is complete it is called a "*Fréchet Space.*"
- A subset B of a quasi-normed space $\mathbf{B}$ is "*bounded*" if for any $\varepsilon > 0$ there is a $\gamma > 0$ such that

$$d_{\mathbf{B}}(\gamma B, 0) \leq \varepsilon.$$

Theorem 15.3. *Uniform Boundedness Principle*

(I) For each a in a set A, let T_a be a continuous linear operator from a Fréchet space $\mathbf{B}_1$ into a quasi-normed space $\mathbf{B}_2$. If, for each $f \in \mathbf{B}_1$, the set $\{T_a f : a \in A\}$ is bounded in $\mathbf{B}_2$, then

$$\lim_{f \to 0} T_a f = 0 \quad \textit{uniformly for } a \in A\,.$$

(II) If, in addition to the assumptions of Part (I), $\mathbf{B}_1$ is a Banach space with norm $\|\cdot\|_{\mathbf{B}_1}$ and $\mathbf{B}_2$ is a normed vector space with norm $\|\cdot\|_{\mathbf{B}_2}$, then the boundedness of $\{\|T_a f\|_{\mathbf{B}_2} : a \in A\}$ at each $f \in \mathbf{B}_1$ implies the boundedness of $\{\|T_a\|_{\mathcal{L}(\mathbf{B}_1,\mathbf{B}_2)}\}$ where

$$\|T_a\|_{\mathcal{L}(\mathbf{B}_1,\mathbf{B}_2)} := \sup_{\|f\|_{\mathbf{B}_1} \leq 1} \|T_a f\|_{\mathbf{B}_2}$$

is the usual operator norm. $\square$

15.1.2 Some Classical Inequalities

Let $(\tilde{\Omega}, \mathcal{F}, \mu)$ be some measure space and for $p \in [1, \infty)$ let $L_p(\tilde{\Omega}, \mathcal{F}, \mu)$ the space of measurable real (or complex)-valued functions f such that

$$\tilde{\|} f \tilde{\|}_p := \left\{ \int_{\tilde{\Omega}} |f|^p(\omega)\mu(\mathrm{d}\omega) \right\}^{\frac{1}{p}} < \infty.$$

[1] The mapping Ψ is called a "*contraction*" and f^* a "*fixed point.*"

[2] This means that $(\mathbf{B}, d_{\mathbf{B}})$ is a vector space such that addition of vectors and multiplication by scalars are continuous operations.

Proposition 15.4. *Hölder's Inequality*[3]
Let $p \in (1, \infty)$ and $\bar{p} := \frac{p}{p-1}$ and f and g be measurable functions on $\tilde{\Omega}$. Then

$$\tilde{\|} fg \tilde{\|}_1 \leq \tilde{\|} f \tilde{\|}_p \tilde{\|} g \tilde{\|}_{\bar{p}}. \tag{15.1}$$

Proof.[4]

(i) We first show

$$ab \leq \frac{a^p}{p} + \frac{b^{\bar{p}}}{\bar{p}} \quad \text{for nonnegative numbers } a \text{ and } b \text{ and } p > 1 \text{ and } \bar{p} := \frac{p}{p-1}. \tag{15.2}$$

Indeed, the function $h(t) := \frac{t^p}{p} + \frac{1}{\bar{p}} - t$ has for $t \geq 0$ the minimum value 0, and this minimum is attained only at $t = 1$. Setting $t = ab^{-\frac{\bar{p}}{p}}$, we obtain (15.2).

(ii) The result is trivial if $\tilde{\|} f \tilde{\|}_p = 0$ or $\tilde{\|} g \tilde{\|}_{\bar{p}|} = 0$ (since then $f = 0$ $\mu-$a.e. or $g = 0$ $\mu-$a.e.) It is also trivial if $\tilde{\|} f \tilde{\|}_p = \infty$ or $\tilde{\|} g \tilde{\|}_{\bar{p}} = \infty$. Otherwise, we apply (15.2) with

$$a := \frac{|f|(\omega)}{\tilde{\|} f \tilde{\|}_p}, \quad b := \frac{|g|(\omega)}{\tilde{\|} g \tilde{\|}_{\bar{p}}}.$$

Then

$$\frac{|f|(\omega)|g|(\omega)}{\tilde{\|} f \tilde{\|}_p \tilde{\|} g \tilde{\|}_{\bar{p}}} \leq \frac{|f|^p(\omega)}{p \tilde{\|} f \tilde{\|}_p^p} + \frac{|g|^{\bar{p}}(\omega)}{\bar{p} \tilde{\|} g \tilde{\|}_{\bar{p}}^{\bar{p}}}.$$

Integrating both sides yields

$$\frac{\tilde{\|} fg \tilde{\|}_1}{\tilde{\|} f \tilde{\|}_p \tilde{\|} g \tilde{\|}_{\bar{p}}} \leq \frac{1}{p} + \frac{1}{\bar{p}}.$$

□

Next, we derive Gronwall's inequality.[5]

Proposition 15.5. *Gronwall's Inequality*
Let φ, ψ, χ be real-valued continuous (or piecewise continuous) functions on a real interval $a \leq t \leq b$. Let $\chi(t) > 0$ on $[a, b]$, and suppose that for $t \in [a.b]$

$$\varphi(t) \leq \psi(t) + \int_a^t \chi(s)\varphi(s)\mathrm{d}s. \tag{15.3}$$

Then

$$\varphi(t) \leq \psi(t) + \int_a^t \chi(s)\psi(s)\mathrm{d}s \, \exp\left[\int_s^t \chi(u)\mathrm{d}u\right] \mathrm{d}s. \tag{15.4}$$

---

[3] For $p = \bar{p} = 2$ we have the Cauchy-Schwarz inequality.

[4] We follow the proofs, provided by Adams (1975), Chap. II, Theorem 2.3, and Folland (1984), Sect. 6, Theorem 6.2.

[5] Cf. Coddington and Levinson (1955), Chap. I.8, Problem 1.

Proof. Set $x(t) := \int_a^t \chi(s)\varphi(s)\mathrm{d}s$. We immediately verify that (15.3) implies

$$x'(t) - \chi(t)x(t) \le \chi(t)\psi(t) \quad \forall t \in [a,b], \quad x(a) = 0.$$

Set

$$f(t) := x'(t) - \chi(t)x(t) \ \forall t \in [a,b].$$

Obviously, $x(\cdot)$ solves the ODE initial value problem

$$x'(t) = \chi(t)x(t) + f(t), \quad x(a) = 0.$$

Employing variation of constants, $x(\cdot)$ has the following representation

$$x(t) = \int_a^t \exp\left[\int_s^t \chi(u)\mathrm{d}u\right] f(s)\mathrm{d}s.$$

Since $f(t) \le \chi(t)\psi(t) \quad \forall t \in [a,b]$ we obtain

$$x(t) \le \int_a^t \exp\left[\int_s^t \chi(u)\mathrm{d}u\right] \chi(s)\psi(s)\mathrm{d}s \ \ \forall t \in [a,b],$$

whence

$$\chi(t)\varphi(t) = x'(t) \le \chi(t)\int_a^t \exp\left[\int_s^t \chi(u)\mathrm{d}u\right] \chi(s)\psi(s)\mathrm{d}s + \chi(t)\psi(t) \ \ \forall t \in [a,b].$$

Dividing both sides of the last inequality by $\chi(t)$ yields (15.4). □

Let $(\Omega_1, \mathcal{F}_1, \mu_1)$ and $(\Omega_2, \mathcal{F}_2, \mu_2)$ be σ-finite measure spaces. $L_p((\mathbf{B}_i, \mathcal{F}_i, \mu_i))$ are the spaces of p-integrable functions on the measure space Ω_i, and the corresponding L_p-norms will be denoted $\|f\|_{\Omega_i,p}$, $i = 1, 2$, where $1 \le p \le \infty$.

Proposition 15.6. *Inequality for Integral Operators*
Let K be an $\mathcal{F}_1 \otimes \mathcal{F}_2$-measurable function on $\Omega_1 \times \Omega_2$ and suppose there exists a finite $C > 0$ such that

$$\int |K(\omega_1,\omega_2)|\mu(\mathrm{d}\omega_1) \le C \quad \mu_2\text{-a.e. and} \int |K(\omega_1,\omega_2)|\mu(\mathrm{d}\omega_2) \le C \ \ \mu_1\text{-a.e.} \tag{15.5}$$

Then, $f \in L_p((\Omega_2, \mathcal{F}_2, \mu_2))$ for $1 \le p \le \infty$ implies that

$$(\bar{K}f)(\omega_1) := \int K(\omega_1,\omega_2) f(\omega_2)\mu_2(\mathrm{d}\omega_2) \in L_p((\Omega_1, \mathcal{F}_1, \mu_1)), \tag{15.6}$$

such that

$$\|\bar{K}f\|_{\Omega_1,p} \le C\|f\|_{\Omega_2,p}. \tag{15.7}$$

Proof.[6] Suppose $p \in (1, \infty)$. Then

$$
\begin{aligned}
&\int |K(\omega_1, \omega_2) f(\omega_2)| \mu_2(d\omega_2) \\
&= \int |K(\omega_1, \omega_2)|^{\frac{p-1}{p} + \frac{1}{p}} |f(\omega_2)| \mu_2(d\omega_2) \\
&\le \left[\int |K(\omega_1, \omega_2)| \mu_2(d\omega_2) \right]^{\frac{p-1}{p}} \left[\int |K(\omega_1, \omega_2)| |f(\omega_2)|^p \mu_2(d\omega_2) \right]^{\frac{1}{p}} \\
&\qquad \text{(by Hölder's inequality (Proposition 15.4))} \\
&\le C^{\frac{p-1}{p}} \left[\int |K(\omega_1, \omega_2)| |f(\omega_2)|^p \mu_2(d\omega_2) \right]^{\frac{1}{p}} \quad \mu_1\text{-a.e.}
\end{aligned}
$$

Hence, by the Fubini–Tonelli theorem,

$$
\begin{aligned}
&\int \left[\int |K(\omega_1, \omega_2) f(\omega_2)| \mu_2(d\omega_2) \right]^p \mu_1(d\omega_1) \\
&\le C^{p-1} \int \int |K(\omega_1, \omega_2)| |f(\omega_2)|^p \mu_2(d\omega_2) \mu_1(d\omega_1) \\
&\le C^p \int |f(\omega_2)|^p \mu_2(d\omega_2) < \infty.
\end{aligned}
$$

We obtain from the last estimate that, by Fubini's theorem,

$$
K(\omega_1, \cdot) f(\cdot) \in L_1((\Omega_2, \mathcal{F}_2, \mu_2)) \ \ \mu_1\text{-a.e..}
$$

Therefore, $\bar{K} f$ is well-defined μ_1-a.e., and

$$
\int |(\bar{K} f)(\omega_1)|^p \mu_1(d\omega_1) \le C^p \|f\|^p_{\Omega_2, p}.
$$

Taking the pth root on both sides, we obtain (15.7).

For $p = 1$ we do not require Hölder's inequality, and other than that the proof is similar. For $p = \infty$ the proof is trivial. □

Proposition 15.6 immediately implies the classical Young inequality.[7]

Proposition 15.7. *Young's Inequality*

Let $f \in \mathbf{W}_{0,1,1}$ and $g \in \mathbf{W}_{0,p,1}$, where $1 \le p \le \infty$. Then the convolution $f * g$ exists for almost every $r \in \mathbf{R}^d$ such that $f * g \in \mathbf{W}_{0,p,1}$ and

$$
\|f * g\|_{0,p,1} \le \|f\|_{0,1,1} \|g\|_{0,p,1}. \tag{15.8}
$$

Proof. We set $K(r, q) := f(r - q)$ and apply Proposition 15.6.[8] □

---

[6] The proof has been adopted from Folland (1984), Sect. 6.3, (6.18) Theorem.

[7] We remark that in Tanabe (1979), Sect. 1.2, Lemma 1.2.3, this inequality is called the "*Haussdorf-Young inequality.*"

[8] Cf. Folland (loc.cit), Sect. 8.1, (8.7).

15.1.3 The Schwarz Space

We define the system of normalized Hermite functions on $\mathbf{H}_0 := L_2(\mathbf{R}^d, \mathrm{d}r)$ *and provide a proof of the completeness of the system. Further, we prove that the normalized Hermite functions are eigenfunctions of* $-\triangle + |\cdot|^2$*, where* $|\cdot|^2$ *acts as a multiplication operator. We then introduce the Schwarz space and its strong dual as well as the chain of associated Hilbert spaces. Finally, we define those Hilbert distribution spaces,* $\mathcal{H}_{-\gamma}$*, for which the imbedding* $\mathbf{H}_0 \subset \mathcal{H}_{-\gamma}$ *is Hilbert-Schmidt.*

As in Chap. 3 we use the abbreviation

$$\hat{\mathbf{N}} := \mathbf{N} \cup \{\mathbf{0}\}.$$

First, we define the usual Hermite functions (i.e,. the Hermite functions that are not normalized!) and the normalized Hermite functions for $d = 1$ and then for $d > 1$:

(i) Case $d = 1$.

$$\left.\begin{aligned} g_k(x) &:= (-1)^k \exp\left(\frac{x^2}{2}\right) \frac{\mathrm{d}^k}{\mathrm{d}x^k} \exp(-x^2), \\ \phi_k &:= \frac{1}{\sqrt{2^k k! \sqrt{\pi}}} g_k, \quad k \in \hat{\mathbf{N}}. \end{aligned}\right\} \tag{15.9}$$

(ii) Case $d > 1$.

Denote the multiindices with a single letter in bold face, i.e., set

$$\begin{aligned} \mathbf{k} &:= (k_1, \ldots, k_d) \\ |\mathbf{k}| &:= k_1 + \ldots + k_d. \end{aligned}$$

Define:

$$g_{\mathbf{k}}(q_1, \ldots, q_d) := \prod_{i=1}^{d} g_{k_i}(q_i), \quad k_i \in \hat{\mathbf{N}}, i = 1, \ldots, d. \tag{15.10}$$

We then obtain the normalized Hermite functions[9]

$$\Phi_{\mathbf{k}} := \frac{1}{\sqrt{\prod_{i=1}^{d} 2^{k_i} k_i! \pi^{\frac{d}{4}}}} g_{\mathbf{k}} = \prod_{i=1}^{d} \phi_{k_i}. \tag{15.11}$$

Further, let $\triangle$ denote the self-adjoint closure of the Laplace operator on $\mathbf{H}_0$ with $\mathrm{Dom}(\triangle) = \mathbf{H}_2$. $|\cdot|^2$ denotes the multiplication operator acting on functions f from a suitable subspace of $\mathbf{H}_0$.

[9] Cf. (15.28) in the proof of the following Proposition 15.8 which establishes that the normalized Hermite functions have norm 1 in $\mathbf{H}_0$.

Proposition 15.8. *The system* $\{\phi_{\mathbf{k}}\}$ *is a complete orthonormal system for* $\mathbf{H}_0 = L_2(\mathbf{R}^d, \mathrm{d}r)$.

Further,

$$((-\Delta + |r|^2)\phi_{\mathbf{k}})(r) \equiv (2|\mathbf{k}| + d)\phi_{\mathbf{k}}(r) \ \ \forall \mathbf{k}, \tag{15.12}$$

i.e., the $\phi_{\mathbf{k}}$ *are eigenfunctions of the self-adjoint unbounded operator* $-\Delta + |\cdot|^2$ *on* $\mathbf{H}_0$.[10]

Proof.

(i) Suppose first $d = 1$. To show the orthogonality we set

$$J_{k,\ell} := \int_{-\infty}^{\infty} g_k(x) g_\ell(x)\mathrm{d}x$$

and suppose $k \geq \ell$. Define the ℓth Hermite polynomial by

$$h_\ell(x) := (-1)^\ell \exp(x^2)\frac{\mathrm{d}^\ell}{\mathrm{d}x^\ell}\exp(-x^2) \tag{15.13}$$

Observe that

$$h_\ell(x)\exp\left(-\frac{x^2}{2}\right) \equiv g_\ell(x),$$

and we verify that

$$h_\ell(x) = 2^\ell x^\ell + p_{\ell-2}(x), \tag{15.14}$$

where $p_{\ell-2}(x)$ is a polynomial of degree $\ell - 2$. Therefore,

$$\begin{aligned} J_{k,\ell} &= (-1)^{k+\ell}\int_{-\infty}^{\infty}\exp(x^2)\left(\frac{\mathrm{d}^k}{\mathrm{d}x^k}\exp(-x^2)\right)\left(\frac{\mathrm{d}^\ell}{\mathrm{d}x^\ell}\exp(-x^2)\right)\mathrm{d}x \\ &= (-1)^k\int_{-\infty}^{\infty}\frac{\mathrm{d}^k}{\mathrm{d}x^k}\exp(-x^2)h_\ell(x)\mathrm{d}x. \end{aligned}$$

Integrating by parts k times we obtain

$$J_{k,\ell} = (-1)^{2k}\int_{-\infty}^{\infty}\exp(-x^2)\left(\frac{\mathrm{d}^k}{\mathrm{d}x^k}h_\ell\right)(x)\mathrm{d}x.$$

Observing that the degree of $h_\ell(\cdot)$ is ℓ we have

$$\left(\frac{\mathrm{d}^k}{\mathrm{d}x^k}h_\ell\right)(x) := \begin{cases} 2^k k!, & \text{if } \ell = k, \\ 0, & \text{if } \ell < k. \end{cases}$$

Hence,

$$J_{k,\ell} := \begin{cases} 2^k k!\sqrt{\pi}, & \text{if } \ell = k, \\ 0, & \text{if } \ell < k. \end{cases} \tag{15.15}$$

Equation (15.15) establishes the orthogonality of the system $\{g_k(\cdot)\}$ and the orthonormality of the system $\{\phi_k(\cdot)\}$.

---

[10] Most of the steps in the following proof are taken from Suetin (1979), Chap. V, Sects. 1 and 3 as well as from Akhiezer and Glasman (1950), Chap. 1, Sect. 11. Cf. also the appendix in the paper by Holley and Stroock (1978) and Kotelenez (1985).

(ii) Next, we show completeness of the one-dimensional system.
Suppose there is an $f \in L_2(\mathbf{R}, \mathrm{d}x)$ such that

$$\int_{-\infty}^{\infty} f(x)\phi_k(x)\mathrm{d}x = 0 \;\; \forall k \in \hat{\mathbf{N}}. \tag{15.16}$$

Setting $\tilde{\phi}_k(x) := \exp(-\frac{x^2}{2})x^k, \;\; k \in \hat{\mathbf{N}}$, we observe that, for any $n \in \hat{\mathbf{N}}$, the linear spans of $\{\phi_0, \ldots, \phi_n\}$ and of $\{\tilde{\phi}_0, \ldots, \tilde{\phi}_n\}$ coincide. Thus, the assumption (15.16) is equivalent to

$$\int_{-\infty}^{\infty} f(x) \exp\left(-\frac{x^2}{2}\right) x^k \,\mathrm{d}x = 0 \;\; \forall k \in \hat{\mathbf{N}}. \tag{15.17}$$

Setting $\mathbf{i} := \sqrt{-1}$ and for $z \in \mathbf{C}$

$$\psi(z) := \int_{-\infty}^{\infty} f(x) \exp\left(-\frac{x^2}{2}\right) \exp(\mathbf{i}xz)\mathrm{d}x, \tag{15.18}$$

we obtain (employing Lebesgue's dominated convergence theorem) that $\psi(\cdot)$ is analytic in the complex plane. By (15.17) Taylor's expansion at $z = 0$ yields

$$\psi^{(k)}(0) = \int_{-\infty}^{\infty} f(x) \exp\left(-\frac{x^2}{2}\right) (\mathbf{i}x)^k \mathrm{d}x = 0 \;\; \forall k \in \hat{\mathbf{N}}, \tag{15.19}$$

where $\psi^{(k)}(0)$ are the kth derivatives of ψ at $z = 0$. Hence,

$$\psi(z) \equiv 0.$$

In particular, we have

$$\psi(\xi) = \int_{-\infty}^{\infty} f(x) \exp\left(-\frac{x^2}{2}\right) \exp(\mathbf{i}x\xi)\mathrm{d}x = 0 \;\; \forall \xi \in \mathbf{R}. \tag{15.20}$$

Multiplying both sides by $\exp(-\mathbf{i}\xi y)$, $y \in \mathbf{R}$, and integrating with respect to $\mathrm{d}\xi$ from $-\gamma$ to γ, yields

$$\int_{-\infty}^{\infty} f(x) \exp\left(-\frac{x^2}{2}\right) \frac{\sin(\gamma(x-y))}{x-y} \mathrm{d}x = 0 \;\; \forall \gamma \text{ and } \forall y\,. \tag{15.21}$$

For fixed γ $\frac{\sin(\gamma(x-y))}{x-y}$ can be extended to a bounded and continuous function in both variables, where the value at 0 equals γ. Let $-\infty < a < b < \infty$ and integrate (15.21) with respect to $\mathrm{d}y$ from a to b. Employing the Fubini–Tonelli theorem we obtain from (15.21)

$$\int_{-\infty}^{\infty} f(x) \exp\left(-\frac{x^2}{2}\right) \int_a^b \frac{\sin(\gamma(x-y))}{x-y} \mathrm{d}y \,\mathrm{d}x = 0 \;\; \forall \gamma\,. \tag{15.22}$$

The following relation is well known and proven in classical analysis books:[11]

$$\int_0^\infty \frac{\sin(u)}{u}\mathrm{d}u = \frac{\pi}{2}. \tag{15.23}$$

Let $c > 0$. Then

$$\int_0^c \frac{\sin(u)}{u}\mathrm{d}u = \sum_{k=0}^{\infty}\int_{c\wedge(k\pi)}^{c\wedge((k+1)\pi)} \frac{\sin(u)}{u}\mathrm{d}u.$$

The series on the right is alternating, whose terms decrease to 0 in absolute value. The first term is in $(0, \pi)$. Therefore,

$$\int_0^c \frac{\sin(u)}{u}\mathrm{d}u \in (0, 2\pi)\ \forall c > 0.$$

Since $\dfrac{\sin(u)}{u}$ is even,

$$\left|\int_a^b \frac{\sin(\gamma(x-y))}{x-y}\mathrm{d}y\right| \le 4\pi \quad \forall\gamma \text{ and } \forall x$$

Changing variables,

$$\int_a^b \frac{\sin(\gamma(x-y))}{x-y}\mathrm{d}y = \int_{\gamma(a-x)}^{\gamma(b-x)} \frac{\sin(\gamma(u))}{u}\mathrm{d}u.$$

Therefore, by (15.23),

$$\lim_{\gamma\to\infty}\int_a^b \frac{\sin(\gamma(x-y))}{x-y}\mathrm{d}y = 1_{(a,b)}(x)\pi + 1_{\{a,b\}}(x)\frac{\pi}{2}, \tag{15.24}$$

where $1_{\{a,b\}}(x)$ is the indicator function of the set $\{a, b\}$, containing the two elements a, b. By Lebesgue's dominated convergence theorem,

$$\lim_{\gamma\to\infty}\int_{-\infty}^{\infty} f(x)\exp\left(-\frac{x^2}{2}\right)\int_a^b \frac{\sin(\gamma(x-y))}{x-y}\mathrm{d}y\,\mathrm{d}x = \pi\int_a^b f(x)\exp\left(-\frac{x^2}{2}\right)\mathrm{d}x.$$

Hence, by (15.22),

$$\int_a^b f(x)\exp\left(-\frac{x^2}{2}\right)\mathrm{d}x = 0\,, \quad -\infty < a < b < \infty. \tag{15.25}$$

The integral in (15.25) is an absolutely continuous function of b, whence

$$f(x)\exp\left(-\tfrac{x^2}{2}\right) = 0 \quad \mathrm{d}x \text{ a.e.,} \quad \text{which is equivalent to } f(x) = 0 \quad \mathrm{d}x \text{ a.e.} \tag{15.26}$$

This establishes the completeness of the system $\{\phi_k\}$ for $d = 1$.

[11] Cf., e.g., Erwe (1968), Chap. VI.8, (289).

(iii) The completeness of the system $\{\phi_{\mathbf{k}}\}$ in $\mathbf{H}_0 = L_2(\mathbf{R}^d, \mathrm{d}r)$ follows from the completeness of the one-dimensional system employing in the following relation Fubini's and the monotone convergence theorems:

$$\left.\begin{aligned}
|f|_0^2 &= \int_{-\infty}^{\infty}\cdots\int_{-\infty}^{\infty}\left[\int_{-\infty}^{\infty} f^2(r_1,\ldots,r_d)\mathrm{d}r_1\right]\mathrm{d}r_2\cdots\mathrm{d}r_d\\
&= \int_{-\infty}^{\infty}\cdots\cdots\int_{-\infty}^{\infty}\left[\sum_{k_1}\left(\int_{-\infty}^{\infty} f(r_1,\ldots,r_d)\phi_{k_1}(r_1)\mathrm{d}r_1\right)^2\right]\mathrm{d}r_2\cdots\mathrm{d}r_d\\
&= \sum_{k_1}\int_{-\infty}^{\infty}\cdots\int_{-\infty}^{\infty}\left[\sum_{k_2}\left(\int_{-\infty}^{\infty}\left(\int_{-\infty}^{\infty} f(r_1,\ldots,r_d)\phi_{k_1}(r_1)\mathrm{d}r_1\right)\right.\right.\\
&\qquad\qquad \left.\left.\phi_{k_2}(r_2)\mathrm{d}r_2\right)^2\right]\mathrm{d}r_3\cdots\mathrm{d}r_d\\
&\ \vdots\\
&= \sum_{k_1}\cdots\sum_{k_d}\left(\int_{-\infty}^{\infty}\cdots\int_{-\infty}^{\infty} f(r_1,\ldots,r_d)\phi_{k_1}(r_1)\cdot\cdots\cdot\phi_{k_d}(r_d)\mathrm{d}r_1\cdots\mathrm{d}r_d\right)^2.
\end{aligned}\right\}
\tag{15.27}$$

Further, the orthonormality follows from the one-dimensional case where (15.15) provides the correct normalizing factor also for the d-dimensional case, i.e.,

$$\|g_{\mathbf{k}}\|_0 = \sqrt{\prod_{i=1}^{d} 2^{k_i} k_i!\pi^{\frac{d}{4}}}\ \ \forall \mathbf{k}. \tag{15.28}$$

(iv) We next prove (15.12) for $d = 1$. Observe that

$$\frac{\mathrm{d}}{\mathrm{d}x}\exp(-x^2) \equiv -2x\ \exp(-x^2),$$

whence we recursively establish

$$\frac{\mathrm{d}^{k+2}}{\mathrm{d}x^{k+2}}\exp(-x^2) + 2x\frac{\mathrm{d}^{k+1}}{\mathrm{d}x^{k+1}}\exp(-x^2) + 2(k+1)\frac{\mathrm{d}^k}{\mathrm{d}x^k}\exp(-x^2) \equiv 0.$$

Recalling (15.13), the last equation may be rewritten as

$$\frac{\mathrm{d}^2}{\mathrm{d}x^2}\left[\exp(-x^2)h_k(x)\right]+2x\frac{\mathrm{d}}{\mathrm{d}x}\left[\exp(-x^2)h_k(x)\right]+2(k+1)\left[\exp(-x^2)h_k(x)\right] \equiv 0.$$

Abbreviating $f(x) := \exp(-x^2)$ we obtain from the previous equation and the equations for the first and second derivatives of f

$$\left(\frac{\mathrm{d}^2}{\mathrm{d}x^2} f(x) + 2x\frac{\mathrm{d}}{\mathrm{d}x} f(x) + 2(k+1) f(x)\right) h_k(x)$$
$$+\left(2\frac{\mathrm{d}}{\mathrm{d}x} f(x) + 2x f(x)\right) \frac{\mathrm{d}}{\mathrm{d}x} h_k(x) + f(x)\frac{\mathrm{d}^2}{\mathrm{d}x^2} h_k(x)$$
$$= f(x)\left(\frac{\mathrm{d}^2}{\mathrm{d}x^2} h_k(x) - 2x\frac{\mathrm{d}}{\mathrm{d}x} h_k(x) + 2k h_k(x)\right) \equiv 0$$

Dividing the last equation by $f(x) = \exp(-x^2)$ yields the differential equations for the Hermite polynomials

$$\left(\frac{\mathrm{d}^2}{\mathrm{d}x^2} h_k\right)(x) - 2x\left(\frac{\mathrm{d}}{\mathrm{d}x} h_k\right)(x) + 2k h_k(x) \equiv 0. \tag{15.29}$$

We differentiate $g_k(x) = \exp\big(-\frac{x^2}{2}\big) h_k(x)$ and obtain

$$\frac{\mathrm{d}}{\mathrm{d}x} g_k(x) = -x\exp\left(-\frac{x^2}{2}\right) h_k(x) + \exp\left(-\frac{x^2}{2}\right)\frac{\mathrm{d}}{\mathrm{d}x} h_k(x)$$

and

$$\frac{\mathrm{d}^2}{\mathrm{d}x^2} g_k(x) = \exp\left(-\frac{x^2}{2}\right)\left[-h_k(x) + x^2 h_k(x) - 2x\frac{\mathrm{d}}{\mathrm{d}x} h_k(x) + \frac{\mathrm{d}^2}{\mathrm{d}x^2} h_k(x)\right].$$

Thus,

$$\frac{\mathrm{d}^2}{\mathrm{d}x^2} g_k(x) + (2k+1-x^2) g_k(x)$$
$$\equiv \exp\left(-\frac{x^2}{2}\right)\left[\frac{\mathrm{d}^2}{\mathrm{d}x^2} h_k(x) - 2x\frac{\mathrm{d}}{\mathrm{d}x} h_k(x)\right.$$
$$\left. + x^2 h_k(x) - h_k(x) + (2k+1-x^2) h_k(x)\right]$$
$$\equiv \exp\left(-\frac{x^2}{2}\right)\left[\frac{\mathrm{d}^2}{\mathrm{d}x^2} h_k(x) - 2x\frac{\mathrm{d}}{\mathrm{d}x} h_k(x) + 2k h_k(x)\right] \equiv 0$$

by (15.29). As a result we obtain the following differential equation for the Hermite functions:

$$\left(\frac{\mathrm{d}^2}{\mathrm{d}x^2} g_k\right)(x) + (2k+1-x^2) g_k(x) \equiv 0. \tag{15.30}$$

Obviously, the normalized Hermite functions $\Phi_k(\cdot)$ are solutions of the same differential equation (15.30), whence we obtain (15.12) for $d = 1$.

Expression (15.11) in addition to the definition of $\triangle = \sum_{i=1}^{d} \partial_{ii}^2$ and (15.30) proves (15.12) also for $d > 1$. □

Let us introduce suitable spaces of distributions. The following set-up is standard and details can be found in Itô (1984), Kotelenez (1985), and Holley and Stroock (loc.cit) and the references therein.

For $\gamma \geq 0$ and $f, g \in C_c^\infty(\mathbf{R}^d; \mathbf{R})$ (the infinitely often continuously differentiable real valued functions with compact support in $\mathbf{R}^d$) we define scalar products and Hilbert norms by

$$\left.\begin{array}{l} \langle f, g\rangle_\gamma := \sum_{\mathbf{k}\in\hat{\mathbf{N}}^d} (d + |\mathbf{k}|)^\gamma \langle f, \phi_\mathbf{k}\rangle_0 \langle g, \phi_\mathbf{k}\rangle_0; \\ \|f\|_\gamma := \sqrt{\langle f, f\rangle_\gamma}. \end{array}\right\} \tag{15.31}$$

Define Hilbert spaces for $\gamma \geq 0$ by

$$\mathcal{H}_\gamma := \{f \in \mathbf{H_0} : \|f\|_\gamma < \infty\},$$

where $\mathcal{H}_0 = \mathbf{H}_0$. Set

$$\mathcal{S} := \cap_{\gamma \geq 0} \mathcal{H}_\gamma .$$

The topology on $\mathcal{S}$ is defined by the countable Hilbert norms $\|\cdot\|_\gamma$, $\gamma \in \hat{\mathbf{N}}$. This topology is equivalent to the topology generated by the metric

$$d_{\mathcal{S},H}(f, g) := \sum_{\gamma=0}^{\infty} (\|f - g\|_\gamma \wedge 1) 2^{-\gamma} .$$

Another definition of $\mathcal{S}$ is the following:[12] Let $f \in C^\infty(\mathbf{R}^d; \mathbf{R})$ and define semi-norms

$$|||f|||_{N,\mathbf{n}} := \sup_{r\in\mathbf{R}^d} (1 + |r|)^N |(\partial^\mathbf{n} f)(r)|.$$

Then

$$\mathcal{S} := \left\{f \in C^\infty(\mathbf{R}^d; \mathbf{R}) : |||f|||_{N,\mathbf{n}} < \infty\right\} \ \forall N \in \mathbf{N}, \mathbf{n} \in \hat{\mathbf{N}}^d.$$

One may show that the metric $d_{\mathcal{S},H}$ is equivalent to the metric

$$d_{\mathcal{S},\sup}(f, g) := \sum_{N,\mathbf{n}} (|||f - g|||_{N,\mathbf{n}} \wedge 1) 2^{-(N+|\mathbf{n}|)},$$

where $\mathbf{n} = (n_1, \ldots, n_d)$ and $|\mathbf{n}| = \sum_{i=1}^{d} n_i$. Therefore, both definitions of $\mathcal{S}$ coincide.

Next, identify $\mathbf{H}_0$ with its strong dual $\mathbf{H}_0'$ and denote by $\mathcal{S}'(\mathbf{R}^d)$ the dual of $\mathcal{S}(\mathbf{R}^d)$. $\mathcal{S}'(\mathbf{R}^d)$ is called the *Schwarz space of tempered distributions on* $\mathbf{R}^d$. For $\gamma \geq 0$ we set

$$\mathcal{H}_{-\gamma} := \left\{f \in \mathcal{S}'(\mathbf{R}^d) : f \in \mathcal{L}(\mathcal{H}_\gamma; \mathbf{R})\right\}$$

[12] Cf., e.g. Treves (1967), Part II, Chap. 25.

(the bounded linear functionals on $\mathcal{H}_\gamma$). We extend the scalar product $\langle f, g\rangle_0$ to a duality $\langle \cdot, \cdot \rangle$ between $\mathcal{S}(\mathbf{R}^d)$ and $\mathcal{S}'(\mathbf{R}^d)$. We verify that the functional norm on $\mathcal{H}_{-\gamma}$, $\|\cdot\|_{-\gamma}$, may be defined as the norm associated with the following scalar product on $\mathcal{H}_{-\gamma}$:

$$\langle f, g\rangle_{-\gamma} = \sum_{\mathbf{k}\in\hat{\mathbf{N}}^d} (d + |\mathbf{k}|)^{-\gamma} \langle f, \phi_\mathbf{k}\rangle\langle g, \phi_\mathbf{k}\rangle.$$

Hence, if $\gamma_1, \gamma_2 > 0$, we obtain the chain of spaces with dense and continuous imbeddings

$$\mathcal{S} := \mathcal{S}(\mathbf{R}^d) \subset \mathcal{H}_{\gamma_1} \subset \mathbf{H_0} = \mathbf{H}_0' \subset \mathcal{H}_{-\gamma_2} \subset \mathcal{S}' := \mathcal{S}'(\mathbf{R^d}) \tag{15.32}$$

Next, for all all $\gamma \in \mathbf{R}$

$$\{\phi_{\mathbf{k},\gamma}\} := \left\{(d + |\mathbf{k}|)^{-\frac{\gamma}{2}}\phi_\mathbf{k}\right\} \tag{15.33}$$

is a CONS in $\mathcal{H}_\gamma$.

For applications to linear PDEs and SPDEs (cf. Sect. 13.1.1) the following characterization of the above Hilbert spaces may be useful. Let "*Dom*" denote the domain of an operator. Consider the fractional powers $(-\triangle + |r|^2)^{\frac{\gamma}{2}}$ of the self-adjoint operator $(-\triangle + |r|^2)$[13] where $\gamma \in \mathbf{R}$. For suitable restrictions or extensions of $(-\triangle + |r|^2)^{\frac{\gamma}{2}}$[14], denoting all restrictions and extensions with the same symbol, (15.12) and the definition of the Hilbert norms $\|\cdot\|_\gamma$ imply

$$\left.\begin{aligned} &\mathcal{H}_\gamma := \mathrm{Dom}((-\triangle + |r|^2)^{\frac{\gamma}{2}}) \quad \text{and} \\ &\left\langle(-\triangle + |r|^2)^{\frac{\gamma}{2}} f, (-\triangle + |r|^2)^{\frac{\gamma}{2}} g\right\rangle_0 \\ = &\sum_{\mathbf{k}\in\hat{\mathbf{N}}^d} (d + |\mathbf{k}|)^{\gamma} \langle f, \phi_\mathbf{k}\rangle\langle g, \phi_\mathbf{k}\rangle = \langle f, g\rangle_\gamma \quad \forall f, g \in \mathcal{H}_\gamma . \end{aligned}\right\} \tag{15.34}$$

In what follows assume $-\infty < \gamma_2 < \gamma_1 < \infty$. We verify that the imbedding

$$\mathcal{H}_{\gamma_1} \subset \mathcal{H}_{\gamma_2}$$

is compact if $\gamma_1 - \gamma_2 > 0$. A finitely additive Gauss measure on $\mathcal{H}_{\gamma_1}$ whose covariance operator is the identity on $\mathcal{H}_{\gamma_1}$ can be extended to a countably additive Gaussian measure on $\mathcal{H}_{\gamma_2}$ if, and only if, the imbedding is "*Hilbert-Schmidt*," i.e., if[15]

$$\sum_{\mathbf{k}} \|\phi_{\mathbf{k},\gamma_1}\|^2_{\gamma_2} < \infty. \tag{15.35}$$

---

[13] Cf., e.g., Pazy (1983), Sect. 2.6, or Tanabe (1979), Sect. 2.3.

[14] Cf. Kotelenez (1985).

[15] Cf. Itô (1984), Kotelenez (1985), and Walsh (1986) in addition to Kuo (1975), Chap. I, Sect. 4, Theorem 4.3, and our previous discussion of the abstract case.

Obviously, the imbedding is Hilbert-Schmidt if, and only if, $\gamma_1 - \gamma_2 > d$. The previous definitions and results imply the following: The imbedding

$$\mathbf{H}_0 \subset \mathcal{H}_{-\gamma} \text{ is Hilbert-Schmidt, i.e., if and only if } \gamma > d. \tag{15.36}$$

15.1.4 Metrics on Spaces of Measures

Properties of various versions of the Wasserstein metric for $\mathbf{M}_f$ and $\mathbf{M}_{\infty,\varpi}$ as well as properties of the weight function ϖ are derived. At the end we compare the Wasserstein metric with the metric in total variation.

In Chap. 4 we introduced $\mathbf{M}_f$ as the space of finite Borel measures on $\mathcal{B}^d$ (the Borel sets in $\mathbf{R}^d$). We define a Wasserstein metric on $\mathbf{M}_f$ as follows:

The space of all continuous Lipschitz functions f from $\mathbf{R}^d$ into $\mathbf{R}$ will be denoted $C_L(\mathbf{R}^d;\mathbf{R})$. Further, $C_{L,\infty}(\mathbf{R}^d;\mathbf{R})$ is the space of all uniformly bounded Lipschitz functions f from $\mathbf{R}^d$ into $\mathbf{R}$. Abbreviate[16]

$$|||f||| := \sup_q |f(q)|; \quad \|f\|_L := \sup_{\{r\neq q, |r-q|\le 1\}} \frac{|f(r)-f(q)|}{\rho(r-q)}; \quad \|f\|_{L,\infty} := \|f\|_L \vee |||f|||. \tag{15.37}$$

For $\mu, \nu \in \mathbf{M}_f$, we set

$$\gamma_f(\mu - \nu) := \sup_{\|f\|_{L,\infty}\le 1} \left| \int f(q)(\mu(\mathrm{d}q) - \nu(\mathrm{d}q)) \right|. \tag{15.38}$$

Obviously $(\mathbf{M}_f, \gamma_f)$ is a metric space where $\gamma_f(\mu-\nu)$ is actually a norm. Endowed with the functional norm the space of all continuous linear bounded functionals from $C_{L,\infty}(\mathbf{R}^d;\mathbf{R})$ into $\mathbf{R}$ (denoted $C^*_{L,1}(\mathbf{R}^d;\mathbf{R})$) is the strong dual of $C_{L,\infty}(\mathbf{R}^d;\mathbf{R})$ and, therefore, a Banach space. $\mathbf{M}_f$ is the cone of (nonnegative) measures in $C^*_{L,1}(\mathbf{R}^d;\mathbf{R})$, and convergence of a sequence $\mu_n \in \mathbf{M}_f$ in the norm-topology implies convergence in the weak-* topology. It follows from the Riesz theorem that the limit μ is a Borel (or Baire) measure. Thus, it is in $\mathbf{M}_f$.[17] Hence, $\mathbf{M}_f$ is closed in the norm-topology and, therefore, it is also complete.

Set

$$\mathbf{M}_d := \left\{ \mu := \sum_{i=1}^N a_i \delta_{r_i}, N \in \mathbf{N}, r_i \in \mathbf{R}^d, a_i \in \mathbf{R} \right\},$$

i.e., $\mathbf{M}_d$ is the space of finite sums of point measures with nonnegative weights. Obviously,

$$\mathbf{M}_d \subset C^*_{L,1}(\mathbf{R}^d;\mathbf{R}).$$

[16] In the definition of the Lipschitz norm (15.37) we may, without loss of generality, restrict the quotient to $|r-q| \le 1$, since for values $|r-q| > 1$ the quotient is dominated by $2|||f|||$.

[17] Cf. Bauer (1968), Sect. VII, 45.

Note that $\mathbf{M}_d$ is dense in $(\mathbf{M}_f, \gamma_f)$ in the topology generated by γ_f.[18] Further, we verify that the sums of point measures with rational nonnegative weights and supports in rational vectors are also dense in $(\mathbf{M}_f, \gamma_f)$. Therefore, we have

Proposition 15.9. $(\mathbf{M}_f, \gamma_f)$ *is a complete separable metric space.*[19]

Choose $\nu = \mathbf{0}$ (the Borel measure that assigns 0 to all Borel sets) in (15.38). For positive measures μ we have

$$\gamma_f(\mu) = < 1, \mu > = \mu(\mathbf{R}^d),$$

i.e., the norm $\gamma_f(\mu)$ of $\mu \in \mathbf{M}$ is the total mass of μ.

Let us now make some comments on the relation of γ_f to the Wasserstein (or Monge-Wasserstein) metric.[20] If μ, ν have equal total finite mass $\bar{m} > 0$ we will call positive Borel measures Q on $\mathbf{R}^{2d}$ joint representations of μ and ν, if $Q(A \times \mathbf{R}^d) = \mu(A)\bar{m}$ and $Q(\mathbf{R}^d \times B) = \tilde{\mu}(B)\bar{m}$ for arbitrary Borel sets $A, B \subset \mathbf{R}^d$. The set of all joint representations of $(\mu, \tilde{\mu})$ will be denoted by $C(\mu, \tilde{\mu})$. For $\mu, \tilde{\mu}$ with mass $\bar{m}$ and $p \geq 1$ define the pth Wasserstein metric by

$$\tilde{\gamma_p}(\mu, \tilde{\mu}) := \left[\inf_{Q \in C(\mu, \tilde{\mu})} \int \int Q\,(\mathrm{d}r, \mathrm{d}q) \rho^p(r - q) \right]^{\frac{1}{p}}. \tag{15.39}$$

For the case of two probability measures μ, ν (or finite measures of equal mass) the Kantorovich-Rubinstein theorem asserts[21]

$$\sup_{\|f\|_L \leq 1} \int f(q)(\mu(\mathrm{d}q) - \nu(\mathrm{d}q)) = \tilde{\gamma}_1(\mu, \nu) = \inf_{Q \in C(\mu, \nu)} \int \int Q\,(\mathrm{d}r, \mathrm{d}q) \rho(r - q) \tag{15.40}$$

The first observation is that, both in (15.38) and (15.40), the sets of f are invariant with respect to multiplication by -1, which allows us to drop the absolute value bars in the definition of the metrics. Further, for two probability measures μ and ν

$$\sup_{\|f\|_L \leq 1} \int f(q)(\mu(\mathrm{d}q) - \nu(\mathrm{d}q)) \geq \sup_{\|f\|_{L,\infty} \leq 1} \int f(q)(\mu(\mathrm{d}q) - \nu(\mathrm{d}q)).$$

Next, for arbitrary constants c_f

$$\begin{aligned} \sup_{\|f\|_L \leq 1} \int (f(q) - c_f)(\mu(\mathrm{d}q) - \nu(\mathrm{d}q)) &= \sup_{\|f\|_L \leq 1} \Big[\int f(q)(\mu(\mathrm{d}q) - \nu(\mathrm{d}q)) \\ &\qquad - c_f(\mu(\mathbf{R}^d) - \nu(\mathbf{R}^d)) \Big] \\ &= \sup_{\|f\|_L \leq 1} \int f(q)(\mu(\mathrm{d}q) - \nu(\mathrm{d}q)), \end{aligned}$$

---

[18] Cf. De Acosta, A. (1982).

[19] By Definition 15.2 and (15.38) $(\mathbf{M}_f, \gamma_f)$ is a separable Fréchet space.

[20] Cf. Dudley (1989), Chap. 11.8).

[21] Cf. Dudley (loc.cit.), Chap. 11, Theorem 11.8.2.

since $\mu(\mathbf{R}^d) - \nu(\mathbf{R}^d) = 0$. On the left-hand side of (15.40) we may therefore, without loss of generality, assume $f(0) = 0$. If we choose two point measures δ_r, δ_0 the (only) joint representation is $\delta_{(r,0)}$. (15.37), in addition to $f(0) = 0$, implies

$$\sup_{\|f\|_L \le 1} f(r) = \rho(r) \le 1.$$

We obtain

$$|||f||| \le 1 \quad \text{if } \|f\|_L \le 1 \text{ and if } f(0) = 0.$$

Thus, we have for two probability measures μ and ν

$$\tilde{\gamma}_1(\mu,\nu) = \sup_{\|f\|_L \le 1, f(0)=0} \int f(q)(\mu(dq) - \nu(dq)) = \sup_{\|f\|_{L,\infty} \le 1} \left[\int f(q)(\mu(dq) - \nu(dq))\right]$$

and therefore, for two probability measures μ and ν

$$\tilde{\gamma}_1(\mu,\nu) = \gamma_f(\mu - \nu).$$

As a consequence, we may in the Kantorovich–Rubinstein theorem replace the $\sup\limits_{\|f\|_L \le 1}$ by the $\sup\limits_{\|f\|_{L,\infty} \le 1}$. "Normalizing" two measures $\mu, \nu \in \mathbf{M}_{\bar{m}}$ with $\bar{m} > 0$ to $\frac{\mu}{\bar{m}}$ and $\frac{\nu}{\bar{m}}$, respectively, we obtain

$$\gamma_f(\mu - \nu) = \bar{m}\tilde{\gamma}_1\left(\frac{\mu}{\bar{m}} - \frac{\nu}{\bar{m}}\right). \tag{15.41}$$

Next, consider two arbitrary measures $\mu, \nu \in \mathbf{M}_f^+$ with total masses $\bar{m}$ and $\bar{n}$, respectively. Suppose $\bar{m} > \bar{n} > 0$. Then,

$$\begin{aligned}\gamma_f(\mu - \nu) &= \sup_{\|f\|_{L,\infty} \le 1} \left|\int f(q)(\mu(dq) - \nu(dq))\right| \\ &= \sup_{\|f\|_{L,\infty} \le 1} \int f(q)(\mu(dq) - \nu(dq))\end{aligned}$$

(as the set of f in question is invariant if the elements are multiplied by (-1))

$$\begin{aligned}&= \sup_{\|f\|_{L,\infty} \le 1} \left[\int f(q)\left(\frac{\bar{n}}{\bar{m}}\mu(dq) - \nu(dq)\right) + \int f(q)\frac{\bar{m} - \bar{n}}{\bar{m}}\mu(dq)\right] \\ &\le \sup_{\|f\|_{L,\infty} \le 1} \int f(q)\left(\frac{\bar{n}}{\bar{m}}\mu(dq) - \nu(dq)\right) + \sup_{\|f\|_{L,\infty} \le 1} \int f(q)\frac{\bar{m} - \bar{n}}{\bar{m}}\mu(dq) \\ &= \gamma_f\left(\frac{\bar{n}}{\bar{m}}\mu - \nu\right) + \int 1(q)\frac{\bar{m} - \bar{n}}{\bar{m}}\mu(dq)\end{aligned}$$

(where $1(q) \equiv 1$)

$$
= \gamma_f \left(\frac{\bar{n}}{\bar{m}} \mu - \nu \right) + \bar{m} - \bar{n}
$$

$$
= \bar{n} \tilde{\gamma}_1 \left(\frac{\mu}{\bar{m}} - \frac{\nu}{\bar{n}} \right) + \bar{m} - \bar{n}
$$

(by (15.41).)

If $\bar{n} = 0$, then $\nu = \mathbf{0}$. We recall that $\gamma_f(\mu - \nu) = \gamma_f(\mu - \mathbf{0}) = \bar{m}$.

For $\bar{m} > \bar{n} > 0$ we now suppose $\frac{\bar{n}}{\bar{m}} \mu \neq \nu$ which is equivalent to $\gamma_f \left(\frac{\bar{n}}{\bar{m}} \mu - \nu \right) =: c > 0$. Let f_n be a sequence with $\|f_n\|_{L,\infty} \leq 1$ and $\gamma \left(\frac{\bar{n}}{\bar{m}} \mu - \nu \right) = c = \lim\limits_{n \to \infty} \int f_n(q) \left(\frac{\bar{n}}{\bar{m}} \mu(dq) - \nu(dq) \right)$. Note that

$$
\lim_{n \to \infty} \int f_n(q) \left(\frac{\bar{n}}{\bar{m}} \mu(\mathrm{d}q) - \nu(\mathrm{d}q) \right) = \lim_{n \to \infty} \left[\int f_n(q) \frac{\bar{n}}{\bar{m}} \mu(\mathrm{d}q) - \int f_n(q) \nu(\mathrm{d}q) \right].
$$

Both integrals in the right-hand side of the last identity are bounded sequences α_n and β_n, respectively. By compactness, they contain convergent subsequences, and we may assume, without loss of generality, that both α_n and β_n converge to real numbers α and β, respectively.

Consider first the case where $-\beta < c$. This implies $\alpha > 0$, whence we may assume that $\alpha_n > 0$ for all n. Since $\alpha_n \longrightarrow \alpha$

$$
\int f_n(q) \frac{\bar{m} - \bar{n}}{\bar{m}} \mu(\mathrm{d}q) = \alpha_n \frac{\bar{m}}{\bar{n}} \frac{\bar{m} - \bar{n}}{\bar{m}} \longrightarrow \alpha \frac{\bar{m}}{\bar{n}} \frac{\bar{m} - \bar{n}}{\bar{m}} = \alpha \left(\frac{\bar{m}}{\bar{n}} - 1 \right) > 0.
$$

Hence,

$$
\gamma_f \left(\frac{\bar{n}}{\bar{m}} \mu - \nu \right) = c = \lim_{n \to \infty} \int f_n(q) \left(\frac{\bar{n}}{\bar{m}} \mu(\mathrm{d}q) - \nu(\mathrm{d}q) \right)
$$

$$
\ldots \bigg) + \lim_{n \to \infty} \int f_n(q) \frac{\bar{m} - \bar{n}}{\bar{m}} \mu(\mathrm{d}q)
$$

$$
\leq \sup_{\|f\|_{L,\infty} \leq 1} \int f(q)(\mu(\mathrm{d}q) - \nu(\mathrm{d}q))
$$

$\ldots \Longleftrightarrow \beta \leq -c$. Since $c > 0$, we may assume

So

$$
\begin{aligned}
\gamma_f\left(\frac{\bar{n}}{\bar{m}}\mu - \nu\right) &= c = \lim_{n\to\infty}\int f_n(q)\left(\frac{\bar{n}}{\bar{m}}\mu(dq) - \nu(dq)\right)\\
&= \frac{\bar{n}}{\bar{m}}\lim_{n\to\infty}\int f_n(q)\left(\mu(dq) - \frac{\bar{m}}{\bar{n}}\nu(dq)\right)\\
&= \frac{\bar{n}}{\bar{m}}\lim_{n\to\infty}\int f_n(q)(\mu(dq) - \nu(dq)) + \frac{\bar{m}-\bar{n}}{\bar{n}}\int f_n(q)\nu(dq)\\
&\le \frac{\bar{n}}{\bar{m}}\lim_{n\to\infty}\int f_n(q)(\mu(dq) - \nu(dq))\\
&\le \frac{\bar{n}}{\bar{m}}\sup_{\|f\|_{L,\infty}\le 1}\int f(q)(\mu(dq) - \nu(dq))\\
&= \frac{\bar{n}}{\bar{m}}\gamma_f(\mu-\nu) < \gamma_f(\mu-\nu).
\end{aligned}
$$

Note that for $\bar{m} > \bar{n}$

$$
(\bar{n}\wedge\bar{m})\gamma_f\left(\frac{\mu}{\bar{m}} - \frac{\nu}{\bar{n}}\right) = \gamma_f\left(\frac{\bar{n}}{\bar{m}}\mu - \nu\right).
$$

The left-hand side is symmetric with respect to $\bar{m}$ and $\bar{n}$. Therefore, we obtain in both cases

$$
(\bar{n}\wedge\bar{m})\gamma_f\left(\frac{\mu}{\bar{m}} - \frac{\nu}{\bar{n}}\right) \le \gamma_f(\mu-\nu). \tag{15.42}
$$

Further, note that

$$
\bar{m}-\bar{n} = \int 1(q)(\mu(dq)-\nu(dq)) \le \sup_{\|f\|_{L,\infty}\le 1}\int f(q)(\mu(dq)-\nu(dq)) = \gamma_f(\mu-\nu)
$$

Altogether,

$$
\left.
\begin{array}{c}
\gamma_f(\mu-\nu) \le (\bar{n}\wedge\bar{m})\gamma_f\left(\frac{\mu}{\bar{m}} - \frac{\nu}{\bar{n}}\right) + |\bar{m}-\bar{n}| \le 2\gamma_f(\mu-\nu)\\
\Longleftrightarrow\\
\gamma_f(\mu-\nu) \le (\bar{n}\wedge\bar{m})\tilde{\gamma}_1\left(\frac{\mu}{\bar{m}} - \frac{\nu}{\bar{n}}\right) + |\bar{m}-\bar{n}| \le 2\gamma_f(\mu-\nu).
\end{array}
\right\} \tag{15.43}
$$

Set

$$
\hat{\gamma}_f(\mu,\nu) := (\bar{n}\wedge\bar{m})\tilde{\gamma}_1\left(\frac{\mu}{\bar{m}} - \frac{\nu}{\bar{n}}\right) + |\bar{m}-\bar{n}| \tag{15.44}
$$

We see that $\hat{\gamma}_f(\mu,\nu)$ defines a metric on $\mathbf{M}_f$. An equivalent version was introduced in Kotelenez (1996). Summarizing the previous results we have shown:

Proposition 15.10. *$\gamma_f(\mu-\nu)$ and $\hat{\gamma}_f(\mu,\nu)$ are equivalent metrics on $\mathbf{M}_f$.* □

Recall that $\mathbf{M}_{\infty,\varpi}$ was defined in (4.5) as the space of all σ-finite Borel measures μ on $\mathbf{R}^d$ such that

$$\mathbf{M}_{\infty,\varpi} := \left\{\mu \in \mathbf{M}_\infty : \int \varpi(q)\mu(\mathrm{d}q) < \infty\right\}$$

where $\varpi(r) = (1+|r|^2)^{-\gamma}$, $\gamma > \frac{d}{2}$, is the weight function from (4.4). $\varpi\mu$ etc. denotes the finite measure that can be represented as μ with density ϖ. Further, defining for $\mu, \nu \in \mathbf{M}_{\infty,\varpi}$

$$\gamma_\varpi(\mu - \nu) := \gamma_f(\varpi(\mu - \nu)), \tag{15.45}$$

we obtain that $(\mathbf{M}_{\infty,\varpi}, \gamma_\varpi)$ is isometrically isomorphic to $(\mathbf{M}_f, \gamma_f)$. Recalling Proposition 15.9, we conclude that $(\mathbf{M}_{\infty,\varpi}, \gamma_\varpi)$ is also a complete separable metric space.

Next, we derive some properties of ϖ. For $k, \ell = 1, \ldots, d$ and f twice continuously differentiable we recall the notation

$$\partial_k f := \frac{\partial}{\partial r_k} f, \quad \partial^2_{k\ell} f := \frac{\partial^2}{\partial r_k \partial r_\ell} f.$$

Let $\beta > 0$ and consider $\varpi_\beta(r) := \varpi^\beta(r)$. We easily verify

$$\left.\begin{aligned} \sum_{k=1}^d |\partial_k \varpi_\beta(q)| + \sum_{k,\ell=1}^d |\partial^2_{k\ell} \varpi_\beta(q)| \le c\varpi_\beta(q), \\ |\varpi_\beta(r) - \varpi_\beta(q)| \le \gamma\beta\varrho(r,q)[\varpi_\beta(r) + \varpi_\beta(q)]. \end{aligned}\right\} \tag{15.46}$$

Indeed, a simple calculation yields

$$\partial_k \varpi(r) = -\gamma(1+|r|^2)^{-\gamma-1} 2r_k$$

$$\partial^2_{k,\ell} \varpi(r) = -\gamma(-\gamma-1)(1+|r|^2)^{-\gamma-2} 2r_k 2r_\ell - \gamma(1+|r|^2)^{-\gamma-1} 2\delta_{k,\ell}.$$

This implies the first inequality. The second inequality is obvious if $|r-q| \ge 1$. For the case $|r-q| < 1$ we may assume $|r| < |q|$ and $\beta = 1$. Then

$$\begin{aligned} &|\varpi(r) - \varpi(q)| = \varpi(r) - \varpi(q) \\ &= \int_{|r|}^{|q|} 2\gamma u(1+u^2)^{-(\gamma+1)} \mathrm{d}u \\ &\le \int_{|r|}^{|q|} \gamma(1+u^2)^{-\gamma}\, \mathrm{d}u \\ &\text{(as } 2u \le (1+u^2)) \\ &\le \int_{|r|}^{|q|} \gamma(1+|r|^2)^{-\gamma}\, \mathrm{d}u \\ &\text{(as } (1+u^2) \text{ is monotone decreasing for } u \ge 0\text{)} \\ &= (|q| - |r|)\varpi(r) \le |r-q|\varpi(r) \\ &\le |r-q|[\varpi(r) + \varpi(q)]. \end{aligned}$$

The following was used in (11.9):

$$\varpi^{-1}(q)\varpi(r) = \left(\frac{1+|q|^2}{1+|r|^2}\right)^{\gamma} \le 2^{\gamma}(1+|r-q|^2)^{\gamma}. \tag{15.47}$$

(15.47) follows from

$$\begin{aligned}\varpi^{-1}(q) &= 1+|q|^2 = 1+|q-r+r|^2 \le 1+2|q-r|^2+2|r|^2\\ &\le 2(1+|q-r|^2+|r|^2) \le 2(1+|r-q|^2)(1+|r|^2).\end{aligned}$$

We have the continuous inclusion

$$(\mathbf{M}_f, \gamma_f) \subset (\mathbf{M}_{\infty,\varpi}, \gamma_\varpi).$$

In what follows we compare the topologies of $(\mathbf{M}_f, \gamma_\varpi)$ and of $(\mathbf{M}_{\infty,\varpi}, \gamma_\varpi)$, respectively, with the topologies of $\mathbf{W}_{m,2,\Phi}$. As a preliminary, we derive a relation between weak and strong convergence in separable Hilbert spaces, using the Fourier expansion of the square of the Hilbert space norm.[22]

Lemma 15.11. *Let* $\mathbf{H}$ *be a separable Hilbert space with scalar product* $\langle\cdot,\cdot\rangle_{\mathbf{H}}$ *and associated Hilbert space norm* $\|\cdot\|_{\mathbf{H}}$ *and let* $\{\varphi_k\}_{k\in\mathbf{N}}$ *be a CONS in* $\mathbf{H}$. *Let* $f_n, f \in \mathbf{H}$ *such that*

(i) $\|f_n\|_{\mathbf{H}} \longrightarrow \|f\|_{\mathbf{H}}$, *as* $n \longrightarrow \infty$,
(ii) $\langle f_n, \varphi_k\rangle_{\mathbf{H}} \longrightarrow \langle f, \varphi_k\rangle_{\mathbf{H}}$ $\forall k \in \mathbf{N}$, *as* $n \longrightarrow \infty$.

Then, $\|f_n - f\|_{\mathbf{H}} \longrightarrow 0$, *as* $n \longrightarrow \infty$.

Proof.

$$\langle f_n, f\rangle_{\mathbf{H}} = \sum_{k \in bfN} \langle f_n, \varphi_k\rangle_{\mathbf{H}}\langle\varphi_k, f\rangle_{\mathbf{H}}.$$

By assumption, the kth term in the above Fourier series converges to $\langle f, \varphi_k\rangle_{\mathbf{H}}\langle\varphi_k, f\rangle_{\mathbf{H}}$ $\forall k$. To conclude that the the left-hand side also converges to $\langle f, f\rangle_{\mathbf{H}}$, we must show that the terms are uniformly integrable with respect to the counting measure. Note that

$$|\langle f_n, \varphi_k\rangle_{\mathbf{H}}\langle\varphi_k, f\rangle_{\mathbf{H}}| \le \frac{1}{2}[\langle f_n, \varphi_k\rangle_{\mathbf{H}}\langle\varphi_k, f_n\rangle_{\mathbf{H}} + \langle f, \varphi_k\rangle_{\mathbf{H}}\langle\varphi_k, f\rangle_{\mathbf{H}}].$$

Again our assumption implies that the first term $\langle f_n, \varphi_k\rangle_{\mathbf{H}}\langle\varphi_k, f_n\rangle_{\mathbf{H}}$ converges for all k to $\langle f, \varphi_k\rangle_{\mathbf{H}}\langle\varphi_k, f\rangle_{\mathbf{H}}$ and its sum, being equal to $\|f_n\|^2_{\mathbf{H}}$, converges by assumption to $\|f\|^2_{\mathbf{H}}$. Consequently,

$$\sum_{k\in\mathbf{N}} \frac{1}{2}[\langle f_n, \varphi_k\rangle_{\mathbf{H}}\langle\varphi_k, f_n\rangle_{\mathbf{H}} + \langle f, \varphi_k\rangle_{\mathbf{H}}\langle\varphi_k, f\rangle_{\mathbf{H}}] \longrightarrow \|f\|^2_{\mathbf{H}}, \ \text{as } n \longrightarrow \infty\,.$$

[22] The result of Lemma 15.11 is actually well known, except that we replaced weak convergence by convergence of the Fourier coefficients (cf. Yosida (1968), Sect. V.1, Theorem 8).

If follows that the terms $\frac{1}{2}[\langle f_n, \varphi_k\rangle_{\mathbf{H}}\langle \varphi_k, f_n\rangle_{\mathbf{H}} + \langle f, \varphi_k\rangle_{\mathbf{H}}\langle \varphi_k, f\rangle_{\mathbf{H}}]$ are uniformly integrable with respect to the counting measure (cf. Bauer (1968), Sect. 20, Korollar 20.5). Since these terms dominate $|\langle f_n, \varphi_k\rangle_{\mathbf{H}}\langle \varphi_k, f\rangle_{\mathbf{H}}|$ we have that the terms $|\langle f_n, \varphi_k\rangle_{\mathbf{H}}\langle \varphi_k, f\rangle_{\mathbf{H}}|$ are also uniformly integrable with respect to the counting measure. Hence, we may apply Lebesgue's dominated convergence theorem and obtain

$$\sum_{k\in\mathbf{N}} |\langle f_n, \varphi_k\rangle_{\mathbf{H}}\langle \varphi_k, f\rangle_{\mathbf{H}} - \langle f, \varphi_k\rangle_{\mathbf{H}}\langle \varphi_k, f\rangle_{\mathbf{H}}| \longrightarrow 0, \quad \text{as } n \longrightarrow \infty\,.$$

This implies

$$\langle f_n, f\rangle_{\mathbf{H}} \longrightarrow \langle f, f\rangle_{\mathbf{H}}, \quad \text{as } n \longrightarrow \infty\,,$$

and similarly for $\langle f, f_n\rangle_{\mathbf{H}}$. Now the statement follows as in Yosida (loc.cit.) from

$$\|f_n - f\|_{\mathbf{H}}^2 = \|f_n|_{\mathbf{H}}^2 + \|f\|_{\mathbf{H}}^2 - \langle f_n, f\rangle_{\mathbf{H}} - \langle f, f_n\rangle_{\mathbf{H}} \longrightarrow 0, \quad \text{as } n \longrightarrow \infty\,. \qquad \square$$

Proposition 15.12. *Let* $\alpha \geq 0$, $(\mathbf{M}, \gamma) \in \{(\mathbf{M}_f, \gamma_f), (\mathbf{M}_{\varpi}, \gamma_{\varpi})\}$, *and let* $\{\varphi_k\}_{k\in\mathbf{N}} \subset \mathcal{S}$ *be a CONS in* $\mathbf{H}_{\alpha,\Phi}$, $\Phi \in \{1, \varpi\}$. *Suppose* $f_n, f \in \mathbf{H}_{\alpha,\Phi} \cap \mathbf{M}$ *such that*

(i) $\|f_n\|_{\alpha,\Phi} \longrightarrow \|f\|_{\alpha,\Phi}$, *as* $n \longrightarrow \infty$,
(ii) $\gamma(f_n - f) \longrightarrow 0$, *as* $n \longrightarrow \infty$.

Then, $\|f_n - f\|_{\alpha,\Phi} \longrightarrow 0$, *as* $n \longrightarrow \infty$.

Proof. Recalling the definitions of γ_f from (15.38) and γ_{ϖ} from (15.45), respectively, we note that for all k we find a suitable constant c_k such that $\|c_k\varphi_k\|_{L,\infty} = 1$ (cf. also the proof of Theorem 8.5). Hence, convergence of f_n to f in $(\mathbf{M}, \gamma)$ implies

$$\langle f_n - f, \varphi_k\rangle_{\alpha,\Phi} \longrightarrow 0 \ \forall k \in \mathbf{N}, \quad \text{as } n \longrightarrow \infty.$$

Employing Lemma 15.11 finishes the proof. $\square$

For our purposes we need to mention two other metrics. The first is the Prohorov metric, $d_{p,\mathbf{B}}$. This is a metric on probability measures μ, ν, defined on the Borel sets of some separable metric space $\mathbf{B}$. The Prohorov metric will be described in Sect. 15.2., (15.37). The second metric is the "*metric of total variation*" on $\mathbf{M}_f$, defined as follows:

$$\|\mu - \nu\|_f := \sup_{\{A\in B^d\}} |\langle \mu - \nu, 1_A\rangle| \tag{15.48}$$

Apparently, $\|\cdot\|_f$ is a norm (on the finite signed measures), restricted to $\mathbf{M}_f$.

Proposition 15.13.

$$\gamma_f(\mu - \nu) \leq \|\mu - \nu\|_f \ \forall \mu, \nu \in \mathbf{M}_f.$$

Proof.

(i) Since $\mathbf{M}_f$ are finite Borel measures on $\mathcal{B}^d$ they are regular. The Riesz representation theorem[23] implies that

[23] Cf. Bauer (1968), Sects. 40, 41, and Folland (1984), Sect. 7.3.

$$\|\mu - \nu\|_f = \sup_{\{|||\varphi|||\le 1, \varphi\in C_0(\mathbf{R}^d;\mathbf{R})\}} |\langle \mu - \nu, \varphi\rangle|. \tag{15.49}$$

Further, the regularity implies that for every $\delta > 0$ there exists a closed ball $B_{L_\delta}(0)$ with finite radius L_δ and center at the origin such that

$$\mu\big(\mathbf{R}^d \setminus B_{L_\delta}(0)\big) + \nu\big(\mathbf{R}^d \setminus B_{L_\delta}(0)\big) \le \delta.$$

Hence,

$$\gamma_f(\mu - \nu) \le \sup_{\{\|\varphi\|_{L,\infty}\le 1\}} \left|\int \varphi(r) 1_{B_{L_\delta}(0)}(r)(\mu - \nu)(dr)\right| + \delta$$

and

$$\|\mu - \nu\|_f \ge \sup_{\{|||\varphi|||\le 1, \varphi\in C_0(\mathbf{R}^d;\mathbf{R})\}} \left|\int \varphi(r) 1_{B_{L_\delta}(0)}(r)(\mu - \nu)(dr)\right| - \delta.$$

(ii) Set

$$\tilde{\eta}(x) := \begin{cases} 1, & \text{if} \quad 0 \le x \le L_\delta\,, \\ 1 + \delta(L_\delta - x), & \text{if} \quad x \in \left(L_\delta + \frac{1}{\delta}\right), \\ 0, & \text{otherwise,} \end{cases}$$

and

$$\eta(r) := \tilde{\eta}(|r|).$$

If $\|\varphi\|_{L,\infty} \le 1$ we verify that the pointwise product of the two functions, φ and η, satisfies the following relations:

$$\varphi\eta \in C_0(\mathbf{R}^d;\mathbf{R}) \text{ and } \|\varphi\eta\|_{L,\infty} \le (1+\delta).$$

(iii) From (ii)

$$\begin{aligned}
&\sup_{\{\|\varphi\|_{L,\infty}\le 1\}} \left|\int \varphi(r) 1_{B_{L_\delta}(0)}(r)(\mu - \nu)(dr)\right| \\
&= \sup_{\{\|\varphi\|_{L,\infty}\le 1\}} \left|\int \varphi(r)\eta(r) 1_{B_{L_\delta}(0)}(r)(\mu - \nu)(dr)\right| \\
&\le \sup_{\{\|\varphi\eta\|_{L,\infty}\le 1+\delta, \varphi\in C_{L,\infty}(\mathbf{R}^d;\mathbf{R})\}} \left|\int \varphi(r)\eta(r) 1_{B_{L_\delta}(0)}(r)(\mu - \nu)(dr)\right| \\
&\le \sup_{\{|||\varphi|||\le 1+\delta, \varphi\in C_0(\mathbf{R}^d;\mathbf{R})\}} \left|\int \varphi(r) 1_{B_{L_\delta}(0)}(r)(\mu - \nu)(dr)\right| \\
&= \sup_{\{|||\varphi|||\le 1, \varphi\in C_0(\mathbf{R}^d;\mathbf{R})\}} \left|\int \varphi(r) 1_{B_{L_\delta}(0)}(r)(\mu - \nu)(dr)\right| (1+\delta).
\end{aligned}$$

Therefore, from (i) we obtain

$$\gamma_f(\mu - \nu) \le \|\mu - \nu\|_f(1+\delta) + \delta(2+\delta).$$

Since $\delta > 0$ was arbitrary the proof is complete. □

15.1.5 Riemann Stieltjes Integrals

Next, we provide some facts about Riemann-Stieltjes integrals, following Natanson (1974).

Definition 15.14. *Let* f *and* g *be real-valued functions, defined on a one-dimensional interval* $[a,b]$ *and* $\{a = t_0^n \le t_1^n \le \ldots . \le t_n^n = b\}$ *be a sequence of partitions such that* $\max_{1\le k\le n}\left(t_k^n - t_{k-1}^n\right) \longrightarrow 0$*, as* $n \longrightarrow \infty$.

(I) Suppose that $V_n(f) := \sum_{k=1}^{n} \left|f(t_k^n) - f(t_{k-1}^n)\right|$ *converges, as* $n \longrightarrow \infty$*. Then,* f *is called to be of* "bounded variation" *and*

$$V_a^b(f) := \lim_{n\to\infty} \sum_{k=1}^{n} \left|f(t_k^n) - f(t_{k-1}^n\right|$$

the variation of f *on* $[a,b]$.

(II) Let $\xi_k^n \in \left[t_{k-1}^n, t_k^n\right]$ *arbitrary. Suppose* $\sum_{k=1}^{n} g\left(\xi_k^n\right)\left[f(t_k^n) - f(t_{k-1}^n)\right]$ *converges, as* $n \longrightarrow \infty$*. Then* g *is said to be* "Riemann-Stieltjes integrable with respect to f" *and*

$$\int_a^b g(t)f(\mathrm{d}t) := \lim_{n\to\infty} \sum_{k=1}^{n} g(\xi_k^n)[f(t_k^n) - f(t_{k-1}^n)] \tag{15.50}$$

the "Riemann–Stieltjes integral" *on* $[a,b]$ *of* g *with respect to* f. □

The following result is well known. The proof is the same as for the classical Riemann integral.[24]

Theorem 15.15. *If* g *is continuous on* $[a,b]$ *and* f *is of bounded variation on* $[a,b]$*, then* g *is Riemann–Stieltjes integrable with respect to* f *and*

$$\left|\int_a^b g(t)f(\mathrm{d}t)\right| \le \max_{a\le t\le b} |g(t)| V_a^b(f)$$ □

The Riemann–Stieltjes integral may be employed to define the Wiener integral (cf. (15.107)) with smooth integrands. This application is based on the following integration-by-parts formula:

Theorem 15.16. g *is Riemann–Stieltjes integrable with respect to* f *if, and only if,* f *is Riemann–Stieltjes integrable with respect to* g *and the following integration-by-parts formula holds:*

$$\int_a^b g(t)f(\mathrm{d}t) + \int_a^b f(t)g(\mathrm{d}t) = f(b)g(b) - f(a)g(a). \tag{15.51}$$

[24] Cf.,e.g., Natanson, (loc.cit.), Chap. 8.7, Theorem 1.

Proof.[25] Without loss of generality, suppose g is Riemann–Stieltjes integrable with respect to f. Let $\{a = t_0^n \le t_1^n \le \ldots . \le t_n^n = b\}$ be a sequence of partitions such that $\max_{1\le k\le n}\left(t_k^n - t_{k-1}^n\right) \longrightarrow 0$, as $n \longrightarrow \infty$, let $\xi_k^n \in \left[t_{k-1}^n, t_k^n\right]$ arbitrary. Set

$$S_n := \sum_{k=1}^{n} f\left(\xi_k^n\right)\left[g\left(t_k^n\right) - g\left(t_{k-1}^n\right)\right].$$

Obviously,

$$S_n = \sum_{k=1}^{n} f\left(\xi_k^n\right) g\left(t_k^n\right) - \sum_{k=1}^{n} f\left(\xi_k^n\right) g\left(t_{k-1}^n\right),$$

whence

$$S_n = -\sum_{k=1}^{n-1} g\left(t_k^n\right)\left[f\left(\xi_{k+1}^n\right) - f\left(\xi_k^n\right)\right] + f\left(\xi_k^n\right) g(b) - f\left(\xi_1^n\right) g(a).$$

Adding and subtracting the right-hand side of (15.51), we obtain

$$S_n = -\left\{\sum_{k=1}^{n-1} g\left(t_k^n\right)\left[f\left(\xi_{k+1}^n\right) - f\left(\xi_k^n\right)\right] + \left[f(b) - f\left(\xi_k^n\right)\right] g(b) - \left[f\left(\xi_1^n\right) - f(a)\right] g(a)\right\}$$
$$+ f(b)g(b) - f(a)g(a).$$

By assumption

$$\sum_{k=1}^{n-1} g\left(t_k^n\right)\left[f\left(\xi_{k+1}^n\right) - f\left(\xi_k^n\right)\right] + \left[f(b) - f\left(\xi_k^n\right)\right] g(b)$$
$$-\left[f\left(\xi_1^n\right) - f(a)\right] g(a) \longrightarrow \int_a^b f(t) g(\mathrm{d}t), \quad \text{as } n \longrightarrow \infty$$

since $\{a =: \xi_0^n \le \xi_1^n \le \ldots . \le \xi_n^n \le \xi_{n+1}^n = b\}$ is a sequence of partitions such that $\max_{1\le k\le n+1}\left(\xi_k^n - \xi_{k-1}^n\right) \longrightarrow 0$, as $n \longrightarrow \infty$ and $t_k^n \in \left[\xi_k^n, \xi_{k+1}^n\right]$ $\forall k, n$. □

Observe that the Riemann–Stieltjes integral is constructed by partitioning the time axis into small intervals and then passing to the limit. This construction may be achieved in two steps. Fixing the integrator $g(\cdot)$, we first define the Riemann–Stieltjes integral for step function integrands $f(\cdot)$.[26] In the second step we may or may not follow the classical procedure outlined above. An alternative method is to approximate other possible integrands "suitably" by step function integrands and extend the integral by continuity (with respect to some norm or metric) to a larger

[25] The proof is adopted from Natanson (loc.cit.), Chap. 8.6.

[26] As we will see in Sect. 15.2.5, the stochastic Itô integral necessarily starts with step function type processes as elementary integrands because the integrator increments in time have to be "orthogonal" to the integrator.

class of integrands. This begs the question of what is "suitable," how large is the class of possible integrands, and what are the properties of the integral. If we wish to have a dominated convergence theorem for the resulting integral, we may first interpret the Riemann–Stieltjes integral for step functions as the Lebesgue–Stieltjes integral with respect to the (signed) Stieltjes measure associated with the integrator $g(\cdot)$. Representing g as the difference of two nondecreasing functions, it suffices to consider only integrators that are nondecreasing. In this case the integrator $g(\cdot)$ must be right continuous to entail the continuity from above of the associated Stieltjes measure and, therefore, the dominated convergence theorem. An answer to our question what is "suitable" etc. is provided by Bichteler (2002) who provides a detailed account of the extension of the Lebesgue–Stieltjes integral from step function integrands to a large class such that the dominated convergence theorem is valid. It is particularly important that, in the stochastic case, this extension leads to the stochastic Itô integral.

In the next section we review some of the analytic properties of right continuous functions with limits from the left.

15.1.6 The Skorokhod Space $D([0, \infty); \mathbf{B})$

We provide some basic definitions and properties of the Skorokhod space of cadlag functions with values in some metric space.

Let $(\mathbf{B}, d_{\mathbf{B}}(\cdot, \cdot))$ be a metric space. The space $C([0, \infty); \mathbf{B})$ is the natural state space for $\mathbf{B}$-valued stochastic processes with continuous sample paths. However, there are many classes of stochastic processes with jumps,[27] and it is desirable to have a state space of $\mathbf{B}$-valued functions that contain both continuous functions and suitably defined functions with jumps. Skorokhod (1956) introduces such a function space and derives both analytic properties of the function space and weak convergence properties of corresponding $\mathbf{B}$-valued stochastic processes. In this section we restrict the presentation to the analytic properties of the function space.

Let $D([0, \infty); \mathbf{B})$ denote the space of $\mathbf{B}$-valued cadlag functions on $[0, \infty)$ (i.e., of functions that are continuous from the right and have limits from the left).[28] As for continuous functions, a suitable metric on $D([0, \infty); \mathbf{B})$ will be defined on compact intervals first and then extended to a metric on $[0, \infty)$. Therefore, let us for the moment focus on $D([0, 1]; \mathbf{B})$ and $C([0, 1]; \mathbf{B})$ (the space of continuous functions on $[0, 1]$ with values in $\mathbf{B}$.

Remark 15.17. Let us endow $C([0, 1]; \mathbf{B})$ with the uniform metric

[27] E.g., Poisson processes or processes arising in a time discretization scheme as our Theorem 2.4 in Chap. 3.

[28] Skorokhod (1956) considers $D([0, 1]; \mathbf{B})$. For more references on our presentation, cf. Kolmogorov (1956), Billingsley (1968), Chap. 3, Jacod (1985), and Ethier and Kurtz (1986), Sect. 3.5.

$$d_{u,1,\mathbf{B}}(f,g) := \sup_{t\in[0,1]} d_{\mathbf{B}}(f(t),g(t)). \tag{15.52}$$

For the theory of stochastic properties the uniform metric has the following two important properties:

(i) Completeness and separability of **B** implies completeness and separability of $(C([0,1];\mathbf{B}), d_{u,1,\mathbf{B}})$. (The completeness is obvious and to see the separability it suffices to take the family of all piecewise linear continuous **B**-valued functions with finitely many rational endpoints and taking values at those endpoints in a countable dense set of **B**.)

(ii) $d_{u,1,\mathbf{B}}(f_n,f) \to 0$ is equivalent to $d_{\mathbf{B}}(f_n(t_n), f(t)) \to 0$ whenever $t_n \to t$. The first property is needed to obtain relative compactness criteria of families of (continuous) random processes with values in a complete and separable metric space **B**.[29] The second property is obviously desirable for many models and their numerical approximations.

Let us now (temporarily!) endow $D([0,1];\mathbf{B})$ with the uniform metric d_u.

(iii) We still obtain completeness of $(D([0,1];\mathbf{B}), d_{u,1,\mathbf{B}})$ if **B** is complete.[30] However, $(D([0,1];\mathbf{B}), d_{u,1,\mathbf{B}})$ is not separable, even if **B** is separable. To verify this statement it suffices to take an arbitrary $b \in \mathbf{B}$ such that $b \neq 0$ and to consider the family

$$\{f(\xi,\cdot) \in D([0,1];\mathbf{B}) : f(\xi,t) := 1_{[\xi,1]}(t)b,\ \xi \text{ irrational}\}.$$

If $\xi_1 \neq \xi_2$ then

$$d_u(f(\xi_1,\cdot), f(\xi_2,\cdot) = d_{\mathbf{B}}(b,0).$$

Hence, there is an uncountable set of nonintersecting nonempty open balls of radius $\frac{1}{2}d_{\mathbf{B}}(b,0)$, which implies the nonseparability of $(D([0,1];\mathbf{B}), d_{u,1,\mathbf{B}})$.

(iv) Property (ii) of $(C([0,1];\mathbf{B}), d_{u,1,\mathbf{B}})$ does not hold either for $(D([0,1];\mathbf{B}), d_{u,1,\mathbf{B}})$. Consider $1 > t_n > t$ and $t_n \downarrow t$. Setting $f_n(t) :\equiv 1_{[t_n,1]}b$ and $f(t) :\equiv 1_{[t,1]}b$, we have

$$d_{u,1,\mathbf{B}}(f_n,f) \equiv d_{\mathbf{B}}(b,0),$$

although $d_{\mathbf{B}}(f_n(t_n), f(t)) \equiv 0$ and, by assumption, $t_n \downarrow t$. □

We have convinced ourselves that the uniform metric $d_{u,1,\mathbf{B}}$ is not a good choice for $D([0,1];\mathbf{B})$. On the basis of the work of Skorokhod (1956), Billingsley (1968) introduces a metric on $D([0,1];\mathbf{B})$, yielding properties analogous to properties (i) and (ii) of $(C([0,1];\mathbf{B}), d_{u,1,\mathbf{B}})$ in Remark 15.17. We will now discuss a generalization of Billingley's metric, $d_{D,\mathbf{B}}$ on $D([0,\infty);\mathbf{B})$, as provided by Ethier and Kurtz (1986), Chap. 3.5, (5.1)–(5.3).

Let Λ be the set of continuous strictly increasing Lipschitz functions $\lambda(\cdot)$ from $[0,\infty)$ into $[0,\infty)$ such that $\lambda(0) = 0$ and

[29] Cf. the following Theorems 15.22 and 15.23.

[30] This assertion follows from Theorem 15.19 and the fact that the uniform metric (extended to $[0,\infty)$) is stronger than the the metric $d_{D,M}$, defined by (15.53).

$$\gamma(\lambda) := \sup_{0 \le s < t} \left| \log \frac{\lambda(t) - \lambda(s)}{t - s} \right| < \infty.$$

Then for $f, g \in D([0, \infty); \mathbf{B})$ we define

$$d_{D,\mathbf{B}}(f, g) := \inf_{\lambda \in \Lambda} \left[\gamma(\lambda) \vee \int_0^\infty \exp(-u) \mathrm{d}(f, g, \lambda, u) du \right], \tag{15.53}$$

where

$$\mathrm{d}(f, g, \lambda, u) := \sup_{t \ge 0} d_{\mathbf{B}}(f(t \wedge u), g(\lambda(t) \wedge u)).$$

Remark 15.18. Choosing $\gamma(t) \equiv t$ in (15.53), it follows immediately that uniform convergence on all compact intervals $[0, T]$ implies convergence in the metric $d_{D,\mathbf{B}}$, i.e., that the metric of uniform convergence on compact intervals is stronger than the metric $d_{D,\mathbf{B}}$. □

Equipped with the metric $d_{D,\mathbf{B}}$, the resulting topology of $D([0, \infty); \mathbf{B})$ is called the Skorokhod topology. If restricted to $C([0, \infty); \mathbf{B})$, the metric topology, generated by $d_{D,\mathbf{B}}$ is the same as the topology, generated by uniform convergence on bounded intervals $[0, T]$ $\forall T$ (cf. Ethier and Kurtz (loc.cit.), Chap. 3.11, Problem 25,b). The usefulness of the metric $d_{D,\mathbf{B}}$ in comparison to some other metrics appears to follow from the following theorem whose proof may be found in Ethier and Kurtz (loc.cit.), Theorem 5.6:

Theorem 15.19. *If* $(\mathbf{B}, d_{\mathbf{B}})$ *is separable, then* $(D([0, \infty); \mathbf{B}), d_{D,\mathbf{B}})$ *is separable. If* $(B, d_{\mathbf{B}})$ *is complete, then* $(D([0, \infty); \mathbf{B}), d_{D,\mathbf{B}})$ *is complete.* □

To analyze convergence of stochastic processes with values in **B** we need criteria of relative compactness. Such criteria may be provided by following the pattern of Ascoli's theorem (or "Arzela-Ascoli theorem") in classical analysis. We remind ourselves that Ascoli's theorem characterizes compactness of continuous functions from a compact metric space **K** into a Banach space in terms of equi-continuity of the family of functions $f(\cdot)$ and compactness of the evaluations $f(x), x \in \mathbf{K}$.[31] The corresponding characterization of compact subsets in $(D([0, \infty); \mathbf{B}), d_{D,\mathbf{B}})$ requires an appropriate definition of a "*modulus of continuity*."

Let $\delta > 0$ and $\{t_0^\delta < t_1^\delta < \cdots < t_{k-1}^\delta < T \le t_k^\delta\}$ be a family of partitions of $[0, T]$ such that for every $T > 0$ $\min_{\{1 \le k \le t_k^\delta \in [0,T]\}} (t_k^\delta - t_{k-1}^\delta) > \delta$. Ethier and Kurtz (loc.cit.), Sect. 6.3, (6.2), characterize the "equi-continuous" subsets of $(D([0, \infty); \mathbf{B}), d_{D,\mathbf{B}})$ in terms of the following "*modulus of continuity*"

$$w'(a, \delta, T) := \inf_{t_k^\delta} \max_k \sup_{s,t \in \left[t_{k-1}^\delta, t_k^\delta\right)} d_{\mathbf{M}}(a(s), a(t)). \tag{15.54}$$

The following theorem, proved by Ethier and Kurtz (loc.cit.), Sect. 3.6, Theorem 6.3, is a generalization of Ascoli's theorem to $(D([0, \infty); \mathbf{B}), d_{D,\mathbf{B}})$ and a somewhat more general form of a theorem proved by Billingsley (1968), Chap. 3.14, Theorem 14.3:

[31] Cf. Dieudonné (1969), Theorem 7.5.7 and, for a typical application in ODEs (the Cauchy-Peano existence theorem), Coddington and Levinson (1955), Chap. 1, Theorem 1.2.

Theorem 15.20. *The closure of $A \subset D([0,\infty);\mathbf{B})$ is compact if and only if the following two conditions hold:*

(i) For every rational $t \geq 0$, there exists a compact set $K_t \subset \mathbf{B}$ such that $a(t) \in K_t$ for all $a \in A$.
(ii) For each $T > 0$,

$$\lim_{\delta \to 0} \sup_{a \in A} w'(a, \delta, T) = 0. \tag{15.55}$$

□

15.2 Stochastics

15.2.1 Relative Compactness and Weak Convergence

We state the basic definitions and theorems regarding weak convergence and relative compactness of random variables and stochastic processes.

Starting with random variables, let $(\Omega, \mathcal{F}, P)$ be a probability space and suppose that all our random variables are defined on $(\Omega, \mathcal{F}, P)$ and take values in a metric space $(\mathbf{B}, d_{\mathbf{B}}(\cdot,\cdot))$. $\mathcal{P}(\mathbf{B})$ denotes the family of Borel probability measures on $\mathbf{B}$. A sequence of $\mu_n \in \mathcal{P}(\mathbf{B})$ is said to "*converge weakly*"[32] to $\mu \in \mathcal{P}(\mathbf{B})$ if

$$\lim_{n\to\infty} \int_{\mathbf{B}} f(a)\mu_n(\mathrm{d}a) = \int_{\mathbf{B}} f(a)\mu(\mathrm{d}a) \quad \forall f \in C_b(\mathbf{B};\mathbf{R}) \tag{15.56}$$

For many estimates it is useful to have a metric on $\mathcal{P}(\mathbf{B})$ such that weak convergence is equivalent to convergence in this metric. To this end, Prohorov (1956), (1.6), defines a metric on $\mathcal{P}(\mathbf{B})$ through

$$d_{p,\mathbf{B}}(\mu,\nu) := \inf\{\eta > 0 : \mu(B) \leq \nu(B^\eta) + \eta \quad \forall B \in \mathcal{C}\}, \tag{15.57}$$

where $\mathcal{C}$ is the collection of closed subsets of $\mathbf{B}$ and

$$B^\eta := \{a \in \mathbf{B} : \inf_{b \in B} d_{\mathbf{B}}(a,b) < \eta\}. \tag{15.58}$$

The metric (15.57) is called the "*Prohorov metric*" on $\mathcal{P}(\mathbf{B})$. Prohorov (loc.cit., Theorem 1.11) proves the following

Theorem 15.21. $(\mathcal{P}(\mathbf{B}), d_{p,\mathbf{B}})$ *is a complete and separable metric space and convergence in* $(\mathcal{P}(\mathbf{B}), d_{p,\mathbf{B}})$ *is equivalent to weak convergence.* □

Ethier and Kurtz (loc.cit.), Chap. 3.1, Corollary 1.9, prove the following.[33]

[32] The reader will notice that in functional analysis this type of convergence would be called "*weak* convergence*."

[33] For a different version, employing the metric in total variation, cf. Dudley (1989), Chap. 9, Theorem 9.3.7.

Theorem 15.22. *Continuous Mapping Theorem*

Let $(\mathbf{B}_i, d_{\mathbf{B}_i}(\cdot,\cdot))$, $i = 1,2$, *be two separable metric spaces, and let* $\Psi : \mathbf{B}_1 \to \mathbf{B}_2$ *be Borel measurable. Suppose that* $\mu_n, \mu \in \mathcal{P}(\mathbf{B}_1)\ \forall n$ *satisfy*

$$\lim_{n\to\infty} d_{p,\mathbf{B}_1}(\mu_n, \mu) = 0.$$

Define probability measures $\nu_n, \nu \in \mathcal{P}(\mathbf{B}_2)$ *as the images of* μ_n *and* μ *under the mapping* Ψ*:*

$$\nu_n := \mu_n \Psi^{-1}, \quad \nu := \mu \Psi^{-1}.$$

Let C_Ψ *be the set of points of* $\mathbf{B}_1$ *at which* Ψ *is continuous. If* $\mu(C_\Psi) = 1$, *then*

$$\lim_{n\to\infty} d_{p,\mathbf{B}_2}(\nu_n, \nu) = 0. \qquad \square$$

A family of probability measures $\mathcal{Q} \subset \mathcal{P}(\mathbf{B})$ is called "*tight*" if for each $\epsilon > 0$ there is a compact set $K \subset S$ such that

$$\inf_{\mu\in\mathcal{Q}} \mu(K) \geq 1 - \epsilon.$$

The following theorem was proved by Prohorov (loc.cit.), Theorem 1.12. In modern literature it is called the "*Prohorov Theorem.*" We adopt the formulation from Ethier and Kurtz, loc.cit., Sect. 3.2, Theorem 2.2.

Theorem 15.23. *Prohorov Theorem*

The following statements are equivalent:

(i) $\mathcal{Q} \subset \mathcal{P}(\mathbf{B})$ *is tight.*

(ii) For every $\eta > 0$, *there exists a compact set* $K \subset \mathbf{B}$ *such that, defining* K^η *is as in (15.58),*

$$\inf_{\mu\in\mathcal{Q}} \mu(K^\eta) \geq 1 - \eta. \tag{15.59}$$

(iii) $\mathcal{Q}$ *is relatively compact, i.e., its closure in* $(\mathcal{P}(\mathbf{M}, d_{p,\mathbf{M}})$ *is compact.* $\square$

We next provide well-known criteria that are equivalent to weak convergence.[34] Let ∂A denote the boundary of $A \subset \mathbf{B}$. A is said to be a P-continuity set if $A \in \mathcal{B}_\mathbf{B}$ and $P(\partial A) = 0$.

Theorem 15.24. *Let* $(\mathbf{B}, d_\mathbf{B})$ *be a separable metric space and let* $\{P_n, P\}$ *be a family of Borel probability measures on the* $(B, d_\mathbf{B})$. *The following conditions are equivalent:*

(a) $P_n \Longrightarrow P$

(b) $\limsup_{n\to\infty} P_n(F) \leq P(F)$ *for all closed sets* $F \subset \mathbf{B}$

(c) $\liminf_{n\to\infty} P_n(G) \geq P(G)$ *for all open sets* $G \subset \mathbf{B}$

(d) $\lim_{n\to\infty} P_n(A) = P(A)$ *for all* $P-$*continuity sets* $A \subset \mathbf{B}$

[34] Cf. Ethier and Kurtz (loc.cit. Theorem 3.1) and Dudley (1989), Chap. 11, Theorem 11.3.3.

(e) $\lim_{n\to\infty} d_{p,\mathbf{B}}(P_n, P) = 0$
(f) $\lim_{n\to\infty} \gamma_f(P_n - P) = 0$ □

Ethier and Kurtz (loc.cit.), Corollary 3.3, prove the following

Corollary 15.25. *Let* $(\mathbf{B}, d_{\mathbf{B}})$ *be a metric space and let* (X_n, Y_n), $n \in \mathbf{N}$, *and* X *be* $\mathbf{B} \times \mathbf{B}$- *and* $\mathbf{B}$-*valued random variables. If* $X_n \Longrightarrow X$ *and* $d_{\mathbf{B}}(X_n, Y_n) \longrightarrow 0$ *in probability, then* $Y_n \Longrightarrow X$. □

We have defined weak convergence, relative compactness, etc. for $\mathbf{B}$-valued random variables in terms of their probability distributions, where all the previous assertions hold provided $(\mathbf{B}, d_{\mathbf{B}})$ is separable and complete. We next extend some of those definitions and results to stochastic processes.

$(\Omega, \mathcal{F}, \mathcal{F}_t, P)$ is called a "*stochastic basis*" if the set Ω is equipped with a σ-algebra $\mathcal{F}$ and a probability measure P and if, in addition, there is an increasing family of σ-algebras $\mathcal{F}_t$, $t \leq 0$ such that

$$\mathcal{F}_s \subset \mathcal{F}_t \subset \mathcal{F}, \quad 0 \leq s \leq t < \infty.$$

$\mathcal{F}_t$, $t \geq 0$, is called a "*filtration*" and we will assume that it is right continuous, i.e., that

$$\cap_{s>0}\mathcal{F}_{t+s} = \mathcal{F}_t.$$

Henceforth we assume that $(\Omega, \mathcal{F}, \mathcal{F}_t, P), t \geq 0$, is a stochastic basis with a right continuous filtration $\mathcal{F}_t$ of σ-algebras and that all $\mathbf{B}$-valued stochastic processes $a(\cdot)$ to be jointly measurable in (t, ω) and adapted to the filtration $\mathcal{F}_t$, where the latter means that $a(t)$ is $\mathcal{F}_t$-measurable for all $t \geq 0$.

Two $\mathbf{B}$-valued stochastic processes $a_i(\cdot)$, $i = 1, 2$, are "*P-equivalent*" if

$$P\{\exists t : a_1(t) \neq a_2(t)\} = 0.$$

This statement is equivalent to the property that these two processes are indistinguishable as elements of a metric space as follows: Let

$$L_{0,\mathcal{F}}([0, T] \times \Omega; \mathbf{B})$$

be the space of $\mathbf{B}$-valued (adapted and jointly measurable) stochastic processes. We can endow $L_{0,\mathcal{F}}([0, T] \times \Omega; \mathbf{B}))$ with the metric

$$d_{(T,\Omega);\mathbf{B}}(a_1, a_2) := \int_0^T \int_\Omega d_{\mathbf{B}}\{(a_1(t, \omega), a_2(t, \omega)) \wedge 1\} P(\mathrm{d}\omega)\mathrm{d}t. \tag{15.60}$$

Then P-equivalence of $a_i(\cdot)$, $i = 1, 2$, just means that

$$d_{(T,\Omega);\mathbf{B}}(a_1, a_2) = 0 \quad \forall T > 0. \tag{15.61}$$

If $a_2(\cdot)$ is P-equivalent to $a_1(\cdot)$, $a_2(\cdot)$ is called a "*modification*" of $a_1(\cdot)$.

For many problems the space of jointly measurable adapted processes is unnecessarily large and, in what follows, we restrict ourselves to $(D([0,\infty);\mathbf{B}), d_{D,\mathbf{B}})$. By Theorem 15.19 $(D([0,\infty);\mathbf{B}), d_{D,\mathbf{B}})$ is a complete and separable metric space if $(\mathbf{B}, d_{\mathbf{B}})$ is complete and separable, and the previous results for random variables could be carried over directly to **B**-valued cadlag stochastic processes. However, for stochastic processes $X(\cdot)$ many properties are shown in terms of the "marginals" $X(t)$. Therefore, the next step is to derive some stochastic Ascoli-type characterization in terms of relative compactness of the marginals and the modulus of continuity (15.54).[35] Recall the notation from (15.58).

Theorem 15.26. *Stochastic Arzela–Ascoli Theorem*

Let $\{a_\alpha(\cdot)\}$ be a family of **B***-valued processes with sample paths in $D([0,\infty);\mathbf{B})$. Then $\{a_\alpha(\cdot)\}$ is relatively compact if, and only if, the following two conditions hold:*

(i) For every $\eta > 0$ and rational $t \ge 0$, there exists a compact set $K_{\eta,t} \subset \mathbf{M}$ such that

$$\inf_{\alpha} \left\{a_\alpha(t) \in K^{\eta}_{\eta,t}\right\} \ge 1-\eta. \tag{15.62}$$

(ii) For every $\eta > 0$ and rational $t \ge 0$, there exists a $\delta > 0$ such that

$$\sup_{\alpha} P\{w'(a_\alpha,\delta,T) \ge \eta\} \le \eta. \tag{15.63}$$

□

The criterion (ii) of the above theorem is difficult to apply. The following criterion, equivalent to condition (ii) of Theorem 15.26, has been obtained by Kurtz (1975).[36] It is especially useful when applying to families of square integrable martingales and if we may choose $\beta = 2$ as the exponent of the following (15.64).[37] We denote by $E[\cdot|\mathcal{F}_t]$ the conditional expectation with respect to $\mathcal{F}_t$.

Theorem 15.27. *Let $\{a_\alpha(\cdot)\}$ be a family of* **B***-valued processes with sample paths in $D([0,\infty);\mathbf{B})$. Then $\{a_\alpha(\cdot)\}$ is relatively compact if, and only if, condition (i) of Theorem 15.26 holds and if the following condition holds:*

For each $T > 0$, there exist $\beta > 0$ and a family $\{\gamma_\alpha(\delta) : 0 < \delta < 1\}$ of nonnegative random variables such that

$$E\Big[d_{\mathbf{M}}(a^{\beta}_{\alpha}(t+u), a_\alpha(t))|\mathcal{F}_t\Big] \le E\Big[\gamma_\alpha(\delta)|\mathcal{F}_t\Big] \tag{15.64}$$

for $0 \le t \le T,\ 0 \le u \le \delta$ in addition to

$$\lim_{\delta\to 0}\sup_{\alpha} E\Big[\gamma_\alpha(\delta)\Big] = 0. \tag{15.65}$$

□

[35] Cf. Ethier and Kurtz (loc.cit), Sect. 3.7, Theorem 7.2. Cf. also Billingsley (1968), Chap. 3.15.

[36] A proof is also found in Ethier and Kurtz (loc.cit.), Sect. 3.8, Theorem 8.6, and Remark 8.7.

[37] Cf. the next subsection for the definition of martingales.

15.2.2 Regular and Cylindrical Hilbert Space-Valued Brownian Motions

Regular and cylindrical Hilbert space-valued Brownian motions are defined in terms of their covariance operators. The $\mathbf{H}_0$*-valued Brownian motion, defined through correlated Brownian noise, is shown to be cylindrical. Imbedded into a weighted* L_2*-space it becomes a regular Brownian motion.*

Let $\mathbf{H}$ be separable Hilbert space with scalar product $< \cdot, \cdot >_{\mathbf{H}}$ and norm $\| \cdot \|_{\mathbf{H}}$. A system of independent vectors $\{\phi_n\} \subset \mathbf{H}$ is called a *complete orthonormal system (CONS)* in $\mathbf{H}$ if for every $f \in \mathbf{H}$:

$$f = \sum_{n \in \mathbf{N}} \langle f, \phi_n \rangle_{\mathbf{H}} \phi_n.$$

Further, a necessary and sufficient condition for completeness of $\{\phi_n\}$ is Parseval's identity:

$$\|f\|_{\mathbf{H}}^2 = \sum_{n \in \mathbf{N}} \langle f, \phi_n \rangle_{\mathbf{H}}^2 \quad \forall f \in \mathbf{H}.$$

A *bounded*, *nonnegative*, and *symmetric* linear operator Q on $\mathbf{H}$ is called "*nuclear*" if its nonnegative square root, $Q^{\frac{1}{2}}$, is "*Hilbert-Schmidt*", i.e., if

$$\sum_{n \in \mathbf{N}} \|Q^{\frac{1}{2}} \phi_n\|_{\mathbf{H}}^2 < \infty. \tag{15.66}$$

As is customary, we denote by $\mathcal{N}(a, b^2)$ the distribution of a real-valued Gaussian random variable with mean a and variance b^2.

Definition 15.28. *Brownian Motion*

(1) A real-valued stochastic process $\beta_1(\cdot)$ *is called a Brownian motion if*

(i) it has a.s. continuous sample paths
(ii) there is a $\sigma > 0$ *such that for* $0 \le s \le t$

$$\beta_1(t) - \beta_1(s) \sim \mathcal{N}(0, \sigma^2(t-s)) \text{ and } \beta_1(t) - \beta_1(s) \text{ is independent of } \sigma(\beta_u : 0 \le u \le s),$$

where $\sigma(\beta_u : 0 \le u \le s)$ *is the* σ*-algebra generated by all* β_u *with* $0 \le u \le s$*. The real-valued Brownian motion is called* "standard" *if* $\sigma = 1$.

(2) Let

(i) $\{\beta_{1,n}\}$ *be a sequence of i.i.d.* $\mathbf{R}$*-valued standard Brownian motions*
(ii) Q_W *a nonnegative and self-adjoint linear operator on* $\mathbf{H}$
(iii) $\{\phi_n\}$ *a complete orthonormal system (CONS) in* $Dom(Q_W) \subset \mathbf{H}$[38]

[38] $Dom(Q_W)$ denotes the domain of the operator Q_W.

The Gaussian **R**-*valued random field* $W(\cdot)$ *defined by*

$$\langle W(t), \varphi\rangle_{\mathbf{H}} := \sum_{n=1}^{\infty} \beta_{1,n} \left\langle \varphi, Q_W^{\frac{1}{2}} \phi_n \right\rangle_{\mathbf{H}}, \quad \varphi \in \mathbf{H}, \tag{15.67}$$

is called an **H**-*valued Brownian motion, where* $Q_W^{\frac{1}{2}}$ *is the nonnegative self-adjoint square root of* Q_W.[39]

The **H**-*valued Brownian motion is called* "regular" *if* $Q_W^{\frac{1}{2}}$ *is Hilbert-Schmidt. Otherwise, the* **H**-*valued Brownian motion is called* "cylindrical." *If* $Q_W^{\frac{1}{2}} = I_{\mathbf{H}}$, *the identity operator on* **H**, *the* **H**-*valued Brownian motion is called* "standard cylindrical". Q_W *is called the* covariance operator *of the* **H**-*valued Brownian motion.* □

In what follows, we restrict ourselves to bounded self-adjoint covariance operators, which covers most infinite dimensional Brownian motions used in the SPDE literature.[40] Observe that an **H**-valued cylindrical Brownian motion $W(\cdot)$, evaluated at fixed t, does not define a countably additive measure on **H**. This implies that $W(\cdot)$ must be treated as a generalized random field.[41] If a stochastic equation is driven by a cylindrical Brownian motion $W(\cdot)$, it is quite natural to ask whether the cylindrical Brownian motion can be imbedded into a larger (separable) Banach space $\mathbf{B} \supset \mathbf{H}$ such that in **B** $W(t)$ defines a countably additive measure. This is the approach taken by the theory of abstract Wiener spaces.[42] The calculations may become easier if, instead of a Banach space, we consider a separable Hilbert space $\tilde{\mathbf{H}}$ such that **H** is continuously and densely imbedded in $\tilde{\mathbf{H}}$. Consider, e.g., an **H**-valued cylindrical Brownian motion with bounded covariance operator Q_W. In this case, we need the imbedding $\mathbf{H} \subset \tilde{\mathbf{H}}$ to be "*Hilbert-Schmidt*," which means

$$\sum_{n=1}^{\infty} \|\phi_n\|_{\tilde{\mathbf{H}}}^2 < \infty,$$

where $\{\phi_n\}$ is a CONS in **H** and the norm in the right-hand side is the norm in $\tilde{\mathbf{H}}$.[43] Such a space $\tilde{\mathbf{H}}$ can be obtained by the completion of **H** with respect to the norm, generated by scalar product

$$\langle f, g\rangle_{\tilde{\mathbf{H}}} := \sum_{n=1}^{\infty} \frac{1}{n^2} \langle f, \phi_n\rangle_{\mathbf{H}} \langle g, \phi_n\rangle_{\mathbf{H}}.$$

The quadratic form on the left-hand side determines a symmetric positive definite Hilbert-Schmidt operator $Q^{\frac{1}{2}}$ from **H** into **H** by setting

---

[39] Recall that the nonnegative square root of a nonnegative self-adjoint operator is uniquely defined and self-adjoint (cf., e.g., Kato (1976), Chap. V.3.11, (3.45)).

[40] Cf. also the analysis of linear SPDEs in the space of distributions, reviewed in Chap. 13.

[41] Cf. Gel'fand and Vilenkin (1964).

[42] Cf. Gross (1965), Kuo (1975) and the references therein.

[43] Cf. Kuo (loc.cit.), Chap. 1, Sect. 4, Theorem 4.3.

$$Q^{\frac{1}{2}}\phi_n = \frac{1}{n}\phi_n \ \forall n \quad \text{or, equivalently,} \quad \left\langle Q^{\frac{1}{2}} f, Q^{\frac{1}{2}} g \right\rangle_{\mathbf{H}} := \langle f, g \rangle_{\tilde{\mathbf{H}}} \tag{15.68}$$

In $\tilde{\mathbf{H}}$ $W(t)$ defines a countably additive Gauss measure. If $W(\cdot)$ is the perturbation of some differential equation, the original differential equation must be redefined on the larger space (which may or may not be possible). An example is pointwise multiplication, which is usually well defined on function space, but not defined on the space of distributions.[44] To be more precise, consider an $\mathbf{H}_0$-valued standard cylindrical Brownian motion $W(\cdot)$ which has been undoubtedly the most popular one among all cylindrical Brownian motions. The reason for this popularity is that it may be represented as

$$\langle W(t), \varphi \rangle_0 = \int_0^t \int \varphi(r) w(\mathrm{d}r, \mathrm{d}s), \tag{15.69}$$

where $w(\mathrm{d}q, \mathrm{d}s)$ is a real-valued Gaussian space–time white noise.[45] Hence, it may serve to model space–time white noise perturbations.[46] By (15.36) the $\mathbf{H}_0$-valued standard cylindrical Brownian motion becomes an $\mathcal{H}_{-\gamma}$-valued regular Brownian motion if $\gamma > d$. It is well known that elements in $\mathcal{H}_{-\gamma}$ cannot be multiplied with each other.

There are other cylindrical Brownian motions of interest in applications. In fact, the perturbation by correlated Brownian motions used in this book give rise to cylindrical Brownian motions. Let us explain.

Consider the kernel the $\mathbf{R}^d$-valued kernel $G(r),\ r \in \mathbf{R}$ from the previous sections such that its one-dimensional components G_k are square integrable with respect to the Lebesgue measure. Define Gaussian $\mathbf{R}$-valued random fields $W_k(\cdot, \cdot)$ by

$$W_k(t, r) := \int_0^t \int G_k(r - q) w(\mathrm{d}q, \mathrm{d}s), \quad k = 1, \ldots, d. \tag{15.70}$$

Proposition 15.29. *The Gaussian random fields $W_k(\cdot, \cdot)$ from (15.70) define cylindrical $\mathbf{H}_0$-valued Brownian motions unless $G_k(r) = 0$ a.e. with respect to the Lebesgue measure.*

Proof.

(i) Employing the series representation (4.14) for the scalar field $w(\mathrm{d}q, \mathrm{d}s)$ yields

$$\langle W_k(t), \varphi \rangle_0 := \sum_{n=1}^{\infty} \int \varphi(r) \int G_k(r - q)\phi_n(q)\mathrm{d}q\, \mathrm{d}r \beta_{1,n}(t), \quad \varphi \in \mathbf{H}, \tag{15.71}$$

[44] Cf. Schwarz (1954). Cf. also our discussion of the bilinear SPDE in Chap. 13.

[45] Jetschke (1986) shows that the right-hand side of (15.69) defines a standard cylindrical Brownian motion. Using distributional calculus, Schaumlöffel (1986) shows also that every standard cylindrical Brownian motion can be represented by the right-hand side of (15.69).

[46] Cf. (4.29) and Definition 2.2. Cf. also our Chap. 13 and the SPDEs driven by cylindrical Brownian motions, which, with a few exceptions, require space dimension $d = 1$, unless the SPDE is linear.

Consequently, the square root of the covariance operator $Q_{W,k}$ is the following integral operator:

$$(Q_{W,k}^{\frac{1}{2}}\phi)(r) := \int G_k(r-q)\phi(q)\,\mathrm{d}q. \tag{15.72}$$

For $W_k(\cdot,\cdot)$ to be regular, $Q_{W,k}^{\frac{1}{2}}$ must be Hilbert-Schmidt. In particular it must be compact (such that the squares of the eigenvalues are summable).

(ii) Korotkov (1983), Chap. 2, Sect. 4, shows that such an integral operator is compact if, and only if, $G_k(r) = 0$ a.e. with respect to the Lebesgue measure dr. We adjust Korotkov's proof to our notation:

Suppose $Q_{W,k}^{\frac{1}{2}}$ is compact. Let $B \subset \mathbf{R}^d$ be a bounded Borel set and d be its diameter. Set $\tilde{r}^m := (md, 0, \ldots, 0)^T \in \mathbf{R}^d$ and define functions $f_m(q) := (U_{\tilde{r}^m} 1_B)(q) := 1_B(q + \tilde{r}^m)$, where $m \in \mathbf{N}$ (and $U_{\tilde{r}^m}$ is the shift operator). The functions f_m are bounded, integrable, and have disjoint supports. So $f_m \longrightarrow 0$ weakly, as $m \longrightarrow \infty$. Since we assumed $Q_{W,k}^{\frac{1}{2}}$ is compact, it follows that $|Q_{W,k}^{\frac{1}{2}} f_m|_0^2 \longrightarrow 0$, as $m \longrightarrow \infty$. Note that

$$\left|Q_{W,k}^{\frac{1}{2}} f_m\right|_0^2 = \int \left(\int G_k(r-q) 1_B(q+\tilde{r}^m)\mathrm{d}q\right)^2 \mathrm{d}r$$
$$= \int \left(\int G_k(r-q) 1_B(q)\mathrm{d}q\right)^2 \mathrm{d}r = \left|Q_{W,k}^{\frac{1}{2}} 1_B\right|_0^2$$

(by the homogeneity of the kernel G_k and the shift invariance of the Lebesgue measure.)

It follows that $|Q_{W,k}^{\frac{1}{2}} 1_B|_0^2 = 0$ for all bounded Borel sets B, whence $G_k(r) = 0$ a.e. □

Although the Brownian motions, defined through (15.70), are cylindrical, they are obviously not standard cylindrical. We now show that $W_k(\cdot)$ from (15.70) become regular Brownian motions if imbedded into a (larger) suitably weighted L_2-space such that the constants are integrable with respect to the weighted measure. A popular choice for the weight function was defined in (4.4):

$$\varpi(r) = (1 + |r|^2)^{-\gamma},$$

where $\gamma > \frac{d}{2}$.[47] We obtain that $\int \varpi(r)\mathrm{d}r < \infty$, i.e., the constants are integrable with respect to the measure $\varpi(r)(r)\mathrm{d}r$. Define the separable Hilbert space

$$\left.\begin{aligned} (\mathbf{H}_{0,\varpi}, \langle\cdot,\cdot\rangle_{0,\varpi}) &:= (L_2(\mathbf{R}^d, \varpi(r)\mathrm{d}r), \langle\cdot,\cdot\rangle_{0,\varpi}), \\ \langle f, g\rangle_{0,\varpi} &:= \int f(r)g(r)\varpi(r)\mathrm{d}r. \end{aligned}\right\} \tag{15.73}$$

[47] Cf., e.g., Triebel (1978).

Clearly, the norm $\|\cdot\|_{0,\varpi}$ is weaker than the norm $|\cdot|_0$ and we have

$$\mathbf{H}_0 \subset \mathbf{H}_{0,\varpi} \tag{15.74}$$

with dense continuous imbedding. It is therefore natural to consider the cylindrical $\mathbf{H}_0$-valued Brownian motions from (15.70) as $\mathbf{H}_{0,\varpi}$-valued.

Proposition 15.30. *The Brownian motions from (15.70) define regular* $\mathbf{H}_{0,\varpi}$*-valued Brownian motions.*

Proof. Let $\phi, \psi \in \mathbf{H}_{0,\varpi}$. The covariance operator $Q_{W,k,\varpi}$ for $W_k(t)$ in $\mathbf{H}_{0,\varpi}$ is given by

$$\left.\begin{aligned} &\frac{\mathrm{d}}{\mathrm{d}t} E \int\int_0^t \int G_k(r-q)w(\mathrm{d}q,\mathrm{d}s)\phi(r)\varpi(r)\mathrm{d}r \\ &\quad\times \int\int_0^t \int G_k(r-q)w(\mathrm{d}q,\mathrm{d}s)\psi(r)\varpi(r)\mathrm{d}r \\ &= \int\int\int G_k(r-q)\phi(r)\varpi(r)G_k(\tilde{r}-q)\psi(\tilde{r})\varpi(\tilde{r})\mathrm{d}\tilde{r}\,\mathrm{d}r\,\mathrm{d}q \\ &\qquad =: \langle Q_{W,k,\varpi}\phi, \psi\rangle_{0,\varpi}. \end{aligned}\right\} \tag{15.75}$$

We show that the square root of $Q_{W,k,\varpi}$ is Hilbert-Schmidt. Let $\{\phi_{n,\varpi}\}$ be a CONS in $\mathbf{H}_{0,\varpi}$. By the dominated convergence and Fubini's theorems, in addition to the shift invariance of the Lebesgue measure,

$$\left.\begin{aligned} &\sum_{n=1}^{\infty}\left\langle Q^{\frac{1}{2}}_{W,k,\varpi}\phi_{n,\varpi}, Q^{\frac{1}{2}}_{W,k,\varpi}\phi_{n,\varpi}\right\rangle_{0,\varpi} \\ &= \sum_{n=1}^{\infty}\langle Q_{W,k,\varpi}\phi_{n,\varpi}, \phi_{n,\varpi}\rangle_{0,\varpi} \\ &= \sum_{n=1}^{\infty}\int\int\int G_k(r-q)\phi_{n,\varpi}(r)\varpi(r)G_k(\tilde{r}-q)\phi_{n,\varpi}(\tilde{r})\varpi(\tilde{r})\mathrm{d}\tilde{r}\,\mathrm{d}r\,\mathrm{d}q \\ &= \int\left[\sum_{n=1}^{\infty}\int\int G_k(r-q)\phi_{n,\varpi}(r)\varpi(r)G_k(\tilde{r}-q)\phi_{n,\varpi}(\tilde{r})\varpi(\tilde{r})\mathrm{d}\tilde{r}\,\mathrm{d}r\right]\mathrm{d}q \\ &= \int\left[\int G_k^2(r-q)\varpi(r)\mathrm{d}r\right]\mathrm{d}q \\ &= \int\left[\int G_k^2(r-q)\mathrm{d}q\right]\varpi(r)\mathrm{d}r\Big] \\ &= |G_k|_0^2\int\varpi(r)\mathrm{d}r < \infty. \end{aligned}\right\} \tag{15.76}$$

□

Since the larger space $\mathbf{H}_{0,\varpi}$ is itself a space of functions, a number of operations, like pointwise multiplication, originally defined on $\mathbf{H}_0$, can be extended to operations on $\mathbf{H}_{0,\varpi}$.

The procedure leading to (15.68) shows that imbeddings from one Hilbert space $\mathbf{H}$ into a possibly larger Hilbert space $\tilde{\mathbf{H}}$ can always be defined through some positive definite symmetric bounded operator on $Q^{\frac{1}{2}}$ from $\mathbf{H}$ into $\mathbf{H}$. The imbedding $\mathbf{H}_0 \subset \mathbf{H}_{0,\varpi}$ from (15.74) defines the multiplication operator $Q^{\frac{1}{2}}$

$$(Q^{\frac{1}{2}} f)(r) := \sqrt{\varpi(r)} f(r) \quad r \in \mathbf{R}^d, \, f \in \mathbf{H}_0. \tag{15.77}$$

We have

$$\partial_k \varpi(r) = -\gamma (1 + |r|^2)^{-\gamma - 1} 2 r_k.$$

So $\nabla \varpi(r) \neq 0$ Lebesgue-a.e. This implies that the spectrum of the multiplication operator $Q^{\frac{1}{2}}$ from (15.77) is (absolutely) continuous (cf. Kato (1976), Chap. 10, Sect. 1, Example 1.9). Therefore, it cannot be compact and, a fortiori, $Q^{\frac{1}{2}}$ cannot be a Hilbert-Schmidt operator $Q^{\frac{1}{2}}$ from $\mathbf{H}$ into $\mathbf{H}$. We conclude that the imbedding (15.77) is not a Hilbert-Schmidt imbedding. The same result holds for any (smooth) and integrable weight function $\lambda(r)$ as long as the $\nabla \lambda(r) \neq 0$ Lebesgue-a.e.

15.2.3 Martingales, Quadratic Variation, and Inequalities

We collect some definitions and properties related to martingales, semi-martingales, and quadratic variation, which are standard in stochastic analysis but may cause problems to the reader from a different background. We also state Levy's characterization of a Brownian motion, the martingale central limit theorem as well as three maximal inequalities. The first two inequalities are Doob's and the Burkholder–Davis–Gundy inequalities. The last is a maximal inequality for Hilbert-valued stochastic convolution integrals.

In this section $(\mathbf{H}, \| \cdot \|_{\mathbf{H}})$ is a separable real Hilbert space[48] with scalar product $< \cdot, \cdot >_{\mathbf{H}}$ and norm $\| \cdot \|_{\mathbf{H}}$.

Definition 15.31. *Martingale*

An $\mathbf{H}$*-valued integrable stochastic process*[49] $m(\cdot)$ *is called an "*$\mathbf{H}$*-valued mean* 0 *martingale" with respect to the filtration* $\mathcal{F}_t$*, if for all* $0 \le t < \infty$

$$E(m(t)|\mathcal{F}_s) = m(s) \ \text{a.s. and} \ m(0) = 0 \ \text{a.s.} \qquad \square$$

An $\mathbf{H}$-valued martingale $m(\cdot)$ always has a cadlag modification, i.e., there is an $\mathbf{H}$-valued martingale $\bar{m}(\cdot)$ that is P-equivalent to $m(\cdot)$. It is customary to work with the "smoothest" modification of a stochastic process, which, in case of an arbitrary $\mathbf{H}$-valued martingale, is the cadlag modification.[50]

---

[48] $\mathbf{H}$ can also be finite dimensional in this setting. In particular, it can be $\mathbf{R}^d$, $d \ge 1$, endowed with the usual scalar product or for $d = 1$ with the usual distance.

[49] Integrability here means that $E\|m(t)\|_{\mathbf{H}} < \infty \ \forall t \ge 0$.

[50] Cf. Metivier and Pellaumail (1980), Sects. 1.16 and 1.17. We remind the reader of the analogy to Sobolev spaces over $\mathbf{R}^d$, $\mathbf{H}_m$, defined earlier. If $m > \frac{d}{2}$, the equivalence classes of $\mathbf{H}_m$ always contain one continuous element. We may call this a continuous modification and assume that all elements of $\mathbf{H}_m$ are continuous.

Henceforth, we will always assume that our martingales are cadlag. Since for many applications (such as integration) it suffices to consider $m(\cdot) - m(0)$, we may also, without loss of generality, assume that our martingales are mean 0.

Within the class of (cadlag) martingales, a special role is assigned to those martingales that have a continuous modification (i.e., where almost all sample paths are continuous). These martingales are called "*continuous martingales*." If, for all t $m(t)$ is square integrable, $m(\cdot)$ is called an "**H**-*valued square integrable martingale*." If there is a sequence of localizing stopping times $\tau_n \longrightarrow \infty$ with probability 1 such that $m(\cdot \wedge \tau_n)$ is a square integrable martingale, $m(\cdot)$ is called a "*locally square integrable martingale*."

Further, we recall the following terminology: $b(\cdot)$ is an "**H**-*valued process of bounded variation*," if with probability 1 the sample paths $b(\cdot, \omega)$ have bounded variation on any finite interval $[0, T]$. If $b(\cdot)$ has a cadlag (continuous) modification, then $b(\cdot)$ is called an "**H**-*valued cadlag (continuous) process of bounded variation*." The sum of an **H**-valued process of bounded variation and an **H**-valued martingale is called an "**H**-*valued semi-martingale*." It is obvious how to extend the local square integrability (or, more generally, p-integrability for $p \geq 1$) to semimartingales.

Probably the most important inequalities in martingale theory and stochastic Itô integration are the following.

Theorem 15.32. *Submartingale and Doob Inequalities*

*Let $m(\cdot)$ be an **H**-valued martingale.*[51] *Then for any stopping time τ*

$$\left.\begin{array}{lll} (i) & P\{\sup\limits_{0\leq t\leq \tau} \|m(t)\|_{\mathbf{H}} > L\} \leq \frac{1}{L} E\|m(\tau)\|_{\mathbf{H}} & (\textit{submartingale inequality}); \\ (ii) & E \sup\limits_{0\leq t\leq \tau} \|m(t)\|_{\mathbf{H}}^2 \leq 4E\|m(\tau)\|_{\mathbf{H}}^2 & (\textit{Doob's inequality}). \end{array}\right\} \tag{15.78}$$

The **proof** *is found in Metivier and Pellaumail (loc.cit.), Sects. 4.8.2 and 4.10.4.* □

Next we define processes of finite quadratic variation (cf. Metivier and Pellaumail (loc. cit.), Chap. 2.3).

Definition 15.33. *Let $\{t_0^n < t_1^n < \cdots < t_k^n < \cdots\}$ be a sequence of partitions of $[0, \infty)$ such that for every $T > 0$ $\max_{\{1\leq k\leq T\}}(t_k^n - t_{k-1}^n) \longrightarrow 0$, as $n \longrightarrow \infty$ and for all n $t_k^n \longrightarrow \infty$, as $k \longrightarrow \infty$. An **H**-valued process $a(\cdot)$ is said to be of* "finite quadratic variation" *if there exists a monotone increasing real-valued process $[a(\cdot)]$ such that for every $t > 0$*

$$\sum_{k\geq 0} \|a(t_k^n \wedge t) - a(t_{k-1}^n \wedge t)\|_{\mathbf{H}}^2 \longrightarrow [a(t)] \text{ in probability, as } n \longrightarrow \infty . \tag{15.79}$$

□

[51] A real-valued cadlag process, $x(\cdot)$, is called a "*submartingale*" if $E(x(t)|\mathcal{F}_s) \geq x(s)$ a.s. for $0 \leq s \leq t$. The sum of submartingales is a submartingale. If $m(\cdot)$ is an **H**-valued martingale then $\|m(\cdot)\|_{\mathbf{H}}$ is a submartingale.

In what follows, let $m(\cdot)$ be an $\mathbf{H}$-valued square integrable martingale. Our goal is to show that $m(\cdot)$ has finite quadratic variation. Recall the space $L_{0,\mathcal{F}}(D([0,T];\mathbf{H}))$, introduced before Definition 4.1. We endow $L_{0,\mathcal{F}}(D([0,T];\mathbf{H}))$ with the metric of convergence in probability

$$d_{\text{prob},D,\mathbf{H}}(f,g) := \inf_{\varepsilon>0} P\{d_{D,\mathbf{H}}(f,g) \geq \varepsilon\} \leq \varepsilon. \tag{15.80}$$

Proposition 15.34. *Suppose $m(\cdot)$ is an $\mathbf{H}$-valued square integrable martingale. Then there is a unique quadratic variation $[m](\cdot)$ such that (15.79) holds with $a(\cdot) := m(\cdot)$. If $m(\cdot)$ is continuous, then its quadratic variation $[m](\cdot)$ is also continuous.*

Proof.

(i) To avoid cumbersome notation describing the jumps in cadlag processes, we provide the proof only for continuous martingales and refer the reader for the more general case to Metivier and Pellaumail, loc.cit.

(ii) Let $\{t_0^n < t_1^n < \cdots < t_k^n < \cdots\}$ be a sequence of partitions of $[0,\infty)$ as in Definition 15.33. We easily verify

$$\begin{aligned}&\|m(t_k^n \wedge t)\|_{\mathbf{H}}^2 - \|m(t_{k-1}^n \wedge t)\|_{\mathbf{H}}^2\\&\quad= \|m(t_k^n \wedge t) - m(t_{k-1}^n \wedge t)\|_{\mathbf{H}}^2 + 2\langle m(t_{k-1}^n \wedge t), m(t_k^n \wedge t) - m(t_{k-1}^n \wedge t)\rangle_{\mathbf{H}}.\end{aligned}$$

Therefore,

$$\left.\begin{aligned}\|m(t)\|_{\mathbf{H}}^2 &= \sum_k \|m(t_k^n \wedge t)\|_{\mathbf{H}}^2 - \|m(t_{k-1}^n \wedge t)\|_{\mathbf{H}}^2\\&= \sum_k \|m(t_k^n \wedge t) - m(t_{k-1}^n \wedge t)\|_{\mathbf{H}}^2 + 2\Big\langle m(t_{k-1}^n \wedge t), m(t_k^n \wedge t) - m(t_{k-1}^n \wedge t)\Big\rangle_{\mathbf{H}}\\&=: S_{n,1}(t) + 2S_{n,2}(t).\end{aligned}\right\} \tag{15.81}$$

Hence, for fixed t, $S_{n,1}(t)$ will converge, if $S_{n,2}(t)$ does and, of course, vice versa. However, $S_{n,2}(t)$ turns out to be a continuous martingale, whence we may use orthogonality of increments, working with squares of the sequence. Therefore, we show the convergence of $S_{n,2}(t)$.

(iii) Observe that for $0 \leq s \leq t$

$$E(\langle m(s), m(t)\rangle_{\mathbf{H}} | \mathcal{F}_s) = \|m(s)\|_{\mathbf{H}}^2. \text{ a.s.} \tag{15.82}$$

Indeed, with probability 1

$$E(\langle m(s), m(t)\rangle_{\mathbf{H}} | \mathcal{F}_s) = E(\langle f, m(t)\rangle_{\mathbf{H}} | \mathcal{F}_s)_{|\{m(s)=f\}}$$

(cf. Gikhman and Skorokhod (1971), Chap. 1.3)

$$= \langle f, E(m(t)|\mathcal{F}_s)\rangle_{\mathbf{H}} = \langle f, m(s)\rangle \mathbf{H}_{|\{m(s)=f\}} = \|m(s)\|_{\mathbf{H}}^2,$$

where the interchangeability of the scalar product and the conditional expectation follows from the linearity and continuity of the scalar product.

(iv) Let $s < t$. Then

$$
\begin{aligned}
&E(S_{n,2}(t)|\mathcal{F}_s)\\
&= \sum_k E(\langle m(t_{k-1}^n \wedge t), m(t_k^n \wedge t) - m(t_{k-1}^n \wedge t)\rangle_{\mathbf{H}}|\mathcal{F}_s)\\
&= \sum_k 1_{\{t_k^n \wedge t \le s\}} E(\langle m(t_{k-1}^n \wedge t), m(t_k^n \wedge t) - m(t_{k-1}^n \wedge t)\rangle_{\mathbf{H}}|\mathcal{F}_s)\\
&\quad + \sum_{k\ge 0} 1_{\{t_k^n \wedge t > s\}} E(\langle m(t_{k-1}^n \wedge t), m(t_k^n \wedge t) - m(t_{k-1}^n \wedge t)\rangle_{\mathbf{H}}|\mathcal{F}_s)\\
&=: E(S_{n,2,1}(t)|\mathcal{F}_s) + E(S_{n,2,2}(t)|\mathcal{F}_s).
\end{aligned}
$$

Obviously,

$$S_{n,2,1}(t) \equiv S_{n,2}(s) \quad \text{a.s.},$$

which is $\mathcal{F}_s$-measurable. Using standard properties of the conditional expectation in addition to (15.82), we obtain with probability 1

$$
\begin{aligned}
&E(S_{n,2,2}(t)|\mathcal{F}_s)\\
&= \sum_{k\ge 0} 1_{\{t_k^n \wedge t > s\}} E(\langle m(t_{k-1}^n \wedge t) E\{(m(t_k^n \wedge t) - m(t_{k-1}^n \wedge t))\mathcal{F}_{s\vee(t_{k-1}^n \wedge t)}\}\rangle_{\mathbf{H}}|\mathcal{F}_s)\\
&= \sum_{k\ge 0} 1_{\{t_k^n \wedge t > s\}} E(\langle m(t_{k-1}^n \wedge t), m(s \vee (t_{k-1}^n \wedge t)) - m(s \vee (t_{k-1}^n \wedge t))\rangle_{\mathbf{H}}|\mathcal{F}_s)\\
&= 0.
\end{aligned}
$$

Combining the previous calculations, we have shown that for $0 \le s \le t < \infty$

$$E(S_{n,2}(t)|\mathcal{F}_s) = S_{n,2}(s) \quad \text{a.s.} \tag{15.83}$$

As all quantities involved in the definition of $S_{n,2}(\cdot)$ are continuous, $S_{n,2}(\cdot)$ is continuous and, obviously, square integrable and $S_{n,2}(0) = 0$ by construction. Thus, we have shown that $S_{n,2}(\cdot)$ is a real-valued continuous square integrable mean zero martingale.

(v) We must show that $S_{n,2}(\cdot)$ is a Cauchy sequence in $L_{0,\mathcal{F}}(C([0,T];\mathbf{H}))$ $\forall T>0$ (cf. before Definition 4.1), where the metric on $L_{0,\mathcal{F}}(C([0,T];\mathbf{H}))$ is now the metric of convergence in probability

$$d_{\text{prob},u,T,\mathbf{H}}(f,g) := \inf_{\varepsilon>0} P\left\{\sup_{0\le t\le T} \|f(t) - g(t)\|_{\mathbf{H}} \ge \varepsilon\right\} \le \varepsilon. \tag{15.84}$$

Analyzing the distance between $S_{n,2}(\cdot \wedge \tau_{n,\delta,N})$ and $S_{m,2}(\cdot \wedge \tau_{n,\delta,N})$ for two partitions $\{t_0^n < t_1^n < \cdots < t_k^n < \cdots\}$ and $\{t_0^m < t_1^m < \cdots < t_k^m < \cdots\}$, respectively, we may, without loss of generality, assume that[52]

$$\{t_0^n < t_1^n < \cdots < t_k^n < \cdots\} \supset \{t_0^m < t_1^m < \cdots < t_k^m < \cdots\}.$$

[52] Otherwise we could compare both $S_{n,2}(\cdot)$ and $S_{m,2}(\cdot)$ with the corresponding sum, based on the union of the partitions, $\{t_0^n < t_1^n < \cdots < t_k^n < \cdots\} \cup \{t_0^m < t_1^m < \cdots < t_k^m < \cdots\}$.

We then rewrite $S_m(\cdot)$ by adding the missing partition points from $\{t_0^n < t_1^n < \cdots < t_k^n < \cdots\}$ and setting

$$\underline{t}_k^n := \max_\ell\{t_\ell^m : t_\ell^m \le t_k^n\}, \quad \forall k.$$

As in the derivation of (15.83), we then see that

$$\begin{aligned}&S_{n,2}(t) - S_{m,2}(t)\\&= \sum_k \langle m(t_{k-1}^n \wedge t) - m(\underline{t}_{k-1}^n \wedge t), m(t_k^n \wedge t) - m(t_{k-1}^n \wedge t)\rangle_{\mathbf{H}}\end{aligned}$$

is itself a real-valued continuous square integrable martingale. For $\delta > 0$ define the following stopping times

$$\tau_{n,\delta} := \inf\left\{0 \le s \le 2T : \sup_k \|m(t_{k-1}^n \wedge s) - m(\underline{t}_{k-1}^n \wedge s)\|_{\mathbf{H}}^2 \ge \frac{\delta}{E\|m(T)\|_{\mathbf{H}}^2 + 1}\right\}$$

and for $N > 0$

$$\tau_{n,N} := \inf\{0 \le s \le 2T : \sum_k \|m(t_k^n \wedge s) - m(t_{k-1}^n \wedge s)\|_{\mathbf{H}}^2 \ge N\}.$$

Setting

$$\tau_{n,\delta,N} := \tau_{n,\delta} \wedge \tau_{n,N},$$

the continuity of $m(\cdot)$ implies,

$$\tau_{n,\delta,N} \longrightarrow 2T, \quad \text{a.s., as } n \longrightarrow \infty\,. \tag{15.85}$$

Further, also $S_{n,2}(t \wedge \tau_{n,\delta,N}) - S_{m,2}(t \wedge \tau_{n,\delta,N})$ is a real-valued continuous square integrable martingale. By (15.85) it suffices to show that $|S_{n,2}(\cdot \wedge \tau_{n,\delta,N}) - S_{m,2}(\cdot \wedge \tau_{n,\delta,N})|$ tends to 0 uniformly on compact intervals in the metric of convergence in probability. Employing Doob's inequality (Theorem 15.32), we obtain that for arbitrary $T > 0$

$$\begin{aligned}E_{0\le t\le T}(S_{n,2}(t \wedge \tau_{n,\delta,N}) - S_{m,2}(t \wedge \tau_{n,\delta,N}))^2 &\le 4E(S_{n,2}(T \wedge \tau_{n,\delta,N})\\&\quad -S_{m,2}(T \wedge \tau_{n,\delta,N}))^2\end{aligned} \tag{15.86}$$

Note that, as in (15.82), for $k < \ell$:

$$\begin{aligned}&E\langle m(t_{k-1}^n \wedge t \wedge \tau_{n,\delta,N}) - m(\underline{t}_{k-1}^n \wedge t \wedge \tau_{n,\delta,N}), m(t_k^n \wedge t \wedge \tau_{n,\delta,N})\\&\quad -m(t_{k-1}^n \wedge t \wedge \tau_{n,\delta,N})\rangle_{\mathbf{H}}\\&\quad \times\langle m(t_{\ell-1}^n \wedge t \wedge \tau_{n,\delta,N}) - m(\underline{t}_{\ell-1}^n \wedge t \wedge \tau_{n,\delta,N}), m(t_\ell^n \wedge t \wedge \tau_{n,\delta,N})\\&\quad -m(t_{\ell-1}^n \wedge t \wedge \tau_{n,\delta,N})\rangle_{\mathbf{H}}\\&= E\langle m(t_{k-1}^n \wedge t \wedge \tau_{n,\delta,N}) - m(\underline{t}_{k-1}^n \wedge t \wedge \tau_{n,\delta,N}), m(t_k^n \wedge t \wedge \tau_{n,\delta,N})\\&\quad -m(t_{k-1}^n \wedge t \wedge \tau_{n,\delta,N})\rangle_{\mathbf{H}} \times \langle m(t_{\ell-1}^n \wedge t \wedge \tau_{n,\delta,N}) - m(\underline{t}_{\ell-1}^n \wedge t \wedge \tau_{n,\delta,N}),\\&\quad \times E(m(t_\ell^n \wedge t \wedge \tau_{n,\delta,N}) - m(t_{\ell-1}^n \wedge t \wedge \tau_{n,\delta,N})|\mathcal{F}_{t_{\ell-1}^n \wedge t \wedge \tau_{n,\delta,N}})\rangle_{\mathbf{H}} = 0\end{aligned}$$

a.s., since $E(m(t_\ell^n \wedge t \wedge \tau_{n,\delta,N}) - m(t_{\ell-1}^n \wedge t \wedge \tau_{n,\delta,N})|\mathcal{F}_{t_{\ell-1}^n \wedge t \wedge \tau_{n,\delta,N}}) = 0$

Hence,

$$
\begin{aligned}
&E(S_{n,2}(T \wedge \tau_{n,\delta,N}) - S_{m,2}(T \wedge \tau_{n,\delta,N}))^2 \\
&= E\Bigg\{ \sum_k \langle m(t^n_{k-1} \wedge T \wedge \tau_{n,\delta,N}) - m(\underline{t}^n_{k-1} \wedge T \wedge \tau_{n,\delta,N}), m(t^n_k \wedge T \wedge \tau_{n,\delta,N}) \\
&\qquad -m(t^n_{k-1} \wedge T \wedge \tau_{n,\delta,N})\rangle_{\mathbf{H}} \Bigg\}^2 \\
&= \sum_k E\langle m(t^n_{k-1} \wedge T \wedge \tau_{n,\delta,N}) - m(\underline{t}^n_{k-1} \wedge T \wedge \tau_{n,\delta,N}), m(t^n_k \wedge T \wedge \tau_{n,\delta,N}) \\
&\quad -m(t^n_{k-1} \wedge T \wedge \tau_{n,\delta,N})\rangle^2_{\mathbf{H}} + \sum_{k\neq\ell} E\langle m(t^n_{k-1} \wedge T \wedge \tau_{n,\delta,N}) \\
&\quad -m(\underline{t}^n_{k-1} \wedge T \wedge \tau_{n,\delta,N}), m(t^n_k \wedge T \wedge \tau_{n,\delta,N}) - m(t^n_{k-1} \wedge T \wedge \tau_{n,\delta,N})\rangle_{\mathbf{H}} \\
&\quad \times \langle m(t^n_{\ell-1} \wedge T \wedge \tau_{n,\delta,N}) - m(\underline{t}^n_{\ell-1} \wedge T \wedge \tau_{n,\delta,N}), m(t^n_\ell \wedge T \wedge \tau_{n,\delta,N}) \\
&\quad -m(t^n_{\ell-1} \wedge T \wedge \tau_{n,\delta,N})\rangle_{\mathbf{H}} \\
&= \sum_k E\langle m(t^n_{k-1} \wedge T \wedge \tau_{n,\delta,N}) - m(\underline{t}^n_{k-1} \wedge T \wedge \tau_{n,\delta,N}), m(t^n_k \wedge T \wedge \tau_{n,\delta,N}) \\
&\quad -m(t^n_{k-1} \wedge T \wedge \tau_{n,\delta,N})\rangle^2_{\mathbf{H}}
\end{aligned}
$$

(since the mixed terms integrate to 0, as shown in the previous formula)

$$
\begin{aligned}
&\le \sum_k E\|m(t^n_{k-1} \wedge T \wedge \tau_{n,\delta,N}) - m(\underline{t}^n_{k-1} \wedge T \wedge \tau_{n,\delta,N})\|^2_{\mathbf{H}} \|m(t^n_k \wedge T \wedge \tau_{n,\delta,N}) \\
&\quad -m(t^n_{k-1} \wedge T \wedge \tau_{n,\delta,N})\|^2_{\mathbf{H}}
\end{aligned}
$$

(by the Cauchy-Schwarz inequality)

$$
\le \sum_k E\|m(t^n_k \wedge T \wedge \tau_{n,\delta,N}) - m(t^n_{k-1} \wedge T \wedge \tau_{n,\delta,N})\|^2_{\mathbf{H}} \frac{\delta}{E\|m(T)\|^2_{\mathbf{H}} + 1}
$$

by the definition of $\tau_{n,\delta,N}$.

By (15.81) and (15.82)

$$
E\|m(t^n_k \wedge T \wedge \tau_{n,\delta,N}) - m(t^n_{k-1} \wedge T \wedge \tau_{n,\delta,N})\|^2_{\mathbf{H}} = E\|m(T \wedge \tau_{n,\delta,N})\|^2_{\mathbf{H}} \le E\|m(T)\|^2_{\mathbf{H}}.
$$

Altogether,

$$
E(S_{n,2}(T \wedge \tau_{n,\delta,N}) - S_{m,2}(T \wedge \tau_{n,\delta,N}))^2 \le \delta. \tag{15.87}
$$

Employing the Chebyshev inequality in addition to (15.86), we have shown that $S_{n,2}(\cdot)$ is a Cauchy sequence in the metric (15.84), and it follows that the limit, $S_2(\cdot)$, is also a continuous real-valued square integrable martingale. We conclude that $S_{n,1}(\cdot)$ converges uniformly in $L_{0,\mathcal{F}}(C([0,T];\mathbf{R})$ for all $T > 0$ as well. □

Returning to the more general case of cadlag martingales we denote the limit of $S_{n,2}(\cdot)$ by

$$S_2(t) =: \int_0^t \langle m(s-), m(\mathrm{d}s)\rangle_{\mathbf{H}}, \qquad (15.88)$$

where $m(s-)$ is the limit from the left of $m(s)$ (cf. Metivier and Pellaumail (loc. cit.)). The integral on the right-hand side of (15.88) is called the "*stochastic Itô integral*" of the linear operator $\langle m(s-), \cdot\rangle_{\mathbf{H}}$ with respect to $m(ds)$ (cf. also the following Sect. 15.2.5 for more details). Summarizing, it follows that

$$\|m\|_{\mathbf{H}}^2(t) \equiv 2\int_0^t \langle m(s-), m(\mathrm{d}s)\rangle_{\mathbf{H}} + [m](t). \qquad (15.89)$$

Remark 15.35. We emphasize that, in general, the quadratic variation is not trivial, i.e., it is not identically 0. (Cf. the following example.) However, we easily see that for a continuous process of bounded variation $b(\cdot)$ its quadratic variation must be identically 0 a.s. On the other hand, comparing the definitions of the variation of a function and the quadratic variation of processes, it is an easy exercise to verify the following statement: If the square integrable martingale $m(\cdot)$ is continuous and $[m](\cdot,\omega)$ is nontrivial, then $m(\cdot,\omega)$ is of unbounded variation a.s. □

Example 15.36. Let $\beta(\cdot)$ be an $\mathbf{R}^d$-valued Brownian motion with covariance matrix $\sigma^2 I_d$ (cf. Definition 15.28, applied to the finite dimensional Hilbert space $\mathbf{R}^d$), i.e., for $0 \le s \le t$ $\beta(t) - \beta(s) \sim \mathcal{N}(0, \sigma^2(t-s)I_d)$. Further, the increments $\beta(t) - \beta(s)$ are independent of $\beta(u)$ for $0 \le u \le s \le t$. Therefore, $\beta(\cdot)$ is an $\mathbf{R}^d$-valued continuous martingale. For fixed t we can compute its quadratic variation directly, employing the strong law of large numbers (LLN) for i.i.d. random variables. With the notation of Definition 15.33 the law of large numbers (LLN) implies

$$\left|\sum_{k\ge 0}\left|\beta(t_k^n \wedge t) - \beta(t_{k-1}^n \wedge t)\right|^2 - t\mathrm{d}\sigma^2\right| \longrightarrow 0, \text{ as } n \longrightarrow \infty, \qquad (15.90)$$

whence,

$$[\beta](t,\omega) \equiv t\mathrm{d}\sigma^2 \quad \text{a.s.} \qquad (15.91)$$

Observe that, by (15.79), the quadratic variation defines a bilinear form as follows: Let $a_1(\cdot)$ and $a_2(\cdot)$ be two $\mathbf{H}$-valued processes with finite quadratic variations $[a_1](\cdot)$ and $[a_2](\cdot)$, respectively. We easily verify that the existence of $[a_1](\cdot)$ and $[a_2](\cdot)$ implies the existence of both $[a_1 + a_2](\cdot)$ and $[a_1 - a_2](\cdot)$. Consequently, the following bilinear form is well defined:

$$[a_1, a_2](\cdot) := \tfrac{1}{4}[a_1 + a_2](\cdot) - \tfrac{1}{4}[a_1 - a_2](\cdot). \qquad (15.92)$$

The left-hand side of (15.92) is called the "*mutual quadratic variation of* $a_1(\cdot)$ *and* $a_2(\cdot)$. If $b(\cdot)$ is an $\mathbf{H}$-valued process of bounded variation and $m(\cdot)$ is an $\mathbf{H}$-valued square integrable martingale

$$[b, m](\cdot) \equiv 0 \text{ a.s.} \qquad (15.93)$$

Expression (15.93) follows in the continuous case relatively easily from the preceding calculations. We refer to Metivier and Pellaumail (loc.cit.), Sect. 2.4.2, for the general case.

We call two **H**-valued square integrable martingales, $m_1(\cdot)$ and $m_2(\cdot)$ "*uncorrelated*" if

$$E\langle m_1(t), m_2(t)\rangle_{\mathbf{H}} = 0 \; \forall t \geq 0. \tag{15.94}$$

For uncorrelated **H**-valued square integrable martingales $m_i(\cdot)$ $i = 1, 2$,

$$[m_1, m_2](\cdot) \equiv 0 \text{ a.s.} \tag{15.95}$$

It follows immediately that (15.95) holds if $m_i(\cdot)$ are independent. If $m_i(\cdot)$ are two **H**-valued square integrable martingales and φ_i are two elements of $\mathbf{H}^*$, the strong dual of **H**,[53] we verify that $(m_i(\cdot), \varphi_i)$ are real-valued square integrable martingales, where $i = 1, 2$, and $(\cdot, \cdot)$ is the duality which extends the scalar product $\langle\cdot, \cdot\rangle_{\mathbf{H}}$. Hence, following Metivier and Pellaumail (loc.cit.), Sec. 2.3.6 and 2.4.4, we define the "*mutual tensor quadratic variation*" of the two martingales $m_i(\cdot)$ as a random bilinear functional, acting on the tensor product $\mathbf{H}^* \otimes \mathbf{H}^*$, by

$$|[m_1, m_2]|(\varphi_1 \otimes \varphi_2)(\cdot) := [(m_1, \varphi_1), (m_2, \varphi_2)](\cdot). \tag{15.96}$$

By (15.95)

$$\left.\begin{array}{c} |[m_1, m_2]|(\cdot) \equiv 0 \text{ a.s. if } m_i(\cdot)\; i = 1, 2, \\ \text{are uncorrelated } \mathbf{H}\text{-valued square integrable martingales.} \end{array}\right\} \tag{15.97}$$

Set for the tensor quadratic variation of $m(\cdot)$

$$|[m]|(\cdot) := |[m, m]|(\cdot).$$

Take a CONS $\{\varphi_n\}$ in $\mathbf{H}^*$. The trace of the bilinear functional $|[m]|(t)$ is defined as

$$\text{Trace}(|[m]|(t) = \sum_{n=1}^{\infty} [(m(t), \varphi_n), (m(t), \varphi_n)],$$

whence,

$$\text{Trace}(|[m]|(\cdot) = [m](\cdot) \text{ a.s.} \tag{15.98}$$

The following characterization of Brownian motion is a special case of Theorem 1.1 of Ethier and Kurtz (loc.cit.), Chap. 7.1, where the covariance matrix may be time dependent.

Theorem 15.37. *Levy's characterization of Brownian motion*

(i) Suppose $m(\cdot)$ is an $\mathbf{R}^d$-valued continuous square integrable martingale and there is a nonnegative definite symmetric matrix $C \in \mathcal{M}_{d\times d}$ such that the tensor quadratic variation satisfies

$$|[m]|(t) \equiv tC \text{ a.s.} \tag{15.99}$$

Then $m(\cdot)$ is an $\mathbf{R}^d$-valued Brownian motion such that for all $t > 0$ $m(t) \sim \mathcal{N}(0, tC)$.

---

[53] Most of the time we may choose $\varphi_i \in \mathbf{H}$ since **H** and $\mathbf{H}^*$ are isometrically isomorphic. However, working with chains of Hilbert spaces as in (15.32) a natural choice could be $\mathbf{H} := \mathcal{H}_{-\gamma}$ for some $\gamma > 0$. In this case $\mathbf{H}^* = \mathcal{H}_{\gamma}$.

The *proof* uses the Itô formula and will be given at the end of Sect. 15.6.4. □

Remark 15.38. We remark that an **H**-valued Brownian motion $W(\cdot)$ is **H**-valued square integrable martingale if, and only if, it is regular. In this case the tensor quadratic variation is given by

$$\|[W]\|(t) \equiv t Q_W \quad \text{a.s.} \tag{15.100}$$

□

A somewhat more general version of the following martingale central limit theorem has been proved by Ethier and Kurtz (loc.cit.), Chap. 7.1, Theorem 1.4.

Theorem 15.39. *Martingale Central Limit Theorem*

(i) Let $m(\cdot)$ be a continuous $\mathbf{R}^d$-valued Gaussian mean-zero martingale with deterministic continuous tensor quadratic variation $\|[m]\|(\cdot)$ and components $\|[m]\|_{k\ell}(\cdot),\ k,\ell = 1,\dots,d$.
(ii) For $n = 1,2,\dots$ let $\{\mathcal{F}_t^n\}$ be a filtration and $m_n(\cdot)$ a locally square integrable $\mathcal{F}_t^n$ martingale with sample paths in $D([0,\infty);\mathbf{R}^d)$ and $m_n(0)$. Let $\|[m_n]\|(\cdot)$ denote the tensor quadratic variation of $m_n(\cdot)$ with components $\|[m_n]\|_{k\ell}(\cdot),\ k,\ell = 1,\dots,d$.

Suppose that the following three conditions hold:
For each $T > 0$ and $k,\ell = 1,\dots,d$

$$\lim_{n\to\infty} E\big[\sup_{t\le T} |\|[m_n]\|_{k\ell}(t) - \|[m]\|_{k\ell}(t-)\big] = 0. \tag{15.101}$$

$$\lim_{n\to\infty} E\big[\sup_{t\le T} |m_n(t) - m_n(t-)\big] = 0. \tag{15.102}$$

For each $t \ge 0$ and $k,\ell = 1,\dots,d$

$$\|[m_n]\|_{k\ell}(t) \longrightarrow c_{k\ell}(t) \text{ in probability.} \tag{15.103}$$

Then

$$m_n(\cdot) \Longrightarrow m(\cdot) \ \text{ in } D([0,\infty);\mathbf{R}^d). \tag{15.104}$$

□

We conclude this section with two additional maximal inequalities.

For real-valued locally square integrable martingales we have the following improvement of Doob's inequality:

Theorem 15.40. *Burkholder-Davis-Gundy Inequality*

There exist universal positive constants c_p, C_p, where $0 < p < \infty$ such that for every real-valued locally integrable martingale $m(\cdot)$ and stopping time $\tau \ge 0$

$$c_p E \sup_{0\le s\le\tau} |m|^{2p}(s) \le E[m]^p(\tau) \le C_p E \sup_{0\le s\le\tau} |m|^{2p}(s). \tag{15.105}$$

The **proof** *for continuous locally square integrable martingales is found in Ikeda and Watanabe (1981), Chap. III.3, Theorem 3.1. the proof of the more general version of just locally integrable martingales is provided by Liptser and Shiryayev (1986), Chap. 1.9, Theorems 6 and 7.*[54] □

We present a special case of an inequality proved by Kotelenez (1984), Theorem 2.1. For its formulation we require the definition of a strongly continuous (two-parameter) semigroup on a real separable Hilbert space with scalar product $\langle \cdot, \cdot \rangle_{\mathbf{H}}$ and norm $\| \cdot \|_{\mathbf{H}}$. Let $I_{\mathbf{H}}$ denote the identity operator on $\mathbf{H}$.

Definition 15.41. *A family of bounded linear operators on* $\mathbf{H}$, $U(t,s)$ *with* $0 \le s \le t < \infty$, *is called a* "strongly continuous two-parameter semi-group" *on* $\mathbf{H}$ *if the following conditions hold:*

(i) $U(t,t) \equiv I_{\mathbf{H}}$
(ii) $U(t,u)U(u,s) = U(t,s)$ *for* $0 \le s \le u \le t < \infty$
(iii) $U(t,s)$ *is strongly continuous in both s and t*[55]

The uniform boundedness principle (Theorem 15.3, Part (II)) implies

$$\sup_{0 \le s \le t \le \bar{t}} \|U(t,s)\|_{\mathcal{L}(\mathbf{H})} < \infty \ \forall \bar{t} < \infty.$$ □

Theorem 15.42. *Maximal Inequality for Stochastic Convolution Integrals*

Let $m(\cdot)$ *be an* $\mathbf{H}$*-valued square integrable cadlag martingale and* $U(t,s)$ *a strongly continuous two-parameter semigroup of bounded linear operators. Suppose there is an* $\eta \ge 0$ *such that*

$$\|U(t,s)\|_{\mathcal{L}(\mathbf{H})} \le e^{\eta(t-s)} \ \ \forall \ 0 \le t < \infty.$$

Then the $\mathbf{H}$*-valued convolution integral* $\int_0^{\cdot} U(\cdot, s)m(ds)$ *has a cadlag modification and for any bounded stopping time* $\tau \le T < \infty$

$$E \sup_{0 \le t \le \tau} \| \int_0^t U(t,s)m(\mathrm{d}s)\|_{\mathbf{H}}^2 \le e^{4T\eta} E[m](\tau). \tag{15.106}$$

□

15.2.4 Random Covariance and Space–time Correlations for Correlated Brownian Motions

The random covariance of both generalized and classical random processes and random fields, related to $\mathbf{R}^d$*-valued correlated Brownian motions, are analyzed. For*

---

[54] Both Liptser and Shiryayev (loc.cit.) and Metivier and Pellaumail (loc.cit.) prove additional inequalities for martingales.

[55] The definition is a straightforward generalization of strongly continuous one-parameter semi-groups. For the definition and properties of strongly continuous one-parameter and two-parameter groups we refer to Kato (1976) and Curtain and Pritchard (1978). However, we do not make assumptions about the existence of a generator $A(t)$. Curtain and Pritchard (1978) call $U(t,s)$ called a "*mild evolution operator*."

the time derivative of correlated and uncorrelated Brownian motions the Wiener integral is employed. We compare the results with the covariance of uncorrelated Brownian motions.

We start with the time correlations of a family of square integrable $\mathbf{R}^d$-valued continuous martingales martingales $m^1(\cdot), m^2(\cdot)$, which are adapted to the same filtration $\mathcal{F}_t$. Note that the mutual tensor quadratic variation $[m^i, m^j]$, $i, j = 1, 2$, is absolutely continuous and, therefore, differentiable in the generalized sense. Similar to the case of an $\mathbf{R}^2$-valued Brownian motion $(\beta_1(\cdot), \beta_2(\cdot))$, we treat the time derivative of $m^i(\cdot)$ as a random distribution over $\mathbf{R}$ (also called "*generalized random processes*") and follow the analysis provided by Gel'fand and Vilenkin[56] for a one-dimensional Brownian motion. The main difference from the case of $(\beta_1(\cdot), \beta_2(\cdot))$ is that $m^1(\cdot), m^2(\cdot)$ is not necessarily Gaussian. Therefore, instead of computing the covariance directly, we may in a first step work with the mutual quadratic variation and then determine the covariance, taking the mathematical expectation of the mutual quadratic variation. Let η, φ, and ψ be test functions from $C_c^\infty(\mathbf{R}; \mathbf{R})$, i.e., from the space infinitely often differentiable real-valued functions on $\mathbf{R}$ with compact support. Denoting by $\langle \varphi, F\rangle$ the duality between a distribution and a test function, the duality between the random distribution (or generalized random process) $\frac{\mathrm{d}}{\mathrm{d}t}m^i(\cdot)$ and the test function η is given by the Wiener integral, i.e.,

$$\left\langle \eta, \frac{\mathrm{d}}{\mathrm{d}t}m^i(\cdot)\right\rangle := \int_0^\infty \eta(t)\mathrm{d}m^i(t), \quad i = 1, 2. \tag{15.107}$$

Integration by parts, in the sense of Riemann–Stieltjes integrals,[57] yields

$$\int_0^T \eta(t)\mathrm{d}m_k^i(t) = -\int_0^T \left(\frac{\mathrm{d}}{\mathrm{d}t}\eta(t)\right)m_k^i(t)\mathrm{d}t + \eta(T)m_k^i(T),$$
$$k = 1, \ldots, d, \quad i = 1, 2, \quad \forall T > 0,$$

where we used $m_k^i(0) = 0$ to simplify the boundary condition.[58] Letting $T \longrightarrow \infty$, the fact that η has finite support implies:

$$\int_0^\infty \eta(t)\mathrm{d}m_k^i(t) = -\int_0^\infty (\frac{\mathrm{d}}{\mathrm{d}t}\eta(t))m_k^i(t)\mathrm{d}t, \quad k = 1, \ldots, d, \quad i = 1, 2. \tag{15.108}$$

Since the test functions have finite support we obtain the existence of the limit of the mutual quadratic variation $[\int_0^T \varphi(t)\mathrm{d}m_k^i(t), \int_0^T \psi(t)\mathrm{d}m_\ell^j(t)]$, as $T \longrightarrow \infty$.

Definition 15.43.

$$Cov_\omega\left[\left(\varphi, \frac{\mathrm{d}}{\mathrm{d}t}m^i\right)\left(\psi, \frac{\mathrm{d}}{\mathrm{d}t}m^j\right)\right]_{k\ell} := \left[\int_0^\infty \varphi(t)dm_k^i(t),\right.$$
$$\left.\int_0^\infty \psi(t)\mathrm{d}m_\ell^j(t)\right] \quad i, j = 1, 2, k, \ell = 1, \ldots, d, \tag{15.109}$$

---

[56] Cf. Gel'fand and Vilenkin (1964), Chap. III.2.5.

[57] Cf. our Theorem 15.16.

[58] The Riemann–Stieltjes interpretation remains obviously valid, if we just assume $\eta(\cdot)$ to be continuous and of bounded variation on $[0, T]$.

will be called the "random covariance" of the generalized random processes $\frac{\mathrm{d}}{\mathrm{d}t}m^i(\cdot)$, $i = 1, 2$, *where the right-hand side of (15.109) is, by definition, equal to* $\lim_{T\to\infty} [\int_0^T \varphi(t)\mathrm{d}m_k^i(t), \int_0^T \psi(t)\mathrm{d}m_\ell^j(t)]$

Generalizing Gel'fand and Vilenkin (loc.cit), we compare the random covariance of $\left(\frac{\mathrm{d}}{\mathrm{d}t}m^1(\cdot), \frac{\mathrm{d}}{\mathrm{d}t}m^2(\cdot)\right)$ with random covariance of $(m^1(\cdot), m^2(\cdot))$, where

$$
\begin{aligned}
&Cov_\omega\left[\left(\frac{\mathrm{d}}{\mathrm{d}t}\varphi, m^i\right)\left(\frac{\mathrm{d}}{\mathrm{d}t}\psi, m^j\right)\right]_{k\ell} \\
&\quad := \int_0^\infty \int_0^\infty \left(\frac{\mathrm{d}}{\mathrm{d}t}\varphi\right)(t)\left(\frac{\mathrm{d}}{\mathrm{d}s}\psi\right)(s)[m_k^i, m_\ell^j](t\wedge s)\mathrm{d}t\,\mathrm{d}s \;\; i, j = 1, 2,
\end{aligned}
\tag{15.110}
$$

By the orthogonality of the martingale increments the right-hand side of (15.110) equals

$$
\begin{aligned}
&\int_0^\infty \int_0^\infty \left(\frac{\mathrm{d}}{\mathrm{d}t}\varphi\right)(t)\left(\frac{\mathrm{d}}{\mathrm{d}s}\psi\right)(s)[m_k^i, m_\ell^j](t\wedge s)\mathrm{d}t\,\mathrm{d}s \\
&\quad = \left[\int_0^\infty \left(\frac{\mathrm{d}}{\mathrm{d}t}\varphi\right)(t)m_k^i(t)\mathrm{d}t, \int_0^\infty \left(\frac{\mathrm{d}}{\mathrm{d}t}\psi\right)(s)m_\ell^j(s)\mathrm{d}s\right].
\end{aligned}
\tag{15.111}
$$

Theorem 15.44.

$$
\left.
\begin{aligned}
Cov_\omega\left[\left(\varphi, \frac{\mathrm{d}}{\mathrm{d}t}m^i\right)\left(\psi, \frac{\mathrm{d}}{\mathrm{d}t}m^j\right)\right] &= Cov_\omega\left[\left(\frac{\mathrm{d}}{\mathrm{d}t}\varphi, m^i\right)\left(\frac{\mathrm{d}}{\mathrm{d}t}\psi, m^j\right)\right] \\
&= \int_0^\infty \varphi(t)\psi(t)\frac{\mathrm{d}}{\mathrm{d}t}[m^i, m^j](t)\mathrm{d}t, \;\; i, j = 1, 2.
\end{aligned}
\right\}
\tag{15.112}
$$

Proof. The first equality in (15.112) follows from integration by parts in the generalized sense. To show the second equality we set

$$
f(t) := [m^i, m^j](t)
$$

and follow the procedure of Gel'fand and Vilenkin (loc.cit) who consider for Brownian motion the case $f(t) \equiv t$. We obtain

$$
\begin{aligned}
&\int_0^\infty \int_0^\infty \left(\frac{\mathrm{d}}{\mathrm{d}t}\varphi\right)(t)\left(\frac{\mathrm{d}}{\mathrm{d}t}\psi\right)(s)f(t\wedge s)\mathrm{d}t\,\mathrm{d}t \\
&= \int_0^\infty \int_0^\infty 1_{\{s\le t\}}\left(\frac{\mathrm{d}}{\mathrm{d}t}\varphi\right)(t)\left(\frac{\mathrm{d}}{\mathrm{d}t}\psi\right)(s)f(t\wedge s)\mathrm{d}t\,\mathrm{d}t \\
&+ \int_0^\infty \int_0^\infty 1_{\{s>t\}}\left(\frac{\mathrm{d}}{\mathrm{d}t}\varphi\right)(t)\left(\frac{\mathrm{d}}{\mathrm{d}t}\psi\right)(s)f(t\wedge s)\mathrm{d}t\,\mathrm{d}t \\
&=: I + II.
\end{aligned}
$$

By Fubini's theorem

$$
I = -\int_0^\infty \varphi(s)\left(\frac{\mathrm{d}}{\mathrm{d}t}\psi\right)(s)f(s)\mathrm{d}s, \;\; II = -\int_0^\infty \psi(t)\left(\frac{\mathrm{d}}{\mathrm{d}t}\varphi\right)(t)f(t)\mathrm{d}t.
$$

Taking into account the boundary conditions

$$f(t)\varphi(t) = 0, \quad t \in \{0, \infty\},$$

we integrate by parts and obtain

$$II = \int_0^\infty \varphi(t)\left(\frac{\mathrm{d}}{\mathrm{d}t} f\right)(t)\psi(t))\mathrm{d}t + \int_0^\infty \varphi(t) f(t) \frac{\mathrm{d}}{\mathrm{d}t}\psi(t)\mathrm{d}t$$

Adding up the representations for I and II we obtain

$$\int_0^\infty \int_0^\infty (\frac{\mathrm{d}}{\mathrm{d}t}\varphi)(t)\left(\frac{\mathrm{d}}{\mathrm{d}t}\psi\right)(s) f(t \wedge s)\mathrm{d}t\,\mathrm{d}t = \int_0^\infty \varphi(t)\psi(t)\frac{\mathrm{d}}{\mathrm{d}t} f(t)\mathrm{d}t.$$

We may rewrite (15.112) in terms of the δ-function with support in $t = 0$ as follows:

$$\left.\begin{aligned} &\mathrm{Cov}_\omega\left[\left(\varphi, \frac{\mathrm{d}}{\mathrm{d}t} m^i\right)\left(\psi, \frac{\mathrm{d}}{\mathrm{d}t} m^j\right)\right] = \int_0^\infty \varphi(t)\psi(t)\frac{\mathrm{d}}{\mathrm{d}t}[m^i, m^j](t)\mathrm{d}t, \\ &= \int_0^\infty \int_0^\infty \varphi(t)\psi(s)\delta_0(t-s)\left[\frac{\mathrm{d}}{\mathrm{d}t} m^i(t), \frac{\mathrm{d}}{\mathrm{d}s} m^j\right]\mathrm{d}s\,\mathrm{d}t, \quad i, j = 1, 2. \end{aligned}\right\} \tag{15.113}$$

Consider now the random covariance for the family of square integrable $\mathbf{R}^d$-valued continuous martingales martingales $m_\varepsilon(\cdot, r_0^i), m_\varepsilon(\cdot, r_0^j)$, defined in (5.2)

$$m_\varepsilon(t, r_0^i) := \int_0^t \int \Gamma_\varepsilon(r(u, r_0^i), q) w(\mathrm{d}q, \mathrm{d}u), \quad i = 1, \ldots N,$$

where the $r(\cdot, r_0^i)$ are the solutions of (5.1) such that (4.11) holds. We assume that the associated diffusion matrix, $D_{\varepsilon,k\ell}(\cdot,\cdot)$, is spatially homogeneous, i.e., that (5.12) holds. The mutual quadratic variations are continuously differentiable and the derivatives are bounded uniformly in all variables, i.e., there is a finite $c > 0$ such that

$$\operatorname*{ess\,sup}_{t,\omega}\left|\frac{\mathrm{d}}{\mathrm{d}t}[m_{\varepsilon,k}(t, r_0^i), m_{\varepsilon,\ell}(t, r_0^j)]\right| \le c \;\; k, \ell = 1, \ldots, d, \;\; i, j = 1, \ldots, N. \tag{15.114}$$

In this case we can take $\mathcal{S}(\mathbf{R})$ instead of $C_c^\infty(\mathbf{R}; \mathbf{R})$. We must show that $|m_\varepsilon|(t, r_0^i)$ at most grows "slowly," as $t \longrightarrow \infty$. To this end we employ a version of the asymptotic *law of the iterated logarithm*, whose proof is found in Loève (1978), Sect. 41.

Lemma 15.45. *Let $\beta(\cdot)$ be a real-valued standard Brownian motion. Then*

$$\limsup_{t\to\infty} \frac{|\beta(t)|}{\sqrt{2t \log\log(t)}} = 1 \;\; a.s. \tag{15.115}$$

□

Corollary 15.46. *There is a finite constant $\bar{c}$ such that for all $i = 1, \ldots, N$ and $k = 1, \ldots, d$*

$$\limsup_{t\to\infty} \frac{|m_{\varepsilon,k}(t, r_0^i)|}{\sqrt{2t \log\log(t)}} \le \bar{c} \;\; a.s. \tag{15.116}$$

Proof. By Proposition 5.2, $m_{\varepsilon,k}(\cdot, r_0^i)$ are one-dimensional Brownian motions with variance $D_{\varepsilon,kk}(0)$. Hence (15.116) follows from (15.115) after a simple deterministic time change, provided $D_{\varepsilon,kk}(0) > 0$. If $D_{\varepsilon,kk}(0) = 0$, then $m_{\varepsilon,k}(\cdot, r_0^i) \equiv 0$. □

As a consequence we obtain "slow" growth of $|m_\varepsilon|(t, r_0^i)$ and we may choose the test functions from $\mathcal{S}(\mathbf{R})$ in the derivation of Theorem 15.44.

We are ready to define the space–time correlations of our driving correlated Brownian motions. Recall that

$$\frac{\mathrm{d}}{\mathrm{d}t} m_\varepsilon(t, r_0) = \int \Gamma_\varepsilon(r(t, r_0), q) w(\mathrm{d}q, \mathrm{d}t), \tag{15.117}$$

interpreted as a generalized random process (in t) and indexed by the spatial initial condition r_0. Considering both t and r_0 to be variables, the left-hand side defines a space–time random field, which is generalized in the variable t. Let $\varphi_d, \psi_d \in \mathcal{S}(\mathbf{R}^d)$ and $\varphi, \psi \in \mathcal{S}(\mathbf{R})$. If F_{d+1} is a function of (r, t) such that $F_{d+1} \in \mathcal{S}'(\mathbf{R}^{d+1})$, the duality between F and $\varphi_d(r) \cdot \varphi(t)$ is defined by

$$(F_{d+1}, \varphi_d \varphi) := \int \int_{-\infty}^{\infty} F_{d+1}(r, t) \varphi_d(r) \varphi(t) \mathrm{d}r \, \mathrm{d}t. \tag{15.118}$$

The duality between arbitrary elements from $\mathcal{S}'(\mathbf{R}^{d+1})$ and test functions is an extension of this duality. Let us focus on two initial conditions, r_0 and $\tilde{r}_0$, respectively.

Definition 15.47.

$$\left.\begin{aligned} &\mathrm{Cov}_{d+1,\omega}\left[\left(\varphi_d \varphi, \frac{\mathrm{d}}{\mathrm{d}t} m_\varepsilon(\cdot, \cdot)\right)\left(\psi_d \psi, \frac{\mathrm{d}}{\mathrm{d}t} m_\varepsilon(\cdot, \cdot)\right)\right]_{k\ell} \\ &= \int\int\int_0^\infty \int_0^\infty \varphi_d(r_0)\psi_d(\tilde{r}_0)\varphi(t)\psi(s)\delta_0(t-s) \\ &\quad \times \int \sum_{m=1}^d \Gamma_{\varepsilon,km}(r(t, r_0), q)\Gamma_{\varepsilon,\ell m}(r(s, \tilde{r}_0), q) \mathrm{d}q \, \mathrm{d}s \, \mathrm{d}t \, \mathrm{d}r_0 \, \mathrm{d}\tilde{r}_0 \end{aligned}\right\} \tag{15.119}$$

will be called the "random covariance" *of the random field* $\frac{\mathrm{d}}{\mathrm{d}t} m_\varepsilon(\cdot, \cdot)$. □

The definition of the δ-function implies

$$\left.\begin{aligned} &\mathrm{Cov}_{d+1,\omega}\left[\left(\varphi_d \varphi, \frac{\mathrm{d}}{\mathrm{d}t} m_{\varepsilon,k}(\cdot, \cdot)\right)\left(\psi_d \psi, \frac{\mathrm{d}}{\mathrm{d}t} m_{\varepsilon,\ell}(\cdot, \cdot)\right)\right]_{k\ell} \\ &= \int\int\int_0^\infty \varphi_d(r_0)\psi_d(\tilde{r}_0)\varphi(t)\psi(t) \\ &\quad \times \int \sum_{m=1}^d \Gamma_{\varepsilon,km}(r(t, r_0), q)\Gamma_{\varepsilon,\ell m}(r(t, \tilde{r}_0), q) \, \mathrm{d}q \, \mathrm{d}t \, \mathrm{d}r_0 \, d\tilde{r}_0. \end{aligned}\right\} \tag{15.120}$$

It follows that the space–time random field of correlated Brownian motions is generalized in the time variable and "classical" in the space variable. Taking the mathematical expectation in (15.120) (or (15.119)) yields the usual covariance as a bilinear functional on the test functions:

$$\left.\begin{aligned}&E\mathrm{Cov}_{d+1,\omega}\left[\left(\varphi_d\varphi,\frac{\mathrm{d}}{\mathrm{d}t}m_{\varepsilon,k}(\cdot,\cdot)\right)\left(\psi_d\psi,\frac{\mathrm{d}}{\mathrm{d}t}m_{\varepsilon,\ell}(\cdot,\cdot)\right)\right]_{k\ell}\\&=\int\int\int_0^\infty\varphi_d(r_0)\psi_d(\tilde{r}_0)\varphi(t)\psi(t)\\&\quad\times\int E\left(\sum_{m=1}^d\Gamma_{\varepsilon,km}(r(t,r_0),q)\Gamma_{\varepsilon,\ell m}(r(t,\tilde{r}_0),q)\right)\mathrm{d}q\,\mathrm{d}t\,\mathrm{d}r_0\,d\tilde{r}_0.\end{aligned}\right\}\quad(15.121)$$

Recall the proof of Theorem 14.2 and suppose that the diffusion matrices are spatially homogeneous and independent of t and μ. Let $\beta(\cdot,\tilde{r}_0)$ be a family of $\mathbf{R}^d$-valued Brownian motions with starts in r_0 and covariance matrices $D_0 I_d$ Brownian motions such that $\beta(\cdot,r_0)$ and $\beta(\cdot,\tilde{r}_0)$ are independent whenever $r_0\neq\tilde{r}_0$. It follows from (15.113) that its covariance is[59]

$$D_0\delta_0(t-s)\otimes\delta_{0,d}(r_0-\tilde{r}_0)\delta_{k\ell}.$$

By the proof of Theorem 14.2 (cf. (14.48)) we have

$$(r_\varepsilon(\cdot,r_0),r_\varepsilon(\cdot,\tilde{r}_0))\Rightarrow(\beta(\cdot,r_0),\beta(\cdot,\tilde{r}_0)),\ \text{as }\varepsilon\to 0\,.$$

Thus, we obtain

$$\begin{aligned}&E\mathrm{Cov}_{d+1,\omega}\left[\left(\cdot\cdot,\frac{\mathrm{d}}{\mathrm{d}t}m_{\varepsilon,k}(\cdot,\cdot)\right)\left(\cdot\cdot,\frac{\mathrm{d}}{\mathrm{d}t}m_{\varepsilon,\ell}(\cdot,\cdot)\right)\right]_{k\ell}\\&\quad\approx D_0\delta_0(t-s)\otimes\delta_{0,d}(r_0-\tilde{r}_0)(\cdot\cdot)\delta_{k\ell}\quad\text{for small }\varepsilon.\end{aligned}\qquad(15.122)$$

"$\cdot\cdot$" in (15.122) is the space for the variables from $\mathcal{S}(\mathbf{R}^d)$ and $\mathcal{S}(\mathbf{R})$, respectively. However, we do not claim that $\beta(\cdot,\cdot)$ is a generalized d-dimensional random field with covariance from the right-hand side of (15.122), because we did not specify the distribution in the spatial variable. Instead of constructing $\beta(\cdot,\cdot)$ as a generalized random field recall the definition of the $\mathbf{R}^d$-valued space–time standard Gaussian white noise $w(\mathrm{d}q,\mathrm{d}t)=(w_1(\mathrm{d}q,\mathrm{d}t),\ldots,w_d(\mathrm{d}q,\mathrm{d}t))^T$. Let us focus on the first component of $w(\mathrm{d}q,\mathrm{d}t)$. Following Walsh (1986), we defined in Definition 2.2 $w_1(\mathrm{d}q,\mathrm{d}t,\omega)$ as a finitely additive signed measure on the Borel sets A of $\mathcal{B}^{d+1}$ of finite Lebesgue measure and as a family of Gaussian random variables, indexed by $(d+1)$-dimensional Borel sets. We also pointed out that the definition of $w_1(\mathrm{d}q,\mathrm{d}t)$ is a multiparameter generalization of the properties of a scalar-valued (standard) Brownian motion $\beta(\cdot)$ as a finitely additive signed random measure, $\beta(\mathrm{d}t,\omega)$, on one-dimensional Borel sets and as a family of Gaussian random variables indexed by one-dimensional Borel sets. The analogy can be extended further. Recall that the generalized derivative of $\beta(\cdot)$ is a Schwarz distribution, i.e.,

[59] Cf. also (5.6).

$$\frac{\mathrm{d}}{\mathrm{d}t}\beta(\cdot,\omega) \in \mathcal{S}'(\mathbf{R}).$$

If we now fix t, then $\int_0^t w(\mathrm{d}q, \mathrm{d}s, \omega)$ defines a finitely additive signed measure on the Borel sets B of $\mathcal{B}^{\mathrm{d}}$ of finite Lebesgue measure. By (15.69), in addition to (15.32) and (15.36), we can interpret the finitely additive signed measure on $\int_0^t w_1(\mathrm{d}\cdot, \mathrm{d}s, \omega)$ as an element from $\mathcal{S}'(\mathbf{R}^d)$. Set

$$\hat{w}_1(r,t,\omega) := \int_0^t \int 1_{Br} w_1(\mathrm{d}q, \mathrm{d}s, \omega), \tag{15.123}$$

where B_r is a closed rectangular domain in $\mathbf{R}^d$ containing 0 and with side lengths $r_1, \ldots, r_d$ and whose sides are parallel to the axes. If all endpoints of B_r have non-negative coordinates, $\hat{w}_1(\cdot,\cdot)$ has a continuous modification, which is called the Brownian sheet (cf. Walsh (loc. cit.)). We will assume that the left hand side of (15.123) is already the Brownian sheet. Hence, we may interpret the integration of a Borel set in $\mathbf{R}_+^{d+1}$ as a multiparameter integration against the Brownian sheet $\hat{w}_1(r,t)$.

We extend the notion of a Brownian sheet to $\hat{w}_1(r,t,\omega)$ for $r \in \mathbf{R}^d$ by taking 2^d i.i.d. Brownian sheets and patching the finitely additive measure $\int_0^\infty \int 1_A(q,t) w_1(\mathrm{d}q, \mathrm{d}t, \omega)$ together as the sum of 2^d measures, determined by the multiparameter integration against 2^d i.i.d. Brownian sheets, defined on each of the 2^d domains D_k, $k = 1, \ldots, 2^d$, where any of the coordinates is either non-negative or negative. Denote these Brownian sheets by $\hat{w}_{1,k}$, $k = 1, \ldots, 2^d$, and observe that the sum of independent normal random variables is normal and that the mean and variance of the sum is the sum of the means and variances. Therefore, we obtain

$$\int_0^\infty \int 1_A(q,t) w_1(\mathrm{d}q, \mathrm{d}t) \sim \sum_{k=1}^{2^d} \int_0^\infty \int 1_{A\cap D_k} 1_A(q,t) \hat{w}_{1,k}(\mathrm{d}q, \mathrm{d}t), \tag{15.124}$$

i.e., both sides in (15.124) are equivalent in distribution. This construction is relatively simple and, setting $\hat{w}_{1,k}(r,t) \equiv 0$ for $r \notin D_k$, yields a multiparameter continuous random field[60]

$$\bar{\hat{w}}_1(r,t) := \sum_{k=1}^{2^d} \hat{w}_{1,k}(r,t) 1_{D_k}(r). \tag{15.125}$$

Hence, we can differentiate $\bar{\hat{w}}_1(r,t,\omega)$ in the generalized sense with respect to all coordinates. Choosing $\varphi(\cdot) \in \mathcal{S}(\mathbf{R})$ and $\varphi_d(\cdot) \in \mathcal{S}(\mathbf{R}^d)$ we obtain[61]

[60] Notice that the extension of the Brownian sheet to negative coordinates is analogous to the construction of a Brownian motion on $\mathbf{R}$, which can be done by taking two i.i.d. Brownian motions $\beta_+(\cdot)$ and $\beta_-(\cdot)$ for the Borel sets in $[0,\infty)$ and $(-\infty, 0]$, respectively.

[61] The time derivative leads to a Schwarz distribution by the previous arguments, because $\int_0^t \int \varphi_d(q) w(\mathrm{d}q, \mathrm{d}s)$ is a one-dimensional Brownian motion.

$$\int_0^\infty \int \varphi_d(q)\varphi(s)\frac{\partial^{d+1}}{\partial s\partial q_1\ldots\partial q_d}\bar{\hat{w}}_1(q,s,\omega)\mathrm{d}q\,\mathrm{d}s = \int_0^\infty \int \varphi(s)\varphi_d(q)w_1(\mathrm{d}q,\mathrm{d}s,\omega).$$

We conclude[62]

$$\frac{\partial^{d+1}}{\partial s\partial q_1\ldots\partial q_d}\bar{\hat{w}}_1(\cdot,\cdot,\omega) \equiv w_1(\mathrm{d}\cdot,\mathrm{d}\cdot,\omega) \;\text{ in } \mathcal{S}'(\mathbf{R}^{d+1}). \tag{15.126}$$

The generalization to $w(\mathrm{d}q,\mathrm{d}t) = (w_1(\mathrm{d}q,\mathrm{d}t),\ldots w_d(\mathrm{d}q,\mathrm{d}t))^T$ is carried out componentwise. Altogether, we obtain that

$$\frac{\partial^{d+1}}{\partial s\partial q_1\ldots\partial q_d}\bar{\hat{w}}(\cdot,\cdot) \equiv w(d\cdot,d\cdot)$$

is a generalized $\mathbf{R}^d$-valued random field whose components take values $\mathcal{S}'(\mathbf{R}^{d+1})$ where $\bar{\hat{w}}(\cdot,\cdot)$ is an appropriately defined $\mathbf{R}^d$-valued Brownian sheet with parameter domain $\mathbf{R}^d \times [0,\infty)$. Its covariance is given by

$$\left.\begin{aligned}
&E\mathrm{Cov}_{d+1,\omega}[(\varphi_d\varphi,\bar{\hat{w}}_k(\cdot,\cdot))(\psi_d\psi,\bar{\hat{w}}_\ell(\cdot,\cdot))]_{k\ell}\\
&= E\mathrm{Cov}_{d+1,\omega}\int_0^\infty \int \varphi_d(q)\varphi(s)\frac{\partial^{d+1}}{\partial s\partial q_1\ldots\partial q_d}\bar{\hat{w}}_k(q,s)\mathrm{d}q\,\mathrm{d}s\\
&\quad\times \int_0^\infty \int \psi_d(r)\psi(t)\frac{\partial^{d+1}}{\partial t\partial r_1\ldots\partial r_d}\bar{\hat{w}}_\ell(r,t)\mathrm{d}r\,\mathrm{d}t\\
&= \int_0^\infty\int_0^\infty\int\int \varphi_d(q)\varphi(s)\psi_d(r)\psi(t)\delta_0(t-s)\otimes\delta_{0,d}(q-r)\delta_{k\ell}\,\mathrm{d}q\,\mathrm{d}s\,\mathrm{d}r\,\mathrm{d}t\\
&= \int\int_0^\infty \varphi_d(q)\psi_d(q)\varphi(t)\psi(t)\mathrm{d}t\,\mathrm{d}q\,\delta_{k\ell}.
\end{aligned}\right\} \tag{15.127}$$

We conclude that the covariance operator of $\mathbf{R}^d$-valued standard Gaussian space–time white noise on $\mathbf{R}^d \times \mathbf{R}_+$ is $\delta_0(t-s)\otimes\delta_{0,d}(q-r)\delta_{k\ell}$. Apart from the factor D_0 this is the same covariance as for the "random field" $\beta(\cdot,\cdot)$. That this coincidence is not accidental follows from Remark 3.10 at the end of Chap. 3.

15.2.5 Stochastic Itô Integrals

We summarize the construction of the Itô integral[63] *in* $\mathbf{R}^d$ *driven by finitely and infinitely many i.i.d Brownian motions. The Itô and the Itô-Wentzell formulas are*

---

[62] In the identification of a measure with a formal Radon–Nikodym derivative of that measure with respect to the Lebesgue measure as as generalized function or generalized field the "Radon–Nikodym derivative" is usually written as a suitable (partial) derivative in the distributional sense. We remind the reader of the identification the Dirac delta function with support in some point $a \in \mathbf{R}^d$ and the point measure $\delta_a(\mathrm{d}r)$. Cf. also Schaumlöffel (1986).

[63] Cf. Itô (1944).

presented. Finally, we comment on the functional analytic interpretation of the Itô integral and provide a sufficient condition for uniform tightness of SODEs.

Let $\beta_1(\cdot)$ a one-dimensional standard Brownian motion and $\phi(\cdot)$ a real-valued $dt \otimes dP$ square integrable adapted process on $[0, T] \times \Omega$. We have seen in (15.115) of the previous section that, under the assumption $\phi(\cdot)$ is a.s. continuous and of bounded variation, we may define the stochastic integral as a Stieltjes integral plus a boundary term:

$$\int_0^t \phi(s)\beta_1(ds) := -\int_0^t \beta_1(s)\phi(ds) + \phi(t)\beta(t) \quad \in [0, T] \tag{15.128}$$

The problem of stochastic integration, however, is that most interesting integrands will not be of bounded variation. An easy way to understand this claim is considering a stochastic ordinary differential equation (SODE):

$$dx = b(x)\beta_1(dt), \quad x(0) = x_0, \tag{15.129}$$

where we may, for the time being, assume that the initial condition is deterministic. Following the procedure of ordinary differential equations (ODEs),[64] we convert (15.129) into an equivalent stochastic integral equation:

$$x(t) = x_0 + \int_0^t b(x(s))\beta_1(ds). \tag{15.130}$$

Assuming that the coefficient $b(x) = bx$ with $b \neq 0 \ \ \forall x$, we again resort to the methods of ODE, apply the Picard-Lindelöf procedure (as $b(x) = bx$ is obviously Lipschitz) and try to solve (15.130) through iteration, defining recursively an approximating sequence as follows:

$$x_1(t) = x_0 + \int_0^t bx_0\beta_1(ds) = x_0 + bx_0\beta_1(t), \tag{15.131}$$

The definition of the first step $x_1(\cdot)$ is trivial. However, we see that $x_1(\cdot) - x_0 \equiv b\beta_1(t)$, which is itself of unbounded variation and with quadratic variation $[\beta_1](t) \equiv t$ by (15.91). Therefore, we cannot explain the right-hand side of the second step, presented in the following (15.132), pathwise as a Stieltjes integral through integration by parts as in (15.108):

$$x_2(t) = x_0 + \int_0^t bx_1(s)\beta_1(ds) = \int_0^t b\{b\beta_1(s) + x_0\}\beta_1(ds). \tag{15.132}$$

Being unable to define the stochastic integral on the right-hand side of (15.132) through integration by parts need not prevent us from trying the Riemann–Stieltjes idea directly. To this end, as in Definition 15.33, consider the sequence of partitions $\{t_0^n < t_1^n < \cdots < t_k^n < \cdots\}$ and let $\gamma \in [0, 1]$. Set

[64] Cf., e.g., Coddington and Levinson (loc.cit.).

$$
\left.\begin{aligned}
S_n(t,\gamma) := \sum_{k\geq 0} b\Big\{\beta_1(t^n_{k-1}\wedge t) + x_0 + \gamma\,(\beta_1(t^n_k\wedge t) - \beta_1(t^n_{k-1}\wedge t))\Big\} \\
\times(\beta_1(t^n_k\wedge t) - \beta_1(t^n_{k-1}\wedge t)),
\end{aligned}\right. \tag{15.133}
$$

i.e., we choose the integrand at a fixed linear combination of the endpoints of the interval. By linearity of the summation,

$$
\left.\begin{aligned}
&S_n(t,\gamma) \\
&= \sum_{k\geq 0} b\gamma\,(\beta_1(t^n_k\wedge t) - \beta_1(t^n_{k-1}\wedge t))^2 \\
&\quad + \sum_{k\geq 0} b\{\beta_1(t^n_{k-1}\wedge t) + x_0\}(\beta_1(t^n_k\wedge t) - \beta_1(t^n_{k-1}\wedge t)) \\
&=: S_{n,1,\gamma}(t) + S_{n,2,\gamma}(t).
\end{aligned}\right\} \tag{15.134}
$$

By (15.90) for every $t \geq 0$

$$
S_{n,1,\gamma}(t) \longrightarrow b\gamma\,[\beta_1](t) \equiv b\gamma\, t \ \text{ a.s., as } n \longrightarrow \infty\,. \tag{15.135}
$$

By the proof of Proposition 15.34,

$$
S_{n,2,\gamma}(t) \longrightarrow b\int_0^t (\beta_1(s) + x_0)\beta_1(\mathrm{d}s) \ \text{ in probability, as } n \longrightarrow \infty\,. \tag{15.136}
$$

Altogether,

$$
S_n(t,\gamma) \longrightarrow b\int_0^t (\beta_1(s) + x_0)\beta_1(\mathrm{d}s) + b\gamma\, t \ \text{ in probability, as } n \longrightarrow \infty, \tag{15.137}
$$

i.e., the limit depends on the choice of the evaluation of the integrand in the approximating sum. We conclude that we cannot use the Stieltjes integral in the Picard-Lindelöf approximation of (15.130).

If we choose the left endpoint in the above approximation, i.e., $\gamma = 0$, we obtain the stochastic Itô integral. In what follows, we provide more details about Itô integration. To avoid a complicated notation, we will only sketch the construction of the Itô integral, driven by Brownian motions, and leave it to the reader to consult the literature about the more general case of martingale-driven stochastic integrals. Set

$$
L_{2,\mathcal{F},\mathrm{loc}}([0,\infty)\times\Omega : \mathcal{M}_{d\times d}) := \left\{\phi : \sum_{k,\ell=1}^{d} E\int_0^T \phi^2_{k\ell}(t,\cdot)\langle\infty\ \forall T\rangle 0\right\}, \tag{15.138}
$$

where $\phi(\cdot,\cdot)$ is an $\mathcal{F}_t$-adapted $\mathcal{M}_{d\times d}$-valued process, jointly measurable in (t,ω) with respect to $dt \otimes P(d\omega)$. Further, let $\beta(\cdot)$ be an $\mathbf{R}^d$-valued $\mathcal{F}_t$-adapted standard Brownian motion. Generalizing the idea of Riemann and Stieltjes, we first define the stochastic Itô integral for step function type processes such that the resulting integral becomes a square integrable martingale. Suppose for $\phi \in L_{2,\mathcal{F},\mathrm{loc}}([0,\infty)\times\Omega : \mathcal{M}_{d\times d})$ there is a sequence of points $\{0 = t_0^n < t_1^n < \cdots < t_k^n < \cdots\}$ and $\tilde{\phi}(t_k^n)$ $\mathcal{F}_{t_k^n}$-adapted $\mathcal{M}_{d\times d}$-valued random variables, $i = 0, 1, \ldots$, such that

$$\phi_n(t,\omega) = \tilde{\phi}(t_0^n) + \sum_{k=1}^{\infty} \tilde{\phi}(t_{k-1}^n) 1_{(t_{k-1}^n, t_k^n]}(t) \ \forall t \ \text{ a.s.} \tag{15.139}$$

$\phi_n(\cdot,\cdot)$ is called "*simple*." The stochastic (Itô) integral of a simple process with respect to $\beta(\cdot)$ is defined by

$$\int_0^t \phi_n(s)\beta(ds) := \tilde{\phi}(0) + \sum_{k=1}^{\infty} \tilde{\phi}(t_{k-1}^n)(\beta(t_k^n \vee t) - \beta(t_{k-1}^n \vee t)) \ \forall t \ \text{ a.s.} \tag{15.140}$$

As in Sect. 15.6.2, we see that $\int_0^{\cdot} \phi_n(s)\beta(ds)$ is a continuous square integrable martingale. Further, it is proved that the class of simple processes is dense in $L_{2,\mathcal{F},\mathrm{loc}}([0,\infty)\times\Omega : \mathcal{M}_{d\times d})$ (cf., e.g., Ikeda and Watanabe (loc.cit.), Chap.II.1 or Liptser and Shiryayev (1974), Chap. 4.2 as well as Itô (1944)). A simple application of Doob's inequality (Theorem 15.32), as for $S_{n,2}(\cdot)$ in the proof of Proposition 15.34, entails that we may construct a Cauchy sequence $\int_0^{\cdot} \phi_n(s)\beta(ds)$ in $L_{0,\mathcal{F}}(C([0,T];\mathbf{R}^d))$, as $\phi_n(\cdot)$ approaches $\phi(\cdot)$ in $L_{2,\mathcal{F},\mathrm{loc}}([0,\infty)\times\Omega : \mathcal{M}_{d\times d})$. The resulting limit is itself a square continuous $\mathbf{R}^d$-valued martingale and is defined as the limit in probability of the approximating stochastic (Itô) integrals.[65]

Definition 15.48. *The stochastic Itô integral of $\phi(\cdot)$ driven by $\beta(\cdot)$ is defined as the stochastic limit in $L_{0,\mathcal{F}}(C([0,T];\mathbf{R}^d))$*

$$\int_0^t \phi(s)\beta(ds) = \lim_{n\to\infty} \int_0^{\cdot} \phi_n(s)\beta(ds), \tag{15.141}$$

where the ϕ_n are simple processes. □

We verify that $\sum_{\ell=1}^{d} \int_0^t \phi_{k,\ell}(s)\beta_\ell(ds)$ are continuous square integrable real-valued martingales for $k = 1,\ldots,d$, where $\phi_{k,\ell}$ are the entries of the $d\times d$ matrix Φ and $\beta_\ell(\cdot)$ are the one-dimensional components of $\beta(\cdot)$. Further, the mutual quadratic variation of these martingales satisfies the following relation:

$$\left[\int_0^t \phi_{k,\ell}(s)\beta_\ell(ds), \int_0^t \phi_{k,\tilde{\ell}}(s)\beta_{\tilde{\ell}}(ds)\right] := \left\{\begin{matrix} \int_0^t \phi_{k,\ell}^2(s)ds, & \text{if} \quad \ell = \tilde{\ell}\,, \\ 0, & \text{if}\, \ell \neq \tilde{\ell}. \end{matrix}\right\} \tag{15.142}$$

[65] Since $\beta(\cdot)$ is continuous, we do not need to take the left hand limit in the integrand.

Expression (15.142) can be shown through approximation with stochastic Itô integrals of simple processes, employing (15.95), since the one-dimensional Brownian motions $\beta_\ell(\cdot)$ are independent and, therefore, also uncorrelated. The latter fact implies that the approximating Itô integrals are uncorrelated. Consequently, the quadratic variation of $\int_0^t \phi(s)\beta(ds)$ is given by

$$\left[\int_0^\cdot \phi(s)\beta(\mathrm{d}s)\right](t) = \sum_{k,\ell=1}^{d} \int_0^t \phi_{k,\ell}(s)\mathrm{d}s, \tag{15.143}$$

We generalize the above construction to an infinite sequence of stochastic Itô integrals. Let $\phi_n(\cdot,\cdot)$ be a sequence of $\mathcal{F}_t$-adapted jointly measurable $\mathcal{M}_{d\times d}$-valued processes and set

$$L_{2,\mathcal{F}}([0,T]\times\Omega\times\mathbf{N}:\mathcal{M}_{d\times d}) := \{(\phi_1,\ldots\phi_n\ldots):\sum_{n=1}^{\infty}\sum_{k,\ell=1}^{d}\int_0^T E\phi^2_{n,k\ell}(t,\cdot) < \infty\}. \tag{15.144}$$

Further, as in (4.14), let $\beta^n(\cdot)$ a sequence of $\mathbf{R}^d$-valued $\mathcal{F}_t$-adapted i.i.d. standard Brownian motions.

As in (15.142), we see that

$$\left[\int_0^t \phi_n(s)\beta^n(\mathrm{d}s), \int_0^t \phi_m(s)\beta^m(\mathrm{d}s)\right] = 0, \quad \text{if } n \neq m \tag{15.145}$$

Hence,

$$m(t) := \sum_n \int_0^t \phi_n(s)\beta^n(\mathrm{d}s) \tag{15.146}$$

is a continuous square integrable $\mathbf{R}^d$-valued martingale with quadratic variation

$$\left[\sum_n \int_0^\cdot \phi_n(s)\beta^n(\mathrm{d}s)\right](t) \equiv \sum_n \sum_{k\ell=1}^{d} \int_0^t \phi^2_{n,k\ell}(s)\mathrm{d}s \tag{15.147}$$

and, by Doob's inequality,

$$E\sup_{0\le t\le T}\left|\sum_n \int_0^t \phi_n(s)\beta^n(\mathrm{d}s)\right|^2 \le 4\sum_n\sum_{k\ell=1}^{d}\int_0^T E\phi^2_{n,k\ell}(s)\mathrm{d}s \;\; \forall T>0. \tag{15.148}$$

Remark 15.49.

(i) The construction of Itô integrals with respect to a series of uncorrelated square integrable martingales follows exactly the same pattern. Similarly, the theory immediately generalizes to semimartingales of the following form

$$a(t) := b(t) + m(t), \tag{15.149}$$

where $b(\cdot)$ is a process of bounded variation and $m(\cdot)$ is a square integrable martingale.

(ii) The following generalization of (15.147) can be derived and is often quite useful. Suppose $\phi(\cdot) \in L_{2,\mathcal{F},\mathrm{loc}}([0,\infty) \times \Omega; \mathcal{M}_{d\times d})$ satisfies

$$\sum_{i,j=1}^{d} \int_0^T E\phi_{ij}^2(s)[m_i](\mathrm{d}s) < \infty \;\forall T, \tag{15.150}$$

where $m(\cdot) = (m_1(\cdot), \ldots, m_d(\cdot))^T$ is a continuous square integrable $\mathbf{R}^d$-valued martingale. The Itô integral

$$\int_0^t \phi(s)m(\mathrm{d}s)$$

is defined similarly to the Itô integral, driven by Brownian motions. It is also a continuous square integrable $\mathbf{R}^d$-valued martingale and

$$\left[\int_0^{\cdot} \phi(s)m(\mathrm{d}s)\right](t) \equiv \sum_{i,j,k=1}^{d} \int_0^t \phi_{ij}(s)\phi_{ik}(s)[m_j, m_k](\mathrm{d}s). \tag{15.151}$$

(iii) Using stopping times (15.150) can be relaxed to

$$\sum_{i,j=1}^{d} \int_0^T \phi_{ij}^2(s)[m_i](\mathrm{d}s) < \infty \quad \text{a.s. } \forall T, \tag{15.152}$$

We obtain that $\int_0^t \phi(s)m(\mathrm{d}s)$ exists and is a continuous locally square integrable $\mathbf{R}^d$-valued martingale. The extension of this statement to stochastic integrals, driven by continuous semimartingales, is obvious.

(iv) Apart from the necessary integrability assumptions on the integrand the most important assumption in stochastic Itô integration is that the integrand $\phi(t_{k-1}^n)$ must be $\mathcal{F}_{t_{k-1}^n}$-adapted and that the integrator increments $m(t_k^n) - m(t_{k-1}^n)$ are orthogonal (i.e. uncorrelated) to $\phi(t_{k-1}^n)$. We abbreviate this property by[66]

$$\phi(t_{k-1}^n) \perp m(t_k^n) - m(t_{k-1}^n) \quad \forall n, k \quad \text{and, for the limits,} \phi(s) \perp m(\mathrm{d}s) \quad \forall s. \tag{15.153}$$

□

The Itô integration has been extended to integrals driven by local (cadlag) martingales.[67] In fact, the Itô integral may be defined for general "good integrators," which are stochastic processes and allow a version of Lebesgue's dominated convergence

[66] For discontinuous martingales we need to evaluate the integrand at $s-$ in (15.153).

[67] Cf., e.g., Metivier and Pellaumail (1980), Liptser and Shiryayev (1974, 1986) as well as Protter (2004).

theorem.[68] One then shows that these good integrators are semimartingales, i.e., can be represented as in (15.149), where $m(\cdot)$ only needs to be integrable. Such a representation is called the "Doob-Meyer decomposition" of a semimartingale. We will return to the "good integrator" approach at the end of this subsection.

Next, we present the Itô formula, which is the most important formula in stochastic analysis and is the key in the transition from a Brownian particle movement to second-order parabolic PDEs and, as shown in Chaps. 8 and 14, also to second-order parabolic SPDEs and their macroscopic limits.[69] Again, we want to avoid a cumbersome notation, and will only present a special case, which was used in the derivation of SPDEs in this volume.

Theorem 15.50. *Itô's Formula: First Extended Chain Rule*

Let $\varphi(r,t)$ be a function from $\mathbf{R}^{d+1}$ into $\mathbf{R}$. Suppose φ is twice continuously differentiable function with respect to the spatial variables and once continuously differentiable with respect to t such that all partial derivatives are bounded. Let $m(\cdot)$ be a continuous square integrable $\mathbf{R}^d$-valued martingale and $b(\cdot)$ a continuous process of bounded variation. We set

$$a(t) := b(t) + m(t),$$

(cf. (15.149)). $\varphi(a(\cdot),t)$ is a continuous, locally square integrable semimartingale and the following formula holds:

$$\left.\begin{aligned} \varphi(a(t),t) = \varphi(a(0),0) &+ \int_0^t \left(\frac{\partial}{\partial s}\varphi\right)(a(s),s)\mathrm{d}s \\ &+ \int_0^t (\nabla\varphi)(a(s),s)\cdot(b(ds)+m(ds)) \\ &+ \frac{1}{2}\sum_{i,j=1}^d \int_0^t (\partial^2_{i,j}\varphi)(a(s),s)[m_i,m_j](\mathrm{d}s), \end{aligned}\right\} \qquad (15.154)$$

where $[m_i,m_j](\cdot)$ are the mutual quadratic variations of the one-dimensional components of $m(\cdot)$.

Proof. (Sketch)

We only sketch the proof and refer the reader for more details to any book on stochastic analysis.[70]

Let $\{t_0^n < t_1^n < \cdots < t_k^n < \cdots\}$ be a sequence of partitions of $[0,\infty)$ as in Definition 15.33. Then,

[68] Cf. Protter (loc.cit.), Chap. IV, Theorem 32.

[69] Recall from Chap. 14 that these macroscopic limits are themselves solutions of second-order parabolic PDEs.

[70] Cf., e.g., Ikeda and Watanabe (loc.cit.), Chap. II.5 or Gikhman and Skorokhod (1982), Chap.3.2.

$$
\left.\begin{aligned}
&\varphi(a(t),t)-\varphi(a(0),0)\\
&=\sum_k \varphi(a(t_k^n\wedge t),t_k^n\wedge t)-\varphi(a(t_{k-1}^n\wedge t),t_{k-1}^n\wedge t)\\
&=\sum_k \varphi(a(t_{k-1}^n\wedge t),t_k^n\wedge t)-\varphi(a(t_{k-1}^n\wedge t),t_{k-1}^n\wedge t)\\
&\quad+\sum_k \varphi(a(t_k^n\wedge t),t_k^n\wedge t)-\varphi(a(t_{k-1}^n\wedge t),t_k^n\wedge t)\\
&=: S_{n,1}(t)+S_{n,2}(t).
\end{aligned}\right\} \tag{15.155}
$$

Apparently,

$$
S_{n,1}(t)\longrightarrow \int_0^t \frac{\partial}{\partial t}\varphi(a(s),s)\mathrm{d}s, \quad \text{as } n\longrightarrow\infty, \text{ uniformly on compact intervals } [0,T]. \tag{15.156}
$$

By Taylor's formula,

$$
\left.\begin{aligned}
S_{n,2}(t)&=\sum_k \varphi(a(t_k^n\wedge t),t_k^n\wedge t)-\varphi(a(t_{k-1}^n\wedge t),t_k^n\wedge t)\\
&=\sum_k \nabla\varphi(a(t_{k-1}^n\wedge t),t_k^n\wedge t)\cdot(a(t_k^n\wedge t)-a(t_{k-1}^n\wedge t))\\
&\quad+\sum_k \frac{1}{2}\sum_{i,j=1}^d \partial_{i,j}^2\varphi(a(t_{k-1}^n\wedge t),t_{k1}^n\wedge t)(a_i(t_k^n\wedge t)\\
&\quad -a_i(t_{k-1}^n\wedge t))(a_j(t_k^n\wedge t)-a_j(t_{k-1}^n\wedge t))\\
&\quad +R_n(t,\varphi,a),
\end{aligned}\right\} \tag{15.157}
$$

where the remainder term $R_n(t,\varphi,a)$ tends to 0 in probability.[71] Next, we note that

$$
\nabla\varphi(a(t_{k-1}^n\wedge t),t_k^n)\ \perp m(t_k^n\wedge t)-m(t_{k-1}^n\wedge t)\ \ \forall n,k. \tag{15.158}
$$

Therefore, $\nabla\varphi(a(t_{k-1}^n\wedge t),t_k^n)$ satisfies the assumptions of the construction of the Itô integral with respect to the semimartingale increments $a(t_k^n\wedge t)-a(t_{k-1}^n\wedge t)$. Hence, the integrability assumptions on $\nabla\varphi(a(t_{k-1}^n\wedge t),t_k^n)$ allow us to conclude

[71] Assuming more differentiability on φ, we may show that the remainder term is dominated by $\sum_n |(a_i(t_k^n\wedge t)-a_i(t_{k-1}^n\wedge t))(a_j(t_k^n\wedge t)-a_j(t_{k-1}^n\wedge t))(a_\ell(t_k^n\wedge t)-a_\ell(t_{k-1}^n\wedge t))|$. As our martingale has finite quadratic variation, this sum has to tend to zero. This is very similar to showing that the quadratic variation of a continuous process of bounded variation equals 0.

$$\left.\begin{array}{l}\sum_k \nabla\varphi(a(t_{k-1}^n \wedge t), t_k^n \wedge t)\cdot(a(t_k^n \wedge t) - a(t_{k-1}^n \wedge t))\\ \longrightarrow \int_0^t \nabla\varphi(a(s), s)\cdot(b(\mathrm{d}s) + m(\mathrm{d}s)) \ \text{ as } n \longrightarrow \infty,\\ \text{uniformly on compact intervals } [0, T],\end{array}\right\} \quad (15.159)$$

Finally, recall that, by (15.93) we have for the mutual quadratic variations

$$[a_i, a_j](t) = [m_i, m_j](t). \quad (15.160)$$

Therefore, the integrability assumptions imply

$$\left.\begin{array}{l}\sum_k \frac{1}{2}\sum_{i,j=1}^d \partial_{i,j}^2\varphi(a(t_{k-1}^n \wedge t), t_{k-1}^n \wedge t)(a_i(t_k^n \wedge t)\\ -a_i(t_{k-1}^n \wedge t))(a_j(t_k^n \wedge t) - a_j(t_{k-1}^n \wedge t))\\ \longrightarrow \frac{1}{2}\sum_{i,j=1}^d \int_0^t \partial_{i,j}^2\varphi(a(s), s)[m_i, m_j](\mathrm{d}s), \ \text{ as } n \longrightarrow \infty,\\ \text{uniformly on compact intervals } [0, T],\end{array}\right\} \quad (15.161)$$

where the right-hand side is the usual (Lebesgue) Stieltjes integral. □

As an application of the Itô formula we now give the proof of Theorem 15.37 (Levy's characterization of Brownian motion), following Ethier and Kurtz (loc.cit.), Chap. 7.1, Theorem 1.1.

Proof of Theorem 15.37
Let $\theta \in \mathbf{R}^d$ be arbitrary and set

$$\left.\varphi(r, t) := \exp[i\theta\cdot r + \frac{1}{2}\theta\cdot(C\theta)t].\right\} \quad (15.162)$$

where $i := \sqrt{-1}$. Applying the Itô formula we obtain

$$\begin{aligned}&\exp\left[i\theta\cdot m(t) + \frac{1}{2}\theta\cdot(C\theta)t\right]\\ &= 1 + \int_0^t \exp\left[i\theta\cdot m(s) + \frac{1}{2}\theta\cdot(C\theta)s\right]\frac{1}{2}\theta\cdot(C\theta)\mathrm{d}s\\ &\quad + \int_0^t \exp\left[i\theta\cdot m(s) + \frac{1}{2}\theta\cdot(C\theta)s\right]i\theta\cdot m(\mathrm{d}s)\\ &\quad - \frac{1}{2}\int_0^t \exp\left[i\theta\cdot m(s) + \frac{1}{2}\theta\cdot(C\theta)s\right]\theta\cdot(C\theta)\mathrm{d}s.\end{aligned}$$

Hence,

$$\left.\begin{aligned}&\exp\left[i\theta\cdot m(t)+\frac{1}{2}\theta\cdot(C\theta)t\right]\\&=1+\int_0^t\exp\left[i\theta\cdot m(s)+\frac{1}{2}\theta\cdot(C\theta)s\right]i\theta\cdot m(\mathrm{d}s).\end{aligned}\right\}\tag{15.163}$$

The right-hand side is (an $\mathcal{F}_t-$) continuous square integrable complex-valued martingale. Therefore, for $0\le s<t$

$$E\left\{\exp\left[i\theta\cdot m(t)+\frac{1}{2}\theta\cdot(C\theta)t\right]\Big|\mathcal{F}_s\right\}=\exp\left[i\theta\cdot m(s)+\frac{1}{2}\theta\cdot(C\theta)s\right]$$

or, equivalently,

$$E\left\{\exp\left[i\theta\cdot(m(t)-m(s))+\frac{1}{2}\theta\cdot(C\theta)t\right]\Big|\mathcal{F}_s\right\}=\exp\left[\frac{1}{2}\theta\cdot(C\theta)(t-s)\right].\tag{15.164}$$

Consequently, $m(\cdot)$ is an $\mathbf{R}^d$-valued Brownian motion with covariance Ct. □

There are many extensions of Itô's formula in finite dimensions as well as extensions to Hilbert space valued semimartingales.[72] In fact, (15.89) is a special case of the Itô's formula in Hilbert space. For finite dimensional more general semimartingales we refer the reader to Ikeda and Watanabe (loc.cit.) and Gikhman and Skorokhod (loc.cit.) Chap. 4.4. Krylov (1977), Chap. II.10, proves an Itô's formula in finite dimensions where the function φ only needs to have second derivatives in the generalized sense in addition to some other conditions. The semimartingales in Krylov's Itô formula are continuous and represented as solutions of Itôs SODEs driven by Brownian motions. We now present a generalization in finite dimensions, which applies to continuous semimartingales but where the function φ is replaced by a semimartingale that depends on a spatial parameter.[73]

Theorem 15.51 *Itô-Wentzell Formula – Second Extended Chain Rule*

Consider the semi–martingale from (7.5)

$$\hat{S}(r,t):=\int_0^t F(r,u)\mathrm{d}u+\int_0^t\int\mathcal{J}(r,p,u)w(\mathrm{d}p,\mathrm{d}u).\tag{15.165}$$

Suppose F is continuously differentiable in r and $\mathcal{J}(r,p,u)$ is twice continuously differentiable in r. Let $a(\cdot)$ be a continuous locally square integrable $\mathbf{R}^d$-valued semimartingale with representation (15.152), $a(\cdot)=b(\cdot)+m(\cdot)$. Denoting by $\hat{S}_\ell$ the ℓ-th coordinate of $\hat{S}$, then the following formula holds for $\ell=1,\ldots,d$:

[72] Cf. for the latter, Metivier and Pellaumail, loc.cit. Chap. 2.

[73] Cf. Kunita, loc.cit., Chap. 3.3, Thereom 3.3.1, *Generalized Itô formula*. In view of the comparison of our approach with Kunita's formalism from our Chap. 7, we use our notation.

$$\left.\begin{aligned}&\hat{S}_\ell(a(t),t)=\hat{S}_\ell(a(0),0)+\int_0^t \hat{S}_\ell(a(s),\mathrm{d}s)\\&+\int_0^t(\nabla\hat{S}_\ell)(a(s),s)\cdot(b(\mathrm{d}s)+m(\mathrm{d}s))\\&+\frac{1}{2}\sum_{i,j=1}^{d}\int_0^t(\partial^2_{i,j}\hat{S}_\ell)(a(s),s)[m_i,m_j](\mathrm{d}s)\\&+\int_0^t\sum_{i,j=1}^{\mathrm{d}}\left[\int\left(\frac{\partial}{\partial r_i}\mathcal{J}_{\ell,j}\right)(a(s),p,s)w_j(\mathrm{d}p,\mathrm{d}s),m_i(\mathrm{d}s)\right].\end{aligned}\right\}\tag{15.166}$$

Proof. (Sketch)

We again only sketch the proof and refer the reader for more details to Kunita (loc.cit.) for a short proof and to Rozovsky (1983), Chap. 1.4, Theorem 9, for a detailed proof, where the semimartingale $\hat{S}(r,t)$ has a somewhat different representation.

We basically repeat the proof of the Itô formula with $\hat{S}_\ell$ instead of φ. The second term on the right-hand side of (15.166) is formally equivalent to $\int_0^t(\frac{\partial}{\partial s}\hat{S}_\ell)(a(s),s)\mathrm{d}s$, which corresponds to the second term on the right-hand side of (15.154). We need to be more careful when applying Taylor's formula to $S_{n,2}(t)$. The problem is that, unlike in (15.158),[74] $\nabla\hat{S}_\ell(a(t^n_{k-1}\wedge t),t^n_k\wedge t)$ is anticipating with respect to the increments $(a(t^n_k\wedge t)-a(t^n_{k-1}\wedge t))$ and we cannot expect the sum over all k to converge to an Itô integral driven by the semimartingale increments $a(\mathrm{d}s)$. Therefore, we must "correct" this term first before we employ martingale sums and Doob's inequality as in Proposition 15.34. We do this as follows:

$$\left.\begin{aligned}&\nabla\hat{S}_\ell(a(t^n_{k-1}\wedge t),t^n_k\wedge t)\cdot(a(t^n_k\wedge t)-a(t^n_{k-1}\wedge t))\\&=(\nabla\hat{S}_\ell(a(t^n_{k-1}\wedge t),t^n_k\wedge t)-\nabla\hat{S}_\ell(a(t^n_{k-1}\wedge t),t^n_{k-1}\wedge t))\cdot(a(t^n_k\wedge t)-a(t^n_{k-1}\wedge t))\\&+\nabla\hat{S}_\ell(a(t^n_{k-1}\wedge t),t^n_{k-1}\wedge t)\cdot(a(t^n_k\wedge t)-a(t^n_{k-1}\wedge t)).\end{aligned}\right\}\tag{15.167}$$

We now have the analogue of (15.161), namely,

$$\nabla\hat{S}_\ell(a(t^n_{k-1}\wedge t),t^n_{k-1}\wedge t)\ \perp\ m(t^n_k\wedge t)-m(t^n_{k-1}\wedge t),\tag{15.168}$$

whence summing up over all k, the integrability assumptions imply

$$\left.\begin{aligned}&\sum_k\nabla\hat{S}_\ell(a(t^n_{k-1}\wedge t),t^n_{k-1}\wedge t)\cdot(a(t^n_k\wedge t)-a(t^n_{k-1}\wedge t))\\&\longrightarrow\int_0^t\nabla\hat{S}_\ell(a(s),s)\cdot(b(\mathrm{d}s)+m(\mathrm{d}s))\\&\text{as } n\longrightarrow\infty,\text{ uniformly on compact intervals } [0,T].\end{aligned}\right\}\tag{15.169}$$

[74] $\varphi(a(s),t)$ is $\mathcal{F}_s$-adapted for all t, because $\varphi(r,t)$ is deterministic.

Next, our assumptions imply that the partial derivatives of $\hat{S}(r,\cdot)$ are themselves continuous semimartingales. Therefore,

$$\sum_k \nabla \hat{S}_\ell(a(t^n_{k-1}\wedge t), t^n_k \wedge t) - \nabla \hat{S}_\ell(a(t^n_{k-1}\wedge t), t^n_{k-1}\wedge t)\cdot (a(t^n_k\wedge t) - a(t^n_{k-1}\wedge t)$$

converges toward the mutual quadratic variation of the partial derivatives of $\hat{S}(r,\cdot)$ and the semimartingale $a(\cdot)$. Further, observe that, by (15.193), the terms of continuous bounded variation do not contribute to the limit. Thus, we obtain the "correction term" to the classical Itô formula

$$\left.\begin{aligned} &\int_0^t \sum_{i,j=1}^d \left[\int \left(\frac{\partial}{\partial r_i}\mathcal{J}_{\ell,j}\right)(a(s),p,s)w_j(\mathrm{d}p,\mathrm{d}s), m_i(\mathrm{d}s)\right] \\ &= \sum_{n=1}^\infty \int_0^t \sum_{i,j=1}^{\mathrm{d}} \left[\left(\frac{\partial}{\partial r_i}\sigma_{n,\ell,j}\right)(a(s),s)\beta^n_j(\mathrm{d}s), m_i(\mathrm{d}s)\right], \end{aligned}\right\} \tag{15.170}$$

employing for the second equality (4.15) with $\{\beta^n(\cdot)\}$ a system of i.i.d. standard $\mathbf{R}^d$-valued Brownian motions and the representation

$$\frac{\partial}{\partial r_i}\sigma_n(r,t) := \int \left(\frac{\partial}{\partial r_i}\mathcal{J}\right)(r,p,t)\widehat{\phi}_n(p)\mathrm{d}p. \qquad \square$$

We remark that the second representation of the correction formula in (15.170) is an extension (from a finite sum to an infinite series) of the formula proved in Rozovski (loc.cit.)

We now return to the "good integrator" approach, following Protter (loc.cit.) and restricting both integrators and integrands to $\mathcal{M}_{d\times d}$- and $\mathbf{R}^d$-valued processes, respectively. In the generality of cadlag integrators we need to have some restrictions on the integrands. The first step is to consider processes that are continuous from the left with limits from the right as integrands, called by its French acronym "*caglad*." If $f(\cdot)$ is cadlag then the process $f(\cdot-)$, defined by

$$f(t-) := \lim_{s\uparrow t} f(s) \;\; \forall t$$

is caglad. Hence, we may work with cadlag processes both for integrators and integrands, taking always the left-hand limits of the integrand. The class of integrands will be denoted $D([0,\infty);\mathcal{M}_{d\times d})$.[75] Our integrators will be cadlag semimartingales with decomposition (15.149). The construction of the stochastic integral

$$J(\phi,a) := \int_0^\cdot f(s-)a(\mathrm{d}s) \in L_0(\Omega; D([0,\infty);\mathbf{R}^d))) \tag{15.171}$$

for $f(\cdot) \in L_0(\Omega; D([0,\infty);\mathcal{M}_{d\times d}))$ and the semimartingale $a(\cdot) \in L_0(\Omega; D([0,\infty);\mathbf{R}^d))$ follows the pattern of the continuous case, which we outlined earlier.[76] It is important to note that for stochastic integral driven by the martingale term, $m(\cdot)$, we employed Doob's inequality (Theorem 15.32), which implies that

[75] We endow $\mathcal{M}_{d\times d}$ with the usual matrix norm $\|\cdot\|_{\mathcal{M}_{d\times d}}$.

[76] Cf. Proposition 15.34 and Definition 15.48.

the stochastic integral is the limit in probability uniformly on compact intervals. This fact is independent of whether or not $m(\cdot)$ is continuous or merely cadlag. For the integral driven by the process of bounded variation, $b(\cdot)$, we also employ the uniform metric and the total variation of the one-dimensional components of $b(\cdot)$ to obtain similar estimates. Therefore, we need to endow both $D([0,\infty); \mathcal{M}_{d\times d})$ and $D([0,\infty); \mathbf{R}^d)$ with the uniform metric $d_{u,\mathcal{M}_{d\times d}}$ and $d_{u,\mathbf{R}^d}$, respectively. The latter is defined by

$$d_{u,\mathbf{R}^d}(f,g) := \sum_{n=1}^{\infty} \sup_{0\le t\le n} \rho(f(t)-g(t))2^{-n} \tag{15.172}$$

and similarly for $d_{u,\mathcal{M}_{d\times d}}$. Further, $L_0(\Omega; D([0,\infty); \mathcal{M}_{d\times d}))$ and $L_0(\Omega; D([0,\infty); \mathcal{M}_{d\times d}))$ are the corresponding classes of processes, endowed with the metric of convergence in probability, $d_{\text{prob},u,\infty,\mathcal{M}_{d\times d}}$ and $d_{\text{prob},u,\infty,\mathbf{R}^d}$, respectively.[77] Protter proves the following[78]

Theorem 15.52. *The mapping*

$$\begin{aligned} J((\cdot),a) : (L_0(\Omega; D([0,\infty); \mathcal{M}_{d\times d})), d_{\text{prob},u,\infty,\mathcal{M}_{d\times d}}) \\ \longrightarrow (L_0(\Omega; D([0,\infty); \mathbf{R}^d)), d_{\text{prob},u,\infty,\mathbf{R}^d}) \end{aligned}$$

is continuous.

We next state a convergence condition for stochastic integrals due to Kurtz and Protter (1996), Sects. 6 and 7. The semimartingale integrators take values in separable Banach space, and most properties are formulated with respect to the weak topology of the Banach space. We formulate a special case of Theorem 7.5 in Kurtz and Protter (loc.cit.), which will be sufficient for the application in the mesoscopic limit theorem.

Let $\mathbf{H}$ be a separable Hilbert space with norm $\|\cdot\|_{\mathbf{H}}$ and scalar product $\langle\cdot,\cdot\rangle_{\mathbf{H}}$. Suppose $m_k^n(\cdot)$ is for each $k\in\mathbf{N}$ a sequence a real-valued square integrable mean zero martingales, $n\in\mathbf{N}\cup\{\infty\}$. Let ϕ_k be a CONS for $\mathbf{H}$ and suppose that for each $n\in\mathbf{N}\cup\{\infty\}$ and $\varphi\in\mathbf{H}$ the series

$$\sum_{k\in\mathbf{N}} m_k^n(\cdot)\langle\phi_k,\varphi\rangle_{\mathbf{H}}$$

converges in mean square uniformly on compact intervals. Set

$$M^n(\cdot) := \sum_{k\in\mathbf{N}} m_k^n(\cdot)\phi_k, \quad n\in\mathbf{N}\cup\{\infty\}. \tag{15.173}$$

---

[77] Cf. (15.84) with $\mathbf{R}^d$ or $\mathcal{M}_{d\times d}$ instead of $\mathbf{H}$.

[78] Protter (loc.cit.), Chap. II, Theorem 11. Protter's definition and his Theorem 11 are stated only for the case where both integrands and integrators are real valued. Since the stochastic integral is constructed for each component in the product between a matrix and a vector we may, without loss of generality, state the theorem in the appropriate multi-dimensional setting.

We call $M^n(\cdot)$ an **H**-*valued weak (square integrable) martingale*. Generalizing the considerations of Sect. 15.2.3 and 15.2.4 to the sequence $\sum_{k\in\mathbf{N}} m_k^n(\cdot)\phi_k$, we define the *tensor quadratic variation of* $M^n(t)$ by

$$|[M^n]|(t) := [m_k^n, m_\ell^n](t)\phi_k \otimes \phi_\ell. \ k, \ell \in \mathbf{N}. \tag{15.174}$$

$\phi_k \otimes \phi_\ell$ in (15.174) is the tensor product of ϕ_k and ϕ_ℓ. We verify that the above convergence condition is equivalent to the statement that for each t $E|[M^n]|(t)$ is a bounded operator on **H**. If $E|[M^n]|(t)$ is nuclear for each t (i.e., it is the product of two Hilbert-Schmidt operators), we call $M^n(\cdot)$ "*regular*." Otherwise we call $M^n(\cdot)$ "*cylindrical*."[79]

Next, let $\mathcal{J}$ be the collection of **H**-valued cadlag processes $\Phi(\cdot)$ which is represented by

$$\Phi(\cdot) = \sum_{k=1}^{L} f_k(\cdot)\tilde{\phi}_k, \quad L \in \mathbf{N}, \tilde{\phi}_k \in \mathbf{H}, \tag{15.175}$$

where $f_k(\cdot)$ are real-valued cadlag processes. The stochastic integral for $\Phi(\cdot) \in \mathcal{A}$ and $M^n(\cdot)$ is defined by

$$\int_0^t \langle \Phi(s-), M^n(ds)\rangle := \sum_{k=1}^{L}\sum_{\ell\in\mathbf{N}} \int_0^t f_k(s-)m_\ell(ds)\langle\tilde{\phi}_k, \phi_\ell\rangle_{\mathbf{H}} \tag{15.176}$$

The quadratic variation of $\int_0\langle\Phi(s-), M^n(ds)\rangle$ can be represented by

$$\left[\int_0^t \langle\phi(s-), M^n(ds)\rangle\right] := \int_0^t \left\langle |[M^n]|(ds)\left(\sum_{k=1}^{L} f_k(s-)\tilde{\phi}_k\right), \sum_{k=1}^{L} f_k(s-)\tilde{\phi}_k\right\rangle_{\mathbf{H}}. \tag{15.177}$$

Hence, by the boundedness of $|[M^n]|(t)$, we extend this definition to all **H**-valued square integrable cadlag processes $\Phi(\cdot)$, using a Fourier expansion of $\Phi(\cdot)$ with respect to some CONS and obtain for the quadratic variation

$$\left[\int_0^t \langle\Phi(s-), M^n(ds)\rangle\right] := \int_0^t \langle|[M^n]|(ds)(\Phi(s-), \Phi(s-)\rangle_{\mathbf{H}}. \tag{15.178}$$

Needless to say that, employing stopping times, we can extend the definition of the stochastic integral and (15.178) to the case where the tensor quadratic variation $|[M^n|](\cdot)$ is only locally integrable.

Definition 15.53. *The sequence* $M^n(\cdot)$ *of* **H**-*valued (possibly cylindrical) martingales,* $n \in \mathbf{N}$ *is called* "uniformly tight" *if for each* T *and* $\delta > 0$ *there is a constant* $K(T,\delta)$ *such that*

$$P\left\{\sup_{0\le t\le T} \frac{1}{K(T,\delta)}\left|\int_0^t \langle\Phi(s-), M^n(ds)\rangle\right| \ge \delta\right\} \le \delta \tag{15.179}$$

for each n *and all* $\mathcal{F}_{n,t}$*-adapted* $\Phi(\cdot) \in \mathcal{A}$ *satisfying* $\sup_{0\le t\le T} \|\Phi_n(t)\|_{\mathbf{H}} \le 1$. □

[79] Cf. Definition 15.28 of regular and cylindrical **H**-valued Brownian motions.

Next, we consider a sequence of stochastic integral equations with valued in $\mathbf{R}^{\bar{k}}$. The integrands are $\hat{F}^n(\hat{r}, s-) = (F_1^n(\hat{r}, s-), \ldots, F_{\bar{k}}^n(\hat{r}, s-))^T$ such that $F_i^n(\hat{r}, \cdot) \in \mathbf{H}$, $n \in \mathbf{N} \cup \{\infty\}$. Further, instead of the usual initial condition there is an $\mathbf{R}^{\bar{k}}$-valued sequence of cadlag processes $U^n(\cdot)$, $n \in \mathbf{N} \cup \{\infty\}$. Hence the sequence of stochastic integral equations can be written as

$$\hat{r}^n(t) = U^n(t) + \int_0^t \left\langle \hat{F}^n(\hat{r}(s-), s-), M^n(\mathrm{d}s) \right\rangle, \ n \in \mathbf{N} \cup \{\infty\}, \tag{15.180}$$

where

$$\langle \hat{F}^n(\hat{r}(s-), s-), M^n(\mathrm{d}s) \rangle := (\langle \hat{F}_1^n(\hat{r}(s-), s-), M^n(\mathrm{d}s) \rangle, \ldots, \langle \hat{F}_{\bar{k}}^n(\hat{r}(s-), s-), M^n(\mathrm{d}s) \rangle)^T.$$

Kurtz and Protter provide conditions for the existence of solutions and uniqueness, and we will just assume the existence of solutions of (15.180) for $n \in \mathbf{N} \cup \{\infty\}$ and uniqueness for the limiting case $n = \infty$. Further, suppose that the following condition holds on the coefficients $\hat{F}^n$:

$$\forall c > 0, t > 0 \quad \sup_{|\hat{r}| \le c} \sup 0 \le s \le t \| F_i^n(\hat{r}, s) - F_i^\infty(\hat{r}, s) \|_{\mathbf{H}} \to 0, \quad \text{as } n \to \infty \,, i = 1, \ldots, \bar{k}. \tag{15.181}$$

Theorem 15.54.

(i) Suppose $\mathbf{H} = \mathbf{H}_0$ and $M^n(\cdot)$, given by (15.173), is uniformly tight and that (15.181) holds. Further, suppose that for any finite $\{\tilde{\phi}_k,\ k = 1, \ldots, L\} \subset \mathbf{H}_0$

$$(U^n(\cdot), \langle M^n(\cdot), \tilde{\phi}_1 \rangle, \ldots, \langle M^n(\cdot), \tilde{\phi}_L \rangle) \Longrightarrow (U^\infty(\cdot), \langle M^\infty(\cdot), \tilde{\Phi}_1 \rangle, \ldots, \langle M^\infty(\cdot), \tilde{\phi}_L \rangle)$$

$$in\ D([0, \infty); \mathbf{R}^{\bar{k}+L}).$$

Then for any finite $\{\tilde{\phi}_k,\ k = 1, \ldots, L\} \subset \mathbf{H}_0$

$$\begin{aligned} &(U^n(\cdot), \hat{r}^n(\cdot), \langle M^n(\cdot), \tilde{\phi}_1 \rangle, \ldots, \langle M^n(\cdot), \tilde{\phi}_L \rangle) \\ &\Longrightarrow (U^\infty(\cdot), \hat{r}^\infty(\cdot), \langle M^\infty(\cdot), \tilde{\phi}_1 \rangle, \ldots, \langle M^\infty(\cdot), \tilde{\phi}_L \rangle) \end{aligned} \tag{15.182}$$

in $D([0, \infty); \mathbf{R}^{2\bar{k}+L})$.

(ii) Fix $\gamma > d$.[80] *Suppose that, in addition to the conditions of part (i), for all $\hat{t}$, $\epsilon > 0$ and $L > 0$ there exists a $\delta > 0$ such that for all n*

$$P\{\sup_{t\hat{t}} |\langle M^n(\cdot), \varphi > |\rangle L\} \le \epsilon \ \ whenever \|\varphi\|_\gamma \le \delta.$$

(ii) Then,

$$(U^n(\cdot), \hat{r}^n(\cdot), M^n(\cdot)) \Longrightarrow (U^\infty(\cdot), \hat{r}^\infty(\cdot), M^\infty(\cdot) \quad in\ D([0, \infty); \mathbf{R}^{2\bar{k}} \times \mathcal{H}_{-\gamma}). \tag{15.183}$$

[80] Recall (15.32) and (15.36).

Proof. Part (i) is a special case of Theorem 7.5 in Kurtz and Protter (loc.cit.) using the uniqueness of the the solution of the limiting equation.

Part (ii) follows from (i) and Walsh (loc.cit.), Corollary 6.16. □

Remark 15.55. To better understand the uniform tightness condition in the paper by Kurtz and Protter, we now restrict the definitions and consequences to the finite-dimensional setting of Theorem 15.51. To this end let $a_n(\cdot)$ be a sequence of $\mathbf{R}^d$-valued cadlag $\mathcal{F}_{n,t}$ semimartingales and $f_n(\cdot)$ families of $\mathcal{M}_{d\times d}$-valued $\mathcal{F}_{n,t}$-adapted cadlag simple processes.

(i) The sequence $a_n(\cdot)$ is called "*uniformly tight*" if for each T and $\delta > 0$ there is a constant $K(T,\delta)$ such that for each n and all $\mathcal{M}_{d\times d}$-valued $\mathcal{F}_{n,t}$-adapted cadlag processes $f_n(\cdot)$

$$P\left\{\sup_{0\le t\le T}\frac{1}{K(T,\delta)}\left|\int_0^t f_n(s-)a_n(ds)\right|\ge\delta\right\}\le\delta \text{ provided } \sup_{0\le t\le T}\|f_n(t)\|_{\mathcal{M}_{d\times d}}\le 1\,.$$

(ii) It follows from Sect. 15.1.6 that $(D([0,\infty);\mathbf{R}^d), d_{u,\mathbf{R}^d})$ is a complete metric space and, by the vector space structure of $D([0,\infty);\mathbf{R}^d)$, it is a Frechét space.[81] Hence, $(L_0(\Omega;D([0,\infty);\mathbf{R}^d)), d_{\text{prob},u,\infty,\mathbf{R}^d})$ is also a Frechét space space. We obtain the same property for $(D([0,\infty);\mathcal{M}_{d\times d}), d_{u,\mathcal{M}_{d\times d}})$ and $(L_0(\Omega;D([0,\infty);\mathcal{M}_{d\times d})), d_{\text{prob},u,\infty,\mathcal{M}_{d\times d}})$. Since the mapping $J(\cdot,a)$ for fixed $a(\cdot)$ is linear Theorem 15.52 asserts

$$J(\cdot,a)\in\mathcal{L}(L_0(\Omega;D([0,\infty);\mathcal{M}_{d\times d})),L_0(\Omega;D([0,\infty);\mathbf{R}^d))), \quad (15.184)$$

i.e., $J(\cdot,a)$ is a bounded linear operator from the Frechét space space $L_0(\Omega;D([0,\infty);\mathcal{M}_{d\times d}))$ into the Frechét space space $L_0(\Omega;D([0,\infty);\mathbf{R}^d))$.

(iii) The subindex n at $f_n(\cdot)$ in Definition 15.53 merely signifies adaptedness to the family $\mathcal{F}_{n,t}$. Therefore, we may drop the subindex at $f_n(\cdot)$, and we may also use the metrics on $[0,\infty)$ in the above definition. Further, we may obviously drop the requirement that $f(\cdot)$ be in the unit sphere (in the uniform metric), incorporating the norm of $f(\cdot)$ into the constants as long as the norm of $f(\cdot)$ does not change with n. The assumption that $f(t,\omega)$ be bounded in t and ω implies boundedness of $(\cdot)$ in the uniform metric in probability. Having accomplished all these cosmetic changes, the property that $a_n(\cdot)$ be uniformly tight implies the following:
For each $f(\cdot)\in L_0(\Omega;D([0,\infty);\mathcal{M}_{d\times d}))$, the set $\{J(f,a_n):n\in\mathbf{N}\}$ is bounded in $L_0(\Omega;D([0,\infty);\mathbf{R}^d))$, i.e., it implies the assumptions of the uniform boundedness principle (Theorem 15.3).

(iv) We expect that a similar observation also holds in the Hilbert space case if, as in the proof of Theorem 15.54, we can find, uniformly in t, a Hilbert-Schmidt imbedding for the integrators. □

---

[81] Cf. Definition 15.2. $D([0,\infty);\mathbf{R}^d)$, endowed with the Skorokhod metric $d_{D,\mathbf{R}^d}$, $(D([0,\infty);\mathbf{R}^d), d_{D,\mathbf{R}^d})$, is both complete and separable. Hence, $(D([0,\infty);\mathbf{R}^d), d_{D,\mathbf{R}^d})$ is a separable Frechét space.

15.2.6 Stochastic Stratonovich Integrals

We briefly describe the Stratonovich integral, following Ikeda-Watanabe, and provide the transformation rule by which one may represent a Stratonovich integral as a sum of a stochastic Itô integral and a Stieltjes integral. Finally, we apply the formulas to a special case of the SODE (4.9).

Recall that by (15.139) and (15.81) the Itô integral for continuous semimartingale integrators is constructed as in a formal Stieltjes approximation, but taking the integrand at the left endpoint of a partition interval and the integrator as the difference of both endpoints. Stratonovich (1964) takes the midpoint.

Let $\tilde{a}(\cdot) = \tilde{b}(\cdot) + \tilde{m}(\cdot)$ and $a(\cdot) = b(\cdot) + m(\cdot)$ be continuous locally square integrable real-valued semimartingales (cf. (15.149)), adapted (as always in this book) to the filtration $\mathcal{F}_t$. Further, let $\{t_0^n < t_1^n < \cdots < t_k^n < \cdots\}$ be a sequence of partitions of $[0, \infty)$ as in Definition 15.33. Set

$$S_n(t, \tilde{a}, a) := \sum_k \frac{1}{2}\{\tilde{a}(t_k^n) + \tilde{a}(t_{k-1}^n)\}(a(t_k^n) - a(t_{k-1}^n)). \tag{15.185}$$

Ikeda and Watanabe show that $S_n(t, \tilde{a}, a)$ converges in probability, uniformly on compact intervals $[0, T]$. The proof of this statement is very similar to the proof of the Itô formula. Thus, the following is well defined:

Definition 15.56.

$$\int_0^t \tilde{a}(s) \circ a(ds) := \lim_{n\to\infty} S_n(t, \tilde{a}, a) \tag{15.186}$$

is called the "Stratonovich integral of $\tilde{a}(\cdot)$ with respect to $a(ds)$."[82] □

The representation of the approximating sequence in (15.185) immediately implies the following transformation rule:

Theorem 15.57. *Under the above assumptions*

$$\int_0^t \tilde{a}(s) \circ a(ds) \equiv \int_0^t \tilde{a}(s) a(ds) + [\tilde{m}, m](t) \; a.s., \tag{15.187}$$

where the stochastic integral in the right-hand side is the Itô integral.

Proof.

$$\left.\begin{aligned} &\sum_k \frac{1}{2}\{\tilde{a}(t_k^n) + \tilde{a}(t_{k-1}^n)\}(a(t_k^n) - a(t_{k-1}^n)) \\ &= \sum_k \tilde{a}(t_{k-1}^n)(a(t_k^n) - a(t_{k-1}^n)) + \sum_k \frac{1}{2}\{\tilde{a}(t_k^n) + \tilde{a}(t_{k-1}^n)\}(a(t_k^n) - a(t_{k-1}^n)). \end{aligned}\right\} \tag{15.188}$$

[82] Cf. with our attempt to define a Stieltjes approximation for (15.132), choosing $\{\beta_1(t_{k-1}^n \wedge t) + x_0 + \gamma(\beta_1(t_k^n \wedge t) - \beta_1(t_{k-1}^n \wedge t))\}$ as the evaluation point for the integrand $\beta_1(\cdot) + x_0$. We saw that Itô's choice is $\gamma = 0$, whereas, by (15.185), Stratonovich's choice is $\gamma = \frac{1}{2}$.

Clearly, the first sum tends to the stochastic Itô integral, whereas the second sum tends to the mutual quadratic variation of $\tilde{a}(\cdot)$ and $a(\cdot)$. By (15.93) this process reduces to the mutual quadratic variation of $\tilde{m}(\cdot)$ and $m(\cdot)$. $\square$

The mutual quadratic variation $[\tilde{m}, m](\cdot)$ in (15.187) is usually called the "*correction term*", which we must add to the Itô integral to obtain the corresponding Stratonovich integral. The generalization to multidimensional semimartingales follows from the real case componentwise.

Although the Stratonovich integral requires more regularity on the integrands and integrators than the Itô integral, transformations of Stratonovich integrals follow the usual chain rule, very much in difference from the Itô integral where the usual chain rule becomes the Itô formula. We adopt the following theorem from Ikeda and Watanabe (loc.cit.), Chap. III.1, Theorem 1.3.

Theorem 15.58. *The Chain Rule*

Suppose $a(\cdot) := (a_1(\cdot), \ldots, a_d(\cdot))$ is an $\mathbf{R}^d$-valued continuous locally square integrable semimartingale and $\varphi \in C^3(\mathbf{R}^d; \mathbf{R})$. Then,

$$\tilde{a}(\cdot) := \varphi(a(\cdot))$$

is a real-valued continuous square integrable semimartingale and the following representation holds:

$$\tilde{a}(t) \equiv \varphi(a(0)) + \sum_{i=1}^{d} \int_0^t (\partial_i \varphi)(a(s)) \circ a_i(\mathrm{d}s). \tag{15.189}$$

$\square$

In what follows, we will apply the Stratonovich integral to the study of a special case of semimartingales, appearing in the definition of our SODEs (4.9) with $\tilde{\mathcal{Y}}(\cdot) \in \mathcal{M}_{f,\mathrm{loc},2,(0,T]}$ $\forall T > 0$. More precisely, we consider

$$\left.\begin{aligned} \mathrm{d}r(t) &= F(r(t), \tilde{\mathcal{Y}}(t), t)\mathrm{d}t + \int \mathcal{J}(r(t), p, t) w(\mathrm{d}p, \mathrm{d}t) \\ r(0) &= r_0 \in L_{2,\mathcal{F}_0}(\mathbf{R}^d), \quad (\tilde{\mathcal{Y}}^+, \tilde{\mathcal{Y}}^-) \in L_{\mathrm{loc},2,\mathcal{F}}(C((0,T]; \mathbf{M} \times \mathbf{M})), \end{aligned}\right\} \tag{15.190}$$

assuming the conditions of Hypothesis 4.2, (i).[83] In the notation of Kunita (cf. also our Sect. 7) (15.190) defines the following space–time field, which is a family of semi-martingales, depending on a spatial parameter:

$$L(r,t) := \int_0^t F(r, \tilde{\mathcal{Y}}(s))\mathrm{d}s + \int_0^t \int \mathcal{J}(r, q, s) w(\mathrm{d}q, \mathrm{d}s). \tag{15.191}$$

Note that $\mathcal{J}$ does not depend on some measure $\tilde{\mathcal{Y}}(\cdot)$. If, in addition F does not depend on the measure process $\tilde{\mathcal{Y}}(\cdot)$, the field is Gaussian. The assumption of independence of the diffusion coefficient $\mathcal{J}$ of measure processes $\tilde{\mathcal{Y}}(t)$ allows us to define the Stratonovich integrals with respect to $w(\mathrm{d}q, \mathrm{d}s)$ in terms of Itô integrals.

[83] (15.190) is a special case of the SODE (4.9).

Proposition 15.59. *Let $z(\cdot)$ be a continuous square integrable $\mathbf{R}^d$-valued semimartingale. Further, suppose that, in addition to Hypothesis 4.2, $\mathcal{J}(r,q,s)$ is twice continuously differentiable with respect to r such that*

$$\sup_{r\in\mathbf{R}^d}\sum_{i,j,k,\ell=1}^{d}\int_0^T\int(\partial^2_{ij}\mathcal{J}_{k,\ell})^2(r,q,s)\mathrm{d}q\,\mathrm{d}s<\infty. \tag{15.192}$$

Then, employing the representation (4.15), for all n the entries of $\int\mathcal{J}(z(\cdot),q,\cdot)$ $\hat{\phi}_n(q)\mathrm{d}q$ are continuous square integrable $\mathbf{R}^d$-valued semimartingales.

Proof. We apply Itô's formula (Theorem 15.50) to the functions

$$\varphi(r,t):=\int\mathcal{J}_{k\ell}(r,q,t)\Phi_n(q)\mathrm{d}q,\quad k,\ell=1,\ldots,d.$$

□

Consequently, employing the series representation (4.15) and assuming the conditions of Proposition 15.59, we can define the Stratonovich integral of $\mathcal{J}(z(s),q,s)$ with respect to $w(\mathrm{d}q,\mathrm{d}s)$ as a series of Stratonovich integrals from Definition 15.56 by[84]

$$\int_0^t\int\mathcal{J}(z(s),q,s)w(\mathrm{d}q,\circ\mathrm{d}s):=\sum_{n=1}^{\infty}\int_0^t\int\mathcal{J}(z(s),q,s)\hat{\phi}_n(q)\mathrm{d}q\circ\beta^n(\mathrm{d}s), \tag{15.193}$$

where the ith component of the stochastic integral is defined by

$$\begin{aligned}&\left\{\int_0^t\int\mathcal{J}(z(s),q,s)\hat{\phi}_n(q)\mathrm{d}q\circ\beta^n(\mathrm{d}s)\right\}_i\\&:=\sum_{j=1}^{d}\int_0^t\int\mathcal{J}_{ij}(z(s),q,s)\phi_n(q)\mathrm{d}q\circ\beta^n_j(\mathrm{d}s).\end{aligned} \tag{15.194}$$

Next, we consider the solution of (15.190). Since the coefficients are bounded this solution is a continuous square integrable $\mathbf{R}$-valued semimartingale.

Proposition 15.60. *Let $r(\cdot)$ be the solution of (15.190). Suppose that, in addition to the conditions of Proposition 15.59, the diffusion matrix, $D_{k\ell}$ associated with the diffusion kernel $\mathcal{J}_{k,\ell}(r,q,s)$ is spatially homogeneous, i.e., assume (8.33), and that the divergence of the diffusion matrix equals* 0 *at* $(0,t)$ $\forall t$, *i.e.,*

$$\sum_{k=1}^{d}(\partial_k\tilde{\mathrm{D}})_{k\ell}(0,t)\equiv 0\ \ \forall\ell. \tag{15.195}$$

[84] Our definition is equivalent to Kunita's definition of the Stratonovich integral (cf. Kunita, loc.cit. Chap. 3.2, and our Chap. 7).

Then

$$\int_0^t \int \mathcal{J}(r(s), q, s) w(\mathrm{d}q, \circ \mathrm{d}s) = \int_0^t \int \mathcal{J}(r(s), q, s) w(\mathrm{d}q, \mathrm{d}s), \tag{15.196}$$

i.e., the Stratonovich and the Itô integrals coincide in this particular case.

Proof. [85]

(i) Set

$$\sigma_{n,ij}(r,t) := \int \mathcal{J}_{ij}(r,q,s)\phi_n(q)\mathrm{d}q. \tag{15.197}$$

By Proposition 15.59 $\sigma_{n,ij}(r(\cdot),\cdot)$ are continuous square integrable semimartingales. Therefore, by (15.190)

$$\int_0^t \sigma_{n,ij}(r(s),s) \circ \beta_j^n(\mathrm{d}s) = \int_0^t \sigma_{n,ij}(r(s),s)\beta_j^n(\mathrm{d}s) + \left[\sigma_{n,ij}(r(t),t), \int_0^t \beta_j^n(ds)\right]$$

and all we need is to compute the correction term $[\sigma_{n,ij}(r(t),t), \int_0^t \beta_j^n(\mathrm{d}s)]$. To this end, we must find the semimartingale representation of $\sigma_{n,ij}(r(t))$. By Itô's formula

$$\begin{aligned}\sigma_{n,ij}(r(t),t) &= \sigma_{n,ij}(r(0),0) + \int_0^t \left(\frac{\partial}{\partial s}\sigma_{n,ij}\right)(r(s),s)\mathrm{d}s \\ &\quad + \int_0^t (\nabla\sigma_{n,ij})(r(s),s)\cdot r(\mathrm{d}s) + \frac{1}{2}\sum_{k\ell=1}^{d}\left(\partial_{k\ell}^2\sigma_{n,ij}\right)(r(s),s)[r_k,r_\ell](\mathrm{d}s),\end{aligned}$$

where the assumptions allow us to interchange differentiation and integration. By (15.93) the processes of bounded variation do not contribute to the calculation of $[\sigma_{n,ij}(r(t),t), \int_0^t \beta_j^n(\mathrm{d}s)]$. Hence,

$$\begin{aligned}&\left[\sigma_{n,ij}(r(t),t), \int_0^t \beta_j^n(\mathrm{d}s)\right] \equiv \sum_{k=1}^{d}\left[\int_0^t (\partial_k\sigma_{n,ij})(r(s),s)r_k(\mathrm{d}s), \int_0^t \beta_j^n(\mathrm{d}s)\right] \\ &\equiv \sum_{k=1}^{d}\left[\int_0^t (\partial_k\sigma_{n,ij})(r(s),s)\left\{\sum_m\sum_{\ell=1}^{d}\sigma_{m,k\ell}(r(s),s)\beta_\ell^m(\mathrm{d}s)\right\}, \int_0^t \beta_j^n(\mathrm{d}s)\right].\end{aligned} \tag{15.198}$$

Observe that β_j^n and β_ℓ^m are independent if $(n,j) \neq (m,\ell)$. Employing (15.151), we obtain

$$\left[\int_0^t \frac{\partial}{\partial r_k}\int \mathcal{J}_{ij}(r(t),q,t)\phi_n(q)\mathrm{d}q, \beta_j^n(t)\right] \equiv \int_0^t (\partial_k\sigma_{n,ij}(r(s),s))\sigma_{n,kj}(r(s),s)\mathrm{d}s. \tag{15.199}$$

[85] The proof is adopted from Kotelenez (2007).

Adding up

$$\sum_n \sum_{j=1}^d \sigma_{n,ij}(r,t)\sigma_{n,kj}(q,t) = \int (\mathcal{J}(r,p,t)\mathcal{J}(q,p,t))_{ik}\,\mathrm{d}p.$$

Hence, summing up the correction terms,

$$\sum_{k=1}^d \left(\frac{\partial}{\partial r_k}\sum_n \sum_{j=1}^d \sigma_{n,ij}(r,t)\sigma_{n,kj}(q,t)_{|q=r}\right)$$
$$= \sum_{k=1}^d \left(\frac{\partial}{\partial r_k}\int (\mathcal{J}(r,p,t)\mathcal{J}(q,p,t))_{ik}\,\mathrm{d}p)_{|q=r}\right).$$

However,

$$\int (\mathcal{J}(r,p,t)\mathcal{J}(q,p,t))_{ik}\,\mathrm{d}p = D_{ik}(r-q,t).$$

Hence,

$$\sum_{k=1}^d \sum_n \sum_{j=1}^d (\partial_k \sigma_{n,ij})(r,t)\sigma_{n,kj}(q,t)_{|q=r} \equiv \sum_{k=1}^d \frac{\partial}{\partial r_k} D_{ik}(r-q,t)_{|q=r} \equiv 0 \tag{15.200}$$

by assumption (15.198). We conclude that the sum of all correction terms equals 0, which implies (15.199). □

Lemma 15.61. *Let $\varphi \in C_c^3(\mathbf{R}^d;\mathbf{R})$ and assume the conditions and notation of Proposition 15.60. Then*

$$\sum_n \int_0^t (\nabla\varphi)(r(s)) \cdot (\circ\, \sigma_n(r(s),s)\beta^n(\mathrm{d}s))$$
$$= \sum_n \int_0^t (\nabla\varphi)(r(s)) \cdot (\sigma_n(r(s),s) \circ \beta^n(\mathrm{d}s)). \tag{15.201}$$

Proof.

(i) As in the proof of Proposition 15.60, we first analyze the one-dimensional coordinates of (15.201). Since the deterministic integral $\int_0^t F(r(s),\tilde{\mathcal{Y}},s)\mathrm{d}s$ is the same for Itô and Stratonovich integrals, we may in what follows assume, without loss of generality, $F \equiv 0$.

(ii) By (15.187),

$$
\left.\begin{aligned}
&\sum_{j=1}^{d}\int_0^t (\partial_k\varphi)(r(s))\circ\sigma_{n,kj}(r(s),s)\beta_j^n(\mathrm{d}s)\\
&=\sum_{j=1}^{d}\int_0^t (\partial_k\varphi)(r(s))\sigma_{n,kj}(r(s),s)\beta_j^n(\mathrm{d}s)\\
&+\frac{1}{2}\left[(\partial_k\varphi)(r(s)),\sum_{j=1}^{d}\int_0^t \sigma_{n,kj}(r(s),s)\beta_j^n(\mathrm{d}s)\right].
\end{aligned}\right\} \qquad (15.202)
$$

To compute the correction term we apply Itô's formula to $(\partial_k\varphi)(r(t))$. Recalling that, by (15.93), only the martingale part in Itô's formula contributes to the mutual quadratic variation, we obtain

$$
\begin{aligned}
&\frac{1}{2}\left[((\partial_k\varphi)(r(t)),\sum_{j=1}^{d}\int_0^t \sigma_{n,kj}(R(s),s)\beta_j^n(\mathrm{d}s)\right]\\
&=\frac{1}{2}\Bigg[\int_0^t\left(\sum_{\ell=1}^{d}(\partial_{k\ell}^2\varphi)(r(s))\right)\left\{\sum_m\sum_{\ell=1}^{d}\sigma_{m,\ell i}(r(s),s)\beta_i^m(\mathrm{d}s)\right\}\\
&\qquad\times\sum_{j=1}^{d}\int_0^t \sigma_{n,kj}(r(s),s)\beta_j^n(\mathrm{d}s)\Bigg]\\
&=\frac{1}{2}\int_0^t\left(\sum_{\ell=1}^{d}(\partial_{k\ell}^2\varphi)(r(s))\right)\left\{\sum_{j=1}^{d}\sigma_{n,\ell j}(r(s),s)\sigma_{n,kj}(r(s),s)\mathrm{d}s\right\}.
\end{aligned}
$$

i.e.,

$$
\left.\begin{aligned}
&\frac{1}{2}\left[(\partial_k\varphi)(r(t)),\sum_{j=1}^{d}\int_0^t \sigma_{n,kj}(r(s),s)\beta_j^n(\mathrm{d}s)\right]\\
&=\frac{1}{2}\sum_{j,\ell=1}^{d}\int_0^t(\partial_{k\ell}^2\varphi)(r(s))\sigma_{n,\ell j}(r(s),s)\sigma_{n,kj}(r(s),s)\mathrm{d}s.
\end{aligned}\right\} \qquad (15.203)
$$

Summing up over all n yields:

$$
\left.\begin{aligned}
&\frac{1}{2}\left[\left(\frac{\partial}{\partial r_k}\varphi\right)(r(t)),\sum_n\sum_{j=1}^{d}\int_0^t \sigma_{n,kj}(r(s),s)\beta_j^n(\mathrm{d}s)\right]\\
&=\frac{1}{2}\sum_{\ell=1}^{d}\int_0^t(\partial_{k\ell}^2\varphi)(r(s))D_{\ell k}(0,s)\mathrm{d}s.
\end{aligned}\right\} \qquad (15.204)
$$

Summing up (15.202) over all n and changing the notation for the indices we obtain from the previous formulas

$$\left.\begin{aligned}&\int_0^t (\partial_k\varphi)(r(s))\circ\sum_n\sum_{\ell=1}^d \sigma_{n,k\ell}(r(s),s)\beta_\ell^n(\mathrm{d}s)\\&=\int_0^t (\partial_k\varphi)(r(s))\int\sum_{\ell=1}^d \mathcal{J}_{k\ell}(r(s),p,s)w_\ell(\mathrm{d}p,\mathrm{d}s)\\&\quad+\frac{1}{2}\sum_{\ell=1}^d\int_0^t(\partial^2_{k\ell}\varphi)(r(s))D_{\ell k}(0,s)\mathrm{d}s.\end{aligned}\right\}\tag{15.205}$$

(iii) Again by (15.187),

$$\left.\begin{aligned}&\sum_{j=1}^d\int_0^t(\partial_k\varphi)(r(s))\sigma_{n,kj}(r(s),s)\circ\beta_j^n(\mathrm{d}s)\\&=\sum_{j=1}^d\int_0^t(\partial_k\varphi)(r(s))\sigma_{n,kj}(r(s),s)\beta_j^n(\mathrm{d}s)\\&\quad+\frac{1}{2}\sum_{j=1}^d\left[(\partial_k\varphi)(r(t))\sigma_{n,kj}(r(t),t),\int_0^t\beta_j^n(\mathrm{d}s)\right].\end{aligned}\right\}\tag{15.206}$$

We now must compute the martingale term of $(\partial_k\varphi)(r(t))\sigma_{n,kj}(r(t),t)$, using Itô's formula. We verify that this term equals

$$\begin{aligned}&\int_0^t\sigma_{n,kj}(r(s),s)(\nabla\partial_k\varphi)(r(s))\cdot\int\mathcal{J}(r(s),p,s)w(\mathrm{d}p,\mathrm{d}s)\\&\quad+\int_0^t(\partial_k\varphi)(r(s))(\nabla\sigma_{n,kj}(r(s),s))\cdot\int\mathcal{J}(r(s),p,s)w(\mathrm{d}p,\mathrm{d}s).\end{aligned}$$

Employing the series representation (4.15) with the notation from (15.197), we obtain for the correction term

$$\begin{aligned}&\frac{1}{2}\sum_{j=1}^d\left[(\partial_k\varphi)(r(t))\sigma_{n,kj}(r(t),t),\int_0^t\beta_j^n(\mathrm{d}s)\right]\\&=\frac{1}{2}\sum_{j=1}^d\Bigg[\int_0^t\sigma_{n,kj}(r(s),s)\left(\sum_{\ell=1}^d\partial^2_{k\ell}\varphi\right)(r(s))\\&\quad\times\left\{\sum_m\sum_{i=1}^d\sigma_{m,\ell i}(r(s),s)\beta_i^m(\mathrm{d}s)\right\},\int_0^t\beta_j^n(\mathrm{d}s)\Bigg]\\&\quad+\frac{1}{2}\sum_{j=1}^d\Bigg[\int_0^t\sum_{\ell=1}^d(\partial_k\varphi)(r(s))(\partial_\ell\sigma_{n,kj})(r(s),s))\\&\quad\times\left\{\sum_m\sum_{i=1}^d\sigma_{m,\ell i}(r(s),s)\beta_i^m(\mathrm{d}s)\right\},\int_0^t\beta_j^n(\mathrm{d}s)\Bigg].\end{aligned}$$

Observing the orthogonality of the Brownian motions the correction term simplifies

$$
\begin{aligned}
&\frac{1}{2}\sum_{j=1}^{d}\left[(\partial_k\varphi)(r(t))\sigma_{n,kj}(r(t),t),\int_0^t \beta_j^n(\mathrm{d}s)\right]\\
&=\frac{1}{2}\sum_{j=1}^{d}\left[\int_0^t \sigma_{n,kj}(r(s),s)\left(\sum_{\ell=1}^{d}\partial_{k\ell}^2\varphi)(r(s))\sigma_{n,\ell j}(r(s),s\right)\beta_j^n(\mathrm{d}s),\int_0^t \beta_j^n(\mathrm{d}s)\right]\\
&+\frac{1}{2}\sum_{j=1}^{d}\left[\int_0^t\left(\sum_{\ell=1}^{d}(\partial_k\varphi)(r(s))\right)(\partial_\ell\sigma_{n,kj})(r(s),s))\sigma_{n,\ell j}(r(s),s)\beta_j^n(\mathrm{d}s),\right.\\
&\qquad\left.\times\int_0^t \beta_j^n(\mathrm{d}s)\right].
\end{aligned}
$$

whence by (15.151) the correction term satisfies the following equation:

$$
\left.
\begin{aligned}
&\frac{1}{2}\sum_{j=1}^{d}\left[(\partial_k\varphi)(r(t))\sigma_{n,kj}(r(t),t),\int_0^t \beta_j^n(\mathrm{d}s)\right]\\
&=\frac{1}{2}\sum_{\ell,j=1}^{d}\int_0^t(\partial_{k\ell}^2\varphi)(r(s))\sigma_{n,kj}(r(s),s)\sigma_{n,\ell j}(r(s),s)\mathrm{d}s\\
&+\frac{1}{2}\sum_{\ell,j=1}^{d}\int_0^t(\partial_k\varphi)(r(s))(\partial_\ell\sigma_{n,kj})(r(s),s))\sigma_{n,\ell j}(r(s),s)\mathrm{d}s.
\end{aligned}
\right\}
\tag{15.207}
$$

It remains to sum up the correction terms over all n. First, we compute the sum of the second terms. Basically repeating the calculations of (15.200),

$$
\left.
\begin{aligned}
&\frac{1}{2}\sum_{\ell,j=1}^{d}\int_0^t((\partial_k\varphi)(r(s))\sum_n(\partial_\ell\sum_n\sigma_{n,kj})(r(s),s))\sigma_{n,\ell j}(r(s),s)\,\mathrm{d}s\\
&=\frac{1}{2}\sum_{\ell,j=1}^{d}\int_0^t(\partial_k\varphi)(r(s))\frac{\partial}{\partial r_\ell}\int\mathcal{J}_{kj}(r,p,s)\mathcal{J}_{\ell j}(q,p,s)\mathrm{d}p_{|r=q=r(s)}\,\mathrm{d}s\\
&=\frac{1}{2}\sum_{\ell=1}^{d}\int_0^t(\partial_k\varphi)(r(s))\frac{\partial}{\partial r_\ell}D_{k,\ell}(r-q,s)_{|r=q=r(s)}\,\mathrm{d}s\\
&=0\quad\text{by assumption (15.195).}
\end{aligned}
\right\}
\tag{15.208}
$$

Summing up over the first terms in the right-hand side of (15.207) yields

$$
\left.
\begin{aligned}
&\frac{1}{2}\sum_{\ell,j=1}^{d}\int_0^t(\partial_{k\ell}^2\varphi)(r(s))\sum_n\sigma_{n,kj}(r(s),s)\sigma_{n,\ell j}(r(s),s)\mathrm{d}s\\
&=\frac{1}{2}\sum_{\ell}^{d}\int_0^t(\partial_{k\ell}^2\varphi)(r(s))D_{k\ell}(0,s)\mathrm{d}s.
\end{aligned}
\right\}
\tag{15.209}
$$

Since the diffusion matrix $D_{k\ell}$ is symmetric, we obtain by (15.204) that the correction terms for both sides of (15.201) are equal. □

We conclude this subsection with the following important observation:[86]

Theorem 15.62. *Suppose that the conditions of Proposition 15.60 hold and let $\varphi \in C_c^3(\mathbf{R}^d;\mathbf{R})$. Denoting by $r(\cdot)$ the solution of the (Itô) SODE (15.190) the following holds:*

$$
\left.
\begin{aligned}
\varphi(r(t)) &\equiv \frac{1}{2}\sum_{k,\ell=1}^{d}(\partial^2_{k\ell}\varphi)(r(s))D_{k\ell}(r(s),s)\mathrm{d}s\\
&+\sum_{k=1}^{d}\int_0^t(\nabla\varphi(r(s))\cdot\int\mathcal{J}_{k,\ell}(r(s),q,s)w_\ell(\mathrm{d}q,\mathrm{d}s)\\
&+\int_0^t(\nabla\varphi)(r(s))\cdot F(r(s),\tilde{\mathcal{Y}}(s),s)\mathrm{d}s\\
&=\sum_{k=1}^{d}\int_0^t(\nabla\varphi)(r(s))\cdot\int\mathcal{J}_{k,\ell}(r(s),q,s)w_\ell(\mathrm{d}q,\circ\mathrm{d}s)\\
&+\int_0^t(\nabla\varphi)(r(s))\cdot F(r(s),\tilde{\mathcal{Y}}(s),s)\mathrm{d}s.
\end{aligned}
\right\}
\tag{15.210}
$$

Proof. The Itô formula (Theorem 15.50) yields the the first part of (15.210). The chain rule for Stratonovich integrals (Theorem 15.58) in addition to Lemma 15.61 implies that $\varphi(r(t))$ equals the right-hand side of (15.210). □

15.2.7 Markov-Diffusion Processes

We review the definition of Markov-diffusion processes and prove the Markov property for a class of semilinar stochastic evolution equations in Hilbert space. We then derive Kolmogorov's backward and forward equations for $\mathbf{R}^d$-valued Markov-diffusion processes. The latter is also called the "Fokker–Planck equation." It follows that the SPDEs considered in this volume are stochastic Fokker-Planck equations for (generalized) densities (or "number densities").

The general description and analysis of Markov processes is found in Dynkin (1965). We will restrict ourselves only to some minimal properties of continuous (nonterminating) Markov-diffusion processes (i.e., where the sample paths are a.s. continuous) with values in $\mathbf{R}^d$. However, to allow applications to SPDEs and other infinite dimensional stochastic differential equations, we will initially consider continuous nonterminating processes with values in a separable metric space $\mathbf{B}$. For the basic definitions we choose the "canonical" probability space $\Omega := C([0,\infty);\mathbf{B})$, endowed with the Borel σ-algebra $\mathcal{F} := \mathcal{B}_{C([0,\infty);\mathbf{B})}$.

[86] Cf. Theorem 5.3 in Kotelenez (2007).

The first object is a family of transition probabilities[87] $P(s, \xi, t, B)$, $0 \le s \le t < \infty, \xi \in \mathbf{B}$, $B \in \mathcal{B}_{\mathbf{B}}$ (the Borel sets on $\mathbf{B}$). By assumption, this family satisfies the following properties:

Definition 15.63.

(I) A family of probability measures on $\mathcal{B}_{\mathbf{B}}$, $P(s, \xi, t, \cdot), 0 \le s \le t < \infty, \xi \in \mathbf{B}$, is called a family of transition probabilities if

(i) $P(s, \cdot, t, B)$ is $\mathcal{B}_{\mathbf{B}}$-measurable for all $0 \le s \le t$, $B \in \mathcal{B}_{\mathbf{B}}$ such that $P(s, \cdot, s, B) = 1_B(\cdot)$ for all $s \ge 0$,

(ii) and if for $0 \le s < u < t$, $\xi \in \mathbf{B}$ and $B \in \mathcal{B}_{\mathbf{B}}$ the "*Chapman-Kolmogorov equation*" holds:

$$P(s, \xi, t, B) = \int_{\mathbf{B}} P(u, \eta, t, B) P(s, \xi, u, \mathrm{d}\eta). \tag{15.211}$$

(II) Suppose, in addition to the family of transition probabilities, there is a probability measure μ on $\mathcal{B}_{\mathbf{B}}$. A probability measure P on $(\Omega, \mathcal{B}_{C([0,\infty);\mathbf{B})})$ is called a "*Markov process*" with transition probabilities $P(s, \xi, t, B)$ and initial distribution μ if

$$P(\omega_{\mathbf{B}}(0) \in B) = \mu(B)\ \forall B \in \mathcal{B}_{\mathbf{B}}. \tag{15.212}$$

and if for all $0 \le s < u < t$ and $B \in \mathcal{B}_{\mathbf{B}}$

$$P(\omega_{\mathbf{B}}(t) \in B | \sigma\{\omega_{\mathbf{B}}(u), 0 \le u \le s\}) = P(s, \omega_{\mathbf{B}}(s), t, B) \text{ a.s.,} \tag{15.213}$$

where the quantity in the left-hand side of (15.213) is the conditional probability, conditioned on the "*past*" up to time s and $\omega_{\mathbf{B}}(\cdot) \in \Omega$.

If $P(s, \xi, t, \cdot) \equiv P_h(t - s, \xi, \cdot)$, where $P_h(t - s, \xi, \cdot)$ is some time-homogeneous family of probability measures with similar properties as $P(s, \xi, t, \cdot)$, then the corresponding Markov process is said to be "*time homogeneous.*" □

Property (15.213) is called the "*Markov property.*" It signifies that knowing the state of the system at time s is sufficient to determine the future.

Obviously, a Markov process is not a stochastic process with a fixed distribution on the sample path but rather a family of processes, indexed by the initial distributions and having the same transition probabilities. It is similar to an ODE or SODE without specified initial value, where the "forcing function" describes the transition from a given state to the next state in an infinitesimal time-step. This begs the question of whether there is a deeper relation between SODEs (or ODEs)[88] and Markov processes. Indeed, it has been Itô's program to represent a large class of Markov processes by solutions of SODEs. To better understand this relation, let us look at a fairly general example of infinite dimensional stochastic differential equations related to semilinear SPDEs.

[87] We follow the definition provided by Stroock and Varadhan (1979), Chap. 2.2.

[88] Notice that ODEs can be considered special SODEs, namely where the diffusion coefficient equals 0.

Let $\mathbf{H}$ and $\mathbf{K}$ be a real separable Hilbert spaces with scalar products $\langle\cdot,\cdot\rangle_{\mathbf{H}}$, $\langle\cdot,\cdot\rangle_{\mathbf{K}}$ and norms $\|\cdot\|_{\mathbf{H}}$, $\|\cdot\|_{\mathbf{K}}$. Suppose there is a strongly continuous two-parameter semigroup $U(t,s)$ on $\mathbf{H}$.[89] In addition to $U(t,s)$, let $B(s,\cdot)$ be a family of (possibly) nonlinear operators on $\mathbf{H}$, $Z\in\mathbf{H}$, $s\geq 0$ such that

$U(t,s)B(s,\cdot)$ is bounded on $\mathbf{H}$ and measurable in $(s,Z)\in[0,t]\times\mathbf{H}$ for all $t\geq 0$.

Further, let $W(\cdot)$ be a $\mathbf{K}$-valued Brownian motion and $C(s,\cdot)$ a family of (possibly) nonlinear operators on $\mathbf{H}$ with values in the space of linear operators from $\mathbf{K}$ into $\mathbf{H}$. Suppose that

$(s,Z)\mapsto U(t,s)C(s,Z)$ is a jointly measurable map from $[0,t]\times\mathbf{H}$ into the bounded linear operators from $\mathbf{K}$ into $\mathbf{H}$ for all $t\geq 0$ such that for all $Z\in\mathbf{H}$ and $t\geq 0$

$$E\|\int_0^t U(t,s)C(s,Z)W(\mathrm{d}s)\|_{\mathbf{H}}^2<\infty.^{90}$$

Consider the following stochastic evolution equation on $\mathbf{H}$:

$$X(t)=X_s+\int_s^t U(t,u)B(u,X(u))\mathrm{d}u+\int_s^t U(t,u)C(u,X(u))W(\mathrm{d}u). \tag{15.214}$$

where X_s is an adapted square integrable initial condition. We assume in what follows that (15.217) has a unique solution $X(\cdot,s,X_s)$, which is adapted, square integrable and such that for any two adapted square integrable conditions X_s and Y_s the following continuous dependence on the initial conditions holds

$$\sup_{s\leq t\leq\bar{t}} E\|X(t,s,X_s)-X(t,s,Y_s)\|_H^2\leq c_{\bar{t}}E\|X_s-Y_s\|_H^2 \quad \forall\bar{t}<\infty, \tag{15.215}$$

where $c(\bar{t})<\infty$. If there is a generator $A(t)$ of $U(t,s)$ the solution of (15.214) is called a "*mild solution*." We adopt this term, whether or not there is a generator $A(t)$. Recall that in the finite dimensional setting of (4.10), as a consequence of global Lipschitz assumptions, (4.17) provides an estimate as in (15.215) if we replace the input processes $\tilde{\mathcal{Y}}$ by the empirical processes with N coordinates. Similar global (or even local) Lipschitz assumptions and linear growth assumptions on (15.214) guarantee the validity of estimate (15.215) (cf. Arnold et al. (1980) or DaPrato and Zabczyk (1992)).

Proposition 15.64. *Under the preceding assumption the solution of (15.214) is a Markov process with transition probabilities*

$$P(s,Z,t,B):=E1_B(X(t,s,Z)),\quad Z\in\mathbf{H},\ B\in\mathcal{B}_{\mathbf{H}}. \tag{15.216}$$

---

[89] Cf. Definition 15.41.

[90] This condition implies that $\int_0^t U(t,s)C(s,Z)W(\mathrm{d}s)$ is a regular $\mathbf{H}$-valued Gaussian random variable, i.e., it is a generalization an $\mathbf{H}$-valued regular Brownian motion whose covariance depends on t. Arnold et al. (loc.cit.) actually assumed $W(\cdot)$ to be regular on $\mathbf{K}$, etc. However, DaPrato and Zabczyk (1992) provide many examples where $W(\cdot)$ can be cylindrical and the "*smoothing effect*" of the semigroup "regularizes" the Brownian motion.

Proof.[91]

(i) Let $\mathcal{G}_s^t := \sigma\{W(v) - W(u) : s \le u \le v \le t\}$ and $\bar{\mathcal{G}}_s^t$ the completed σ-algebra. The existence of a solution $X(t, s, Z, \omega)$ of (15.214), which is measurable in $(s, Z, \omega) \in [0, t] \times \mathbf{H} \times \Omega$ with respect to the σ-algebra $\mathcal{B}_{[0,t]} \otimes \mathcal{B}_{\mathbf{H}} \otimes \bar{\mathcal{G}}_s^t$ follows verbatim as the derivation of property 3) of Theorem 4.5.[92] Hence the left-hand side of (15.216) defines a family of probability measures that is measurable in Z.

(ii) Next, we show the Markov property (15.213) with $X(\cdot)$ replacing $\omega_{\mathbf{B}}(\cdot)$. The following function

$$f_B(Z, \omega) := 1_B(X(t, s, Z, \omega)$$

is bounded and $\mathcal{B}_{\mathbf{H}} \otimes \bar{\mathcal{G}}_s^t$-measurable. Consider functions

$$g(Z, \omega) := \sum_{i=1}^{n} h_i(Z)\phi_i(\omega)$$

where all $h_i(\cdot)$ are bounded $\mathcal{B}_{\mathbf{H}}$-measurable and all $\phi_i(\cdot)$ are bounded $\bar{\mathcal{G}}_s^t$-measurable. Note that $\phi_i(\cdot)$ are independent of $\sigma\{X(u), 0 \le u \le s\}$, whence

$$E[\phi_i(\omega)|\sigma\{X(u), 0 \le u \le s\}] = E\phi_i(\cdot) = E[\phi_i(\omega)|\sigma\{X(s)\}].$$

Therefore,

$$\begin{aligned} E[g(X(s), \omega)|\sigma\{X(u), 0 \le u \le s\} &= \sum_{i=1}^{n} h_i(X(s))E[\phi_i(\omega)|\sigma\{X(s)\}] \\ &= E[g(X(s), \omega)|\sigma\{X(s)\}]. \end{aligned} \tag{15.217}$$

A variant of the monotone class theorem (cf. Dynkin (1961), Sect. 1.1, Lemma 1.2) implies that (15.217) can be extended to all bounded $\mathcal{B}_{\mathbf{H}} \otimes \bar{\mathcal{G}}_s^t$-measurable functions. Consequently, it holds for $f_B(Z, \omega)$ and the Markov property (15.213) follows.

(iii) For $0 \le s \le u \le t < \infty$, by the uniqueness of the solutions of (15.214),

$$X(t, s, Z) = X(t, u, X(u, s, Z)) \text{ a.s.}$$

Thus,

$$\left.\begin{aligned} P(s, Z, t, B) &= E1_B(X(t, s, Z)) = E1_B(X(t, u, X(u, s, Z))) \\ &= E[E[1_B(X(t, u, X(u, s, Z)))|\sigma\{X(u, s, Z)\}]] = E[P(u, X(u, s, Z), t, B)] \\ &= \int_{\mathbf{H}} P(u, \tilde{Z}, t, B)P(s, Z, u, d\tilde{Z}). \end{aligned}\right\} \tag{15.218}$$

This is the Chapman–Kolmogorov equation (15.211) and the proof of Proposition 15.64 is complete. □

[91] We follow the proof provided by Arnold et al. (1980), which itself is a generalization of a proof by Dynkin (1965), Vol. I, Chap. 11, for finite-dimensional SODEs.

[92] Cf. Chap. 6.

Corollary 15.65. *The unique solution* $(r^1(\cdot, s, q^1), \ldots, r^N(\cdot, s, q^N))$ *of (4.10) is an* $\mathbf{R}^{Nd}$*-valued continuous Markov process.* □

In what follows, we restrict ourselves to Markov-diffusion processes with values in $\mathbf{R}^d$. Accordingly, the canonical probability space is $(\Omega, \mathcal{F}) := (C([0,\infty); \mathbf{R}^d), \mathcal{B}_{C([0,\infty);\mathbf{R}^d)})$. We now write $r(\cdot)$ for an element of $C([0,\infty); \mathbf{R}^d)$.

Example 15.66.

(I)

$$P(s,q,t,B) := \int 1_B(r)(2\pi(t-s))^{-\frac{d}{2}} \exp\big(-\frac{|r-q|^2}{2(t-s)}\big)\mathrm{d}r, \tag{15.219}$$

which is the transition function of an $\mathbf{R}^d$-valued standard Brownian motion $\beta(\cdot)$. The most frequently used initial distribution is $\mu := \delta_0$.
A Markov process, generated by the transition function (15.219), is best represented by the solution of the following (simple!) SODE

$$\mathrm{d}r = \mathrm{d}\beta \;\; r_0 = q, \tag{15.220}$$

(II) Consider the ODE

$$\frac{\mathrm{d}}{\mathrm{d}t} r = F(r,t), \;\; r(s) = q. \tag{15.221}$$

Suppose the forcing function F is nice so that the ODE has a unique solution for all initial values (s,q), which is measurable in the initial conditions (s,q). Denote this solution $r(t,s,q)$. Set

$$P(s,q,t,B) := 1_B(r(t,s,q)). \tag{15.222}$$

(III) A more general example of a Markov process are the solutions of (4.10), considered in $\mathbf{R}^{dN}$.[93] In this case, we set

$$P(s,q,t,B) := E1_B(r_N(t,s,q)). \tag{15.223}$$

□

Let $P(s,q,t,B)$ be a family transition probabilities $P(s,q,t,B)$ satisfies the requirements of Definition 15.62. From the Chapman–Kolmogorov equation we obtain a two-parameter semigroup $U'(t,s)$ on the space of probability measures through

$$(U'(t,s)\mu)(\cdot) = \int P(s,q,t,\cdot)\mu(\mathrm{d}q). \tag{15.224}$$

Suppose that $U'(t,s)$ is strongly continuous on the space of probability measures, endowed with the restriction of the norm of the total variation of a finite signed measures. By the Riesz representation theorem[94] the transition probabilities also define in a canonical way a strongly continuous two-parameter semigroup $U(t,s)$ of bounded operators on the space $C_0(\mathbf{R}^d; \mathbf{R})$, setting for $f \in C_0(\mathbf{R}^d; \mathbf{R})$

[93] Cf. Corollary 15.65.

[94] Cf., e.g., Folland (1984), Chap. 7.3.

$$(U(t,s)f)(q) := \int P(s,q,t,\mathrm{d}\tilde{q})f(\tilde{q}). \tag{15.225}$$

If the transition probabilities are time-homogeneous, the corresponding semigroups, $U'(t,s)$ and $U(t,s)$, depend only on $t-s$ and have time independent generators, denoted A' and A, respectively. To better understand the relation between Kolmogorov's backward and forward equations and the general theory of semigroups let us first look at the time-homogeneous case. In this case we will write $P(t,q,B)$ for the transition probabilities. $C_0(\mathbf{R}^d;\mathbf{R})$ is the domain of the semigroup $U(t)$, endowed with the supremum norm $|||\cdot|||$. If $f \in \mathrm{Dom}(A)$ we obtain

$$(Af)(q) := \lim_{t\to 0}((U(t)f)(q) - f(q)) = \lim_{t\to 0}\frac{1}{t}\int P(t,q,\mathrm{d}r)(f(r)-f(q)), \tag{15.226}$$

where the limit is taken with respect to the norm $|||\cdot|||$. (15.226) is called "*Kolmogorov's backward equation*" in weak form, where "backward" refers to the fact that the evaluation of the generator (for Markov-diffusions the closure of a differential operator) is computed with respect to the initial variable q.[95]

We have seen in Chap. 14, (14.38)–(14.39) that a time-inhomogeneous Markov process, represented by an SODE, can be transformed into a time-homogeneous Markov process by adding time as an additional dimension. However, the analysis of backward and forward equations in the original time-inhomogeneous setting provides more insight into the structure of the time evolution of the Markov process. Therefore, we will now derive the backward equation for the time-inhomogeneous case (with time-dependent generator $A(s)$). Note that for the time-homogeneous case $P(t-s,q,B)$

$$\frac{\partial}{\partial t}P(t-s,q,B)_{|t-s=u} = -\frac{\partial}{\partial s}P(t-s,q,B)_{|t-s=u}.$$

Thus, we obtain in weak form the following "backward" equations for the two-parameter semigroups, $U'(t,s)$ and $U(t,s)$:[96]

$$-\frac{\partial}{\partial s}U'(t,s) = U'(t,s)A'(s), \quad -\frac{\partial}{\partial s}U(t,s) = U(t,s)A(s). \tag{15.227}$$

We denote the duality between measures and test functions by $\langle\cdot,\cdot\rangle$. Assuming $f \in \mathrm{Dom}(A(u))\ \forall u$, we then have

$$\left.\begin{aligned}&\langle (U'(t,s-h) - U'(t,s))\mu, f\rangle = \langle \mu, (U(t,s-h) - U(t,s))f\rangle \\ &= \int_{s-h}^{s}\langle \mu, U(t,u)A(u)f\rangle \mathrm{d}u = \int_{s-h}^{s}\langle A'(u)U'(t,u)\mu, f\rangle \mathrm{d}u\end{aligned}\right\} \tag{15.228}$$

[95] The reader, familiar with the semigroups of operators on a Banach space and their generators, will recognize that the left-hand side of (15.226) is the definition of the generator of $U(t)$ (cf., e.g., Pazy (1983)).

[96] For the analytic treatment of time-dependent generators and the existence of associated two-parameter semigroups we refer the reader to Tanabe (1979), Chap. 5.2, where $U(t,s)$ is called "*fundamental solution*." Note that, in our setting, Tanabe considers two-parameter semigroups on the space of test functions with $-A(t)$ being a generator.

We divide the quantities in (15.228) by h and pass to the limit. Choosing as the initial distribution at time s the Dirac measure δ_q yields "*Kolmogorov's backward equation*" for the two-parameter flow of transition probabilities

$$-\frac{\partial}{\partial s}P(s,q,t,\cdot) = (A'(s)P(s,\cdot,t,\cdot))(q), \quad P(s,q,s,\cdot) = \delta_q(\cdot). \tag{15.229}$$

For a solution of an SODE (or ODE) we compute the generator $A(s)$ of $U(t,s)$ through Itô's formula (or the simple chain rule). To simplify the notation, suppose that $N = 1$ in (4.10) and that the coefficients do not depend on the empirical process. Let $f \in C_c^2(\mathbf{R}^d;\mathbf{R})$. Then

$$\begin{aligned}(A(s)f)(q) &:= (F(q,s)\cdot\nabla f)(q)\\ &+\frac{1}{2}\sum_{k,\ell=1}^{d}\left\{\sum_{j=1}^{d}\int \mathcal{J}_{kj}(q,\tilde{q},s)\mathcal{J}_{\ell j}(q,\tilde{q},s)\mathrm{d}\tilde{q} + \sigma_{kj}^{\perp}(q,s)\sigma_{\ell j}^{\perp}(q,s)\right\}(\partial_{k\ell}^2 f)(q),\end{aligned} \tag{15.230}$$

where $\sigma^{\perp} := \sigma_1^{\perp}$. The only term missing from Itô's formula is the stochastic term, because it has mean 0 (15.230) and its generalizations and specializations are called the "*Itô-Dynkin formula*" for the generator.

Expression (15.228) lead to the backward equation, by varying the backward time argument s. Let us present the analogous argument by varying the forward time t. Then,

$$\left.\begin{aligned}&\langle (U'(t+h,s) - U'(t,s))\mu, f\rangle = \langle \mu, (U(t+h,s) - U(t,s))f\rangle\\ &= \int_t^{t+h}\langle \mu, A(u)U(u,t)f\langle \mathrm{d}u = \int_t^{t+h}\langle A'(u)U'(u,t)\mu, f\rangle \mathrm{d}u.\end{aligned}\right\} \tag{15.231}$$

We divide the quantities in (15.231) by h and pass to the limit. Choosing as the initial distribution at time s the Dirac measure δ_q yields "*Kolmogorov's forward equation*" for the two-parameter flow of transition probabilities

$$\frac{\partial}{\partial t}P(s,q,t,\cdot) = (A'(t)P(s,q,t,\cdot)), \quad P(s,q,s,\cdot) = \delta_q(\cdot). \tag{15.232}$$

Next, suppose that $P(s,q,t,B)$ has a density, $p(s,q,t,r)$, with respect to the Lebesgue measure so that

$$\int 1_B P(s,q,t,\mathrm{d}r) = \int 1_B p(s,q,t,r)\mathrm{d}r \;\; \forall t > s \ge 0, q \in \mathbf{R}^d, B \in \mathcal{B}^d. \tag{15.233}$$

Computing the duality of the right-hand side of (15.227) with start in μ against a smooth test function f we obtain

$$\langle (A'(t)\int p(s,q,t,\cdot)\mu(\mathrm{d}q), f\rangle = \int \mu(\mathrm{d}q)\int p(s,q,t,r)(A(t)f)(r)\mathrm{d}r. \tag{15.234}$$

We are mainly interested in the case when the generator A is the closure of a second-order differential operator as in (15.230), which defines a Markov-diffusion process. If, for such an operator, the coefficients are sufficiently smooth, we can integrate by parts in (15.234) and obtain adjoint generator, A^*, as an unbounded operator in $\mathbf{H}_0 = L_2(\mathbf{R}^d, \mathrm{d}r)$

$$\int p(s,q,t,r)(A(t)f)(r)\mathrm{d}r = \int (A^*(t)p(s,q,t,\cdot))(r)f(r)\mathrm{d}r. \tag{15.235}$$

Put (15.232) and (15.235) together and suppose that the domain of $A(t)$ is dense in $C_0(\mathbf{R}^d;\mathbf{R})$ for all $t \geq 0$. We then obtain "*Kolmogorov's forward equation*" for the density or, as it is called in the physics and engineering literature, the "*Fokker–Planck equation*":

$$\left(\frac{\partial}{\partial t}p\right)(s,q,t,r) \equiv (A^*p(s,q,t,\cdot))(r), \quad p(s,q,s,r) = \delta_q(r). \tag{15.236}$$

For the simplified version of (4.10), considered in (15.230), the Fokker–Planck equation is cast in the form

$$\left.\begin{array}{l}\left(\frac{\partial}{\partial t}p\right)(s,q,t,r) = -(\nabla \cdot F(\cdot,t)p(s,q,,t\cdot))(r)\\ +\frac{1}{2}\sum_{k,\ell=1}^{d}\partial^2_{k\ell}\left\{p(s,q,t,\cdot)\sum_{j=1}^{d}\int \mathcal{J}_{kj}(\cdot,\tilde{q},t)\mathcal{J}_{\ell j}(\cdot,\tilde{q},t)\mathrm{d}\tilde{q} + \sigma^{\perp}_{kj}(\cdot,t)\sigma^{\perp}_{k\ell}(\cdot,t)\right\}(r).\end{array}\right\} \tag{15.237}$$

Remark 15.67. As we have mentioned previously, the independence of the coefficients of the empirical process and $N = 1$ was assumed for notational convenience. We now compare (15.237) with the SPDEs in (8.26) or, more generally, in (8.55). Replacing the deterministic density $p(s,q,\cdot)$ by $X(t,\cdot)$ and adding the stochastic terms, we conclude that the SPDEs derived in this book are stochastic Fokker–Planck equations for the (generalized) density $X(t,\cdot)$. The generalized density corresponds to what is called "*number density*" in physics and engineering. □

15.2.8 Measure-Valued Flows: Proof of Proposition 4.3

(i) The definition of $\mu(\cdot)$ is equivalent to

$$\mu(t,\mathrm{d}r) = \int \delta_{\varphi(t,q)}(\mathrm{d}r)\mu_0(\mathrm{d}q).$$

So,

$$\gamma_{\varpi}(\mu(t)) \leq \int \varpi(r)\mu(t,\mathrm{d}r) = \int \varpi(\varphi(t,q))\mu_0(\mathrm{d}q).$$

Let $\beta > 0$ and consider $\varpi_\beta(r) := \varpi^\beta(r)$. By Itô's formula $\forall q$

$$\varpi_\beta(\varphi(t,q)) = \varpi_\beta(q) + \int_0^t (triangledown \varpi_\beta)(\varphi(s,q)) \cdot [\mathrm{d}b(s,q) + \mathrm{d}m(s,q)]$$
$$+ \frac{1}{2} \sum_{k,\ell=1}^d (\partial^2_{k\ell} \varpi_\beta)(\varphi(s,q)) \mathrm{d}[m_k(s,q), m_\ell(s,q)].$$

Hence, by the Burkholder-Davies-Gundy inequality and the usual inequality for increasing processes,

$$\left.\begin{aligned}
& E \sup_{0 \le t \le T} \varpi_\beta(\varphi(t,q) \\
& \le \varpi_\beta(q) + c \int_0^T \sum_{k=1}^d E|\partial_k \varpi_\beta(\varphi(s,q))| \mathrm{d}|b_k(s,q)| \\
& + cE \sqrt{\int_0^T |\sum_{k,\ell}^d \partial_k \varpi_\beta(\varphi(s,q)) \partial_\ell \varpi_\beta(\varphi(s,q))| \mathrm{d}[m_k(s,q), m_\ell(s,q)]} \\
& + cE \int_0^T \sum_{k,\ell}^d |\partial^2_{k,\ell} \varpi_\beta(\varphi(s,q))| \mathrm{d}[m_k(s,q), m_\ell(s,q)] \\
& \le \varpi_\beta(q) + c \int_0^T \sum_{k=1}^d E \varpi_\beta(\varphi(s,q)) \mathrm{d}|b_k(s,q)| \\
& + cE \sqrt{\int_0^T \sum_{k,\ell}^d \varpi^2_\beta(\varphi(s,q)) \mathrm{d}\{[m_k(s,q)] + [m_\ell(s,q)]\}} \\
& + cE \int_0^T \sum_{k,\ell}^d \varpi_\beta(\varphi(s,q)) \mathrm{d}\{[m_k(s,q)] + [m_\ell(s,q)]\} \\
& \text{(by (15.46))} \\
& \le \varpi_\beta(q) + c \int_0^T \sum_{k=1}^d E \varpi_\beta(\varphi(s,q)) \mathrm{d}|b_k(s,q)| \\
& + c \sqrt{\int_0^T \sum_{k,\ell}^d E \varpi^2_\beta(\varphi(s,q)) \mathrm{d}\{[m_k(s,q)] + [m_\ell(s,q)]\}} \\
& + cE \int_0^T \sum_{k,\ell}^d \varpi_\beta(\varphi(s,q)) \mathrm{d}\{[m_k(s,q)] + [m_\ell(s,q)]\}
\end{aligned}\right\} \quad (15.238)$$

From Itô's formula and (15.46) we obtain for $\tilde{\beta} > 0$

$$\begin{aligned}
& E \varpi_{\tilde{\beta}}(\varphi(t,q) \\
& \quad \le \varpi_{\tilde{\beta}}(q) + c \int_0^T \sum_{k=1}^d E \varpi_{\tilde{\beta}}(\varphi(s,q)) \mathrm{d}|b_k(s,q)| \\
& \quad + cE \int_0^T \sum_{k,\ell}^d \varpi^2_{\tilde{\beta}}(\varphi(s,q)) \mathrm{d}\{[m_k(s,q)] + [m_\ell(s,q)]\}.
\end{aligned}$$

Gronwall's inequality[97] implies

$$E\varpi_{\tilde\beta}(\varphi(t,q) \le c_{T,b,[m]}\varpi_{\tilde\beta}(q), \tag{15.239}$$

where the finite constant $c_{T,b,[m]}$ depends on T, $\tilde\beta$, the bounds of the derivatives of b and $\{[m_k(s,q)]+[m_\ell(s,q)]\}$. We employ (15.239) (with $\tilde\beta := 2\beta$) to estimate the square root in the last inequality of (15.238):

$$\begin{aligned} &c\sqrt{\int_0^T \sum_{k,\ell}^d E\varpi_\beta^2(s,q)\mathrm{d}\{[m_k(s,q)]+[m_\ell(s,q)]\}} \\ &\le c\sqrt{\int_0^T \sum_{k,\ell}^d E\varpi_\beta^2(s,q)c_{T,b,[m]}\mathrm{d}s]\}} \le \hat c_{T,b,[m]}\varpi_\beta(q) \end{aligned} \tag{15.240}$$

To estimate the remaining integrals in (15.238) we again use Gronwall's inequality and obtain altogether

$$E\sup_{0\le t\le T}\varpi_\beta(\varphi(t,q)) \le \bar c_{T,b,[m]}\varpi_\beta(q), \tag{15.241}$$

where $\bar c_{T,b,[m]}$ depends only upon T, β and upon the bounds of the characteristics of φ, but not upon φ itself. In what follows we choose $\beta = 1$.

We may change the initial measure μ_0 on $\mathcal{F}_0$-measurable set of arbitrary small probability such that

$$\operatorname{ess\,sup}_\omega \int \varpi(q)\mu_0(\mathrm{d}q) < \infty.$$

Therefore, $E\int\varpi(q)\mu_0(\mathrm{d}q) < \infty$. We may use conditional expectations instead of absolute ones in the preceding calculations[98] and assume in the first step that μ_0 is deterministic. Hence $E\sup_{0\le t\le T}\varpi(\varphi(t,q))$ is integrable against the initial measure μ_0, and we obtain

$$\begin{aligned} E\sup_{0\le t\le T}\int\varpi(\varphi(t,q))\mu_0(\mathrm{d}q) &\le \int E\sup_{0\le t\le T}\varpi(\varphi(t,q))\mu_0(\mathrm{d}q) \\ &\le c_{T,b}E\int\varpi(q)\mu_0(\mathrm{d}q) < \infty. \end{aligned} \tag{15.242}$$

As a result we have a.s.

$$\begin{aligned} \sup_{0\le t\le T}\gamma_\varpi(\mu(t)) &\le \sup_{0\le t\le T}\int\varpi(\varphi(t,q))\mu_0(\mathrm{d}q) \\ &\le \int\sup_{0\le t\le T}\varpi(\varphi(t,q))\mu_0(\mathrm{d}q) < \infty. \end{aligned} \tag{15.243}$$

---

[97] Cf. Proposition 15.5.

[98] Cf. (15.238).

(ii) We show that $\mu(\cdot)$ is continuous (a.s.) in $\mathbf{M}_{\infty,\varpi}$. Let $t > s \geq 0$. Then,

$$\begin{aligned}
\gamma_{\varpi}(\mu(t)-\mu(s)) = & \sup_{\|f\|_{L,\infty}\leq 1}\left|\int [f(\varphi(t,q))\varpi(\varphi(t,q)) \right.\\
& \left. -f(\varphi(s,q))\varpi(\varphi(s,q))]\mu_0(dq)\right| \\
\leq & \sup_{\|f\|_{L,\infty}\leq 1}\int \left|f(\varphi(t,q))-f(\varphi(s,q))\right|\varpi(\varphi(t,q))]\mu_0(dq) \\
& + \sup_{\|f\|_{L,\infty}\leq 1}\left|\int f(\varphi(s,q))[\varpi(\varphi(t,q))-\varpi(\varphi(s,q))]\mu_0(dq)\right| \\
=: & \, I(t,s)+II(t,s).
\end{aligned}$$

Since the Lipschitz constant for all f in the supremum is 1

$$I(t,s) \leq \int \rho(\varphi(t,q)-\varphi(s,q))\varpi(\varphi(t,q))\mu_0(dq).$$

Further, $\varphi(\cdot,q)$ is uniformly continuous in $[0,T]$ and $\rho(\varphi(t,q)-\varphi(s,q)) \leq 1$. Therefore, we conclude by (15.243) and Lebesgue's dominated convergence theorem that a.s.

$$\lim_{\delta\to 0}\sup_{0\leq s<t\leq T,|t-s|\leq\delta} I(t,s)=0.$$

Similarly,

$$|f(\varphi(s,q)(\varpi(\varphi(t,q))-\varpi(\varphi(s,q)))| \leq 2\sup_{0\leq t\leq T}\varpi(\varphi(t,q)),$$

where by (15.243) the right-hand side is a.s. integrable with respect to $\mu_0(dq)$. Therefore,

$$\begin{aligned}
II(t,s) = & \sup_{\|f\|_{L,\infty}\leq 1} |\int f(\varphi(s,q))[\varpi(\varphi(t,q))-\varpi(\varphi(s,q))]\mu_0(dq)| \\
\leq & \int |\varpi(\varphi(t,q))-\varpi(\varphi(s,q))|\mu_0(dq) \longrightarrow 0, \text{ as } |t-s|\to 0
\end{aligned}$$

by the continuity of the integrand in t and by Lebesgue's dominated convergence theorem. Thus, we have shown that with probability 1

$$\lim_{\delta\to 0}\sup_{0\leq s<t\leq T,|t-s|\leq\delta}\gamma_{\varpi}(\mu(t)-\mu(s))=0, \tag{15.244}$$

i.e., $\mu(\cdot)$ has continuous sample paths in $\mathbf{M}_{\infty,\varpi}$.

(iii) The adaptedness of $\mu(\cdot)$ follows directly from the construction. This finishes the proof of Part (a).

To show Part (b) we may again assume that the initial distribution is deterministic. The proof for $p=1$ follows immediately from (15.241). We may, therefore, without loss of generality, assume $p>1$. Hence, we obtain

$$E \sup_{0\le t\le T} \gamma_{\varpi}^{p}(\mu(t)) \le E\Big[\int \sup_{0\le t\le T} \varpi(\varphi(t,q))\mu_0(\mathrm{d}q)\Big]^p$$

$$= E\Big[\int \sup_{0\le t\le T} \varpi(\varphi(t,q))\varpi^{-\frac{p-1}{p}}(q)\varpi^{\frac{p-1}{p}}(q)\mu_0(\mathrm{d}q)\Big]^p$$

$$\le E\Big[\int \sup_{0\le t\le T} \varpi^{p}(\varphi(t,q))\varpi^{-(p-1)}(q)\mu_0(\mathrm{d}q)\Big]\Big[\int \varpi(q)\mu_0(\mathrm{d}q)\Big]^{p-1}$$

(by Hölder's inequality)

$$= \Big[\int E\{\sup_{0\le t\le T} \varpi^{p}(\varphi(t,q))\}\varpi^{-(p-1)}(q)\mu_0(\mathrm{d}q)\Big]\Big[\int \varpi(q)\mu_0(\mathrm{d}q)\Big]^{p-1}$$

(by Fubini's theorem)

$$\le c_T\Big[\int \varpi(q)^p\varpi^{-p+1}\mu_0(\mathrm{d}q)\Big]\Big[\int \varpi(q)\mu_0(\mathrm{d}q)\Big]^{p-1}$$

(by (15.241) for $\beta = p$)

$$= c_T\Big[\int \varpi(q)\mu_0(\mathrm{d}q)\Big]^p$$

$= c_T\gamma_{\varpi}^{p}(\mu_0)$ (by the definition of γ_{ϖ}). □

15.3 The Fractional Step Method

We sketch the fractional step method, following Goncharuk and Kotelenez (1998).[99]

Let $\mathbf{H}$ be some finite or infinite dimensional Hilbert space and consider the following stochastic evolution equation

$$\left.\begin{aligned} \mathrm{d}X &= A(t,X)\mathrm{d}t + B(t,X)\mathrm{d}M + \tilde{A}(t,X)\mathrm{d}t + \tilde{B}(t,X)\mathrm{d}\tilde{M}, \\ X(0) &= X_0, \end{aligned}\right\} \tag{15.245}$$

where $M(t)$ and $\tilde{M}(t)$ are Hilbert-valued (possibly cylindrical) martingales and the other coefficients are suitably defined operators (possibly nonlinear, unbounded, and random – adapted to some filtration). A solution of (15.245), if it exists, is an element of $\mathbf{H}$ for fixed ω and t. To analyze (15.245) we decompose it into two equations:

[99] The method is widely used in numerical analysis. Another term for the method is "*time splitting method*." For an application of this method to the decomposition of a coercive SPDEs driven by a regular Brownian motion into a deterministic and a purely stochastic equation, we refer to Bensoussan et al. (1992). The author wants to thank J. Duan for informing him about the existence of this paper.

$$\left.\begin{aligned} \mathrm{d}Y &= A(t,Y)\mathrm{d}t + B(t,Y)\mathrm{d}M, \\ Y(0) &= Y_0, \end{aligned}\right\} \tag{15.246}$$

and

$$\left.\begin{aligned} \mathrm{d}Z &= \tilde{A}(t,Z)\mathrm{d}t + \tilde{B}(t,Z)\mathrm{d}\tilde{M}, \\ Z(0) &= Z_0. \end{aligned}\right\} \tag{15.247}$$

Of course, other formal decompositions of (15.245) are possible as well.[100] Given the above decomposition of (15.245), we need to make the following assumptions:

- Assume the mutual tensor quadratic variation of noise processes, $M(t)$, $\tilde{M}(t)$, equals 0 for all t.
- Assume solvability of (15.246) and (15.247) for suitable, adapted initial conditions such that the increments of $\tilde{M}$ are orthogonal to the initial conditions of (15.246) and that the increments of M are orthogonal to the initial conditions of (15.247).

The fractional step method works as follows:

First solve (15.246) on a small time interval $[0, t_1]$ with initial condition $Y(0) = X_0$. Take the solution at the end point of this interval, denoted by $Y(t_1, Y_0)$, as the initial condition for (15.247) at time $t_0 = 0$(!). The solution of (15.247) at t_1, denoted $Z(t_1, Y(t_1, Y_0)$, serves as the initial condition of (15.246) on a small time interval $[t_1, t_2]$ etc. Decomposing a finite interval $[0, t]$ into n small intervals $[t_{i-1}, t_i]$ the above conditions imply the existence of a process $X_n(\cdot)$ and, under some additional assumptions the process $X_n(\cdot)$ may converge to the solution $X(\cdot, X_0)$, the fractional step product of the equations (15.246) and (15.247).

Clearly the assumption that the two martingales are uncorrelated is necessary for the fractional step method if we want to avoid working with anticipating solutions. To appreciate this method, recall that the mass in the SPDEs (8.25) and (8.26) is conserved. Goncharuk and Kotelenez employ this method to extend these noncoercive SPDEs with mass conservation to SPDEs that include a creation and annihilation of mass (as in reaction phenomena). Another application is qualitative insight by decomposing an equation of type (15.245) into simpler equations for which it may be easier to prove certain properties. If these properties hold for both equations (15.246) and (15.247) they may be preserved in the fractional step product of both equations and therefore also hold for (15.245). An example is positivity or comparison of solutions.[101]

The assumption that the two martingales be uncorrelated is void if one of the equations does not contain a stochastic term, i.e., $M(\cdot) \equiv 0$ or $\tilde{M}(\cdot) \equiv 0$, as in (5.68), (5.134), and (5.135) and similar decompositions in Chap. 5.

[100] Cf., in particular, the decomposition of (5.68) into (5.134) and (5.135).

[101] Cf. Goncharuk and Kotelenez (loc.cit.).

15.4 Mechanics: Frame-Indifference

Following Kotelenez et al. (2007) we derive the representation of frame-indifferent functions and matrices, which we used in Sect. 5.[102]

Definition 15.68. *A scalar function* $\phi(r^1, \ldots, r^m)$, $r^i \in \mathbf{R}^d$, $i = 1, \ldots, m$ *is* "isotropic" *whenever*

$$\phi(r^1, \ldots, r^m) = \phi(Qr^1, \ldots, Qr^m) \tag{15.248}$$

for every $d \times d$*-orthogonal matrix* Q. □

Theorem 15.69. *Cauchy's Representation Theorem* $\phi(r^1, \ldots, r^m)$ *is an isotropic function of* m *vectors in* $\mathbf{R}^d$ *if and only if it can be expressed as a scalar function* φ *of the* $\frac{m(m+1)}{2}$ *inner products* $\{r^i \cdot^j : \; i, j = 1, \ldots, m\}$.

The proof may be found in Truesdell and Noll (1965), Chap. B.II, Section 11). □

The following proposition is a simple exercise in linear algebra. For the convenience of the reader we include a proof.

Proposition 15.70. *Let* $A \in \mathcal{M}_{d\times d}$ *and assume*

$$AB = BA \quad \forall B \in \mathcal{O}(d), \tag{15.249}$$

where $\mathcal{O}(d)$ *are the orthogonal matrices over* $\mathbf{R}$. *Then there is an* $\eta \in \mathbf{R}$ *such that*

$$A = \eta I_d, \tag{15.250}$$

where I_d *is the identity matrix over* $\mathbf{R}^d$.[103]

Proof.

(i) Let e^i be the ith unit column vector in (the Euclidean) $\mathbf{R}^d$, i.e., if e^i_k is the kth coordinate then

$$e^i_k := \begin{cases} 1, & \text{if} \quad k = i\,, \\ 0, & \text{if} \quad k \neq i\,. \end{cases}$$

Consider the following elements from $\mathcal{O}$:

$$B_{(1,\ldots,1,\pm 1,1,\ldots,1),j} := (e_n, \ldots, e_{j+1}, \pm e_j, e_{j-1}, \ldots, e_1),$$

i.e., one matrix has the sign "plus" at all column vectors and the other matrix has a "minus" sign exactly at the jth column vector and "plus" at all other column vectors. Suppose that A has the following representation

[102] The proofs are essentially due to Leitman.

[103] If we assume A symmetric, we can restrict the class of orthogonal matrix to the subgroup $\mathcal{O}_+(d)$, which are the orthogonal matrices B with $\det(B) = 1$. For this case, an alternative proof has been provided by Kotelenez et al. (loc.cit., Appendix A).

$$A = \begin{pmatrix} a_{1,1} & a_{1,2} & \dots & a_{1,d} \\ a_{2,1} & a_{2,2} & \dots & a_{2,d} \\ & & \vdots & \\ a_{d,1} & a_{d,2} & \dots & a_{d,d} \end{pmatrix}$$

Let R_i denote the ith row vector of A and C_j the jth column vector. We note that

$$B_{(1,\dots,1,\pm1,1,\dots,1),j}A = \begin{pmatrix} R_d \\ \vdots \\ \pm R_{d+1-j} \\ \vdots \\ R_1 \end{pmatrix},$$

i.e., all row vectors of the matrix A appear in reverse order and all of them but the $(d+1-j)$th one have the sign "plus." R_{d+1-j} has $\pm$, depending on the sign at the jth column vector in $B_{(1,..1,\pm1,1,..,1),j}$. In this reverse order, $\pm R_{d+1-j}$ is the jth row of the matrix product. Further, we have

$$AB_{(1,\dots,1,\pm1,1,\dots,1),j} = (C_d, \dots, \pm C_j, \dots, C_1).$$

By condition (15.249) we must have

$$\begin{pmatrix} R_d \\ \vdots \\ \pm R_{d+1-j} \\ \vdots \\ R_1 \end{pmatrix} = (C_d, \dots, \pm C_j, \dots, C_1).$$

Comparing the jth rows in the above identity, we obtain for $k = 1, \dots, d$

$$\pm a_{d+1-j,k} = \begin{cases} a_{j,d+1-k}, & \text{if} \quad k \neq d+1-j, \\ \pm a_{j,j} & \text{if} \quad k = d+1-j. \end{cases}$$

Hence,

$$a_{i,j} = 0, \quad \text{if } i \neq j$$

and for $j = 1, \dots, d$

$$a_{d+1-j,d+1-j} = a_{j,j}.$$

It follows, in particular, A is a diagonal matrix.[104]

(ii) [105]We next switch columns in $B(1, \dots, 1)$, setting for $i < j$,

$$B^{ij}(1, \dots, 1) := (e_d, \dots, e_{d+1-j}, \dots, e_{d+1-i}, \dots, e_1).$$

[104] This finishes the proof of (15.250) for $d = 2$.

[105] Alternatively, we could now adopt the proof of Kotelenez et al. (loc.cit.), since diagonal matrices are symmetric.

Then,

$$B^{ij}(1,\ldots,1)A = \begin{pmatrix} R_d \\ \vdots \\ R_{d+1-j} \\ \cdot \\ R_{d+1-i} \\ \vdots \\ R_1 \end{pmatrix},$$

and

$$AB^{ij}(1,\ldots,1) = (C_d,\ldots,C_{d+1-j},\ldots,C_{d+1-i},\ldots,C_1).$$

The entry at the ith row and $(d+1-j)$th column of $B^{ij}(1,\ldots,1)A$ equals $a_{d+1-j,d+1-j}$, and the entry at the ith row and $(d+1-j)$-column of $AB^{ij}(1,\ldots,1)$ equals $a_{i,i}$. Hence, by condition (15.249) and step (i),

$$a_{d+1-j,d+1-j} = a_{i,i}, \quad i,j = 1,\ldots,d.$$

□

Recall the definitions of

$$P(r) := \frac{rr^T}{|r|^2},\ P^{\perp}(r) := I_d - P(r),$$

from (5.28) which are the projections on the subspace spanned by r, denoted $\{r\}$, and the subspace orthogonal to $\{r\}$, respectively. We denote the latter subspace by $\{r\}^{\perp}$.

Theorem 15.71. *Representation for Frame-Indifferent Functions*

(i) A scalar function, $\vartheta : \mathbf{R}^d \longrightarrow \mathbf{R}$ is frame-indifferent if and only if there is a scalar-valued function $\alpha : \mathbf{R}_+ \longrightarrow \mathbf{R}$ such that

$$\vartheta(r) \equiv \alpha\big(|r|^2\big). \tag{15.251}$$

(ii) A vector function, $G : \mathbf{R}^d \longrightarrow \mathbf{R}^d$ is frame-indifferent if, and only if, there is a scalar-valued function $\beta : \mathbf{R}_+ \longrightarrow \mathbf{R}$ such that

$$G(r) \equiv \beta\big(|r|^2\big)r. \tag{15.252}$$

(iii) A symmetric matrix-valued function, $A : \mathbf{R}^d \longrightarrow \mathcal{M}_{d\times d}$ is frame-indifferent if, and only if, there are two scalar-valued functions $\lambda, \lambda^{\perp} : \mathbf{R}_+ \longrightarrow \mathbf{R}$ such that

$$A(r) \equiv \lambda\big(|r|^2\big)P(r) + \lambda^{\perp}\big(|r|^2\big)P^{\perp}(r). \tag{15.253}$$

Proof.[106] Note that for $d = 1$ (i) asserts that the function is even, and the result clearly holds. Henceforth we suppose that $d \geq 2$.

Part (i) follows from Cauchy's representation theorem with $m = 1$.

For Part (ii) observe that by the frame-indifference G must be odd, whence $G(0) = 0$. We may suppose that $r \neq 0$. For an arbitrary constant vector a set

$$\phi(r, a) := G(r) \cdot a. \tag{15.254}$$

We verify that $\phi(r, a)$ is an isotropic scalar function in the two vectors r and a. Further, it follows that $\phi(r, a)$ is linear in a for arbitrary r. Hence, by Cauchy's Representation Theorem

$$G(r) \cdot a = \varphi(|r|^2, |a|^2, r \cdot a) \tag{15.255}$$

for some scalar-valued function φ os three scalar variables. Let r be fixed. φ as a function of a must also be linear. Consider the restriction of $\varphi(|r|^2, |a|^2, r \cdot a)$ to $a \in \{r\}^{\perp}$, which maps a onto $\varphi(|r|^2, |a|^2, 0)$. By the linearity in a,

$$\varphi\big(|r|^2, |-a|^2, 0\big) = \varphi\big(|r|^2, |a|^2, 0\big) = -\varphi\big(|r|^2, |a|^2, 0\big) = 0. \tag{15.256}$$

Hence,

$$G(r) \cdot a_{|a \in \{r\}^{\perp}} \equiv 0. \tag{15.257}$$

Thus,

$$G(r) \in \big\{\{r\}^{\perp}\big\} = \{r\}. \tag{15.258}$$

Equivalently there must be a scalar function $\hat{\varphi}(r)$ such that

$$G(r) = \hat{\varphi}(r)r.$$

As $G(0) = 0$ we may define $\hat{\varphi}(0) = 0$. We see that the frame-indifference of G implies the frame-indifference of $\hat{\varphi}$. By Part (i) there is a scalar function β, defined on $\mathbf{R}_+$ such that

$$\hat{\varphi}(r) \equiv \beta\big(|r|^2\big).$$

Hence,

$$G(r) \equiv \beta\big(|r|^2\big)r,$$

which proves Part (ii).

To prove Part (iii) observe that $A(0)$ commutes with all orthogonal matrices. By Proposition 15.70 this implies

$$A(0) = \eta I_d \tag{15.259}$$

for some $\eta \in \mathbf{R}$. Thus, Part (iii) holds if $r = 0$, and we may, in what follows, suppose $r \neq 0$. Note that the frame-indifference for $A(r)$ implies

$$A(r) = A(-r).$$

---

[106] The proof has been provided by M. Leitman. Cf. Kotelenez et al. (2007).

Set $G(r) := A(r)r$. We verify that $G(r)$ is frame-indifferent, whence by Part (ii)

$$G(r) = \lambda\big(|r|^2\big)r$$

for some real valued function λ, defined on $\mathbf{R}_+$. Hence, $(\lambda(|r|^2), r)$ is an eigenpair for $A(r)$. By symmetry of A, $\{r\}$ is also an eigenspace of $A^T(r)$ with eigenvalue $\lambda(|r|^2)$. Let $a \in \{r\}^\perp$. Then

$$A(r)a \cdot r = a \cdot A^T(r)r = \lambda\big(|r|^2\big)a \cdot r = 0. \tag{15.260}$$

Hence, $A(r)a \in \{r\}^\perp$, i.e., the subspace $\{r\}^\perp$ is also invariant under the action of $A(r)$. Employing the projections $P(r)$ and $P^\perp(r)$ $A(r)$ must have the decomposition

$$\left.\begin{aligned} A(r) &= P(r)A(r)P(r) + P^\perp(r)A(r)P^\perp(r) \\ &= \lambda(|r|^2)P(r) + P^\perp(r)A(r)P^\perp(r). \end{aligned}\right\} \tag{15.261}$$

Set

$$A^\perp(r) := P^\perp(r)A(r)P^\perp(r).$$

$A^\perp(r)$ is obviously frame-indifferent. Choose an orthogonal matrix $\hat{Q}$ that leaves the subspace $\{r\}$ invariant. Then, $\hat{Q}r = \pm r$. Further, $A(r)$ is even and so is $A^\perp$. Hence,

$$A^\perp(\hat{Q}r) = A^\perp(\pm r) = A^\perp(r) = \hat{Q}A^\perp(r)\hat{Q}^T. \tag{15.262}$$

Therefore, $A^\perp(r)$ commutes with every orthogonal matrix $\hat{Q}$ that leaves $\{r\}$ invariant. Note that the set of all orthogonal transformations that leave the subspace $\{r\}$ invariant is a subgroup equivalent to the group of all orthogonal transformations on $\{r\}^\perp$. Invoking Proposition 15.70, we conclude that the action of $A^\perp(r)$ on $\{r\}^\perp$ is a multiple of the identity on $\{r\}^\perp$ where the factor depends on r. Thus,

$$A^\perp(r) = \psi(r)P^\perp(r) \tag{15.263}$$

for some scalar function $\psi(r)$, where again the frame-indifference of $A^\perp(r)$ implies that $\psi(r)$ is also frame-indifferent. Therefore,

$$\psi(r) \equiv \lambda^\perp\big(|r|^2\big),$$

whence we obtain (15.253) from (15.253) and (15.255). □

Finally, we show that the diffusion matrix, defined in (5.33) is symmetric, as a result of the frame-indifference of the kernel G_ε.

Proposition 15.72. *For fixed i, j*

$$D_{\varepsilon,k\ell,ij} = D^T_{\varepsilon,k\ell,ij}. \tag{15.264}$$

Proof. Note that the frame-indifference of G_ε implies that G_ε is odd. Then,

$$D_{\varepsilon,k\ell,ij}(\hat{r}) = \int G_\varepsilon(r^i - q)G_\varepsilon^T(r^j - q)\,\mathrm{d}q$$

$$= \int G_\varepsilon\left(\frac{1}{2}(r^i - r^j) - q\right) G_\varepsilon^T\left(-\frac{1}{2}(r^i - r^j) - q\right)\mathrm{d}q$$

(by shift invariance)

$$= \int G_\varepsilon\left(-\frac{1}{2}(r^i - r^j) + q\right) G_\varepsilon^T\left(\frac{1}{2}(r^i - r^j) + q\right)\mathrm{d}q$$

(since G_ε is odd)

$$= \int G_\varepsilon\left(-\frac{1}{2}(r^i - r^j) - q\right) G_\varepsilon^T\left(\frac{1}{2}(r^i - r^j) - q\right)dq$$

(changing variables $\tilde{q} = -q$)

$$= D_{\varepsilon,k\ell,ij}^T(\hat{r}).$$

□

Subject Index

Symbols

$A_{\varepsilon,\mathbf{R}}$ - (5.75)
$\alpha_{\varepsilon}(|r|^2)$ - (5.61)
$\alpha_{\perp,\varepsilon}(|r|^2)$ - (5.61)
$\Longrightarrow$ denotes weak convergence

B is some metric space or function space
$\mathbf{B}_{+}$ is the cone of functions with nonnegative values if
$\mathbf{B}_{d,1,m} := \{f : \mathbf{R}^d \to \mathbf{R}$: f is $\mathcal{B}^d - \mathcal{B}^1$ measurable and $\exists\ \partial^{\mathbf{j}} f,\ |\mathbf{j}| \le \mathbf{m}\}$, where $m \in \mathbf{N} \cup \{0\}$ - after (8.32)
$\mathcal{B}^d$ - Borel σ−algebra of $\mathbf{R}^d$

$C_b(q)$ denotes the closed cube in $\mathbf{R}^d$, parallel to the axes, centered at q and with side length b.
$\circ$ denotes the Stratonovich differential
$\mathbf{C}_n$ – (3.6)
$C([s, T]; \mathbf{K})$ is the space of continuous **K**-valued functions on $[s, T]$, where **K** is a metric space.
$C_b^m(\mathbf{R}^d, \mathbf{R})$ is the space of m times continuously differentiable bounded real valued functions on $\mathbf{R}^d$, where all derivatives up to order m are bounded
$C_0^m(\mathbf{R}^d, \mathbf{R})$ is the subspace of $C_b^m(\mathbf{R}^d, \mathbf{R})$ whose elements and derivatives vanish at infinity
$C_c^m(\mathbf{R}^d, \mathbf{R})$ is the subspace of $C_0^m(\mathbf{R}^d, \mathbf{R})$, whose elements have compact support
$C_c^{\infty}(\mathbf{R}; \mathbf{R})$ is the subspace of $C_c^m(\mathbf{R}^d, \mathbf{R})$ whose elements are infinitely often differentiable
$C_{L,\infty}(\mathbf{R}^d; \mathbf{R})$ is the space of all uniformly bounded Lipschitz functions f from $\mathbf{R}^d$ into $\mathbf{R}$.
Cov_{ω} - random covariance of two processes, Definition 15.43
$Cov_{d+1,\omega}$ - random covariance of a random field, Definition 15.47

"$\cdot$" denotes also the scalar product in $\mathbf{R}^k$
d_p, $d_{p,\mathbf{B}}$ - the Prohorov metric

$d_\infty(r^{(\cdot)}, q^{(\cdot)}) := \sum_{k=1}^{\infty} 2^{-k}\rho(r_k - q_k)$
$D([0,\infty); \mathbf{B})$ - The (Skorokhod) space of $\mathbf{B}$-valued cadlag functions with domain $[0,\infty)$, where $\mathbf{B}$ be some topological space.
$D_{\varepsilon,k\ell,ij}(\hat{r})$ - (5.33)
$(D([0,\infty); \mathbf{M}), d_{D,M})$ - (15.53)
$\bar{D}_\varepsilon(\sqrt{2}r)$ - (5.45)
$\tilde{D}_{k\ell}(\mu, r, q, t) := \frac{d}{dt}[M_k(t, r, \mu), M_\ell(t, q, \mu)]$ - 175
$\hat{D}_{k\ell}(s, r) := -\tilde{D}_{k\ell}(s, r) + D_{k\ell}(s)$ - (11.64)
$D_{k\ell}(\mu, r, t) := \tilde{D}_{k\ell}(\mu, r, r, t)$ - 175
$\underline{D}_{k\ell}(\mu, r, t) \equiv \tilde{D}_{k\ell}(\mu, 0, t)$ - (8.34)
$\overline{\overline{D}}(\mu, r, t) = D(\mu, r, t) + (\sigma^\perp(\mu, r, t))^2$ - (8.55)
δ denotes the Fréchet derivative
$\delta R \approx \frac{1}{n}$ - mesoscopic length unit -
δs - correlation time, microscopic time unit -
$\delta\sigma$ - mesoscopic time unit -
$diam\ (A) := \sup_{b,\tilde{b}\in A} d_B(b, \tilde{b})$ - diameter of a Borel set $A \subset \mathbf{B}$
$\diamond$ denotes this empty state, $\hat{\mathbf{R}}^d := \mathbf{R}^d \cup \{\diamond\}$.
$d_{prob,u,T,\mathbf{H}}$ - metric of convergence in probability, uniformly on $[0, T]$, values in $\mathbf{H}$
$\partial^{\mathbf{n}} = \frac{\partial^{\mathbf{n}}}{(\partial r_{n_1})^{n_1}\ldots(\partial r_{n_d})^{n_d}}$ is a partial differential operator
$\partial_r^{\mathbf{n}}$ is a partial differential operator, acting on the variable r
$\partial_k, \partial_{k\ell}^2$ - partial derivatives of first and second order
$\partial_{k,r}, \partial_{k\ell,r}^2$ - partial derivatives of first and second order, acting on the variable r
$div F := \sum_{k=1}^{d} \partial_k F_k$

$\sim$ - equivalence in distribution
$\sqrt{\varepsilon}$ - correlation length
η_n - friction coefficient.
$\mathcal{E}(f, g) = \mathcal{E}_1(f, g) + \mathcal{E}_2(f, g)$ - Dirichlet forms

$Fl(r)$ is the $d-$dimensional flux
$\bar{F}l(x)$ is the radial flux
$\mathcal{F}_{\mathbf{M},t}$, - the cylinder set filtrations on $\mathbf{M}_{[0,\infty)}$
$\mathcal{F}_{\mathbf{M},s,t}$ - the cylinder set filtrations on $\mathbf{M}_{[s,\infty)}$
$\mathcal{F}_{n,s-}$ - (3.19)
$\hat{\mathcal{F}}_{n,s-}$
$\{\phi_\mathbf{k}\}_{\mathbf{k}\in\hat{\mathbf{N}}^d}$ - the normalized Hermite functions
$\Phi \in \{1, \varpi\}$ - weight function

$\gamma_f(\mu - \nu)$ - Wasserstein metric on finite measures -
$\gamma_\varpi(\mu - \nu)$ - metric on a class of $\sigma-$finite measures
$\gamma \in \{\gamma_f, \gamma_\varpi\}$

$\tilde{\gamma_p}(\mu, \tilde{\mu})$ - (15.39)
$\hat{\gamma}_f(\mu, \nu)$, - (15.44)
$G(r) = G_\varepsilon(r)$ - kernel, governing the interaction between large and small particles - Chap. 2
$G_{\varepsilon,M}$ - Maxwellian kernel
$G_n(r)$ - (2.1)
$\mathcal{G}_t$ - the σ-algebra generated by $w(dp, du)$ between 0 and t
$\mathcal{G}_{s,t}$ - the σ-algebra generated by $w(dp, du)$ between s and t
$\mathcal{G}_{s,t}^{\perp}$ - σ-algebra generated by $\{d\beta^{\perp,n}(u)\}_{n\in\mathbf{N}}$ between s and t
$\mathcal{G}_t^{\perp}$ - σ-algebras generated by $\{d\beta^{\perp,n}(u)\}_{n\in\mathbf{N}}$ between 0 and t
$\bar{\mathcal{G}}_{s,t}$ - completed $\sigma-$algebra $\mathcal{G}_{s,t}^{\perp}$
$\mathcal{G}_{n,s}$ - (3.19)
$G(u, r) := (2\pi u)^{-\frac{d}{2}} \exp\left(\frac{-|r|^2}{2u}\right), u > 0$

$\mathbf{H}_0 := L_2(\mathbf{R}^d, dr)$ -
$(\mathbf{H}_0, \|\cdot\|_0) = (\mathbf{W}_{0,2,1}, \|\cdot\|_{0,2,1})$ - 182
$\mathbf{H}_m := \mathbf{W}_{m,2,1}$.
$\mathbf{H}_w$ - (4.31), (15.74)

I_d is the identity matrix in $\mathbf{R}^d$
$I_A^n(t)$ - occupation measure
$\tilde{I}_A^n(t)$ - (3.7)
$\tilde{I}_{A,\mathbf{J}_n}^n(t)$ - (3.9)
$\tilde{I}_A^{n,\perp}(t)$ - (3.9)
$\tilde{I}_A^{n,\perp,c}(t)$ - (3.15)

$(\mathbf{K}, d_K)$ is some metric space with a norm d_K,

"l.s.t." means "localizing stopping time"
$L_{p,loc}(\mathbf{R}^d \times \Omega; \mathbf{B})$ - Definition 6.2
$L_{0,\mathcal{F}_s}(\mathbf{K})$ is the space of $\mathbf{K}$-valued $\mathcal{F}_s$-measurable random variables
$L_{0,\mathcal{F}}(C([s,T];\mathbf{K}))$ - $\mathbf{K}$-valued continuous adapted and $dt \otimes dP$-measurable processes
$L_{p,\mathcal{F}_s}(\mathbf{K})$ - those elements in $L_{0,\mathcal{F}_s}(\mathbf{K})$ with finite $p-$th moments of $d_{\mathbf{K}}(\xi, e)$
$L_{p,\mathcal{F}}(C([s,T];\mathbf{K}))$ are those elements in $L_{0,\mathcal{F}}(C([s,T];\mathbf{K}))$ such that $E\sup_{s\le t\le T} d_{\mathbf{K}}^p(\xi(t), e) < \infty$
$L_{p,\mathcal{F}}(C([s,\infty);\mathbf{K}))$ - space of processes $\xi(\cdot)$ such that $\xi(\cdot\wedge T)\in L_{p,\mathcal{F}}(C([s,T];\mathbf{K}))\forall T > s$.
$L_{loc,p,\mathcal{F}}(C([s,T];\mathbf{K}))$ - 61
$L_{loc,p,\mathcal{F}}(C((s,T];\mathbf{K}))$ - 61
$L_{1,\mathcal{F}}([0,T]\times\Omega)$ is the space of $dt\otimes dP$-measurable and adapted integrable processes
$L_{p,\mathcal{F}}([0,T]\times\Omega;\mathbf{B})$ is the space of $dt\otimes dP$-measurable and adapted p integrable $\mathbf{B}$-valued processes - Section 8.1

$\mathcal{L}_{k,\ell,\mathbf{n}}(s) := \partial^{\mathbf{n}} \tilde{D}_{k\ell}(s,r)_{|r=0}$ - (8.40)
$\mathcal{L}(\mathbf{B})$ - the set of linear bounded operators from a Banach space $\mathbf{B}$ into itself
$\Lambda_m := \{(p^1, \ldots, p^m) \in \mathbf{R}^{d \cdot m} : \exists i \neq j,\ i, j \in \{1, \ldots, m\}, \quad \text{with } p^i = p^j\}$
$\Lambda(t, \mathbf{D}_n, R, J^n)$ - (3.1)

m - mass of a small particle.
$\hat{m}$ - mass of a large particle.
$m(dx) := A(S^{d-1})x^{d-1}dx$, where $A(S^{d-1})$ is the surface measure of the unit sphere in $\mathbf{R}^d$
$\overline{m} = \overline{m}(m)$ - the smallest even integer with $\overline{m} > d + m + 2$
$\mathcal{M}_{d \times d}$ - the $d \times d$ matrices
$\mathcal{M}_{f,s,d} := \{\mathcal{X}_s : \mathcal{X}_s = \mathcal{X}_{s,N} := \sum_{i=1}^{N} m_i \delta_{r_s^i}$
$\mathbf{m}_{+k} := \mathbf{m} + 1_k$
$m(dx)$ - (5.141)
$\mathbf{M}_f$ - finite Borel measures on $\mathbf{R}^d$
$\mathbf{M}_{\infty,\varpi} := \{\mu \in \mathbf{M}_\infty : \int \varpi(q)\mu(dq) < \infty\}$
$\mu_b := \mu$, if $\gamma_\varpi(\mu) < b$, and = $\frac{\mu}{\gamma_\varpi(\mu)} b$, if $\gamma_\varpi(\mu) \geq b$. - (9.2)
$\wedge$ - denotes "minimum"
$\vee$ - denotes "maximum"

$n \in \mathbf{N}$ - index characterizing the discretization
$\mathbf{n} = (n_1, \ldots n_d) \in (\mathbf{N} \cup \{0\})^d$
$|\mathbf{n}| := n_1 + \ldots + n_d$.
$\mathbf{n} \leq \mathbf{m}$ iff $n_i \leq m_i$ for $i = 1, \ldots, d$. $\mathbf{n} < \mathbf{m}$ iff $\mathbf{n} \leq \mathbf{m}$ and $|\mathbf{n}| < |\mathbf{m}|$.
$|\cdot|$ denoted both the absolute value and the Euclidean norm on $\mathbf{R}^d$.
$\|\cdot\|$ denotes the maximum norm on $\mathbf{R}^d$.
$|\|F_\ell\|| := \sup_q |F_\ell(q)|$ denotes the sup-norm.
$\|\cdot\|_{L,\infty}$ denotes the bounded Lipschitz norm - (15.37)
$|\|f\||_m := \max_{|\mathbf{j}| \leq m} \sup_{r \in \mathbf{R}^d} |\partial^{\mathbf{j}} f(r)|$
$\|f\|_{m,p,\Phi}^p := \sum_{|\mathbf{j}| \leq m} \int |\partial^{\mathbf{j}} f|^p(r)\Phi(r)dr,$
$\|f\|_m := \|f\|_{m,2,1}$ - before (8.33)
$\|\cdot\|_f$ - dual norm on the finite Borel measures w.r.t. $C_0(\mathbf{R}^d, \mathbf{R})$
$\|\cdot\|_{\mathcal{L}(\mathbf{B})}$ - the norm of the set of linear bounded operators from a Banach space $\mathbf{B}$ into itself
$\mathbf{0}$, i.e., the "null" element.
$\mathcal{O}(d)$ are the orthogonal $d \times d$-matrices over $\mathbf{R}^d$
$\Omega := \{\hat{\mathbf{R}}^{\mathbf{d}} \times \mathbf{R}^{\mathbf{d}}\}^{\mathbf{N}}$.
$(\Omega, \mathcal{F}, \mathcal{F}_t, P)$ is a stochastic basis with right continuous filtration
$\mathbf{1} = (1, \ldots, 1)$
$1_k := (0, \ldots, 0, 1, 0, \ldots, 0)$

$(\pi_{s,t} f)(u) := f(u \wedge t), (u \geq s)$.
$\pi_2 : \hat{\mathbf{M}}_f \longrightarrow \mathbf{M}_f \quad \pi_2((\{r^i(t)\}^\infty, \mathcal{X}(t))) = \mathcal{X}(t))$ - (8.69)
$\pi_N : \mathbf{M}_{[0,T]} \rightarrow \mathbf{M}_{[0,T],N} \subset \hat{\mathbf{M}}_{[0,T]}$ - $\eta(\cdot) \mapsto \pi_N \eta(\cdot) := \eta(g_N(\cdot))$ - (6.2)

$\pi_{\mathbf{n}}(r) := \Pi_{i=1}^{d} r_i^{n_i}$, if for all $i = 1, \ldots, d$ $n_i \geq 0$ and $\pi_{\mathbf{n}}(r) = 0$ otherwise.
$P(r) := \frac{rr^T}{|r|^2}$ - projection
$P^{\perp}(r) := I_d - P(r)$ - projection
$\bar{P}(t, a, B)$ - marginal transition probability distribution - (5.25)
$\psi(x) \in \{\log(x + e), 1(x)\}$
Ψ - the set of measurable continuous semi-martingale flows φ from $\mathbf{R}^d$ into itself - cf. Proposition 4.3
$\partial^{\mathbf{j}} f = \frac{\partial^{j_1, \ldots, j_d}}{\partial r^{j_1} \ldots \partial r^{j_d}} f$

$q_n(t, \lambda, \iota)$ - the position of a small particle with start in $(\bar{R}^{\lambda}]$ and velocity from B_{ι}.

$\hat{\mathbf{R}}^d := \mathbf{R}^d \cup \{\diamond\}$.
$(\bar{R}^{\lambda}]$ - small cube, $\bar{R}^{\lambda}$ is the center of the cube
$r_n^i(t)$ - position of i-th large particle
$(\mu - \frac{D}{2}\triangle)^{-\gamma} = \frac{1}{\Gamma(\gamma)} \int_0^{\infty} u^{\gamma - 1} e^{-\mu u} T(Du) du$ - the resolvent of $D\triangle$
$R_{\mu}^{\gamma} := \mu^{\gamma} (\mu - \frac{1}{2}\triangle)^{-\gamma}, \quad R_{\mu,D}^{\gamma} := \mu^{\gamma} (\mu - \frac{D}{2}\triangle)^{-\gamma}$
$\hat{r} := (r^1, r^2) \in \mathbf{R}^{2d}$
$\{r\}$ - subspace spanned by r (Chap. 5)
$\{r\}^{\perp}$ - subspace orthogonal to $\{r\}$ (Section 5)
$\rho(r - q) := |r - q| \wedge 1$ - (4.1)
$\bar{\rho}(r, q) = \rho(r - q)$, if $r, q \in \mathbf{R}^d$ and $= 1$, if $r \in \mathbf{R}^d$ and $q = \diamond$.
$\rho_N(r_N, q_N) := \max_{1 \leq i \leq N} \rho(r_i, q_i)$

$(S_{\sigma} f)(r)$ - rotation operator - Chap. 10
$\mathcal{S}'$ - Schwarz space of tempered distributions over $\mathbf{R}^d$
$\sigma\{\cdot\}$ denotes the $\sigma-$algebra generated by the the quantities in the braces
$[\cdot, \cdot]$ - the mutual quadratic variation of square integrable martingales.
$\hat{\sum}_A$ - summation, restricted to a set $\mathbf{C}_n$ - after (3.20)

τ_n is a sequence of localizing stopping times
$\tau(c, b, \omega) := \inf_n \tau_n(c, b, \omega)$ - stopping time - (9.8)

U_h = shift operator - Chap. 10

$\varpi(r) = (1 + |r|^2)^{-\gamma}$ - weight function
$\varpi_{\beta}(r) := \varpi^{\beta}(r)$ $(\beta > 0)$

$\bar{w}$ - average speed of small particles
$w_{\ell}(dr, dt)$ be i.i.d. real valued space-time white noises on $\mathbf{R}^d \times \mathbf{R}_+$, $\ell = 1, \ldots, d$
$\check{w}(dp, t) := w(dp, T - t) - w(dp, T)$
$\mathbf{W}_{m,p,\Phi} := \{f \in \mathbf{B}_{d,1,m} : \|f\|_{m,p,\Phi} < \infty\}$.

$\mathcal{X}_N(t)$ - empirical (measure) process for N large particles, associated with (4.10).

$\tilde{\mathcal{X}}_N(t)$ - (3.7)

$\mathcal{X}(t)$ - (9.55)

$\mathbf{0} \in \mathcal{M}_{d\times d}$ - the matrix with all entries being equal to 0

References

Adams, R.A. (1975), *Sobolev Spaces.* Academic Press, New York.

Akhiezer, N.I. and Glasman, I.M. (1950), *Theory of Linear Operators in Hilbert space.* State Publisher of Technical-Theoretical Literature, Moscow, Leningrad (in Russian).

Albeverio, S., Haba, Z. and Russo, F. (2001), *A Two-Space Dimensional Semilinear Heat Equations Perturbed by (Gaussian) White Noise.* Probab. Th. Rel. Fields 121, 319–366.

Albeverio, S. and Röckner, M. (1991), *Stochastic Differential Equations in Infinite Dimensions: Solutions via Dirichlet Forms.* Probab. Th. Rel. Fields 89, 347–386.

Aldous, D.J. (1985), *Exchangeability and Related Topics.* Ecole d'Ete de Probabilites de Saint-Flour XIII—1985, Lecture Notes in Math., 1117, Springer, Berlin Heidelberg New York, pp. 1–198.

Arnold, L. (1973), *Stochastische Differentialgleichungen-Theorie und Anwendungen.* R. Oldenburg, München-Wien (English translation: Stochastic differential equations and applications, Wiley, New York).

Arnold, L., Curtain, R.F. and Kotelenez, P. (1980), *Nonlinear Stochastic Evolution Equations in Hilbert Space.* Universität Bremen, Forschungsschwerpunkt Dynamische Systeme, Report # 17.

Asakura, S. and Oosawa, F. (1954), *On Interactions between Two Bodies Immersed in a Solution of Macromolecules.* J. Chem. Phys. 22, 1255–1256.

Balakrishnan, A.V. (1983), *On Abstract Bilinear Equations with White Noise Inputs.* Appl. Math. Optim. 10, 359–366.

Baxendale, P. and Rozovsky, B. (1993), *Kinematic Dynamo and Intermittence in a Turbulent Flow.* Geophys. Astrophys. Fluid Dyn. 73, 33–60.

Bauer, H. (1968), Wahrscheinlichkeitstheorie und Grundzűge der Maßtheorie. de Gruyter & Co., Berlin (in German).

Bensoussan, A., Glowinski, R. and Răşcanu, A. (1992), *Approximation of Some Stochastic Differential Equations by the Splitting Up Method.* Appl. Math. Optim. 25, 81–106.

Bhatt, A.G., Kallianpur, G. and Karandikar, R.I. (1995), *Uniqueness and Robustness of Solutions of Measure-Valued Equations of Nonlinear Filtering.* Ann. Probab. 23(4), 1895–1938.

Bichteler, K. (2002), *Stochastic Integration with Jumps.* Encyclopedia of Mathematics and its Applications 89, Cambridge University Press, Cambridge, New York.

Billingsley, P. (1968), *Convergence of Probability Measures.* Wiley, New York.

Blömker, D. (2000), *Stochastic Partial Differential Equations and Surface Growth.* Wißner Verlag, Augsburg.

Blount, D. (1996), *Diffusion Limits for a Nonlinear Density Dependent Space–Time Population Model.* Ann. Probab. 24(2), 639–659.

Bojdecki, T. and Gorostiza, L.G. (1986), *Langevin Equations for $\mathcal{S}'$-Valued Gaussian Processes and Fluctuation Limits of Infinite Particle Systems.* Probab. Th. Rel. Fields 73, 227–244.

Bojdecki, T. and Gorostiza, L.G. (1991), *Gaussian and Non-Gaussian Distribution Valued Ornstein-Uhlenbeck Processes.* Can. J. Math. 43(6), 1136–1149.

Bogachev, V.I. (1997), *Gaussian Measures.* Nauka, Moscow (in Russian).

Borkar, V.S. (1984), *Evolution of Interacting Particles in a Brownian Medium.* Stochastics 14, 33–79.

Bourbaki, N. (1977), *Éléments de Mathématique – Livre IV: Intégration.* Nauka, Moscow (Russian Translation).

Brzeźniak, Z. and Li, Y. (2006), *Asymptotic Compactness and Absorbing Sets for 2D Stochastic Navier-Stokes Equations on Some Unbounded Domains.* Transactions of the American Mathematical Society S-0002-9947(06)03923-7.

Carmona, R. and Nualart, D. (1988a), *Random Nonlinear Wave Equations: Smoothness of the Solutions.* Probab. Th. Rel. Fields 79, 469–508.

Carmona, R. and Nualart, D. (1988b), *Random Nonlinear Wave Equations: Propagation of Singularities.* Ann. Probab. 16(2), 730–751.

Carmona, R. and Rozovsky, B. (1999), *Stochastic Partial Differential Equations: Six Perspectives.* Mathematical Surveys and Monographs, Vol. 64, American Mathematical Society.

Chojnowska-Michalik, A. (1976), *Stochastic Differential Equations in Hilbert Spaces.* Ph.D. Thesis. Institute of Mathematics, Polish Academy of Science.

Chorin, A.J. (1973), *Numerical Study of a Slightly Viscous Flow.* J. Fluid Mech. 57, 785–796.

Chow, P.L. (1976), *Function Space Differential Equations Associated with a Stochastic Partial Differential Equation.* Indiana University Mathematical J. 25(7), 609–627.

Chow, P.L. (1978), Stochastic Partial Differential Equations in Turbulence Related Problems. In: Bharucha-Reid, A.T. (ed.) *Probability Analysis and Related Topics, Vol. I,* Academic Press, New York, pp. 1–43.

Chow, P.L. (2002), *Stochastic Wave Equations with Polynomial Nonlinearity.* Ann. Appl. Probab. 12(1), 361–381.

Chow, P.L. and Jiang, J.-L. (1994), *Stochastic Partial Differential Equations in Hölder Spaces.* Probab. Th. Rel. Fields 99, 1–27.

Coddington, E.A. and Levinson, N. (1955), *Theory of Ordinary Differential Equations.* McGraw-Hill, New York.

Crauel, H., Debussche, A. and Flandoli, F. (1997), *Random Attractors.* J. Dyn. Differ. Eqns. 9(2), 307–341.

Crisan, D. (2006), *Particle Approximations for a Class of Stochastic Partial Differential Equations.* Appl. Math. Optim. 54, 293–314.

Curtain, R.F. (1981), *Markov Processes Generated by Linear Stochastic Evolution Equations.* Stochastics 5, 135–165.

Curtain, R.F. and Falb, P.L. (1971), *Stochastic Differential Equations in Hilbert Space.* J. Differ. Eqns. 10, 412–430.

Curtain, R.F. and Kotelenez, P. (1987), *Stochastic Bilinear Spectral Systems.* Stochastics 20, 3–15.

Curtain, R.F. and Pritchard, A.J. (1978), *Infinite Dimensional Linear Systems Theory.* Springer-Verlag, Lecture Notes in Control and Information Sciences, Vol. 8, Berlin Heidelberg New York.

Dalang, R. (1999), *Extending Martingale Measure Stochastic Integral with Applications to Spatially Homogeneous SPDEs.* EJP 4, 1–29.

Dalang, R. and Mueller, C. (2003), *Some Nonlinear SPDEs that are Second Order in Time.* EJP 8(1), 1–21.

Dalang, R., Mueller, C. and Tribe, R. (2006), *A Feynman-Kac-Type Formula for the Deterministic and Stochastic Wave Equations and Other SPDEs.* Preprint.

Dalang, R. and Nualart, D. (2004), *Potential Theory for Hyperbolic SPDEs* J. Funct. Anal. 227, 304–337.

Dalang, R. and Sanz-Sol, M. (2005), *Regularity of the Sample Paths of a Class of Second Order SPDEs* J. Funct. Anal. 227, 304–337.

Dalang, R. and Walsh, J. (2002), *Time-Reversal in Hyperbolic SPDEs.* Ann. Probab. 30(1), 213–252.

Daletskii, Yu. L. and Goncharuk, N. Yu. (1994), *On a Quasilinear Stochastic Differential Equation of Parabolic Type.* Stochastic Anal. Appl. 12(1), 103–129.

Da Prato, G. (1982), *Regularity Results of a Convolution Stochastic Integral and Applications to Parabolic Stochastic Equations in a Hilbert Space.* Conferenze del Seminario Matematico dell'Universtitá di Bari, Nor. 182, Laterza.

Da Prato, G. (2004), *Kolmogorov Equations for Stochastic PDEs.* Birkhäuser, Boston Basel Berlin.

Da Prato, G. and Debussche, A. (2002), *2D Navier-Stokes Equations Driven by a space–time White Noise.* J. Funct. Anal., 196(1), 180–210.

Da Prato, G. and Debussche, A. (2003), *Strong Solutions to the Stochastic Quantization Equations.* Ann. Probab. 31(4), 1900–1916.

Da Prato, G., Iannelli, M. and Tubaro, L. (1982), *Some Results on Linear Stochastic Differential Equations in Hilbert Spaces.* Stochastics 23, 1–23.

Da Prato, G., Kwapien, S. and Zabczyk, J. (1987), *Regularity of Solutions of Linear Stochastic Equations in Hilbert Spaces.* Stochastics, 23, 1–23.

Da Prato, G. and Zabczyk, J. (1991), *Smoothing Properties of the Kolmogorov Semigroups in Hilbert Space.* Stochast. Stochast. Rep. 35, 63–77.

Da Prato, G. and Zabczyk, J. (1992), *Stochastic Equations in Infinite Dimensions.* Cambridge University Press, Cambridge.

Da Prato, G. and Zabczyk, J. (1996), *Ergodicity for Infinite Dimensional Systems.* Cambridge University Press, London Mathematical Society, Lecture Note Series 229, Cambridge.

Davies, E.B. (1980), *One-Parameter Semigroups.* Academic Press, New York.

Dawson, D.A. (1972), *Stochastic Evolution Equations.* Math. Biosci. 15, 287–316.

Dawson, D.A. (1975), *Stochastic Evolution Equations and Related Measure Processes.* J. Multivariate Anal. 5, 1–52.

Dawson, D.A. (1993), Private Communication.

Dawson, D.A. (1993), *Measure-Valued Markov Processes.* Ecole d'Ete de Probabilites de Saint-Flour XXI—1991, Lecture Notes in Math., 1541, Springer, Berlin, pp. 1–260.

Dawson, D.A. and Gorostiza, L. (1990), *Generalized Solutions of a Class of Nuclear Space Valued Stochastic Evolution Equations.* Appl. Math. Optim. 22, 241–263.

Dawson, D.A. and Hochberg, K.L. (1979), *The Carrying Dimension of a Stochastic Measure Diffusion.* Ann. Probab. 7, 693–703.

Dawson, D.A., Li, Z. and Wang, H. (2003), *A Degenerate Stochastic Partial Differential Equation for the Putely Atomic Superprocess with Dependent Spatial Motion.* Infinite Dimensional Analysis, Quantum Probability and Related Topics 6(4), 597–607.

Dawson, D.A. and Vaillancourt, J. (1995), *Stochastic McKean-Vlasov Equations.* No. DEA 2. 199–229.

Dawson, D.A., Vaillancourt, J. and Wang, H. (2000). *Stochastic Partial Differential Equations for a Class of Interacting Measure-Valued Diffusions.* Ann. Inst. Henri Poincare, Probabilites et Statistiques 36(2), 167–180.

De Acosta, A. (1982), *Invariance Principles in Probability for Triangular Arrays of B-Valued Random Vectors and Some Applications.* Ann. Probab. 2, 346–373.

Dellacherie, C. (1975), *Capacités et processus stochastiques.* Mir, Moscow (in Russian – Translation from French (1972), Springer, Berlin Heidelberg New York).

Dieudonné, J. (1969), *Foundations of Modern Analysis.* Academic Press, New York.

Doering, C.R. (1987), *Nonlinear Parabolic Stochastic Differential Equations with Additive Colored Noise on* $\mathbf{R}^d \times \mathbf{R}_+$*: A Regulated Stochastic Quantization.* Commun. Math. Phys. 109, 537–561.

Donati-Martin, C. and Pardoux, E. (1993), *White Noise Driven SPDEs with Reflection.* Probab. Th. Rel. Fields 95, 1–24.

Donnelly, P. and Kurtz, T.G. (1999), *Particle Representations for Measure-Valued Population Models.* Ann. Probab. 27, 166–205.

Dorogovtsev, A. (2004a), Private Communication.

Dorogovtsev, A. (2004b), *One Brownian Stochastic Flow.* Th. Stochast. Process. 10(26), 3–4, 21–25.

Dorogovtsev, A. and Kotelenez, P. (2006), *Stationary Solutions of Quasilinear Stochastic Partial Differential Equations.* Preprint, Department of Mathematics, CWRU.

Duan, J., Lu, K. and Schmalfuss, B. (2003), *Invariant Manifolds for Stochastic Partial Differential Equations.* Ann. Probab. 31(4), 2109–2135.

Duan, J., Lu, K. and Schmalfuss, B. (2004), *Smooth Stable and Unstable Manifolds for Stochastic Evolutionary Equations.* J. Dyn. and Differ. Eqns. 16(4), 949–972.

Dudley, R. (1989), *Real Analysis and Probability.* Wadsworth and Brooks, Belmont, California.

Dunford, N. and Schwartz, J. (1958), *Linear Operators, Part I.* Interscience, New York.

Dürr, D., Goldstein, S. and Lebowitz, J.L. (1981), *A Mechanical Model of Brownian Motion.* Commun. Math. Phys. 78, 507–530.

Dürr, D., Goldstein, S. and Lebowitz, J.L. (1983), *A Mechanical Model for the Brownian Motion of a Convex Body.* Z.Wahrscheinlichkeitstheorie verw. Gebiete. 62, 427–448.

Dynkin, E.B. (1961), *Die Grundlagen der Theorie der Markoffschen Prozesse.* Springer Verlag, Berlin Göttingen Heidelberg.

Dynkin, E.B. (1965), *Markov Processes.* Vol. I and II. Springer Verlag, Berlin Göttingen Heidelberg.

Ehm, W., Gneiting, T. and Richards, D. (2004), *Convolution Roots of Radial Positive Functions with Compact Support.* Transactions of the American Mathematical Society 356, 4655–4685.

Einstein, A. (1905), *Über die von der molekularkinetischen Theorie der Wärme gefordete Bewegung von in ruhenden Flüssigkeiten suspendierten Teilchen.* Ann.d.Phys. 17 (quoted from the English translation: (1956) *Investigation on the Theory of Brownian Movement*), Dover Publications, New York.

Erwe, F. (1968), *Differential- und Integralrechnung II.* Bibliographisches Institut, Mannheim.

Ethier, S.N. and Kurtz, T.G. (1986), *Markov Processes – Characterization and Convergence.* Wiley, New York.

Faris, W.G. and Jona-Lasinio, G. (1982), *Large Fluctuations for a Nonlinear Heat Equation with Noise.* J. Phys. A. Math. Gen. 15, 3025–3055.

Flandoli, F. (1995), *Regularity Theory and Stochastic Flows for Parabolic SPDEs.* Gordon and Breach, London.

Flandoli, F. and Maslowski, B. (1995), *Ergodicity of the 2-D Navier-Stokes Equation Under Random Perturbations.* Comm. Math. Phys. 172(1), 119–141.

Fleming, W.H. and Viot, M. (1979), *Some Measure-Valued Markov Processes in Population Genetics Theory.* Indian University Mathematics Journal, 28(5), 817–843.

Folland, G.B. (1984), *Real Analysis – Modern Techniques and their Applications.* Wiley & Sons, New York.

Fouque, J.-P. (1994), Private Communication.

Friedman, A. (1975), *Stochastic Differential Equations and Applications, Vol. 1.* Academic Press, New York San Francisco London.

Friedman, A. (1976), *Stochastic differential equations and applications, Vol. 2.* Academic Press, New York San Francisco London.

Fukushima, M. (1980), *Dirichlet Forms and Markov Processes.* North-Holland/ Kodansha. Amsterdam Oxford New York.

Funaki, T. (1983), *Random Motion of Strings and Related Stochastic Evolution Equations.* Nagoya Math. J. 89, 129–193.

Funaki, T. and Spohn, H. (1997), *Motion by Mean Curvature from the Ginzburg-Landau $\nabla\phi$ Interface Model.* Commun. Math. Phys. 185, 1–36.

Gärtner, J. (1988): On the McKean-Vlasov limit for interacting diffusions. Math. Nachr. 137, 197–248.

Gel'fand I.M. and Vilenkin, N.Ya. (1964), *Generalized Functions, Vol 4.* Academic Press, New York.

Giacomin, G., Lebowitz, J.L. and Presutti, E. (1999), Deterministic and Stochastic Hydrodynamic Equations Arising from Simple Microscopic Model Systems. In: *Stochastic Partial Differential Equations: Six Perspectives,* Carmona R.A. and Rozovskii B. (eds.), Mathematical Surveys and Monographs, Vol. 64, American Mathematical Society, pp. 107–152.

Gikhman, I.I. and Mestechkina, T.M. (1983), *A Cauchy Problem for Stochastic First Order Partial Differential Equations and Their Applications.* Preprint (in Russian).

Gikhman, I.I. and Skorokhod, A.V. (1968), *Stochastic Differential Equations.* Naukova Dumka, Kiev (in Russian – English Translation (1972): *Stochastic Differential Equations.* Springer, Berlin Heidelberg New York).

Gikhman, I.I. and Skorokhod, A.V. (1971), *Theory of Random Processes. Vol. I.* Nauka, Moscow (in Russian – English Translation (1974): *The Theory of Stochastic Processes I.* Springer, Berlin Heidelberg New York).

Gikhman, I.I. and Skorokhod, A.V. (1982), *Stochastic Differential Equations and Their Applications.* Naukova Dumka, Kiev (in Russian).

Goetzelmann, B., Evans, R. and Dietrich, S. (1998), *Depletion Forces in Fluids.* Phys. Rev. E 57(6), 6785–6800.

Goncharuk, N. and Kotelenez, P. (1998), *Fractional Step Method for Stochastic Evolution Equations.* Stoch. Proc. Appl. 73, 1–45.

Gorostiza, L.G. (1983), *High Density Limit Theorems for Infinite Systems of Unscaled Branching Brownian Motions.* Ann. Probab. 11(2), 374–392.

Gorostiza, L.G. and León (1990), *Solutions of Stochastic Evolution Equations in Hilbert Space.* Preprint.

Greksch, W. and Tudor, C. (1995), *Stochastic Evolution Equations. A Hilbert Space Approach.* Akademie Verlag, Berlin.

Gyöngy, I. (1982), *On Stochastic Equations with Respect to Semimartingales III.* Stochastics 7, 231–254.

Gross, L. (1965), *Abstract Wiener Spaces.* Proc. 5th. Berkeley Sym. Math. Stat. Prob. 2, 31–42.

Gurtin, M.E. (1981), *An Introduction to Continuum Mechanics.* Mathematics in Science and Engineering, Vol. 158, Academic Press, New York.

Haken, H. (1983), *Advanced Synergetics.* Springer, Berlin Heidelberg New York.

Heisenberg, W. (1958), The Copenhagen Interpretation of Quantum Physics. In *Physics and Philosophy, Volume Nineteen of World Perspectives*, Harper and Row New York.

Holden H., Øksendal, B., Ubøe, J. and Zhang, T. ((1996), *Stochastic Partial Differential Equations. A Modeling, White Noise Functional Approach.* Birkhäuser, Boston.

Holley, R. (1971), *The Motion of a Heavy Particle in an Infinite One-Dimensional Gas of Hard Spheres.* Z. Wahrscheinlickeitstheor. Verw. Geb. 17, 181-219.

Holley, R. and Stroock, D.W. (1978), *Generalized Ornstein-Uhlenbeck Processes and Infinite Particle Branching Brownian Motions.* Publ. RIMS, Kyoto Univ. 14, 741–788.

Ibragimov, I.A. (1983), *On Smoothness Conditions for Trajectories of Random Functions*. Theory Probab. Appl. **28**(2), 240–262.

Ichikawa, A. (1982), *Stability of Semilinear Stochastic Evolution Eqautions.* J. Math. Anal. Appl. 90, 12–44.

Ikeda, N. and Watanabe, S. (1981), *Stochastic Differential Equations and Diffusion Processes*. North Holland, New York.

Iscoe, I. (1988), *On the Supports of Measure-Valued Critical Branching Brownian Motion.* Ann. Probab. 16(1), 200–221.

Iscoe, I. and McDonald, D. (1989), *Large Deviations for ℓ^2-Valued Ornstein-Uhlenbeck Processes.* Ann. Probab. 17, 58–73.

I'lin, A.M. and Khasminskii, R.Z. (1964), *On equations of Brownian motions.* Probab. Th. Appl. IX(3), 466–491 (in Russian).

Itô, K. (1944), *Stochastic Integral.* Proc. Japan. Acad. Tokyo, 20, 519–524.

Itô, K. (1983), *Distribution-Valued Processes Arising from Independent Brownian Motions.* Math. Z 182, 17–33.

Itô, K. (1984), *Foundation of Stochastic Differential Equations in Infinite Dimensional Spaces.* CBMS-NSF Regional Conference Series, SIAM.

Jacod, J. (1985), *Theoremes limite pour les processus.* Ecole d'Ete de Probabilites de Saint-Flour XIII—1985, Lecture Notes in Math., 1117, Springer, Berlin Heidelberg New York, pp. 298–409

Jetschke, G. (1986), *On the Equivalence of Different Approaches to Stochastic Partial Differential Equations.* Math. Nachr. 128, 315–329.

Kallianpur, G. and Wolpert, R. (1984), *Infinite Dimensional Stochastic Diffrential Equation Models for Spatially Distributed Neurons.* Appl. Math. Optim. 12, 125–172.

Kallianpur, G. and Xiong, J. (1995), *Stochastic Differential Equations in Infinite Dimensional Spaces*. Institute of Mathematical Statistics, Lecture Notes – Monograph Series, Hayward, California.

Kantorovich, L.V. and Akilov, G.P. (1977), *Functional Analysis.* Nauka, Moscow (in Russian).

Kato, T. (1976), *Perturbation Theory for Linear Operators.* Springer, Berlin Heidelberg New York.

Khasminskii, R.Z. (1969), *Stability of Systems of Differential Equations Under Random Perturbations of their Parameters.* Moscow, Nauka (in Russian).

Kim, K.-H. (2005), *L_p-Estimates for SPDE with Discontinuous Coefficients in Domains.* EJP 10(1), 1–20.

Knoche-Prevot, C. and Röckner, M. (2006), *A Concise Course on Stochastic Partial Differential Equations.* Lecture Notes, Preprint.

Kolmogorov, A.N. (1956), *On Skorokhod's Convergence.* Theory Probab. Appl. 1(2), 239–247 (in Russian).

Konno, N. and Shiga, T (1988), *Stochastic Differential Equations for Some Measure Valued Diffusions.* Prob. Th. Rel. Fields 79, 201–225.

Korotkov, V.B. (1983), *Integral Operators.* Nauka, Moscow (in Russian).

Kotelenez, P. (1982), *A Submartingale Type Inequality with Applications to Stochastic Evolution Equations.* Stochastics 8. 139–151.

Kotelenez, P. (1984), *A Stopped Doob Inequality for Stochastic Convolution Integrals and Stochastic Evolution Equations.* Stoch. Anal. Appl. 2(3), 245–265.

Kotelenez, P. (1985), *On the Semigroup Approach to Stochastic Evolution Equations.* In Arnold, L. and Kotelenez, P. (eds.) *Stochastic space–time Models and Limit Theorems*, 95–139.D. Reidel, Dordrecht.

Kotelenez, P. (1986), *Law of Large Numbers and Central Limit Theorem for Linear Chemical Reactions with Diffusion.* Probab. Ann. Probab. 14(1), 173–193.

Kotelenez, P. (1987), *A Maximal Inequality for Stochastic Convolution Integrals on Hilbert Space and space–time Regularity of Linear Stochastic Partial Differential Equations.* Stochastics 21, 345–458.

Kotelenez, P. (1988), *High Density Limit Theorems for Nonlinear Reactions with Diffusion.* Probab. Th. Rel. Fields 78, 11–37.

Kotelenez, P. (1992a), *Existence, Uniqueness and Smoothness for a Class of Function Valued Stochastic Partial Differential Equations.* Stochast. Stochast. Rep. 41, 177–199.

Kotelenez, P. (1992b), *Comparison Methods for a Class of Function Valued Stochastic Partial Differential Equations.* Probab. Th. Rel. Fields 93, 1–19.

Kotelenez, P. (1995a), *A Stochastic Navier Stokes Equation for the Vorticity of a Two-dimensional Fluid.* Ann. Appl. Probab. 5(4) 1126–1160.

Kotelenez, P. (1995b), *A Class of Quasilinear Stochastic Partial Differential Equations of McKean-Vlasov Type with Mass Conservation.* Probab. Th. Rel. Fields 102, 159–188

Kotelenez, P. (1995c), Particles, Vortex Dynamics and Stochastic Partial Differential Equations. In: Robert J. Adler et al. (eds.), *Stochastic Modelling in Physical Oceanography*. Birkhäuser, Boston, pp. 271–294.

Kotelenez, P. (1996), *Stochastic Partial Differential Equations in the Construction of Random Fields from Particle Systems. Part I: Mass Conservation.* CWRU, Department of Mathematics, Preprint, pp. 96–143.

Kotelenez, P. (1999), Microscopic and Mesoscopic Models for Mass Distributions. In: *Stochastic Dynamics*, Crauel, H. and Gundlach, M. (eds.), (in honor of Ludwig Arnold), Springer, Berlin Heidelberg New York.

Kotelenez, P. (2000), Smooth Solutions of Quasilinear Stochastic Partial Differential Equations of McKean-Vlasov Type. In: *Skorokhod's Ideas in Probability Theory,*

Korolyuk, V. Portenko, and N. Syta, H. (eds.), (in honor of A.V. Skorokhod), National Academy of Sciences of Ukraine, Institute of Mathematics, Kyiv.

Kotelenez, P. (2002), *Derivation of Brownian Motions from Deterministic Dynamics of Two Types of Particles.* CWRU, Department of Mathematics, Technical Report No. 02–149

Kotelenez, P. (2005a), *From Discrete Deterministic Dynamics to Stochastic Kinematics – A Derivation of Brownian Motions.* Stochast. Dyn., 5(3), 343–384.

Kotelenez, P. (2005b), *Correlated Brownian Motions as an Approximation to Deterministic Mean-Field Dynamics.* Ukrainian Math. J. T 57(6), 757–769.

Kotelenez, P. (2007), *Itô and Stratonovich Stochastic Partial Differential Equations. Transition from Microscopic to Macroscopic Equations.* Quarterly of Applied Mathematics. (to appear).

Kotelenez, P. and Kurtz, T.G. (2006), *Macroscopic Limit for Stochastic Partial Differential Equations of McKean-Vlasov Type.* Preprint.

Kotelenez, P., Leitman M. and Mann, J.A., Jr. (2007), *On the Depletion Effect in Colloids.* Preprint.

Kotelenez, P. and Wang, K. (1994), Newtonian Particle Mechanics and Stochastic Partial Differential Equations. In: Dawson, D.A. (ed.), *Measure Valued Processes, Stochastic Partial Differential Equations and Interacting Systems*, Centre de Recherche Mathematiques, CRM Proceedings and Lecture Notes, Vol. 5, 130–149.

Krylov, N.V. (1977), *Controlled Diffusion Processes.* Nauka, Moscow (in Russian).

Krylov N.V. (1999), An Analytical Approach to SPDEs. In: *Stochastic Partial Differential Equations: Six Perspectives*, Carmona, R.A. and Rozovskii, B. (eds.), Mathematical Surveys and Monographs, Vol. 64, American Mathematical Society, pp. 185–242.

Krylov, N.V. (2005), *Private Communication.*

Krylov, N.V. and Rozovsky, B.L. (1979), *On stochastic evolution equations.* Itogi Nauki i tehniki, VINITI, 71–146 (in Russian).

Kunita, H. (1990), *Stochastic Flows and Stochastic Differential Equations.* Cambridge University Press, Cambridge, New York.

Kuo, H.H. (1975), *Gaussian measures in Banach spaces.* Springer, Berlin Heidelberg New York.

Kupiainen, A. (2004), *Statistical Theories of Turbulence.* In: Random Media (J. Wehr ed.) Wydawnictwa ICM, Warszawa.

Kurtz, T.G. (1975), *Semigroups of Conditioned Shifts and Approximation of Markov Processes.* Ann. Probab. 3, 618–642.

Kurtz, T.G. and Protter, P.E. (1996), *Weak Convergence of Stochastic Integrals and Differential Equations II: Infinite Dimensional Case.* Probabilistic Models for Nonlinear Partial Differential Equations, Lecture Notes in Mathematics, Vol. 1627, Springer, Berlin Heidelberg New York, 197–285.

Kurtz, T.G. and Xiong, J. (1999), *Particle Representations for a Class of Nonlinear SPDEs.* Stochast. Process Appl. 83, 103–126.

Kurtz, T.G. and Xiong, J. (2001), *Numerical Solutions for a Class of SPDEs with Application to Filtering.* Stochastics in Finite/Infinite Dimensions (in honor of Gopinath Kallianpur) , 233–258. Birkhäuser, Boston.

Kurtz, T.G. and Xiong, J. (2004), *A stochastic evolution equation arising from the fluctuation of a class of interacting particle systems.* Comm. Math. Sci. 2, 325–358.

Kushner, H.J. (1967), *Dynamical Equations for Optimal Nonlinear Filtering.* J. Differ. Eqns., 3(2), 179–190.

Ladyženskaja, O.A., Solonnikov, V.A. and Ural'ceva, N.N. (1967)), *Linear adn Quasilinear Equations of Parabolic Type.* Nauka, Moscow, 1967 (in Russian)(English Translation (1968), American Mathematical Society).

Lifshits, E.M. and Pitayevskii, L.P. (1979), *Physical Kinetics. Theoretical Physics X.* Nauka, Moscow (in Russian).

Lions, J.L. and Magenes, E. (1972), *Non-Homogeneous Boundary Value Problems and Applications I.* Springer, Berlin Heidelberg New York.

Lions, P.L. and Souganidis, P. F. (2000), *Uniqueness of Weak Solutions of Fully Nonlinear Stochastic Partial Differential Equations.* C.R. Acad. Sci. Paris. t. 331, Sèrie I, 783–790.

Liptser, P.Sh. and Shiryayev, A.N. (1986), *Theory of Martingales.* Nauka, Moscow (in Russian).

Liptser, P.Sh. and Shiryayev, A.N. (1974), *Statistics of Random Processes.* Nauka, Moscow (in Russian).

Loève, M. (1978), *Probability Theory II.* Springer, Berlin Heidelberg New York.

Ma, Z.-M. and Röckner, M. (1992), *Introduction to the Theory of (Non-Symmetric) Dirichlet Forms.* Springer, Berlin Heidelberg New York.

Marchioro, C. and Pulvirenti, M. (1982), *Hydrodynamics and Vortex Theory..* Comm. Math. Phys. 84, 483–503.

Marcus, M. and Mizel, V. J. (1991), *Stochastic Hyperbolic Systems and the Wave Equations.* Stochast. Stochast. Rep., 36, 225–244.

Marcus, R. (1974), *Parabolic Itô Equations.* Trans. Am. Math. Soc. 198, 177–190.

Martin-Löf, A. (1976), *Limit Theorems for the Motion of a Poisson System of Independent Markovian Particles with High Density.* Z. Wahrsch.verw. Gebiete 34, 205–223.

Metivier, M. (1988), *Stochastic Partial Differential Equations in Infinite Dimensional Spaces.* Scuola Normale Superiore, Pisa.

Metivier, M. and Pellaumail, J. (1980), *Stochastic Integration.* Adademic Press, New York.

Mikulevicius, R. and Rozovsky, B.L. (1999), Martingale Problems for Stochastic PDE's. In: *Stochastic Partial Differential Equations: Six Perspectives*, Carmona, R. A. and Rozovskii, B. (eds.), Mathematical Surveys and Monographs, Vol. 64, American Mathematical Society, pp. 243–325.

Mikulevicius, R. and Rozovsky, B.L. (2004), *Stochastic Navier-Stokes Equations for Turbulent Flows.* Siam J. Math. Anal. 35(5), 1250–1310.

Mikulevicius, R. and Rozovsky, B.L., *Global L_2-Solutions of Stochastic Navier-Stokes Equations.* Ann. Probab. 33, No 1, 137–176.

Monin, A.S. and Yaglom, A.M. (1965), *Statistical Fluid Mechanics: Mechanics of Turbulence.* Nauka, Moscow (English Translation, Second Printing: (1973), The MIT Press, Vols. 1 and 2.)

Mueller, C. (2000), *The Critical Parameter for the Heat Equation with a Noise Term to Blow up in Finite Time.* Ann. Probab. 28(4), 1735–1746.

Mueller, C. and Perkins, E.A. (1992) *The Compact Support Property for Solutions to the Heat Equation with Noise.* Probab. Th. Rel. Fields 93, 325–358.

Mueller, C. and Sowers, R. (1995) *Travelling Waves for the KPP Equation with Noise.* J. Funct. Anal., 128, 439–498.

Mueller, C. and Tribe, R. (1995) *Stochastic P.D.E.'s Arising from the Long Range Contact and Long Range Voter Models.* Probab. Th. Rel. Fields 102(4), 519–546.

Mueller, C. and Tribe, R. (2002) *Hitting Properties of a Random String.* EJP 7, 1–29.

Natanson, I.P. (1974), *Theory of Functions of a Real Variable.* Nauka, Moscow (in Russian).

Neate, A.D. and Truman, A. (2006), *A One-Dimensional Analysis of Turbulence and its Intermittence for the d-Dimensional Stochastic Burgers Equations.* Preprint.

Nelson, E. (1972), *Dynamical Theories of Brownian Motions.* Princeton University Press, Princeton, N.J.

Nualart, D. and Zakai, M. (1989), *Generalized Brownian functionals and the solution to a stochastic partial differential equation.* J. Funct. Anal. 84(2), 279–296.

Oelschläger, K. (1984), *A Martingale Approach to the Law of Large Numbers for Weakly Interacting Stochastic Processes.* Ann. Probab. 12, 458–479.

Oelschläger, K. (1985), *A Law of Large Numbers for Moderately Interacting Diffusions.* Z. Wahrscheinlichkeitstheorie verw. Gebiete 69, 279–322.

Pardoux, E. (1975), *Equations aux dérivées partielles stochastique non linéaires monotones.* Etude de solutions fortes de type Itô. These.

Pardoux, E. (1979), *Stochastic Partial Differential Equations and Filtering of Diffusion Processes.* Stochastics 3, 127–167.

Pazy, A. (1983), *Semigroups of linear operators and applications to partial differential equations* (Applied Mathematical Sciences) 44, Springer, Berlin Heidelberg New York.

Perkins, E. (1992), *Measure-Valued Branching Measure Diffusions with Spatial Interactions.* Probab. Th. Rel. Fields 94, 189–245.

Perkins, E. (1995), *On the Martingale Problem for Interactive Measure-Valued Branching Diffusions.* Mem. AMS 115, No. 549, 1–89.

Peszat, S. and Zabczyk, J. (2000), *NonLinear Stochastic Wave and Heat Equations.* Probab. Th. Rel. Fields 116, 421–443.

Peszat, S. and Zabczyk, J. (2006), *Stochastic Partial Differential Equations Driven by Lévy Processes.* Book. Preprint.

Prohorov, Yu.V. (1956), *Convergence of Random Measures and Limit Theorems in Probability Theory.* Th. Probab. Appl. 1(2), 157–214.

Protter, P.E. (2004), *Stochastic Integration and Differential Equations.* Applications of Mathematics, Springer, Berlin Heidelberg New York.

Reimers, M. (1989), *One Dimensional Stochastic Partial Differential Equations and the Branching Measure Diffusion.* Probab. Th. Rel. Fields 81, 319–340.

Roukes, M. (2001), *Plenty of Room, Indeed.* Scientific American, Special Issue on Nanotechnology, September 2001.

Rozovsky, B.L. (1983), *Stochastic Evolution Systems.* Nauka, Moscow (in Russian – English Translation (1990), Kluwer Academic, Dordrecht).

Schaumlöffel, K.-U. (1986), *Verallgemeinerte Zufällige Felder und Lineare Stochastische Partielle Differentialgleichungen.* Master Thesis, University of Bremen.

Schmalfuss, B. (1991), *Long-Time Behavior of the Stochastic Navier-Stokes Equations.* Math. Nachr. 152, 7–20.

Schmuland, B. (1987), *Dirichlet Forms and Infinite Dimensional Ornstein-Uhlenbeck Processes.* Ph.D. Thesis, Carleton University, Ottawa, Canada.

Schwarz, L. (1954), *Sur l'impossibilité de la multiplication de distributions.* Comptes Rendus Hebdomadaires de Séances de l'Academie de Sciences (Paris), 239, 847–848.

Shiga, T. (1994), *Two Contrasting Properties of Solutions for One-Dimensional Stochastic Partial Differential Equations.* Can. J. Math. 46(2), 415–437.

Sinai, Ya.G. and Soloveichik, M.R. (1986), *One-Dimensional Classical Massive Particle in the Ideal Gas.* Commun. Math. Phys. 104, 423–443.

Skorokhod, A.V. (1956), *Limit Theorems for Stochastic Processes.* Theor. Probab. Appl.1 (269–290).

Skorokhod, A.V. (1987), *Asymptotic Methods of the Theory of Stochastic Differential Equations.* Kiev, Naukova Dumka (in Russian).

Skorokhod, A.V. (1996), *On the Regularity of Many-Particle Dynamical Systems Perturbed by White Noise.* Preprint.

Souganidis, P.F. and Yip, N.K. (2004), *Uniqueness of Motion by Mean Curvature Perturbed by Stochastic Noise.* Ann. I. H. Poincaré - AN 21, 1–23.

Spohn, H. (1991), *Large Scale Dynamics of Interacting Particles.* Springer, Berlin Heidelberg New York.

Sritharan, S. and Sundar, P., *Large Deviations for the Two-Dimensional Navier Stokes Equations with Multiplicative Noise.* Stoch. Proc. Appl. 116, 1636–1659.

Stratonovich, R.L. (1964), *A New Representation of Stochastic Integrals and Equations.* Vestnik MGU, Ser. 1, 1, 3–12 (in Russian).

Stroock, D.W. and Varadhan, S.R.S. (1979), *Multidimensional Diffusion Processes.* Springer, Berlin Heidelberg New York.

Suetin, P.K. (1979), *Classical Orthogonal Polynomials.*, 2nd edition, Nauka, Moscow (in Russian).

Szász, D. and Tóth, B. (1986a), *Bounds for the Limiting Variance of the "Heavy Particle" in* **R**. Commun. Math. Phys. 104, 445–455.

Szász, D. and Tóth, B. (1986b), *Towards a Unified Dynamical Theory of the Brownian Particle in an Ideal Gas.* Commun. Math. Phys. 111, 41–62.

Tanabe, H. (1979), *Equations of Evolution.* Pitman, London.

Treves, F. (1967), *Topological Vector Spaces, Distributions and Kernels.* Academic Press, New York.

Triebel, H. (1978), *Interpolation theory, function spaces, differential operators. VEB Deutscher Verlag der Wissenschaften.* Berlin.

Truesdell, C. and Noll, W. (1965), *Encyclopedia of Physics, Vol. III/3 - The Nonlinear Field Theories of Mechanics.* Springer, Berlin Heidelberg New York.

Tubaro, L. (1984), *An Estimate of Burkholder Type for Stochastic Processes Defined by a Stochastic Integral.* Stoch. Anal. Appl. 187–192.

Tulpar, A., Van Tassel, P.R. and Walz, J.Y. (2006), *Structuring of Macro-Ions Confined between Like-Charged Surfaces.* Langmuir 22(6): 2876–2883.

Uhlenbeck, G.E. and Ornstein, L.S. (1930), *On the Theory of the Brownian Motion.* Phys. Rev., 36, 823–841.

Üstünel, A.S. (1985), *On the Hypoellipticity of Stochastic Partial Differential Equations.* Proceedings of the IFIP-WG 7/1 Working Conference, LN in Control and Information Sciences, Vol. 69, Springer, Berlin Heidelberg New York.

Üstünel, A.S. (1995), *An Introduction to Analysis on Wiener Space.* LN in Mathematics 1610, Springer, Berlin Heidelberg New York.

Van Kampen, N.G. (1983), *Stochastic Processes in Physics and Chemistry.* North Holland, Amsterdam, New York.

Vaillancourt, J. (1988), *On the Existence of Random McKean-Vlasov limits for Triangular Arrays of Exchangeable Diffusions.* Stoch. Anal. Appl. 6(4), 431–446.

Viot, M. (1975), *Solutions faibles d'équations aux dérivées partielles stochastique non linéaires.* These.

Vishik, M.J. and Fursikov A.V. (1980), *Mathematical Problems of Statistical Hydromechanics.* Nauka, Moscow (Tranlation into Enlish (1988), Kluwer, Dordrecht.,)

Walsh, J.B. (1981), *A Stochastic Model of Neural Response.* Adv. Appl. Prob. 13, 231–281.

Walsh, J.B. (1986), *An Introduction to Stochastic Partial Differential Equations.* Ecole d'Eté de Probabilité de Saint Fleur XIV. Lecture Notes in Math. 1180. Springer, Berlin Heidelberg New York, 265–439.

Wang, H.(1995), *Interacting Branching Particles System and Superprocesses.* Ph.D. Thesis. Carleton University, Ottawa.

Wang, W., Cao, D. and Duan J. (2005), *Effective Macroscopic Dynamics of Stochastic Partial Differential Equations in Perforated Domains.* SIAM. J. Math. Ana Vol 38, No 5, 1508–1527.

Watanbe, S. (1984), *Lectures on Stochastic Differential Equations and Malliavin Calculus.* Tata Institute of Fundamental Research, Bombay.

Weinan, E. Mattingly, J. and Sinai, Ya. (2001), *Gibbsian Dynamics and Ergodicity for the Stochastically Forced Navier-Stokes Equation.* Comm. Math. Phys. 224, 83–106.

Wong, E. and Zakai, M. (1965), *On the Relation between Ordinary and Stochastic Differential Equations,* Int. J. Eng. Sci. 3, 213–229.

Yaglom, A.M. (1957), *Some Classes of Random Fields in n-Dimensional Space, Related to Stationary Random Processes,* Theor. Probab. Appl. II(3), 273–320.

Yip, N.K. (1998), *Stochastic Motion by Mean Curvature.* Arch. Rational Mech. Anal. 144, 313–355.

Yosida, K. (1968), *Functional Analysis*, Second Edition. Springer, Berlin Heidelberg New York.

Zabczyk, J. (1977), *Linear Stochastic Systems in Hilbert Spaces: Spectral Properties and Limit Behavior.* Institute of Mathematics, Polish Academy of Sciences, Report No. 236.

Zakai, M. (1969), *On the Optimal Filtering of Diffusion Processes.* Z. Wahrschein. Verw. Geb.11, 230–243.

Printed in the United States of America